Malaria Methods and Protocols

METHODS IN MOLECULAR MEDICINE™

John M. Walker, Series Editor

77. **Psychiatric Genetics:** *Methods and Reviews,* edited by *Marion Leboyer and Frank Bellivier,* 2003

76. **Viral Vectors for Gene Therapy:** *Methods and Protocols,* edited by *Curtis A. Machida,* 2003

75. **Lung Cancer:** *Volume 2, Diagnostic and Therapeutic Methods and Reviews,* edited by *Barbara Driscoll,* 2003

74. **Lung Cancer:** *Volume 1, Molecular Pathology Methods and Reviews,* edited by *Barbara Driscoll,* 2003

73. **E. coli:** *Shiga Toxin Methods and Protocols,* edited by *Dana Philpott and Frank Ebel,* 2003

72. **Malaria Methods and Protocols,** edited by *Denise L. Doolan,* 2002

71. *Hemophilus influenzae* **Protocols,** edited by *Mark A. Herbert, E. Richard Moxon, and Derek Hood,* 2002

70. **Cystic Fibrosis Methods and Protocols,** edited by *William R. Skach,* 2002

69. **Gene Therapy Protocols, 2nd ed.,** edited by *Jeffrey R. Morgan,* 2002

68. **Molecular Analysis of Cancer,** edited by *Jacqueline Boultwood and Carrie Fidler,* 2002

67. **Meningococcal Disease:** *Methods and Protocols,* edited by *Andrew J. Pollard and Martin C. J. Maiden,* 2001

66. **Meningococcal Vaccines:** *Methods and Protocols,* edited by *Andrew J. Pollard and Martin C. J. Maiden,* 2001

65. **Nonviral Vectors for Gene Therapy:** *Methods and Protocols,* edited by *Mark A. Findeis,* 2001

64. **Dendritic Cell Protocols,** edited by *Stephen P. Robinson and Andrew J. Stagg,* 2001

63. **Hematopoietic Stem Cell Protocols,** edited by *Christopher A. Klug and Craig T. Jordan,* 2002

62. **Parkinson's Disease:** *Methods and Protocols,* edited by *M. Maral Mouradian,* 2001

61. **Melanoma Techniques and Protocols:** *Molecular Diagnosis, Treatment, and Monitoring,* edited by *Brian J. Nickoloff,* 2001

60. **Interleukin Protocols,** edited by *Luke A. J. O'Neill and Andrew Bowie,* 2001

59. **Molecular Pathology of the Prions,** edited by *Harry F. Baker,* 2001

58. **Metastasis Research Protocols:** *Volume 2, Cell Behavior In Vitro and In Vivo,* edited by *Susan A. Brooks and Udo Schumacher,* 2001

57. **Metastasis Research Protocols:** *Volume 1, Analysis of Cells and Tissues,* edited by *Susan A. Brooks and Udo Schumacher,* 2001

56. **Human Airway Inflammation:** *Sampling Techniques and Analytical Protocols,* edited by *Duncan F. Rogers and Louise E. Donnelly,* 2001

55. **Hematologic Malignancies:** *Methods and Protocols,* edited by *Guy B. Faguet,* 2001

54. *Mycobacterium tuberculosis* **Protocols,** edited by *Tanya Parish and Neil G. Stoker,* 2001

53. **Renal Cancer:** *Methods and Protocols,* edited by *Jack H. Mydlo,* 2001

52. **Atherosclerosis:** *Experimental Methods and Protocols,* edited by *Angela F. Drew,* 2001

51. **Angiotensin Protocols,** edited by *Donna H. Wang,* 2001

50. **Colorectal Cancer:** *Methods and Protocols,* edited by *Steven M. Powell,* 2001

49. **Molecular Pathology Protocols,** edited by *Anthony A. Killeen,* 2001

48. **Antibiotic Resistance Methods and Protocols,** edited by *Stephen H. Gillespie,* 2001

47. **Vision Research Protocols,** edited by *P. Elizabeth Rakoczy,* 2001

46. **Angiogenesis Protocols,** edited by *J. Clifford Murray,* 2001

45. **Hepatocellular Carcinoma:** *Methods and Protocols,* edited by *Nagy A. Habib,* 2000

44. **Asthma:** *Mechanisms and Protocols,* edited by *K. Fan Chung and Ian Adcock,* 2001

43. **Muscular Dystrophy:** *Methods and Protocols,* edited by *Katherine B. Bushby and Louise Anderson,* 2001

42. **Vaccine Adjuvants:** *Preparation Methods and Research Protocols,* edited by *Derek T. O'Hagan,* 2000

41. **Celiac Disease:** *Methods and Protocols,* edited by *Michael N. Marsh,* 2000

40. **Diagnostic and Therapeutic Antibodies,** edited by *Andrew J. T. George and Catherine E. Urch,* 2000

39. **Ovarian Cancer:** *Methods and Protocols,* edited by *John M. S. Bartlett,* 2000

38. **Aging Methods and Protocols,** edited by *Yvonne A. Barnett and Christopher R. Barnett,* 2000

METHODS IN MOLECULAR MEDICINE™

Malaria Methods and Protocols

Edited by

Denise L. Doolan

*Malaria Program, Naval Medical Research Center,
Silver Spring, MD;
Department of Molecular Microbiology and Immunology,
School of Hygiene and Public Health, Johns Hopkins University,
Baltimore, MD*

Humana Press ✳ Totowa, New Jersey

© 2002 Humana Press Inc.
999 Riverview Drive, Suite 208
Totowa, New Jersey 07512

www.humanapress.com

All rights reserved. No part of this book may be reproduced, stored in a retrieval system, or transmitted in any form or by any means, electronic, mechanical, photocopying, microfilming, recording, or otherwise without written permission from the Publisher. Methods in Molecular Medicine™ is a trademark of The Humana Press Inc.

The content and opinions expressed in this book are the sole work of the authors and editors, who have warranted due diligence in the creation and issuance of their work. The publisher, editors, and authors are not responsible for errors or omissions or for any consequences arising from the information or opinions presented in this book and make no warranty, express or implied, with respect to its contents.

The opinions and assertions contained herein are the private ones of the authors and are not to be construed as official or reflecting the views of the Department of the Navy, the Department of the Army, the Department of Defense or the U.S. Government.

This publication is printed on acid-free paper. ∞
ANSI Z39.48-1984 (American Standards Institute) Permanence of Paper for Printed Library Materials.

Production Editor: Adrienne Howell

Cover design by Patricia F. Cleary
Cover illustration: Digitized phase-contrast images of parasitized red blood cells adhering to purified CD36 in a microslide under flow in both a typical medium-power (background on cover) field of view and at a higher magnification (inset on cover). *See* full caption and discussion on pp. 566, 567 (Fig. 3 A, B in Chapter 53).

For additional copies, pricing for bulk purchases, and/or information about other Humana titles, contact Humana at the above address or at any of the following numbers: Tel.: 973-256-1699; Fax: 973-256-8341; E-mail: humana@humanapr.com; or visit our Website: http://humanapress.com

Photocopy Authorization Policy:
Authorization to photocopy items for internal or personal use, or the internal or personal use of specific clients, is granted by Humana Press Inc., provided that the base fee of US $10.00 per copy, plus US $00.25 per page, is paid directly to the Copyright Clearance Center at 222 Rosewood Drive, Danvers, MA 01923. For those organizations that have been granted a photocopy license from the CCC, a separate system of payment has been arranged and is acceptable to Humana Press Inc. The fee code for users of the Transactional Reporting Service is: [0-89603-823-8/02 $10.00 + $00.25].

Printed in the United States of America. 10 9 8 7 6 5 4 3 2 1

Library of Congress Cataloging-in-Publication Data

Malaria methods and protocols/ edited by Denise L. Doolan.
  p. cm. -- (Methods in molecular medicine; 72)
  Includes bibliographical references and index.
  ISBN 0-89603-823-8 (alk. paper)
  1. Malaria--Laboratory manuals. I. Doolan, Denise L. II. Series

QR201.M3 M337 2002
616.9'362'0072--dc21

2001051656
CIP

Preface

The *Plasmodium* spp. parasite was identified as the causative agent of malaria in 1880, and the mosquito was identified as the vector in 1897. Despite subsequent efforts focused on the epidemiology, cell biology, immunology, molecular biology, and clinical manifestations of malaria and the *Plasmodium* parasite, there is still no licensed vaccine for the prevention of malaria. Physical barriers (bed nets, window screens) and chemical prevention methods (insecticides and mosquito repellents) intended to interfere with the transmission of the disease are not highly effective, and the profile of resistance of the parasite to chemoprophylactic and chemotherapeutic agents is increasing.

The dawn of the new millennium has seen a resurgence of interest in the disease by government and philanthropic organizations, but we are still faced with complexities of the parasite, the host, and the vector, and the interactions among them. *Malaria Methods and Protocols* offers a comprehensive collection of protocols describing conventional and state-of-the-art techniques for the study of malaria, as well as associated theory and potential problems, written by experts in the field. The major themes reflected here include assessing the risk of infection and severity of disease, laboratory models, diagnosis and typing, molecular biology techniques, immunological techniques, cell biology techniques, and field applications.

The contributors and I hope that this manual will assist researchers at all stages of their careers in carrying out high caliber research to address the difficult problems with which we are faced. I express my deepest gratitude to all the many authors who have contributed to *Malaria Methods and Protocols*, without whose efforts this volume would not have been possible, and to my husband and friends for their support and understanding.

Denise L. Doolan

Contents

Preface v
Contributors xi

PART I. ASSESSING RISK OF INFECTION AND SEVERITY OF DISEASE

1 Vector Incrimination and Entomological Inoculation Rates
  John C. Beier 3

2 Epidemiological Measures of Risk of Malaria
  *J. Kevin Baird, Michael J. Bangs, Jason D. Maguire,
  and Mazie J. Barcus* 13

PART II. LABORATORY MODELS

3 Maintenance of the *Plasmodium berghei* Life Cycle
  Robert E. Sinden, Geoff A. Butcher, and A. L. Beetsma 25

4 Mouse Models for Pre-Erythrocytic–Stage Malaria
  Laurent Rénia, Elodie Belnoue, and Irène Landau 41

5 Mouse Models for Erythrocytic-Stage Malaria
  Latifu A. Sanni, Luis F. Fonseca, and Jean Langhorne 57

6 Nonhuman Primate Models : *I.* Nonhuman Primate Host–Parasite
  Combinations
  William E. Collins 77

7 Nonhuman Primate Models : *II.* Infection of Saimiri *and* Aotus *Monkeys
  with* Plasmodium vivax
  William E. Collins 85

PART III. DIAGNOSIS AND TYPING

8 Vector Analysis
  John C. Beier 95

9 Genotyping of *Plasmodium* spp.: *Nested PCR*
  Georges Snounou 103

10 Genotyping of *Plasmodium falciparum* : *PCR-RFLP Analysis*
  Ingrid Felger and Hans-Peter Beck 117

11 Microsatellite Analysis in *Plasmodium falciparum*
  Xin-zhuan Su and Michael T. Ferdig 131

12 Quantitation of Liver-Stage Parasites
  by Automated TaqMan® Real-Time PCR
  Adam A. Witney, Robert M. Anthony, and Daniel J. Carucci 137

13 Quantitation of Liver-Stage Parasites by Competitive RT-PCR
  Kyle C. McKenna and Marcelo R. S. Briones 141

PART IV. MOLECULAR BIOLOGY TECHNIQUES

14 Extraction and Purification of *Plasmodium* Total RNA
    Till Voss 151

15 Extraction and Purification of *Plasmodium* Parasite DNA
    Hans-Peter Beck 159

16 Southern Blotting of Parasite DNA
    Tobias Spielmann 165

17 SDS-PAGE and Western Blotting of *Plasmodium falciparum* Proteins
    Roland A. Cooper 177

18 Nested PCR Analysis of *Plasmodium* Parasites
    Georges Snounou and Balbir Singh 189

19 RFLP Analysis
    Hans-Peter Beck 205

20 Analysis of Gene Expression by RT-PCR
    Peter Preiser 213

21 *In Situ* Detection of RNA in Blood- and Mosquito-Stage Malaria Parasites
    Joanne Thompson 225

22 Purification of Chromosomes from *Plasmodium falciparum*
    Daniel J. Carucci, Paul Horrocks, and Malcolm J. Gardner 235

23 Construction of Genomic Libraries from the DNA of *Plasmodium* Species
    **Leda M. Cummings, Dharmendar Rathore,
    and Thomas F. McCutchan** 241

24 Construction of a Gene Library with Mung Bean
    Nuclease-Treated Genomic DNA
    Dharmendar Rathore and Thomas F. McCutchan 253

25 Construction of *Plasmodium falciparum* λ cDNA Libraries
    David A. Fidock, Dharmendar Rathore, and Thomas F. McCutchan 265

26 Production of Stage-Specific *Plasmodium falciparum* cDNA Libraries
    Using Subtractive Hybridization
    **David A. Fidock, Thanh V. Nguyen, Brenda T. Beerntsen,
    and Anthony A. James** 277

27 Construction and Screening of YAC Libraries
    Cecilia P. Sanchez, Martin Preuss, and Michael Lanzer 291

28 Episomal Transformation of *Plasmodium berghei*
    Chris J. Janse and Andrew P. Waters 305

29 Gene Targeting in *Plasmodium berghei*
    Vandana Thathy and Robert Ménard 317

PART V. IMMUNOLOGICAL TECHNIQUES

30 Peptide Vaccination
    Valentin Meraldi, Jackeline F. Romero, and Giampietro Corradin 335

31 DNA Vaccination
    Richard C. Hedstrom and Denise L. Doolan 347
32 Assessing Antigen-Specific CD8+ and CD4+ T-Cell Responses in Mice After Immunization with Recombinant Viruses
    Moriya Tsuji 361
33 Assessing CD4+ Helper T-Lymphocyte Responses by Lymphoproliferation
    Isabella A. Quakyi and Jeffrey D. Ahlers 369
34 Limiting Dilution Analysis of Antigen-Specific CD4+ T-Cell Responses in Mice
    Elsa Seixas, Jean Langhorne, and Stuart Quin 385
35 Cell Trafficking: *Malaria Blood-Stage Parasite-Specific CD4+ T Cells After Adoptive Transfer into Mice*
    Chakrit Hirunpetcharat and Michael F. Good 401
36 Assessing Antigen-Specific Proliferation and Cytokine Responses Using Flow Cytometry
    Catherine E. M. Allsopp and Jean Langhorne 409
37 Cytokine Analysis by Intracellular Staining
    Aftab A. Ansari and Ann E. Mayne 423
38 Assessment of Antigen-Specific CTL- and CD8+-Dependent IFN-γ Responses in Mice
    Katrin Peter, Régine Audran, Anilza Bonelo, Giampietro Corradin, and José Alejandro López 437
39 Assessing Antigen-Specific CD8+ CTL Responses in Humans
    Denise L. Doolan 445
40 Human Antibody Subclass ELISA
    Pierre Druilhe and Hasnaa Bouharoun-Tayoun 457
41 Systemic Nitric Oxide Production in Human Malaria: *I. Analysis of NO Metabolites in Biological Fluids*
    Nicholas M. Anstey, Craig S. Boutlis, and Jocelyn R. Saunders 461
42 Systemic Nitric Oxide Production in Human Malaria: *II. Analysis of Mononuclear Cell Nitric Oxide Synthase Type 2 Antigen Expression*
    Jocelyn R. Saunders, Mary A. Misukonis, J. Brice Weinberg, and Nicholas M. Anstey 469

PART VI. CELL BIOLOGY TECHNIQUES

43 In Vitro Culture of *Plasmodium* Parasites
    James B. Jensen 477
44 Automated Synchronization of *Plasmodium falciparum* Parasites by Culture in a Temperature-Cycling Incubator
    J. David Haynes and J. Kathleen Moch 489
45 Hepatic Portal Branch Inoculation
    John B. Sacci, Jr. 499

46	Hepatocyte Perfusion, Isolation, and Culture ***John B. Sacci, Jr.***	*503*
47	Inhibition of Sporozoite Invasion: *The Double-Staining Assay* ***Laurent Rénia, Ana Margarida Vigário, and Elodie Belnoue***	*507*
48	Inhibition of Liver-Stage Development Assay ***John B. Sacci, Jr.***	*517*
49	T-Cell Mediated Inhibition of Liver-Stage Development Assay ***Laurent Rénia, Elodie Belnoue, and Ana Margarida Vigário***	*521*
50	Antibody-Dependent Cellular Inhibition Assay ***Pierre Druilhe and Hasnaa Bouharoun-Tayoun***	*529*
51	Erythrocytic Malaria Growth or Invasion Inhibition Assays with Emphasis on Suspension Culture GIA ***J. David Haynes, J. Kathleen Moch, and Douglas S. Smoot***	*535*
52	Analysis of CSA-Binding Parasites and Antiadhesion Antibodies ***Michal Fried and Patrick E. Duffy***	*555*
53	Analysis of the Adhesive Properties of *Plasmodium falciparum*-Infected Red Blood Cells Under Conditions of Flow ***Brian M. Cooke, Ross L. Coppel, and Gerard B. Nash***	*561*
54	Preparation of Adhesive Targets for Flow-Based Cytoadhesion Assays ***Brian M. Cooke, Ross L. Coppel, and Gerard B. Nash***	*571*
55	Triton X-114 Phase Partitioning for Antigen Characterization ***Lina Wang and Ross L. Coppel***	*581*
56	Immunoprecipitation for Antigen Localization ***John G. T. Menting and Ross L. Coppel***	*587*

PART VII. TESTING INTERVENTIONS FOR MALARIA CONTROL

57	Field Trials ***Pedro L. Alonso and John J. Aponte***	*607*
	Index	*617*

Contributors

JEFFREY D. AHLERS • *Molecular Immunogenetics and Vaccine Research Section, National Cancer Institute, National Institutes of Health, Bethesda, MD*
CATHERINE E. M. ALLSOPP • *Molecular Parasitology, Institute of Molecular Medicine, University of Oxford; John Radcliffe Hospital, Oxford, UK*
PEDRO L. ALONSO • *Epidemiology and Biostatistics Unit, Hospital Clinic, University of Barcelona, Barcelona, Spain; Centro de Investigacao em Saude da Manhica; Hospital Clinic Collaborative Programme, Maputo, Mozambique*
AFTAB A. ANSARI • *Department of Pathology and Laboratory Medicine, The Robert W. Woodruff Health Sciences Center, Emory University School of Medicine, Atlanta, GA*
NICHOLAS M. ANSTEY • *Tropical Medicine and International Health Unit, Menzies School of Health Research; Division of Medicine, Royal Darwin Hospital, Darwin, Australia*
ROBERT M. ANTHONY • *Malaria Program, Naval Medical Research Center, Silver Spring, MD*
JOHN J. APONTE • *Centro de Investigacao em Saude de Manhica, Manhica, Maputo, Mozambique; and Unidad de Epidemiología, Hospital Clinic, Universidad de Barcelona, Barcelona, Spain*
RÉGINE AUDRAN • *Institute of Biochemistry, University of Lausanne, Epalinges, Switzerland*
J. KEVIN BAIRD • *Parasitic Diseases Program, US Naval Medical Research Unit #2, Jakarta, Indonesia*
MICHAEL J. BANGS • *Parasitic Diseases Program, US Naval Medical Research Unit #2, Jakarta, Indonesia*
MAZIE J. BARCUS • *Parasitic Diseases Program, US Naval Medical Research Unit #2, Jakarta, Indonesia*
HANS-PETER BECK • *Swiss Tropical Institute, Basel, Switzerland*
BRENDA T. BEERNTSEN • *Department of Veterinary Pathobiology, University of Missouri, Columbia, MO*
A. L. BEETSMA • *Department of Biology, Imperial College of Science, Technology and Medicine, London, UK*
JOHN C. BEIER • *Department of Tropical Medicine, School of Public Health and Tropical Medicine, Tulane University, New Orleans, LA*
ELODIE BELNOUE • *Département d'Immunologie, Institut Cochin, INSERM/CNRS, Université René Déscartes; Hôpital Cochin, Paris, France*
ANILZA BONELO • *Institute of Biochemistry, University of Lausanne, Epalinges, Switzerland*

HASNAA BOUHAROUN-TAYOUN • *Bio-medical Parasitology Unit, Institut Pasteur, Paris, France*

CRAIG S. BOUTLIS • *Tropical Medicine and International Health Unit, Menzies School of Health Research, Darwin, Australia*

MARCELO R. S. BRIONES • *Disciplina de Microbiologia, Imunologia, e Parasitologia, Escola Paulista de Medicina, Universidade Federal de Sao Paulo, Sao Paulo, Brazil*

GEOFF A. BUTCHER • *Department of Biology, Imperial College of Science Technology and Medicine, London, UK*

DANIEL J. CARUCCI • *Malaria Program, Naval Medical Research Center, Silver Spring, MD*

WILLIAM E. COLLINS • *Division of Parasitic Diseases, National Center for Infectious Diseases, Centers for Disease Control and Prevention, Chamblee, GA*

BRIAN M. COOKE • *Department of Microbiology, Monash University, Clayton, Victoria, Australia*

ROLAND A. COOPER • *Laboratory of Malaria and Vector Research, National Institute of Allergy and Infectious Diseases, National Institutes of Health, Bethesda, MD*

ROSS L. COPPEL • *Department of Microbiology and Victorian Bioinformatics Consortium, Monash University, Clayton, Victoria, Australia*

GIAMPIETRO CORRADIN • *Department of Biochemistry, University of Lausanne, Epalinges, Switzerland*

LEDA M. CUMMINGS • *The Institute for Genome Research, Rockville, MD*

DENISE L. DOOLAN • *Malaria Program, Naval Medical Research Center, Silver Spring, MD; Department of Molecular Microbiology and Immunology, School of Hygiene and Public Health, Johns Hopkins University, Baltimore, MD*

PIERRE DRUILHE • *Bio-medical Parasitology Unit, Institut Pasteur, Paris, France*

PATRICK E. DUFFY • *Seattle Biomedical Research Institute, Seattle, WA; Walter Reed Army Institute of Research, Silver Spring, MD*

INGRID FELGER • *Swiss Tropical Institute, Basel, Switzerland*

MICHAEL T. FERDIG • *Laboratory of Parasitic Diseases, National Institute of Allergy and Infectious Disease, National Institutes of Health, Bethesda, MD*

DAVID A. FIDOCK • *Department of Microbiology and Immunology, Albert Einstein College of Medicine of Yeshiva University, The Bronx, NY*

LUIS F. FONSECA • *Division of Parasitology, National Institute for Medical Research, London, UK*

MICHAL FRIED • *Seattle Biomedical Research Institute, Seattle, WA*

MALCOLM J. GARDNER • *The Institute for Genome Research, Rockville, MD*

MICHAEL F. GOOD • *Queensland Institute of Medical Research, Royal Brisbane Hospital, Brisbane, Queensland, Australia*

J. DAVID HAYNES • *Department of Immunology, Walter Reed Army Institute of Research; Malaria Program, Naval Medical Research Center, Silver Spring, MD*

RICHARD C. HEDSTROM • *Regulatory Affairs Section, DynPort Vaccine Company, Frederick, MD*

CHAKRIT HIRUNPETCHARAT • *Department of Microbiology, Faculty of Public Health, Mahidol University, Bangkok, Thailand*

PAUL HORROCKS • *Institute of Molecular Medicine, University of Oxford; John Radcliffe Hospital, Oxford, UK*

ANTHONY A. JAMES • *Department of Molecular Biology and Biochemistry, University of California, Irvine, CA*

CHRIS J. JANSE • *Department of Parasitology, Leiden University Medical Centre, Leiden, The Netherlands*

JAMES B. JENSEN • *Brigham Young University, Provo, UT*

IRÈNE LANDAU • *Laboratoire de Biologie Parasitaire, Musée National d'Histoire Naturelle, Paris, France*

JEAN LANGHORNE • *Division of Parasitology, National Institute for Medical Research, London, UK*

MICHAEL LANZER • *Abteilung Parasitologie, Hygiene Institut, Universität Heidelberg, Heidelberg, Germany*

JOSÉ ALEJANDRO LÓPEZ • *Institute of Biochemistry, University of Lausanne, Epalinges, Switzerland*

JASON D. MAGUIRE • *Department of Medical Parasitology, US Naval Medical Research Unit #2, Jakarta, Indonesia*

ANN E. MAYNE • *Department of Pathology and Laboratory Medicine, The Robert W. Woodruff Health Sciences Center, Emory University School of Medicine, Atlanta, GA*

THOMAS F. MCCUTCHAN • *Growth and Development Section, Laboratory of Malaria and Vector Research, National Institute of Allergy and Infectious Diseases, National Institutes of Health, Bethesda, MD*

KYLE C. MCKENNA • *Department of Ophthalmology, Emory University, Atlanta, GA*

ROBERT MÉNARD • *Unité de Biologie et Génétique du Paludisme, Départment de Parasitologie, Institut Pasteur, Paris, France*

JOHN G. T. MENTING • *Department of Microbiology, Monash University, Clayton, Victoria, Australia*

VALENTIN MERALDI • *Institute of Biochemistry, University of Lausanne, Epalinges, Switzerland*

MARY A. MISUKONIS • *Tropical Medicine and International Health Unit, Menzies School of Health Research, Darwin, Australia*

J. KATHLEEN MOCH • *Malaria Program, Naval Medical Research Center; Department of Immunology, Walter Reed Army Institute of Research, Silver Spring, MD*

GERARD B. NASH • *Department of Physiology, University of Birmingham Medical School, Birmingham, UK*

THANH V. NGUYEN • *Department of Molecular Biology and Biochemistry, University of California, Irvine, CA*

KATRIN PETER • *Institute of Biochemistry, University of Lausanne, Epalinges, Switzerland*

PETER R. PREISER • *Division of Parasitology, National Institute for Medical Research, London, UK*
MARTIN PREUSS • *Abteilung Parasitologie, Hygiene Institut, Universität Heidelberg, Heidelberg, Germany*
ISABELLA A. QUAKYI • *Department of Biology, Georgetown University, Washington, DC*
STUART QUIN • *Division of Parasitology, National Institute for Medical Research, London, UK*
DHARMENDAR RATHORE • *Growth and Development Section, Laboratory of Malaria and Vector Research, National Institute of Allergy and Infectious Diseases, National Institutes of Health, Bethesda, MD*
LAURENT RÉNIA • *Département d'Immunologie, Institut Cochin, INSERM/CNRS, Université René Déscartes; Hôpital Cochin, Paris, France*
JACKELINE F. ROMERO • *Institute of Biochemistry, University of Lausanne, Epalinges, Switzerland*
JOHN B. SACCI, JR. • *Department of Microbiology and Immunology, University of Maryland School of Medicine, Baltimore, MD*
CECILIA P. SANCHEZ • *Abteilung Parasitologie, Hygiene Institut, Universität Heidelberg, Heidelberg, Germany*
LATIFU A. SANNI • *Division of Parasitology, National Institute for Medical Research, London, UK*
JOCELYN R. SAUNDERS • *Tropical Medicine and International Health Unit, Menzies School of Health Research, Darwin, Australia*
ELSA SEIXAS • *Department of Biology, Imperial College of Science, Technology and Medicine, London, UK*
ROBERT E. SINDEN • *Department of Biology, Imperial College of Science, Technology and Medicine, London, UK*
BALBIR SINGH • *Faculty of Medicine and Health Sciences, Universiti Malaysia Sarawak (UNIMAS), Sarawak, Malaysia*
DOUGLAS S. SMOOT • *Combat Casualty Care Department, Naval Medical Research Center, Bethesda, MD*
GEORGES SNOUNOU • *Bio-medical Parasitology, Institut Pasteur, Paris, France*
TOBIAS SPIELMANN • *Swiss Tropical Institute, Basel, Switzerland*
XIN-ZHUAN SU • *Laboratory of Parasitic Diseases, National Institute of Allergy and Infectious Diseases, National Institutes of Health, Bethesda, MD*
VANDANA THATHY • *Departments of Medicine and of Microbiology and Immunology, Albert Einstein College of Medicine, Bronx, NY*
JOANNE THOMPSON • *Ashworth Laboratories, Institute of Cell, Animal, and Population Biology, University of Edinburgh, Edinburgh, UK*
MORIYA TSUJI • *Department of Medical and Molecular Parasitology, New York University School of Medicine, New York, NY*
ANA MARGARIDA VIGÁRIO • *Département d'Immunologie, Institut Cochin, INSERM/CNRS, Université René Déscartes; Hôpital Cochin, Paris, France*
TILL VOSS • *Swiss Tropical Institute, Basel, Switzerland*

Contributors

LINA WANG • *Department of Microbiology, Monash University, Clayton, Victoria, Australia*

ANDREW P. WATERS • *Department of Parasitology, Leiden University Medical Centre, Leiden, The Netherlands*

J. BRICE WEINBERG • *Division of Hematology, Department of Medicine, Duke University, Durham, NC*

ADAM A. WITNEY • *Malaria Program, Naval Medical Research Center, Silver Spring, MD; Henry M. Jackson Foundation, Rockville, MD*

I

Assessing Risk of Infection and Severity of Disease

1

Vector Incrimination and Entomological Inoculation Rates

John C. Beier

1. Introduction

This chapter provides standard methods for the incrimination of *Anopheles* mosquito species serving as malaria vectors and associated methods for measuring the intensity of transmission. In any malaria-endemic area, one or more species of *Anopheles* mosquitoes serve as malaria vectors. To show that an *Anopheles* mosquito species serves as a malaria vector in nature, it is necessary to demonstrate:

1. An association in time and space between the Anopheles species of mosquito and cases of malaria in humans. After study sites are selected, longitudinal field studies are established to sample mosquito populations. Adult mosquitoes are sampled by using trapping techniques such as landing/biting collections, light traps, pyrethrum spray catches inside houses, and outdoor aspiration collections. Larval mosquitoes developing in aquatic habitats normally are sampled by dipping methods. Mosquitoes are identified by standard taxonomic methods and also by molecular methods if mosquitoes belong to a species complex. The standard methods for performing landing/biting collections are described in this chapter; other types of mosquito trapping methods are described in **refs.** *1* and *2*.
2. Evidence of direct contact between the Anopheles species and humans. Catching a mosquito biting humans through landing/biting catches conclusively establishes contact between that mosquito species and humans. A second method involves immunologically identifying human blood in the abdomen of field-captured *Anopheles* mosquitoes. A direct enzyme-linked immunosorbent assay (ELISA) suitable for bloodmeal identification of African malaria vectors is described in this chapter *(3)*.
3. Evidence that the Anopheles species harbors malaria sporozoites in the salivary glands. Sporozoites may be detected in mosquitoes through the dissection and microscopic examination of mosquito salivary glands *(4)* or through ELISA methods *(5)*. Both methods are described in this chapter.

The process of vector incrimination is often done in conjunction with longitudinal field studies to measure the intensity of transmission. In an endemic area, the intensity of transmission is determined by calculating the entomological inoculation rate (EIR), which is the product of the mosquito biting rate times the proportion of mosquitoes with sporozoites. EIRs, calculated as the sum total for each individual vector species of mosquito, are expressed in terms of average numbers of infective bites per person per unit time. For example, EIRs in endemic areas of Africa generally range from 1 to >1000 infective bites per year *(6)*. In this chapter, methods are described for calculating EIRs from landing/biting catches and determinations of sporozoite rates. Further details and references are provided on the use of EIRs for epidemiological studies and for determining levels of control necessary for achieving reductions in malaria prevalence and the incidence of severe disease.

2. Materials
2.1. Equipment

1. Mouth aspirators for collecting mosquitoes.
2. Flashlights.
3. Hand-held global position satellite (GPS) system receiver.
4. Low-intensity kerosene lantern.
5. Screened paper pint cups for holding live mosquitoes.
6. Labels and/or permanent marker pens.
7. Glass microscope slides.
8. Phase-contrast compound microscope and dissecting microscope with light source.
9. Surgical scalpel blades.
10. Glass rods or plastic pestles for grinding mosquitoes.
11. 1.8-mL plastic tubes with snap-on caps for holding mosquito samples.
12. 15- and 50-mL tubes for mixing ELISA reagents.
13. Freezer for storing mosquito samples at –20 or –70°C.
14. Refrigerator.
15. 8-Channel manifold attached to 60-mL plastic syringe.
16. Polyvinyl chloride (PVC) microtiter plates.
17. Absorbent tissue paper.
18. ELISA plate reader.

2.2. Reagents

1. Chloroform, ether, and/or 70% ethanol for killing mosquitoes.
2. Physiological saline or medium-199 (M-199) for dissecting mosquitoes.
3. Phosphate-buffered saline (PBS).
4. PBS–Tween-20 (PBS–Tw20) wash solution for ELISA: Add 500 µL of Tween-20 to 1 L of PBS, mix, and store in a refrigerator.
5. Boiled casein blocking buffer (BB) for ELISA: Suspend 5.0 g casein in 100 mL of 0.1 N sodium hydroxide and bring to a boil while stirring on a hot plate. After casein has dissolved, slowly add 900 mL of PBS, allow to cool, and adjust the pH to 7.4 with hydrochloric acid (HCl). Add 0.1 g thimerosal and 0.02 g phenol red. Mix well using a magnetic stirrer and store in a refrigerator; shelf life is 7 to 10 d.
6. Blocking buffer Nonidet P-40 (BB–NP-40) for grinding mosquitoes: prepare by adding 5 µL of NP-40 to each 100 µL BB and mixing. Make fresh daily.
7. Capture and conjugated monoclonal antibodies for sporozoite ELISA tests.
8. Peroxidase substrate (ABTS) and phosphatase substrate.
9. Recombinant proteins as positive controls for sporozoite ELISA tests.
10. Host-specific peroxidase conjugates (anti-host IgG, H&L) and phosphatase-labeled anti-bovine IgG (H&L) for bloodmeal ELISA.
11. Host sera as controls for the bloodmeal ELISA.

3. Methods
3.1. Site Selection and Mosquito Sampling Stations

1. Study sites are selected based on study objectives that may be related to epidemiological studies, malaria control operations, vaccine or drug testing under natural conditions, or an

abundance of vector species of mosquitoes. Study sites may range in size from a cluster of a few houses to whole communities. Prior to selecting and working in sites, it is advisable to discuss study objectives and operations with community leaders and residents and to obtain their consent for the field studies (*see* **Note 1**).
2. Normally, it is necessary to develop study site maps based on either traditional mapping methods or through geographic information systems (GIS) using GPS receivers for determining latitudes and longitudes of houses and other landmark features within sites. The maps are used to facilitate field studies logistically and serve as a foundation for the analysis of spatial data on mosquito populations and transmission.
3. Sampling stations for mosquito trapping are selected within sites. For highly endophilic and anthropophilic mosquitoes like the African malaria vectors, sampling stations are normally houses or homesteads comprising family units of houses. For exophilic or zoophilic *Anopheles* species, sampling stations can be either outdoor areas or animal sheds. The number of sampling stations depends upon logistical capabilities such as the number of mosquito collectors available and the expected frequency of sampling within the study sites. Sampling stations are normally fixed and used repeatedly throughout the duration of field studies. Alternatively, sampling stations can be selected randomly during each sampling period. While this is sometimes useful from a statistical perspective, we have found that this approach makes it more difficult to obtain good cooperation from communities.

3.2. Landing/Biting Catches of Anopheles Mosquitoes

Landing/biting catches of *Anopheles* mosquitoes on human volunteers (*see* **Note 2**) are performed either by individual human collectors, by pairs of collectors, or by up to four collectors working simultaneously. Collections are normally performed at night, during the biting cycles of the *Anopheles* mosquitoes. Each collector is responsible for catching, by mouth aspirator with the aid of a flashlight, mosquitoes that are attracted to and in the process of biting humans (i.e., host-seeking mosquitoes). The trapping technique simulates the natural situation whereby mosquitoes contact and bite humans, and so it is regarded as the gold standard. For each malaria vector field study, it is necessary to evaluate all other trapping methods for evaluating host contact against the gold standard. Procedurally, landing/biting catches are performed as follows:

1. Collectors with their arms and legs exposed are seated in chairs or on mats on the ground at sampling stations. It is common to perform indoor and outdoor biting catches simultaneously at the same sampling stations, with the outdoor collectors positioned at least 5 m from surrounding houses. Trapping inside houses provides information on the numbers of mosquitoes biting inside, while performing the sampling outdoors provides comparable information on outdoor biting rates.
2. Collectors catch landing/biting mosquitoes from themselves and from their partners with a hand-held mouth aspirator (or mechanical aspirator). Each collector uses a flashlight to locate landing/biting mosquitoes. Additional background light from a low-intensity lantern is advisable.
3. Each aspirated mosquito is placed in screened pint cups, labeled according to sampling station. Some studies also segregate mosquito collections by hour of capture, and this requires additional cups labeled by hour.
4. Landing/biting collections are normally performed throughout the night as dictated by the natural biting habits of the target mosquitoes. Logistically, it is feasible for individuals or teams of collectors to work one-half hour every hour throughout the night. Alternatively, it is feasible for half the team to work continuously during the first half of the night and the rest of the team to work the second half of the night.

5. After collections, mosquitoes in cups are normally killed either by freezing or by exposure to chloroform or ether. For immediate processing, mosquitoes may be aspirated out of cups and blown into 70% ethanol followed by transfer to PBS or M-199. Mosquitoes may also be stored in Carnoy's solution for cytogenetic studies or for longer-term storage before processing (*see* Chapter 8).
6. Mosquitoes are identified according to taxonomic methods or by molecular techniques (*see* Chapter 8).
7. The biting rate for each mosquito species is calculated as the number of mosquitoes per person per unit of sampling effort. For example, a biting rate of 2 per day is derived from one collector who catches one mosquito while working throughout the night in half-hour shifts. A biting rate of 40 per day is derived from a team of two collectors working in half-hour shifts during the whole night and catching 40 mosquitoes.

3.3. Determination of Sporozoite Rates in Anopheles Mosquitoes

3.3.1. Dissection and Microscopic Examination of Mosquito Salivary Glands

1. Salivary gland dissections are performed on mosquitoes freshly killed by freezing, exposure to ether, or by blowing into 70% ethanol.
2. Place an individual mosquito on a glass slide, with head in contact with a small drop of physiological saline or PBS or M-199.
3. View slide containing mosquito on a dissecting microscope at ×10 to ×30.
4. Hold two dissecting needles, which can be made conveniently by placing the 27-gage needle from a 1-mL tuberculin syringe on the end of the movable shaft (rubber stopper removed), between your thumb and forefinger. Place one needle (bevel down) on the thorax of the mosquito while placing the other needle against the mosquito head (bevel facing toward the head). Simultaneously place pressure on the thorax while pulling the head away from the thorax. As the head moves away from the thorax, observe the salivary glands and cut them with the needle controlling the head. The cut should be made in one continuous motion as soon as the glands are seen; otherwise, it is necessary to reposition the head and try again. Sometimes the glands become stuck in the mosquito thorax, and it is necessary to tease apart the tissue to locate and cut the glands.
5. After severing the salivary glands, remove the head and thorax and any other extraneous tissue.
6. Place a glass cover slip over the salivary glands, now lying in the dissection media.
7. Transfer the slide to a compound microscope and observe the preparation at ×100 to locate the salivary glands.
8. Apply gentle pressure to the cover slip to disrupt the glands and then search at ×400 the entire area of the salivary glands for sporozoites (which normally measure about 1×10 μm). Experienced dissectors can typically dissect a mosquito within 1 min and reliably examine the preparation within 2 min.
9. Sporozoite infections are normally scored according to the number of sporozoites observed: 1+ (1–10 sporozoites), 2+ (11–100 sporozoites), 3+ (101–1000), and 4+ (>1000 sporozoites).
10. Record results. Normally, each field-collected mosquito is given a unique identifier (*see* **Note 3**).
11. Standard procedures are also available for removing sporozoite material from slides and testing the sporozoites by ELISA to determine *Plasmodium* species *(7)*. Various additional procedures are available for determining sporozoite loads, the number of sporozoites found in the salivary glands of individual mosquitoes *(8,9)*.

3.3.2. Sporozoite ELISA Methods

ELISA methods exist for testing field-collected mosquitoes for sporozoites representing each of the four species of *Plasmodium* affecting humans *(5,10,11)*. The sporozoite ELISA detects circumsporozoite protein that is either from intact sporozoites or in soluble form within the mosquito. Based on comparisons with the gold standard dissection method, the sporozoite ELISA provides a reasonable estimate of the true sporozoite rate in wild-caught mosquitoes *(12)*.

1. Prepare the mosquito sample for ELISA testing. Label sets of 1.8-mL tubes with the corresponding mosquito sample numbers. Add 50 µL of BB–NP-40 to each vial. Using a sharp clean surgical blade, cut the mosquito between the thorax and the abdomen (normally done on a filter paper). Transfer the head–thorax with forceps to the labeled tube, and transfer the abdomen to the corresponding tube for bloodmeal identification if the mosquito is blood-fed. If the mosquito is not blood-fed or no bloodmeal analysis is required, discard abdomen. Grind the mosquito in the tube using a nonabsorbent glass rod or plastic pestle. Add 200 µL of the BB to bring the total sample volume to 250 µL. To avoid contamination, clean the pestle and wipe it dry before grinding the next sample. Repeat the procedure until all samples are prepared. Arrange samples in numbered order within storage boxes and keep samples in a freezer at –20 or –70°C until testing.
2. Coat number-coded ELISA plates with monoclonal antibody (MAb). In each well, add 50 µL of the diluted capture MAb. Cover the plates with another clean ELISA plate and incubate for 30 min at room temperature in subdued light.
3. Block the plates. Using an 8-channel manifold attached to a vacuum pump, aspirate the capture MAb from the microtiter plate. Bang the plate hard on an absorbent tissue paper or gauze to ensure complete dryness. Fill each well with BB using a manifold attached to a 60-mL syringe. Incubate for 1 h at room temperature in subdued light.
4. Load the plates with mosquito samples. Aspirate the blocking buffer from the wells using the manifold attached to a vacuum pump and bang plate to complete dryness. Place 50 µL of 100, 50, 25, 12, 6, 3, 1.5, 0 pg of positive control recombinant protein in the first column wells. Into the second column, add 50 µL per well of the negative controls; normally, field-collected male *Anopheles* mosquitoes or culicine mosquitoes are used as negative controls. Load 50 µL of each mosquito sample to the remaining wells of the plate, checking carefully that numbered mosquito samples are placed in the wells according to the completed ELISA data form. Cover the plate and incubate for 2 h at room temperature in subdued light.
5. Add peroxidase-conjugated monoclonal antibody. After 2 h, aspirate the triturate from the wells and wash the plate two times with PBS-Tw20. Add 50 µL of the peroxidase-labeled enzyme and incubate for 1 h at room temperature.
6. Add the substrate. Aspirate the enzyme conjugate from the wells and wash three times with PBS-Twn 20. Using a multichannel pipet, add 100 µL of ABTS substrate and incubate for 30 min. Positive reactions, which appear green, can be determined by reading plates at 414 nm using an ELISA plate reader; absorbance values two times the mean of negative controls provides a valid cutoff for sample positivity *(13)*. Alternatively, results can be read visually with a high degree of accuracy *(14)*. Record results for each tested mosquito.

3.4. Bloodmeal ELISA Methods

Both direct and indirect ELISA procedures are routinely used to identify bloodmeals of wild-caught mosquitoes. Strategies for using ELISAs for bloodmeals depend upon study objectives. For example, Edrissian et al. *(15)* used a direct ELISA to screen over 5000 *Anopheles* for human blood; they reported that an experienced technician could easily screen over 1000 samples per week. Burkot and DeFoliart *(16)* used an indirect ELISA to identify 16 host sources, including wild animals. Their studies involved producing antisera for each host tested. To bypass extensive production of antisera and to shorten the overall testing time, we developed a simple direct ELISA that uses only commercially available reagents *(3)*. The test reliably detects bloodmeals from humans and from a spectrum of domestic animals for which conjugated antisera are commercially available. The test employs a two-step screening for human and cow bloodmeals, making it particularly useful in Africa where major malaria vectors feed primarily on humans and cows. The assay is performed as follows:

1. Prepare wild-caught half-gravid to freshly fed mosquitoes by cutting them transversely at the thorax between the first and third pairs of legs (under a dissecting microscope, ×10–20). In a labeled tube, place the posterior part of the mosquito containing the bloodmeal in 50 µL PBS and grind with a pestle or pipet repeatedly. Dilute sample 1:50 with PBS and freeze samples at –20°C until testing.
2. Load 96-well polyvinyl microtiter plates with mosquito bloodmeal samples by adding 50 µL of each sample per well. On the same plate, add 50-µL samples of positive control antisera for human and cow (diluted 1:500 in PBS), and four or more negative control unfed female mosquitoes or male mosquitoes obtained from the same field collections and handled as above. Cover and incubate at room temperature for 3 h (or overnight).
3. Wash each well twice with PBS-Tw20.
4. Add 50 µL of host-specific conjugate (anti-host IgG, H&L) diluted 1:2,000 (or as determined in control tests) in 0.5% BB containing 0.025% Tween-20, and incubate 1 h at room temperature.
5. Wash wells three times with PBS–Tw-20.
6. Add 100 µL of ABTS peroxidase substrate to each well.
7. After 30 min, read each well with an ELISA reader. Samples are considered positive if absorbance values exceed the mean plus three standard deviations of four negative control, unfed female, or male mosquitoes. The dark green positive reactions for peroxidase (or the dark yellow reactions for phosphatase) may also be determined visually *(14)*.
8. The following modification is used in the two-step procedure for determining a second host source in the same microtiter plate well where mosquito samples were screened for human blood *(3)*. A second conjugate, phosphatase-labeled anti-bovine IgG (1:250 dilution of a 0.5 mg/mL stock solution) is added to the peroxidase-labeled anti-human IgG solution (**step 4**). Screen bloodmeals first for human IgG by adding peroxidase substrate, and after reading absorbance at 30 min, wash the wells three times with PBS–Tw20. Add 100 µL of phosphatase substrate and read plates after 1 h to determine positive cow reactions.
9. Each laboratory should initially establish the sensitivity and specificity of the assay for each conjugated antisera and different lots of reagents. The assay can detect sera diluted to around 1:10,000,000. The degree to which commercially available antisera crossreact with sera from different hosts varies according to manufacturers. For standardization and to reduce levels of nonspecific reactivity, it is sometimes necessary to add 1:500 dilutions of heterologous sera to the conjugate solutions (*see* **step 4**). It is important to note that the assay works equally well with frozen, dried, or Carnoy's fixed mosquito samples, and that each 1-mg vial of conjugated antisera can be aliquoted, frozen, and used to test up to 20,000 mosquito samples.

3.5. Entomological Inoculation Rates (see Note 4)

3.5.1. Calculation of EIRs

The EIR is calculated as the product of the mosquito biting rate and the sporozoite rate. **Table 1** provides an example for the calculation of the EIR from site-specific data on mosquito biting rates and sporozoite rates. Beyond calculating daily EIRs, it is also

Table 1
Entomological Inoculation Rate

Species	Biting rate[a]	Sporozoite rate (%)[b]	Daily EIR
A	10	5.00	0.50
B	4	2.00	0.08
C	2	1.00	0.02
Total	16	3.75	0.60

The EIR is calculated as the product of the biting rate times the sporozoite rate. In this hypothetical example, there are three vector species of *Anopheles* mosquitoes. The daily EIR of 0.60 infective bites per person per night is calculated as the sum total of EIRs for each of the three species.
[a]Number of mosquitoes per person per night.
[b]Percentage of mosquitoes with salivary gland sporozoites by dissection or ELISA.

useful to calculate monthly or annual EIRs based on averaged values of biting rates and sporozoite rates. For the example given in **Table 1**, if these were the only sample data available, then an estimate of the monthly EIR could be obtained by multiplying the daily EIR of 0.60 by 30 d to yield an estimated monthly EIR of 18. For field studies, it is advisable to have two or more point determinations of biting rates per month and statistically reliable estimates of sporozoite rates.

3.5.2. Relationships Between EIRs and Measures of Human Malaria

Time series data on EIRs for given sites can be related directly to measures of human malaria in several simple ways. Graphically, measures of the EIR and human malaria such as prevalence or incidence can be graphed along the y-axis and related to sampling time points on the x-axis. In addition, such temporal changes in EIR and infection or disease can be related to environmental parameters such as temperature and rainfall. The same data can be graphed with EIRs on the x-axis and human malaria data on the y-axis. This approach, when combined with regression analysis, provides both a graphical and a statistical account of the variation in human infection or disease explained by EIRs.

Several studies in Africa provide good examples of how EIRs can be related to the following:

1. Incidence of *P. falciparum* infection in children *(17)*.
2. Incidence of severe life-threatening cases of *P. falciparum* in children *(18,19)*.
3. Prevalence of *P. falciparum* *(6)*.

It is important to note that malaria prevalence data, which is often used as the basis for guiding control operations, is not a sensitive indicator of the intensity of malaria transmission by vector populations. Malaria prevalence rates from 40 to >90% can occur at any EIR exceeding one infective bite per person per year. Control operations therefore need to be guided by both entomological data on EIRs and traditional measures of human malaria infection and disease.

4. Notes

1. Ethical concerns must be addressed for each malaria vector field study. Mosquito trapping in malaria endemic areas normally involves local mosquito collectors who are recruited from study communities. For landing/biting mosquito sampling methods, local mosquito

collectors normally do not face any excess risks for malaria infection beyond what they would normally experience sleeping in their own homes. Some studies have traditionally offered malaria prophylactic drugs to collectors. However, in highly endemic areas, this practice goes beyond normal protective measures of the community and may even be detrimental to the long-term health of the collectors. It is advisable that malaria field studies make sure that mosquito collectors have proper access to curative antimalarial drugs and health facilities whenever they develop symptomatic malaria infections.

2. The landing/biting catch method is the gold standard for determining human contact with mosquitoes. Other methods such as pyrethrum spray catches or CDC light traps may be used to estimate rates of human biting for each *Anopheles* species. However, for each study area, it is necessary to establish quantitative relations between the sampling method proposed and the gold standard method. Estimates of correction factors from regression analysis based on data from comparative sampling for one site do not generally hold universally throughout the ranges of mosquitoes *(20)*.

3. Care must be taken with data management. Malaria vector studies normally yield large numbers of mosquitoes from different trapping methods that are processed by a variety of different methods ranging from taxonomic identifications to sporozoite ELISA testing. Primary data sets can be established in matrix format with each mosquito represented by rows and columns represented by collection and processing data. Prospects of having >15 variables of data for each individual mosquito and over 50,000 mosquitoes in a single dataset demand careful attention in terms of data entry, data management, and analysis.

4. The EIR is a direct measure of the intensity of malaria transmission by vector populations. In malaria endemic areas outside Africa and Papua New Guinea, annual EIRs may be lower than one infective bite per person per year *(21)*. The low EIRs are often due to sporozoite rates substantially less than 1%. In some areas of Central and South America, for example, it is not uncommon to find fewer than 1 in 1000 mosquitoes infected with malaria parasites. Under such conditions, investigators may alternatively calculate the vectorial capacity (VC), an indirect measure of the potential for transmission that considers the biting rate of the vector population, the human blood-feeding rate, the vector survival rate, and the extrinsic incubation period of the malaria parasite *(22)*. Some of the practical limitations and errors associated with the use of the VC are discussed by Dye *(23)*.

Acknowledgments

This work was supported by the National Institutes of Health grants AI29000, AI45511, and TW01142.

References

1. Service, M. W. (1976) *Mosquito Ecology*. Wiley, New York.
2. Service, M. (1993) *Mosquito Ecology: Field Sampling Methods*. Elsevier Applied Science, New York.
3. Beier, J. C., Perkins, P. V., Wirtz, R. A., Koros, J., Diggs, D., Gargan, T. P., and Koech, D. K. (1988) Bloodmeal identification by direct enzyme-linked immunosorbent assay (ELISA), tested on *Anopheles* (Diptera: Culicidae) in Kenya. *J. Med. Entomol.* **25,** 9–16.
4. WHO (1975) *Manual on Practical Entomology in Malaria. Part II. Methods and Techniques.* WHO Offset Publication 13, Geneva.
5. Wirtz, R. A. and Burkot, T. R. (1991) Detection of malaria parasites in mosquitoes. Chap 4 in *Advances in Disease Vector Research,* vol. 8, Springer-Verlag, New York, pp. 77–106.
6. Beier, J. C., Killeen, G. F., and Githure, J. I. (1999) Short report: Entomological inoculation rates and *Plasmodium falciparum* malaria prevalence in Africa. *Am. J. Trop. Med. Hyg.* **61,** 109–113.
7. Beier, J. C., Copeland, R. S., Onyango, F. K., Asiago, C. M., Ramadhan, M., Koech, D. K., and Roberts, C. R. (1991) *Plasmodium* species identification by ELISA for sporozoites removed from dried dissection slides. *J. Med. Entomol.* **28,** 533–536.
8. Kabiru, E. W., Mbogo, C. M., Muiruri, S. K., Ouma, J. H., Githure, J. I., and Beier, J. C. (1997)

Sporozoite loads of naturally infected *Anopheles* in Kilifi District, Kenya. *J. Am. Mosq. Control Assoc.* **13,** 259–262.
9. Beier, J. C., Onyango, F. K., Ramadhan, M., Koros, J. K., Asiago, C. M., Wirtz, R. A., et al. (1991) Quantitation of malaria sporozoites in the salivary glands of wild Afrotropical *Anopheles. Med. Vet. Entomol.* **5,** 63–70.
10. Wirtz, R. A., Burkot, T., Graves, P. M., and Andre, R. G. (1987) Field evaluation of enzyme-linked immunosorbent assays for *Plasmodium falciparum* and *Plasmodium vivax* sporozoites in mosquitoes (Diptera: Culicidae) from Papua New Guinea. *J. Med. Entomol.* **24,** 433–437.
11. Beier, J. C., Perkins, P. V., Wirtz, R. A., Whitmire, R. E., Mugambi, M., and Hockmeyer, W. T. (1987) Field evaluation of an enzyme-linked immunosorbent assay (ELISA) for *Plasmodium falciparum* sporozoite detection in anopheline mosquitoes from Kenya. *Am. J. Trop. Med. Hyg.* **36,** 459–468.
12. Beier, J. C., Perkins, P. V., Koros, J. K., Onyango, F. K., Gargan, T. P., Wirtz, R. A., et al. (1990) Malaria sporozoite detection by dissection and ELISA to assess infectivity of Afrotropical *Anopheles* (Diptera: Culicidae). *J. Med. Entomol.* **27,** 377–384.
13. Beier, J. C., Asiago, C. M., Onyango, F. K., and Koros, J. K. (1988) ELISA absorbance cut-off method affects malaria sporozoite rate determination in wild Afrotropical *Anopheles. Med. Vet. Entomol.* **2,** 259–264.
14. Beier, J. C. and Koros, J. K. (1991) Visual assessment of sporozoite and bloodmeal ELISA samples in malaria field studies. *J. Med. Entomol.* **28,** 805–808.
15. Edrissian, G. H., Manouchehry, A. V., and Hafizi, A. (1985) Application of an enzyme-linked immunosorbent assay (ELISA) for determination of the human blood index in anopheline mosquitoes collected in Iran. *J. Am. Mosq. Control Assoc.* **1,** 349–352.
16. Burkot, T. R. and DeFoliart, G. R. (1982) Bloodmeal sources of *Aedes triseriatus* and *Aedes vexans* in a southern Wisconsin forest endemic for LaCrosse encephalitis virus. *Am. J. Trop. Med. Hyg.* **31,** 376–381.
17. Beier, J. C., Oster, C. N., Onyango, F. K., Bales, J. D., Sherwood, J. A., Perkins, P. V., et al. (1994) *Plasmodium falciparum* incidence relative to entomologic inoculation rates at a site proposed for testing malaria vaccines in western Kenya. *Am. J. Trop. Med. Hyg.* **50,** 529–536.
18. Mbogo, C. N., Snow, R. W., Kabiru, E. W., Ouma, J. H., Githure, J. I., Marsh, K., and Beier, J. C. (1993) Low-level *Plasmodium falciparum* transmission and the incidence of severe malaria infections on the Kenyan coast. *Am. J. Trop. Med. Hyg.* **49,** 245–253.
19. Mbogo, C. N., Snow, R. W., Khamala, C. P., Kabiru, E. W., Ouma, J. H., Githure, J. I., et al. (1995) Relationships between *Plasmodium falciparum* transmission by vector populations and the incidence of severe disease at nine sites on the Kenyan coast. *Am. J. Trop. Med. Hyg.* **52,** 201–206.
20. Mbogo, C. N., Glass, G. E., Forster, D., Kabiru, E. W., Githure, J. I., Ouma, J. H., and Beier, J. C. (1993) Evaluation of light traps for sampling anopheline mosquitoes in Kilifi, Kenya. *J. Am. Mosq. Control Assoc.* **9,** 260–263.
21. Alles, H. K., Mendis, K. N., and Carter, R. (1998) Malaria mortality rates in South Asia and in Africa: implications for malaria control. *Parasitol. Today* **14,** 369-375.
22. Garrett-Jones, C. and Shidrawi, G. R. (1969) Malaria vectorial capacity of a population of *Anopheles gambiae. Bull. World Health Organ.* **40,** 531–545.
23. Dye, C. (1986) Vectorial capacity: must we measure all its components? *Parasitol. Today* **2,** 203–209.

2

Epidemiological Measures of Risk of Malaria

J. Kevin Baird, Michael J. Bangs, Jason D. Maguire, and Mazie J. Barcus

1. Introduction

Estimates of the risk of infection by the parasites that cause malaria govern decisions regarding vector control, chemoprophylaxis, therapeutic management, and clinical classifications of immunological susceptibility to infection. Gauging the risk of malaria represents a critical step in its management and the investigation of its consequences. The term malariometry is applied to the numerical measure of risk of malaria in communities *(9)*. Many approaches have been developed and applied to malariometry, but no single method stands out as universally applicable. Instead, individual measures of risk must be suitable for specific questions posed in the context of what may be practically measured. For example, passive surveillance provides a superior measure of risk where the infrastructure of diagnosis and reporting is well developed and the risk of infection relatively low, e.g., in the United States, where conducting active cross-sectional surveys would yield little useful information at great cost. Active surveillance for cases is suited to areas with relatively high risk, unreliable diagnostic capabilities and inadequate reporting infrastructure. This chapter strives to catalog measures of risk of malaria and define their utility in the context of local parameters of endemicity, infrastructure, and intent of inquiry.

The risk of malaria is highly dependent on interactions between the host, parasite, mosquito vector, and environment, a relationship known as the epidemiologic triad of disease. Changes in any one of these elements may profoundly impact risk of infection. Measures of risk of malaria may be broadly classified as either indirect or direct. Indirect measures gauge risk through surrogate markers of risk of infection such as rainfall, altitude, temperature, entomological parameters, spleen rates, antibody titers, or patterns of antimalarial drug use in a community. Direct measures of risk depend on diagnoses of malaria (clinical or microscopic) and their relationship to a variety of denominators representing classes of persons at risk over some unit of time. In general, indirect measures apply data conveniently at hand to estimate risk of malaria. By contrast, direct estimates of risk often require deliberate effort to collect data for the sole purpose of gauging risk of malaria.

An area supporting active malaria transmission is termed endemic. Transmission of infection may be unstable or stable, the primary difference being a fluctuating low to high incidence versus a consistently high incidence over successive years. Malariologists have long graded endemic malaria according to risk of infection as reflected in the proportion of children and adults having enlarged spleens (spleen rate, *see* below). However, these terms have evolved into a more general use and are routinely applied in the absence of supporting spleen rate measures. The following terms have been used to empirically gauge regional risk according to criteria described by Bruce-Chwatt *(9)*:

1. Hypoendemic: Little transmission, and the effects of malaria on the community are unimportant.
2. Mesoendemic: Variable transmission that fluctuates with changes in one or many local conditions, e.g., weather or disturbance to the environment.
3. Hyperendemic: Seasonally intense malaria transmission with disease in all age groups.
4. Holoendemic: Perennial intense transmission with protective clinical immunity among adults.

2. Indirect Estimates of Risk of Malaria
2.1. Environmental (Rainfall, Altitude, Temperature)

Transmission of malaria requires mosquito vectors in the genus *Anopheles*. These insects exhibit exquisite sensitivity to the environmental parameters of temperature and humidity. Thus, rainfall, altitude, and temperature govern the activity and abundance of anopheline mosquitoes and the transmission of malaria. Within ranges of temperature (20–30°C) and humidity (>60%) that vary for each vector species, the mosquito survives and is capable of transmitting malaria. When the limits of temperature and humidity tolerance are exceeded, the vectors die, and the risk of malaria evaporates. Variations of temperature and humidity within the viable range for mosquitoes can also affect the duration of sporogony, the time required for development of the parasite in the mosquito after taking a bloodmeal from an infected human so that a new host can be infected. Cooler temperatures generally prolong sporogony, decreasing the period of infectivity. High relative humidity increases mosquito life-span, so that each infective mosquito can infect more hosts. The risk of transmission by anophelines, however, depends on an available pool of infectious humans so that even in a favorable environment with the appropriate vector, transmission cannot be sustained without adequate numbers of already infected humans. Conditions perfectly suitable for anopheline survival allow seasonably abundant mosquito populations in the United States, but infection is rare because of the lack of infectious humans in the region.

2.2. Entomological

Chapter 1 details the use of the entomological inoculation rate as a measure of risk of infection. The following terms have been used to describe risk of malaria according to entomological criteria:

1. Human landing rate = anophelines captured/person-night.
2. Infected mosquito = oocysts in stomach wall by dissection.
3. Infective mosquito = sporozoites in salivary gland by dissection.
4. Sporozoite rate = infective anophelines/anophelines captured.
5. Entomological inoculation rate = human landing rate × sporozoite rate.
   = infective mosquito bites/person-night.

2.3. Clinical
2.3.1. Spleen Rate

The spleen rate is the proportion of people in a given population having enlarged spleens expressed as a percentage. The relationship between malaria and the spleen

Table 1
WHO Criteria for Classification of Endemicity by Spleen Rates

Endemicity	Children aged 2–9yr (%)	Adults (>16 yr)
Hypoendemic	0–10	No measure
Mesoendemic	11–50	No measure
Hyperendemic	>50	"High" (≥25%)
Holoendemic	>75	"Low" (<25%)

Table 2
Criteria for Scoring Spleen Size[a]

Score	Description of spleen as measured with subject in recumbent position
0	Normal, not palpable
1	Palpable below costal margin on deep inspiration
2	Palpable below costal margin but not beyond midpoint between costal margin and umbilicus
3	Palpable below limits for score of 2 but not beyond umbilicus
4	Palpable below umbilicus but not beyond midpoint between umbilicus and symphisis pubis
5	Palpable beyond limits for score of 4

[a]AES = $\Sigma(\text{Hackett score}_i \times n_i)/N$, where Hackett scores of 1 through 5 are included. n_i = number of spleens measured with a given Hackett score. $N = n_1 + n_2 + n_3 + n_4 + n_5$ (total number of non-Hackett 0 spleens measured)

rate has served as an indirect marker of risk of malaria long before plasmodia were known as the cause (Dempster introduced the method in India in 1848). Chronic exposure to malaria causes spleen enlargement. Thus, the percentage of people having enlarged spleens and the degree of enlargement correlate with risk of infection in a community. The WHO has classified endemicity as gauged by spleen rate as shown in **Table 1**.

Important caveats complicate this convenient estimate of risk. Under conditions of epidemic malaria, where risk of infection may be very high, spleen rates may be close to nil because exposure is acute rather than chronic. The spleen rate is useful as a relative measure of risk only where stable malaria prevails. The distinction between hyperendemic and holoendemic on the basis of spleen rate in adults, developed on the basis of observations in sub-Saharan Africa, does not appear to hold true on the island of New Guinea where adults consistently have enlarged spleens in the face of holoendemic malaria.

A secondary estimate of risk utilizing spleen measurements is the average enlarged spleen (AES). The AES represents the mean Hackett score derived from a sample of enlarged spleens. **Table 2** describes grading of spleen enlargement on the basis of palpation.

Spleens with Hackett scores of zero are not included in the calculation of AES. A higher AES between two sites with comparable spleen rates may be interpreted as consistent with higher risk. A recent malariometric survey in Papua New Guinea found that, although spleen rates did not vary with altitude for children less than 10 yr of age, AES decreased with increasing altitude, closely paralleling altitude specific prevalence of malaria *(4)*.

2.3.2. Serology

Serological tests have been applied to demonstrate exposure to infection among individual subjects. Unfortunately, no serological test reliably assesses either degree of naturally acquired immunity or the extent of exposure in individuals. Risk of exposure among groups, however, may be assessed by serological analyses. Detection of antibodies to the circumsporozoite protein (CSP) in European travelers returning from malarious areas suggests that anti-CSP antibodies can serve as an indicator of the relative risk of infection in travelers to specific regions *(5–7)*. The prevalence of antibodies to malaria antigens such as CSP, merozoite surface antigen (MSA), and erythrocytic stage antigens has also served as an indirect indicator of risk of infection *(2,8)* and in some studies correlates with spleen rates *(8)*, prevalence of positive smears *(1,8)*, and parasite density *(2)*. Serological assays may also be useful in assessing risk in regions with recent increases in transmission during epidemics or with declining transmission during eradication efforts *(9–11)*. However, no currently available serological assay system can reliably serve as a quantitative measurement of exposure.

3. Direct Estimates of Risk of Malaria
3.1. Passive Surveillance
3.1.1. Passive Case Detection

Passive case detection (PCD) is the detection and reporting of malaria cases restricted to people seeking treatment for illness at a health post, clinic, or hospital. The PCD is reported as simply the number of cases treated. The PCD proportion, or ratio of malaria cases to some defined total (hospitalizations, febrile illnesses, deaths), is often used to gauge the burden of illness caused by malaria relative to all other causes definable in the setting from which those data were collected. This is often called the PCD rate, although it is actually a proportion. The quantitative value of the PCD rate varies according to the rigor of diagnostic procedures and the likelihood of malaria parasitemia masking some other underlying cause of illness (*see* **Subheading 3.2.8.**). The availability of reporting health-care delivery facilities also impacts interpretation of PCD data. PCD data are used to gauge risk in communities with the assumption that virtually all infections prompt seeking of treatment at a reporting facility.

In a setting where malaria is reported on the basis of microscopically confirmed infection, and the proportion of fevers caused by malaria is relatively low, the PCD rate may serve as a reliable estimate of risk of malaria relative to other febrile illnesses in the community. In general, this scenario is true where malaria is hypo- to mesoendemic and health-care providers are more likely to report malaria on the basis of diagnostic criteria that exclude other more common causes of febrile illness. In this scenario, the reported PCD proportion may carry good sensitivity and specificity.

The PCD proportion as a measure of risk is less reliable where malaria is hyper- to holoendemic. This problem stems from three important features specific to heavily endemic areas. First, because malaria dominates as a cause of febrile illness, health-care providers tend to make presumptive diagnoses of malaria in patients presenting with fever. Thus, other causes of febrile illness are often erroneously classified as malaria so that the specificity of the PCD proportion may be extremely low. Second, because self-treatment often relieves symptoms, the true burden of malaria in the community may be underreported. Finally, naturally acquired immunity in hyper- to holoendemic regions leaves most infections unnoticed at treatment facilities.

The PCD totals may be analytically applied in a variety of ways. If the PCD total is believed to have captured most infections in the community, then monthly or annual PCD case totals may be divided by the mid-interval population of the areas served by the reporting facilities to define incidence of infection.

Incidence of malaria = malaria PCD total/mid-interval population/unit time

More often, a PCD proportion is applied as a direct measure of the contribution of malaria to febrile disease in communities expressed as a percentage.

PCD proportion = (malaria cases/patients seeking care for febrile illness) × 100

3.1.2. Annual Parasite Incidence

The number of malaria cases per 1000 population per year is called the annual parasite incidence, or API. The API represents the most broadly applied measure of risk of infection. Many health authorities rely upon the API as the core measure of risk of infection. The statistic is often used as the basis for comparing risk between communities, districts, provinces, and nations. The means of deriving the numerator for the API, "cases of malaria", varies a great deal. Comparisons of risk based on the API demand consideration of the sources and case definitions; for example, clinical diagnosis versus smear-confirmed diagnosis for reported total cases. The API numerator often represents a hybrid of PCD and active surveillance methods (*see* below), and the relative contributions of each impact interpretation of the API.

Annual parasite incidence = reported infections/1000 person (mid-year)/year

A statistic used to help interpret the API is the annual blood examination rate, or ABER. This is the number of blood films examined per 1000 population. The ABER reflects the degree of diagnostic effort made to identify malaria. For example, between two locations having comparable API estimates, the location having the lower ABER may be considered higher risk because less effort produced an equal density of infections.

Annual blood examination rate = blood film exams/1000 person (mid-year)/year

3.2. Active Surveillance

3.2.1. Active Case Detection

Systematic screening of communities for people with fever and examination of blood films collected from them is called active case detection (ACD). The analytical application of infections discovered by ACD varies. The data collected by ACD may be applied as follows:

Fever rate = (people with fever/people examined for fever) × 100

Malaria among fevers rate = (people with malaria/people with fever) × 100

These estimates serve as measures of the prevalence of fever in the community at sampling and the proportion of fevers likely caused by malaria, respectively. Health officers often use ACD to monitor the progress of control efforts within specific areas, but also include these detected cases into the numerator for API. Thus, the vigor of case detection impacts on the API. In this sense, the importance of interpreting the API in the context of ABER may be appreciated.

3.2.2. Active Case Survey

The primary distinction between active case survey (ACS) and ACD is sampling that does not exclude people without fever. Whereas ACD estimates the prevalence of fever and the proportion of fevers caused by malaria, ACS measures the prevalence of parasitemia independently of fever. The ACS may be the best approach to assessing risk of infection in communities where parasitemia often occurs without fever, that is, where malaria is hyper- to holoendemic. The survey includes recording both febrile and afebrile (or apyrexic) malaria cases. The analytical application of ACS may include the following:

Point prevalence of malaria = (people with parasitemia/people examined) × 100

Febrile malaria rate = (people with febrile malaria/people with malaria) × 100

Point prevalence refers to the ratio of parasitemic individuals to the total number of individuals examined at a single point in time. This provides an estimate of risk by indicating the number of people infected at any given time with higher prevalence indicating higher rates of transmission. However, for diseases like malaria that may be seasonal in some locations, the point prevalence in June might not be a good estimator of risk for an individual traveling to the area in November.

3.2.3. Gametocyte Rate

Analysis of point prevalence data often includes specific designation of the prevalence of gametocytemia in study subjects, the carriage of sexual stage parasites that can be transferred to feeding mosquitoes to complete the life cycle and allow new infections of human hosts. In highly endemic areas where immunity to disease often develops, such individuals are frequently asymptomatic and remain undiagnosed and untreated, serving as reservoirs for new infection in the community. Because gametocytemia declines with increasing age in malaria endemic regions *(1,12)*, gametocyte rates in young children may serve as an indicator of risk, particularly when comparing gametocyte rates between two different locations.

3.2.4. Period Prevalence

Compared to point prevalence, period prevalence may better reflect risk for persons who will be exposed to infection over an extended period. This may be especially true where risk fluctuates appreciably. Whereas point prevalence addresses the question of whether a person currently has malaria, period prevalence addresses the question of whether that subject had malaria at any time during the period under investigation. This serves as a measure of the probability that an individual in a defined population will be a case at any given time over a defined period. The numerator of this estimate includes both newly incident cases (*see* below) and cases that may have developed before the survey was initiated. The time of onset of infection may not be known, especially in endemic regions where some subjects may be asymptomatic and remain undiagnosed. Thus, incident cases and prevalence may not be distinguishable. Assuming a stable population over time, period prevalence is calculated as follows:

Period prevalence = (people with malaria at start of period + new cases of malaria during period)/population under study) × 100

Period prevalence requires identifying a sample of individuals from a population, screening each of them for malaria on enrollment, and systematically repeating malaria smears during the defined period in order to identify new cases. Multiple infections among any given individual allows the possibility of a period prevalence of >100%. The greatest limitation with period prevalence is the fact that many population totals are dynamic. The migration of people, and other important changes like mass drug administration, chemoprophylaxis, or any other intervention that alters the number of people at risk, makes the drawing of meaningful statistical inference from measures of period prevalence difficult. In many instances, the effort required to define period prevalence would be better spent measuring incidence, which provides much less ambiguous measures of risk (see **Subheading 3.3.**).

3.2.5. Cumulative Incidence

Cumulative incidence (CI) represents the probability or proportion having malaria or a specific outcome of infection, for example, cerebral malaria, relapse, death, or chemotherapeutic failure over defined intervals. Cumulative incidence is often referred to as an "attack rate," even though the estimate represents a proportion rather than a true rate. The measurement of CI requires a prospective study of people free of infection at the outset. These may be uninfected people newly arriving in a malarious area or people cured of malaria immediately before the observation period. The number of new infections is divided by the number of people at risk, expressed in the context of the period of observation. For example, if one follows 100 people for 10 weeks and 25 get malaria, then the 10-week cumulative incidence of malaria is 25%. More often, larger populations are followed for longer periods, and the estimate may be complicated by losses to follow-up, migration, or death by other causes.

When individual follow-up times vary for study subjects, a convenient means of calculating cumulative incidence is the actuarial method (also called life table). This approach takes into account the loss of individuals from the study population in the denominator by presuming that the mean withdrawal time occurred at the midpoint of the surveillance interval. For example, in a 1-yr study of malaria with monthly intervals of observation, all subjects lost to follow-up would be assumed to have done so at the midinterval, that is, 2 wk. This approach conveniently averages person-time losses across the interval. This is especially useful when such losses are likely, for example, the occurrence of vivax malaria among subjects recruited to gauge the incidence density of falciparum malaria. The actuarial method for estimating CI is as follows:

Cumulative incidence = attack rate (%) over a defined period

Cumulative incidence (actuarial) = incident cases/[population at start of study − (number of withdrawals/2)]

3.2.6. Incidence Density

Incidence density of malaria estimates the risk of infection in a population expressed as a true rate; that is, the number of new infections per unit person- time. This estimate requires a cohort of people who are free of infection and prospectively followed over a defined period. This is most often accomplished by giving radical curative therapy before the follow-up phase, but newly arrived migrants into malarious areas may also provide a suitable cohort without radical cure. Incidence density is calculated simply as the number of new infections divided by the sum of person-time at risk. People lost to follow-up contribute person-time to the denominator up to the point of loss. Their contribution should not be counted for the full period of the study. For example, if one follows 100 subjects for 52 wk and 25 become infected, the incidence density is 25 infections per 100 person-years, or 0.25 infections per person-year, assuming each individual contributed 52 wk of follow-up time. This overly simplistic example assumes no losses to follow-up. However, a subject infected at wk 2 contributes only 2 wk of person-time to the denominator, not 52. Assume that infections occur evenly over the 52 wk and that approximately one infection occurs every 2 wk. In this scenario, losses in person-time at risk due to infection outcomes amount to 650 person-weeks, or 12.5 yr. Taking these losses into account, the incidence density would be estimated at 25 infections/87.5 person-years, or 0.29 infections/person-year. Further, assume that 20 people were lost follow-up at anywhere from wk 1 to wk 51 of the observation period, yielding a total loss of 12.5 person-years at risk. Thus, the true incidence density in the hypothetical cohort would be estimated as 25 infections per 75 person-years, or 0.33 infections/person-year (or everyone experiencing, on average, an infection once every 3 yr).

$$\text{Incidence density} = \text{infections/person-year at risk}$$

3.2.7. Attributable Risk

Attributable risk (or risk difference) represents an estimate of the risk of disease that may be attributed to a specific exposure. In its simplest form, it is the additional amount of disease in those exposed over the background amount of disease in the unexposed population and is given by:

$$AR = I_e - I_u$$

where I_e is the incidence in the exposed population and I_u is the incidence in the unexposed population.

In conducting malaria studies in transmission areas, it is difficult to reliably differentiate between reinfection and recurrent parasitemia following therapy. The attributable risk statistic, calculated by subtracting the coincident incidence rate for a given population from the rate of recurrent parasitemia, estimates the rate of therapeutic failure. The efficacy of standard mefloquine therapy against uncomplicated *Plasmodium falciparum* infections was evaluated in children aged 6 to 24 mo in the Kassena-Nankana District of northern Ghana, West Africa. The incidence of late recrudescence, or therapeutic failure, was calculated as the difference between the incidence of recurrent parasitemia during wk 3 and 4 after mefloquine therapy and the known attack rate of malaria in the region for the cohort. The incidence of recurrent parasitemia at d 28 was 6.3 infections/person-year at risk. However, this incidence rate approximated the known reinfection rate in this cohort (5.7 infections/person-year). Thus, the observed parasitemia in the treatment group could be almost wholly attributed to the measured reinfection rate in this cohort. This method has been reported previously in comparing the efficacy of antimalarial drug regimens *(13,14)*.

3.2.8. Attributable Fraction

Attributable fraction represents an estimate of the risk of disease in a community that is attributable to a particular risk factor. Because many symptoms of malaria are non-specific, attributable fraction is a useful measure in looking at clinical markers or case definitions for malaria in a community. The statistic assigns a probability of a single fever episode being due to malaria. For a single exposure variable, the overall attributable fraction is given by

$$AF = p(R-1)/R$$

where p is the exposure prevalence among cases and R is the relative risk of disease associated with the exposure *(15)*. The probability that any individual case is attributable to malaria is calculated without multiplying by p *(16)*.

Probabilities derived from logistic regression models can increase the precision of the attributable fraction estimates by allowing fever risk as a continuous function of parasite density *(16)* and can be further extended to include other covariates *(17)*.

Schellenberg et al. *(18)* used the fraction of fever cases in a population attributable to malaria at each level of parasite density to evaluate the sensitivity and specificity of alternative case definitions for malaria, and to provide a direct estimate of malaria-attributable fever. The attributable fraction statistic also has been used to determine the specificity and sensitivity of case-definition thresholds *(19)*.

References

1. Genton, B., al-Yaman, F., Beck, H. P., Hii, J., Mellor, S., Narara, A., et al. (1995) The epidemiology of malaria in the Wosera area, East Sepik Province, Papua New Guinea, in preparation for vaccine trials. I. Malariometric indices and immunity. *Ann. Trop. Med. Parasitol.* **89,** 359–376.
2. May J., Mockenhaupt, F. P., Ademowo, O. G., Falusi, A. G., Olumese, P. E., Bienzle, U., and Meyer, C. G. (1999) High rate of mixed and subpatent malaria infections in Southwest Nigeria. *Am. J. Trop. Med. Hyg.* **61,** 339–343.
3. Pribadi, W., Sutanto, I., Atmosoedjono, S., Rasidi, R., Surya, L. K., and Susanto, L. (1998) Malaria situation in several villages around Timika South Central Irian Jaya, Indonesia. *Southeast Asian J. Trop. Med. Public Health* **29**, 228–235.
4. Hii, J., Dyke, T., Dagoro, H., and Sanders, R. C. (1997) Health impact assessments of malaria and Ross River virus infection in the Southern Highlands Province of Papua New Guinea. *P.N.G. Med. J.* **40,** 14–25.
5. Nothdurft, H. D., Jelinek, T, Bluml, A., von Sonnenburg, F., and Loscher, T. (1999) Seroconversion to circumsporozoite antigen of *Plasmodium falciparum* demonstrates a high risk of malaria transmission in travelers to East Africa. *Clin. Infect. Dis.* **28,** 641,642.
6. Jelinek, T., Bluml, A., Loscher, T., and Northdurft, H. D. (1998) Assessing the incidence of infection with *Plasmodium falciparum* among international travelers. *Am. J. Trop. Med. Hyg.* **59,** 35–37.
7. Cobelens, F. G., Verhave, J. P., Leentvaar-Kuijpers, A., and Kager, P. A. (1998) Testing for anti-circumsporozoite and anti-blood-stage antibodies for epidemiologic assessment of *Plasmodium falciparum* infection in travelers. *Am. J. Trop. Med. Hyg.* **58,** 75–80.
8. Al-Yaman, F., Genton, B., Kramer, K.J., Taraika, J., Chang, S. P., Hui, G. S., and Alpers, M. P. (1995) Acquired antibody levels to *Plasmodium falciparum* merozoite surface antigen 1 in residents of a highly endemic area of Papua New Guinea. *Trans. R. Soc. Trop. Med. Hyg.* **89,** 555–559.
9. Bruce-Chwatt, L. J., Draper, C. C., and Konfortion, P. (1973) Seroepidemiologic evidence of eradication of malaria from Mauritius. *Lancet* **2,** 547–551.
10. Lobel, H. O., Najera, A. J., Ch'en, W. I., Munore, P., and Mathews, H. M. (1976) Seroepidemiologic investigations of malaria in Guyana. *J. Trop. Med. Hyg.* **79,** 275–284.
11. Tikasingh, E., Edwards, C., Hamilton, P. J. S., Commissiong, L. M., and Draper, C. C. (1980) A malaria outbreak due to *Plasmodium malariae* on the island of Grenada. *Am. J. Trop. Med. Hyg.* **29,** 715–719.
12. Gilles, H. M. (1993) Epidemiology of malaria, in *Bruce-Chwatt's Essential Malariology*, 3rd ed., Arnold, London.
13. ter Kuile, F. O., Dolan, G., Nosten, F., Edstein, M. D., Luxemburger, C., Phaipun, L., Chongsuphajaisiddhi, T., Webster, H. K., and White, N. J. (1993) Halofantrine versus mefloquine in treatment of multidrug-resistant falciparum malaria. *Lancet.* **341,** 1044–1049.

14. von Seidlein, L., Milligan, P., Pinder, M., Bojang, K., Anyalebechi, C., Gosling, R., et al. (2000) Efficacy of artesunate plus pyrimethamine-sulphadoxine for uncomplicated malaria in Gambian children: a double-blind, randomised, controlled trial. *Lancet* **355,** 352–357.
15. Rothman, K. J., Adami, H. O., and Trichopoulos D. (1998) Should the mission of epidemiology include the eradication of poverty? *Lancet* **352,** 810–813.
16. Smith, T., Genton, B., Baea, K., Gibson, N., Taime, J., Narara, A., et al. (1994) Relationships between *Plasmodium falciparum* infection and morbidity in a highly endemic area. *Parasitology* **109,** 539–549.
17. Prybylski, D., Khaliq, A., Fox, E., Sarwari, A. R., and Strickland, G. T. (1999) Parasite density and malaria morbidity in the Pakistani Punjab. *Am. J .Trop. Med. Hyg.* **61,** 791–801.
18. Smith, T., Schellenberg, J. A., and Hayes, R. (1994) Attributable fraction estimates and case definitions for malaria in endemic areas. *Stat. Med.* **13,** 2345–2358.
19. McGuinness, D., Koram, K., Bennett, S., Wagner, G., Nkrumah, F., and Riley, E. (1998) Clinical case definitions for malaria: clinical malaria associated with very low parasite densities in African infants. *Trans. R. Soc. Trop. Med. Hyg.* **92,** 527–531.

II

LABORATORY MODELS

3

Maintenance of the *Plasmodium berghei* Life Cycle

Robert E. Sinden, Geoff A. Butcher, and A. L. Beetsma

1. Introduction

Plasmodium berghei was probably first described in 1946 by Vincke in blood films of the stomach contents of *Anopheles dureni*. In 1948, it was subsequently found in blood films of *Grammomys surdaster* collected in Kisanga, Katanga; blood was passaged to white rats and became the K173 strain made widely available by the Institute for Tropical Medicine in Antwerp. A trio of fascinating papers describing the discovery and early analysis of the biology of *P. berghei (1–3)* give an early indication of the natural and catholic host range of this parasite (**Table 1**), a property that possibly underlies the successful transmission of the parasite to a variety of laboratory hosts. Although the natural vector of *P. berghei* is *A. dureni*, early laboratory studies *(3,4)* showed that a wide range of colonized mosquitoes would successfully transmit the parasite (**Table 2**).

Since their introduction to the laboratory, rodent malarias have made an enormous contribution to our understanding of the biology, cell biology, and immunology of the malaria parasites *(5)*. A search of the more recent literature available on electronic databases indicates that there are 10,617 publications on *P. falciparum*, compared to 2655 on *P. berghei*, 413 on *P. chabaudi*, and 152 on *P. vinckei*.

Notwithstanding this large knowledge base, other important factors currently contributing to the enormous potential of research on rodent malarias, and *P. berghei* in particular, include the following:

1. The wide availability of susceptible, genetically defined or knockout mouse strains.
2. An extensive range of well-characterized clones of *P. berghei,* some with important biological phenotypes, e.g., an inability to produce mature sexual-stage parasites.
3. The facility to make and analyze in vivo-defined gene knockouts in this parasite species currently surpasses that in other *Plasmodium* species (e.g., *P. falciparum* and *P. knowlesi*).
4. The parasite can be transmitted through major vector species from Africa, India, and South America (*see* **Table 2**).
5. Infections can readily be synchronized.
6. All the life cycle stages can be grown in vitro, thus permitting direct comparison of in vivo and in vitro data.
7. The genome is now being analyzed, and expressed sequence tags (EST) and gene sequence survey (GSS) databases are under construction.

Table 1
List of Natural and Laboratory Hosts of *P. berghei*

Natural hosts	Laboratory hosts
Muridae	
Thamnomys surdaster	Brown Norway rat (<4 wk)
Praomys jacksoni	White rat (<4 wk)
Leggada belle	Mouse
Saccotomus campestris	*Thamnomys surdaster*
Mastomys coucha	*Grammamys rutilans*
Æthomys sp.	*Ondatra zibethica* (musk rat)
Pelomys frater	Hamster (good for gametocytes)
Lophyromys aquilus	
Crocidura turba	
Sciuridae	
Unnamed	
Carnassieridae?	
Unnamed	

See **refs. *1–3***.

Table 2
Vectorial Capacity of Mosquitoes for P.berghei

Vector	Nonvector
Anopheles dureni **(*1*)**	*Culex bitaeneorhynchus* **(*1*)**
A. stephensi **(*1*)** (var mysorensis)	*Aedes aegypti* **(*1,2*)**[a]
A. gambiae **(*1*)**	*Anopheles albimanus* **(*1*)**
A. atroparvus **(*1*)**	*A. aztecus* **(*1*)**
A. quadrumaculotus **(*1*)**	*A. concolor* **(*1*)**
A. maculipennis **(*2*)**	*A. coustani* **(*1*)** var ziemanni
A. albimanus **(*3*)**	*A. fluviatilis* **(*1*)**
	A. funestus **(*1*)**
	A. hyrcanus **(*1*)**
	A. jamesi **(*1*)**
	A. pulcherrimus **(*1*)**
	A. splendidus **(*1*)**
	A. subpictus **(*1*)**

[a]Produced oocysts (SP28 strain S. Keiberg) (*see* **refs. *3*** and ***4***, and M.-C. Rodriguez, personal communication).

2. Materials

2.1. Anesthetics for Rodents

When it is not important whether the host-body temperature falls, for example, for the routine blood-feeding of stock mosquitoes, the anesthetics Nembutal or Saggatal (60 mg/mL pentobarbitone sodium [BP]; Rhone Meriuex) are very satisfactory. For mice, rats or rabbits, 0.125 mL Saggatal per100 g body weight is a routine dose and is administered intraperitoneally. However, animals rapidly accommodate to these anesthetics. Alternatively, for invasive procedures, rats may be anesthetized with halothane–oxygen.

For mice, we prefer to use a mixture of Rompun™ (2-(2,6-xylidino)-5,6-dihydro-4*H*-1,3-thiazine hydrochloride, 2% stock solution; Bayer), and Vetalar™ (100 mg/mL ketamine; Parke-Davis)—both kept at 4°C. In a sterile tube, mix 1 vol of Rompun with 2 vol of Vetalar and 3 vol of sterile phosphate-buffered saline (PBS). This will keep at 4°C for up to 2 wk without loss of activity. Use this diluted mixture at 0.05 mL/10 g body weight. Deliver intramuscularly into the thigh; there may be local bleeding, the animal may become transitorily hyperactive, but thereafter anesthesia is deep and thus suitable for cardiac bleeds or for exposure to mosquitoes. Anesthesia may persist up to 45 min.

2.2. Phenylhydrazine for the Induction of Reticulocytosis

Prepare a sterile stock solution of 1.2 mg/mL phenylhydrazine (phenylhydrazinium chloride; BDH 10189) in PBS. Use at a rate of 10 μL/g body weight, inoculated intraperitoneally to induce reticulocytosis in mice. Inoculation should be 3 d before infection of the host by blood transfer.

2.3. Heparin Anticoagulant

Prepare a stock solution of 1 mg/mL (ca. 300 U/mL) preservative-free heparin (Sigma 9133) in PBS. Use at a nominal final dilution of 1:10 in blood.

2.4. Giemsa Staining of Blood Films

Dilute concentrated Giemsa stain (BDH R66) to 20% or 10% in buffer (0.7 g anhydrous KH_2PO_4 plus 1.0 g anhydrous Na_2HPO_4 per liter distilled H_2O). Stain air-dried, methanol-fixed cells for 10 or 45 min, respectively, rinse very briefly in tap water or buffer, and air-dry.

2.5. Fructose/PABA Feed for Mosquitoes

Combine 8 g fructose and 0.05 g *p*-aminobenzoic acid (PABA). Make up to 100 mL in distilled water. Filter-sterilize, or autoclave. Store at 4°C.

2.6. Membrane Feeders

2.6.1. Feeder Sources

1. Discovery Workshops, 516A Burnley Road, Accrington, Lancashire BB5 6JZ, UK; e-mail: discovery@mail.internexus.co.uk. Volume: 1 mL. Has an integral electrical heater.
2. Department of Aeronautics, Imperial College. Volume: 1 mL. Perspex construction with demountable feeder units. Requires a circulating water bath.
3. All-glass feeders following design of Wade are available from various sources (e.g., Bioquip Products, 17803 LaSalle Avenue, Gardena, CA 90248-3602).

2.6.2. Membranes

1. For the majority of cases, 2-way stretch Parafilm-M available from Merck Ltd. (Merck House, Poole, Dorset BH15 1TD, UK) is perfectly satisfactory.
2. For the fastidious mosquito (or experimenter), Baudruche membrane is available from John Long, New Jersey.

2.7. Mercurochrome Staining of Oocysts

Prepare a 1% solution of mercurochrome (BDH-Merck 29177-4Y) in PBS; this can be stored for 1 mo at 4°C. To prolong the observation time of stained oocysts on midguts, postfix in 1% formaldehyde or 1% glutaraldehyde. The two solutions can be mixed if required, but the guts become more resistant to flattening under pressure of the cover slip, and as a consequence, the observation of the gut wall is often more difficult.

2.8. Fixation and Staining of Exoerythrocytic Stage Cultures

Rinse cultures in multiwell slides or on coverslips briefly in PBS and fix in Bouin's Fixative (85 mL of saturated picric acid, 10 mL of 40% formaldehyde, 5 mL of acetic acid) for 10–30 min. Then stain cultures in 10% Giemsa stain (*see* **Subheading 2.4.**) overnight. Wash briefly in Giemsa buffer, then treat with 60% acetone in water to enhance differentiation, then for 20 s each in 100% acetone, Histoclear R, and Euparal essence. Mount preparations in Euparal Vert.

2.9. Culture Medium for HepG2 Cells and EE Cultures

HepG2 cells are available from a wide range of sources including ATCC (Rockville, MD) and the European Collection of Cell Cultures (CAMR, Porton Down, Salisbury, Wilt., SP4 OJ6, UK). Although some authors have suggested subclone HepG2 A16 is a preferred option, we have found no differences in the susceptibility of this cell line from a variety of sources. Stock cells are maintained in 25-cm^2 plastic Falcon flasks in minimal essential medium (MEM) supplemented with 10% fetal calf serum (FCS), 50 µg/mL penicillin, 100 µg/mL streptomycin, 50 µg/mL neomycin, 1 mM L-glutamine, and nonessential amino acids (1× Flow mixture). Maintain cells at 37°C in 5% CO$_2$ in air. For subculture, inoculate trypsinized stock cultures at 1×10^5 cells per well onto cover slips in 4-well or 24-well Nunc tissue culture plates, or Lab-Tek 8-chamber sides. When near confluence, irradiate the cells at 3–3.5 krad from a cobalt-60 source.

2.10. Blood-Stage Culture Medium

Dissolve 10.41 g of RPMI 1640 powder (Sigma, Poole, Dorset, UK) in 960 mL deionized water, and add 5.94 g of HEPES. Check that pH is 7.4. Sterilize by filtration and store for up to 4 wk at –20°C. Immediately before use, add 4.2 mL of sterile 5% NaHCO$_3$ solution, 11 mL of FCS, and 5.5 mg neomycin to 96 mL medium.

2.11. Ookinete Culture Medium

Dissolve powdered RPM1 1640 medium (Sigma R 4130) containing 0.025 M HEPES in 900 mL deionized water. Add 0.05 g of hypoxanthine, 2 g of NaHCO$_3$, 50,000 IU penicillin, and 50 mg streptomycin, and make up to 1 L final volume. Adjust pH to 8.3 with 1 M NaOH. Then add either FCS to a final concentration of 20%, or (less satisfactorily) add Ultroser-G (Gibco) to a final concentration of 0.4%. Store complete medium in appropriate volumes at –20°C.

2.12. Purification of Asexual Stages

2.12.1. Equipment

1. Slides.
2. Scissors, forceps.
3. 10- to 50-mL syringes.
4. 1-mL syringes.
5. Needles (26-gage 1/2, 0,45 × 13).
6. Nylon wool.

2.12.2. Reagents

1. Mice (strain not critical).
2. *P. berghei* clone 2.33.
3. Phenylhydrazine-HCl.
4. Rompun™/Vetelar™/PBS.
5. Heparin (300 U/mL).
6. 70% Ethanol.
7. Blood-stage culture medium (*see* **Subheading 2.10.**).
8. Whatman CF11 cellulose powder.
9. Nycodenz (Nycomed Pharma AS, Oslo, Norway).
10. PBS.

2.13. Purification of Schizonts

2.13.1. Equipment

1. Slides.
2. Scissors, forceps.
3. 10- to 50-mL syringes.
4. 1-mL syringes.
5. Needles (26*G* A1/2, 0,45 × 13).
6. Nylon wool.

2.13.2. Reagents

1. Wistar rat (180–250 g).
2. *P. berghei* clone 2.34 or 2.33.
3. Phenylhydrazine-HCl.
4. Rompun/Vetelar/PBS.
5. Heparin (300 U/mL).
6. 70% Ethanol.
7. Blood-stage culture medium (*see* **Subheading 2.10.**).
8. Whatman CF11 cellulose powder.
9. Nycodenz (Nycomed Pharma AS).
10. PBS.

2.14. Purification of Asexual-Free Gametocytes

2.14.1. Equipment

1. Slides.
2. Scissors, forceps.
3. 10- to 50-mL syringes.
4. 1-mL syringes.
5. Needles (26-gage A1/2, 0,45 × 13).
6. Nylon wool.

2.14.2. Reagents

1. Theilers Original (TO) mice.
2. *P. berghei* clone 2.34.
3. Phenylhydrazine-HCl.
4. Rompun/Vetelar/PBS.
5. Heparin (300 U/mL).

6. 70% Ethanol.
7. Blood-stage culture medium (*see* **Subheading 2.10.**).
8. Whatman CF11 cellulose powder.
9. Nycodenz (Nycomed Pharma AS).
10. PBS.

2.15. Enrichment of Ookinetes

2.15.1. Equipment

1. Slides.
2. Scissors, forceps.
3. 10- to 50-mL syringes.
4. 1-mL syringes.
5. Needles (26-gage A1/2, 0,45 × 13).
6. Nylon wool.

2.15.2. Reagents

1. Theilers Original (TO) mice.
2. *P. berghei* clone 2.34.
3. Phenylhydrazine-HCl.
4. Rompun/Vetelar/PBS.
5. Heparin (300 U/mL).
6. 70% Ethanol.
7. Ookinete culture medium (*see* **Subheading 2.11.**).
8. Whatman CF11 cellulose powder.
9. Nycodenz (Nycomed Pharma AS).
10. PBS.

2.16. Purification of Sporozoites

2.16.1. Equipment

1. Slides.
2. Scissors, forceps.
3. 1-mL syringes.
4. Needles (26-gage A1/2, 0,45 × 13).
5. Nylon wool.
6. 0.5- and 1.5-mL Eppendorf tubes.

2.16.2. Reagents

1. Theilers Original (TO) mice.
2. *P. berghei* clone 2.34.
3. Phenylhydrazine-HCl.
4. Rompun/Vetelar/PBS.
5. Heparin (300 U/mL).
6. 70% Ethanol.
7. Ookinete culture medium (*see* **Subheading 2.11.**).
8. Whatman CF11 cellulose powder.
9. Nycodenz (Nycomed Pharma AS).
10. PBS.

P. berghei Life Cycle

3. Methods

The practical aspects of running the *P. berghei* life cycle are divided here as follows: parasite storage; maintenance in the rodent host; maintenance in the mosquito vector; maintenance in vitro, and methods for the purification of life-cycle stages.

3.1. Parasite Storage

The malarial genome is very "fragile," suffering chromosomal changes within very short periods—most notably during sustained asexual erythrocytic schizogony (i.e., repeated mechanical passage). To overcome this, it is necessary to select against biologically incompetent genotypes (genetic deletions or rearrangements) either by regular transmission through the mosquito (normally within eight serial blood passages), and/or by the establishment of large stocks of cryopreserved blood samples, preferably from a sporozoite-infected mouse.

3.1.1. Cryopreservation of P. berghei

3.1.1.1. BLOOD-STAGES (SEE NOTE 1)

1. Bleed anesthetized rodent into heparinized syringe (final heparin concentration, 30 U/mL blood ca. 100 µg/mL). Place blood on ice immediately.
2. If possible, arrange for the blood to be stirred slowly and continuously on ice while adding an equal volume of cold (0–4°C) 20% dimethyl sulfoxide (DMSO) in blood-stage culture medium (*see* **Subheading 2.10.**) dropwise over a period of not less than 10 min. Make numerous small aliquots of the final mixture (0.1 mL is adequate) on ice in polythene cryotubes.
3. Freeze samples in a controlled manner (ca. 1.0°C/min) to –80°C. This can be achieved simply by placing the ampules in a closed, thin-walled polystyrene box in a –80°C freezer or in dry ice, but to achieve maximal yields of viable parasites it is preferable to use a programmable cell freezer. A suitable program is shown in **Table 3**.
4. Store at –196°C indefinitely.
5. Thawing: Take care if ampules have been submerged in liquid N_2, since they may have liquid N_2 inside and will explode when thawed rapidly (if not before). Thus: loosen cap of ampule immediately upon removal from liquid nitrogen; thaw at 37°C in a water bath (note cap is "open") until ice is almost gone; and store on ice until inoculated by the intraperitoneal route into new host. In our hands, inocula of up to 0.5 mL are readily accepted by a 25-g mouse.

3.1.1.2. SPOROZOITES

In laboratories that have mosquito colonies, the need to preserve the parasite at the sporozoite stage is rare; however, this has been done with limited success, that is, 5% successful transmissions *(6)*. Greater success has been recorded more recently *(7,8)*.

1. Take 21-d -infected *Anopheles*, and immobilize with ice or CO_2.
2. Sterilize by rinsing briefly in 70% ethanol then dry on sterile filter paper.
3. Dissect salivary glands as cleanly as possible (*see* **Subheading 3.3.2.3.**), and rinse individually in sterile medium.
4. Transfer the salivary glands from ca. 50 mosquitoes into a loose-fitting Teflon-glass homogenizer. Gently rupture the salivary glands by two or three passes of the pestle without rotation. Treat this sporozoite suspension as described above (*see* **Subheading 3.1.1.1., steps 2–5**). NOTE: for sporozoite inoculation the intraperitoneal route is more reproducible, but has a lower "take" rate than intravenous inoculation.

**Table 3
Controlled Freezing of *P. berghei***

Start temperature (°C)	Rate of cooling	End temperature (°C)
+4	−0.5°C/min	−2
−2	0°C/min for 17 min	−2
−2	−0.5°C/min	−9
−9	−0.3°C/min (but add heat sink)	−12
−12	−1°C/min	−25
−25	−2°C/min	−70
−70	−5°C/min	−196
−196	Hold 2 h	−196

3.2. Maintenance in the Rodent Host

3.2.1. Parasite Strains

The most frequently used strains of *P. berghei* are: K173, ANKA, NK65, SP11, and LUKA *(9)*. *P. berghei* ANKA clone 2.34 generates gametocytes and is commonly used for maintaining the infection in mosquitoes. *P. berghei* clone 2.33, derived in the same cloning study as 2.34, produces no mature gametocytes but otherwise gives similar parasitemias to clone 2.34, which can only be maintained by mechanical passage from mouse to mouse with the inherent genetic risks described in **Subheading 3.1**.

3.2.2. Host Strains

Laboratory colonies of the thicket rat, *Thamnomys surdaster* are rare; consequently, the blood-stages of *P. berghei* have more often been passaged in a wide variety of mouse strains. Successful transmission to mosquitoes is reported from a narrower selection, including Theiler's Original (TO–outbred), BALB/c, C57BL/6, Swiss, and A/J. The peak period of infectivity for mosquitoes (usually 2–4 d after blood inoculation) is similar in BALB/c and severe combined immunodeficient (SCID) mice.

The most susceptible rat strain is the Brown Norway. *P. berghei* ANKA will usually not go above 5% parasitemia in adult laboratory rats (e.g., Wistar and Sprague-Dawley) but will give high parasitemias in weanling rats, comparable to those in mice. The pattern of infectivity to mosquitoes is the same for rats (adult and weanling) as mice but peak infectivity is delayed a day or two, depending on the number of parasites initially inoculated.

Hamsters reportedly harbor good gametocyte infections and are used as an alternative to mice or rats to maintain *P. berghei* infectious to mosquitoes. They are relatively more difficult to infect intravenously and are not available as genetically defined inbred strains.

3.2.3. Infection of Mice with P. berghei ANKA

Mice are maintained on food and water *ad libitum* in a 12-h alternating light–dark cycle. Mice are infected by intraperitoneal or intravenous (tail vein) injections of blood or sporozoites, or by mosquito bite. Avoid passage from mouse to mouse more than eight times.

For blood passage, infect mice from a frozen stabilate or with heparinized blood. Routinely, 0.05 mL of blood with a 10% parasitemia (i.e., 10^7–10^8 parasitized red blood

P. berghei Life Cycle

cells [RBC]) will give a 10% parasitemia in 3 d when injected intraperitoneally into TO mice (*see* **Note 2**).

To infect mice by mosquito bite, restrain them mechanically or anaesthetize them with Rompun/Vetalar (*see* **Subheading 2.1.**) and lay them for more than 15 min on a cage of up to 50 starved (minimum of 3 h without fructose) infected mosquitoes (with > d 14 infection). To obtain high parasitemias, and especially gametocyte infections, within a short time, for example, for ookinete cultures—pretreat mice with phenylhydrazine (*see* **Subheading 2.2.**) and infect with higher doses of parasitized RBCs, up to 0.2 mL ip of high parasitemia (50% infected RBC) blood (*see* **Note 3**).

3.2.4. Monitoring Parasitemias

1. Hold the mouse by the tail, make a small incision in the tip of the tail and place a small drop of blood on a microscope slide. Smear this by using a second slide, dry the film in stream of warm air, fix in methanol for 30 s and stain in Giemsa stain for 10–45 min. Dry the film in air and observe with a ×50–×100 oil immersion objective. Make films at least on alternate days. Make sure you can distinguish parasites from inclusions in RBC (e.g., Howell-Jolly bodies), platelets on top of RBC, or from precipitated stain.
2. To obtain accurate counts it is convenient to use an eyepiece graticule that subdivides the field into squares. Count an appropriate number of RBC and parasitized RBC (*see* **Notes 4** and **5**) and calculate the percentage infected. The number of RBC you need to count for accurate counts can be calculated from:

$$N = 45.5 \left[(I - P)/P\right]$$

where N = number of RBC that must be counted; P = number of parasites per unit number of RBCs; I = selected unit (e.g., 2000 RBCs).

3.2.5. Drug Cure of Asexual Infections of P. berghei in Mice

Cure of erythrocytic stages malaria in mice can be achieved providing the parasitemia has not risen too high (over 10%). Use any of the regimes listed below as appropriate.

1. Chloroquine treatment: 5–10 mg/kg/d over 4 d (but *P. berghei* is often resistant to it).
2. Sulfadiazine: 10 mg/L in drinking water for 2 d.
3. Combination of atovaquone (0.8 mg/mL) and pyrimethamine (1.0 mg/mL): 0.1–0.2 mL of each over 4 d, given orally, ip as an aqueous suspension, or ip dissolved in olive oil *(10)*.
4. Artemether: 138 mg/kg given orally or im is reported to cure *P. berghei* *(11)*.
5. Pyrimethamine: 0.2 mL of 1 mg/mL given ip on 2 successive days.

3.3. Maintenance of P. berghei in the Mosquito

3.3.1. Infection of Mosquitoes

3.3.1.1. DIRECT FEEDS ON MICE

1. At 6 d prior to the feeding of the mosquitoes, treat a mouse intraperitoneally with phenylhydrazine (*see* **Subheading 2.2.**).
2. At 3 d prior to feeding, infect mouse intraperitoneally with 10^7–10^8 parasitized RBC.
3. At 1 d prior to feeding, place mosquitoes in an appropriate container. Paper ice cream containers are excellent mosquito cages; mosquitoes are happy at densities from 0.5 to 1/mL. Do not feed mosquitoes with sugar in the 24 h preceding the blood-feed.
4. The day of the feed, you may wish to record the parasitemia or gametocytemia of the mouse; by using a Giemsa-stained blood film (*see* **Note 4**).

5. An additional useful assay is to check whether the male gametocytes in the blood are potentially infectious by testing for male gamete formation—the exflagellation test: take one drop of ookinete medium and add directly to it an equal volume of tail blood, mix thoroughly and cover immediately with a cover slip and seal with petroleum jelly. Keep at 20–25°C and observe after 10–20 min. Observe a cell monolayer with a ×40 objective preferably with phase or interference contrast. Exflagellation is noted by the violent lashing movements of the male flagella, which can be clustered (early) or widespread (late). Often red blood cells aggregate around the exflagellating microgametocyte forming characteristic clusters (*see* **Note 6**).
6. Having selected an appropriate mouse, give an anesthetic that is known not to reduce host body temperature, nor to inhibit parasite infectivity (e.g., Rompun/Vetelar) (*see* **Subheading 2.1.**). Shave the belly of the rodent to permit rapid feeding of the mosquitoes. Place the anesthetized rodent on the netting of the cage—but take care not to restrict its breathing. Keep the rodent warm (particularly if small) (*see* **Note 7**). Maintain this cage at 19–21°C, in a draft-free, darkened environment for a minimum of 15 min (or until the appropriate number of mosquitoes have fed). A dark red light can be used if it is necessary to observe the feeding process (although we often feed under normal laboratory lighting, which does not inhibit the voracious mosquito).
7. After feeding, engorged females are very delicate, avoid handling them if at all possible for 24 h. For many purposes, the day after feeding you may wish to remove unfed or partially fed mosquitoes (i.e. those unlikely to be infected). A variety of methods may be used:
   a. Mosquitoes previously starved for 24 h can be further starved for 48 h (but give water). Unfeds will usually die.
   b. Cool the mosquitoes in a refrigerator or cold room, open cage, and remove unfeds (do not touch the "feds" if possible). Never do this immediately after the feed because it will stop the infectious process.
   c. Knock the mosquitoes out with CO_2; while anesthetized remove the unfeds. NOTE: mosquitoes recover very quickly from CO_2.
8. After "sorting" return fructose/PABA feed to the mosquitoes and maintain at 19–21°C 70–80% relative humidity (RH) for the desired period. (24 h for ookinete studies; 8–10 d for oocyst counting; 18–21 d for sporozoite infection of salivary glands).
9. At 4–6 d after the infectious feed give the mosquitoes a second (uninfected) bloodmeal. This raises the number of surviving mosquitoes very significantly, and within them a much higher yield of viable parasites. We do not permit infected mosquitoes to lay eggs.

3.3.1.2. Membrane Feeding of Mosquitoes

Instead of feeding directly on an infected mouse, there are numerous reasons why one would need to feed on artificial mixtures of infected blood and added reagents. To do this, a membrane feeder is used.

1. All mosquito manipulation techniques are as given in **Subheading 3.3.1., steps 1–5**.
2. Prepare a membrane feeder. Stretch the membrane (Baudruche membrane or 2-way-stretch Parafilm—the latter stretched to the point of breaking in both directions) over the feeder and secure well. Not more than 5 min before adding blood, raise feeder temperature to 37–39°C.
3. Anesthetize the rodent deeply (*see* **Subheading 2.1.**) and collect blood as rapidly as possible. Keep blood at 37°C (for short periods) or rapidly cool to 0°C (for greater periods, e.g., >30 min of manipulation).
4. Introduce gametocyte-infected blood, or ookinete culture (*see* **Subheading 3.4.3.**), at 30–50% hematocrit into the feeder.
5. Feed to mosquitoes as above. There is no effective time limit on the feeding of ookinete cultures, but beware as these infections can be so heavy as to induce significant mosquito mortality over the succeeding 24 h.

P. berghei Life Cycle

3.3.2. Monitoring P. berghei Infections in Mosquitoes

3.3.2.1. Ookinetes

1. Anesthetize mosquitoes (*see* **Subheading 3.3.1.1., step 7**).
2. Dissect midgut from mosquito onto a microscope slide that has a drop (2 µL) of ookinete medium. To do this, cut off the mosquito head (the "side" of a 20*G* needle is a good miniscalpel).
3. Then holding the thorax with a blunt needle, pull on the penultimate abdominal segment gently to remove blood-filled midgut (which is delicate!).
4. Rupture midgut into the drop of ookinete medium. Make a homogeneous suspension: if thick, make a blood smear; if dilute, observe under phase-contrast microscopy and/or in a hemocytometer.

3.3.2.2. Oocysts

Dissect mosquito as detailed in **Subheading 3.3.2.1.** into either PBS (if oocysts to be observed immediately by phase or interference contrast microscopy), or into 0.1% mercurochrome (*see* **Subheading 2.7.**) (poisonous) in PBS. After staining in the latter for 5–60 min, transfer to 1% glutaraldehyde or formaldehyde if permanent preparations are needed, otherwise observe by brightfield microscopy.

3.3.2.3. Sporozoites

1. Anesthetize >18 d infected mosquitoes as in **Subheading 3.3.1.1., step 7**.
2. Transfer mosquitoes to microscope slide and add a drop (2 µL) of appropriate medium (PBS for observation, exoerythrocytic (EE) culture medium (*see* **Subheading 2.9.**) if being used for infection studies.
3. Cut head off mosquito cleanly using the side of a 20*G* hypodermic needle.
4. Press gently on the thorax with a blunt needle.
5. Cut away the very first drop of tissue that emerges from the neck and transfer to the drop of liquid. This will contain both sets of salivary glands that appear as opalescent or transparent trilobed "bunches-of-grapes."
6. Gently tease the glands apart with needles, or homogenize gently (*see* **Subheading 3.4.5.**) to rupture the glands.
7. Observe under a ×40 objective lens, preferably using phase-contrast microscopy. Sporozoites are straight or sinuous wormlike cells 12-20 µm in length.

3.4. Maintenance of P. berghei In Vitro

It is not now essential to maintain rodents to transmit *P. berghei* in the laboratory. It must be emphasized, however, that it is considerably more expensive and time-consuming to maintain the parasite in culture. At present it is also essential to obtain fresh reticulocytes from rats regularly to culture the blood-stage parasite, thus the use of animals cannot be avoided. It is now possible to culture the entire sporogonic cycle of *P. berghei* (*23*); however, a good mosquito colony remains essential. The ability to compare parasite development in vitro and in vivo permits penetrating insights into the parasites' biology.

3.4.1. Exoerythrocytic Stage Culture (13)

3.4.1.1. Host Cells

1. HepG2 cells are cultured by standard methods in HepG2–EE culture medium (*see* **Subheading 2.9.**) at 37°C under a gas mixture of 5% CO_2 in air.

2. To facilitate microscopic analysis of the parasites, overgrowth of the HepG2 cells is prevented by the irradiation of preconfluent cultures with 3000–3500 rad from a cobalt-60 (gamma) source. Replenish the medium following irradiation and infect cells only after a delay of 2–24 h.

3.4.1.2. PARASITES

1. Add salivary-gland sporozoites (*see* **Subheading 3.3.2.3.**) in the culture medium in as small a volume as possible, then leave for 3 h to promote contact between sporozoites and HepG2 cells and enhance subsequent invasion.
2. Change the medium 3 h after "infection," and at 24-h intervals thereafter.
3. At 48 h later, the blood-stage culture medium (*see* **Subheading 2.10.**) can be added for the released merozoites to invade, and blood-stage cultures can then be initiated. To assist in the rupture of hepatocytes and merozoite release, add a small sterile stirrer bar and stir vigorously for an extended period.
4. Cultures are best monitored by Giemsa staining of preparations wet-fixed in Bouin's fluid or by indirect fluorescent antibody test (IFAT) using an appropriate antibody (young stages: anti-CSP; mid-stages: anti Pbl-1; late stages: anti-MSP-1).

3.4.2. Blood-Stage Culture

See **ref. 14** for further details.

1. Prepare erythrocytes: induce reticulocylosis in a Wistar rat with phenylhydrazine (in this case at 60 mg/kg body wt).
2. Collect blood into a heparinized syringe (*see* **Subheading 2.3.**) 4–6 d after treatment.
3. Remove leukocytes by passing blood down a Whatman CF11 column in blood-stage culture medium (*see* **Subheading 2.10.**). Provided the medium is changed at intervals of 48 h (*see* **Note 8**), the eluted blood may be maintained in complete culture medium for up to 1 wk at 4°C.
4. The initial culture is established at a hematocrit of 8–12% and a parasitemia of 0.5–2.0%. Maintain cultures at 37°C under a constant flow of 5% CO_2 in 10% O_2 and 85% N_2 in cylindrical glass culture vessels with a surface:volume ratio of 0.5/mm. This culture equipment is described in great detail by Mons *(14)* and may be purchased from Glass Instruments BV (Arnemuiden, The Netherlands).
5. Stir the cultures constantly at 50 rpm; once every 24 h, stir the culture at 400 rpm to assist in the rupture of the accumulated schizonts.
6. Following rapid stirring, remove 20–50% of the infected RBC and replace with fresh uninfected RBC.
7. Change medium twice daily by stopping the magnetic stirrer, allowing the RBC to sediment, and adding fresh medium.
8. Feed gametocytes from such cultures by membrane feeding to mosquitoes, to continue the life cycle (*see* **Subheading 3.3.1.2.**). Concentrate culture to 40% hematocrit by centrifugation at 200*g* for 10 min at 37°C. Resuspend in heat-inactivated fetal calf serum and place in a membrane feeder. Alternatively, the gametocytes can be introduced into ookinete culture (*see* **Subheading 3.4.3.**).

3.4.3. Ookinete Culture

See **refs. 15** and **16**.

1. Add heparinized infected blood (from infections between 3 and 10 d old) to 10 vol of ookinete medium at 19–21°C. Maintain at this temperature in air for 24 h, after which time mature ookinetes will be found.
2. Phagocytic white blood cells can occasionally be very active in ookinete cultures, and severely deplete parasite numbers *(17,18)*. It is thus advisable to remove these with

P. berghei Life Cycle

Whatman CF11 cellulose powder columns, or Plasmodipure filters *(18)* (*see* **Note 9**). Prepare in a cold room a short, fat column containing 2.5 mL of packed CF11 for every 10 mL of blood to be treated. Wash the CF11 with complete ookinete medium until the eluate is the same color (purple) as the medium applied (usually 5–10 column volumes eluted). Immediately pass heparinized blood through the column at 4°C. Elute the blood with ookinete medium, saving only the most concentrated fractions (judged by eye).

3. Make the volume of the collected blood up to ×10 the original volume of blood collected and culture at 19–21°C as in **step 1** (*see* **Note 8**).

3.4.4. Culture of Sporogonic Stages

See **ref. 23** for full details.

1. Prepare culture vessels: For 8-well slide chambers add 90 µL cold Matrigel, then warm to 37°C for 30 min to harden. Add ookinetes (*see* **Subheading 3.4.3.**) and *Drosophilia* S2 cells (10:1 ratio; 10^4 ookinetes/well) and incubate in modified Schneider's medium at 19°C±1°C for 18–25 d. Medium and S2 cells are refreshed every 48–72 h.

Modified Schneider's medium is made up as follows. 83.48 mL Schneider's medium; 15 mL heat-inactivated fetal bovine serum; 23.8 mM sodium bicarbonate; 36.7 mM hypoxanthine; 200 µL lipoprotein and cholesterol; 44 mM paraamino benzoic acid; 10,000 U penicillin; 10 mg streptomycin; 20 mg gentamicin.

3.5. Purification of Life-Cycle Stages

3.5.1. Purification of Mixed Asexual Blood Stages

1. Infect a phenylhydrazine-treated mouse with *P. berghei* clone 2.33. Note parasitemia and bleed mouse when this is approx 20%. Collect blood into 10 vol of blood-stage culture medium (*see* **Subheading 2.10.**).
2. Centrifuge 5 min at 1300g. Take off supernatant and resuspend pellet in 5 mL of PBS.
3. Gently load cell suspension onto a 55% Nycodenz/blood-stage culture medium cushion. Centrifuge for 30 min at 1300g (without brake). Collect interface and wash twice in 10 mL of PBS (as in **step 2**).
4. Count cells using a hemocytometer.

3.5.2. Purification of Schizonts (see Notes 10 and 11)

1. Infect rat(s) (intraperitoneally) with 5×10^7 *P. berghei* 2.33 in 1 mL of PBS; by d 4 the parasitemia should be about 1%.
2. Anesthetize the rat. Collect heparinized blood in a 50-mL tube containing 20 mL of blood-stage culture medium.
3. Centrifuge blood for 8 min at 340g at room temperature. Discard supernatant, including the buffy coat.
4. Mix red blood cell pellet with 40 mL of blood-stage culture medium and add cell suspension to 1-L Erlenmeyer flask containing 80 mL of blood-stage culture medium. Gas the flask(s) with 10% O_2, 5% CO_2, and 85% N_2 for 5 min (200 cm^3/min). Close flask with a rubber stopper.
5. Place flask(s) on a shaker table at 37°C (temperature should not exceed 38°C) overnight (or for 20 h).
6. Make a Giemsa-stained blood smear from the culture and determine the number (abundance) of schizonts. If appropriate, gently load culture on top of a 55% Nycodenz–RPMI culture medium cushion (35 mL culture per 10 mL 55% Nycodenz cushion).
7. Spin cells at 1300g for 30 min without brake. Take off interface, which contains infected schizonts and leukocytes. Spin cells down at 1200g for 5 min.

3.5.3. Purification of Asexual-Free Gametocytes

This procedure is modified from **ref. 19**.

1. Infect phenylhydrazine-treated mice with strain 2.34 as above (*see* **Subheading 3.3.1.1.**).
2. On d 2 and 3 postinfection, treat mice with sulfadiazine (*see* **Subheading 3.2.5., step 2**), or on d 5–6 treat with pyrimethamine (*see* **Subheading 3.2.5., step 5**).
3. On d 4 (sulfadiazine) or 7 (pyrimethamine), collect blood (*see* **Subheading 1.4.1.**) in blood-stage culture medium (*see* **Subheading 2.10.**) at 37°C.
4. Preferably work in 37°C room from now on. Remove white blood cells using CF11 or Plasmodipur filters (*see* **Subheading 3.4.3.**). Spin cells down at 1200*g* for 5 min. Remove supernatant and pellet cells in 5 mL of PBS.
5. Gently load cell suspension on top of a 48% Nycodenz/blood-stage culture medium cushion.
6. Spin cells at 1300*g* for 30 min without brake. Take off the interface and wash twice in 10 mL of PBS.
7. Count cells using a hemocytometer.

3.5.4. Enrichment of Ookinetes

1. Prepare gametocytes as detailed in **Subheading 3.5.3.**
2. Set up ookinete culture as detailed in **Subheading 3.4.3.** After 24 h, check for presence of ookinetes by Giemsa-stained blood films.
3. Spin down culture at 1300*g* for 5 min. Remove supernatant and resuspend pellet in 10 mL of PBS.
4. Gently load cell suspension on top of a 20 mL of 55% Nycodenz/ookinete culture medium cushion. Spin cells at 1300*g* for 30 min without brake. Remove interface and wash twice in 10 mL of PBS.
5. Count ookinetes in a hemocytometer.

3.5.5. Purification of Thoracic Sporozoites

See **ref. 20** for further details.

1. Remove heads and abdomens from alcohol-sterilized mosquitoes with a sharp blade. Place thoraces into a minimal volume of HepG2–EE culture medium (*see* **Subheading 2.9.**).
2. Take a 0.5-mL Eppendorf tube, make a hole with a heated 19*G* needle in the tip of the tube and plug the tube with a small ball of glass wool.
3. Place thoraces in 200 µL of medium in the tube. Put this tube inside a 1.5-mL Eppendorf tube such that it is retained by the neck of the outer tube and spin at 3000*g* for 4 min.
4. Collect sporozoite suspension from the larger tube.

4. Notes

1. Mons *(14)* has suggested that the passage from cryopreserved blood as opposed to direct mouse-mouse passage, results in a higher gametocytemia in the ensuing infection. It has been speculated that the cell debris resulting from cells killed by the cryopreservation protocol is responsible for inducing an unknown response in the host that in turn promotes gametocytogenesis. Noting that in poorly cryopreserved preparations considerable parasite death will occur, it is not unexpected that the rate of parasite growth in the infected host is slower following inoculation of cryopreserved parasites. It is not unusual for the parasitemia in these infections to lag 1 or 2 d behind the normal pattern of infection when using passaged blood inoculation.
2. Infections of mice with low doses of malaria (10^4–10^5 parasites), may result in some—especially TO or CBA strains—dying of cerebral malaria on d 7 or 8 with low parasitemias. BALB/c mice are unlikely to do this. Mice that ultimately develop cerebral malaria may

P. berghei Life Cycle

initially show hind leg paralysis the day before death. Such animals should be humanely killed to avoid unnecessary suffering.

3. *P. berghei* ANKA 2.34 infections in mice (TO, CBA, and BALB/C strains) should rise rapidly and normally kill the mice within 1–2 wk. With this clone, parasitemias should reach high levels (i.e., over 50%), though sometimes there is a lag phase when parasitemias remain almost constant for several days at about 5–10%. Other infections, such as *Haemobartonella*, will interfere with the malaria, so any abnormal patterns of infection should be investigated.

4. The distribution of parasites is not random within a thin blood film. Large parasites (schizonts and mature gametocytes) are carried to the "tail" of the smear. We have found if two simple transects oriented in a St. Andrew's Cross across the whole smear are viewed, the parasites encountered approximate their real densities counted by other methods.

5. Parasite counts on malaria infections are routinely done on air-dried methanol-fixed Giemsa-stained thin films. For finding parasites in blood at the lowest detectable parasitemias, thick films can be used: make a puddle of blood instead of a smear, dry the film at 37°C for 20 min, and stain without fixing for about 5 min; wash thick films very cautiously as they have a tendency to float off the slide.

6. In contaminated preparations, it is easy for an inexperienced observer to misidentify spirochetes or other motile bacteria as male gametes. The distinguishing characteristic is that malarial male gametes move with alternate periods of fast and slow undulation *(21)*.

7. Remember that each anopheline mosquito will ingest 1–2 µL of blood. A 25-g mouse can lose no more than 500 µL of blood without a risk of cardiac failure. Therefore, feed only the appropriate number of mosquitoes for the body weight of the rodent.

8. Speed of working is critical to the success of use of CF11 or Plasmodipur filters (*see* **Note 9**) for ookinete culture. It is essential that the parasites are passed through the column before gametogenesis is induced, otherwise the flagellate male gametes become enmeshed in the column, thereby reducing or eliminating fertilization of the female gametes that will pass through the column (L. Grimes and R. Sinden, unpublished observations).

9. Instead of CF11 columns, Plasmdipur filters (Euro-Diagnostica B.V., Beijerinckweg 18, P.O. Box 5005, 6802 EA Arnhem, The Netherlands, tel: +31-26-3630364/ Fax: +31-26-3645111) can be used. Simply screw the filter onto a 50-mL syringe and gently push the blood through the filter. A maximum of 50 mL of diluted blood can be used.

10. The method described here can also be used to synchronize erythrocytic schizogony and gametocytogenesis *(14,22)*.

11. Two different strains that do not produce gametocytes have been described. The first is clone 2.33, the other is K173. The protocol described here can be used with both strains.

References

1. van den Berghe, L. (1954) The history of the discovery of *Plasmodium berghei*. *Indian J. Malariol.* **8**, 241–243.
2. Vincke, I. H. (1954) Natural history of *Plasmodium berghei*. *Indian J. Malariol.* **8**, 245–256.
3. Yoeli, M. (1965) Studies on *Plasmodium berghei* in nature and under experimental conditions. *Trans. R. Soc. Trop. Med. Hyg.* **59**, 255–276.
4. Bray, R. S. (1954) The mosquito transmission of *Plasmodium berghei*. *Indian J. Malariol.* **8**, 263–274.
5. Killick-Kendrick, R. (1978) Taxonomy, zoogeography and evolution, in *Rodent Malaria* (Killick-Kendrick, R. and Peters, W., eds.), Academic Press, London, pp. 1–52.
6. Bafort, J. (1968) The effects of low temperature preservation on the viability of sporozoites of *Plasmodium berghei*. *Ann. Trop. Med. Parasitol.* **62**, 301–304.
7. Collins, W. E., Morris, C. L., Richardson, B. B., Sullivan, J. S., and Galland, G. G. (1994) Further studies on the sporozoite transmission of the Salvador I strain of *Plasmodium vivax*. *J. Parasitol.* **80**, 512–517.
8. Hollingdale, M. R., Leland, P., Sigler, C. I., and Leef, J. L. (1985). *In vitro* infectivity of cryopreserved *Plasmodium berghei* sporozoites to cultured cells. *Trans. R. Soc.Trop. Med. Hyg.* **79**, 206-208.

9. Walliker, D. and Beale, G. (1993) Synchronization and cloning of malaria parasites. *Methods Mol. Biol.* **21,** 57-66.
10. Fowler, R. E., Sinden, R. E., and Pudney, M. (1995) Inhibitory activity of the anti-malarial Atovaquone (566C80) against ookinetes, oocysts, and sporozoites of *Plasmodium berghei. J. Parasitol.* **81,** 452-458.
11. Shmuklarsky, M. J., Klayman, D. L., Milhous, W. K., Kyle, D. E., Rossan, R. N., Ager, A. L., et al. (1993) Comparison of beta-Artemether and beta-Arteether against malaria parasites *in vitro* and *in vivo. Am. J. Trop. Med. Hyg.* **48,** 377–384.
12. Sinden, R. E. (1996) Infection of mosquitoes with rodent malaria, in *Molecular Biology of Insect Disease Vectors: A Methods Manual* (Crampton, J. M., Beard, C. B., and Louis, C., eds.), Chapman and Hall, London, pp. 67–91.
13. Sinden, R. E., Suhrbier, A., Davies, C. S., Fleck, S. L., Hodivala, K., and Nicholas, J. C. (1990) The development and routine application of high-density exoerythrocytic-stage cultures of *Plasmodium berghei. Bull.WHO* **68,** 115–125.
14. Mons, B. (1986) Intraerythrocytic differentiation of *Plasmodium berghei. Acta Leidensia* **54,** 1–83.
15. Sinden, R. E., Hartley, R. H., and Winger, L. (1985) The development of *Plasmodium* ookinetes *in vitro*: an ultrastructural study including a description of meiotic division. *Parasitol.* **91,** 227–244.
16. Janse, C. J., Mons, B., Rouwenhorst, R. J., Klooster Van Der, P. F. J., Overdulve, J. P., and Kaay Van Der, H. J. (1985) *In vitro* formation of ookinetes and functional maturity of *Plasmodium berghei* gametocytes. *Parasitol.* **91,** 19–29.
17. Sinden, R. E. and Smalley, M. E. (1976) Gametocytes of *Plasmodium falciparum:* phagocytosis by leucocytes *in vivo* and *in vitro. Trans. Roy. Soc. Trop. Med. Hyg.* **70,** 344,345.
18. Janse, C. J., Camargo, A., Delportillo, H. A., Herrera, S., Waters, A. P., Kumlien, S., Mons, B., and Thomas, A. (1994). Removal of leucocytes from *Plasmodium vivax* infected blood. *Ann.Trop.Med.Parasitol.* **88,** 213–216.
19. Beetsma, A. L., Vandewiel, T. J. J. M., Sauerwein, R. W., and Eling, W. M. C. (1998). *Plasmodium berghei* ANKA: purification of large numbers of infectious gametocytes. *Exp. Parasitol.* **88,** 69–72.
20. Ozaki, L. S., Gwadz, R. W., and Godson, G. N. (1984) Simple centrifugation method for rapid separation of sporozoites from mosquitoes. *J. Parasitol.* **70,** 831–833.
21. Sinden, R. E. and Croll, N. A. (1975) Cytology and kinetics of microgametogenesis and fertilization of *Plasmodium yoelii nigeriensis. Parasitol.* **70,** 53–65.
22. Janse, C. J. and Waters, A. P. (1995) *Plasmodium berghei*: the application of cultivation and purification techniques to molecular studies of malaria parasites. *Parasitol. Today* **11,** 138–143.
23. Al-Olayan, E., Beetsma, A. L., Butcher, G. A., Sinden, R. E., and Hurd, H. (2002) Complete development of mosquito phases of the malaria parasite in vitro. *Science,* **295,** 677–679.

4

Mouse Models for Pre-Erythrocytic–Stage Malaria

Laurent Rénia, Elodie Belnoue, and Irène Landau

1. Introduction

The erythrocytic stages of the malarial parasites were first observed in man by Laveran in 1880 *(1)* and in birds by Danilevsky in 1884 *(2)*. This was followed by the discovery and elucidation of the sporogonous cycle by Ross (1897) *(3)* and Grassi (1900) *(4)*. Schaudinn produced an apparent link between these two phases, as he claimed to have observed the penetration of a sporozoite into an erythrocyte *(5)*. However, this observation could never be repeated, and scientists remained puzzled by the latency between sporozoite injection and the appearance of parasites in the blood. They, therefore, suspected that there was a tissue phase. This phase was not identified until 40 yr later, first in birds *(6,7)*, and then in monkeys and men *(8,9)*. It was demonstrated that subcutaneous macrophages were the target of avian malaria sporozoites and that hepatocytes were the target of mammalian sporozoites.

Most of the discoveries about the nature and the biology of the life cycle of *Plasmodium* parasites have been intimately linked to the use of models. *P. gallinaceum* was described by Brumpt in 1935 *(10)*. It was very popular because there was a need to develop new antimalarial drugs during and after World War II. This organism was capable of infecting domestic fowls, and its most common experimental vector (*Aedes aegypti*) was easily maintained in the laboratory. This model (and also *P. lophurae*) provided a wealth of data *(11)*. Extrapolations from these studies paved the way for studies with human or other animal malaria parasites. Nevertheless, strict extrapolation to human malaria was impossible because *P. gallinaceum* exo-erythrocytic stages (EES) occur inside macrophages after sporozoite injection and consequent generations of EES are produced from cryptozoites or from merozoites from blood-stage infections *(11)*. This phenomenon has never been observed with mammalian parasites because EES occurs in hepatocytes only after sporozoite injection; hence, mammalian sporozoites and EES have been grouped under the denomination of pre-erythrocytic stage (PE).

Understanding this phase of the malaria parasite required new and more accessible tools. The introduction of rodent plasmodia, *P. berghei* in 1948 *(12)*, *P. vinckei* in 1952 *(13)*, *P. chabaudi* in 1965 *(14)*, and *P. yoelii* in 1965 *(15)*, each with its specific morphology and red cell tropism and suitable for a wide range of small laboratory animals, brought a new dimension to experimental malariology. The liver stages of mouse malarias have been extensively studied now and do not differ from one another or from those of the species that infect primates, except for their rapid development (48–68 h) *(16–19)*.

Liver stages are initiated in the laboratory after injection of sporozoites directly into the blood or by mosquito bites. Sporozoites injected intravenously reach the liver and disappear from the blood after 30–60 min *(20)*. Mosquito bites inject sporozoites into the avascular skin tissue *(21)* before blood ingestion by the mosquito. The sporozoites migrate to the broken vessel or may travel to the liver through the lymphatic system *(22)*. It is now clear that the sporozoite enters the hepatocyte through a complex series of molecular interactions involving at least two of its proteins *(23)*. The subsequent development takes place in a vacuole, where the sporozoite changes into a rounded form, the trophozoite, that grows and gives rise to the formation of 4000–40,000 merozoites. The schizonts burst at the end of schizogony and release merozoites into the circulation. These then initiate the blood infection.

Hamsters, rats, laboratory-bred tree rats, *Thamnomys rutilans* (the natural host of *P. yoelii*, *P. chabaudi*, and *P. vinckei*) *(14–16)*, *Grammomys surdaster* (the natural host of *P. berghei,* and possibly *P. vinckei*) *(12)*, and mice are suitable hosts in which to study rodent malaria parasites. Although they are not always as susceptible to sporozoite infection as rats, mice have been extensively used because they are easy to handle; there is now a wide range of genetically defined strains, and many immunological and genetic tools are available.

Mouse models of PE have been used to study the biology and the chemotherapy and to elucidate the immune mechanisms involved in protection against PE. Such models also permit studies that would be otherwise impossible—for example, the investigation of host or parasite genetic factors in susceptibility to the infection and the development of model(s) of protective immunity by immunization with irradiated sporozoites. The relative ease with which infected mosquitoes can be reliably obtained and the development of hepatic culture systems have also allowed studies on several aspects of PE. Two of the four murine species, *P. yoelii* and *P. berghei,* are used routinely in a few laboratories. Most of the data have been obtained with a limited number of lines and clones of these two parasites infecting a limited number of mouse strains. Because every species and every line or clone of a particular species has its particular characteristics, the resulting infection may vary considerably from host to host, or even within strain of the same host. It may, therefore, be difficult to extrapolate from one rodent model to another and from rodent models to man. Nevertheless, these models have provided an enormous amount of information about PE. Most of the work has been done with sporozoites and liver schizonts, but a few reports have provided data on liver merozoites.

The complete life cycles of *P. berghei* and *P. yoelii* have been established in many laboratories and the production of infective sporozoites is more or less standardized. These two species have a number of common characteristics:

1. Their livers stages are of short duration in vivo (under 50 h).
2. Rat, hamster, and mouse hosts are spontaneously receptive to the blood stages
3. The majority of their lines or clones have a predilection for reticulocytes in the circulation.

The following descriptions of PE of *P. berghei* and *P. yoelii* are based on those of Killick-Kendrick *(24)* and Landau and Boulard *(25)*.

P. berghei (12). Two gametocyte-producing lines of the strain ANKA and NK65 are used. The best temperature for schizogony is 19–21°C and sporozoites appears in the glands from d 14 after mosquito blood feeding. Liver stage development takes less than 50 h *(16–18,26)*. The estimated number of merozoites produced is 5000–8000.

P. yoelii yoelii (15). The gametocyte-producing lines used routinely are 17X and 265 BY. The best temperature for schizogony is 24°C and sporozoites appear in the glands from d 9 after mosquito blood feeding. Liver stage development takes less than 50 h *(16)*. The estimated number of merozoites produced is 7500–8000.

The following sections give detailed descriptions of the procedures required to induce and study PE in vivo, establish a model of protection by immunization with irradiated sporozoites, and study PE in vitro.

2. Materials
2.1. Induction and Study of PE
2.1.1. Animals

1. Mice (any strain may be used) (*see* **Note 1**). They should be housed in an animal facility free from any endemic pathogens (e.g., mouse hepatitis virus) that interfere with PE development.

2.1.2. Mosquito Dissection

1. *Plasmodium yoelii* 265 BY or 17X-infected *Anopheles stephensi* mosquitoes. For details, *see* Chapter 3 and **Note 2**)
2. Neubauer or Malassez chamber.
3. Sterilin tubes.
4. Ice bucket and ice.
5. 1-mL Syringes.
6. 26-gage needles.
7. A glass homogenizer (10 mL) (Thomas, Philadelphia, PA).
8. Sterile microscope slides.
9. Sterile tips for 200-µL pipet.
10. Small Petri dishes (60-mm diameter, Corning).
11. 70% ethanol.
12. 0.9% NaCl with 1% penicillin/streptomycin (Life Technologies Inc.): prepare a 100X stock solution, aliquot and store at –20°C.
13. Washing medium A: William's medium containing 5 µg/mL of Fungizone (Sigma).
14. Washing medium B: William's medium containing 2% penicillin–streptomycin.

2.2. Sporozoite Infection In Vivo

1. 1 mL Syringes.
2. Sterile 0.9% NaCl (passed through a 0.22-µm filter).
3. Beaker with warm tap water.
4. Holder for restraining mice (Polylabo, Paris, France).

2.3. Histology
2.3.1. Classical Histology (see Note 3)

1. Carnoy's fixative: 60% absolute ethanol, 30% chloroform, and 10% acetic acid; mix immediately before use under a chemical hood.
2. Butanol.
3. Paraffin solution.
4. Histoplast (Shandon).

5. Rehydration kit: toluene, 100, 90, and 70% ethanol.
6. Phosphate buffer: Prepare 10X solution by dissolving 5 g of $Na_2HPO_4 \cdot 2H_2O$, and 2 g of KH_2PO_4 in 1 L of distilled water; adjust to pH 7.4 and filter through a 0.45-µm filter. Prepare 1X working stock in distilled water.
7. Giemsa solution: Dilute Giemsa (Merck) 1 in 8 in 1X phosphate buffer.
8. Acetone.
9. Colophonium resin: dissolve in acetone at 150 mg/mL.
10. Toluene.
11. Mounting medium: Eukit (O. Kindler GmbH and Co).

2.3.2. Immunohistology

1. OCT embedding compound (Miles Labs).
2. Liquid nitrogen in a container covered with aluminum foil.
3. Tweezers.
4. Moist chamber.
5. Cryostat.
6. Cork oak container: a 0.3-cm-wide slice of cork oak surrounded by a transparent paper, to make a flexible container.
7. Pap pen (Research Products International or Shandon/Lipshaw).
8. PBS-azide: 0.001% (w/v) sodium azide (Sigma) in 1X phosphate-buffered saline (PBS) (pH 7.5). Take great care when handling sodium azide, as this chemical is highly toxic. Use gloves and a mask when manipulating the product in powder form. Store at 4°C.
9. Acetone (reagent grade), absolute methanol, and formaldehyde (Sigma).
10. Silanized slides: Clean slides by washing in acetic acid and in 70% ethanol. Soak slides in a bath containing 5 mL of TESPA (3-aminopropyltriethoxysilane, Sigma) in 200 mL of acetone. Wash slides twice in acetone, then twice in deionized water. Drain off excess water by touching one side of the slide on a tissue paper. Dry the slides at room temperature under a ventilated hood.
11. Blocking solution: 5% goat serum (Jackson laboratories) or 1% bovine serum albumin (Sigma) in PBS.
12. Mouse anti-PfHSP70.1 *(27)* serum: make aliquots and store at –20°C.
13. Fluorescein isothiocyanate (FITC)-labeled goat anti-mouse IgG (lyophilized, Biosys, Compiègne, France). Store lyophilized product at 4°C. Reconstitute with distilled sterile water at 1 mg/mL, make aliquots, and store at –20°C.
14. Evan's Blue (Sigma): Prepare a 100X stock of 5 % (w/v). Store at 4°C.
15. Aquamount (Shandon/Lipshaw).
16. Cover slips, 20 × 60 mm.

2.4. Model for PE Immunity: Irradiated Sporozoites

1. X-ray machine.
2. 1-mL Syringes.
3. Sterile 0.9% NaCl.

2.5. PE Studies In Vitro

2.5.1. Isolation of Hepatocyte

For details regarding reagents for hepatocyte perfusion, *see* Chapter 46. For hepatocyte isolation, the following materials are needed:

1. Female mice, 7–12 wk-old.
2. Eight-chamber plastic Lab-Teck slides (Nunc Inc., IL).

3. Dissecting materials (tweezers, scissors).
4. Complete medium: William's medium (Life Technologies, Edinburgh, Scotland) supplemented with 10% fetal calf serum (make aliquots and store at –20°C; Dutscher), 1% penicillin/streptomycin, and 0.5 µg/mL amphotericine B. Prepare fresh and store at 4°C.

2.5.2. Parasite Staining

1. Moist chamber.
2. PBS-azide.
3. Absolute methanol. Store at room temperature in a special cabinet. Store a 10-mL aliquot at –20°C before staining parasites.
4. Mouse anti-PfHSP70.1 serum.
5. FITC-labeled goat anti-mouse IgG.
6. Evan's Blue (Sigma): prepare a 100X stock of 5% (w/v). Store at 4°C.
7. PBS/glycerol.
8. Cover slips, 20 × 60 mm.

3. Methods

3.1. Induction and Study of PE In Vivo

3.1.1. Isolation of Sporozoites

1. Collect infected mosquitoes with sporozoites in their salivary glands in a sterilin vial.
2. Shake the vial vigorously to stun the mosquitoes.
3. Catch the stunned mosquitoes with tweezers and immerse them in 70% ethanol.
4. Transfer them to a small Petri dish containing 3 mL of 0.9% NaCl supplemented with 1% penicillin–streptomycin.
5. Place mosquitoes on a slide and dissect them under a microscope using sterile 24-gage needles. Cut off the head of mosquitoes; use light pressure on the thorax to release the salivary glands (which appear translucid), pull the salivary glands away and immerse them in a drop of medium (50 µL) on the opposite side of the dissecting slide. Take care to avoid bringing the legs or other part of the mosquito's body into contact with the salivary glands.
6. At the end of the dissection of a series of mosquitoes (10–15 per slide), collect salivary glands and place them in the glass homogenizer (kept on ice).
7. Homogenize the glands and dilute the resulting sporozoite suspension with complete medium in preparation for counting.
8. Place a suspension of sporozoites in a Neubauer or Malassez counting chamber and allow to stand for 5–10 min before counting. The sporozoites sink slowly, especially in a small volume and in a thick suspension, so counting them before they sink will be difficult and inaccurate. Phase-contrast illumination facilitates rapid, accurate counting. The most convenient magnification for counting sporozoites is 300X.

3.1.2. Injection of Sporozoites

1. Dilate the mouse tail vein by immersing it in warm water for 1–3 min.
2. Inject sporozoites at the desired concentration in a 0.1-mL volume into one of the tail veins. Any resistance when pushing the syringe plunger means that the needle is not in the vein but rather in the skin. Remove the needle and inject into another site or into another vein.

3.2. Detection of Liver Stages In Vivo

3.2.1. Histology (see Note 3)

1. Euthanize the mouse and immediately remove small (1-cm) pieces of liver. Immerse them into Carnoy's fixative. Handle with fingers, never with forceps.

2. Leave pieces of tissue for 6–12 h in the fixative and then place them in butanol (twice). They can be stored in butanol indefinitely.
3. Transfer the tissue pieces directly from butanol to the embedding medium (paraffin).
4. Cut 5-μm-thick sections.
5. Rehydrate the sections by washing them for 10 min in acetone, 20 min in 100% ethanol, 15 min in 90% ethanol, 20 min in 70% ethanol, and, finally, in water.
6. Stain the sections with a 10% Giemsa solution in phosphate buffer for 1 h.
7. Immerse the slides in tap water for an additional 1 h.
8. Differentiate with 2–5 drops of the acetone–Colophonium solution. The number of drops applied to the sections depends on the color obtained. The sections should appear blue, not pink.
9. When the desired color is obtained, rinse the slide with 60% acetone and wash in toluene 100% and then 40% toluene. A white depot means that the Colophonium has not been removed; in that case, rinse again with acetone and toluene.
10. Mount slides in Eukit.
11. The cytoplasm of a successful preparation should appear blue and the nuclei red.

3.2.2. Immunohistology

3.2.2.1. Freezing of Tissue Samples

1. Remove the liver and place in a Petri dish containing PBS. Cut into pieces of 0.5 cm^3.
2. Place each piece into the center of the cork oak container.
3. Add enough OCT embedding compound to the fragment to cover the tissue.
4. Immerse the container in liquid nitrogen for 5–10 min.
5. Store the container at –80°C.

3.2.2.2. Cutting Cryosections of Tissue

1. Remove the container from dry ice or liquid nitrogen and place in the cryostat. The cryostat should always be at approx –20°C.
2. Cover the cryostat chuck with a thin layer of OCT embedding compound and place the tissue on it, correctly orientated.
3. Allow the tissue preparation to freeze solid (10–15 min at –20°C) inside the cryostat. If the cryostat is warmer or colder than the recommended temperature (–20°C), cutting will be difficult and may damage the tissue.
4. Set the cryostat to cut 5-μm-thick sections.
5. Consecutive sections (maximum of three) of tissues are collected on the silanized slide. The difference in temperature will flatten the section so that they stick to the slide during contact.
6. Allow the sections to air-dry overnight at room temperature.

3.2.2.3. Section Fixation and Staining

1. Immerse the slides in a staining dish containing acetone for 5 min and then air-dry for 10 min at room temperature. Acetone should be tried first because it preserves antigenic determinants, but it may distort the liver tissue. If distortion is observed, use methanol; alternatively, use 4% formaldehyde in PBS but note that some antigen determinants may be lost.
2. Encircle the tissue sections on the glass slides using a Pap pen. The Pap pen (water repellent wax) creates a hydrophobic barrier that prevents the reagents spreading over the slide. This also allows the use of small quantities of reagent with minimal risk of drying. Stain slides as soon as possible. If they must be stored, store at –70°C but note that prolonged storage can lead to the denaturation of certain epitopes.
3. Place the slides in a moist chamber and add PBS-azide to the slides. PBS should be added carefully because the force of the liquid could remove or alter the tissue.

Mouse Models for Pre-Erythrocytic Stages

4. Incubate the slides for 5 min at room temperature. Avoid letting the slides dry. Carry out all incubations in the moist chamber.
5. Remove PBS-azide by capillary action, by placing a corner of the slide on a paper towel.
6. Add anti-PfHSP70.1 serum (diluted 1/100 in PBS-azide) to the tissue within the area marked by the Pap pen, and incubate for 30 min at 37°C.
7. Wash the sections twice carefully with PBS-azide and incubate with FITC-labeled anti-mouse antibody diluted 1/100 in PBS-azide with 0.5% Evan's Blue for 30 min at 37°C. The solution should appear light blue and not dark blue (which indicates that too much Evan's blue was added).
8. Wash the sections twice carefully with PBS-azide and place 1–2 drops of aquamount on the tissue.
9. Place a cover slip on the tissue. Slides can be stored at 4°C for no more than 1 mo.

3.3. Model for PE Immunity: Irradiated Sporozoites (see Note 4)

1. Keep sporozoites on ice.
2. Irradiated at 12,000 rads. The time of irradiation depends on the radiation source.
3. Primary immunization: Adjust sporozoite concentration to 75,000 in 0.1 mL of 0.9% NaCl and inject into the mice via the tail vein.
4. First boost (2 wk after primary immunization): Inject mice with 25,000 sporozoites in 0.1 mL of 0.9% NaCl.
5. Second boost (1 wk after first boost): Inject mice with 25,000 sporozoites in 0.1 mL of 0.9% NaCl.
6. Challenge: Isolate live sporozoites from infected mosquitoes and dilute in sterile 0.9% NaCl to a concentration of 4000 sporozoites in 0.1 mL. Inject the mice intravenously in the tail vein.
7. Monitor the blood parasitemia from d 3 to d 12 after challenge by Giemsa staining of thin blood smears (for a detailed protocol, *see* Chapter 5). Protected mice will have no blood-stage parasites (*see* **Note 5**).

3.4. In Vitro Model of PE (see Notes 6–8)

3.4.1. Isolation of Mouse Hepatocytes

See Chapter 46 for details.

1. Prepare hepatocytes by collagenase perfusion of a liver biopsy from BALB/c mice.
2. Seed hepatocytes (60×10^3 in complete medium) in Lab-Tek wells.
3. Incubate for 24 h in 3.5% CO_2 at 37°C before use. This time is normally sufficient to allow hepatocytes to adhere to the Lab-Teck slides and become confluent.

3.4.2. Sporozoite Preparation

1. Prepare mosquitoes (*see* **Subheading 3.1.**) and wash them in 70% ethanol.
2. Transfer washed mosquitoes to a small Petri dish containing 3 mL of washing solution A. Wash twice in two other Petri dishes containing the same medium. Then, wash twice in Petri dishes containing washing solution B.
3. Place the mosquitoes on a sterile slide and dissect out the salivary glands under a microscope using sterile 24-gage needles on syringes. Legs or other part of the mosquito's body should not be mixed with the salivary glands.
4. Collect the salivary glands are with a 200-µL pipet using a sterile tip and transfer them to a small tissue grinder, kept on ice.
5. Homogenize the glands and dilute the resulting sporozoite suspension. Count the sporozoites.

3.4.3. Sporozoite Infection In Vitro

1. Add 6×10^4 sporozoites in 50 µL of complete medium to each well.
2. Incubate cultures for 3 h in 3.5% CO_2 at 37°C.
3. Wash cultures three times with complete medium to remove sporozoites, which will not have completed their entry. These washes also remove most fungi or bacteria in the sporozoite suspension; omission of this step may lead to culture contamination. For all washing steps, the volume of the well should be removed using a tip placed in the corner of each well and the new solution should be added slowly. Lab-Teks should be handled carefully to avoid damaging hepatocytes.
4. Change the medium 21–24 h after sporozoite infection.
5. Incubate cultures for an additional 24 h.

3.4.4. Parasite Staining (see Note 7)

1. Stop cultures 45–48 h after sporozoite inoculation. Remove the medium completely removed using a 1-mL pipet tip.
2. Wash cultures twice with PBS-azide.
3. Add cold methanol (0.3 mL) to each well and incubate at room temperature for 10 min.
4. Remove the methanol and wash culture twice with PBS-azide.
5. Add anti-PfHSP70.1 serum (1/100 dilution in PBS-azide; see **Note 7**) to the hepatocyte culture and incubate for 30 min at 37°C.
6. Wash cultures twice with PBS-azide and incubate for 30 min at 37°C with FITC-labeled anti-mouse antibody diluted 1/100 in PBS-azide with 0.5% Evan's Blue.
7. Wash Lab-Tecks twice with PBS-azide and remove the wells.
8. Mount the remaining slides with PBS-azide/glycerol and store at 4°C before observation.
9. Read slides with an epifluorescence microscope at 250X to 400X magnification.

4. Notes

1. Rodent models of PE. Despite their obvious advantages (see **Subheading 1.**), mouse models for PE might not be the best suited to answer particular questions. Keep in mind that the sporozoites of *P. yoelii* or *P. berghei* are more infectious and develop more rapidly and completely in their respective natural host (*Thamnomys* and *Grammomys*) than they do in mice *(17,28,29)*. The immune mechanisms in different species may also vary. Remember that nonspecific resistance is frequent in unnatural hosts (rats or mice) *(30–34)*, whereas it is rare in the natural host *Thamnomys (28)*. This innate immune response of unnatural hosts is also characterized by an infiltrate of mononuclear cells, neutrophils, and eosinophils around late-stage schizonts and emerging merozoites and is likely to contribute to the lower susceptibility of mice to *P. berghei* infection, as compared with *Thamnomys*, and of mice to *P. berghei* as compared with *P. yoelii (31)*.

   When a new strain of mice has to be used, preliminary experiments should be performed to determine the minimum number of sporozoites required for infection. This number could vary between strain of parasites and even between lines of the same parasite infecting a given strain of mouse. The best route of sporozoite inoculation for inducing a blood infection is intravenous (either through the tail vein or through the retro-ocular sinus). Mice injected intravenously with sporozoites have more schizonts than animals given the same number intraperitoneally. However, the density of infection varies less in mice given sporozoites intraperitoneally *(26,35)*. Nevertheless, it should be kept in mind that the natural infection is via mosquito bite, and, because the details of the sporozoite journey to the liver are still uncertain, it may be important to confirm results by doing challenge by mosquito bites.

Many liver schizonts in vivo may be needed for certain experiments (i.e., for histological studies). They can be obtained by injecting a very large number of sporozoites intravenously (20–50 times more than normally required to induce a blood-stage infection). Alternatively, they may be obtained by injecting sporozoites intraportally (*see* Chapter 46) *(36,37)* or directly into the liver *(38)*. The major advantage of the latter method is that it can be performed with a very low number of sporozoites and this could be very important with lines or clones of parasites grown in mosquitoes that are not efficient vectors.

Mosquitoes are stunned to facilitate dissection. They may be also anesthetized using CO_2. Mass dissections of salivary glands is time-consuming and labor-intensive. Alternative techniques based on the grinding of thoraces have been proposed *(39–41)*. However, this seems to save little time and the suspension so produced contains debris that may kill the host animal when given intravenously *(39)*. A purifying step may provide a clean suspension of *P. berghei* sporozoites *(41)*.

2. Sporozoite production. Regular production of sporozoites is the major limiting factor for studies on PE. Details of sporozoite production are provided in Chapter 3, but we will comment on a few points here. Regular cyclical transmission requires gametocyte-producing strains. Parasite strains maintained by blood passage usually lose the capacity to produce gametocytes. Even in the laboratory, where a routine cyclical transmission of murine malaria parasites has been established, sporozoites sometimes become unobtainable or noninfective for no apparent reasons. This could be the result of a very small variation of temperature or humidity in the insectary, but often no explanations can be found. Keep in mind that the development in the mosquito takes 10–18 d because these limitations affect the planning and the number of experiments that can be performed in a relatively short period of time.

Only a few clones and lines of *P. berghei* and *P. yoelii* are propagated in the laboratory through cyclical transmission. Lines contain different clones, and clones may undergo phenotype variation *(42,43)*. Passage through mosquitoes may result in a large number of new clones being created by recombination, which can occur during the sexual stage when the parasite undergoes meiosis *(44,45)*. This can alter the production of sporozoites, modify the growth characteristics of PE and subsequent liver stages, and lead to conflicting results when studying immune mechanisms against PE. In such circumstances, cyclical transmission should be reinitiated with deep-frozen batches or from parasites obtained from another laboratory.

Cyclical transmission of *P. vinckei* and *P. chabaudi* has never been performed routinely in the laboratory because it is more difficult with these species than with P. berghei and P. Yoelli, although it has been done successfully *(16,26,46,47)*. Despite the widespread use of *P. chabaudi* in mice as a model for blood stage malaria (*see* Chapter 5), only a few studies have been done on PE of this stage *(16,26,46–48)* or have used sporozoites to induce a blood-stage infection *(49,50)*.

3. Detecting the parasite. The best staining of hepatic schizonts is obtained with the Giemsa–Colophonium method *(51)*, not with fixatives containing paraformaldehyde. It allows the detection of parasites in the liver as soon as they possess four to five nuclei (15–25 h, depending on the species). Morphological alterations, such as wrong division of nuclei containing large and condensed aggregates, retarded or immature schizonts, basophilic aggregates in the cytoplasm, and vacuolization, are readily seen with this staining. These morphological alterations provide a useful indication in inhibition studies.

Histological analysis also provides important information on *in situ* events. It can reveal the presence of granulomas or cellular infiltrates in normal or immunized animals. The cellular composition of the infiltrates can be determined with antibodies to different cell subsets and/or to different lymphokines produced by these cells. For example, this type of analysis has revealed that infected hepatocytes, from a rat immunized with irradiated sporozoites and subsequently challenged with live sporozoites, expressed the nitric oxide synthase. The product of this enzyme, nitric oxide, is thought to be involved in the protection conferred by irradiated sporozoites *(52)*.

Immunofluorescence studies on cryosections or Carnoy-fixed sections can be used to study the production of proteins during the liver stages. Anti-Pfhsp70.1 antibodies recognize all four rodent malaria species (*see* **Note 7**). Antibodies against other antigens may be used, but a negative result does not necessarily mean that the antigen is absent. An antigenic site may be denatured by the fixatives. It may be better to test antibodies on in vitro-infected hepatocytes to ascertain the presence of antigen(s) during the liver stage.

4. Model for PE immunity: irradiated sporozoites. Protection occurs only when irradiated sporozoites are injected by a route enabling liver infection *(48)*. They give very little or no protection when injected intramuscularly, intraperitoneally, subcutaneously, or even orally, either with or without adjuvants *(53–56)*. X-ray-irradiated sporozoites need to be alive: too low an irradiation dose leads to blood-stage infection, whereas too high a dose leads to the absence of protection *(57,58)*. At the dose (10,000–15,000 rads) that results in protection but not blood infection, sporozoites transform into trophozoites, which remain blocked in their development, unable to divide *(57–62)*. Induction of a protective response depends greatly on the host/parasite combination, the dose of irradiation, the number of sporozoites injected, and the number of immunizations *(61,63–65)*. Thus, a single injection of as few as 1000 irradiated *P. berghei* sporozoites confers protection in A/J mice, and protection is always increased when going from a single injection to three injections of either *P. yoelii* in BALB/c mice or *P. berghei* in A/J mice *(65)*. Thus, protection appears to be inversely correlated with the susceptibility of the host to the sporozoite. For example, C57BL/6 mice are extremely sensitive to the development of *P. berghei* and are the most difficult to protect, requiring three immunizations with 30,000 irradiated sporozoites. On the other hand, BALB/c are quite refractory to *P. berghei* infection and are the easiest to protect *(65)*. The dose of sporozoites and the number of immunizations required to obtain protection should be determined for a strain of mice that has never been tested previously. A suspension of salivary glands corresponding to an equal number of sporozoites should be used as a control in preliminary experiments to rule out any nonspecific effects.

Sporozoites can also be attenuated by γ-irradiation. Both methods produce attenuated *P. berghei* sporozoites able to confer protection against a sporozoite challenge *(59,66)*. Ultraviolet irradiation has been used with *P. gallinaceum* sporozoites and is effective *(67)*, but this method has never been used with rodent parasites and its potentiality remains to be investigated.

Immunization of mice and rats with liver trophozoite-infected hepatocytes produced by injecting irradiated or normal *P. yoelii* or *P. berghei* sporozoites *(58,68)* can also protect against an infectious sporozoite challenge. These models have not been widely used, but they deserve further study.

5. Quantification of parasites in the liver. Successful liver stage development results in blood-stage infection. The effects of drugs or other inhibitory factors can be determined on the basis of the emergence—or not—of a blood infection (or sometimes a delay in patency of infection). However, because only a single schizont is sufficient to result in a blood-stage infection, techniques have been developed to quantify liver-stage development. Microscopic investigation has been rarely used because it is time-consuming and labor-intensive *(37,69)*. Only recently have molecular techniques (polymerase chain reaction, Northern blotting) been developed to quantify the liver stage load, thus simplifying the assessment of protection (*[70–74]*; *see* Chapters 14 and 15).

6. Sporozoite development in vitro. Hepatocytes can be grown in different plastic substrates or on collagen-coated glass wells. A particular substrate should be used to suit the type of

experiment to perform. If few schizonts are needed, for titration of antibodies for example, microcultures containing 2000 hepatocytes in a Petri dish may be used. If more schizonts are needed (i.e., for inhibition assays [see Chapters 47 and 49]), Lab-Teks are the material of choice.

The rate of sporozoite development in vitro is usually low when mouse hepatocytes are used. In a successful experiment, 50–500 schizonts per Lab-Tek well are obtained. Primary hepatocyte cultures need to be produced for each experiment and this requires trained personnel. Cultures also frequently become contaminated with bacteria or fungi present in the solution containing the sporozoites despite the presence of antibiotics and antimycotics. The tumoral hepatoma cell line HepG2-A16 *(75–77)* has, for practical reasons, been much used more often to grow *P. berghei* than has the more problematic mouse primary culture of hepatocyte *(78)*. However, the susceptibility of HepG2-A16 to *P. berghei* may be totally artificial, contrasting with its inability to permit intracellular development of *P. yoelii* and *P. falciparum (79,80)*. This demonstrates critical differences that may also be a concern with regard to their susceptibility to invasion, and results obtained with this cell line should be confirmed with primary hepatocyte culture.

7. Parasite staining. The parasite can be detected with mouse sera containing antibodies that react to other liver-stage antigens. Only a limited number of antigens expressed by the liver parasite have been identified, limiting the variety of antibodies that may be used to detect the parasite. Antibodies to PfHsp70.1 *(27)* are very useful because of the high conservation of malaria Hsp; all malaria liver stage parasites can be recognized. Antibodies against antigens such as MSP1 *(81,82)*, 17kD *(83)*, and SSP2 *(84)* can also be used, but they are mostly species-restricted. The reactivity of the anti-liver-stage antibodies should be checked regularly and the appropriate dilution to be used determined. Avoid repeated cycles of freezing and thawing to prevent antibody denaturation.

8. Inhibition assays. In vitro cultures have been used to determine the effect of antibodies, drugs, and T-cells on the development of the developing liver stage. Assays such as the double-staining assay (*see* Chapter 47) *(85)* or the inhibition of liver-stage development assay (ILSDA) (*see* Chapter 48) *(86)* have been used to measure the inhibitory effect of antibodies drugs and other molecules. Another more specific assay, the TILSA, has been devised to study the action of T-cells against liver stage parasites (*see* Chapter 49) *(87–90)*.

Acknowledgments

Some of the aforementioned work was made possible in part by financial support from Institut Electricité et Santé, UNDP/World Bank/WHO Special Program for Research and Training in Tropical Diseases (TDR), the European Community programs INCO-DC, Fondation pour la Recherche Médicale, and the Junta Nacional de Investigação Cientifica e Tecnologica (JNICT). Elodie Belnoue is supported by a predoctoral fellowship from the Ministère de l'Education Nationale, de la Recherche et de la Technologie. We thank Dr. Jean-Gérard Guillet for his active support. The English text was edited by Dr. Owen Parkes.

References

1. Laveran, A. (1880) Note sur un nouveau parasite trouvé dans le sang de plusieurs malades atteints de fièvre palustre. *Bull. Acad. Med. (Paris)* **9,** 1235,1236.
2. Danilevsky, B. (1884) On the parasites of the blood (Haematozoa). *Russk. Med.* **46–48,** 15–18.
3. Ross, R. (1897) On some peculiar pigmented cells found in two mosquitoes fed on malarial blood. *Br. Med. J.* **2,** 1786–1788.
4. Grassi, B. (1900) Studi di un zoologo sulla malaria. *Accad. Lincei (Roma)* **9,** 215–223.
5. Schaudinn, F. (1902) Studien über krankheitserregende protozoen, II. *Arb. Kais. Ges. Amt. Berlin* **19,** 169–173.

6. Raffaele, G. (1936) Presumibili forme iniziale di evoluzione di *P. relictum*. *Rev. Mal. (Italy)* **15**, 318–324.
7. James, S. P. and Tate, P. (1937) New knowledge of life cycle of malaria parasites. *Nature* **139**, 545.
8. Shortt, H. E. and Garnham, P. C. C. (1948) Demonstration of a persisting exoerythrocytic cycle in *P. cynomolgi* and its bearing on the production of relapse. *Br. Med. J.* **1**, 1225–1228.
9. Shortt, H. E. and Garnham, P. C. C. (1948) Pre-erythrocytic stage in mammalian malaria parasites. *Nature* **161**, 126.
10. Brumpt, E. (1935) Paludisme aviaire: *Plasmodium gallinaceum* n. sp. de la poule domestique. *Comp. Rend. Acad. Sci. (Paris)* **200**, 783–786.
11. Huff, C. G. (1963) Experimental research on avian malaria. *Adv. Parasitol.* **1**, 1–65.
12. Vincke, I. H. and Lips, M. (1948) Un nouveau *Plasmodium* d'un rongeur sauvage du Congo, *Plasmodium berghei* n. sp. *Ann. Soc. belge. Méd. Trop.* **28**, 197–204.
13. Rodhain, J. (1952) *Plasmodium vinckei* n. sp. Un deuxième *Plasmodium* de parasite de rongeurs sauvages au Katanga. *Ann. Soc. belg. Med. Trop.* **32**, 275–279.
14. Landau, I. (1965) Description de *Plasmodium chabaudi* n. sp. Parasite de rongeurs africains. *C. R. Acad. Sci. Paris.* **260**, 3758–3761.
15. Landau, I. and Chabaud, A. G. (1965) Natural infection by 2 plasmodia of the rodent *Thamnomys rutilans* in the Central African Republic. *C. R. Acad. Sci. Paris.* **261**, 230–232.
16. Landau, I. and Killick-Kendrick, R. (1966) Rodent plasmodia of the Republique Centrafricaine: the sporogony and tissue stages of *Plasmodium chabaudi* and *P. berghei yoelii*. *Trans. R. Soc. Trop. Med. Hyg.* **60**, 633–649.
17. Yoeli, M., Vanderberg, J., Upmanis, R. S., and Most, H. (1965) Primary tissue phase of *Plasmodium berghei* in different experimental hosts. *Nature* **208**, 903.
18. Yoeli, M. and Most, H. (1965) Studies on sporozoite induced-infection of rodent malaria. I. The pre-erythrocytic stages of *Plasmodium berghei*. *Am. J. Trop. Med. Hyg.* **14**, 700–714.
19. Bafort, J. (1968) Primary exo-erythrocytic forms of *Plasmodium vinckei*. *Nature* **217**, 1264,1265.
20. Nussenzweig, R. S., Vanderberg, J. P., Sanabria, Y., and Most, H. (1972) *Plasmodium berghei*: accelerated clearance of sporozoites from blood as part of immune-mechanism in mice. *Exp. Parasitol.* **31**, 88–97.
21. Sidjanski, S. and Vanderberg, J. P. (1997) Delayed migration of *Plasmodium* sporozoites from the mosquito bite site to the blood. *Am. J. Trop. Med. Hyg.* **57**, 426–429.
22. Ponnudurai, T., Lensen, A. H., van Gemert, G. J., Bolmer, M. G., and Meuwissen, J. H. (1991) Feeding behavior and sporozoite ejection by infected *Anopheles stephensi*. *Trans. R. Soc. Trop. Med. Hyg.* **85**, 175–180.
23. Frevert, U. and Crisanti, A. (1999) Invasion of vertebrate cells: hepatocytes, in *Malaria: Parasite Biology, Pathogenesis, and Protection* (Sherman, I. W., ed.), ASM Press, Washington, DC, pp. 73–91.
24. Killick-Kendrick, R. (1974) Parasitic protozoa of the blood of rodents: a revision of *Plasmodium berghei*. *Parasitology* **69**, 225–237.
25. Landau, I. and Boulard, Y. (1978) Life cycles and morphology, in *Rodent Malaria* (Killick-Kendrick, R. and Peters, W., eds.), Academic Press, London, pp. 53–84.
26. Wéry, M. (1968) Studies on the sporogony of rodent malaria parasites. *Ann. Soc. belg. Méd. Trop.* **48**, 1–137.
27. Rénia, L., Mattei, D., Goma, J., Pied, S., Dubois, P., Miltgen, F., Nussler, A., Matile, H., Ménégaux, F., Gentilini, M., and Mazier, D. (1990) A malaria heat shock like protein epitope expressed on the infected hepatocyte surface is the target of antibody-dependent cell-mediated cytotoxic mechanisms by non-parechymal liver cells. *Eur. J. Immunol.* **20**, 1445–1449.
28. Bafort, J. (1967) La transmission cyclique du *Plasmodium vinckei*. *Ann. Soc. belg. Méd. Trop.* **47**, 271–276.
29. Yoeli, M., Upmanis, R. S., Vanderberg, J., and Most, H. (1966) Life cycle and patterns of development of *Plasmodium berghei* in normal and experimental hosts. *Mil. Med.* **131(Suppl)**, 900–910.
30. Khan, Z. M., Ng, C., and Vanderberg, J. P. (1992) Early hepatic stages of *Plasmodium berghei*: release of circumsporozoite protein and host cellular inflammatory response. *Infect. Immun.* **60**, 264–270.
31. Khan, Z. M. and Vanderberg, J. P. (1991a) Role of host cellular response in differential susceptibility of nonimmunized BALB/c mice to *Plasmodium berghei* and *yoelii* sporozoites. *Infect. Immun.* **59**, 2529–2534.
32. Khan, Z. M. and Vanderberg, J. P. (1991b) Eosinophil-rich, granulomatous inflammatory response to *Plasmodium berghei* hepatic schizonts in nonimmunized rats is age-related. *Am. J. Trop. Med. Hyg.* **45**, 190–201.
33. Vanderberg, J. P., Khan, Z. M., and Stewart, M. J. (1993) Induction of hepatic inflammatory response by *Plasmodium berghei* sporozoites protects BALB/c mice against challenge with *Plasmodium yoelii* sporozoites. *J. Parasitol.* **79**, 763–767.
34. Meis, J. F. G. M., Jap, P. H. K., Hollingdale, M. R., and Verhave, J. P. (1987) Cellular response against exoerythrocytic forms of *Plasmodium berghei* in rats. *Am. J. Trop. Med. Hyg.* **37**, 506–510.

35. Wéry, M. and Killick-Kendrick, R. (1967) Differences in the numbers of tissues schizonts of *Plasmodium berghei yoelii* in the livers of individual mice, and a comparison of two routes of inoculation of sporozoites. *Trans. R. Soc. Trop. Med. Hyg.* **61,** 447,448.
36. Scheller, L. F., Wirtz, R. A., and Azad, A. F. (1994) Susceptibility of different strains of mice to hepatic infection with *Plasmodium berghei*. *Infect. Immun.* **62,** 4844–4847.
37. Verhave, J. P. (1975) Immunization with sporozoites. 1-123. PhD thesis, Catholic University of Nitmegen.
38. Held, J. R., Contacos, P. G., Jumper, J. R., and Smith, C. S. (1967) Direct hepatic inoculation of sporozoites for the study of the exo-erythrocytic stages of simian malaria. *J. Parasitol.* **53,** 656,657.
39. Vincke, I. H. and Bafort, J. M. (1968). Méthodes de standardization de l'inoculum de sporozoites de *Plasmodium berghei*. *Ann. Soc. belg. Méd. Trop.* **48,** 181–194.
40. Ozaki, L. S., Gwadz, R. W., and Godson, G. N. (1984) Simple centrifugation method for rapid separation of sporozoites from mosquitoes. *J. Parasitol.* **70,** 831–833.
41. Pacheco, N. D., Strome, C. P., Mitchell, F., Bawden, M. P., and Beaudoin, R. L. (1979) Rapid, large-scale isolation of *Plasmodium berghei* sporozoites from infected mosquitoes. *J. Parasitol.* **65,** 414–417.
42. Amani, V., Idrissa-Boubou, M., Pied, S., Marussig, M., Walliker, D., Mazier, D., and Rénia, L. (1998) Cloned lines of *Plasmodium berghei* ANKA differ in their abilities to induce experimental cerebral malaria. *Infect. Immun.* **66,** 4093–4099.
43. Gilks, C. F., Walliker, D., and Newbold, C. I. (1990) Relationships between sequestration, antigenic variation and chronic parasitism in *Plasmodium chabaudi chabaudi*-a rodent malaria model. *Parasite Immunol.* **12,** 45–64.
44. Walliker, D., Carter, R., and Morgan, S. (1971) Genetic recombination in malaria parasites. *Nature* **232,** 561,562.
45. Walliker, D., Carter, R., and Morgan, S. (1973) Genetic recombination in *Plasmodium berghei*. *Parasitology* **66,** 309–320.
46. Wéry, M. (1966) Etude du cycle sporogonique de *Plasmodium chabaudi* en vue de la production massive de sporozoites viables et de formes exo-érythrocytaires. *Ann. Soc. belg. Méd. Trop.* **46,** 755–788.
47. Bafort, J. (1971) The biology of rodent malaria with particular reference to *Plasmodium vinckei vinckei* Rodhain. *Ann. Soc. Belg. Méd. Trop.* **51,** 1–204.
48. Nussenzweig, R. S., Vanderberg, J. P., Spitalny, G. L., Rivera-Ortiz, C., Orton, C., and Most, H. (1972) Sporozoite induced immunity in mammalian malaria. A review. *Am. J. Trop. Med. Hyg.* **21,** 722–728.
49. Brannan, L. R., McLean, S. A., and Phillips, R. S. (1993) Antigenic variants of *Plasmodium chabaudi chabaudi* AS and the effects of mosquito transmission. *Parasite Immunol.* **15,** 135–141.
50. McLean, S. A., Phillips, R. S., Pearson, C. D., and Walliker, D. (1987) The effect of mosquito transmission of antigenic variants of *Plasmodium chabaudi*. *Parasitology* **94,** 443–449.
51. Bray, R. S. and Garnham, P. C. C. (1962) The Giemsa-Colophonium method for staining protozoa in tissue sections. *Indian J. Malariol.* **16,** 153–155.
52. Klotz, F. W., Scheller, L. F., Seguin, M. C., Kumar, N., Marletta, M., Green, S. J., and Azad, A. F. (1994) Co-localization of inducible-nitric oxide synthase and *Plasmodium berghei* in hepatocytes from rats immunized with irradiated sporozoites. *J. Immunol.* **5,** 3391–3395.
53. Spitalny, G. L. and Nussenzweig, R. S. (1972) Effect of various routes of immunization and methods of parasite attenuation on the development of protection against sporozoite-induced rodent malaria. *Proc. Helminthol. Soc. (Wash. DC)* **39,** 506–514.
54. Jakstys, B. P., Alger, N. E., Harant, J. A., and Silverman, P. (1974) Ultrastructural analysis of *Plasmodium berghei* sporozoite antigens prepared by freeze thawing and heat inactivation. *J. Protozool.* **21,** 344–348.
55. Kramer, L. D. and Vanderberg, J. P. (1975) Intramuscular immunization of mice irradiated *Plasmodium berghei* sporozoites. Enhancement of protection with albumin. *Am. J. Trop Med. Hyg.* **24,** 913–916.
56. Alger, N. E. and Harant, J. (1976) *Plasmodium berghei*: heat treated sporozoite vaccination of mice. *Exp. Parasitol.* **40,** 261–268.
57. Vanderberg, J. P., Nussenzweig, R. S., Most, H., and Orton, C. (1968) Protective immunity of X-irradiated sporozoites of *Plasmodium berghei*. II. Effect of radiation on sporozoites. *J. Parasitol.* **54,** 1175–1180.
58. Scheller, L. F. and Azad, A. F. (1995) Maintenance of protective immunity against malaria by persistent hepatic parasites derived from irradiated sporozoites. *Proc. Natl. Acad. Sci. USA* **92,** 4066–4068.
59. Ramsey, J. M., Hollingdale, M. R., and Beaudoin, R. L. (1982) Infection of tissue-stage cells with ^{60}Co gamma-irradiated malaria sporozoites, in *Nuclear Techniques in the Study of Parasitic Infections*. International Atomic Energy, Vienna, pp. 19–25.

60. Siegler, C. I., Leland, P., and Hollingdale, M. R. (1984) *In vitro* infectivity of irradiated *Plasmodium berghei* sporozoites to cultured hepatoma cells. *Am. J. Trop. Med. Hyg.* **33,** 544–547.
61. Nussler, A., Follezou, J. Y., Miltgen, F., and Mazier, D. (1989) Effect of irradiation on *Plasmodium* sporozoites depends on the species of hepatocytes infected. *Trop. Med. Parasitol.* **40,** 468,469.
62. Mellouk, S., Lunel, F., Sedegah, M., Beaudoin, R. L., and Druilhe, P. (1990) Protection against malaria induced by irradiated sporozoites. *Lancet* **335,** 721.
63. Weiss, W. R., Good, M. F., Hollingdale, M. R., Miller, L. H., and Berzovsky, J. A. (1989) Genetic control of immunity to *Plasmodium yoelii* sporozoites. *J. Immunol.* **143,** 4263–4266.
64. Rodrigues, M. M., Nussenzweig, R. S., and Zavala, F. (1993) The relative contribution of antibodies, CD4+ and CD8+ T cells to sporozoite-induced protection against malaria. *Immunology* **80,** 1–5.
65. Jaffe, R. I., Lowell. G. H., and Gordon, D. M. (1990) Differences in susceptibility among mouse strains to infections with *Plasmodium berghei* (ANKA clone) sporozoites and its relationship to protection by gamma-irradiated sporozoites. *Am. J. Trop. Med. Hyg.* **420,** 309–313.
66. Smrkovski, L. L., Mc Connell, E., and Tubergen, T. A. (1983) Effect of ^{60}CO-irradiation on the development and immunogenicity of *Plasmodium berghei* sporozoites in *Anopheles stephensi* mosquitoes. *J. Parasitol.* **69,** 814–817.
67. Mulligan, H. W., Russel, P. F., and Mohan, B. N. (1941) Active immunization of fowls against *Plasmodium gallinaceum* by injections of killed homologous sporozoites. *J. Mal. Inst. India* **4,** 25–34.
68. Rénia, L., Rodrigues, M. M., and Nussenzweig, V. (1994) Intrasplenic immunization with infected hepatocytes: a mouse model for studying protective immunity against malaria pre-erythrocytic stage. *Immunology* **82,** 164–168.
69. Scheller, L. F., Stump, K. C., and Azad, A. F. (1995) *Plasmodium berghei*: Production and quantitation of hepatic stages derived from irradiated sporozoites in rats and mice. *J. Parasitol.* **81,** 58–62.
70. Ferreira, A., Enea, V., Morimoto, T., and Nussenzweig, V. (1986) Infectivity of *Plasmodium berghei* sporozoites measured with a DNA probe. *Mol. Biochem. Parasitol.* **19,** 103–109.
71. Arreaza, G., Corredor, V., and Zavala, F. (1991) *Plasmodium yoelii*: quantification of the exoerythrocytic stages based on the use of ribosomal RNA probes. *Exp. Parasitol.* **72,** 103–105.
72. Li, J., Zhu, J. D., Appiah, A., McCutchan, T. F., Long, G. W., Milhous, W. K., and Hollingdale, M. R. (1991) *Plasmodium berghei*: quantitation of *in vitro* effects of antimalarial drugs on exoerythrocytic development by a ribosomal RNA probe. *Exp. Parasitol.* **72,** 450–458.
73. Briones, M. R., Tsuji, M., and Nussenzweig, V. (1996) The large difference in infectivity for mice of *Plasmodium berghei* and *Plasmodium yoelii* sporozoites cannot be correlated with their ability to enter into hepatocytes. *Mol. Biochem. Parasitol.* **77,** 7–17.
74. Hulier, E., Petour, P., Snounou, G., Nivez, M. P., Miltgen, F., Mazier, D., and Rénia, L. (1996) A method for the quantitative assessment of malaria parasite development in organs of the mammalian host. *Mol. Biochem. Parasitol.* **77,** 127–135.
75. Hollingdale, M. R., Leland, P., Leef, J. L., and Schwartz, A. L. (1983) Entry of *Plasmodium berghei* sporozoites into cultured cells, and their transformation into trophozoites. *Am. J. Trop. Med. Hyg.* **32,** 685–690.
76. Hollingdale, M. R., Leland, P., and Schwartz, A. L. (1983) *In vitro* cultivation of the exoerythrocytic stage of *Plasmodium berghei* in a hepatoma cell line. *Am. J. Trop. Med. Hyg.* **32,** 682–684.
77. Suhrbier, A., Janse, C., Mons, B., Fleck, S. L., Nicholas, J., Davies, C. S., and Sinden, R. E. (1987) The complete development *in vitro* of the vertebrate phase of the mammalian malarial parasite *P. berghei*. *Trans. R. Soc. Trop. Med. Hyg.* **81,** 907–910.
78. Long, G. W., Leath, S., Schuman, R., Hollingdale, M. R., Ballou, W. R., Sim, B. K. L., and Hoffman, S. L. (1990) Cultivation of the exoerythrocytic stage of *Plasmodium berghei* in primary cultures of mouse hepatocytes and continuous mouse cell lines. *In Vitro Cell. Dev. Biol.* **25,** 857–862.
79. Hollingdale, M. R. (1985) Malaria and the liver. *Hepatology* **5,** 327–335.
80. Calvo-Calle, J. M., Moreno, A., Eling, W. M. C., and Nardin, E. H. (1994) *In vitro* development of infectious liver stages of *P. yoelii* and *P. berghei* malaria in human cell lines. *Exp. Parasitol.* **79,** 362–373.
81. Suhrbier, A., Holder, A. A., Wiser, M. F., Nicholas, J., and Sinden, R. E. (1989) Expression of the precursor of the major merozoite surface antigens during the hepatic stage of malaria. *Am. J. Trop. Med. Hyg.* **40,** 19–23.
82. Rénia, L., Ling, I. T., Marussig, M., Miltgen, F., Holder, A. A., and Mazier. D. (1997) Immunization with a recombinant C-terminal fragment of *Plasmodium yoelii* merozoite surface protein 1 protects mice against homologous but not heterologous *P. yoelii* sporozoite challenge. *Infect. Immun.* **65,** 4419–4423.
83. Charoenvit, Y., Mellouk, S., Sedegah, M., Toyoshima, T., Leef, M. F., De la Vega, P., et al. (1995) *Plasmodium yoelii*: 17-kDa hepatic and erythrocytic stage protein is the target of an inhibitory monoclonal antibody. *Exp. Parasitol.* **80,** 419–429.
84. Rogers, W. O., Malik, A., Mellouk, S., Nakamura, K., Rogers, M. D., Szarfman, A., et al. (1992) Characterization of *Plasmodium falciparum* sporozoite surface protein 2. *Proc. Natl. Acad. Sci. USA* **89,** 9176–9180.

85. Rénia, L., Miltgen, F., Charoenvit, Y., Ponnudurai, T., Verhave, J. P., Collins, W. E., and Mazier, D. (1988) Malaria sporozoite penetration: a new approach by double staining. *J. Immunol. Methods* **112,** 201–205.
86. Mellouk, S., Berbiguier, N., Druilhe, P., Sedegah, M., Galey, B., Yuan, L., et al. (1990) Evaluation of an *in vitro* assay aimed at measuring protective antibodies against sporozoites. *Bull. World. Hlth. Organ.* **68(Suppl.),** 52–59.
87. Hoffman, S. L., Isenbarger, D., Long, G. W., Sedegah, M., Szarfman, A., Waters, L., et al. (1989) Sporozoite vaccine induces genetically restricted T cell elimination of malaria from hepatocytes. *Science* **244,** 1078–1081.
88. Weiss, W. R., Mellouk, S., Houghten, R. A., Sedegah, M., Kumar, S., Good, M. F., et al. (1990) Cytotoxic T cells recognize a peptide from the circumsporozoite protein on malaria-infected hepatocytes. *J. Exp. Med.* **171,** 763–773.
89. Rénia, L., Salone-Marussig, M., Grillot, D., Corradin, G., Miltgen, F., Del Giudice, G., et al. (1991) *In vitro* activity of CD4+ and CD8+ T lymphocytes from mice immunized with a malaria synthetic peptide. *Proc. Natl. Acad. Sci. USA* **88,** 7963–7967.
90. Rénia, L., Grillot, D., Marussig, M., Corradin, G., Miltgen, F., Lambert, P.-H., et al. (1993) Effector functions of circumsporozoite peptide-primed CD4$^+$ T cell clones against *Plasmodium yoelii* liver stages. *J. Immunol.* **150,** 1471–1478.

5

Mouse Models for Erythrocytic-Stage Malaria

Latifu A. Sanni, Luis F. Fonseca, and Jean Langhorne

1. Introduction

Mouse malaria infections are useful models with which to study the interactions between the host and the erythrocytic stage of the parasite. These models have several advantages for immunological and related studies. There are inbred and congenic strains with defined major histocompatibility complex (MHC) haplotypes and a number of natural mutants defective in components of the immune response. More recently, an increasing number of gene-targeted mice with defective genes of immunological relevance are being developed. The availability of these mice makes it possible to determine the exact components of the immune response that may be important in protective immunity and pathology.

Infections caused by rodent parasites can vary in virulence depending on the species of *Plasmodium* and the strain of mouse. Parasites such as *Plasmodium berghei and P. vinckei* and some strains of *P. yoelii* and *P. chabaudi* cause lethal infections in mice, whereas infections with *P. yoelii, P. chabaudi chabaudi, P. chabaudi adami*, and *P. vinckei petterei* are cleared after the initial acute parasitemia or after a subsequent low-grade chronic parasitemia *(1–3)* (*see* **Fig. 1**).

Most of the laboratory strains of parasite have been isolated from African Thicket Rats, *Thamnomys rutilans*, and none is a natural pathogen of the laboratory mouse *(1)*. Therefore, unlike the human pathogens, there has been no evolutionary adaptation of parasite and laboratory mouse. For this reason, and because the rodent parasites differ in virulence and in the fine details of immune elimination from the host, it has been argued that mouse models are not very relevant to the human disease. However, it is likely that the principles by which erythrocytic stage parasites can be recognized by the immune system and the possible pathological consequences of infection are similar in mice and humans. Clearly, the more similar the models are among themselves, the more likely it is that they will reflect human malaria infections.

In this chapter, we have compared erythrocytic-stage infections of a number of mouse malaria parasites, and detailed some of the characteristics and information available on mechanisms of pathology and immunity. Parasitological methods and those methods pertaining to malaria-associated disease are described in this chapter, with the exception of maintenance of mosquito colonies (*see* Chapter 3). The immunological and cytokine assays used to determine mechanisms of immunity in the different models here are detailed in **Part 5**.

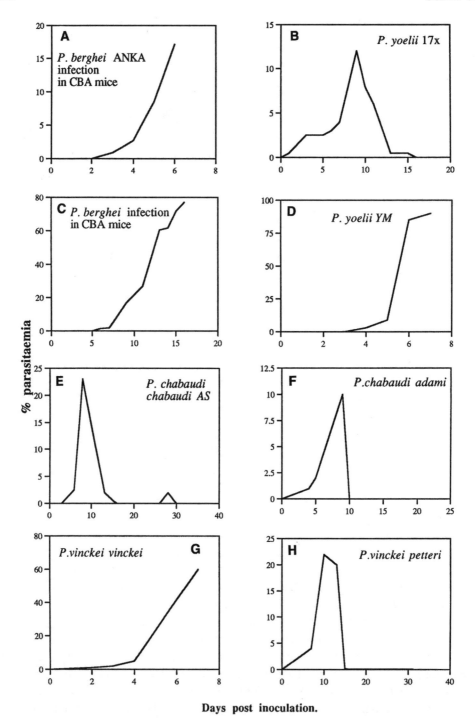

Fig. 1. Diagrammatic representation of the course of primary infection in different mouse malaria models.

Table 1 lists the major rodent parasites that have been used to investigate mechanisms of pathogenesis (including cerebral malaria) and immunity, and also for design of therapeutic drugs. Some of these parasites are uniformly lethal in all strains of mice

Table 1
Experimental Malaria Infection in Different Mouse Strains

Plasmodium	Strain/clone	Mouse strain	RBC infected	Lethal infection	Experimental study
berghei	ANKA	CBA/T6 BALB/c (*see* **Note 1a**) C57BL/6	Mature RBC and reticulocytes	Yes: d 6–8 (*4,5*)	Pathogenesis of CM
	K173	DBA/2J CBA/T6, BALB/c DBA/2J	Mature RBC and reticulocytes Mature RBC and reticulocytes	Yes: d 15–22 (*4*) Yes: d 15–22 (*4*)	Resolving CM model Non CM control for CM study
yoelii		C57BL6	Mature RBC and reticulocytes (*6*)	Yes: d 6–8 (*5*)	Pathogenesis of CM
	17x	Swiss BALB/c	Reticulocyte	Yes: d 7–9 (*7*)	Pathogenesis CM (*see* **Note1b**)
		CBA/Ca BALB/c, C57BL/6, DBA/2J, C3H	Reticulocyte (*8*)	No (*8–10*)	Immune mechanisms Pathogenesis
	YM (lethal)	CBA/T6, BALB/c C57BL/6 DBA/2J SWISS	Mature RBC and reticulocytes	Yes: d 7–8 postinfection (*11*)	Vaccine
vinckei vinckei		BALB/c	Mature RBC	Yes: d 8 (*12*)	Chemotherapy Immune mechanisms Pathogenesis

(*continued*)

Table 1
Experimental Malaria Infection in Different Mouse Strains

Plasmodium	Strain/clone	Mouse strain	RBC infected	Lethal infection	Experimental study
chabaudi chabaudi	AS	C57BL/6 NIH, CBA/Ca B10 series BALB/c A/J DBA/2J	Mature RBC	No: peak d 8–10 *(13) (see* **Note 1***)*	Immune mechanisms Pathogenesis
	CB	CBA/Ca	Mature RBC Mature RBC or all	Yes: d 9 *(13)* Yes: in 20–50%, after d 8 *(14)* *(see* **Note 1c***)*	Immune mechanisms
chabaudi adami	556 KA	BALB/c C3H C57BL/6	Mature RBC	No: peak d 7–11 *(15,16)*	Immune mechanisms Pathogenesis
vinckei petteri	CR	C57BL/6 BALB/c	Mature RBC	No: peak d 10 *(17)*	Chemotherapy Immune mechanisms

and some are lethal in only some strains of mice. Recent studies, particularly those using a variety of gene-targeted mice and natural mutant mice on mixtures of strain backgrounds, have underlined the importance of the genetic background other than MHC in the susceptibility to lethal malaria infections. Therefore, when untested mouse strains or gene-targeted mice on a mixed genetic background are to be used in an immunologically related study, it is advisable to carry out a pilot experiment to determine the course of infection and pathology in the mice of that background.

Lethal infections have often been used to evaluate new chemotherapeutic drugs and potential vaccine candidate molecules. Some of the models—such as *P. berghei* (ANKA), *P. vinckei*, and *P. chabaudi chabaudi*—also allow investigations of malarial disease and immunologically related pathology (**Tables 2–4**). Nonlethal infections have been used predominantly to investigate mechanisms of immunity and regulation of the immune response (**Table 4**). Comparisons between mouse infections are not always easy, as it is rare that the same methods of assessing immunity or pathology have been carried out in each case. In addition, some parasite/host combinations have been more extensively investigated than others.

The various models for pathology have been studied for different purposes, and so far there has been no investigation of a link between parasite sequestration and cerebral malaria (CM) as seen in human *Plasmodium falciparum*. In the most commonly used mouse model for CM, *P. berghei* ANKA, there is no documented evidence of parasite sequestration in the brain. However, the immunological and inflammatory processes involved in CM seen in this model have been defined in detail. By contrast, *P. yoelii* 17X in some mouse strains does sequester and is able to induce a form of CM in some strains of mice (*see* **Note 1**). However, the mechanisms involved have not been determined.

Most of the nonlethal infections exhibit transient disease such as weight loss, hypothermia, anemia, and hypoglycemia (**Table 3, Fig. 2**) and therefore may be suitable models for investigating the mechanisms involved in these different sequelae. Although the role of inflammatory cytokines has not been investigated in all cases, there does seem to be some consensus that tumor necrosis factor-α (TNF-α), interferon-γ (IFN-γ), and possibly interleukin-1 (IL-1) are involved (**Table 4**). Resistance and susceptibility to lethal infections, on the other hand, do not appear to be due to the same mechanisms. Susceptibility to a lethal infection of *P. chabaudi chabaudi* and lethal non-*CM P. yoelii* is accompanied by an early CD4$^+$ Th2–like response and low inflammatory responses.

Investigations of immune mechanisms of parasite clearance and immunity to reinfections have been carried out in detail in only a limited number of parasite–mouse strain combinations (**Table 4**). Although there are differences in the ability of antibody-independent mechanisms to control acute stage parasitemia, experiments in immunologically deficient gene-targeted-, mutant- or antibody-depleted mice in general show that clearance and immunity to reinfection are antibody- and B-cell dependent. Thus immunological study of any of the nonlethal infections is equally likely to furnish useful information. However *P. chabaudi chabaudi*, *P. chabaudi adami*, and *P. vinckei* parasites would be the models of choice for investigating the mechanisms of antibody-independent parasite killing, as they are more susceptible to these types of effector mechanisms.

Table 2
Pathology Associated with Mouse Malaria Infections Used in the Study of the Pathogenesis of Cerebral Malaria

Pathology	PbA/CBA or BALB/c[a]	PbA/DBA[b]	PbK/CBA[c]	PbK/C57BL/6[d]	P. yoelii 17x in susceptible strains of mice (Swiss/CDR1)
Clinical					
Weight loss	+[e]	+	+	+ (18)	?[f]
Hypothermia	+	+	+	+ (4)	?
Convulsions	++++ (4)	+ (4)	−[g] (5)	++++ (5)	+++ (19)
Limb paralysis	++++ (4)	+ (4)	− (5)	++++ (5)	?
Coma	+ (4)	− (4)	− (5)	+ (5)	?
Anatomical					
Cerebral edema	++++ (4)	+ (4)	± (4)	++++ (20)	?
Petechial hemorrhage in the brain	++++ (4,55)	++ (4)	− (4)	++++ (20)	++++ (21)
Sequestration of parasitized RBCs in brain	−	−	−	−	+ (7,21)
Sequestration of monocytes in brain	++++ (4)	++ (4)	+ (4)	++++ (20)	−
Splenomegaly	+	+	+	+	+ (11)
Pulmonary edema	+ (4)	−	−	?	?
Haematological					
Anemia	−	++++	++++	−	++++ (22)
Biochemical					
Hypoglycemia	− (23)	?	?	?	+ (22)
Brain lactic acidosis	+ (24)	?	−	?	+ (19)
Systemic lactic acidosis	−	?	+(24)	?	+ (22)

[a]PbA/CBA, *P. berghei* ANKA infection in CBA mice.
[b]PbA/DBA, *P. berghei* ANKA infection in DBA mice.
[c]PbK/CBA, *P. berghei* K173 infection in CBA mice.
[d]PbK/C57, *P. berghei* K173 infection in C57 mice.
[e]+, present.
[f]?, not known.
[g]−, absent.

Table 3
Pathology Seen in Experimental Mouse Malaria Infections Where Cerebral Malaria is not Thought to Occur

Pathology	P. yoelii	P. yoelii YM	P. chabaudi chabaudi AS	P. chabaudi chabaudi AS	P. chabaudi chabaudi adami	P. vinckei vinckei
	C3H		C57BL/6	A/J	BALB/c	BALB/c
	CBA		CBA/Ca	DBA/2		CBA
	BALB/c		B10.A			
	C57Bl/6		B10.D2			
	BALB/c					
Clinical						
Weight loss	?[a]	?	+[b] (13)	+ (13)	−[c]	+ (26)
Hypothermia	?	?	+ (13)	+ (13)	?	?
Anatomical						
Sequestration of parasitized RBCs	?	?	+ liver (27)	+ liver (28)	?	+ liver (29)
Splenomegaly	+ (9)	+ (11)	+	+	+ (26)	+ (26)
Pulmonary edema	?	?	?	?	?	+ (12)
Hematological						
Anemia	+ (9)	?	+ (30)	+ (30)	+ (26)	+ (26,29)
Biochemical						
Hypoglycemia	− (31)	+ (25)	+ (13,31)	+ (13)	?	+ (29,32)

[a]?, not known.
[b]+, present.
[c]−, absent.

Table 4
Immune Mechanisms Induced During Mouse Malaria Infections

Plasmodium	Mouse strain	Antibody response	T cell response	CM or pathology	Immune response/cytokines associated with Resistance or susceptibility	Parasite clearance Acute phase	Parasite clearance Chronic phase	Resistance to reinfection
berghei ANKA	CBA C57BL/6 BALB/c	N/A[a] Th2 (11)	Th1,	CD4+, CD8+, IFN-γ, TNF-α (5,33–35)	N/A	N/A	N/A	N/A
berghei K173	CBA	N/A	? N/A	N/A	N/A	N/A	N/A	N/A
yoelii YM	BALB/c C57BL/6	IgG1, IgG2a, IgG2b (36) Immunity following vaccination	Low Th1 High Th2 (11)	? TNFα (21)	High IFN-γ High IL-4 (36) Needed for immunity following vaccination	N/A	N/A	N/A
yoelii 17X Nonlethal	C3H CBA BALB/c C57Bl/6	IgM + IgG1 ++ IgG2++ IgG3 + (8)	Th1, Th2 (9,11)	N/A	High early Th1 (11)	?	B cells Ab IgG$_{2a}$ (?) (8,37)	B cells

Erythrocytic-Stage Mouse Models

Species	Mouse strain	Ig isotypes	Th response	Cytokines	Early response	Effectors	Other	Protection
vinckei vinckei	BALB/c CBA/ca	N/A	Th1	TNFα, IL-1 IFNγ (*12,29,32*)				CD4 T cells B cell (*38*)
chabaudi chabaudi AS	C57BL/6 BALB/c	IgM + IgG2a,b +++ IgG1 +/− IgG3 ++	Th1 then Th2 γδ T cells (*40–42*)	TNFα (*43*) High IL-12 (*44*)	High early Th1 response (*44*)	Th1 γδ T cells IFN (*41,42,45,46*)	Th2 B cells Ab (*40,41*)	B cells Ab (*47*)
chabaudi chabaudi AS	A/J DBA/2		Early Th2 (*48*)	Low IL-12 (*44*)	Low early Th1 High early Th2 (*44*)	N/A	N/A	N/A
chabaudi adami	C3H BALB/c	IgM + IgG2a,b +++ IgG3 +++ IgG1 +/− (*8*)	Th1 Little Th2 (*49*)	?	High early Th1 Low Th2 (*49*)	γδ T cells T cells (*15,16,50*)	B cells (*15*)	CD4+ T cells (*39,50–52*)
vinckei petteri	C57BL/6	?				T cells (*17*)	B cells	

[a]N/A, not applicable because the infection is lethal.
See also footnotes to **Table 3**.

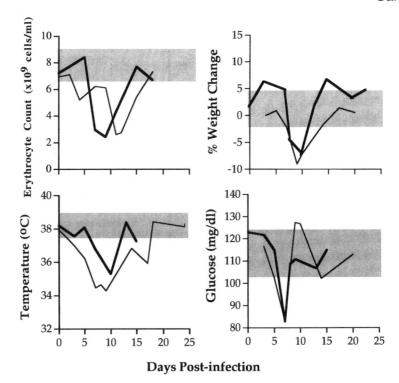

Fig. 2. Diagrammatic representation of the transient clinical signs associated with nonlethal male (thin line) and female (thick line) C55BL/6 mice. The shaded areas represent the normal range of values of uninfected *P. chabaudi chabaudi* (AS) infection in mice.

2. Materials

2.1. Equipment

2.1.1. Measurement of Temperature

1. Rectal thermometer.

2.1.2. Measurement of Weight

1. Balance.

2.1.3. Determination of Parasitemias

1. Microscope with ×100 immersion objective.
2. Eyepiece with reticule.
3. Hemocytometer.

2.1.4. Detection of Anaemia

1. Analytical flow cytometer.

2.1.5. Crude Brain/Organ Extraction

1. Glass homogenizer with teflon pestle.
2. Motor drive for teflon pestle.

2.1.6. Packed Cell Volume (PCV) Hematocrit

1. Heparinized microhematocrit tubes.
2. Microhematocrit centrifuge.

2.1.7. Proton Nuclear Magnetic Resonance Spectrometry (^{1}H-NMR)

1. NMR spectrometer.
2. NMR tubes.
3. Beckman centrifuge.

2.1.8. Spectrophotometry

1. Spectrophotometer.
2. Quartz microcuvettes.

2.1.9. Measurement of Glucose

1. Glucose meter (Boehringer Mannheim, Lewes, Sussex, UK).
2. BM™ test strips (Boehringer Mannheim).

2.1.10. Brain Density

1. Microanalytical balance with a resolution of at least 0.00001 g (10 µg).
2. 2-mL glass pycnometer (Fisher Inc.).

2.2. Reagents

2.2.1. Giemsa Staining of Blood Films

1. Methanol.
2. Giemsa (Giemsa Stain Improved R66, Gurr; BDH).
3. Giemsa buffer: 0.015 M NaCl, 0.001 M phosphate buffer, pH 7.0.

2.2.2. Parasite Freezing Medium

1. Combine 400 mL glycerol (100%), 100 mL of 1.4 M sodium lactate solution (stock 1.4 M = 157 g/L), 10 mL of 0.5 M potassium chloride solution (stock 0.5 M = 37.5 g/L), and 450 mL of distilled water. Adjust pH to 7.4 with NaH_2PO_4.

2.2.3. Drabkin's Solution

1. Combine 0.2 g $K_3Fe(CN)_6$, 0.05 g KCN, and 1 g $NaHCO_3$. Dilute to 1000 mL in distilled water. Drabkin's solution is stable for at least 4 mo when stored in a dark bottle at 4°C. Standard for the reaction is 2-Cyanmethemoglobin, 80 mg/dL *(53)* (*see* **Note 2**).

2.2.4. Evans' Blue Dye

1. 2% (w/v) Evans' Blue dye in phosphate-buffered saline (PBS), pH 7.4.

2.2.5. Histology

1. 2 IU/mL heparin.
2. 10% Neutral buffered formalin.

2.2.6. Brain/Organ Extraction

1. 6.0% (w/v) ice-cold perchloric acid ($HClO_4$).
2. 1 M potassium hydroxide.
3. Liquid nitrogen.

2.2.7. Anesthetic Agent

1. Ether.

2.2.8. ^{1}H-NMR

1. 2,2,3,3-Tetradeuterotrimethylsilyl-1-propionic acid (TSP).
2. ^2H_2O.

2.2.9. Pyruvate

1. Triethanolamine, 0.5 M, pH 7.6; EDTA, 5 mM (TEA).
2. 7 mM βNADH solution.
3. 225 kU/L L-lactate dehydrogenase (LDH).

2.2.10. Fixation Buffer

1. Heat 1 L of distilled water until boiling in a microwave.
2. Pour into a bottle with 8.5 g of NaCl and 10 g of paraformaldehyde and stir well to dissolve.
3. Allow the solution to cool to room temperature and then pH 7.2–7.4.
4. Filter through 0.22-µm filter and store at 4°C.

3. Methods

3.1. Giemsa Staining of Blood Films

1. Make a blood smear on a clean glass slide and let it dry.
2. Fix the blood smear by immersion in methanol for a few seconds and air-dry.
3. Immerse for 10–20 min in 10% (v/v) Giemsa solution in Giemsa buffer.
4. Wash under running water and dry.
5. Examine on microscope at ×100 objective, with immersion oil.
6. Quickly scan through the slide and choose an area where the red blood cells (RBC) are uniformly distributed. There should be no overlapping cells.
7. Count all the RBC present in one microscope field.
8. Count all the parasitized RBC present in that same field.
9. Calculate the total number of parasitized RBC and express this as a percentage of the total RBC counted (*see* **Note 3**).

3.2. Frozen Stock of Parasites

1. Bleed the infected mice using a heparinized syringe and keep the blood on ice (*see* **Note 4**).
2. Centrifuge the blood at 1000*g* for 10–15 min at 4°C.
3. Remove the supernatant and estimate the volume of the pellet (packed cell volume).
4. Add 2 vol of freezing medium to 1 vol pellet by adding 1/5 of the total glycerol volume dropwise with gentle mixing. Leave cells for 5 min and then add the remaining 4/5 of glycerol volume slowly with mixing.
5. Aliquot approx 250 µL of the parasite/glycerol mixture per freezing vial.
6. Place vials in –70°C and transfer them to liquid nitrogen after approx 48 h (*see* **Note 5**).

3.3. Thawing Frozen Samples of Parasites

1. Remove the frozen vial of parasites from liquid nitrogen and keep on ice until immediately before dilution and injection.
2. Thaw rapidly and mix with equal volume of 0.9% saline.
3. Inject mice intraperitoneally or intravenously.

3.4. Infection of Mice by Injection of Parasitized Erythrocytes Obtained from Another Mouse

Mice should always be infected with a known number of parasites as the inoculum dose determines the kinetics of appearance of parasites and the peak parasitemia. The doses used are usually between 10^4 and 10^6 parasites (infected erythrocytes) per mouse.

1. Determine the concentration of parasitized erythrocytes per milliliter in the blood of the mouse:
   a. Take a drop of blood from the mouse tail vein and do a smear.
   b. Proceed as described in **Subheading 3.1.** to Giemsa stain the blood smear and count the parasitemia.
   c. Calculate the number of RBC per milliliter present in the blood of the mouse using a hemocytometer.
   d. Calculate the number of parasitized RBC per millilter.
   e. Make the appropriate dilution in 0.9% saline in order to inject 100 µL per mouse with the required number of parasites.

3.5. Infection of Mice by Feeding Infected Mosquitoes

In nature, malaria parasites are transmitted between hosts, through the bite of the *Anopheles* mosquitoes. This process can be done in the laboratory, but is necessary to have access to a colony of mosquitoes (*see* Chapter 3).

1. Infect mice by injecting parasitized RBC intraperitoneally or intravenously.
2. When gametocytes are present in the circulation, allow 50 mosquitoes contained in paper cups to feed on the anesthetized infected mice.
3. Several days later (depending on the parasite species), when sporozoites appear in the salivary glands of the mosquitoes, allow them to feed on noninfected mice (*see* **Notes 6** and **7**).

3.6. Infection of Mice by Injection of Sporozoites

Mice can also be infected by the artificial injection of sporozoites. Sporozoites can be obtained by using the method described in Chapter 3.

1. Dissect salivary glands of malaria-infected mosquitoes.
2. Rupture the salivary glands using a glass homogenizer.
3. Count the number of sporozoites using a hemocytometer.
4. Adjust to the desired concentration and inject intravenously (*see* **Note 8**).

3.7. Clinical Examination of Infected Mice

1. When mice are becoming sick from malaria infection, they become progressively quiet (i.e., they move less when disturbed) and develop a ruffled fur and a hunched back.
2. With the development of anemia, the limbs become pale.
3. When cerebral malaria develops, the mice start to convulse; this may be clonic or tonic in nature. Limb paralysis may also develop.
4. CBA strains of mice infected with *P. berghei* ANKA go into a nonarousable coma before death on d 6–7 postinoculation (*see* **Fig. 1A**).
5. Cerebral involvement during malaria infection in mice is demonstrated by holding the mouse vertically by the tail. When a mouse suffering from cerebral malaria is held upside-down in this way, it undergoes a spiral motion along the vertical axis.

3.8. Measurement of Temperature

Body temperature is measured using a rectal thermometer at different days during the infection.

3.9. Measurement of Weight

Body weight is measured in grams at various time points during the course of infection. The results can be presented as percentage of the body weight on day 0 (day of infection).

3.10. Detection of Anemia

Anemia can be defined as a reduction in number of RBC or decrease in the concentration of hemoglobin in whole blood. Anemia can be detected clinically in the mouse by examining the limbs for pallor. A definitive diagnosis of anemia lies with the determination of the number of erythrocytes per milliliter of blood, the packed cell volume (PCV) also known as the hematocrit and the hemoglobin concentration. Hemorrhage, hemolysis, and impaired blood formation are the commonest causes of diminished blood hemoglobin concentration. During malaria infection, red blood cell hemolysis is the major cause of reduced blood hemoglobin concentration.

3.10.1. Counting the Number of Erythrocytes by Flow Cytometry

Anemia is measured as the number of RBC/mL by diluting 5 µL of blood from the tail vein in 200 µL of fixation buffer. This volume is after made up to 1 mL in the same buffer. The number of events (RBC) are counted on a flow cytometer (FACscan, Becton-Dickinson) for a defined time of 10 s. The number of events counted is used to calculate the concentration of erythrocytes per milliliter against a standard curve of known erythrocyte concentrations. This standard curve is made by making twofold dilutions from a blood sample and counting the number of RBC/mL using a hemocytometer and also counting the number of events in the flow cytometer. These will give the relation between the number of events counted in the flow cytometer and the number of RBC counted in the hemocytometer.

3.10.2. Determination of Packed Cell Volume or Hematocrit

1. Obtain blood samples from either the tip of the mouse tail or via an axillary artery.
2. Suck up the blood samples into an heparinized microhematocrit tubes by capillary action to approx 75% capacity and seal.
3. Centrifuge the microhematocrit tubes containing the blood sample at 10,000g for 3 min, using a microhematocrit centrifuge.
4. Determine the hematocrit using a reader supplied with the centrifuge.

3.10.3. Measurement of Hemoglobin in Whole Blood

The biochemical method utilized in the determination of blood hemoglobin concentration involves the oxidation of Fe(II) of hemoglobin to Fe(III) by ferricyanide. This is then converted into the stable cyanmethemoglobin by addition of potassium cyanide (KCN).

$$HbFe(II) + Fe(III)(CN) \rightarrow HbFe(III) + Fe(II)(CN)$$

$$HbFe(III) + CN \rightarrow HbFe(III)CN$$

where HbFe(II) is hemoglobin, HbFe(III) is methemoglobin, and HbFe(III)CN is cyanmethemoglobin.

3.10.3.1. Procedure

1. Label cuvets, blank, standard, and samples.
2. Dispense 6 mL of standard into the cuvette labeled standard, and 6 mL of Drabkin's solution into all other cuvettes.
3. Mix blood specimens thoroughly by inversion, and transfer 20 µL of whole blood into the appropriate cuvets.
4. Mix and let stand for approx 5 min.
5. Read absorbance of samples at 540 nm.
6. If the absorbance of standard is S and absorbance of a test sample is T, then:

    hemoglobin concentration $C = 301 \ [(T \times \text{concentration of standard})/S] \times 1/1000$

    $= 0.301 \ [(T \times \text{concentration of standard})/S]$

    Blood dilution in Drabkin's solution is 1/301; 2-cyanmethemoglobin standard is 80 mg/dL *(53)*.

3.11. Measurement of Blood Glucose

Glucose oxidase catalyzes the oxidation of glucose to glucono-δ-lactone and hydrogen peroxide. A color develops in the presence of peroxidase and an oxidizable chromogen.

Glucose oxidase

Glucose + O_2 → Gluconolactone + H_2O_2

Peroxidase

H_2O_2 + chromogen → Oxidized chromogen + H_2O.

Glucose level in the blood is measured by spreading 14 µL of blood from the tail vein onto BM test strips and analyzing them with a glucose meter according to the manufacturer's instructions.

3.12. Measurement of Brain Weight

Dissect whole-brain samples out of the cranial cavity of mice and quickly weigh on a balance.

3.13. Determination of Brain Density

Brain density during malaria infection *(54)* used to be determined using the Percoll gradient method of Tengvar et al. *(55)*. However, it has been replaced by the much simpler pycnometric method of DiResta et al *(56)*.

Use a microanalytical balance with a resolution of at least 0.00001 g (10 µg). Before starting, carefully clean the pycnometer and wipe dry using a lint-free paper.

1. Use a 2-mL glass pycnometer with the central hole of the glass stopper plugged.
2. Place two small alignment marks on the ground glass stopper and on the outside neck of the flask as described previously by DiResta and colleagues *(56)*.
3. Align the marks to each other to ensure that the final volume remains the same between measurements.
4. Calculate brain density from weight values as described below *(56)*.

$$d_s = s^* \ d_w / (m_{pf} - m_t + s)$$

where d_s = density of brain sample (g/mL); d_w = density of water (g/mL); s = weight of sample (g); m_{pf} = weight of pycnometer and water (g); m_t = weight of pycnometer, water, and brain sample (g).

All measurements should be made at the same temperature.

a. Fill pycnometer with nanopure water, and stopper such that the marks align, then wipe the outside of the flask with lint-free wipes, and weigh to obtain m_{pf}.
b. Weigh the brain tissue specimen to obtain s.
c. Remove some water from within the pycnometer, and then transfer the weighed brain tissue specimen into the pycnometer. Refill the pycnometer with additional nanopure water, and stopper as described above. Again, dry the outside of the flask with lint-free paper, and then measure the weight of the pycnometer, water, and brain tissue (m_t).

This method of determining brain density is simple and accurate for specimens with volumes of at least 0.12 mL.

3.14. Determination of Breakdown of the Blood Brain Barrier by Measurement of Evans' Blue Leakage

1. Inject 200 µL of 2% (w/v) Evans' blue dye in PBS into the tail vein.
2. After 1 h, anaesthetize the mice with ether and perfuse with 5 mL of saline via the left ventricle.
3. Dissect out brain and compare the color to that of uninfected mice.

3.15. Preparation of Organs for Histology

1. Anesthetize the mice with ether inhalation.
2. Perfuse mice with saline containing 2 IU/mL heparin via the left ventricle *(54)*.
3. Then perfuse with 100 mL of 10% neutral buffered formalin.
4. Fix brain in 10% neutral buffered formalin for another 24 h.
5. Dehydrate, embed in wax, and section.

3.16. Brain Extraction for Lactate and Alanine Measurement

1. Anesthetize the mice with ether inhalation.
2. Decapitate mice such that heads fall straight into liquid nitrogen.
3. Chisel out whole-brain samples from the frozen heads.
4. Quickly weigh in 5 mL of 6% (w/v) ice-cold perchloric acid ($HClO_4$) and homogenize using a motor-driven Teflon pestle and glass homogenizer.
5. Centrifuge the brain homogenates produced at 2000g for 10 min at 4°C in a Beckman centrifuge.
6. Recover supernatants and neutralize to pH 7.0 with 1 M potassium hydroxide.

3.17. ^{1}H-NMR Analysis of Brain Lactate and Alanine

1. Freeze-dry 2 mL of the neutralized supernatants overnight.
2. Reconstitute the lyophilized samples in 0.65 mL of ^2H_2O, containing 2 mM TSP as an internal chemical shift and concentration reference, and dispense into 5-mm NMR tubes for the acquisition of ^{1}H-NMR spectra.
3. Acquire ^{1}H-NMR spectra at 600.13 MHz on a spectrometer, using a gradient inverse.
4. Acquire spectra across 64 K data points using a spectral width of 6000 Hz and a duty cycle of 30 s for "fully-relaxed" spectra, and consisting of the sum of 32 transients.
5. Transform spectra with 0.5 Hz of exponential multiplication and 1° of zero-filling.
6. Determine concentrations by comparing the size of the integral of the resonance of interest with that from TSP.

3.18. Spectrophotometric Assay for Lactate and Pyruvate

3.18.1. Lactate

Brain and blood lactate can also be measured spectrophotometrically using the Boehringer Mannheim kit that employs a method modified from that of Gutman and Wahlfeld *(57)*.

3.18.2. Pyruvate

1. For brain tissue analyses of pyruvate:
   a. In a quartz microcuvette, successively add 0.5 mL of neutralized sample, 0.25 mL of TEA solution and 0.02 mL of 7 mM βNADH solution.
   b. Mix thoroughly.
   c. Monitor the change in absorbance at 340 nm until a linear change with time is reached.
   d. Record absorbance (A_1).
   e. Then add 0.01 mL of 225 kU/L of LDH solution and mix.
   f. Read the absorbance every 2 min for 10 min, and by extrapolation of these values to the time of addition of LDH the final absorbance (A_2) can be determined.
   g. Calculate ΔA (i.e., $A_1 - A_2$) and use for the determination of brain pyruvate concentration using the formula shown below *(58)*.

   $$C = \Delta A \times 0.78 \times (V_a + V_{KHCO3}) \times (V_{HClO4} + m_t \times 0.75)/(\varepsilon \times 10 \times 0.5 \times V_a \times m_t)$$

   where C = brain concentration of pyruvate (μmol/g wet weight); V_a = volume of brain homogenate supernate (mL), neutralized with 2 M potassium hydrogen carbonate = 1 mL; V_{KHCO3} = volume of KHCO$_3$ used for neutralization; V_{HClO4} = volume of HClO$_4$ in which whole-brain sample was homogenized (mL); m_t = brain tissue weight (g); ε = extinction coefficient of NADH at 340 nm = 6.3 (L × mmol^{-1} × cm^{-1}).

   The above formula assumes that the fluid content of 1 g brain tissue is 0.75 mL.

2. For deproteinized blood:
   a. Successively add into a quartz microcuvette, 1.0 mL of neutralized sample, 0.05 mL of 7 mM NADH and mix thoroughly.
   b. Monitor the change in absorbance at 340 nm as described above until constant, and record absorbance (A_1).
   c. Add 0.01 mL of 225 kU/L L-lactate dehydrogenase and mix.
   d. Monitor the absorbance and determine the final absorbance (A_2).
   e. Calculate ΔA and use for the determination of blood pyruvate concentration using the following formula:

   $$C_b = \Delta A \times 1.06 \times (V_a + V_p) \times (2 \times 1.06 \times 0.8 + 2)/(\varepsilon \times 10 \times 1.0 \times V_a \times 2) \text{ (mmol/L)}$$

   where C_b = blood concentration of pyruvate; $\Delta A = A_2 - A_1$; V_a = volume of protein-free supernate neutralized with KHCO$_3$ = 1 mL; V_p = final volume of neutralized supernate (ml); ε = extinction coefficient of NADH at 340 nm = 6.3 (L × mmol^{-1} × cm^{-1}).

4. Notes

1. (a) BALB/c mice in some laboratories are resistant to the development of CM induced by *P. berghei* ANKA *(59)*. Hence, this infection has also been used as control during CM studies in CBA mice. (b) The course of *P. yoelii 17x* infection is normally mild, with a low level of parasitemia. Certain strains have been reported to become more virulent and pro-

duce a fulminating and fatal infection (from CM) in CF1 and A/J mice within 6-7 d post intraperitoneal inoculation *(22)*. However, studies by Rest to reproduce this mouse model of CM were unsuccessful because the parasite virulence decreased during passage *(60)*. (c) With some strains of mouse malaria parasites, male mice tend to be more susceptible to a lethal infection than female mice.
2. Drabkin's solution contains cyanide, although at a very low concentration that should not pose a dangerous hazard. However, care must be taken when preparing and handling the solution.
3. If the number of parasites is very low (parasitemia less then 0.1%), at least 50 to 100 similar fields should be counted. With higher parasitemias, a total of 10 fields is sufficient.
4. Mice should be bled when parasitemia is still ascending, before the peak. If it is a synchronous parasite, mice should be bled at a time of the day when parasites are in the ring form.
5. Record the strain, number, and gender of mice used, strain of parasite, parasitemia at the moment of bleeding, number of vials frozen, and position in the nitrogen container. It is important to make a stock of parasites with several vials frozen from the same infected blood. The number of passages of parasitized blood from mouse to mouse should be noted. The number should not be greater than four. After four passages, thaw another frozen vial from the liquid nitrogen stock.
6. To detect the presence of sporozoites in the salivary glands of the mosquitoes, dissect some salivary glands on a microscope slide, place a cover slip, and examine microscopically.
7. In this case, it is not possible to be sure how many parasite forms are injected by the feeding mosquito.
8. The number of parasites used to infect is known but it is important to note that some of them may not be viable. Depending on the nature of the study, there should also be a group of mice injected with salivary glands dissected from noninfected mosquitoes to function as a control group.
9. The advantage of this method is that the ^{1}H-NMR spectrum contains resonances from numerous important metabolites and, at high field strengths, resolution of these resonances is possible. Thus, the concentration of various brain metabolites can be measured simultaneously from a single brain sample, therefore allowing a comprehensive assessment of the metabolic status of the brain during the infection of a mouse with a malaria parasite strain that can cause cerebral complications.

References

1. Landau, I. and Boulard, Y. (1978) Life cycles and morphology, in *Rodent Malaria* (Killick-Kendrick, R. and Peters, E., eds.), Academic Press, New York, pp. 53–84.
2. Cox, F. E. G. (1988) Major animal models in malaria research: rodents, in *Malaria: Principles and Practice of Malariology,* Vol. 2 (Wernsdorfer, W. H. and McGregor, I., eds.), Churchill Livingstone, New York, pp. 1503–1543.
3. Langhorne, J. (1994) The immune response to the blood stages of Plasmodium in animal models. *Immunol. Lett.* **41,** 99–102.
4. Neill, A. L. and Hunt, N. H. (1992) Pathology of fatal and resolving *Plasmodium berghei* cerebral malaria in mice. *Parasitology* **105,** 165–175.
5. Sanni, L. A., Thomas, S. R., Tattam, B. N., Moore, D. E., Chaudhri, G., Stocker, R., et al. (1998) Dramatic changes in oxidative tryptophan metabolism along the kynurenine pathway in experimental cerebral and non-cerebral malaria. *Am. J. Pathol.* **152,** 611–619.
6. Curfs, J. H. A. J., Schetters T. P. M., Hermsen, C. C., Jerusalem, C. R., Van Zon, A. A. J. C., and Eling, W. M. C. (1993) Immunological aspects of cerebral lesions in murine malaria. *Clin. Exp. Immunol.* **75,** 136–140.
7. Kaul, D. K., Nagel, R. L., Llena, J. F., and Shear, H. L. (1994) Cerebral malaria in mice: demonstration of cytoadherence of infected red blood cells and microrheologic correlates. *Am. J. Trop. Med. Hyg.* **54,** 512–521.
8. Langhorne, J., Evans, C. B., Asofsky, R., and Taylor, D. W. (1984) Immunoglobulin isotype distribution of malaria-specific antibodies produced during infection with *Plasmodium chabaudi chabaudi* and *Plasmodium yoelii. Cell. Immunol.* **87,** 452–461.

9. Jayawardena, A. N., Targett, G. A. T., Carter, R. L., Leuchars, E., and Davies, A. J. S. (1977) The immunological response of CBA mice to *P. yoelli*. I. General characteristics of the effects of T cell deprivation and reconstitution with thymus grafts. *Immunology* **32**, 849–859.
10. Weinbaum, F. I., Evans, C. B., and Tigelaar, R. E. (1976) Immunity to *Plasmodium berghei yoelii* in mice. I. The course of infection in T cell and B cell deficient mice. *J. Immunol.* **117**, 1999–2005.
11. De Souza, J. B., Williamson, K. H., Otani, T., and Playfair, J. H. (1997) Early gamma interferon responses in lethal and nonlethal murine blood-stage malaria. *Infect. Immun.* **65**, 1593–1598.
12. Kremsner, P. G., Neifer, S., Chaves, M. F., Rudolph, R, and Bienzle, U. (1992) Interferon-g induced lethality in the phase of *Plasmodium vinckei* malaria despite effective parasite clearance by chloroquine. *Eur. J. Immunol.* **22**, 2873–2878.
13. Cross, C. E. and Langhorne, J. (1998) *Plasmodium chabaudi chabaudi* (AS): Inflamatory cytokines and pathology in an erythrocytic-stage infection in mice. *Exp. Parasitol.* **90**, 220–229.
14. Jarra, W. and Brown, K. N. (1985) Protective immunity to malaria: studies with cloned lines of *Plasmodium chabaudi* and *P. berghei* in CBA/Ca mice.I. The effectiveness and inter and intraspecies specificity of immunity induced by infection. *Parasite Immunol.* **7**, 595–606.
15. Grun, J. L. and Weidanz, W. P. (1981) Immunity to *Plasmodium chabaudi chabaudi adami* in the B-cell-deficient mouse. *Nature* **290**, 143–145.
16. Van der Heyde, H. C., Manning, D. D. and Weidanz, W. P. (1993) Role of CD4+ T cells in the expansion of the CD4-, CD8- gd T cell subset in the spleens of mice during blood-stage malaria. *J. Immunol.* **151**, 6311–6317.
17. Cavacini, L.A., Parke, L. A., and Weidanz, W. P. (1990) Resolution of acute malarial infections by T cell-dependent non-antibody-mediated mechanisms of immunity. *Infect. Immun.* **58**, 2946–2950.
18. Curfs, J. H. A., Van der Meide, P. H., Billiau, A., Meuwissen, J. H. E., Eling, W, M. C. (1993)*Plasmodium berghei*: recombinant interferon-g and the development of parasitaemia and cerebral pesions in malaria-infected mice. *Exp. Parasitol.* **77**, 212–223.
19. Sharma, M, C., Tripathi, L. M., Sagar, P., Dutta, G. P., Pandey, V. C. (1992) Cerebral ammonia levels and enzyme changes during *Plasmodium yoelii* infection in mice. *J. Trop. Med. Hyg.* **95**, 410–415.
20. Sanni, L. A., Fu, S., Dean, R. T., Bloomfield, G., Stocker, R., Chaudhri, G. C., et al. (1999) Are reactive oxygen species involved in the pathogenesis of murine cerebral malaria? *J. Infect. Dis.* **179**, 217–222.
21. Yoeli, M. and Hargreaves, B. J. (1974) Brain capillary blockage produced by a virulent strain of rodent malaria. *Science* **184**, 572,573.
22. Krishna, S., Shoubridge, E. A., White, N. J., Weatherall, D. J., and Radda, G. K. (1983) *Plasmodium yoelii*: blood oxygen and brain function in the infected mouse. *Exp. Parasitol.* **56**, 391–396.
23. Thumwood, C. M. (1987) The pathogenesis of murine cerebral malaria. Ph.D. Thesis, Australian National University, Canberra, Australia.
24. Sanni, L. A. (1998) Some neurochemical correlates of murine cerebral malaria. Ph.D. Thesis, University of Sydney, New South Wales, Australia.
25. Elased, K. M., Taverne, J., and Playfair, J. H. (1996) Malaria, blood glucose, and the role of tumour necrosis factor (TNF) in mice. *Clin. Exp. Immunol.* **105**, 443–449.
26. Silverman, P. H., Schooley, J. C., and Mahlmann, L. J. (1987) Murine malaria decreases hemopoietic stem cells. *Blood* **69**, 408–413.
27. Gilks, C. F., Walliker, D., and Newbold, C. I. (1990) Relationships between sequestration, antigenic variation and chronic parasitism in *Plasmodium chabaudi chabaudi*- a rodent model. *Parasite Immunol.* **12**, 45–64.
28. Cox, J., Semoff, S., and Hommel, M. (1987) *Plasmodium chabaudi*: a rodent malaria model for in-vivo and in-vitro cytoadherence of malaria parasites in the absence of knobs. *Parasite Immunol.* **9**, 543–561.
29. Clark, I. A., Cowden, W. B., and Butcher, G. A. (1987) Possible roles of Tumor necrosis factor in the pathology of malaria. *Am. J. Pathol.* **129**, 192–199.
30. Stevenson, M. M., Lyanga, J. J., and Skamene, E. (1982) Murine malaria: genetic control of resistance to *Plasmodium chabaudi*. *Infect. Immun.* **38**, 80–88.
31. Elased, K. and Playfair, J. H. L. (1994) Hypoglycemia and Hyperinsulinemia in Rodent Models of Severe Malaria Infection. *Infect. Immun.* **62**, 5157–5160.
32. Rockett, K. A., Awburn, M. M., Rockett, E. J., and Clark, I. A. (1994) Tumor necrosis factor and Interleukin-1 synergy in the context of malaria pathology. *Am. J. Trop. Hyg.* **50**, 735–742.
33. Grau, G. E., Piguet, P. F., Engers, H. D., Louis, J. A., Vassalli, P., and Lampert, P. H. (1986) L3T4+ T-lymphocytes play a major role in the pathogenesis of murine cerebral malaria. *J. Immunol.* **137**, 2348–2354.
34. Grau, G. E., Piguet, P. F., Vassalli, P., and Lambert, P. H. (1989) Tumor-necrosis factor and other cytokines in cerebral malaria: experimental and clinical data. *Immunol. Rev.* **112**, 49–70.
35. Yanez, D. M., Manning, D. D., Cooley, A. J., Weidanz, W. P., and van der Heyde, H. C. (1996) Participation of lymphocyte subpopulations in the pathogenesis of experimental murine cerebral malaria. *J. Immunol.* **157**, 1620–1624.

36. De Souza, J. B., Ling, I, T., Ogun, S. A., Holder, A. A., and Playfair, J. H. L. (1996) Cytokines and antibody subclass associated with protective immunity against blood-stage malaria in mice vaccinated with the C terminus of merozoite surface protein 1 plus a novel adjuvant. *Infect. Immun.* **64,** 3532–3536.
37. Jayawardena, A. N., Janeway, C. A. Jr., and Kemp, J. D. (1979) Experimental malaria in the CBA/N mouse. *J. Immunol.* **123,** 2532–2539.
38. Kumar, S., Good, M. F., Dontfraid, F., Vinetz, J. M., and Miller, L. (1989) Interdependence of CD4+ T cells and malarial spleen in immunity *to Plasmodium vinckei vinckei. J. Immunol.* **143,** 2017–2023.
39. Grun, J. L. and Weidanz, W. P. (1983) Antibody-independent immunity to reinfection malaria in B-cell-deficient mice. *Infect. Immun.* **41,** 1197–1204.
40. Langhorne, J. (1989) The role of CD4+ T cells in the immune response to *Plasmodium chabaudi. Parasitol. Tod.* **5,** 362–364.
41. Langhorne, J., Gillard, S., Simon, B., Slade, S., and Eickmann, K. (1989) Frequencies of distinct response kinetics for cells with Th1 and Th2 characteristics during infection. *Intl. Immunol.* **1,** 416–424.
42. Seixas, E. M. G. and Langhorne, J. (1999) gd T cells contribute to control of chronic parasitaemia in *Plasmodium chabaudi* infections in mice. *J. Immunol.* **162,** 2837–2841.
43. Bate, C. A. W., Taverne, J., and Playfair, J. H. L. (1988) Malarial parasites induce TNF production by macrophages. *Immunology* **64,** 227–231.
44. Sam, H. and Stevenson, M. M. (1999) In vivo IL-12 production and IL-12 receptors B1 and B2 mRNA expression in the spleen are differentially up-regulated in resistant B6 and susceptible A/J mice during early blood-stage *Plasmodium chabaudi* AS malaria. *J. Immunol.* **162,** 1582–1589.
45. Slade, S. J. and Langhorne, J. (1989) Production of IFN-g during infection of mice with *Plasmodium chabaudi chabaudi. Immunbiology* **179,** 353–365.
46. Stevenson, M. M., Tam, F. F., Belosevic, M., van der Meide, P. H. and Podoba, J. E. (1990) Role of endogenous gamma interferon in host response to infection with blood-stage *Plasmodium chabaudi* AS. *Infect. Immun.* **58,** 3225–3232.
47. Taylor-Robinson, A. W. (1998) Immunoregulation of malarial infection: balancing the vices and virtues. *Int. J. Parasitol.* **28,** 135–148.
48. Stevenson, M. M. and Tam, M.-F. (1993) Differential induction of helper T cells subsets during blood-stage *Plasmodium chabaudi* AS infection in resistant and susceptible mice. *Clin. Exp. Immunol.* **92,** 77–83.
49. Taylor-Robisnson, A. W. and Phillips, R. S. (1992) Functional Characterization of protective CD4+ T-cell clones reactive to the murine malaria parasite *Plasmodium chabaudi. Immunology* **77,** 99–105.
50. Grun, J. L., Long C.A., and Weidanz, W.P. (1985) Effects of splenectomy on antibody-independent immunity to *Plasmodium chabaudi adami* malaria. *Infect. Immun.* **48,** 853–858.
51. Brake, D. A., Long, C. A., and Weidanz, W. P. (1985) Effects of splenectomy on antibody-independent immunity to *Plasmodium chabaudi adami* malaria. *Infect. Immun.* **48,** 853–856.
52. Brake, D. A., Weidanz, W. P., and Long, C. A. (1986) Antigen-specific interleukin 2-propagated T lymphocytes confer resistance to a murine malarial parasite: *Plasmodium chabaudi adami. J. Immunol.* **137,** 347–352.
53. Fairbanks, V. F. and Klee, G. G. (1987) Biochemical aspects of hematology, in *Fundamentals of Clinical Chemistry,* 3rd ed. (Tietz, N.W., ed.), W. B. Saunders, Philadelphia, PA, pp. 789–824.
54. Thumwood, C. M., Hunt, N. H., Clark, I. A., and Cowden, W. B. (1988) Breakdown of the blood-brain barrier in murine cerebral malaria. *Parasitology* **96,** 579–589.
55. Tengvar, C., Forssen, M., Hultstrom, D., Olsson, Y., Pertoft, H., and Petterson, A. (1982) Measurement of edema in the nervous system. *Acta Neuropathologica* (Berlin). **57,** 143–150.
56. DiResta, G. R., Lee, J., Lau, N., Ali, F., Galicich, J. H., and Arbit, E. (1990) Measurement of brain tissue density using pycnometry. *Acta Neurochir.* **51,** 34–36.
57. Gutman, I. and Wahlfeld, A. W. (1974) Lactate, in *Methods of Enzymatic Analysis,* Vol. 3, 2nd ed. (Bergmeyer, H. U., Bergmeyer, J., and Grabi, M., eds.), Academic Press, New York, pp. 1464–1468.
58. Lamprecht, W. and Heinz, F. (1985) Pyruvate, in *Methods of Enzymatic Analysis,* Vol. 6–8, 3rd ed. (Bergmeyer, H. U., Bergmeyer, J., and Grabi, M., eds.), Academic Press, New York, pp. 570–577.
59. De Kossodo, S. and Grau, G. E. (1993) Profiles of cytokine production in relation with susceptibility to cerebral malaria. *J. Immunol.* **151,** 4811–4820.
60. Rest, J. R. (1982) Cerebral malaria in inbred mice. I. A new model and its pathology. *Trans. R. Soc. Trop. Med . Hyg.* **76,** 410–415.

6

Nonhuman Primate Models

I. Nonhuman Primate Host–Parasite Combinations

William E. Collins

1. Introduction

Malaria parasites infect a variety of animals, including reptiles, birds, rodents, nonhuman primates, and humans *(1)*. The most commonly studied hosts for biologic, immunologic, and chemotherapeutic studies are rodents and nonhuman primates. The nonhuman primate models of interest are those that are susceptible to the human-infecting malaria parasites, *Plasmodium falciparum*, *P. vivax*, *P. malariae*, and *P. ovale*, and the malaria parasites naturally infective to monkeys and apes. Presented here are various combinations of parasite species and strains with primate hosts suitable for various immunologic and chemotherapeutic studies. Of particular interest are those models susceptible to the human malaria parasites. However, parasites naturally infective to monkeys and apes have characteristics that make them very suitable for a variety of laboratory-based investigations.

2. Materials

2.1. Animals

Primates require special housing and care as set forth in the *Guide for the Care and Use of Laboratory Animals*. The New World monkeys susceptible to infection with human malaria parasites *P. falciparum*, *P. vivax*, and *P. malariae* are *Aotus nancymai* and *A. vociferans* from Peru, *A. lemurinus lemurinus* and *A. lemurinus griseimembra* from Panama and Colombia, and *A. azarae boliviensis* from Bolivia. *Saimiri boliviensis* and *S. peruviensis* from Bolivia and Peru and *S. sciureus* from Guyana, Columbia, and Panama have been used for many studies with *P. falciparum* and *P. vivax*. Because of difficulties in housing and management of chimpanzees, *Pan troglodytes*, and the need to splenectomize them to increase susceptibility, these animals are only infected in those instances where infections in monkeys are unsatisfactory.

2.2. Parasites

Different strains of *P. falciparum*, *P. vivax*, and *P. malariae* have been adapted to *Aotus* and *Saimiri* monkeys. *Plasmodium ovale* has proven to be noninfectious to these monkeys; chimpanzees are used in studies with *P. ovale*, *P. vivax*, and *P. malariae*. Rarely, chimpanzees are infected with *P. falciparum*. The malaria parasites of Old

World monkeys, *P. knowlesi*, *P. coatneyi*, *P. fragile*, *P. cynomolgi*, *P. inui*, *P. fieldi*, *P. simiovale*, and *P. gonderi*, and the parasites of New World monkeys, *P. brasilianum* and *P. simium*, provide a number of characteristics useful for comparative studies with the human parasites. Species naturally infectious to gibbons, orangutans, gorillas, chimpanzees, and lemurs are seldom studied because their natural hosts are endangered or unavailable. Newly described parasites of African monkeys, *P. petersi* and *P. georgesi*, have not yet been made available for experimental study.

3. Methods

This chapter details specific nonhuman primate host–parasite combinations. The related technical methods employed are presented in Chapter 7. Protocols are approved by institutional Animal Care and Use Committees in accordance with U. S. Public Health Service Policy (1986) (*see* **Notes 1–4**).

3.1. Plasmodium falciparum

Many different isolates/strains of *P. falciparum* have been adapted to develop in *Aotus* and *Saimiri* monkeys. Serial blood passage rapidly results in increased virulence and loss in their ability to produce infective gametocytes. Thus, stabilates are stored frozen over liquid N_2 and, whenever possible, frequent sporozoite passage is made to ensure the continued production of gametocytes. Splenectomy of the host markedly increases parasite count and is often essential for the production of abundant infective gametocytes.

Major *P. falciparum* strains are presented in the following subsections.

3.1.1. Vietnam Oak Knoll (FVO)

The Vietnam Oak Knoll strain of *P. falciparum* was originally isolated from a soldier from Vietnam who was admitted to the Oak Knoll Naval Hospital in June 1968 *(2)*. The parasite was passaged to an *Aotus* monkey and has been maintained either by blood passage in *Aotus* or frozen. In *Aotus nancymai*, parasitemia greater than 10% occurs following intravenous inoculation of 1×10^4 parasites. The parasite is resistant to chloroquine. Infected animals are cured by treatment with quinine and mefloquine. This strain is a standard for blood-stage vaccine trials and chemotherapeutic studies in intact *A. nancymai*, *A. lemurinus lemurinus* and *A. lemurinus griseimembra* monkeys *(3–6)*. Gametocytes are no longer produced.

3.1.2. Santa Lucia

This is a well-established strain of parasite that was isolated from a patient in El Salvador, Central America. It is susceptible to most of the standard antimalarial drugs and is a good infector of mosquitoes. Sporozoite transmission to *A. lemurinus griseimembra* has been demonstrated repeatedly. Gametocyte production has been maintained by frequent sporozoite passage to splenectomized *Aotus* spp. It is a standard parasite for transmission-blocking and antisporozoite vaccine studies in *A. lemurinus griseimembra* and *A. vociferans* monkeys *(7,8)*.

3.1.3. Uganda Palo Alto (FUP)

This strain was adapted initially to develop in *A. lemurinus griseimembra* monkeys. There are many lines of this strain currently available. It has been used for screening of antimalarial drugs and blood-stage vaccine trials in *S. boliviensis*, *S. sciureus*, *A. nancymai*, and *A. azarae boliviensis*. Splenectomy of *A. nancymai* is required to obtain uniformly high-density parasitemia. The Roche strain of FUP has been adapted to produce uniformly high parasitemia in intact *S. boliviensis* monkeys and is thus suitable for blood-stage vaccine trials in this host. The parasite is susceptible to treatment with chloroquine; it no longer produces infective gametocytes.

3.1.4. Indochina I/CDC (INDO-1)

This strain has been used in Australia for vaccine trials in Guyanan *Saimiri* monkeys. The parasite infects different species of *Aotus* as well as *Saimiri* monkeys. The parasite is resistant to chloroquine and sulfadoxine–pyrimethamine and somewhat tolerant of mefloquine. Early stabilates produce infective gametocytes. Following serial passage through *S. sciureus* monkeys, the line produces high-density asexual parasitemia, but gametocytes are no longer produced.

3.1.5. Malayan IV

This is one of the first chloroquine-resistant strains adapted to *Aotus*. It has been used in both monkey and human drug-susceptibility trials. Infective gametocytes are produced in *Aotus* monkeys and sporozoite transmission has been obtained to splenectomized *A. lemurinus griseimembra*.

3.1.6. Other Strains of P. falciparum Adapted to New World Monkeys

Camp: Originally from Malaysia; highly virulent to intact *Aotus*; susceptible to treatment with chloroquine; no longer produces gametocytes.
Geneve: Probably of African origin; highly virulent to *S. boliviensis* monkeys; susceptible to treatment with chloroquine; does not produce infective gametocytes; has been used in vaccine trials in splenectomized monkeys.
Cambodian I: Originally from Cambodia; resistant to chloroquine; produces infective gametocytes in *Aotus* monkeys.
FCH/4: Originally from the Philippines; adapted to *Aotus* monkeys from in vitro culture; does not produce infective gametocytes.
Montagnard S-1: Isolated from a Montagnard refugee living in North Carolina; susceptible to chloroquine; grows in *Aotus* monkeys; produces infective gametocytes.
West Africa I: Isolated from an American who acquired infection in Nigeria; susceptible to chloroquine; adapted to *Aotus* monkeys; produces infective gametocytes.
Haitian I: Originally from Haiti; highly virulent to *Aotus* monkeys; does not produce infective gametocytes.
Haitian III: Originally from Haiti; adapted to *Aotus* monkeys; does not produce infective gametocytes.
Panama II: Originally from Panama; susceptible to chloroquine; adapted to *Aotus* monkeys; produces infective gametocytes.

3.2. Plasmodium vivax

Major *P. vivax* strains are presented in the following subsections.

3.2.1. Salvador I

This is the standard strain for sporozoite vaccine trials in *S. boliviensis* monkeys *(9)*. Challenge with 10,000 sporozoites injected intravenously gives a predictable infection. The parasite was isolated from a patient in El Salvador, Central America. Massive numbers of infected mosquitoes have been obtained by feeding on gametocytes from chimpanzees. Infective gametocytes are also produced in *Aotus* and *Saimiri* monkeys. It is susceptible to chloroquine and primaquine.

3.2.2. Salvador II

This parasite has been extensively used for mosquito infection studies and is suitable for the testing of transmission-blocking vaccines; it is susceptible to chloroquine and primaquine.

3.2.3. Chesson

This is an old strain that was used in human drug trials in the 1950s and 1960s. It has been studied in different species of *Aotus*. The predictability of sporozoite-induced infections in *Saimiri* is less than that of Salvador I or Salvador II. Mosquitoes have been infected by feeding on gametocytes from chimpanzees and monkeys. It is believed to be tolerant to primaquine.

3.2.4. Brazil I/CDC

This is a relatively recent isolate from a patient whose infection was resistant to treatment with primaquine. It is readily infective to mosquitoes and *Aotus* and *Saimiri* monkeys.

3.2.5. AMRU-1

This is the first chloroquine-resistant strain of the parasite adapted to New World monkeys. It is resistant to chloroquine, but susceptible to amodiaquine and mefloquine. It grows well in *Aotus*; it is less well adapted to *Saimiri*.

3.2.6. Indonesia XIX/CDC

This parasite was isolated from blood samples collected in Indonesia. It is resistant to chloroquine, amodiaquine, and somewhat tolerant to mefloquine. It does not readily infect mosquitoes. It grows very well in splenectomized *Saimiri* and *Aotus* monkeys.

3.2.7. Indonesia I

This parasite was isolated from an American returning from Indonesia. It has a low level of resistance to chloroquine. It grows well in splenectomized *Saimiri* monkeys.

3.2.8. Mauritania I and II

These parasites were isolated on two separate occasions from a patient who acquired his infection in Mauritania. It is the only *P. vivax* parasite of African origin that has been adapted to New World monkeys.

3.2.9. Vietnam Palo Alto

This parasite is highly virulent and produces exceptionally high asexual parasite counts in *Aotus* monkeys. It no longer produces infective gametocytes. It is suitable for the testing of blood-stage vaccines in nonsplenectomized *Saimiri* and *Aotus* monkeys.

3.2.10. Thai III

This is a type II strain of *P. vivax* with a variant circumsporozoite surface protein. It develops well in New World monkeys. Massive numbers of sporozoites have been produced in mosquitoes fed on gametocytes from chimpanzees.

3.2.11. Vietnam Ong

This strain was isolated from a refugee who probably acquired the infection in Indonesia. It readily grows in *Aotus* monkeys, readily infects mosquitoes, and is susceptible to standard anti-malarial drugs.

3.2.12. Vietnam Nam

This strain was isolated from a refugee who acquired the infection in Vietnam or Indonesia. It readily grows in *Aotus*, readily infects mosquitoes, and is susceptible to standard antimalarial drugs.

3.2.13. North Korean

This strain is originally from North Korea. The *hibernans* type of *P. vivax* has a greatly extended prepatent period and relapse interval. It is susceptible to treatment with chloroquine.

3.2.14. Other Strains of P. vivax Adapted to New World Monkeys

The strains are classified as follows: Panama; Pakchong (Thailand); Rio Meta (Colombia); Sumatra; Apastepeque (El Salvador); Miami I and II; Honduras; Indochina I; New Guinea (Henderson); Nicaragua (Nica); West Pakistan. All of these strains are adapted to grow in *Aotus* monkeys. All blood-induced infections are apparently susceptible to chloroquine. There is limited information on mosquito infection.

3.3. Plasmodium malariae

3.3.1. Uganda I/CDC Strain

This parasite has been extensively studied in different species of *Aotus* and *Saimiri* monkeys. It requires splenectomy for high-density parasitemia. Mosquito infection by feeding on monkeys is rare. Mosquito infections are readily obtained by feeding on gametocytes from chimpanzees *(10,11)*.

3.3.2. China I/CDC Strain

This parasite was isolated from an infection that had persisted for many years in a patient *(12)*. The parasite develops in *Aotus* monkeys and chimpanzees. It does not produce infective gametocytes.

3.4. Plasmodium ovale

3.4.1. Nigerian I Strain

This strain has been studied extensively in chimpanzees. It readily infects different species of mosquitoes *(13)*. Only exoerythrocytic stages have been demonstrated in New World monkeys.

3.5. Plasmodium knowlesi

3.5.1. H, Malaysian, Philippine, Hackeri, and Nuri Strains

All strains develop well in rhesus monkeys (*Macaca mulatta*). Infections induced by passage of parasitized erythrocytes are usually fatal. Sporozoite-induced infections are approx 60% fatal. Because the parasite has a quotidian asexual life cycle, parasitemia must be monitored very closely. Exoerythrocytic stages are readily demonstrated in liver sections following sporozoite challenge. Infections are readily induced in *Aotus* and *Saimiri* monkeys; humans are also susceptible to infection. *Macaca fascicularis*, a natural host of this parasite, usually survive infection. *Anopheles dirus* is a very susceptible laboratory vector. This is a nonrelapsing parasite, in that resting stages in the liver (hypnozoites) are not produced. The parasite has been used extensively for immunologic and vaccine trials.

3.6. Plasmodium coatneyi

3.6.1. Hackeri Strain

This parasite is highly virulent to rhesus monkeys; approx 30% of sporozoite-induced infections are fatal. It has a tertian erythrocytic cycle. Mature forms sequester in the deep tissue, resulting in cerebral malaria, similar to *P. falciparum* in humans. The parasite also develops well in *M. fascicularis*. It has not been shown to develop in erythrocytes of New World monkeys. *Anopheles dirus* is an excellent vector for laboratory studies. The parasite does not relapse. The parasite is frequently used for pathologic studies and rarely for immunologic study.

3.7. Plasmodium fragile

3.7.1. Nilgiri and Type Strains

The parasite is highly virulent to rhesus monkeys; sporozoite-induced infections are approx 30% fatal. Blood-induced infections in rhesus monkeys are frequently fatal, whereas *M. fascicularis* rarely die. It has a tertian erythrocytic cycle. Mature forms sequester. Both of these strains develop well in New World monkeys. The Nilgiri strain no longer produces infective gametocytes. The type species produces infective gametocytes in macaques; infection of mosquitoes by feeding on New World monkeys has not been successful. It is readily transmitted by *Anopheles dirus* to macaques and New World monkeys. The parasite has been used for vaccine trials in *Saimiri boliviensis*.

3.8. Plasmodium cynomolgi

3.8.1. Berok, Gombak, B (bastianellii), Cambodian, M (mulligan), Rossan, Smithsonian, Pig Tail, and Ceylon (ceylonensis) Strains

This complex of parasites readily infects macaques; not all strains have been adapted to develop in New World monkeys. The parasite has a tertian erythrocytic cycle. *Plasmodium cynomolgi* is readily transmitted by most laboratory-maintained species of *Anopheles*. The parasite has many biological similarities to *P. vivax* and is used as a model for the study of relapsing malaria. The parasite has been experimentally transmitted to humans. Because the circumsporozoite surface protein (CSP) varies markedly between strains, it is a fertile area for molecular evolutionary studies. The parasite has been used extensively in the testing of drugs against exoerythrocytic and erythrocytic stages.

3.9. Plasmodium inui

3.9.1. Celebes I and II, CDC, Hackeri, Hawking, I (leucosphyrus), Leaf Monkey I and II, Mulligan, N-34, OS (shortti), Perak, Perlis, Philippine, Taiwan I and II, and Walter Reed Strains

This complex of parasites readily infects macaques and New World monkeys to varying degrees. The parasite has been transmitted to humans. It has a quartan erythrocytic cycle. There is also marked variation in the CSP between strains and is thus another parasite with marked potential for molecular evolutionary and vaccine studies. The parasite is readily infective to different anopheline mosquitoes. *Anopheles dirus* is the most effective vector.

3.10. Plasmodium simium

3.10.1. Howler Strain

This parasite naturally infects howler and woolly spider monkeys in Brazil and readily infects *Aotus* and *Saimiri* monkeys inoculated with trophozoites or sporozoites. It does not infect macaques. The parasite has a tertian erythrocytic cycle and has many similarities to *P. vivax*. Relapse has not been demonstrated. It is readily transmitted by most laboratory-available anopheline mosquitoes. The parasite has had limited study.

3.11. Plasmodium brasilianum

3.11.1. Ateles and Peruvian I, III, and IV Strains

This parasite naturally infects many species of monkeys throughout South America. This is a quartan parasite closely related, if not identical, to *P. malariae*. It has been transmitted to humans. *Plasmodium brasilianum* can serve to test vaccines directed against the human malaria parasite. The parasite does not relapse. It is readily transmitted by most laboratory-maintained *Anopheles* mosquitoes. Some strains of this parasite have been shown to be resistant to treatment with chloroquine.

3.12. Plasmodium fieldi

3.12.1. Hackeri, N-3, and ABI (Anopheles balabacensis introlatus) Strains

These parasites have characteristics similar to the human malaria parasite *P. ovale*. Infections in macaques seldom produce high-density erythrocytic infections. *Plasmodium fieldi* has a tertian erythrocytic cycle. Exoerythrocytic stages develop in the hepatocytes of *Aotus* monkeys, but erythrocytic stages are rarely seen. Relapses occur at frequent intervals. The most effective laboratory vector is *Anopheles dirus*.

3.13. Plasmodium simiovale

3.13.1. Type Strain

This parasite from Sri Lanka also has characteristics similar to the human malaria parasite *P. ovale* and to *P. fieldi*. The parasite has a tertian erythrocytic cycle; in macaques, the asexual parasite counts rarely exceed 2–3%. It has not been shown to develop in erythrocytes of New World monkeys, but it does develop in hepatocytes. Relapses occur at frequent intervals. The most effective laboratory vector is *Anopheles dirus*.

3.14. Plasmodium gonderi

3.14.1. Mandrill Strain

This parasite is found in African monkeys and readily develops in macaques. It has a tertian periodicity of the erythrocytic cycle and many similarities to *P. cynomolgi*. It apparently does not relapse from residual liver-stage schizonts. Exoerythrocytic stages have been demonstrated in hepatocytes of *Aotus* and *Saimiri* monkeys as well as macaques. It does not develop in erythrocytes of New World monkeys. This parasite is transmissible by most laboratory-maintained anopheline mosquitoes. This parasite has been little studied.

4. Notes

1. Most institutions have ethical standards that will not allow death to be an end point in any study involving nonhuman primates. Supportive care and treatment should be available to prevent painful suffering or risk of death.
2. Different studies in nonhuman primates can often be combined to obtain the maximum information and/or material from each infection.
3. Animals previously infected with one species of *Plasmodium* can frequently be successfully reinfected with a heterologous species. This is particularly true for animals that have been splenectomized.
4. Some parasites of monkeys are infectious to humans. Vigilance must be maintained to protect against infection of mosquitoes that may transmit the infection to humans or other primates.

References

1. Garnham, P. C. C. (1966) *Malaria Parasites and Other Haemosporidia*, Blackwell Scientific Publications, Oxford. pp. 1–1114.
2. Geiman, Q. M. and Meagher, M. J. (1967) Susceptibility of a New World monkey to *Plasmodium falciparum* from man. *Nature* **215**, 437–439.
3. Schmidt, L. H. (1978) *Plasmodium falciparum* and *Plasmodium vivax* infections in the owl monkey (*Aotus trivirgatus*). I. The course of untreated infections. *Am. J. Trop. Med. Hyg.* **27**, 671–702.
4. Schmidt, L. H. (1978) *Plasmodium falciparum* and *Plasmodium vivax* infections in the owl monkey (*Aotus trivirgatus*). II. Responses to chloroquine, quinine, and pyrimethamine. *Am. J. Trop. Med. Hyg.* **27**, 703–717.
5. Schmidt, L. H. (1978) *Plasmodium falciparum* and *Plasmodium vivax* infections in the owl monkey (*Aotus trivirgatus*). III. Methods employed in the search for new blood schizonticidal drugs. *Am. J. Trop. Med. Hyg.* **27**, 718–737.
6. Collins, W. E., Galland, G. G., Sullivan, J. S., and Morris, C. L. (1994) Selection of *Aotus* monkey models for testing *Plasmodium falciparum* blood-stage vaccines. *Am. J. Trop. Med. Hyg.* **51**, 224–232.
7. Collins, W. E., Galland, G. G., Sullivan, J. S., Morris, C. L., Richardson, B. B., and Roberts, J. M. (1996) The Santa Lucia strain of *Plasmodium falciparum* as a model for vaccine studies. I. Development in *Aotus lemurinus griseimembra* monkeys. *Am. J. Trop. Med. Hyg.* **54**, 372–379.
8. Collins, W. E., Galland, G. G., Sullivan, J. S., Morris, C. L., Richardson, B. B., and Roberts, J. M. (1996) The Santa Lucia strain of *Plasmodium falciparum* as a model for vaccine studies. II. Development in *Aotus vociferans* as a model for testing transmission-blocking vaccines. *Am. J. Trop. Med. Hyg.* **54**, 380–385.
9. Sullivan, J. S., Morris, C. L., McClure, H. M., Strobert, S., Richardson, B. B., Galland, G. G., et al. (1996) *Plasmodium vivax* infections in chimpanzees for sporozoite vaccine challenge studies in monkeys. *Am. J. Trop. Med. Hyg.* **55**, 344–349.
10. Collins, W. E., Schwartz, I. K., Skinner, J. C., and Broderson, J. R. (1984) Studies on the Uganda I/CDC strain of *Plasmodium malariae* in Bolivian *Aotus* monkeys and different anophelines. *J. Parasitol.* **70**, 677–681.
11. Collins, W. E., McClure, H. M., Strobert, E., Filipski, V. K., Skinner, J. C., Stanfill, P. S., et al. (1990) Infection of chimpanzees with the Uganda I/CDC strain of *Plasmodium malariae*. *Am. J. Trop. Med. Hyg.* **42**, 99–103.
12. Collins, W. E., Lobel, H. O., McClure, H. M., Strobert, E., Galland, G. G., Taylor, F., et al. (1994) The China I/CDC strain of *Plasmodium malariae* in *Aotus* monkeys and chimpanzees. *Am. J. Trop. Med. Hyg.* **50**, 28–32.
13. Morris, C. L., Sullivan, C. L., McClure, H. M., Strobert, E., Richardson, B. B., Galland, G. G., et al. (1996) The Nigerian I/CDC strain of *Plasmodium ovale* in chimpanzees. *J. Parasitol.* **82**, 444–448.

7

Nonhuman Primate Models

II. Infection of Saimiri *and* Aotus *Monkeys with* Plasmodium vivax

William E. Collins

1. Introduction

Infections with human malaria parasites in New World monkeys offer opportunities to determine host–parasite interactions and relationships, to produce malarial antigens for diagnostic and molecular characterization, to conduct drug and vaccine efficacy trials, and to produce gametocytes for mosquito infectivity studies *(1–7)*. Isolates of *Plasmodium vivax* from different geographic areas vary in their infectivity to monkeys and mosquitoes. Because they are human pathogens, standard precautions must be followed to avoid accidental infection. Additionally, primates require special housing and care as set forth in *Guide for the Care and Use of Laboratory Animals*. Protocols are approved by institutional Animal Care and Use Committees in accordance with U.S. Public Health Service Policy (1986).

2. Materials

2.1. Animals

Different species of New World monkeys are susceptible to infection with different strains and isolates of *P. vivax (8–22)*. The most commonly available species are *Aotus nancymai*, *Aotus vociferans*, *Saimiri boliviensis,* and *Saimiri peruviensis*. Feral and laboratory-born animals are suitable. Animals are imported for research purposes only; laboratory-bred animals are frequently available.

2.2. Parasites

Many different strains of *P. vivax* have been adapted to New World monkeys. These include Panama Achiote, Honduras I, Indochina I, North Korean, Pakchong (Thailand), Salvador I, Salvador II, Vietnam Palo Alto, Chesson (New Guinea), AMRU-1 (Papua New Guinea), India VII, Indonesia XIX, Brazil I, and West Pakistan (*see* Chapter 6). Parasites are maintained frozen over liquid nitrogen or by serial passage from animal to animal (*see* **Note 1**).

2.3. Equipment

1. Centrifuge with a swing-out rotor.
2. 37°C water bath (optional).

3. 37°C oven.
4. Microscope with an eyepiece containing a Howard disk.
5. 19-, 20-, 21-, and 25-gage needles; syringes.
6. Sterile 50-mL conical screw-capped polypropylene tubes.
7. Sterile plastic vials.
8. Sterile pipets, pipetting equipment.
9. Vacutainer tubes with anticoagulent heparin or acid citrate dextrose (ACD).
10. Capillary tubes, 5 µL, equipped with suction bulb.
11. Microscope slides.
12. 500-mL brown bottle.
13. Coplin jars.
14. Glass beads, size 3 mm.
15. Whatman #1 filter paper.
16. Neubauer Cell Counting chamber, cover slips.
17. Autoclaved muslin or sterile paper drape.
18. Surgical soap.
19. Surgical blades, size 40.
20. Ovariohysterectomy hook or allis forceps.
21. Curved Mayo or Metzenbaum scissors.
22. 1/2 circle straumic needle.
23. 3-0 or 4-0 chromic gut.
24. 3-0 Dexon®.
25. Vitafil® type nonabsorbable surtures.
26. 3-in., 18-gage intubation needle.

2.4. Reagents

1. Glycerolyte 57 solution: 57 g Glycerin USP, 1.6 g sodium lactate, 30 mg KCl buffered with 51.7 mg NaH_2PO_4 and 124.2 mg Na_2HPO_4 (Baxter Healthcare, Fenwal Div., Dearfield, IL).
2. Ketamine hydrochloride, 100 mg/mL (Fort Dodge Animal Health, Fort Dodge, Iowa).
3. Acepromazine maleate, 100 mg/mL (Butler, Columbus, OH).
4. Heparin.
5. Methyl alcohol (absolute, acetone-free).
6. Giemsa stain (Fisher, Certified Biological Stain): Prepare by adding 3 g Giemsa, 270 mL absolute methyl alcohol (acetone-free), and 140 mL glycerin to a clean, dry 500-mL brown bottle that contains 15 mL glass beads (size 3 mm) (*see* **Note 2**). Shake daily for at least 30–60 min, for a 2-wk period. The quality of the stain will improve with age. Keep bottle tightly closed. Before use, filter an aliquot of stain through Whatman #1 filter paper into a small, dry, brown glass bottle. Shake bottle before filtering. Keep tightly capped.
7. Chloroquine phosphate (Sanofi Winthrop, New York, NY).
8. Rivoquine Syrup (Rivopharm Labs., Manno, Switzerland).
9. Primaquine phosphate (Sanofi Winthrop, New York, NY).
10. Lariam HCl (mefloquine) (Hoffman-LaRoche & Co., Ltd.).
11. Mercurochrome.
12. Triton X-100 (LabChem., Inc.).
13. Glycerol (glycerin) (Sigma).
14. RPMI 1640 (Gibco-BRL).
15. Phosphate-buffered water (PBW, pH 7.0–7.2).
16. NaCl/saline, 12, 1.6, and 0.9%.
17. Dextrose solution, 0.2%.

3. Methods

3.1 Thawing of Frozen Parasitized Erythrocyte Stocks

Parasites are maintained frozen in Glycerolyte.

1. Thaw vials of frozen cells in a 37°C water bath or at room temperature. Process rapidly after thawing.
2. Transfer to a 50-mL conical centrifuge tube by sterile pipet, noting volume.
3. Poke hole in cap of tube with a 20-gage needle.
4. Take up 0.2X blood volume of 12% NaCl with a small syringe and a 21-gage needle.
5. Add the 12% saline to the parasite pellet, dropwise through the hole while mixing gently.
6. Let stand at room temperature for 5 min without shaking.
7. Take up 10X blood volume of 1.6% NaCl with a larger syringe and 19-gage needle.
8. Add the 1.6% saline dropwise through the hole while mixing gently.
9. Spin at 1400 rpm for 10 min.
10. Remove supernatant by aspiration and resuspend pellet by gentle shaking.
11. Add 10X blood volume of 0.9% NaCl + 0.2% dextrose solution as described in **step 4**.
12. Spin at 1400 rpm for 10 min.
13. Remove supernatant by aspiration and resuspend pellet by gentle shaking.
14. Add culture medium or saline.

3.2. Infection of Monkey with Parasitized Erythrocytes

1. Restrain monkey, either manually or by intramuscular injection with 0.15–0.24 mL ketamine.
2. Remove monkey from cage; on a flat surface, have handler immobilize monkey on its back with head forward.
3. Grasp foot and ankle in one hand and slightly twist leg outward.
4. Wipe groin area with disinfectant.
5. Insert 25-gage needle into femoral vein (vein is usually shallow and to the outside of the groove between the thigh muscles).
6. Slowly inject parasitized erythrocytes.
7. Hold a pledget over site of injection until blood no longer oozes.
8. Return animal to cage.

3.3. Monitoring Parasitemia

1. Remove monkey from cage and restrain manually, belly-down with head facing forward.
2. Shave calf of hind leg.
3. Grasp hind leg and surface sterilize back of calf from thigh to ankle.
4. Prick leg below the knee carefully avoiding the saphenous vein.
5. Gently squeeze leg until drop of blood appears.
6. Collect blood into a 5-µL capillary tube equipped with suction bulb.
7. Blood is expressed onto a prescribed area (15 mm × 5 mm) on a precleaned microscope slide; a second drop of blood is added to the same slide and a thin film prepared *(23)*.
8. Label slide with the animal identification number and the date.
9. Hold gauze pledget to prick site until blood no longer oozes.
10. Return animal to cage.

3.4. Staining Blood Films and Determining Parasitemia

3.4.1. Staining of Blood Smears

1. Air-dry blood smears on slides.
2. Fix thin films by dipping in methyl alcohol; leave thick films unfixed.

3. Add 50 mL of phosphate-buffered water (PBW, pH 7.0–7.2) to coplin jar.
4. Add 1 mL of Giemsa stain and 2 drops of a 1:20 dilution of Triton X-100 to jar. Mix.
5. Immerse slides with blood films in stain for 45 min
6. For thin films, dip slides in PBW to remove excess stain; for thick films, immerse slides in PBW for 3–5 min to destain.
7. Place slides in a 37°C convection oven until dry.

3.4.2. Counting Parasites

Determine the number of parasites per microliter of blood by counting the number of parasites in a band through the width of the rectangular thick blood film, using the method of Earle and Perez *(23)*, with a microscopic eyepiece containing a Howard disk. Determine the number of gametocytes by counting the number of parasites on the thin film and recording per 100 white blood cells (WBC) (*see* **Notes 3–5**).

3.5. Splenectomy

In order to increase asexual parasite count and to greatly increase the production of infective gametocytes, animals are frequently splenectomized, either before infection or soon after the infection is patent (*see* **Note 6**). All surgical procedures are conducted under general anesthesia by a qualified veterinary surgeon. All surgeries should be performed in an AAALAC-approved surgical suite appropriate for aseptic surgery.

1. Fast all animals for 12 h prior to surgery.
2. Anesthetize monkey with ketamine at a dosage of 15–25 mg/kg body weight prior to surgery.
3. For prolonged procedures, administer isoflurane gas by using a standard closed oxygen delivery apparatus with CO_2 absorption; gas is given at a concentration of 1–2%.
4. Monitor anesthesia by respiratory rate, depth, and skeletal muscle tone.
5. Prepare animal for surgery by clipping with a #40 surgical blade from the xyphoid to the pubis over the ventral body region.
6. Scrub area three times with surgical soap, swab three times with 70% alcohol, and allow to dry.
7. Prepare the final surgical field by drapping with an autoclaved muslin or sterile paper drape, fenestrated for the surgical approach.
8. Perform splenectomy through a midline incision above the umbilicus approximately one-third of the distance advancing toward the xyphpoid cartilage.
9. Make an incision of 1–2 cm through the skin and linea alba.
10. Exteriorize spleen by using an ovariohysterectomy hook or allis forceps.
11. After fenestration, ligate the vasculature at intervals using 3-0 or 4-0 chromic gut, to allow for the most effective hemostasis and minimal tissue trauma.
12. Exise the spleen at the lesser curvature using curved Mayo or Metzenbaum scissors.
13. Check the ligatures for hemostasis and replace into the abdominal cavity.
14. Close the abdominal laparotomy incision in the linea with 3-0 Dexon using a 1/2 circle straumic needle.
15. Close skin with nonabsorbable surtures of Vitafil type using simple interrupted placement at 0.5-cm intervals.

3.6. Collection of Blood or Serum

For the continued well-being of the monkey, blood collection is restricted to 4 mL/kg over a 2-wk period. If larger amounts are required, exchange transfusion of up to 10 mL is acceptable.

1. For routine blood collection, collect blood from the femoral vein using a syringe equipped with a 25-gage needle.

Infection of Saimiri *and* Aotus *Monkeys with* Plasmodium vivax

2. For exchange transfusion, collect blood from donor animals into ACD. Up to 10 mL can be removed with a syringe if blood is quickly replaced with new blood or packed erythrocytes. Otherwise, butterflies are inserted into opposite legs. As blood is slowly removed from one side, donor blood or cells are introduced into the other.
3. Hold sterile gauze to needle sites until all bleeding ceases.

3.7. Freezing of Parasitized Erythrocytes

In order to maintain stocks of parasites for subsequent inoculation, parasitized erythrocytes are stored frozen over liquid N_2.

1. Collect blood into heparin.
2. Wash blood twice in serum-free RPMI medium.
3. Spin at 1200 rpm for 10 min. Remove medium.
4. Estimate the RBC packed cell volume. Calculate 2X the RBC packed volume; this is the total Glycerolyte volume to use.
5. Transfer cells to a 50-mL conical centrifuge tube and add one-fifth of the calculated Glycerolyte volume to the blood cells, at a rate of 1–2 drops/s, using a syringe and needle.
6. Let stand 5 min.
7. Aliquot 0.5–1.0 mL parasite suspension per vial, ensuring each tube is labeled with parasite and date.
8. Freeze overnight at –70°C.
9. Transfer to liquid N_2 storage next day.

3.8. Mosquito Infection

1. Anesthetize the monkey. For mosquito infection, monkeys are anesthetized with ketamine. However, some animals rapidly develop a tolerance for ketamine alone. In that case, 1 mL of acepromazine maleate (10 mg/mL) is added to 10 mL of ketamine (100 mg/mL); animals are then injected intramuscularly with 0.15 mL/kg of the mixture.
2. Immobilize the anesthetized monkey face-down to a plastic board using Velcro straps. Ensure that the center of the board has been cut out such that the belly of the monkey can rest directly on the top of a mosquito-containing cage.
3. Allow the mosquitoes to feed directly through the cage mesh onto the animal until the mosquitoes are engorged.
4. Return the monkey to the holding cage.
5. Incubate mosquitoes at $25 \pm 1°C$. Place cotton pledgets moistened with 5–10% sugar solution on top of mosquito cage, and change pledgets daily.
6. Six to 9 d after feeding, anesthetize a sample of the mosquitoes and examine the midguts for the presence of oocysts. Mount the midguts in mercurochrome solution to make identification easier. The number of oocysts per gut $\times$ 100 = Gut Infection Index.
7. Beginning 12 d or more after infection, remove salivary glands of mosquitoes and examine for the presence of sporozoites. Do NOT use mercurochrome for examination of salivary glands.

3.9. Sporozoite Transmission to Other Monkeys

Infections can be induced in monkeys either via the bites of infected mosquitoes or via the intravenous or intrahepatic inoculation of sporozoites dissected from the salivary glands (*see* **Note 7**).

1. Allow caged mosquitoes to feed directly on the monkeys using the same procedure as was used for mosquito infection (*see* **Subheading 3.7.**).
2. After feeding, dissect individual mosquitoes to determine the intensity of infection in those mosquitoes that fed.

3. Determine the number of sporozoites in the salivary gland preparations and grade as 1+ (1–10 sporozoites), 2+ (11–100 sporozoites), 3+ (101–1000 sporozoites), or 4+ (greater than 1000 sporozoites).
4. For intravenous challenge, dissect mosquito salivary glands in 20% serum saline, pool preparations, crush under a cover slip, and wash into a vial.
5. Add a portion of the sporozoite suspension to a Neubauer Cell Counting chamber and determine concentration. Adjust to desired inoculum.
6. Inject sporozoite suspension into the femoral vein of the recipient monkey.

3.10. Treatment

Most *P. vivax* strains are susceptible to treatment with chloroquine (10 mg/kg daily for 3 d) (*see* **Notes 8** and **9**). Chloroquine-resistant strains such as AMRU-1 and Indonesia XIX can be effectively treated with mefloquine (single treatment with 20 mg/kg). Radical cure of sporozoite-induced infections is usually obtained by a combination treatment with chloroquine or mefloquine (as earlier) and primaquine (2.5 mg/kg × 7 d). Drug tablets are crushed in a mortar and combined with water. Alternatively, Chloroquine Syrup (Rivoquine, 10 mg chloroquine/mL) is a convenient formulation for treatment of New World monkeys.

1. Use a 3-in., 18-gage intubation needle for oral administration.
2. Hold the monkey upright with arms behind its back and legs firmly extended down.
3. Tilt the head of the monkey back and allow the needle to gently drop down the throat until the syringe is at the level of the teeth.
4. Administer drug suspension slowly, and then remove the needle.

4. Notes

1. Parasitemia is often slow to develop following the inoculation with parasites that have been stored frozen; some blood-induced infections are slow to rise. Two or 3 wk may elapse before parasites are detected on a thick blood film or the parasitemia rises above 1000/µL.
2. All reagents used for making Giemsa stain should be free of water; only freshly opened bottles of methanol and glycerin should be used. Because Giemsa is photosensitive, brown glass bottles should be used. The brown bottle and glass beads should be clean and completely dry.
3. Not all animals develop high-density parasitemia; maximum parasite counts in splenectomized animals are usually in the range of 10,000–40,000/µL.
4. Primary episodes of parasitemia are usually followed by recrudescences; secondary peak parasite counts are usually lower than primary attacks.
5. Gametocytes are rarely seen; however, when microgametocytes are present early in the infection, mosquito infection is commonly obtained.
6. Mosquito infection is rarely obtained by feeding on animals that have NOT been splenectomized. Mosquito infection is more apt to occur when mosquitoes are fed during the primary episode of ascending parasitemia. As a general rule, infection begins when the asexual parasite count approaches or surpasses 1000/µL and often ceases when the count reaches or exceeds 10,000/µL.
7. Large numbers of sporozoites are required to assure infection of New World monkeys; prepatent periods vary markedly between animals. Transmission rates to *Saimiri* monkeys are normally higher than to *Aotus* monkeys injected with similar numbers of sporozoites.
8. Sporozoite-induced infections in *Aotus* monkeys rarely relapse after treatment with chloroquine; relapse is more common in *Saimiri* monkeys.

9. Heparin is counterindicated for transfusion; hemorrhage often results when large volumes of heparin are injected.

References

1. Schmidt, L. H. (1973) Infections with *Plasmodium falciparum* and *Plasmodium vivax* in the owl monkey - model systems for basic biological and chemotherapeutic studies. *Trans. Roy Soc. Trop. Med. Hyg.* **67**, 446–474.
2. Schmidt, L. H. (1978) *Plasmodium falciparum* and *Plasmodium vivax* infections in the owl monkey (*Aotus trivirgatus*). I. The courses of untreated infections. *Am. J. Trop. Med. Hyg.* **27,** 671–702.
3. Schmidt, L. H. (1978) *Plasmodium falciparum* and *Plasmodium vivax* infections in the owl monkey (*Aotus trivirgatus*). II. Responses to chloroquine, quinine, and pyrimethamine. *Am. J. Trop. Med. Hyg.* **27,** 703–717.
4. Schmidt, L. H. (1978) *Plasmodium falciparum* and *Plasmodium vivax* infections in the owl monkey (*Aotus trivirgatus*). III. Methods employed in the search for new blood schizonticidal drugs. *Am. J. Trop. Med. Hyg.* **27,** 718–737.
5. Collins, W. E., Nussenzweig, R. S., Ballou, W. R., Ruebush, T. K. II, Nardin, E. H., Chulay, J. D., et al. (1989) Immunization of *Saimiri sciureus boliviensis* monkeys with recombinant vaccines based on the circumsporozoite protein of *Plasmodium vivax*. *Am. J. Trop. Med. Hyg.* **40,** 455–464.
6. Collins, W. E., Nussenzweig, R. S., Ruebush, T. K. II, Bathurst, I., Nardin, E. H., Gibson, H. L., et al. (1990) Further studies on the immunization of *Saimiri sciureus boliviensis* monkeys with recombinant vaccines based on the circumsporozoite protein of *Plasmodium vivax*. *Am. J. Trop. Med. Hyg.* **43,** 576–583.
7. Collins, W. E., Sullivan, J. S., Morris, C. L., Galland, G. G., Jue, D. L., Fang, S., et al. (1997) Protective immunity induced in squirrel monkeys with a multiple antigen construct (MAC) against the CS protein of *Plasmodium vivax*. *Am. J. Trop. Med. Hyg.* **56,** 200–210.
8. Collins, W. E., Contacos, P. G., Stanfill, P. S. and Richardson, B. B. (1973) Studies on human malaria in *Aotus* monkeys I. Sporozoite transmission of *Plasmodium vivax* from El Salvador. *J. Parasitol.* **59,** 606–608.
9. Collins, W. E., Skinner, J. C., Richardson, B. B., Stanfill, P. S., and Contacos, P. G. (1974) Studies on human malaria in *Aotus* monkeys V. Blood-induced infections of *Plasmodium vivax*. *J. Parasitol.* **60,** 393–398.
10. Collins, W. E., Warren, McW, Contacos, P. G., Skinner, J. C., Richardson, B. B., and Kearse, T. S. (1980) The Chesson strain of *Plasmodium vivax* in *Aotus* monkeys and anopheline mosquitoes. *J. Parasitol.* **66,** 488–497.
11. Collins, W. E., Warren, McW., Skinner, J. C., and Richardson, B. B. (1980) The West Pakiston strain of *Plasmodium vivax* in *Aotus* monkeys and anopheline mosquitoes. *J. Parasitol.* **66,** 780–785.
12. Campbell, C. C., Collins, W. E., Chin, W., Roberts, J. M., and Broderson, J. R. (1983) Studies on the Sal I strain of *Plasmodium vivax* in the squirrel monkey (*Saimiri sciureus*). *J. Parasitol.* **69,** 689–695.
13. Collins, W. E., Skinner, J. C., Krotoski, W. A., Cogswell, F. B., Gwadz, R. W., Broderson, J. R., et al. (1985) Studies on the North Korean strain of *Plasmodium vivax* in *Aotus* monkeys and different anophelines. *J. Parasitol.* **71,**: 20–27.
14. Collins, W. E., Warren, McW., Huong, A. Y., Skinner, J. C., Sutton, B. B., and Stanfill, P. S. (1986) Studies on the comparative infectivity of fifteen strains of *Plasmodium vivax* to laboratory-reared anophelines, with special reference to *Anopheles culicifacies*. *J. Parasitol.* **72,** 521–524.
15. Collins, W. E., Skinner, J. C., Pappaioanou, M., Broderson, J. R., McClure, H. M., Strobert, E., et al. (1987) Chesson strain *Plasmodium vivax* in *Saimiri sciureus boliviensis* monkeys. *J. Parasitol.* **73,** 929–934.
16. Collins, W. E., Skinner, J. C., Pappaioanou, M. J., Ma, N. S-F., Broderson, J. R., Richardson, B. B., and Stanfill, P. S. (1987) Infection of *Aotus vociferans* (Karyotype V) monkeys with different strains of *Plasmodium vivax*. *J. Parasitol.* **73,**: 536–540.
17. Collins, W. E., Skinner, J. C., Pappaioanou, M., Broderson, J. R., Ma, N. S.-F., Filipski, V., et al. (1988) Infection of Peruvian *Aotus nancymai* monkeys with different strains of *Plasmodium falciparum, P. vivax,* and *P. malariae*. *J. Parasitol.* **74,** 392–398.
18. Collins, W. E., Skinner, J. C., Pappaioanou, M, Broderson, J. R., Filipski, V. K., McClure, H. M., et al. (1988) Sporozoite-induced infections of the Salvador I strain of *Plasmodium vivax* in *Saimiri sciureus boliviensis* monkeys. *J. Parasitol.* **74,** 582–585.
19. Collins, W. E., Schwartz, I. K., Skinner, J. C., Morris, C., and Filipski, V. K. (1992) Studies on the susceptibility of the Indonesia I/CDC strain of *Plasmodium vivax* to chloroquine. *J. Parasitol.* **78,** 344–349.
20. Collins, W. E., Morris, C. L., Richardson, B. B., Sullivan, J. S., and Galland, G. G. (1994) Further studies on the sporozoite transmission of the Salvador I strain of *Plasmodium vivax*. *J. Parasitol.* **80,** 512–517.

21. Nayar, J. K., Baker, R. H., Knight, J. W., Sullivan, J. S., Morris, C. L., Richardson, B. B., et al. (1997) Studies on a primaquine-tolerant strain of *Plasmodium vivax* from Brazil in *Aotus* and *Saimiri* monkeys. *J. Parasitol.* **83,** 1174–1177.
22. Collins, W. E., Sullivan, J. S., Morris, C. L., Galland, G. G., Richardson, B., and Nesby, S. (1998) Adaptation of a strain of *Plasmodium vivax* from Mauritania to New World monkeys and anopheline mosquitoes. *J. Parasitol.* **84,** 619–621.
23. Earle, W. C. and Perez, M. (1932) Enumeration of parasites in the blood of malarial patients. *J. Lab. Clin. Med.* **17,** 1124–1130.

III

DIAGNOSIS AND TYPING

8

Vector Analysis

John C. Beier

1. Introduction

This chapter describes methods relating to the handling and processing of *Anopheles* mosquitoes captured during malaria vector field studies. Areas covered include the following methods

1. Handling mosquitoes in the field.
2. Developing assembly-line approaches for specimen processing.
3. Taxonomically identifying mosquitoes.
4. Classifying blood-feeding stages, parity, and Christophers' stages of ovarian development.
5. Using molecular methods for identifying specific species of mosquitoes belonging to *Anopheles* species complexes.

This chapter complements Chapter 1.

2. Materials
2.1. Equipment

1. Screened pint cups for holding mosquitoes.
2. Paper towels.
3. Filter paper.
4. Petri dishes of different sizes.
5. 1.8-mL, 1.5-mL, and 0.5-mL plastic tubes with snap-on caps.
6. Insulted cooler for transporting mosquitoes.
7. Ice packs.
8. Color-indicator desiccant.
9. Glass wool.
10. Dissecting and compound microscopes.
11. Field and laboratory processing data forms.
12. Glass cover slips.
13. Aspirator for handling mosquitoes.
14. Plastic disposable beakers of different sizes.
15. No. 5 forceps.
16. Hand-held motor with polypropylene pellet pestle.
17. 1.5-mL grinding tube.
18. Freezer and refrigerator.
19. Micropipets of different sizes, ranging from 10 µL to 1 mL.

20. Vacuum dryer and/or a drying oven.
21. Thermocycler.
22. Agarose gel and gel apparatus.
23. Ultraviolet light source.

2.2. Reagents

1. Carnoy's solution: 3 parts absolute ethanol : 1 part glacial acetic acid.
2. Physiological saline, phosphate-buffered saline (PBS), or M-199 media for dissecting mosquitoes.
3. Mercurochrome solution.
4. DNA extraction buffer (DEB): 0.5% SDS, 0.2 M NaCl, 25 mM ethylenediaminetetraacetic acid (EDTA), and 10 nM Tris-HCl, pH 8.0. Prepare fresh. To prepare 10 mL, combine 50 mg of sodium dodecyl sulfate (SDS), 0.4 mL of 5 M NaCl, 0.5 mL of 0.5 M EDTA, and 0.1 mL of 1.0 M Tris-HCl, pH 8.0. Bring the total volume to 10 mL using ddH_2O stored in an incubator at 37°C, and heat to 65°C for 10 min. Swirl until mixed thoroughly.
5. RNAse A and proteinase K (20 µg/mL).
6. Phenol and chloroform.
7. 70% Ethanol.
8. ddH_2O.
9. Primers for mosquito species diagnosis.
10. 1 M MgCl$_2$.
11. *Taq* polymerase.
12. DNA markers such as 100 bp and/or pUC18.
13. Ethidium bromide.

3. Methods

3.1. Handling Mosquitoes in the Field

1. Live-caught mosquitoes from biting/landing catches and daytime resting collections indoors are normally held alive in screened pint cups. If they are to be transported to the field laboratory within hours of collection, then they should be held either indoors or in the shade, with minimal handling. If they are to be held in the field more than a couple of hours, then precautions such as placing water-soaked paper towels over the cups need to be taken to ensure that the mosquitoes do not become too dry or too hot.
2. Freshly killed mosquitoes from any type of collection method require special handling methods, and several options are available. Mosquitoes from pyrethrum spray collections can be held on filter paper in Petri dishes containing water moistened cotton; this will allow the mosquitoes to be held for 3–6 h before further processing either in the field or field laboratory. Alternatively, mosquitoes can be placed in labeled tubes and held in an insulated cooler containing ice packs; this is a standard method for transport when the field laboratory is within a couple of hours drive from the field site.
3. When it is not possible to transport mosquitoes from field sites to the laboratory during the same day, it is necessary to take more elaborate steps for mosquito processing. Sometimes it is feasible to identify and process mosquitoes directly in the field, through setting up a field laboratory containing microscopes and other essential supplies and equipment. Some investigators prefer to dry mosquitoes by placing individual specimens in individual 0.5-mL labeled tubes containing desiccant granules covered by glass wool.
4. Freshly collected mosquitoes can also be preserved in Carnoy's solution. It is quite effective for a variety of purposes:
    a. Mosquitoes from the same collections can be pooled in individual labeled tubes.
    b. Mosquito samples in Carnoy's solution can be held at room temperature for 24 h or more, after which they can either be refrigerated for several days and/or frozen indefinitely.

c. Mosquitoes held in Carnoy's solution can be processed by standard taxonomic identification, cytogenetics, polymerase chain reaction (PCR) methods for species identification, bloodmeal and sporozoite enzyme-linked immunosorbent assay (ELISA), and other procedures, except routine salivary gland dissections for sporozoites where fresh specimens are required.

Carnoy's solution is an especially good preservative for situations where teams spend several days to weeks in the field and specimens must be processed later in the main laboratory (Y. T. Toure, personal communication).

5. Logistically, it is normal for field teams to complete standard field forms for each collection and to turn these forms over to the laboratory teams doing the subsequent processing. In such field studies, it is necessary to give unique identifier numbers to each mosquito, either in the field at the time of collection, or in the laboratory during subsequent processing.

3.2. Assembly-Line Approaches for Specimen Processing

Mosquitoes captured in the field require processing using a variety of methods. For the sake of efficiency, it is standard practice to operate an assembly-line mosquito processing operation using one or more qualified technicians at each station. Below is one possible strategy when mosquito dissections are required:

1. Station No. 1: *Mosquito Identifications*. For each screened cup or tube of mosquitoes from the field, identify individual mosquitoes taxonomically, using a dissecting microscope. Assign each identified mosquito a unique identifier number and place it in a small Petri dish labeled with the number.
2. Station No. 2: *Records*. Record the identifier number for each mosquito on field forms along with specific field data (e.g., study site, type of collection, station or house number, and date of collection) and the taxonomic identification. Normally, data are entered on standard field forms, but it may also be possible to enter the data directly into a computer.
3. Station No. 3: *Mosquito Dissections*. Place each mosquito on a labeled glass slide and examined under a dissecting microscope.
    a. Note blood-feeding stages (*see* **Subheading 3.4.1.**).
    b. Remove the ovaries and note both parity (parous or nulliparous) and Christophers' stages of ovarian development (*see* **Subheadings 3.4.2.** and **3.4.3.**).
    c. Check for malaria oocysts, dissect the midgut in physiological saline, add a small drop of diluted mercurochrome solution, and add a cover slip (*1*).
    d. Dissect the salivary glands to examine them for malaria sporozoites.
    e. Record results on standard forms.
4. Station No. 4: *Examination of Mosquito Midguts and Salivary Glands*. Using a compound microscope, inspect midguts for malaria oocysts (×100) and examine salivary glands for malaria sporozoites (×400). Record results on standard forms.
5. Station No. 5: *Additional Processing*. One or more technicians can be added to the assembly line to help prepare additional mosquito samples. For example, specific parts of the mosquito can be prepared for bloodmeal identification or sporozoite ELISA testing (*see* Chapter 1), determination of sugar-feeding (*2*), measurements of wing-length (*3*), PCR identification of species (*see* **Subheading 3.5.**), or any other type of processing needed for the scope of investigations being conducted.
6. Check data forms to ensure they are complete and accurate. Also ensure that mosquito samples to be frozen are numbered correctly and that each vial number corresponds to the corresponding numbers on the data forms.

3.3. Taxonomic Identification of Mosquitoes (see Note 1)

1. Immobilize live wild-caught mosquitoes by aspirating them from cages, blowing them into disposable beakers containing 70% ethanol, and within 5 min, transferring them by

forceps to other beakers containing either physiological saline, PBS, or M-199 media. This method immobilizes but does not normally kill the mosquitoes, and if the mosquitoes are allowed to dry on the slide, they will recover and fly away.
2. Place mosquitoes individually on a glass slide and examine at ×10 under a dissection microscope.
3. Identify the mosquito using standard taxonomic criteria. There are standard taxonomic keys available for each major geographic region and these need to be consulted. For the African region, one useful key is Gillies and DeMeillon *(4)*.

3.4. Classification of Blood-Feeding Stages and Christophers' Stages of Ovarian Development

Figure 1 provides diagrams of mosquito blood-feeding stages, differences in ovaries between parous and nulliparous mosquitoes, and Christophers' stages of ovarian development; further information is available in a WHO training manual *(1)*. After the mosquito is identified, the following parameters can easily be determined.

3.4.1. Blood-Feeding Stages

1. Examine the mosquito on a labeled glass slide, under a dissecting microscope at ×10.
2. Classify blood-feeding stages as either empty (E), fed (F), half-gravid (HG), or gravid (G), according to the amount and condition of the blood in the mosquito midgut (**Fig. 1A**).
3. Record results on the standard forms.
4. For indoor-resting mosquitoes, data on the mosquito blood-feeding stages provides an account of the proportion of mosquitoes feeding on blood per day. The frequency of blood-feeding defines the duration of gonotrophic cycles, an important parameter in calculating the survival rate based on determinations of mosquito parity rates *(5,6)*.

3.4.2. Parity Status

1. Dissect ovaries from the same mosquito, as above, into a drop of physiological saline or PBS by gently pulling the last two abdominal segments with forceps while securing the mosquito thorax with another pair of forceps.
2. Examine the ovaries to determine parity by observing them at about ×40. Nulliparous mosquitoes contain tightly coiled ovarioles while parous mosquitoes do not. **Figure 1B** shows ovaries of nulliparous and parous females. Record results.
3. If mosquito is parous, determine the Christophers' stage of ovarian development.
4. The parity rate can be used to calculate the daily survival rate of mosquito populations, as described by Service *(5,6)*.

3.4.3. Christophers' Stages of Ovarian Development

1. For parous mosquitoes, gently tease apart one ovary to release the oocytes.
2. Observe the oocytes at ×40 and record the Christophers' stage of oocyte development as stage I, II, III, IV, or V (*see* **Fig. 1C**). Defining characteristics of each stage are as follows *(1)*:

"Stage I, egg follicle round, yolk granules absent; Stage II, egg follicle oval, yolk granules present; Stage II-early, a few fine granules of yolk around the nucleus of the ovum; Stage II-mid, yolk granules easily visible under low power; Stage II-late, yolk granules very abundant occupying about half the follicle; Stage III, yolk occupying about three-quarters of the follicle; Stage IV, egg follicle sausage-shaped; Stage V, ova fully formed with well-developed floats."

Record results on standard forms.
3. For some *Anopheles* species, over at least part of their geographic ranges, a portion of mosquitoes feed on blood multiple times per gonotrophic cycle *(2)*. Using combined data

A Blood-feeding stages

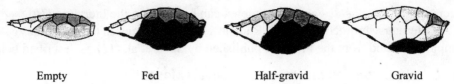

Empty   Fed   Half-gravid   Gravid

B Parity status

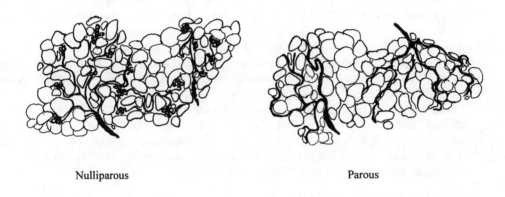

Nulliparous   Parous

C Stages of ovarian development

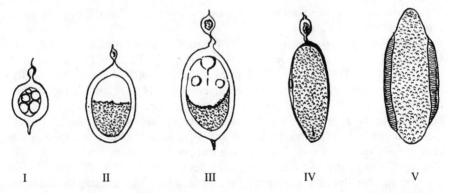

I   II   III   IV   V

Fig. 1. (**A**) Mosquito blood-feeding stages. (**B**) Ovaries of mosquitoes showing differences between parous and nulliparous specimens. (**C**) Christophers' stages of ovarian development. Drawings from freshly dissected mosquitoes by Ross Frohn.

on blood-feeding stages and Christophers' stages, those mosquitoes that are classified as F (i.e., fed) and contain stage IV or V oocytes are clearly refeeding before the completion of normal gonotrophic cycles. Additional information on multiple blood-feeding can be found in refs. *2* and *7–9*.

3.5. Molecular Methods for Identifying Species in Anopheles *Species Complexes*

For *Anopheles* species complexes, it is standard practice to use either cytogenetic or molecular approaches for identifying species. Methods for cytogenetically identifying species in the *Anopheles gambiae* complex can be found in the literature *(10)* (*see* **Note 2**). An alternative protocol for the PCR identification of species in the *Anopheles gambiae* complex, adapted from the methods published by Scott et al. *(11)*, is described below.

3.5.1. DNA Extraction of Ethanol Preserved Mosquito

1. Separate individual mosquitoes with forceps and allow them to dry overnight on filter paper inside a covered Petri dish.
2. In a 1.5-mL grinding tube, place a single dry mosquito. Grind the mosquito with a hand-held motor with a polypropylene pellet pestle for at least 10 s or until the mosquito is powdered. Leave the pellet pestle in the tube and continue for other samples.
3. Add 120 µL of warm DEB and 1.5 µL of RNAse A. Grind for another 30 s to homogenize the solution. Incubate at 37°C for 60 min.
4. Add 3 µL of proteinase K (20 µg/µL). Incubate at 50°C for 60 min.
5. Add 60 µL of each phenol and chloroform. Vortex and microfuge for 10 min at 12,000g.
6. Transfer the supernatant to a 0.5-mL tube and add 300 µL of ice-cold 95% ethanol. Allow to precipitate overnight in the –20°C freezer.
7. Centrifuge for 10 min at 12,000g and discard the supernatant. Add 300 mL of ice-cold 70% ethanol without disturbing the pellet. Let it sit for 2 or 3 min and discard the supernatant.
8. Remove all excess ethanol with a micropipette. Allow pellet to completely dry by either using a vacuum or allowing the pellet to dry at 37°C with the cap open for 10 min.
9. Resuspend the pellet in 20 µL of Tris-EDTA (TE) pH 7.4.

3.5.2. Polymerase Chain Reaction

1. Prepare diluted mosquito DNA. Add 1 µL of mosquito DNA to 999 µL of ddH$_2$O for a 1:1000 dilution.
2. Species diagnostic primers for the *A. gambiae* ribosomal DNA (rDNA) intergenic spacer region are shown in **Table 1**.
3. Prepare a PCR cocktail. A typical 25-µL PCR would include the following: 2.5 µL of 10X buffer, 4 µL of deoxyribonucleoside triphosphate (dNTP) (5 m*M*), 1 µL of each of the primers AR, GA, ME, and UN, 1.5 µL of MgCl$_2$, 0.2 µL of *Taq*, and 0.8 µL of ddH$_2$O. The total volume of the PCR cocktail will be equivalent to 24 µL multiplied by the number of mosquito samples that are being tested. Therefore, aliquot a total volume of 24 µL of the cocktail into the appropriate number of PCR tubes (*see* **Note 3**). Add 1 µL of diluted mosquito DNA (1:1000) into each tube and 1 µL of ddH$_2$O into one tube as a negative control (*see* **Note 4**). Always run negative and positive controls.
4. Run on PCR program: 94°C for 1 min; 35 cycles at 94°C for 15 s; 50°C for 15 s; and 72°C for 30 s; and, finally, 72°C for 3 min. The length of time for each step should be tested for each PCR thermocycler because they may need to be adjusted to optimize amplification.
5. When the PCR program is completed, load 10 µL of PCR product plus 1 µL of agarose dye to each well of a 2% agarose gel. Run one or two lanes of DNA size marker (e.g., 100 bp or pUC18) on the same gel.
6. Run the 2% agarose gel at 70 V for approx 2 h. The front dye should be about half-way through the length of the gel.
7. For species identification, stain the gel with ethidium bromide and observe the gel under ultraviolet light to detect the rDNA bands amplified by the PCR. The size of the band will reveal the species: *A. gambiae,* 390 bp; *A. arabiensis,* 315 bp; *A. melas,* 464 bp; *A. merus,* 466 bp; *A. quadriannulatus,* 153 bp.

4. Notes

1. Personnel who are responsible for taxonomic identifications need to be properly trained and supervised. Because in most areas there will generally be only 2 to 10 common spe-

**Table 1
Species Diagnostic Primers for *A. gambiae* Ribosomal DNA
(rDNA) Intergenic Spacer Region**

Primer name	Primer sequence (5' to 3')	T_m (°C)
UN	GTG TGC CCC TTC CTC GAT GT	58.3
GA	CTG GTT TGG TCG GCA CGT TT	59.3
ME	TGA CCA ACC CAC TCC CTT GA	57.2
AR	AAG TGT CCT TCT CCA TCC TA	47.4
QD	CAG ACC AAG ATG GTT AGT AT	42.7

cies of *Anopheles* mosquitoes, personnel can quickly learn to identify the major species, without continually checking with keys. It must be stressed that proper training is essential because misidentifications are not acceptable.

2. One of the drawbacks of using cytogenetic methods is that identification of adult females is limited only to those in the half-gravid stage; the methods are also technically demanding and require specialized training of personnel. PCR methods are becoming more routine for standard field studies.
3. A smaller volume of PCR (e.g., 15 µL) could be used to reduce usage of the reagents.
4. A small piece of raw mosquito body (e.g., legs) also works.

Acknowledgments

I thank Dr. Guiyun Yan for kindly sharing his laboratory protocol for PCR identification of *Anopheles gambiae* species and for commenting on the article, and Ross Frohn for providing the artwork in Fig. 1. This work was supported by the National Institutes of Health grants AI29000, AI45511, and TW01142.

References

1. WHO (1975) *Manual on Practical Entomology in Malaria. Part II. Methods and Techniques.* WHO Offset Publication 13, Geneva.
2. Beier, J. C. (1996) Frequent blood-feeding and restrictive sugar-feeding behavior enhance the malaria vector potential of *Anopheles gambiae* s.l. and *An. funestus* (Diptera: Culicidae) in western Kenya. *J. Med. Entomol.* **33,** 613–618.
3. Lyimo, E. O. and Takken, W. (1993) Effects of adult body size on fecundity and the pre-gravid rate of *Anopheles gambiae* females in Tanzania. *Med. Vet. Entomol.* **7,** 328–332.
4. Gillies, M. T. and DeMeillon, B. (1968) *The Anophelinae of Africa South of the Sahara.* The South African Institute for Medical Research, Johannesburg.
5. Service, M. W. (1976) *Mosquito Ecology.* Wiley, New York.
6. Service, M. W. (1993) *Mosquito Ecology: Field Sampling Methods.* Elsevier Applied Science, New York.
7. Briegel, H. and Horler, E. (1993) Multiple blood meals as a reproductive strategy in *Anopheles* (Diptera: Culicidae). *J. Med. Entomol.* **30,** 975–985.
8. Klowden, M. J. and Chambers, G. M. (1992) Reproductive and metabolic differences between *Aedes aegypti* and *A. albopictus* (Diptera: Culicidae). *J. Med. Entomol.* **29,** 467–471.
9. Klowden, M. J. and Briegel, H. (1994) Mosquito gonotrophic cycle and multiple feeding potential: contrasts between *Anopheles* and *Aedes* (Diptera: Culicidae). *J. Med. Entomol.* **31,** 618–622.
10. Coluzzi, M., Sabatini, A., Petrarca, V., and Di Deco, M. A. (1979) Chromosomal differentiation and adaptation to human environments in the *Anopheles gambiae* complex.*Trans. R. Soc. Trop. Med. Hyg.* **73,** 483–497.
11. Scott, J. A., Brogdon, W. G., and Collins, F. H. (1993) Identification of single specimens of the *Anopheles gambiae* complex by the polymerase chain reaction. *Am. J. Trop. Med. Hyg.* **49,** 520–529.

9

Genotyping of *Plasmodium* spp.
Nested PCR

Georges Snounou

1. Introduction

It is now established that *Plasmodium falciparum* parasites found in any one area are highly diverse and that individual hosts, vertebrate and insect, are often concurrently infected by multiple parasite lines (for example, *see* **refs. 1–4**). Different parasite lines of the same *Plasmodium* species are also known to vary substantially with respect to parameters such as the parasitological and clinical course of the infection, the pattern of recrudescence, immunological cross-reactivity, susceptibility to drugs, and transmissibility by various vector species. The relationship between the complexity of the parasite populations and such pathological, immunological, and epidemiological factors is poorly understood and clearly merits further investigation (for example, *see* **refs. 4–16**).

Until recently, only phenotypic differences could be used to distinguish between parasite lines: (a) direct observations of the natural and experimental course of the infection, susceptibility to drugs, and infectivity to mosquitoes and/or (b) immunological or biochemical analysis of parasites. The former is excluded for human infections on ethical and practical considerations, whereas the latter is limited by the availability of specific reagents and the requirement for relatively large quantities of parasite materials. Thus, minor parasite populations were overlooked, and low-grade infections and most infected mosquitoes were also excluded from analysis. The discovery of polymorphic genes in the genome of *P. falciparum* provided a means to characterize parasite diversity genotypically. Since the detection of DNA can be achieved with high sensitivity and specificity by polymerase chain reaction (PCR) amplification, this has become the optimal method for the analysis of parasite diversity *(1,16–22)*. In this context, it is important to bear in mind that the suitability of any PCR genotyping strategy, i.e., the choice of genetic marker, the efficiency of amplification, and the capacity to discriminate between variants, depends primarily on the questions being asked.

Any polymorphic sequence can be used as a genetic marker, provided the following conditions are fulfilled: (a) The sequence must be present as a single copy in the genome of the parasite; (b) the variable region must be present in all the parasites, i.e., null variants must not occur naturally; and (c) the polymorphic sequence must be stable, i.e., it must remain unaltered throughout the numerous mitotic divisions of asexual parasites and following the meiotic division that takes place each time the parasite is transmitted by the mosquito. The usefulness of a given polymorphic sequence as a genetic marker will then depend on the degree of polymorphism in the parasite population and on the ease with which the variants can be discriminated.

Single-base differences between parasite lines constitute the simplest polymorphisms. These mutations are often disproportionately associated with non-synonymous amino acid substitutions implying a functional role (for example, *CSP* and the C-terminus of *MSP1*). Nonetheless, point mutations might naturally occur in parasite populations. Other polymorphisms are characterized by the presence of unique sequences that can be used to divide the variants into distinct families (for example, *EBA-175*, *MSP1*, and *MSP2*). The most striking polymorphisms are those resulting from repetitive sequences, where the number of a given repeat unit can vary between parasite lines (for example, *CSP*, *MSP1*, *MSP2*, *GLURP*, and so on). In some cases, two or more types of repeat units can be found and their number and arrangement can characterize a given parasite population. Polymorphisms resulting from repetitive sequences have been deemed the most suitable for rapid genotyping of parasite populations, since most allelic variants can be simply distinguished by size following electrophoresis in agarose or polyacrylamide gels.

Three highly polymorphic repetitive regions from different *P. falciparum* genes (*MSP1*, *MSP2*, and *GLURP*) have become the most frequently used genetic markers for genotyping, and a protocol for their amplification and analysis is described in this chapter.

2. Materials

2.1. Designated Areas

Ideally, the following procedures should be carried out in physically separated rooms (*see* **Note 1**).

1. DNA template preparation from samples and their storage.
2. Storage, preparation and aliquoting of PCR reagents.
3. Addition of DNA template to amplification reaction tubes.
4. Addition of template from the Nest 1 to the Nest 2 reactions.

Analysis and storage of PCR product (*see* **Note 2**).

2.2. Equipment (see Note 3)

1. Micropipettors and tips.
2. Thermal cycler.
3. Apparatus for agarose gel electrophoresis.
4. Photographic equipment and UV transilluminator.
5. Refrigerator (4°C) and freezer (–20°C).
6. Microcentrifuge.

2.3. Reagents (see Note 4)

1. PCR buffer (10X stock): 500 mM KCl, 100 mM Tris-HCl, pH 8.3, 20 mM MgCl$_2$ and 1 mg/mL gelatin. Store at 4°C or –20°C (*see* **Note 5**).
2. dNTP stock solution: 5 mM of each of the four dNTPs: dATP, dCTP, dGTP, and dTTP. Store working stocks at –20°C, and back-up stocks at –70°C.

3. Oligonucleotide primers: A 2.5 μM stock of each (*see* **Note 6**). Store working stocks at –20°C, and back-up stocks at –70°C.
4. AmpliTaq polymerase (Cetus) (*see* **Note 5**). Store at –20°C.
5. Mineral oil. Store at room temperature.
6. Loading buffer (5X stock): 50 mM Tris, pH 8.0, 75 mM EDTA, pH 8.0, 0.5% SDS, 30% w/v sucrose, 10% w/v Ficoll (average mol wt = 400,000), and approx 0.25% w/v of Orange G dye. If Ficoll is not available, use 40% sucrose. Store at room temperature.
7. TBE buffer (10X stock): 1 M Tris, 1 M boric acid, 50 mM EDTA. A pH of approx 8.3 should be obtained without any adjustment. Store at room temperature.
8. High resolution agarose such as NuSieve™ agarose or MetaPhor™ agarose.
9. Ethidium bromide solution: 10 mg/mL in water. Extreme care should be taken when handling this solution as this chemical is highly carcinogenic. Store in the dark at 4°C.
10. Water (sterile).

3. Methods

3.1. Sample Collection and Template Preparation

Samples to be analyzed by PCR can be obtained in different forms either from the vertebrate or the insect hosts. It is beyond the scope of this chapter to provide an exhaustive review of the collection procedure and the various methods used to obtain DNA template suitable for PCR amplification protocols. Nonetheless, the following points should be considered.

1. Correct, consistent, and clear labeling is of the utmost importance. Care must be taken that a unique code/number is assigned to each sample, and that this code/number is legibly written on the label. Labeling of the sample tubes or filter paper spots must be done with water-proof indelible ink. Tubes are best labeled using sticky labels and marker pens suitable for cold storage. Keeping the tubes dry before and after freezing prolongs the life of the label. A further label on the top of the tube, though desirable, is not always practical.
2. The value of the highly sensitive PCR protocols lies in its ability to detect very low numbers of parasites. The potential advantage of PCR over microscopic examination is therefore best realized when a substantial volume of blood (≥5 μL) is screened in each PCR assay. Depending on the method of DNA isolation and the nature of the sample, the final volume of the DNA template might substantially exceed the original volume of the blood sample from which it was purified. Concentrating the DNA template, for example, by ethanol precipitation, might be considered in these cases.
3. For epidemiological studies where sampling is performed in remote areas, over lengthy periods of time, or by untrained staff, the simplest method of collection is strongly recommended. Finger-prick blood spotted and dried on filter paper offers the most practical and robust method of sample collection, storage and transport. Commercially available kits for this type of collection, although relatively costly, provide a standardized sample volume per spot from which the DNA template can be rapidly obtained (for example, ISOCODE™ STIX PCR preparation dipsticks from Schleicher and Schüll, Dassel, Germany).
4. The presence of PCR inhibitors in the DNA template can seriously reduce the sensitivity of the amplification assay. The anticoagulants heparin or EDTA when used in high concentration might be carried over in the template solution. The presence of large quantities of host DNA can also reduce the efficiency of the reaction, a problem particularly encountered if parasites are sought in tissue samples when fixative, such as formalin, can also cause problems. The nature of the support used for dried blood spot collection, the storage, and processing of such samples, can influence the efficiency of the assay *(23,24)*. Finally, PCR inhibitors are often found in templates purified from mosquito samples

(25,26). The choice of DNA template preparation method and/or dissection of the midguts or the salivary glands before template preparation minimizes such inhibition *(27)*.
5. It is advisable to test the efficacy of the PCR protocol for each given combination of samples and template purification method. This is achieved by spiking uninfected samples with a known quantity of parasites or serially diluting blood containing a known quantity of parasites, then subjecting these samples to routine storage before preparing the template and establishing the level of PCR detection sensitivity, and whether inhibition of the PCR assay occurs.

3.2. The Amplification Reactions

The amplification strategy used for genotyping *P. falciparum* parasites is the nested PCR. Two advantages of this method are substantially increased detection sensitivity and a decreased sensitivity to minor variations in the amplification conditions and DNA template quality. The disadvantages are the additional time and materials required, and the substantially increased risk of contamination. The latter can be minimized provided the precautions and methodology described later are assiduously followed.

In the first amplification reaction (Nest 1), oligonucleotide primers pairs which will hybridize to conserved sequences flanking the repeat polymorphic regions of the genes are used (*see* **Note 7**). The product of this first reaction is then used as the DNA template for separate second amplification reactions (Nest 2) in which the oligonucleotide primers used recognize sequences contained within the DNA fragment amplified in the first reaction (**Fig. 1**).

GLURP: The RII repeat region of this gene consists of repeat units of usually 60 bp, whose number can differ between different parasite lines *(28)*. A semi-nested PCR strategy is used to amplify the allelic variants of *GLURP*, where the Nest 2 reaction is carried out with the anti-sense primer G-OR used in Nest 1 and G-NF, which is internal but overlaps with G-OF (**Fig. 1**).

MSP1: Sequence analysis revealed that the polymorphic block 2 of *MSP1* can occur as one of three distinct families (K1, MAD20, or RO33) which derive their name from the parasite lines in which they were first observed *(29)*. Different repeated sequence units characterize the K1 and MAD20 families, whereas the RO33 block 2 regions consists of a unique sequence. The sequences in block 2 that flank the repeated region of K1 or MAD20 are unique to each family and are shared amongst all the allelic variants. Three separate Nest 2 reactions are therefore performed to complete the genotyping of parasites using the *MSP1* marker (**Fig. 1**): one specific for the K1 family [(M1–KF) + (M1–KR)], one specific for the MAD20 family [(M1–MF) + (M1–MR)], and finally one specific for the RO33 family [(M1–RF) + (M1–RR)].

MSP2: Sequence analysis revealed that the polymorphic block 3 of *MSP2* can occur as one of two distinct families (FC27 and 3D7/IC) which derive their name from the parasite lines in which they were first observed *(30)*. Different repeated sequence units characterize the FC27 and 3D7/IC families, and the sequences in block 3 that flank these repeated regions are unique to each family and are shared amongst all the allelic variants. Two separate Nest 2 reactions are therefore performed to complete the genotyping of parasites using the *MSP2* marker (**Fig. 1**): One specific for the FC27 family [(M2–FCF) + (M2–FCR)] and one specific for the 3D7/IC family [(M2–ICF) + (M2–ICR)].

There are distinct advantages to the use of a family-specific amplification strategy for the genotyping of *MSP1* and *MSP2* which requires three additional PCR reactions, as compared to a strategy where the primers of the Nest 2 reaction would target sequences conserved amongst all allelic variants. In the latter case, establishing to

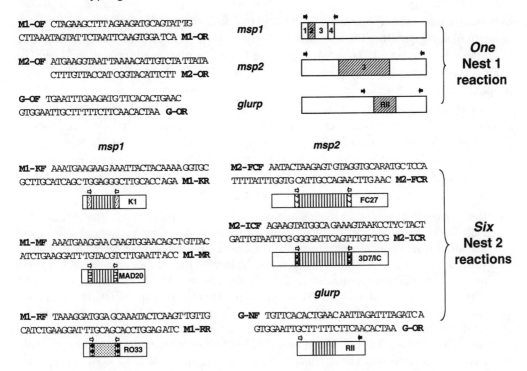

Fig. 1. Schematic representation of *msp1*, *msp2*, and *glurp* genes with the sequences of the different oligonucleotide primer pairs and their approximate positions and that of their products. The name of each oligonucleotide primer is in bold, and is placed before the sequence (presented 5' to 3') for "sense" primers, and after the sequence for "anti-sense" primers.

which family the amplified fragments belong would necessitate Southern blotting of the PCR product and a total of five hybridizations. This can be circumvented for *MSP2* by using restriction fragment length polymorphism (RFLP) analysis where one or more restriction enzyme digestions have to be performed *(20)*. Variants resulting from a recombination event in the repeat regions between lines whose *MSP1* or *MSP2* belong to different families, have so far only been observed within the block 3 repeat region of *MSP2*. Detection of these "hybrid" variants can be theoretically achieved by carrying out two further Nest 2 reactions [(M2–FCF) + (M2–ICR)] and [(M2–ICF) + (M2–FCR)]. Another important advantage of family-specific amplification is the improved sensitivity in detecting minor parasite populations. Samples obtained from areas with relatively high endemicity often contain multiple parasite lines. As a result of competition for reactants, the allelic variants representing minor parasite populations (a ratio of 1 in 50 or less) are not amplified with sufficient efficiency for detection *(21)*. Adopting a family-specific amplification strategy ensures that this type or competition affects only allelic variants of the same family. Thus in a given sample, the parasites of one allelic family will be consistently detected irrespective of the presence of large number of parasites from another allelic family.

3.2.1. Setting up the Amplification Reaction

For all the PCR amplifications, the volume used for each reaction is 20 µL (*see* **Note 8**). A master mix containing all the reagents, except for the DNA, is prepared and aliquoted into the reaction tubes (*see* **Notes 9–11**), and overlaid with mineral oil (*see* **Note 12**). It is worth making sure that the reagents are fully thawed and vortexed before being used to prepare the master mix (*see* **Note 13**). The DNA template is always added last. The risk of cross-contamination and aerosol contamination is minimized if only ONE tube is open at any one time.

In order to minimize contamination, template addition is made after the aliquoted reaction mixtures have been overlaid with oil. Reaction mixtures for Nest 1 and Nest 2, as well as template addition for Nest 1, can be made and aliquoted in the same room. However, template addition for the Nest 2 reaction MUST be performed in a separate room. Product analysis MUST be done in a third room, in which none of the reagents used for PCR should ever be opened or, preferably, stored.

1. Remove the PCR buffer, oligonucleotide primers from the freezer and allow to thaw. The PCR buffer is stable indefinitely at room temperature. The oligonucleotide primers will not suffer from being placed repeatedly at ambient temperature for a few min at a time.
2. Add the appropriate volumes of reagents to the labeled master mix tube, in the following order: water, PCR buffer, oligonucleotide primers. The final $MgCl_2$ concentration is 1 mM; variations in the $MgCl_2$ (2–4 mM) have been found to affect the specificity of the amplification. All the oligonucleotide primers for the Nest 1 and Nest 2 reactions are used at a final concentration of 250 nM, except for the GLURP Nest 1 primers (G-OF and G-OR) which are used at a 125 nM. When lower concentrations are used the efficiency of amplification was decreased. The oligonucleotides pairs used in the Nest 1 and Nest 2 reactions are given in **Fig. 1**.
3. Remove the dNTP aliquot from the freezer, thaw, add appropriate amount to the master mix tube, and replace immediately in the freezer. The final concentration of each dNTP is 125 µM.
4. Remove the *Taq* polymerase from the freezer, add the appropriate amount to the master mix tube, and replace immediately in the freezer. A total of 0.4 U of enzyme is used for each 20 µL reaction (2 U/100 µL).
5. Mix the contents of the master mix tube thoroughly by vortexing (*see* **Note 14**).
6. Aliquot 20 µL of the master mix to each tube. Remember to open only one experimental tube at a time, closing it as soon as the master mix is aliquoted. This will minimize the possibility of contamination. The same tip can be used for all the tubes.
7. Add 50 µL of mineral oil to each tube, taking care to avoid splashes by adding the oil on the side of the wall at the top of the tube and allowing gravity to do the rest. It is very important that the same volume of oil is added to each tube (*see* **Note 15**). The same tip can be used for all the tubes.
8. DNA template must be added last and only after overlaying the master mix with mineral oil, so as to ensure a minimum risk of contamination. The appropriate amount of DNA template (usually 1 µL) is added by immersing the micropipet tip INSIDE the oil, preferably in the middle of the oil layer. The droplet will sink in the oil and merge with the master mix. This procedure will prevent aerosols escaping and avoid contaminating other tubes (*see* **Note 16**).
9. When aliquoting the product of the Nest 1 reaction as a DNA template for the Nest 2 reaction, the following procedure must be used. Remove the DNA template aliquot from the Nest 1 tube from UNDER the oil, taking care that it is the product and not the oil that is removed. This is achieved by placing the tip of the micropipette in the middle of the aqueous layer, just below the oil and without touching the tube walls. Check that it is indeed the DNA template that is added to the Nest 2 tube, by observing the droplet as it is added to the oil layer in the Nest 2 tube. This procedure will prevent aerosols escaping and avoid contaminating other tubes.

10. Do NOT mix the contents by vortexing the tubes as this will result in an increased risk of contamination, and will necessitate a centrifugation step.

3.2.2. The Amplification Reactions

The cycling parameters for the PCR amplifications are as follows (*see* **Notes 15,17,18**).

Step 1: 95°C for 5 min   Initial denaturation
Step 2: X°C for 2 min   Annealing
Step 3: 72°C for 2 min   Extension
Step 4: 94°C for 1 min   Denaturation
Step 5: Repeat **steps 2–4** for a total of 25 cycles (Nest 1) or 30 cycles (Nest 2)
Step 6: X°C for 2 min   Final annealing
Step 7: 72°C for 5 min   Final extension
Step 8: The reaction is completed by reducing the temperature to 25°C (*see* **Note 19**).

X = 58°C for Nest 1 (M1–OF/M1–OR; M2–OF/M2–OR; G-OF/G-OR) and the Nest 2 reactions of *GLURP* (G-NF/G-OR).

X = 61°C for the Nest 2 reactions of *MSP1* (M1–KF/M1–KR; M1–MF/M1–MR; M1–RF/M1–RR) and of *MSP2* (M2–FCF/M2–FCR; M2–ICF/M2–ICR).

These conditions have been determined using a PTC-100 Thermal cycler (MJ Research Inc.).

3.3. PCR Product Analysis

3.3.1. Electrophoresis

As a result of the very high sensitivity of the Nested PCR methodology, the amount of amplification product obtained from one parasite is sufficient for visualization by ethidium bromide staining following electrophoresis. Separation of the PCR product is most simply achieved by agarose gel electrophoresis. The best resolution between variants that differs only slightly in size is achieved when NuSieve™ or MetaPhor™ agarose (FMC Bioproduct Inc.) are used. All of these agarose gels can be re-used a large number of times (*see* **Notes 20** and **21**), although the presence of ethidium bromide might affect the migration of some of the DNA fragments. Gels are made and electrophoresed in TBE Buffer. The dissolved agarose mixture should be allowed to cool to approx 55°C before pouring into the gel cast, and then allowed to set for at least 30 min (with an additional 30 min in the fridge for MetaPhor gels) before removing the combs and loading the PCR product. The approximate size range of the PCR products differs for the different markers: 125–250 bp for *MSP1* variants, 250–450 bp for the FC27 variants of *msp2*, and for these markers a 3% MetaPhor gel gives very good resolution, for the 3D7/IC variants of *MSP2* a range of 450–700 bp is usual and 600–1200 bp for the *GLURP* variants, 2.5% NuSieve™:Agarose (3:1) gels give adequate resolution. It is likely that variants that fall outside this size range exist.

In order to prepare the PCR product for electrophoresis, add 5 µL of loading buffer to the tubes at the end of the Nest 2 amplification. A brief centrifugation will bring the loading buffer under the oil layer. It is very important to mix the PCR reaction with the loading buffer before loading onto the gel. This is best done not by vortexing, which will require a further centrifugation step, but by repeated up-and-down pipetting just prior to loading onto the gel. The positive controls for the PCR amplification will be sufficient as a size marker. It is advisable, and sufficient, to use 12.5–15 µL from the total of 25 µL when analyzing the PCR product, and to store the remainder, until the results have been interpreted and recorded correctly. The tracking dye, Orange G migrates below the level of the smallest PCR product expected, thus electrophoresis is carried out until the Orange G reaches the end of the gel.

DNA is visualized under ultraviolet (UV) illumination, following staining by ethidium bromide. Ethidium bromide is a very carcinogenic chemical. Consequently, ethidium bromide is neither added to the gel nor to the TBE buffer. The gel is better stained following electrophoresis, by immersion in TBE buffer containing ethidium bromide (0.1–0.5 µg/mL) for a period of 30 min. It is then destained by placing in TBE buffer for a minimum of 10 min. Destaining times of up to 3 h are unlikely to result in appreciable loss of the DNA product bands.

3.3.2. Interpretation

The oligonucleotides presented here for the amplification of *MSP1*, *MSP2*, and *GLURP* are highly specific and a product is only observed when *P. falciparum* DNA from the appropriate parasite line is added. The efficiency of amplification by nested PCR is such that the amount of product obtained does not alter with a large range of parasite DNA in the original sample *(31,32)*. When the amount of parasite material present in the sample under analysis is very low, inconsistency in the amplification might be observed. In other words, repeated amplification from the same sample could result in the detection of product in some cases but not others, the "all-or-none" effect *(31)*. This is due to the very high sensitivity of the PCR analysis presented here, in which only a few parasite genomes are required to give a positive result. Thus, if the concentration of DNA template is equivalent to less than one parasite per aliquot used for the analysis, the result of the amplification will depend on the probability of picking the target DNA in a given aliquot from this template solution.

When the template consists of DNA from a cloned parasite line, a single band should be observed following amplification. However the product of the Nest 1 reaction might be observed. This product is approx 120 bp larger than the Nest 2 product for *GLURP*, 230 bp larger for *MSP2*, and 830 bp for *MSP1*. Moreover, carry-over of Nest 1 oligonucleotides into the Nest 2 reaction could initiate amplification of additional products of larger size than the genuine product (e.g., product resulting from [{M2–OF} + {M2–ICR}] or [{M2–ICF} + {M2–OR}]). These specific yet "artifact" bands are more frequently observed if high oligonucleotide concentrations are used, when the number of parasites in the initial sample is high, or if too many Nest 1 cycles are carried out. Generally, the detection of multiple bands in the PCR product is interpreted to reflect the presence of as many different lines in the initial sample. It is important to note that each band does not necessarily represent a parasite clone, since a sample containing many genetically different parasite lines which all share the same variant of a particular genetic marker, will only exhibit one band following PCR genotyping for this marker.

Genotyping of *P. falciparum* populations can be used for different purposes:

1. Comparative studies of the parasites in sequential samples from the same person, or from parasites that develop into oocysts or sporozoites in a mosquito fed on the blood obtained on a particular time point. In these case the PCR product(s) for each molecular marker are best run side-by-side.
2. Determination of frequencies of the various markers in the parasites sampled from the same human or mosquito population but during different seasons, or from different sub-

PCR Genotyping

groups (e.g., child versus adult, symptomatic versus asymptomatic, or different *Anopheles* species), or from populations in different geographical locations. In such cases it is important to classify a particular variant consistently since comparison between different gels is difficult. This is best achieved through the use of agarose gels with high resolution and with respect to defined molecular size markers, which can be commercially obtained or which consist of mixture of defined products from previously characterized parasite samples.

3. Calculation of the multiplicity of infection or MOI *(11)*, which is defined as the minimum number of genetically different parasite lines present in a given sample. For a particular sample, the number of variants (bands) observed for each of the genetic markers is counted (for *msp1*, the bands observed for the K1, MAD20, and RO33 families are added, for *msp2* the bands observed for the FC27 and 3D7/IC families are added). The highest number will then be considered as the MOI of this sample. Estimation of the minimum number of parasite clones present in a sample can also be derived using statistical methods *(33)*.

4. Notes

1. There are two major problems encountered with the use of PCR. The first is the reduction of the amplification efficiency, which is easily detectable through the use of appropriate controls and which is mainly due to human error, problems with the equipment, or the quality of the reagents or the DNA template. The second is contamination, which is invariably the result of human error. Two types of contamination are possible. In the first, parasite material is transferred between samples (cross-contamination), or between the samples and the PCR reagents. In the second, PCR product from either of the two nested amplification reactions, comes into contact with the samples, the reagents or the equipment. The second type of contamination is by far the more serious, and unfortunately the easiest to achieve. The extent of the precautions to be taken ultimately depends on the particular situation of each laboratory, and consequently a compromise will almost always have to be reached. Paramount in the mind of the researcher should be the adoption of procedures in which the probability of contact between the PCR template and the PCR reagents and equipment is minimized, and the PCR product NEVER comes into contact with either the samples, the DNA template, the PCR reagents, or the equipment used for their handling. Thus, physically separated areas for the equipment and reagents, must be assigned to the different procedures.

2. When space is at a premium, the first three areas could be combined, provided that care is taken to prevent cross-contamination between tubes. The addition of template for the Nest 2 reaction however, must be performed in a separate room, and handling of the PCR product requires yet another room. It must be stressed that the setting up of PCR amplifications does not need to be performed under sterile conditions. Ultimately, only one room is required to be dedicated for PCR analysis, namely the room in which the PCR product is handled. Only access to two other rooms, one for procedures (i)–(iii) and one for (iv), will be needed. Since these procedures are relatively short in duration, this should pose little problem even in overcrowded laboratories. The rooms in which procedures (i)–(iii) are carried out are considered as PCR clean, whereas those in which procedures (iv) and (v) are performed are PCR dirty. The risk of contamination in the room assigned to procedure (v) is a certainty, whereas in the room where procedure (iv) is carried out the risk of contamination is minimal provided good technique is used (*see* **Note 4**). It should be noted that any room can be considered as an absolutely PCR clean room for malaria work, provided that the template of the amplification reaction, namely any genetic material from *Plasmodium* species (tissue or blood samples, cultured parasites and lysates, bacterial clones or PCR product), have never been handled within this room.

3. All the equipment (tubes, pipet tips, and containers) to be used for the PCR analysis should be purchased new and dedicated to a specific procedure, and therefore, the appropriate room. Ideally, three sets of micropipets should be purchased, one for sample handling and DNA template preparation, one for PCR reagents handling, and finally one for PCR product handling. When using nested PCR, an additional micropipet must obtained and dedicated for DNA template addition from the Nest 1 to the Nest 2 reaction. Positive displacement micropipets, which reduce the risk of contamination, are only advisable for handling solutions that contain DNA template. Filtered tips that are expensive, might for example be considered for the DNA template addition from the Nest 1 to the Nest 2 reaction, but are not essential to obtain contamination-free amplification. The most efficient method to decontaminate equipment is to soak it for a few hours in a 0.1–0.2 M of HCl, should this be possible. The acid depurinates DNA, and thus renders it refractive to PCR amplification. UV exposure of equipment and surfaces provides an efficient way to neutralize contaminating DNA, whereas autoclaving although of some use, is not as efficient at degrading DNA.
4. All the chemicals to be used for the procedure must be obtained before initiating the project, and dedicated to PCR work. They must only be opened in an absolutely PCR clean room. Aliquoting, weighing, pH adjustments, and storage must be done with absolutely PCR clean materials. During storage, a Parafilm strip should be used as an extra seal around the closed lid of the bottles. Aliquoting from the stocks to obtain working solutions must be done under sterile and PCR clean conditions. The hierarchy of stocks and their back-up should be considered carefully. It is strongly advisable to divide all chemicals and solutions into at least two, or preferably three, aliquots. The first aliquot is used as the working stock, and the second aliquot should be considered as the back-up stock, and only used when the working stock runs out, or becomes "suspect." Finally, a reference stock should be considered. This stock should hopefully never have to be used, but should provide a back-up should a disaster befall the laboratory and should consequently be stored in a separate laboratory/building. Contamination of any one of the chemicals or stocks will render the results of the analysis totally invalid. Once a stock is made, it should be divided up immediately, and the working stock aliquot tested for efficiency and lack of contamination.
5. The composition of the PCR buffer depends on the *Taq* polymerase used. The buffer used should be the one provided with the *Taq* polymerase purchased. These enzymes are currently available from a variety of sources (bacterial strains and companies). The reaction conditions described in this chapter have been optimized using the AmpliTaq polymerase. It is advisable to establish the optimal conditions if other enzyme/buffer combinations are used. This is best achieved using standard templates known to contain a defined number of genomes from standard parasite lines (0.1, 1, 10, 100, 1000, and 10,000 genomes per aliquot). Optimal $MgCl_2$ concentrations should be first established as this component usually has the most impact on the efficiency of the amplification reaction. The effective final concentration of Mg^{2+} might be inadvertently reduced by the addition of EDTA which is often used in the buffer used to resuspend the DNA template or the presence of large concentrations of dNTPs since these molecules chelate Mg^{2+} ions in equimolar proportions.
6. The calculations used to obtain the concentration of the oligonucleotides have been performed assuming that: a) 330 g is the average mol wt of each base, and b) an optical density $OD_{260\ nm} = 1.0$ for a 1-cm path in a quartz cuvette, is equivalent to a solution containing 33 µg of oligonucleotide primer per milliliter of water. However, the above quantity is only a crude estimate. Accurate calculations are made by taking the base composition into account, and thus an extinction coefficient is obtained for each individual primer. This is usually the method employed by companies who supply oligonucleotides. The sequences and the names of the oligonucleotide primers are given in **Fig. 1**.

7. The efficiency of detection of allelic variants was found to be similar whether the Nest 1 reaction is performed when all three primers pairs ([{M1–OF} + {M1–OR}], [{M2–OF} + {M2–OR}] and [{G–OF} + {G–OR}] needed to amplify the targeted polymorphic regions of *MSP1*, *MSP2*, and *GLURP*, respectively) are present simultaneously or singly. The Nest 1 reaction is therefore routinely performed as a multiplex amplification.
8. This small reaction volume has been found sufficient for the purposes of highly sensitive detection. Larger volumes will not improve the efficiency of detection, but will cost more in materials, particularly the expensive *Taq* polymerase. Nonetheless, when the volume of the template aliquot to be analyzed is large, or when dilution of possible PCR inhibitors in this template is sought, larger reaction volumes become necessary, but usually only for the Nest 1 reaction.
9. PCR reactions do NOT need to be set up under sterile conditions. The tubes and pipette tips do NOT require autoclaving, although they must be PCR clean. It should be noted that autoclaving does not destroy DNA totally, and cannot therefore, be used to "clean up" reagents and materials. Since the reagents are stored at –20°C, sterility is unnecessary. Moreover, once the reaction is initiated, the temperature does not fall below 58°C and is often at 94°C, conditions hardly suitable for bacterial growth or endonuclease activity.
10. Despite what is written in most manuals, it is strongly recommended to AVOID the use of gloves when setting up PCR. Most types of gloves generate a large amount of static electricity, which will trap any aerosols that might be formed around the fingers, and promote the dissemination of these possible contaminants. In addition gloves are not only uncomfortable but also expensive. Washing of hands before setting up the reactions ensures minimal contamination.
11. It is preferable to have PCR tubes with a flat lid, on which the label can be written. It is also very wise to place the tubes in racks, in the same order as that of the samples, with this order being kept throughout the PCR analysis.
12. The use of mineral oil is crucial when performing nested PCR since it provides an excellent barrier against cross-contamination. The addition of DNA template into the oil layer drastically reduces the risks of aerosol formation and therefore, contamination with DNA template or the Nest 1 PCR product. The development of thermal cyclers with heated lids has obviated the need of an oil barrier against evaporation. For the reasons above, it is still strongly advisable to add mineral oil to these reactions. This is all the more important to minimize contamination when arrays of tube or 96-well plates are used since multiple samples are simultaneously open to the atmosphere.
13. Provided the freezer where the reagents are stored is close by, and that the scheme described is followed, the requirement for ice during the setting up of the PCR can be avoided.
14. The master mix is stable at room temperature for several hours without loss of efficiency.
15. The heating and cooling rates and the accuracy of the temperature calibration are two parameters that can affect the efficiency of the amplification. The fit of the tube into the machine is extremely important as it affects the rate of heat transfer. The use a drop of oil in the "holes" of the heating block in the thermal cycler can be of help unless the manufacturers advise against such procedure. Tube thickness varies between different suppliers, and will also affect the rate of heat transfer, thus the indiscriminate use of different types of tubes should be avoided. The use of tubes specifically designed for a particular thermal cycler is therefore an advantage. The tube content volume will also alter the rate of heat transfer, thus it is important to keep the volume of the mineral oil overlay constant.
16. When a large number of samples are analyzed, the risks of mis-aliquoting, for example, mistakenly adding two different templates to one tube, are relatively high. A brief loss of concentration can therefore invalidate the results, without this being necessarily noticeable by the researcher. A simple way to avoid this problem is to move each tube along the rack (or to a back or front row) immediately after template addition, remembering to close

17. The optimization of the parameters has been performed using a PTC-100 thermal cycler (MJ Research Inc.), and with the AmpliTaq enzyme/PCR Buffer combination described. It is quite possible that changes to cycle number, step times, annealing temperature and even oligonucleotide and dNTP concentrations, could be made without altering the sensitivity and the resolution of the technique. If a different thermal cycler, or another enzyme/buffer combination are used, the reaction conditions described might have to be. Optimization of the reaction conditions, and confirmation that the sensitivity and specificity of the assay are adequate, are best performed using a series of templates containing DNA derived from known numbers of parasite genomes (1, 10, 100, 1000, etc.) belonging to each of the different allelic families to be analyzed.
18. The efficiency of amplification can be monitored by including a positive control, to which the template added represents the minimum quantity of DNA required for the detection of the PCR product, namely that obtained from 1–10 parasites. A further positive control to which 1000-fold more template is added will monitor the quality of the amplification reagents. A minimum of one negative control (no DNA template or human DNA) must be performed each time a reaction is performed. In order to confirm that insignificant levels of contamination arise when performing the PCR analysis, a run of alternate positive and negative samples should be subjected to the nested PCR protocol on a regular basis.
19. Once the reaction is complete, the DNA product is quite stable, since any potential contaminating nucleases will have been destroyed by the lengthy incubation at high temperatures. Consequently, the use of a soak temperature lower than 20°C at **step 8** is unnecessary, and actually will only result in the thermal cycler working unnecessarily. In fact, in tropical countries where the humidity might be high, temperatures lower than 25°C will result in considerable water condensation on and around the heating block.
20. Following use, gels can be stored indefinitely at room temperature, provided they are submerged in TBE Buffer. It is advisable to change the TBE Buffer once or twice in the first few days of storage, as ethidium bromide and the tracker dye diffuse out of the gels.
21. In order to re-use a gel, break it into small pieces and re-boil it making sure that all the agarose has dissolved. The use of a microwave is by far the most rapid and practical way to re-melt a gel. Despite frequent re-use, up to thirty times in our experience, the resolving power needed for the present purpose and the physical integrity of the gel are retained. Two problems are likely to be encountered. The reduction in gel volume through the occasional loss of gel pieces, or more commonly through evaporation due to frequent re-boiling. The latter will alter the agarose concentration but is easily remedied by the addition of distilled water. The accumulation of dust and other dirt particles in the gel will eventually become troublesome. The commonest source is the powder from the gloves that must be worn when handling the gels. This is minimized by pre-washing the gloved hands and the use of lint-free towels. The amount of money saved when large numbers of gels are required is substantial. An important point to bear in mind, is that re-used gels will always contain small amounts of ethidium bromide and should therefore be handled with care.

References

1. Mercereau-Puijalon, O., Fandeur, T., Bonnefoy, S., Jacquemot, C., and Sarthou, J.-L. (1991) A study of the genomic diversity of *Plasmodium falciparum* in Senegal. 2. Typing by the use of the polymerase chain reaction. *Acta. Trop.* **49**, 293–304.
2. Felger, I., Tavul, L., Kabintik, S., Marshall, V. M., Genton, B., Alpers, M. P., and Beck, H.-P. (1994) *Plasmodium falciparum*: Extensive polymorphism in merozoite surface antigen 2 alleles in an area with endemic malaria in Papua New Guinea. *Exp. Parasitol.* **79**, 106–116.
3. Viriyakosol, S., Siripoon, N., Zhu, X. P., Jarra, W., Seugorn, A., Brown, K. N., and Snounou, G.

(1994) *Plasmodium falciparum*: Selective growth of subpopulations from field samples following *in vitro* culture, as detected by the polymerase chain reaction. *Exp. Parasitol.* **79**, 517–525.
4. Contamin, H., Fandeur, T., Rogier, C., Bonnefoy, S., Konate, L., Trape, J.-F., and Mercereau-Puijalon, O. (1996) Different genetic characteristics of *Plasmodium falciparum* isolates collected during successive clinical malaria episodes in Senegalese children. *Am. J. Trop. Med. Hyg.* **54**, 632–643.
5. Babiker, H. A., Ranford-Cartwright, L. C., Currie, D., Charlwood, J. D., Billingsley, P., Teuscher, T., and Walliker, D. (1994) Random mating in a natural population of the malaria parasite *Plasmodium falciparum*. *Parasitol.* **109**, 413–421.
6. Paul, R. E. L., Packer, M. J., Walmsley, M., Lagog, M., Ranford-Cartwright, L. C., Paru, R., and Day, K. P. (1995) Mating patterns in malaria parasite populations of Papua New Guinea. *Science* **269**, 1709–1711.
7. Engelbrecht, F., Felger, I., Genton, B., Alpers, M., and Beck, H.-P. (1995) *Plasmodium falciparum*: malaria morbidity is associated with specific merozoite surface antigen 2 genotypes. *Exp. Parasitol.* **81**, 90–96.
8. Ntoumi, F., Contamin, H., Rogier, C., Bonnefoy, S., Trape, J.-F., and Mercereau-Puijalon, O. (1995) Age-dependent carriage of multiple *Plasmodium falciparum* merozoite surface antigen-2 alleles in asymptomatic malaria infections. *Am. J. Trop. Med. Hyg.* **52**, 81–88.
9. Arez, A. P., Palsson, K., Pinto, J., Franco, A. S., Dinis, J., Jaenson, T. G. T., Snounou, G., and Do Rosario, V. E. (1997) Transmission of mixed malaria species and strains by mosquitoes, as detected by PCR, in a study area in Guinea-Bissau. *Parassitol.* **39**, 65–70.
10. Färnert, A., Snounou, G., Rooth, I., and Björkman, A. (1997) Daily dynamics of *Plasmodium falciparum* subpopulations in asymptomatic children in a holoendemic area. *Am. J. Trop. Med. Hyg.* **56**, 538–547.
11. Beck, H.-P., Felger, I., Huber, W., Steiger, S., Smith, T. A., Weiss, N. A., Alonso, P. L., and Tanner, M. (1997) Analysis of multiple *Plasmodium falciparum* infections in Tanzanian children during the phase III trial of the malaria vaccine SPf66. *J. Inf. Dis.* **175**, 921–926.
12. Snounou, G. and Beck, H.-P. (1998) The use of PCR-genotyping in the assessment of recrudescence or reinfection after antimalarial treatment. *Parasitol. Today* **14**, 462–467.
13. Haywood, M., Conway, D. J., Weiss, H., Metzger, W., Alessandro, U. d., Snounou, G., Targett, G. A. T., and Greenwood, B. M. (1999) Reduction in the mean number of *Plasmodium falciparum* genotypes in Gambian children immunised with the malaria vaccine Spf66. *Tran. R. Soc. Trop. Med. Hyg.* **93**, S1/65–S1/68.
14. Arez, A. P., Snounou, G., Pinto, J., Sousa, C. A., Modiano, D., Ribeiro, H., et al. (1999) A clonal *Plasmodium falciparum* population in an isolated outbreak of malaria in the Republic of Cabo Verde. *Parasitol.* **118**, 347–355.
15. Smith, T., Felger, I., Tanner, M., and Beck, H. P. (1999) Premunition in Plasmodium falciparum infection: insights from the epidemiology of multiple infections. *Trans. R. Soc. Trop. Med. Hyg.* **93 Suppl 1**, 59–64.
16. Snounou, G., Zhu, X., Viriyakosol, S., Jarra, W., Siripoon, N., Thaithong, S., and Brown, K. N. (1999) Biased distribution of *msp1* and *msp2* allelic variants in *Plasmodium falciparum* populations in Thailand. *Trans. R. Soc. Trop. Med. Hyg.* **93**, 369–374.
17. Ranford-Cartwright, L. C., Balfe, P., Carter, R., and Walliker, D. (1991) Genetic hybrids of *Plasmodium falciparum* identified by amplification of genomic DNA from single oocysts. *Mol. Biochem. Parasitol.* **49**, 239–243.
18. Snewin, V. A., Herrera, M., Sanchez, G., Scherf, A., Langsley, G., and Herrera, S. (1991) Polymorphism of the alleles of the merozoite surface antigens MSA1 and MSA2 in *Plasmodium falciparum* wild isolates from Colombia. *Mol. Biochem. Parasitol.* **49**, 265–275.
19. Wooden, J., Gould, E. E., Paull, A. T., and Sibley, C. H. (1992) *Plasmodium falciparum*: a simple polymerase chain reaction method for differentiating strains. *Exp. Parasitol.* **75**, 207–212.
20. Felger, I., Tavul, L., and Beck, H.-P. (1993) *Plasmodium falciparum*: A rapid technique for genotyping the merozoite surface protein 2. *Exp. Parasitol.* **77**, 372–375.
21. Contamin, H., Fandeur, T., Bonnefoy, S., Skouri, F., Ntoumi, F., and Mercereau-Puijalon, O. (1995) PCR typing of field isolates of *Plasmodium falciparum*. *J. Clin. Microbiol.* **33**, 944–951.
22. Viriyakosol, S., Siripoon, N., Petcharapirat, C., Petcharapirat, P., Jarra, W., Thaithong, S., Brown, K. N., and Snounou, G. (1995) Genotyping of *Plasmodium falciparum* isolates by the polymerase chain reaction and potential uses in epidemiological studies. *Bull. WHO* **73**, 85–95.
23. Cox-Singh, J., Mahayet, S., Abdullah, M. S., and Singh, B. (1997) Increased sensitivity of malaria detection by nested polymerase chain reaction using simple sampling and DNA extraction. *Int. J. Parasitol.* **27**, 1575–1577.
24. Färnert, A., Arez, A. P., Correia, A. T., Björkman, A., Snounou, G., and Do Rosário, V. (1999) Sampling and storage of blood and the detection of malaria parasites by polymerase chain reaction. *Trans. R. Soc. Trop. Med. Hyg.* **93**, 50–53.
25. Schriefer, M. E., Sacci, J. B., Jr., Wirtz, R. A., and Azad, A. F. (1991) Detection of polymerase

chain reaction-amplified malarial DNA in infected blood and individual mosquitoes. *Exp. Parasitol.* **73,** 311–316.
26. Siridewa, K., Karunanayake, E. H., and Chandrasekharan, N. V. (1996) Polymerase chain reaction-based technique for the detection of *Wuchereria bancrofti* in human blood samples, hydrocele fluid, and mosquito vectors. *Am. J. Trop. Med. Hyg.* **54,** 72–76.
27. Arez, A. P., Lopes, D., Pinto, J., Franco, A. S., Snounou, G., and Do Rosário, V. E. (2000) *Plasmodium* sp.: optimal protocols for PCR amplification of low parasite numbers from mosquito (*Anopheles* sp.) samples. *Exp. Parasitol.*, **94,** 269–272.
28. Borre, M. B., Dziegiel, M., Høgh, B., Petersen, E., Rieneck, K., Riley, E., et al. (1991) Primary structure and localization of a conserved immunogenic *Plasmodium falciparum* glutamate rich protein (GLURP) expressed in both the preerythrocytic and erythrocytic stages of the vertebrate life cycle. *Mol. Biochem. Parasitol.* **49,** 119–131.
29. Miller, L. H., Roberts, T., Shahabuddin, M., and McCutchan, T. F. (1993) Analysis of sequence diversity in the *Plasmodium falciparum* merozoite surface protein-1 (MSP-1). *Mol. Biochem. Parasitol.* **59,** 1–14.
30. Smythe, J. A., Peterson, M. G., Coppel, R. L., Saul, A. J., Kemp, D. J., and Anders, R. F. (1990) Structural diversity in the 45-kilodalton merozoite surface antigen of *Plasmodium falciparum*. *Mol. Biochem. Parasitol.* **39,** 227–234.
31. Snounou, G., Viriyakosol, S., Zhu, X. P., Jarra, W., Pinheiro, L., Do Rosario, V. E., et al. (1993) High sensitivity of detection of human malaria parasites by the use of nested polymerase chain reaction. *Mol. Biochem. Parasitol.* **61,** 315–320.
32. Snounou, G., Pinheiro, L., Gonçalves, A., Fonseca, L., Dias, F., Brown, K. N., et al. (1993) The importance of sensitive detection of malaria parasites in the human and insect hosts in epidemiological studies, as shown by the analysis of field samples from Guinea Bissau. *Trans. R. Soc. Trop. Med. Hyg.* **87,** 649–653.
33. Hill, W. G. and Babiker, H. A. (1995) Estimation of numbers of malaria clones in blood samples. *Proc. R. Soc. Lond. B* **262,** 249–257.

10

Genotyping of *Plasmodium falciparum*

PCR–RFLP Analysis

Ingrid Felger and Hans-Peter Beck

1. Introduction

Many studies in molecular epidemiology of malaria require the researcher to enumerate or to characterize multiple *Plasmodium falciparum* infections found concurrently in one carrier. These individual clones can be distinguished by a genotyping scheme based on the different alleles of a polymorphic gene.

The gene for the merozoite surface protein 2 (*MSP2*) of *P. falciparum* (*1*) is highly polymorphic; so far 82 alleles have been found (*2*). This extensive diversity constitutes *MSP2* as an excellent marker gene for single-locus genotyping. Polymorphism of *MSP2* is mainly due to different size or copy numbers of the central repeats, leading to variation in gene size. The repeats are located within a dimorphic nonrepetitive region that defines the allelic family. The primers used for amplifying *MSP2* by polymerase chain reaction (PCR) have been selected from the 5' and 3' conserved region immediately flanking the central polymorphic part of the gene. The resulting PCR products vary in length from 378 bp to approx 740 bp. The discrimination power of this length polymorphism is further increased by digesting the PCR product with one or more restriction enzymes. The digested fragments are separated by polyacrylamide gel electrophoresis. This results in genotype-specific patterns of bands that allow to discriminate a multitude of different *MSP2* alleles, and to determine the number of concurrent genotypes per blood sample (multiplicity of infection). New alleles can be easily determined. This method based on restriction fragment length polymorphism of the *MSP2* PCR product (PCR–RFLP) (*3,4*) provides the highest resolution at this particular locus.

The PCR–RFLP method has a wide range of applications. Because the genotype-specific RFLP patterns are stable in time and space, different samples or entire population genetic studies can easily be compared. This is of particular importance when individual alleles need to be followed in consecutive blood samples of the same individual, for example, in longitudinal studies, drug and vaccine trials, or other intervention studies.

2. Materials

2.1. Starting Material

P. falciparum DNA for PCR is routinely isolated from 10 µL of whole blood or 5 µL of blood cell pellets (packed cells) that have been stored either frozen or dried on filter papers. (**CAUTION:** When handling blood, wear gloves and clean up with 70% ethanol. All samples should be treated as potentially infectious material.) So far, several methods for DNA extraction have been successfully used (*see* **Note 1**). For several recent field studies, we have used with very good results the Isocode™ Stix PCR template preparation dipsticks (Schleicher & Schuell, Dassel, Germany), which combine a convenient sampling, storage and shipment of blood dried on a solid matrix, with a fast and easy DNA preparation. Briefly, one of the perforated blood-impregnated Stix triangles is transferred to a PCR tube, twice washed with 500 µL of dH_2O by vortexing (3 × 10 s). After a short spin and removal of excess liquid, the PCR mix is added directly to the triangle and PCR is started.

2.2. PCR–RFLP

2.2.1. Equipment

1. BRL V15.17 vertical gel electrophoresis apparatus with matching glass plates and 1.5-mm spacers, for electrophoresis of polyacrylamide gels (Biometra).
2. UV transilluminator (312 nm), for visualizing stained gels. **CAUTION:** Wear a protective acrylic glass face shield to avoid eye and skin damage.
3. Polaroid camera with film type 667 or photoimager, for documentation of the gels.

2.2.2. Reagents

Unless otherwise stated, all chemicals are derived from Merck, Darmstadt, Germany.

1. PCR primers (Operon Technologies, Alameda, USA):
   For primary PCR: primers S2 and S3 corresponding to nucleotides 3–23 and 789–811 from the 5' and 3' conserved region of the MAD71 sequence of *MSP2* (*5*):

   S2 5'- GAA GGT AAT TAA AAC ATT GTC -3' (sense)

   S3 5'- GAG GGA TGT TGC TGC TCC ACA G -3' (antisense)

   For nested PCR: primers S1 and S4 corresponding to nucleotides 111–129 and 709–728 of the same sequence:

   S1 5'- GAG TAT AAG GAG AAG TAT G -3' (sense)

   S4 5'- CTA GAA CCA TGC ATA TGT CC -3' (antisense)

   These two primer pairs are specific for *P. falciparum MSP2* and do not give rise to a PCR product if human DNA or DNA of any other *Plasmodium* species is the only template.
2. *Taq* DNA polymerase (Biometra).
3. 10X PCR buffer: 500 m*M* KCl, 100 m*M* Tris-HCl, pH 8.8, 15 m*M* $MgCl_2$. If 10X PCR buffer provided by the supplier of *Taq* DNA polymerase is to be used, the magnesium chloride concentration will need to be adjusted using the magnesium stock solution also provided.
4. 100 m*M* deoxyribonucleoside triphosphate (dNTP) stock solutions (Pharmacia, Dübendorf, Switzerland): prepare 5 mL of a 2 m*M* dNTP mix from commercially available preneutralized stock solutions of all four dNTPs (100 m*M*) by adding 100 µL of each dNTP (dATP, dCTP, dGTP, and dTTP) to 4.6 mL of 10 m*M* Tris-HCl, pH 7.4. Store frozen in aliquots of 500 µL or less, depending on your laboratory's throughput (*see* **Note 2**).
5. Restriction enzymes: *Hin*fI, *Dde*I (and for some applications *Scr*FI) with appropriate 10X incubation buffer solutions (New England Biolabs/Bioconcept, Allschwil, Switzerland).

6. Sample loading buffer: 0.2% bromophenol blue (Merck), 0.2% xylene cyanol (Fluka, Buchs, Switzerland), 30% glycerol (Merck), 10 mM Tris-HCl, pH 7.0, 10 mM EDTA, pH 8.0.
7. Premixed acrylamide/bisacrylamide solution (37.5:1) (cat. no. 10325-025; Gibco-BRL Life Technologies, Basel, Switzerland) (*see* **Note 3**).
8. N,N',N' tetramethylethylenediamine (TEMED) (Serva).
9. Ammonium persulfate (Serva): a 10% solution (1 g ammonium persulfate/10 mL dH$_2$O) can be stored for up to 3 wk at 4°C.
10. 5X Tris-borate-EDTA (TBE) buffer: 5 M Tris-HCl, 4 M boric acid, 10 mM EDTA, pH 8.0. To prepare 1 L of 5X TBE buffer stock, combine 54 g of Tris-HCl, 27.5 g of boric acid, 980 mL of dH$_2$O, and 20 mL of 0.5 M EDTA, pH 8.0.
11. DNA size marker: 1 kb DNA ladder (Gibco-BRL Life technologies AG). Prepare 1 mL of marker solution by combining 600 µL of dH$_2$O, 300 µL of sample loading buffer, and 100 µL of the commercially available 1 kb ladder stock solution (1 mg/mL). Load 10 µL of marker solution (= 1 µg of 1 kb ladder) per lane.
12. Ethidium bromide, for staining of gels. Commercially available 1% stock solution in dH$_2$O (cat. no. 111608; Merck). If only undissolved ethidium bromide is available, prepare a 10 mg/mL (w/v) stock solution in dH$_2$O. Store the stock solution light proof at 4°C. Prepare working solution for staining gels by adding 10 µL of ethidium bromide stock solution per 100 mL of 1X TBE; the staining solution can be stored in a lightproof container at room temperature for about 2 wk. **CAUTION:** Ethidium bromide is a powerful mutagen. Wear gloves when preparing and handling these solutions and stained gels.

3. Methods

3.1. Polymerase Chain Reaction

In order to achieve the high sensitivity that is necessary for detecting very low parasite densities, two consecutive rounds of PCR are performed: a primary and a nested PCR (*see* **Note 4**). For nested PCR, a different set of primers is used. These nested primers are located within the primary PCR product generated by the outer primers.

3.1.1. PCR Profile

The same amplification conditions are used for primary and nested PCR: 5 min at 94°C, followed by 30 cycles of 30 s at 94°C, 2 min at 55°C, 2 min at 70°C.

After the last cycle, a final extension step of 7 min at 72°C is included in order to complete all partial extension products and to allow for annealing of single-stranded complementary products.

This profile was designed for the Perkin Elmer DNA Thermal cycler 480. When using a different thermal cycler, the cycle conditions should be optimized.

3.1.2. PCR

1. While thawing the ingredients of the PCR mix, label 0.5-mL PCR tubes on their lids with the sample number (in epidemiological studies, it is best to use the original bleeding code) and P for primary or N for nested PCR.
2. Prepare a master PCR mix without DNA for all samples to be amplified and for a negative control (with dH$_2$O in place of the DNA template; *see* **Note 5**). Combine 78 µL of dH$_2$O plus DNA template (*see* **Note 6**), 10 µL of 10X PCR buffer, 10 µL dNTPs (2 mM), 1 µL primer S2 (50 µM), 1 µL of primer S3 (50 µM), 0.3 µL of *Taq* polymerase (5 U/µL), to a final volume of 100 µL. For nested PCR, use 2 µL of the primary PCR product and replace primers S2 + S3 by primers S1 + S4. The final volume of 100 µL will allow for sufficient material for several different or repeated restriction digests.

3. Distribute 95 μL (in primary PCR) or 98 μL (nested) of master mix in each 0.5 mL reaction tube and overlay with 3 drops of mineral oil to prevent evaporation during the reaction. If the thermal cycler has a top heater, overlaying with oil is not necessary.
4. Add the DNA template (for primary PCR: 5 μL DNA solution; for nested PCR: 2 μL primary product) through the oil layer into the reaction mixture. It is important to use a new tip for each sample (see **Note 7**).
5. Do not delay unnecessarily the transfer of samples into the thermal cycler. Start the preprogrammed cycler immediately.

3.1.3. Prevention of Cross-Contamination

1. A potential source of contamination is right at the beginning of DNA extraction, when aliquots of blood need to be removed from microtainers. Small particles of dried blood tend to fall off when opening the tube. Therefore, clean up and change gloves frequently. Because of this risk, there is an advantage for using blood collected on filter papers or other solid matrices.
2. Use aerosol protected pipet tips for preparing the PCR master mix. If these tips are not available, follow **step 7**.
3. Make aliquots of dH_2O and mineral oil in 1.5-mL tubes and discard after use.
4. After distributing the master mix and oil into the tubes, close all tubes. When adding the template to one tube, keep all other tubes closed.
5. Carry out all pipetting steps for the PCR reaction mix in a designated "clean" area. Wash this area after every use with a cleansing solution that eliminates DNA from surfaces, e.g., DNA-OFF™ (Eurobio, Les Ulis, France). If no such cleansing solution is available, use dishwashing liquid and water and dry with paper towels.
6. Always wear fresh gloves when working in the PCR area, and change gloves whenever contamination is suspected.
7. Designate one set of pipets for preparation of the PCR mix only. This is essential if no aerosol protected tips are available.
8. Routinely clean the removed shaft of pipetors either by soaking in DNA-OFF or in water and soap (follow the supplier's instructions for cleaning) or by exposing the pipetor to UV irradiation (both sides) in a Stratalinker UV Crosslinker.
9. Ensure that each person in the laboratory has his or her own set of PCR reagents.

3.1.4. Quality Control in Molecular Epidemiological Studies

1. Epidemiological studies with big sample sizes are prone to sample mixups. To estimate this, 10% of all samples are randomly selected and repeated once, starting from the DNA extraction. Results are compared. In a few cases, formerly negative samples will turn positive, or an additional infection may appear in multiple clone infections, or others may disappear. This is probably due to very low densities of the respective parasite clones where the detection limit of PCR may be reached. Only totally discrepant genotyping results should be repeated because this indicates a definite mixup of samples.
2. If only samples that were positive by microscopy were chosen for genotyping, all PCR-negative samples are repeated once. If PCR is repeatedly negative despite a positive microscopic slide, it is likely that this is not due to a PCR-associated technical error (e.g., forgetting to add the template) but rather to microscopical error (e.g., wrong determination of the *Plasmodium* species) or to mislabeling of slides or blood samples.

3.2. Gel Electrophoresis

1. Nested PCR products and restriction digests are run on 10% polyacrylamide (PAA) gels using 1.5 mm spacers. The separation is improved if the gels are prepared at least 1 d before use.

2. To pour a vertical 10% PAA gel: combine 17 mL of dH$_2$O, 10 mL of 5X TBE, 15 mL of acrylamide/bisacrylamide solution, 20 µL of TEMED, and 500 µL of 10% ammonium persulfate. Mix all ingredients well and pour gels without delay; several gels may be poured simultaneously. Insert comb. If the gel leaks during the polymerization process, add a few drops of the above gel mix with a 1-mL pipetor. Store gels at 4°C for up to 2 wk. Older gels show a better resolution.
3. Mix 10% (= 10 µL) of the nested PCR product with 3 µL of sample loading buffer and load onto a 10% PAA gel. Include 1 µg of the 1-kb ladder as a DNA length standard.
4. Run PAA gels in 1X TBE buffer in a BRL V 15.17 vertical gel electrophoresis apparatus for 2.5 h at 200 V.
5. Stain the gel for 15 min in ethidium bromide staining solution, visualize by UV and document.

3.3. PCR-RFLP Analysis of MSP2 Genotypes

In routine genotyping, initially, a *Hin*fI digest of each nested PCR product is made and analyzed. In the case of low multiplicity (1–3) infections and predominantly FC27-type infections, a *Hin*fI digest is usually sufficient to determine multiplicity and individual genotypes. In order to genotype infections of high multiplicity, *Dde*I digests are also made, because this enzyme selectively digests 3D7-type alleles, while FC27-type alleles remain uncut. This fact greatly facilitates the enumeration of 3D7-type alleles in multiple-clone infections.

3.3.1. Hinfl Digest

1. Digest 10 µL of nested PCR product with the restriction enzyme *Hin*fI. In a total volume of 20 µL reaction mix, combine 10 µL of nested PCR product (*see* **Note 8**), 7.7 µL of dH$_2$O, 2 µL of 10X buffer, and 0.3 µL (3 U) of *Hin*fI. In general, several samples are digested simultaneously. For multiple digestions, prepare a master reaction mix (all ingredients except the nested PCR product), calculated for the total number of samples plus one to allow for pipetting inaccuracies. Distribute 10-µL aliquots to prelabeled tubes, and add the respective nested PCR product.
2. Incubate 1.5- to 2-h incubation at 37°C.
3. Add 5 µL of sample loading buffer to stop the reaction.
4. Load the total volume (25 µL) onto a PAA gel. Protocols for gels, electrophoresis, and documentation as described in **Subheading 3.2.**

3.3.2. Analysis of Hinfl Digest

3.3.2.1. DETERMINATION OF THE ALLELIC FAMILY

The alleles of *MSP2* fall into two allelic families: the FC27- and 3D7-type alleles. The *Hin*fI fragments at the 5' and 3' ends are family-specific and conserved. The conserved fragments indicating the allelic family are as follows:

1. FC27-type: A 137 bp *Hin*fI fragment at the 5' end and a 115 bp fragment at the 3' end (**Fig. 1**) (*see* **Note 9**).
2. 3D7-type: A 70- and 108-bp fragment, both from the 3' end (**Fig. 2**) (*see* **Note 9**).

3.3.2.2. DETERMINATION OF INDIVIDUAL FC27-TYPE ALLELES

Genotyping FC27-type alleles is very straightforward because this family has a very conserved repeat structure (and therefore exactly predictable *Hin*fI restriction fragment lengths): a 96-bp tandem repeat unit and a 36-bp repeat unit are conserved in sizes but vary in copy number. The restriction maps and the repeat organizations of the most frequent FC27-type alleles are shown in **Fig. 1**. In all alleles of this family, at least

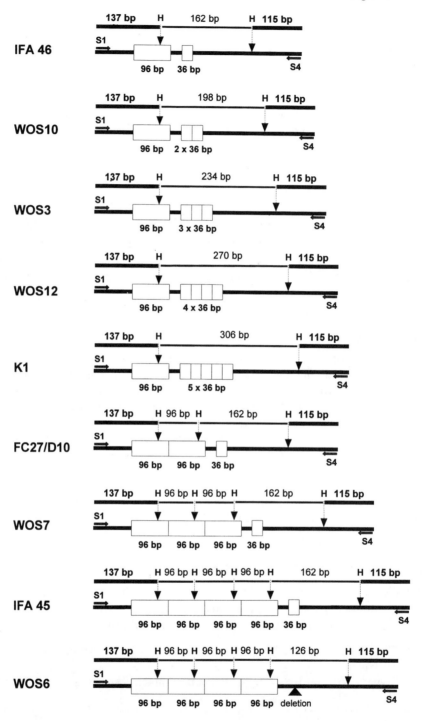

Fig. 1. Organization of the 96 bp and 36 bp repeats and *Hin*fI restriction sites of PCR products of the most frequent FC27-type *MSP2* alleles. The central variable *Hin*fI fragment of FC27-type alleles varies according to the copy number of the 36 bp repeat. Note that if <u>one</u> 96-bp <u>fragment</u> appears in an *Hin*fI digest, in actual fact <u>two</u> 96-bp <u>repeats</u> are present in the allele. H, *Hin*fI restriction site; S1 and S4, location of nested PCR primers.

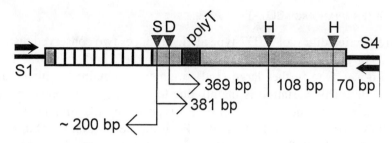

Fig. 2. Structural organization of PCR products of 3D7-type *MSP2* alleles. The repeats, indicated as white boxes, vary in size (4 to 10 amino acids), sequence and copy number (1 to 15 copies), and are often in a scrambled array. The 3D7 family-specific, nonrepetitive region is shown as a gray box. S1 and S4, nested primers, depicted as arrows, located in the 5' and 3' conserved flanking regions of *MSP2*, which are indicated as black line; Poly T, polythreonine stretch, a site of further size variation; S, *Scr*FI restriction site; D, *Dde*I; H, *Hin*fI. Resulting S, D, and H restriction fragments are given in base pairs.

one 96 bp repeat is present. If two repeats are present, a 96-bp *Hin*fI fragment appears on the gel after electrophoresis. If 3 or 4 copies of this repeat are present, the 96-bp fragment shows double intensity of staining. The discrimination between 3 or 4 or more copies of the 96-bp repeat is possible by estimating the size of the uncut nested PCR product, or by running it side by side with an uncut reference allele of the expected size (*see* **Note 10**).

The essential rules of genotyping FC27-type alleles are as follows:

1. Besides the two conserved fragments (137 and 115 bp), either one or two additional allele-specific fragments are seen after running the *Hin*fI-digested PCR product on a gel.
2. For the additional allele-specific fragment, different but exactly defined sizes are possible: 162 bp, or 198 bp, or 234 bp and so on, with the difference always being 36 bp (due to different numbers of 36 bp repeats within this fragment).
3. If a second allele-specific fragment appears, this is always 96 bp in size. However, this second allele-specific fragment is missing in all those FC27-type alleles that have only one 96-bp repeat.

Exceptions from the above rules are as follows:

1. *Deletions.* So far, one partial deletion of a 36-bp repeat (found in several alleles, with Wos6 being the only frequent allele with this deletion) has been observed, thus reducing the size of the 162-bp fragment (D10 and Wos7 in **Fig. 1**) to a 126-bp fragment (Wos6 in **Fig. 1**).
2. *Duplications.* In blood samples from African countries, several alleles belonging to the FC27 family were found with an amplification (up to 23-fold) of a 9-bp unit within the otherwise conserved 137 bp *Hin*fI fragment of the 5' end (*6*). This third tandem repeat unit is found in some alleles in addition to the 96- and 36-bp repeats always present in FC27-type alleles. Such amplifications result in a great variety of fragment sizes and complicate genotyping. But these alleles are all of low allelic frequency. All more frequent alleles show the standard 137 bp *Hin*fI fragment at their 5' end. In conclusion, if occasionally a *Hin*fI fragment of untypical size between 137 bp and 250 bp appears in an African sample, these rare alleles have to be considered. A list of FC27-type alleles from Tanzania, their *Hin*fI fragments and a photograph showing some frequent PCR–RFLP patterns of FC27-type can be found in **ref. 2**.

3.3.2.3. DETERMINATION OF 3D7-TYPE ALLELES

*Hin*fI restriction digest of a 3D7-type allele produces two conserved restriction fragments of 70 bp and 108 bp, which both derive from the 3' end of the *MSP2* nested PCR product (**Fig. 2**). One further large *Hin*fI fragment is obtained, which varies in size between alleles. This variation is mainly due to differences in repeat sizes and copy numbers. The repeat structure of several 3D7-type alleles and alignments can be found in UNDP/World Bank/WHO TDR Programme Malaria Database http://www.wehi.edu.au/MalDB-www/genomeInfo/FTPaccess/htmledFTP/Alignments/msp2.html and in **ref. 7**. Individual alleles were named according to the size of this fragment, for example, $3D7_{370}$. Other, unusual restriction patterns can also, though rarely, be found in this allelic family, mostly due to point mutations, creating new *Hin*fI restriction sites. 3D7-type alleles found in Tanzania with deviations from the conserved *Hin*fI pattern are listed in **ref. 2**.

3.3.2.4. RECOMBINANT FORMS BETWEEN THE TWO ALLELIC FAMILIES

Alleles that represent recombinations between the FC27 and the 3D7 families have also been found. These recombinations were detected by *Hin*fI restriction fragments of unusual fragment lengths. Nucleotide sequencing revealed that in all examples analyzed so far, the site of recombination was located at the 5' end of the 96-bp repeat unit (*6*). In all recombinations for which sequence data is accessible, the 5' end derived from a 3D7-type allele, and the 3' end was of the FC27 type. PCR products of *MSP2*, representing a recombination between the 3D7 and the FC27 allelic families, are grouped with FC27-type alleles, because they mainly contain FC27-type sequences, in particular, the FC27 family-specific 36- and 96-bp repeats.

3.3.2.5. ANALYZING MULTIPLE INFECTIONS

The *Hin*fI restriction patterns of blood samples with a high number of concurrent infections is very complex. Therefore, in order to learn PCR–RFLP genotyping, it is advisable to start with one round of experiments using DNA from cultured material or from blood samples with a known single infection only. **Figure 3** shows an example of a complex restriction pattern and its analysis (*see* **Note 11**).

3.3.3. Additional Restriction Digests with Dde*I*

Because the variable, allele-specific *Hin*fI fragment of 3D7-type alleles is too large (250–550 bp) to identify an allele unequivocally even by PAA gels, a further restriction digest with *Dde*I is sometimes necessary because it produces smaller allele-specific fragment sizes. This allows further differentiation between 3D7-type alleles of similar sizes (detected differences <10 bp). *Dde*I digests produce two fragments (**Fig. 4**), one of which is highly variable and allele-specific. The C-terminal, large *Dde*I fragment is either 369 bp or 387 bp long in most alleles (**Fig. 4A,B**), but occasionally other sizes or a deletion of this site can occur, as shown in **Fig. 4D,E**.

A further advantage of *Dde*I digests of multiple clone infections is that in contrast to the 3D7-type alleles, the alleles of the FC27 allelic family remain uncut (with the exception of a few alleles producing a short 39-bp *Dde*I fragment, which is too small to interfere with the analysis of the 3D7-type alleles).

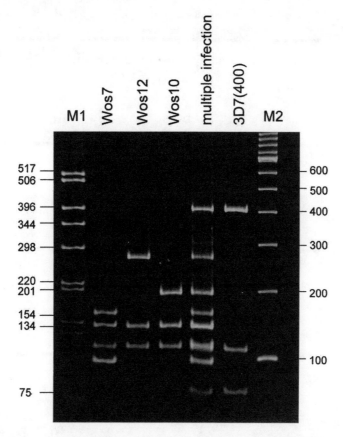

Fig. 3. *Hin*fI digest of *MSP2* alleles of a *Plasmodium falciparum* multiple infection from a Tanzanian blood sample, which shows a multiplicity of four concurrent infections. In order to demonstrate how a complex restriction pattern is analyzed, digests of corresponding alleles (Wos 7 to $3D7_{400}$) were loaded beside the RFLP pattern of the multiple infection. Lane M1: 1 µg of 1 kb ladder (Gibco-BRL Life Technologies); lane M2: 1 µg of 100 bp DNA Marker (Gibco-BRL Life Technologies). Reproduced with permission from **ref. 2**.

3.3.4. Comparison of Consecutive Samples of the Same Patient

If infections from paired blood samples are to be compared (longitudinal samples), both a *Hin*fI and a *Dde*I restriction digest are generally performed. The additional *Dde*I digest increases the discriminatory power of the PCR–RFLP genotyping scheme considerably (*see* **Note 12**). Digests are performed as described in **Subheading 3.3.1.** However, when separating the restriction fragments in a PAA gel, it is important that all consecutive samples of the same patient are loaded side by side. By doing so, it becomes straightforward to identify newly appearing or disappeared alleles in a series of consecutive blood samples. The analysis of individual alleles is performed as described in **Subheadings 3.3.2.** and **3.3.3.**

4. Notes

1. Methods for extraction of *Plasmodium* DNA that have been successfully used for PCR–RFLP genotyping:

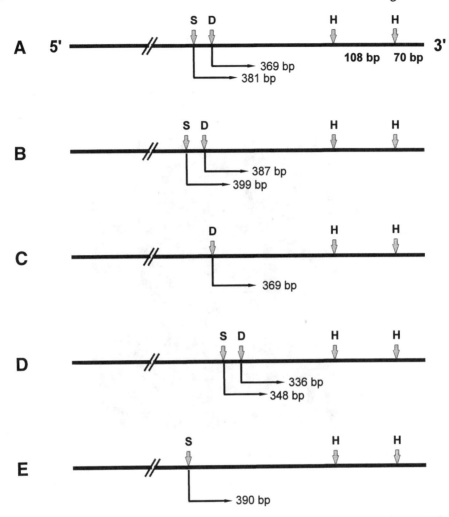

Fig. 4. Restriction maps of five subgroups of 3D7-type *MSP2* PCR products. The region upstream of the *Dde*I and *Scr*FI sites contains repeats and is highly variable and allele-specific. The region downstream of these sites corresponds to the nonrepetitive family-specific part of 3D7-type alleles. *Dde*I and *Scr*FI restriction fragments are indicated by arrows and given in base pairs. The two constant and family-specific *Hinf*I restriction fragments are bolded in a. Variants **a** to **c** are the most frequent ones, variants d. and e. are found less frequently. D, *Dde*I; S, *Scr*FI; H, *Hinf*I.

   a. DNA from frozen blood: rapid boiling method *(5)*; guanidine isothiocyanate (GTC) preparation with subsequent phenol/chloroform extraction *(8)*; commercially available DNA purification kit (QIAamp® Blood Kit; Qiagen, Basel, Switzerland).
   b. DNA from dried blood on solid matrices: Isocode™ Stix PCR Template Preparation Dipstick (Schleicher & Schuell); chelex extraction from blood-blotted filter paper *(9)*; methanol-fixation of blood-blotted filter paper.

Each of these methods has its pros and cons. The choice of the DNA extraction method depends on questions such as: Is long-term storage of the DNA required? Which amount of DNA is required? A comparison of methods used for DNA extraction in studies of molecular epidemiology and a detailed description can be found in **ref. 8**.

2. The quality of the dNTP mix is the most critical factor for the outcome of the PCR amplification. If the yield of a PCR was poorer than that obtained in previous experiments, repeat the experiment with a new aliquot dNTPs. If the dNTP mix had been through more than three freeze/thaw cycles, or if the mix had been left at room temperature for a longer period of time than is required for thawing, the dNTPs may be degraded. Prior to use, thaw the dNTP mix at room temperature or by hand until most of the frozen solution has melted, then keep on ice while removing the amount needed, and freeze again as soon as possible.
3. For optimal resolution in PAA gels, it is essential to use good quality PAA such as the one indicated. PAA from some other suppliers may not give satisfying results.
4. When DNA from parasite culture or from blood samples of high known parasite density (e.g., clinical cases) is used, only a nested PCR needs to be performed. Then the PCR conditions are changed to 45 cycles of 30 s at 94°C, 1 min at 50°C, 2 min at 70°C. However, molecular epidemiological studies generally involve many samples of widely differing densities, and for good comparability and for adequate sensitivity all samples need to be amplified in the same way, by a primary plus nested PCR.
5. A negative control must be included in each batch of PCRs. If the negative control comes out positive after running in parallel with the other samples on a gel, this indicates contamination of the master mix with foreign DNA or products from previous amplification reactions or from other samples of the same batch. If the control is contaminated, all PCRs of the batch need to be discarded. Often the size of the contaminating PCR product gives a hint to the source of contamination, for example, one of the positive samples in the same batch. A *Hin*fI digested can also be performed to find the reason for contamination.

   A positive control is included if samples of unknown parasitemia are to be amplified, and where the possibility exists that all samples of one batch may be parasite-negative. Use as positive control DNA of a tested positive blood sample or material from *P. falciparum* in vitro cultures. This allows control for quality and completeness of the PCR mix. If no amplification product is seen in the positive control, it is very likely that one of the ingredients into the PCR mix was omitted.
6. Using 5 µL of DNA (corresponding to about 1 µL of whole blood in most DNA preparations) in a primary PCR gives sufficient sensitivity even for very low parasite densities below the detection limit of microscopy. Due to specific questions asked or due to availability, more or less DNA may be used. DNA from 5–10 µL of blood can be used, but this requires that the DNA solution is of high purity, for example, DNA from an Isocode™ dipstick is adequate, but the DNA from a rapid-boil preparation is not adequate.

   It is essential to use the same amount of DNA throughout an entire study to ensure comparable sensitivity.

   Excessive input of DNA (e.g., from a sample with very high parasite density) can occasionally result in inhibition of PCR; a visible band appears if an aliquot of the primary PCR product is loaded on a gel. In this case, dilute the primary PCR at least 1:10.000.

   When preparing the master mix for several reactions, subtract the volume of the DNA template, which will be added later to each tube, from 78 µL, which is the volume of template plus water.
7. When processing many reactions at a time, pipetting errors can occur. Great care has to be taken to add the correct template in the corresponding PCR tube. It is helpful to align the template and the PCR tubes in rows (preferentially in numerical order), and to translocate completed tubes to another position in the rack. Do not discriminate tubes with or without template by leaving lids open or closed because of the risk of contaminating open PCR tubes, particularly when performing nested PCRs where already amplified primary PCR products have to be handled on the bench.
8. The nested PCR product is covered with an oil layer. Oil carryover does not harm the digestion, but the sample will be difficult to load into the gel slot. To get rid of the viscous

oil on the outside of pipet tips, pull the tip out of the tube along the wall of the PCR tube. If the yield of the nested PCR is only low (e.g., 10 µL of nested PCR product giving only a faint band on the PAA gel), 15 µL of PCR product or more can be digested with *Hin*fI. However, if the volume of a digest exceeds 25 µL, precipitation with isopropanol is necessary due to the limited capacity of the gel slot.

9. In a 10% PAA gel, the 115 bp *Hin*fI fragment specific for FC27 allelic family runs inseparably together with the 108 bp 3D7 family-specific fragment, if both types are present in one blood sample. The difference in size is evident only in single infections. Therefore, these fragments cannot be used to determine the allelic family. This may be achieved with the other family-specific *Hin*fI fragments.

10. It is helpful to reamplify frequently occurring alleles from single clone infections, once they have been clearly identified, in order to build up a collection of reference alleles, for example, FC27-type reference alleles with the repeat organizations shown in **Fig. 1**. Reference alleles can also be obtained by PCR amplification of *MSP2* from DNA from in vitro-cultured parasites. Use reference alleles (digested or undigested) for side-by-side runs with query genotypes.

11. Genotyping of single clone infections is straightforward, by checking the *Hin*fI restriction fragment pattern for the presence of the family-specific conserved fragments, followed by determination of the allele-specific fragment. If necessary, use digested reference alleles for side by side comparisons.

    In multiple infections, first start with screening for the 137-bp fragment, indicating the presence of at least one FC27-type allele, and for a 70-bp fragment, indicating the presence of at least one or more 3D7-type alleles. Second, check for a 96-bp fragment, which indicates the presence of at least one allele of FC27 type with more than one 96 bp repeat. As a next step, try to determine the fragments between 150 and 300 bp. In most cases, these alleles are of FC27 type, representing the large internal *Hin*fI fragment of FC27-type alleles with several 36 bp repeats (*see* **Fig. 1**). The central *Hin*fI fragments of these alleles have fixed sizes. The remaining high molecular fragments in general are of 3D7 type, and very variable in size. If there are doubts about the number of 3D7-type alleles in a multiple infection, make a *Dde*I digest, which cuts 3D7-type alleles (with only rare exceptions), but not those of FC27 type.

12. *Scr*FI digests can also be performed to further resolve ambiguities. As with the *Dde*I fragment, the N-terminal *Scr*FI fragment is highly variable due to repeats, while the C-terminal fragment occurs in three sizes, which are shown in **Fig. 4**, with a and b being the fragment sizes most frequently found.

References

1. Scythe, J. A., Peterson, M. G., Compel, R. L., Saul, A. J., Kemp, D. J., and Anders, R. F. (1990) Structural diversity in the 45-kilodalton merozoite surface antigen of *Plasmodium falciparum*. *Mol. Biochem. Parasitol.* **39,** 227–234.
2. Felger, I., Irion, A., Steiger, S., and Beck, H. P. (1999) Genotypes of merozoite surface protein 2 of *Plasmodium falciparum* in Tanzania. *Trans. R. Soc. Trop. Med. Hyg.* **93,** 3–9.
3. Felger, I., Tavul, L., and Beck, H. P., (1993) *Plasmodium falciparum:* A rapid technique for genotyping the merozoite surface protein 2. *Exp. Parasitol.* **77,** 372–375.
4. Felger, I., Tavul, L., Kabintik, S., Marshall, V., Genton, B., Alpers, M., and Beck, H. P. (1994) *Plasmodium falciparum*: Extensive polymorphism in merozoite surface antigen 2 alleles in an area with endemic malaria in Papua New Guinea. *Exp. Parasitol.* **79,** 106–116.
5. Foley, M., Ranford-Cartwright, L. C., and Babiker, H.A. (1992) Rapid and simple method for isolating malaria DNA from fingerprick samples of blood. *Mol. Biochem. Parasitol.* **53,** 241–244.
6. Irion, A., Beck, H. P., and Felger, I. (1997) New repeat unit and hot spot of recombination in FC27-type alleles of the gene for *Plasmodium falciparum* merozoite surface protein 2. *Mol. Biochem. Parasitol.* **90,** 367–370.
7. Felger, I., Marshall, V. M., Reeder, J. C., Hunt, J. A., Mgone, C. S., and Beck, H. P. (1997) Sequence diversity and molecular evolution of the merozoite surface antigen 2 of *Plasmodium falciparum*. *J. Mol. Evol.* **45,** 154–160.

8. Henning, L., Felger, I., and Beck, H. P. (1999) Rapid DNA extraction for molecular epidemiological studies of malaria. *Acta Tropica* **72**, 149–155.
9. Plowe, C. V., Djimde, A., Bouare, M., Doumbo, O., and Wellems, T. E. (1995) Pyrimethamine and proguanil resistance-conferring mutations in *Plasmodium falciparum* dihydrofolate reductase: polymerase chain reaction methods for surveillance in Africa. *Am. J. Trop. Med. Hyg.* **52**, 565–568.

11

Microsatellite Analysis in *Plasmodium falciparum*

Xin-zhuan Su and Michael T. Ferdig

1. Introduction

Microsatellites (MS) are simple sequence repeats (SSRs) such as $(CA)_n$, $(TAA)_n$, or $(TA)_n$ that have been found in all eukaryotes studied *(1–3)*. The number of repeated nucleotides per unit is generally limited to between 1 and 5, in contrast to the longer repeats characteristic of minisatellites. Whereas $(CA)_n$ and $(GT)_n$ are the most abundant repeats in mammals, $(TA)_n$ and $(TAA)_n$ are the most common SSRs in plants *(4)*. Similar to many plants, the genome of the human malaria parasite, *Plasmodium falciparum*, is rich in A-T SSRs, particularly $(TA)_n$ and $(TAA)_n$ *(5)*.

SSRs vary in their numbers of tandem repeat units among individuals or strains. These variations, known as simple sequence length polymorphisms (SSLPs), can provide a valuable source of information for genetic studies *(6)*. SSLPs are assayed by polymerase chain reaction (PCR) using primers designed to sequences flanking the repeat regions to amplify DNAs from different individuals or isolates (**Fig. 1**). PCR products of various sizes, representing different alleles from parasite isolates, can be detected and compared by autoradiography after separation of radioactively labeled PCR products on DNA sequencing gels (**Fig. 2**) or by laser, with fluorescently labeled products, using an automated sequencer *(7)*.

SSLPs have been referred to as second-generation genetic markers because of their advantages over traditional restriction fragment length polymorphisms (RFLPs). Genotypes from SSLPs can be generated more rapidly and from much smaller amounts of DNA than is required for RFLP analysis. This is critical in studies for which only limited amounts of DNA samples can be obtained. SSLPs are highly informative, exhibiting multiple alleles in a general population in contrast to the two alleles typical for RFLPs. More important, the entire process of PCR amplification and product detection can be automated using a robotic system and automated sequencer to type and score genotypes *(7)*. Multiple markers can be assayed in a single gel by multiplexing PCR reactions and by labeling PCR products with different fluorescent dyes. Such a high throughput analysis has made it possible to search for genes of interest by genome-wide scanning with hundreds of markers *(8)* and is essential for population studies. Many genetic maps employing SSLPs have been developed in various systems *(9–11)*. These maps have greatly enhanced efforts to map and clone genes involved in diseases and in biological processes.

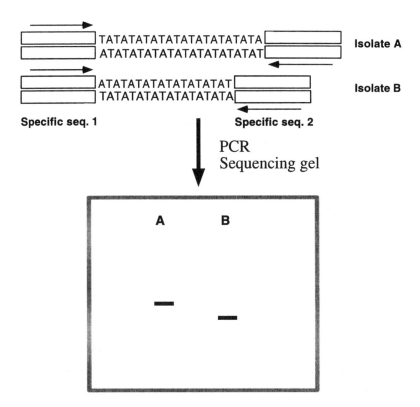

Fig. 1. Diagram illustrating the principle of microsatellite analysis. Parasite isolates A and B have different number of TA repeats, 11 repeats for A and 9 repeats for B, but have the same DNA sequences flanking the repeat region. A single pair of PCR primers is designed to the flanking DNA sequences and used to amplify the variable repeat region from both parasites' DNA. The PCR product from isolate A is 4 base pairs longer than the product from isolate B and, therefore, migrates slower in a polyacrylamide gel.

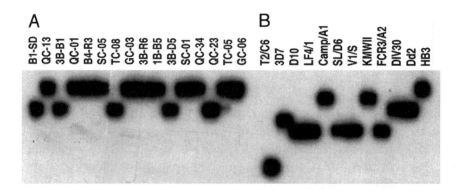

Fig. 2. Autoradiograph of PCR products from 16 progeny DNAs of a Dd2 × HB3 genetic cross (20) and DNAs from 12 P. falciparum isolates. Microsatellite TA81 was amplified as described above. Two alleles were present in the progeny (**A**), but six alleles were detected from the 12 isolates (**B**).

MS typing is a powerful tool in genetic studies of malaria parasites. For example, segregation of polymorphic MS markers in a genetic cross have been used to map gene(s) involved in chloroquine resistant in *P. falciparum* (12) and will play an important role in identifying determinants of a defect in male gamete development (13) and an "X" factor in folate utilization (14). MS markers also have been used to study parasite population structure and dynamics (15) and to "fingerprint" parasite isolates (16). Recently, more than 800 MS markers have been developed from the *P. falciparum* genome and placed in a high-resolution genetic map (17,18), providing an important tool for linkage analysis and positional cloning, allelic association studies (19,20), and parasite population genetics.

In the following sections, we discuss procedures for testing microsatellites from the human malaria parasite *P. falciparum*.

2. Materials

2.1. Equipment

1. Repeating pipet, capable of dispensing 2 and 10 µL (cat. nos. 7683-D10, 7683-D55, 7681-B30, or 7681-B40; Thomas Scientific).
2. Thin-wall 200-µL PCR tubes.
3. Thermal cycler: MJ PTC100 or Perkin Elmer GeneAmp 9700.
4. Heat-block or water bath, adjusted to 37°C.
5. Electrophoresis apparatus: any DNA sequencing apparatus.
6. ABI377 with GeneScan software or equivalent, for fluorescent detection.
7. 3 MM Whatman filter paper.
8. Saran wrap.
9. Gel dryer.
10. X-ray film: BioMax film from Kodak.

2.2. Reagents

For long-term storage, keep all reagents at –20°C.

1. Synthetic oligonucleotides: diluted to 50–100 p*M*/µL in water. Primers are generally 20–22-mers, with between 5 and 7 GCs (about 25–30% GC content).
2. *Taq* polymerase: usually 5 U/µL (from any commercial supplier).
3. PCR buffer (pH 8.3): 10 m*M* Tris-HCl, 50 m*M* KCl, 1.5 m*M* MgCl$_2$, 0.01% gelatin. All commercially available *Taq* polymerases are supplied with an appropriate 10X buffer.
4. Polynucleotide kinase (with 10X buffer): 10 U/µL (Boehringer Mannheim).
5. Radioactive nucleotide: [γ^{32}P]ATP.
6. Fluorescent-dUTPs: R110 (blue), R6G (green), and Temra (yellow) (Perkin Elmer). Dilute the fluorescent-dUTP (1:10) to final concentrations of 10 µ*M* for R110 and R6G and 40 µ*M* for Temra.
7. 10 m*M* deoxyribonucleoside triphosphate (dNTP): combine 100 µL of each of 100 m*M* dGTP, 100 m*M* dATP, 100 m*M* dCTP, 100 m*M* dTTP (Boehringer Mannheim), and 600 µL H$_2$O.
8. Manual sequencing dye: 95% formamide, 20 m*M* EDTA, 0.05% bromophenol blue, and 0.05% xylene cyanol.
9. Automatic sequencing dye: combine 800 µL of deionized formamide and 200 µL of 25 m*M* EDTA (pH 8.0) containing 50 mg/mL blue dextran.
10. 10X Tris-borate-EDTA (TBE) (pH 8.3): combine 108 g of Tris-HCl, 55 g of boric acid, 40 mL of 0.5 *M* EDTA, and make final volume to 1000 mL with H$_2$O. Store at room temperature.

11. 6% bispolyacrylamide (neurotoxic) gel stock: combine 200 mL of 30% bisacrylamide (290 g acrylamide, 10 g NN'-methylene-bisacrylamide, H_2O to 1000 mL), 50 mL of 10X TBE, 460 g of urea, and make final volume to 1000 mL with H_2O. Store at 4°C.
12. 10% ammonium persulfate: dissolve 0.1 g of ammonium persulfate in 1 mL of H_2O. Prepare fresh immediately before pouring the gel.
13. TEMED.
14. Sterile H_2O.

3. Methods

3.1. Microsatellite Assay with Radioactive Labeling

1. DNA preparation: parasite DNA prepared by any established procedure is suitable. For quick preparation of DNA from parasite cultures or blood samples, wash red blood cells with PBS (1×, pH 7.2) and lyse with 0.05% saponin in PBS. Centrifuge the parasites at 10,000*g* for 5 min, wash again with PBS, resuspend in 100 μL of water, and boil at 100°C for 10 min. The samples can be used immediately or stored at –20°C.
2. ^{32}P labeling oligonucleotide: one of the PCR primers is labeled with $\gamma^{32}P$ before amplification. Mix 300 p*M* of oligonucleotide, 1 μL of polynucleotide kinase, 2 μL of 10X kinase buffer, 1 μL of fresh $\gamma^{32}P$-ATP, and make volume to 20 μL with H_2O. Incubate at 37°C for 30 min, and dilute the reaction to 60 μL with H_2O.
3. PCR mixture (*see* **Notes 1** and **2**):
   a. 10X PCR buffer   1.4 μL
   b. 10 m*M* dNTP   0.3 μL
   c. DNA   4 μL (~10 ng)
   d. *Taq* polymerase   0.1 μL (0.5 U)
   e. Primer 1   0.5 μL (10 p*M*/μL)
   f. Primer 2   1 μL (^{32}P labeled)
   g. H_2O to 14 μL

   For multiple PCR tubes, pipet 4 μL of DNA into each tube and combine the remaining reagents (except the ^{32}P-labeled primer) as a cocktail. After addition of the labeled primer, deliver 10 μL of the mix into each tube using a repeating pipet (Thomas Scientific). Overlay the reactions with 10 μL of mineral oil if thermal cycler is not equipped with a heated cover.
4. PCR cycles (*see* **Note 3**):
   a. Initial denaturation   92°C, 2 min (optional)
   b. Cycling (35 cycles)
      i. Denature   92°C for 20 s
      ii. Annealing   45°C for 10 s and 40°C for 10 s
      iii. Extension   60°C for 30 s
   c. Final Extension   60°C, 2 min
5. Polyacrylamide gel: prepare and pour a 6% polyacrylamide gel while waiting for the PCR cycling to finish. Combine 100 mL of 6% bisacrylamide solution, 100 μL of TEMED, and 200 μL of 10% ammonium persulfate, and mix well carefully, but do not introduce air bubbles. Use a 60-mL syringe to pour the gel mixture between gel plates *(20)*. Allow at least 1 h for gel to completely polymerize.
6. Sample loading and electrophoresis: mix 4 μL of PCR product with 6 μL of manual sequencing dye, place the mixture in an 80°C oven for 5 min, and load 2 μL per well onto the gel. Run the gel (in 1X TBE) at constant wattage to keep temperature at approx 50°C until the blue dye reaches the bottom of the gel (2–3 h).
7. Gel removal: remove one glass plate, overlay a 3MM paper (precut to the size of the gel) on the top of the gel, transfer the gel onto the 3MM paper by lifting the paper away from the second glass plate, cover the gel with Saran wrap, and dry the gel in a gel dryer at 80°C for about 30–60 min.

8. Autoradiography: expose the dried gel to an X-ray film with DuPont intensifying screen and place the gel at –70°C for 10 min to a few hours. A typical autoradiograph is shown in **Fig. 2**.

3.2. Microsatellite Assay with Fluorescent Labeling

Fluorescent labeling of PCR products requires equipment such as an ABI377 to read signals from fluorescent dyes. The advantages of fluorescent labeling include not utilizing radioactive materials and the potential for high throughput analysis.

1. PCR mixture:
    a. 10X PCR buffer   1.4 µL
    b. 10 mM dNTP   0.3 µL
    c. Primer 1   0.5 µL (50–100 pM/µL)
    d. Primer 2   0.5 µL (50–100 pM/µL)
    e. DNA   4 µL (~5 ng)
    f. *Taq* polymerase   0.1 µL (0.5 U)
    g. Fluorescent dye   0.1 µL (10 mM/µL)
    h. H$_2$O   7.1 µL
2. PCR cycles: same as **Subheading 3.1., step 4**.
3. Gel and instrument setup: see instructions for ABI377.
4. Sample loading: mix 1–2 µL of PCR products (*see* **Note 4**) with 2 µL of loading dye (*see* **Subheading 2.2., item 9**), denature the samples at 90°C for 2 min, load 1–2 µL/sample onto the gel, and run the GeneScan program for 2 h.
5. Gel images can be printed directly from computer screen for record keeping or analyzed with GeneScan software.

4. Notes

1. The use of aerosol barrier pipet tips is recommended for reducing cross-contamination.
2. The reaction volumes can be scaled up if necessary. For small volumes, care should be taken to avoid variation in pipetting. We always prepare a master reaction mix and combine the mixture with 4 µL of diluted DNA solution. Use of a repeating device and multiple channel pipetors is recommended.
3. We use a single PCR cycling program to amplify hundreds of different primer combinations. All of our primers were designed according to guidelines described above. Not all the primers work equally efficiently in the PCR program. For multiplexing PCRs with different combinations of primers, additional modification may be necessary. Our experience has been that radioactive primer labeling works better in multiplex reactions than does incorporation of fluorescent dUTPs. In multiplex fluorescent reactions, microsatellites with smaller PCR products are usually amplified more efficiently.
4. The PCR products from fluorescent labeling reaction can be precipitated by adding 50 µL of 70% ethanol with 0.5 mM MgCl$_2$ to the reactions. The products are precipitated at room temperature for 15 min and centrifuged at 3000g for 15 min. Dye (4 µL) is added to the DNA pellets, and the samples are processed accordingly.

References

1. Miesfeld, R., Krystal, M., and Arnheim N. (1981) A member of a new repeated sequence family which is conserved throughout eukaryotic evolution is found between the human d and b globin genes. *Nucleic Acids Res.* **9,** 5931–5947.
2. Hamada, H., Petrino, M. G., and Kakunapa, T. (1982) A novel repeated element with Z-DNA-forming potential is widely found in evolutionarily diverse eukaryotic genomes. *Proc. Natl. Acad. Sci. USA* **79,** 6465–6469.
3. Tautz, D. and Renz, M. (1984) Simple sequences are ubiquitous repetitive components of eukaryotic genomes. *Nucl. Acids Res.* **12,** 4127–4138.

4. Lagercrantz, U., Ellegren, H., and Anderson, L. (1993) The abundance of various polymorphic microsatellite motifs differs between plants and vertebrates. *Nucl. Acids Res.* **21,** 1111–1115.
5. Su, X-z. and Wellems, T. E. (1996) Toward a high-resolution *Plasmodium falciparum* linkage map: polymorphic markers from hundreds of simple sequence repeats. *Genomics* **33,** 430–444.
6. Tautz, D. (1989) Hypervariability of simple sequences as a general source for polymorphic DNA markers. *Nucleic Acids Res.* **17,** 6463–6471.
7. Mansfield, D. C., Brown, A. F., Green, D. K., Carothers, A. D., Morris, S. W., Evens, H. J., and Wright, A. F. (1994) Automation of genetic linkage analysis using fluorescent microsatellite markers. *Genomics* **24,** 225–233.
8. Lee, J. H., Reed, D. R., Li, W. D., Xu, W., Joo, E. J., Kilker, R. L., et al. (1999) Genome scan for human obesity and linkage to markers in 20q13. *Am. J. Hum. Genet.* **64,** 196–209.
9. Dib, C., Faure, S., Fixames, C., Samson, D., Drouot, N., Vignal, A., et al. (1996) A comprehensive genetic map of the human genome based on 5264 microsatellites. *Nature* **380,** 152–154.
10. Dietrich, W. F., Miller, J., Steen, R., Merchant, M. A., Damron-Boles, D., Husain, Z., et al. (1996) A comprehensive genetic map of the mouse [published erratum appears in *Nature* (1996) **381,** 172]. *Nature* **380,** 149–152.
11. Ela, W. K., Goodman, A., Eller, M., Chevrette, M., Delgade, J., Neuhauss, S., et al. (1998) A microsatellite genetic linkage map for zebrafish (*Danio rerio*). *Nature Genet.* **18,** 338–343.
12. Su, X-z., Kirkman, L. A., Fujioka, H., and Wellems, T. E. (1997) Complex polymorphisms in an ~300 kDa protein are linked to chloroquine-resistant *P. faciparum* in Southeast Asia and Africa. *Cell* **91,** 593–603.
13. Vaidya, A. B., Muratova, F., Guinet, F., Keister, D., Wellems, T. E., and Kaslow, D. C. (1995) A genetic locus on *Plasmodium falciparum* chromosome 12 linked to a defect in mosquito-infectivity and male gametogenesis. *Mol. Biochem. Parasitol.* **69,** 65–71.
14. Wang, P., Read, M., Sims, P. F. G., and Hyde, J. (1997) Sulfadoxine resistance in the human malaria parasite *Plasmodium falciparum* is determined by mutations in dihydropteroate synthetase and an additional factor associated with folate utilization. *Mol. Microbiol.* **23,** 979–986.
15. Anderson, T. J. C., Su, X-z., Rockarie, M., and D, K. P. (1999) Twelve microsatellite markers for characterization of *Plasmodium falciparum* from finger prick blood samples. *Parasitology* **119, 1**13–125.
16. Su, X.-z., Curcci, D., and Wellems, T. E. (1998) *Plasmodium falciparum*: parasite typing with a multi-copy microsatellite marker pfRRM. *Exp. Parasitol.* **89,** 262–265.
17. Su, X.-z., Ferdig, M. F., Chuong, H. Q., Huang, Y., Liu, A., You, J., Wootton, J. C., and Wellems, T. E. (1999) A comprehensive genetic map for human malaria parasite *Plasmodium falciparum. Science* **286,** 1351–1353. http://www.ncbi.nlm.nih.gov/Malaria/markersNmaps.html
18. Lander, E. S. and Schork, N. J. (1994) Genetic dissection of complex traits. *Science* **265,** 2037–2048.
19. Ferdig, M. T. and Su, X.-z. (2000) Microsatellite markers and genetic mapping in *Plasmodium falciparum. Parasitol. Today* **16,** 307–312.
20. Wellems, T. E., Panton, L. J., Gluzman, I. Y., do Rosario, V. E., Gwadz, R. W., Walker-Jonah, A., and Krosstad, D. J. (1990) Chloroquine resistance not linked to mdr-like genes in a *Plasmodium falciparum* cross. *Nature* **345,** 253–255.

12

Quantitation of Liver-Stage Parasites by Automated TaqMan® Real-Time PCR

Adam A. Witney, Robert M. Anthony, and Daniel J. Carucci

1. Introduction

The liver stages of the malaria parasite have long been a difficult part of the life cycle to study. Very little is known about the localization of the parasite within the liver and no method for the estimation of liver parasite burden is in common use. Differences in liver parasite burden would provide a good measure of preerythrocytic malaria vaccine efficacy, which currently is measured indirectly by the counting of blood-stage parasites that develop following a sporozoite challenge. However, the measurement of blood-stage parasitemia is a slow, painstaking process requiring a high level of technical ability. In addition, a biological amplification of parasites occurs in hepatocytes, with one infected hepatocyte yielding approx 4000–40,000 merozoites, suggesting that counting blood-stage parasitemia may not be an accurate way of quantifying preerythrocytic stage vaccine efficacy. A direct method would, therefore, be preferable. Several methods have been described to attempt to examine liver-stage parasitemia directly, including the use of radioactive oligonucleotides to probe RNA blots *(1)*, high performance liquid chromatography (HPLC) *(2)*, or competitive polymerase chain reaction (PCR) to quantitate PCR products generated to specific parasite targets *(3)*. However, these methods are cumbersome and not conducive to high-throughput screening procedures. The TaqMan® real-time fluorescent PCR method described here is a fast and reproducible method for analyzing liver-stage parasitemia in rodent malaria models. The method has previously been used to investigate pre-erythrocytic malaria vaccine efficacy *(4)*.

2. Materials

2.1. Equipment

1. ABI Prism 7700 Sequence Detector (PE Applied Biosystems, Foster City, CA).
2. Omni EZ Connect Tissue Homogenizer with 10 mm sawtooth generator probe (95 mm length) (Omni International, VA).
3. 96-well optical plates.

2.2. Reagents

1. TaqMan® PCR Core Reagents (PE Applied Biosystems).
2. TaqMan® Rodent GAPDH Control Reagents (PE Applied Biosystems).
3. *P. yoelii* specific PCR primers: Py685F 5'-CTTGGCTCCGCCTCGATAT-3'; Py782R 5'-TCAAAGTAACGAGAGCCCAATG-3'.

4. 6-Carboxyfluorescein (FAM)-labeled 18S rRNA specific probe: 6FAM-CTGGCCCTT TGAGAGCCCACTGATT-TAMRA
5. Trizol (Life Technologies Inc., Gaithersburg, MD).
6. *DNase I* (Life Technologies Inc.).
7. Superscript II Reverse Transcriptase (Life Technologies Inc.).
8. RNAlater (Ambion Inc., Austin, TX).

3. Methods

3.1. Extraction of Total RNA from Whole Mouse Livers

1. Excise the liver and store in 10 mL of RNAlater. Livers can be stored under these conditions at 4°C for up to 1 wk (refer to manufacturer's protocol for additional storage details).
2. Remove liver from RNAlater solution and immediately homogenize for 90 s in 1 mL of Trizol per 50 mg of tissue weight (approx 15 mL). Wash the probe between liver samples by activating in 1 L of dH_2O, followed by 1 L of 0.1 M NaOH, followed by a further wash in 1 L of dH_2O.
3. The homogenate can be stored at –80°C for up to 1 mo.

3.2. Isolation of Total RNA from Liver Homogenates

1. Remove 1 mL of the liver homogenate into a 1.5-mL microtube.
2. Incubate at room temperature for 5 min.
3. Add 200 µL of chloroform.
4. Shake vigorously for 15 s.
5. Incubate at room temperature for 10 min.
6. Spin at 12,000g for 15 min at 4°C.
7. Remove the aqueous phase (~500 µL) to a fresh microtube.
8. Add 500 µL of isopropanol.
9. Incubate at room temperature for 10 min.
10. Spin at 12,000g for 10 min at 4°C.
11. Wash with 1 mL of 75% ethanol.
12. Spin at 7000g for 5 min at 4°C.
13. Air dry pellet (DO NOT spin dry under vacuum) for no longer than 10 min (*see* **Note 1**).
14. Resuspend in 200 µL of dH_2O.
15. Incubate at 55°C for 10 min.
16. Store at –80°C.

3.3. Synthesis of cDNA from Total RNA

1. Remove contaminating genomic DNA by treatment with *DNaseI*. In a 20-µL reaction volume, add approx 10 µg of total RNA and 2 µL of *DNaseI*.
2. Incubate at room temperature for 15 min.
3. Stop the reaction by adding 2 µL of 25 mM EDTA and incubating at 65°C for 10 min.
4. To a 10 µL aliquot of the DNase-treated RNA, add 1 µL of 500 ng/µL random hexamers and incubate at 70°C for 10 min.
5. Chill on ice for at least 1 min.
6. Add 4 µL of 5X first-strand buffer, 2 µL of 0.1M DTT, and 1 µL of 10 mM dNTPs.
7. Incubate at 42°C for 2 min.
8. Add 2 µL of Superscript II enzyme.
9. Incubate at 42°C for 50 min.
10. Inactivate by heating to 70°C for 10 min.
11. Add 180 µL of dH_2O to the sample.
12. Further dilute by adding 180 µL dH_2O to 20 µL of diluted sample, ready for real-time amplification.

3.4. Real-Time Quantitation of Target Material

1. Prepare a master mix containing the following material per well:
   a. 5X TaqMan® buffer A   4.0 µL
   b. 25 mM MgCl$_2$ 8.8 µL
   c. 10 mM dATP   0.8 µL
   d. 10 mM dCTP   0.8 µL
   e. 10 mM dGTP   0.8 µL
   f. 10 mM dUTP   0.8 µL
   g. GAPDH probe   0.2 µL
   h. rRNA probe   0.08 µL
   i. GAPDH for primer   0.32 µL
   j. GAPDH rev primer   0.08 µL
   k. rRNA for primer (Py685F) 0.12 µL
   l. rRNA rev primer (Py782R) 0.12 µL
   m. dH$_2$O   2.48 µL
   n. Taq Gold   0.2 µL
   o. UNG   0.4 µL
2. Set-up a 96-well optical plate containing four plasmid controls in triplicate each for GAPDH and rRNA plasmids, containing 10 pg, 1 pg, 100 fg and 10 fg (see **Note 2**). In the remaining wells, setup 20-µL aliquots of each sample in triplicate.
3. Aliquot 20 µL of the master mix into each well.
4. Set-up the ABI 7700 according to the manufacturer's protocols, with the machine detecting both FAM and VIC dyes in both wells.
5. Run the TaqMan® machine using the standard PCR conditions as stated in the manual (1 cycle of 50°C for 2 min, 95°C for 10 min; 50 cycles of 95°C for 15 s and 60°C for 1 min).

3.5. Analysis of Results

The threshold cycle (C_T) is the cycle at which the amplification curve for a particular target reaches a fluorescence that is greater than 10 standard deviations above the background fluorescence (measured as the mean fluorescence over cycles 3 to 15). Standard curves for each target are generated by plotting the C_T's of the four plasmid controls against the log of the amount of starting plasmid template. Control plasmid equivalents are deduced by measuring the corresponding amount of control plasmid for the C_T measured for each target within each sample. A measure of parasite burden can be calculated as the ratio of the parasite 18S rRNA to mouse GAPDH control plasmid equivalents. The efficacy of test vaccines can be deduced using standard statistical comparisons of test vaccine groups and control vaccine groups.

4. Notes

1. Overdrying the RNA pellets makes resuspension in dH$_2$O very difficult.
2. Plasmid controls are constructed by amplifying the mouse GAPDH and *P. yoelii* 18S rRNA target DNA sequences from their respective genomic DNA, and cloning into pCRScript (Stratagene) by standard protocols.

References

1. Arreaza, G., Corredor, V., and Zavala, F. (1991) *Plasmodium yoelii*: quantitation of the exoerythrocytic stages based on the use of ribosomal probes. *Exp. Parasitol.* **72,** 103–105
2. Hulier, E., Pétour, P., Snounou, G., Nivez, M., Miltgen, F., Mazier, D., and Rénia, L. (1996) A method for the quantitative assessment of malaria parasite development in organs of the mammalian host. *Mol. Biochem. Parasitol.* **77,** 127–135.
3. Briones, M. R. S., Tsuji, M., and Nussenzweig, V. (1996) The large difference in infectivity for mice of *Plasmodium berghei* and *Plasmodium yoelii* sporozoites cannot be correlated with their ability to enter into hepatocytes. *Mol. Biochem. Parasitol.* **77,** 7–17.
4. Witney, A. A., Doolan, D. L., Anthony, R. M., Weiss, W. R., Hoffman, S. L., and Carucci, D. J. (2001) Determining liver stage parasite burden by real time quantitative PCR as a method for evaluating pre-erythrocytic malaria vaccine efficacy. *Mol. Biochem. Parasitol.* **118,** 233–245.

13

Quantitation of Liver-Stage Parasites by Competitive RT-PCR

Kyle C. McKenna and Marcelo R. S. Briones

1. Introduction

1.1. Background

A standard method for evaluating inhibitory effects on preerythrocytic parasites in murine malaria models is to measure the time to patency of a blood-stage infection following a sporozoite challenge. The prevention of a blood-stage infection or a delay in patency suggests that preerythrocytic parasites have been eliminated or that liver-stage development has been inhibited. However, susceptibility to a challenge with blood-stage parasites must also be determined to conclude that preerythrocytic parasites and not blood-stage parasites were targeted. As a more direct method, liver-parasite burden can be measured. Due to the paucity of liver-stage parasites at physiological challenge doses, <100 sporozoites *(1)*, liver-stage parasites are very difficult to detect by conventional methods, such as immunofluorescence or Giemsa staining of parasites in liver sections, unless a very large challenge dose is used ($>1.0 \times 10^6$ sporozoites) *(2)*. Molecular techniques that target *Plasmodium*-specific small subunit rRNA are much more sensitive, requiring less than 5.0×10^4 sporozoites to quantitate liver stages *(3,4)*. *Plasmodium* rRNA has been now measured in total liver RNA by spot-blot hybridization with *Plasmodium*-specific rRNA probes *(3)* and as described here by quantitative-competitive reverse transcriptase–polymerase chain reaction (RT-PCR) *(4)*.

1.2. Principle of Quantitative-Competitive RT-PCR

The principle of quantitative-competitive RT-PCR is to amplify an unknown amount of a target sequence from a fixed amount of cDNA while simultaneously amplifying a sequence from a known amount of an added competitor plasmid that competes for available primers in the PCR reaction *(5)*. Target and competitor plasmid amplicons are resolved on ethidium-bromide stained agarose gels and are distinguished by size (**Fig. 1A**). As both the target sequence and the competitor sequence use the same primer set for PCR amplification, under PCR conditions where primers are limiting, target amplification varies inversely with increasing amounts of competitor plasmid added to the PCR reaction (**Fig. 1A**). By determining the ratio of amplified target to amplified competitor from a series of PCR amplifications with different concentrations of com-

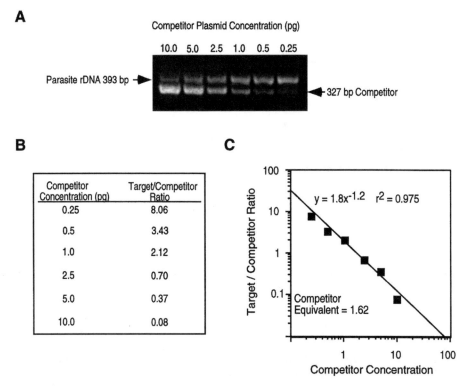

Fig. 1. Measurement of *P. yoelii* small-subunit rRNA from total liver RNA of a mouse challenged with 5.0×10^4 sporozoites by quantitative-competitive RT-PCR. (**A**) PCR amplification of *Plasmodium*-specific rDNA from total liver cDNA with the addition of titrated concentrations of *Plasmodium* rDNA competitor plasmid visualized on an ethidium bromide-stained 2% agarose gel. (**B**) Target to competitor amplicon ratios from individual competitive-PCR reactions with the indicated concentrations of competitor plasmid as determined by quantification of parasite rDNA and competitor plasmid amplicons by NIH Image. (**C**) Linear regression analysis for the determination of competitor equivalent.

petitor plasmid (**Fig. 1B**), a competitor plasmid concentration can be determined where the target and competitor plasmid amplification are equivalent, a target to competitor ratio = 1 (**Fig. 1C**). This concentration, termed competitor equivalent, provides a quantitative value for the target molecule in the cDNA sample.

1.3. Plasmodium *rDNA Competitor Plasmid Construction*

The parasite rDNA competitor sequence consists of a wild-type parasite rDNA template that has been genetically engineered with the removal of 66 bp of internal sequence. To generate the smaller competitor sequence, a wild-type parasite rDNA sequence was first amplified from cDNA of a sporozoite-challenged mouse with *Plasmodium* rRNA-specific primers PB1 and PB2 and then cloned into pUC119 (**Fig. 2A**). Primer PB1 corresponds to position 1563–1587 and primer PB2 corresponds to position 1926–1950 of the *P. berghei* A type 18s rRNA *(6)*. The plasmid containing the wild-type sequence was then used as a template in a second PCR with primers PB2 and PB3 (**Fig. 2B**). Primer PB3 consists of primer PB1 plus 25 bp of parasite rDNA sequence found 66 bp 3' of primer PB1 (position 1653–1678). Consequently, 66 bp of

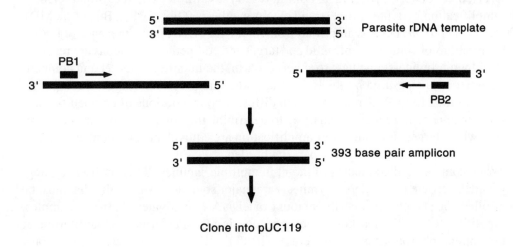

Fig. 2. Schematic of *Plasmodium* rDNA competitor construction.

Plasmodium rDNA sequence are looped-out during primer annealing and a shorter parasite rDNA amplicon is generated during replication of the parasite rDNA template. This competitor amplicon was cloned into pUC119 for use in quantitative-competitive RT-PCR of *Plasmodium berghei* or *Plasmodium yoelii* liver stages.

1.4. The Use of Quantitative-Competitive RT-PCR for the Quantification of Plasmodium *Liver Stages*

Quantification of *Plasmodium* rRNA from total liver RNA by quantitative-competitive RT-PCR involves four major procedures. First, total RNA is isolated from whole liver by a modified Chomczynski and Sacchi technique *(7)*. Second, total liver RNA is converted to cDNA utilizing a commercially available kit, the Superscript™ preamplification system for first strand cDNA synthesis (Gibco-BRL, Bethesda, MD). Third, *Plasmodium* rDNA is amplified from total liver cDNA in the presence of known concentrations of competitor plasmid and target and competitor amplicons are resolved on ethidium bromide-stained agarose gels. Fourth, the intensity of target and competitor amplicons is measured by the NIH Image software program. Target to competitor amplicon ratios from PCR reactions with different concentrations of competitor plasmid are then used in regression analysis to determine the competitor plasmid concentration where target and competitor amplification are equivalent. This concentration is used as a relative measure of liver-parasite burden.

When comparing the amounts of target in multiple samples, RNA purity and integrity, and the efficiency of cDNA synthesis are major sources of variability that must be controlled for. To ensure that the amount of cDNA is equivalent between samples, competitive RT-PCR of a housekeeping gene is also performed. Quantitation of hypoxanthine phosphoribosyl transferase (HPRT) mRNA has been used for this purpose *(4,8)*. The HPRT competitor sequence is part of the polycompetitor plasmid (pPQRS) used to quantitate mouse cytokine expression *(9)*.

2. Materials

2.1. Equipment

1. Tenbroeck tissue grinder (VWR Scientific, S. Plainfield, NJ).
2. 60-mm diameter Petri dishes.
3. Scissors.
4. 2- and 0.5-mL microcentrifuge tubes.
5. Eagle Eye II Still Video System (Stratagene, La Jolla, CA).
6. NIH Image software program.

2.2. Total Liver RNA Isolation Reagents

1. Phosphate-buffered saline (PBS), pH 7.4.
2. Stock denaturing solution: 4 M guanidinium thiocyanate, 25 mM sodium citrate, pH 7.0, 0.5% *n*-lauryl sarcosomine.
3. Working denaturing solution: stock denaturing solution supplemented with 0.1 M 2-mercaptoethanol.
4. Trizol® reagent (Gibco-BRL, Grand Island, NY).
5. Chloroform.
6. Isopropanol.
7. Ethanol.
8. Diethyl pyrocarbonate (DEPC)-treated water (*see* **Note 1**).
9. 2 M sodium acetate, pH 5.5.

2.3. cDNA Synthesis Reagents

1. RNase-free, DNase I, 1 U/µL (Boehringer Mannheim, Mannheim, Germany).
2. DNase I reaction buffer (10X): 200 mM Tris-HCl (pH 8.4), 500 mM KCl, 20 mM MgCl$_2$.
3. RNase H, 2 U/µL.
4. 25 mM EDTA.
5. Random hexamers, 50 ng/µL.
6. PCR buffer (10X): 200 mM Tris-HCl (pH 8.4), 500 mM KCl.

Competitive RT-PCR

7. 25 mM MgCl$_2$.
8. 10 mM dNTP.
9. 0.1 M DTT.
10. Superscript™ preamplification system for first strand cDNA synthesis (Gibco-BRL).
11. DEPC-treated water.
12. 70% ethanol.

2.4. Competitive-PCR Reagents

1. PCR Supermix (Gibco-BRL).
2. *Taq* DNA polymerase (Sigma, St Louis, MO).
3. Parasite rDNA-specific primers:

    PB1: 5' AGG ATG TAT TCG CTT TAT TTA ATG C 3'
    PB2: 5' TCT TGT CCA AAC AAT TCA TCA TAT C 3'
 PB3: 5' AGG ATG TAT TCG CTT TAT TTA ATG CTT AGA TAT ACT AGG CTG CAC GCG TG 3'

4. 6X DNA loading buffer (commercially available).

3. Methods

3.1. Total Liver RNA Isolation

1. Euthanize mouse by cervical dislocation.
2. Remove the liver. Dissect away the gallbladder from the liver and discard the gallbladder.
3. Place the liver in a Petri dish containing PBS, pH 7.4. Gently move the liver with forceps through the PBS buffer to wash the liver. Remove the liver to a second Petri dish containing PBS, pH 7.4 and repeat wash. Remove the liver to a third Petri dish without PBS.
4. Mince the liver with scissors and then add the minced liver to an iced Tenbroeck tissue grinder.
5. Add 4 mL of working denaturing solution (*see* **Note 2**).
6. Homogenize the liver in the working denaturing solution. Keep on ice.
7. Remove 400 µL of the liver homogenate to a 2.0-mL microcentrifuge tube (*see* **Note 3**).
8. Add 1.2 mL of the Trizol reagent. Vortex and then leave at room temperature for 5 min.
9. Centrifuge at 12,000g for 10 min at 4°C.
10. Remove 1.0 mL of the supernatant to a fresh 2.0-mL microcentrifuge tube.
11. Add 320 µL of chloroform to the supernatant. Vortex and then leave at room temperature for 2–15 min.
12. Centrifuge at 12,000g for 10 min at 4°C.
13. Transfer 0.5 mL of the aqueous layer to a new 2.0-mL microcentrifuge tube.
14. Add 800 µL of isopropanol, kept at room temperature, to the aqueous layer to precipitate the RNA from solution. Place on a rocker at room temperature for 5–10 min.
15. Centrifuge at 12,000g for 10 min at 4°C.
16. Aspirate the supernatant and save the pellet.
17. Wash the pellet with 0.5 mL of ice-cold 70% ethanol. Vortex.
18. Centrifuge at 12,000g for 10 min at 4°C.
19. Aspirate the supernatant and then dry the pellet by placing the microcentrifuge tube open and under the hood until the pellet becomes translucent. Drying of the pellet normally takes from 30 min to 1 h.
20. Resuspend the pellet in 250 µL of DEPC-treated water. Keep on ice.
21. Determine the concentration of isolated RNA by absorbance of UV light at 260-nm wavelength. Typically, a 1:100 dilution is made from the RNA solution in DEPC-treated water for determining the concentration (*see* **Note 4**).

22. For storage of RNA, reprecipitate the RNA solution by adding 1/10 vol of 2 M sodium acetate, pH 5.5 (25 µL), and then 2.5 vol of 100% ice-cold ethanol (625 µL). Vortex and store solution at –20°C.

3.2. cDNA Synthesis Reaction

1. Vortex RNA precipitate stored in ethanol vigorously to homogenize precipitate.
2. Aliquot 10 µg of RNA stored in ethanol into a 0.5-mL microcentrifuge tube.
3. Centrifuge at 12,000g for 30 min at 4°C.
4. Aspirate the supernatant and then wash the pellet with 0.5 mL of ice-cold 70% ethanol.
5. Centrifuge at 12,000g for 10 min at 4°C.
6. Aspirate the supernatant and then dry the pellet by placing the microcentrifuge tube open and under the hood until the pellet becomes translucent. Drying of the pellet normally takes 5–10 min.
7. Resuspend the pellet in 10 µL of DEPC-treated water. Vortex and then keep on ice.
8. In a new 0.5-mL microcentrifuge tube, add the following reagents made as a master mix:
    a. 6 µL of DEPC-treated water.
    b. 1 µL of 10X DNase I reaction buffer.
    c. 1 µL of DNase I (1 U/mL).
9. Add 2 µg of RNA (2 µL).
10. Incubate at room temperature for 15 min.
11. Stop the DNase I reaction by adding 1 µL of 25 mM ethylenediaminetetraacetic acid (EDTA) and then incubating the reaction at 65°C for 15 min. Keep the reaction tube on ice after incubation.
12. To the reaction tube, add the following reagents made as a master mix:
    a. 9 µL of DEPC-treated water.
    b. 5 µL of random hexamers (50 ng/mL).
13. Incubate reaction tube at 70°C for 10 min and then immediately place on ice at least 1 min.
14. To the reaction tube, add the following reagents made as a master mix:
    a. 4 µL of 10X PCR buffer.
    b. 4 µL of 25 mM MgCl$_2$.
    c. 2 µL of 10 mM dNTP.
    d. 4 µL of 0.1 M DTT.
15. Incubate at room temperature for 5 min.
16. Add 2 µL of Superscript™ enzyme (200 U/µL).
17. Place in a thermocycler programmed as follows:
    Step 1: 25°C, 10 min.
    Step 2: 42°C, 50 min.
    Step 3: 70°C, 15 min.
18. Remove from the thermocycler and briefly centrifuge the reaction tube.
19. Add 2 µL of *Escherichia coli* RNase H (2 U/µL).
20. Incubate reaction tube at 37°C for 20 min.
21. After incubation, store reaction tube at –20°C until use.

3.3. Competitive-PCR Amplification

1. To a microcentrifuge tube, add the following reagents made as a master mix:
    a. 46 µL of PCR Supermix (Gibco-BRL).
    b. 1 µL of 600 mM PB1 primer (12 mM final concentration).
    c. 1 µL of 600 mM PB2 primer (12 mM final concentration).
    d. 0.2 µL of *Taq* DNA polymerase (5 U/µL).
    e. 1 µL of competitor plasmid of known concentration.

Competitive RT-PCR

2. Add 5 µL of the total liver cDNA mixture to the reaction tube.
3. Place reaction tube in a thermocycler programmed as follows:
   Step 1: 94°C, 2 min.
   Step 2: 94°C, 1 min.
   Step 3: 60°C, 2 min for 35 cycles.
   Step 4: 72°C, 1 min for 35 cycles.
   Step 5: 72°C, 10 min for 35 cycles.
4. Briefly centrifuge the reaction tube and then concentrate the PCR sample to ~10 µL by placing in a speed vacuum at medium heat. Concentrating the PCR sample normally takes 30 min.
5. Add 2.5 µL of 6X DNA loading buffer and then run the entire sample on a 2% 1X Tris-acetate (TAE) agarose gel to resolve the target and competitor amplicons. The target amplicon appears as a 393 base-pair fragment and the competitor amplicon appears as a 327 base-pair fragment.
6. Take an electronic image of the gel using the Eagle Eye II Still Video System (Stratagene) and store the image onto a disk.

3.4. Quantitation of Target and Competitor Amplicons by NIH Image

1. Once the NIH Image program has been accessed (*see* **Note 5**), open the file containing the electronic image of the agarose gel to be quantified by choosing OPEN from the FILE menu on the top of the screen.
2. The electronic image of the gel will now appear on the screen. Invert the image by selecting INVERT from the EDIT menu on the top of the screen. Target and competitor bands on the displayed image will become black on a white background. Now select THRESHOLD from the OPTIONS menu at the top of the screen. Adjust the threshold so that the target and competitor bands can be clearly discerned from one another. This is accomplished by clicking and dragging the mouse on the pixel gradient (LUT) found on the left-hand side of the screen (*see* **Note 6**).
3. With the threshold now set, select ANALYZE PARTICLES from the ANALYZE menu on the top of the screen. The PARTICLE ANALYSIS OPTION window will appear, allowing you to set values to define the number of pixels that distinguish a band from background. After setting these values, select OK. The image will now appear with a number on each band that was quantified.
4. To view the results of the quantified bands, select SHOW RESULTS from the ANALYZE menu on the top of the screen. A window will appear showing the area and the density of the pixels for each numbered band. By multiplying the area by the density, a quantitative value is given to the band. Divide the target and competitor amplicon values to determine a target to competitor ratio for each respective competitor concentration.

4. Notes

1. Care should be taken when working with RNA to prevent RNA degradation by RNases. Microcentrifuge tubes, and pipet tips should be purchased as RNase-free and used only for RNA work. Gloves should be worn and changed routinely. When possible, solutions should be treated with DEPC remove RNases before use.
2. The stock denaturing solution may be stored for up to 3 mo and should be prepared with autoclaved DEPC-treated water under a fume hood. The working solution is made fresh prior to each RNA isolation.
3. This step may be used as a stopping point. The liver homogenate may be stored at –70°C for later use. When RNA is to be isolated, simply thaw the homogenate on ice and then continue with the procedure.
4. The RNA concentration measured in micrograms per microliter is determined as followed:

$$(OD_{260} \times \text{dilution factor} \times 40)/1000$$

5. NIH Image was developed at the United States National Institutes of Health and is a part of the public domain. The program is available at no cost and can downloaded off the Internet in both Macintosh and PC formats at http://rsb.info.nih.gov/nih-image/
6. If while setting the threshold in NIH Image the background is black and the bands are white, the image of the gel must first be inverted prior to being opened in NIH Image. The Eagle Eye II Still Video system will store the electronic image of the gel as a TIFF file. Using Adobe PhotoShop invert the image so that bands appear black on a white background and then store the image as a pict file. Open the PICT file in NIH Image and proceed directly to setting the threshold.

References

1. Rosenberg, R., Wirtz, R. A., Schneider, I., and Burge, R. (1990) An estimation of the number of malaria sporozoites ejected by a feeding mosquito. *Trans. Soc. Trop. Med. Hyg.* **84,** 209–212.
2. Scheller, L. F., Stump, K. C., and Azad, A. F. (1995) *Plasmodium berghei*: production and quantification of hepatic stages derived from irradiated-sporozoites in rats and mice. *J. Parasitol.* **81,** 103–105.
3. Arreaza, G., Corredor, V., and Zavala, F. (1991) *Plasmodium yoelii*: quantification of exoerythrocytic stages based on the use of ribosomal RNA probes. *Exp. Parasitol.* **72,** 103–105.
4. Briones, M. R. S., Tsuji, M., and Nussenzweig, V. (1996) The large difference in infectivity for mice of *P. berghei* and *P. yoelii* sporozoites can not be correlated with their ability to enter hepatocytes. *Mol. Biochem. Parasitol.* **77,** 7–17.
5. Gause, W. C. and Adamovicz, J. (1995) Use of PCR to quantitate relative differences in gene expression, in *PCR Primer: A Laboratory Manual* (Dieffenbach, C. W., and Dveksler, G. S., eds.), Cold Spring Harbor Laboratory Press, Cold Spring Harbor, NY, pp. 308–309.
6. Gunderson, J. H., McCutcheon, T. F., and Sogin, M. L. (1986) Sequence of the small subunit ribosomal RNA gene expressed in the bloodstream stages of *Plasmodium berghei*: evolutionary implications. *J. Protozool.* **33,** 525–529.
7. Chomczynski, P. and Sacchi, N. (1987) Single-step method of RNA isolation by acid guanidinium thiocyanate-phenol chloroform extraction. *Anal. Biochem.* **162,** 156–159.
8. Gantt, S. M., Myung, J. M., Briones, M. R. S., Corey, E. J., Omura, S., Nussenzweig, V., and Sinnis, P. (1998) Proteasome inhibitors block development of *Plasmodium spp. Antimicrob. Agents Chemother.* **42,** 2731–2738.
9. Reiner, S. C., Zheng, S., Corry, D. B., and Locksley, R. M. (1993) Constructing polycompetitor cDNAs for quantitative PCR. *J. Immunol. Meth.* **165,** 37–46.

IV

MOLECULAR BIOLOGY TECHNIQUES

14

Extraction and Purification of *Plasmodium* Total RNA

Till Voss

1. Introduction

The isolation of RNA and its subsequent characterization by various applications, such as the generation of cDNA libraries, Northern blot analysis, and reverse transcriptase polymerase chain reaction studies (RT-PCR), are fundamental for understanding the mechanisms underlying gene expression. In order to obtain starting material of high quality, it is crucial to isolate pure and intact RNA from living cells. Effective cellular lysis and membrane disruption, inhibition of endogenous ribonuclease activity, deproteinization of the lysate, recovery of intact RNA, and storage of RNA are the most important factors to be considered when preparing RNA.

This chapter describes a method of extracting and purifying *Plasmodium falciparum* total RNA meeting these demands. The procedure can be accomplished within 4 or 5 h yielding high-quality RNA. Like most current methods of RNA isolation, the protocol presented here is based on the single-step acid guanidinium thiocyanate (GTC)–phenol–chloroform RNA extraction method described by Chomczynski and Sacchi *(1)*. After harvesting parasites from in vitro cultures, cells are homogenized in a GTC-extraction buffer, which, with its highly denaturing and chaotropic (biologically disruptive) properties, disrupts cells very rapidly while completely inactivating ribonucleases *(2)* (*see* **Note 1**). Subsequent extraction of this lysate with acidic phenol:chloroform followed by centrifugation separates RNA from proteins and genomic DNA. Total RNA (nuclear and cytoplasmic) is precipitated from the aqueous supernatant with isopropanol, and the RNA pellet resuspended in storage buffer. The quality of the isolated RNA, in terms of integrity, purity, and yield, can then be determined by agarose gel electrophoresis and spectrophotometry. The purified RNA is suitable for a variety of RNA applications; reference for further processing of total RNA is provided in **Subheadings 3.6.** and **3.7.**

Separation of DNA from RNA by acidic phenol:chloroform extraction always leaves traces of DNA behind in the RNA fraction. This can be critical in RNA applications such as RT-PCR and nuclease protection assays, and therefore removal of contaminating genomic DNA from total RNA samples is discussed in detail.

2. Materials
2.1. Equipment

1. Bench centrifuge with a swing-out rotor that can be cooled to 4°C. The indications given in this chapter refer to the Jouan CR 4 11 centrifuge using rotor E4. In order to guarantee appropriate speeds using other centrifuges, gravity (g) forces are given.
2. Sterile and *RNase*-free 15-mL and 50-mL screw-capped polypropylene tubes (e.g., Falcon) for use in a bench centrifuge.
3. Table-top microcentrifuge at 4°C.
4. Sterile and *RNase*-free microcentrifuge tubes (*see* **Note 2**).
5. Pasteur pipets, plugged pipet tips, and plastic pipets.
6. Heating block set at 55°C.
7. Automatic pipetor.
8. Agarose gel electrophoresis equipment.
9. Spectrophotometer
10. *RNase*-free bench space dedicated to RNA work.

2.2. Reagents

1. 0.05% (w/v) saponin (Sigma, cat. no. 8047-15-2) dissolved in parasite culture medium (without $NaHCO_3$ and albumax/serum).
2. RNA extraction buffer *(4)*: 4 M guanidine isothiocyanate (GTC), 25 mM sodium citrate (pH 7.0), 0.5% *N*-lauroylsarcosine (Sarkosyl, Fluka, cat. no. 61743 [137-16-6]), 0.1 M 2-mercaptoethanol (2-ME). Prepare a stock solution by mixing 100 mL of dH_2O, 6 mL of 0.75 M sodium citrate (pH 7.0), and 9 mL of 10% (w/v) *N*-lauroylsarcosine. Add 85.3 g of GTC and stir at room temperature to dissolve. Store stock solution at room temperature. Before use, add 70 µL of 2-ME to 10 mL of stock solution to obtain the RNA extraction buffer. Prepared buffer can be stored up to 1 mo at room temperature.
3. 2 M sodium acetate (pH 4.0): Add 8.21 g of sodium acetate (anhydrous) to 20 mL of dH_2O and adjust the pH to 4.0 with glacial acetic acid. Add dH_2O to a final volume of 50 mL.
4. H_2O-saturated acidic phenol (it is not necessary to adjust the pH). Store aliquots of H_2O-saturated phenol in polypropylene tubes at –20°C. After thawing, store phenol aliquots at 4°C in a lightproof container for no longer than 2 wk (*see* **Note 3**).
5. Chloroform:isoamylalcohol (49:1).
6. Isopropanol.
7. Diethyl pyrocarbonate (DEPC)-treated H_2O. Add DEPC to solutions at a concentration of 0.1% (v/v) and incubate for several hours to overnight at room temperature. Note that it is absolutely necessary to autoclave DEPC-treated solutions in order to destroy the reagent that can also react with RNA. **WARNING:** DEPC is a suspected carcinogen, so always wear gloves and work in a fume hood when pipetting DEPC (*see* **Note 4**).
8. 70% ethanol (prepared with DEPC-treated dH_2O).
9. Formamide (e.g., Gibco-BRL)
10. Formamide gel-loading buffer: Mix 8 mL of formamide, 2 mL of Tris-EDTA (TE) and add a tip of a spatula each of bromophenol blue and xylenecyanol.
11. *RNase*-free 5 M ammonium acetate. This solution can either be treated with DEPC and autoclaved, or purchased (e.g., Ambion).

3. Methods
3.1. Harvesting of Parasites

This protocol is designed for the isolation of *P. falciparum* total RNA from a 10 mL culture (5% hematocrit). When working with larger or smaller culture volumes, adjust the reagent volumes accordingly.

1. Transfer the parasite culture from the culture dish or flask into a sterile screw-capped polypropylene tube (e.g., Falcon tube) and spin for 5 min at 800g at room temperature in a bench centrifuge (swing-out rotor).
2. Carefully take off the supernatant, resuspend the erythrocyte pellet in 2.5 mL (5 vol) of ice-cold 0.05% saponin (*see* **Subheading 2.2.**), and allow red blood cell (RBC) lysis to proceed for 10 min on ice. Pellet the released parasites at 4°C for 5 min at 3500g in a bench centrifuge (*see* **Note 5**).
3. Carefully remove the supernatant with a Pasteur pipet, optimally using an automatic pipetor. Make sure not to disturb the parasite pellet; it is advisable to leave behind a small volume of supernatant.

3.2. RNA Extraction

Contamination with exogenous *RNases* often is responsible for obtaining degraded RNA. Therefore, dedicate all reagents solely for RNA extraction and analysis, and always use *RNase*-free plugged pipet tips and pipets when working with these reagents. Furthermore, since *RNases* are present almost everywhere (fingertips, door knobs, centrifuge switches, and so on), always wear gloves when handling RNA samples and reagents. Change gloves frequently.

1. Resuspend parasites in 1 mL of RNA extraction buffer (2 vol of initial RBC pellet) by rapidly but gently pipetting up and down several times until the lysate becomes homogeneous (*see* **Note 6**). Homogenization should be achieved as quickly as possible, since only rapid mixing of the denaturing RNA extraction buffer with cell components will result in an immediate inactivation of endogenous *RNases*.
2. Add 100 µL (1/10 vol) of 2 M Na-acetate pH 4.0 and mix gently by inverting the tube 10 times.
3. Add 1.2 mL (approx 1 vol) of H_2O-saturated acidic phenol and mix thoroughly by inverting the tube 20 times (*see* **Note 7**).
4. Add 800 µL (approx 1/3 vol) of chloroform:isoamylalcohol (49:1) and mix thoroughly by inverting the tube 20 times. Place the tube on ice for 5 min.
5. Spin at 3500g at 4°C for 30 min.
6. Transfer 800 µL of the aqueous supernatant (*see* **Note 8**) to a *RNase*-free microcentrifuge tube containing 700 µL of isopropanol (7/8 vol), and precipitate RNA at –20°C for 1 h (*see* **Notes 9** and **10**).
7. Pellet the RNA at 15,000g at 4°C in a table-top microcentrifuge for 20 min.
8. Gently pour off the supernatant and dry the RNA pellet for 5 min by placing the open tube upside-down on tissue paper (*see* **Note 11**).
9. Resuspend the pellet in 450 µL of RNA extraction buffer (*see* **Notes 12** and **13**).
10. Add 50 µL (1/10 vol) of 2 M sodium acetate (pH 4.0), mix and add 500 µL (1 vol) of acid H_2O-saturated phenol. Mix thoroughly by inverting the tube 20 times.
11. Add 330 µL (1/3 vol) of chloroform:isoamylalcohol (49:1), and mix thoroughly by inverting the tube 20 times.
12. Spin in a microcentrifuge at 15,000g at 4°C for 10 min.
13. Take off the supernatant and precipitate RNA by the addition of approx 7/8 vol of isopropanol. Incubate at –20°C for 1 h.
14. Pellet the RNA at 15,000g at 4°C in a tabletop microcentrifuge for 20 min.
15. Wash RNA pellet once in 70% ethanol (prepared with DEPC-treated dH_2O) and spin again at 15,000g for 5 min at 4°C. Gently pour off the ethanol wash solution and air-dry the RNA pellet for 5 min.
16. Resuspend RNA in an appropriate volume of formamide (approx 0.5–2 µg RNA/µL formamide) (*see* **Note 14**), and incubate at 55°C for 5 min to aid resuspension of RNA (*see* **Note 15**). Briefly vortex the tube gently.

3.3. Quality Control of Total RNA Preparations

3.3.1. Analysis of RNA by Agarose Gel Electrophoresis

1. Prepare a standard 1.2% agarose gel and pre-run the gel at 5–6 V/cm for 15 min before loading the RNA sample.
2. Mix 1 µL of 0.5–2 µg total RNA with 4 µL of RNA loading buffer, and incubate at 65°C for 5 min to denature the RNA. Cool the tube on ice and load the sample on the prepared agarose gel.
3. Electrophorese the RNA sample at 5–6 V/cm until the bromophenol blue dye-front has migrated along half the length of the gel. Include a sample of a DNA size marker of known concentration on the same gel (*see* **Note 16**).
4. Stain the gel for 10 min in an *RNase*-free ethidium bromide solution (1 mg/mL in 1X Tris-borate-EDTA [TBE]). Inspect the gel under UV light for degradation and/or contamination with genomic DNA (*see* **Fig. 1** and **Note 17**).

3.3.2. Spectrophotometric Quantitation of RNA

Dilute an aliquot of total RNA 1:50 in TE and measure the absorbance at 260 nm. Dilute an equal volume of formamide 1:50 in TE and use this as the reference sample. An RNA concentration of 40 µg/mL will give an absorbance value of $OD_{260} = 1.0$. The RNA concentration of the original sample can easily be determined by multiplication of the calculated value by the dilution factor (*see* **Note 18**).

3.4. Storage of RNA

Store RNA in formamide at –20°C. Under these conditions, RNA is stable for many years *(4)* (*see* **Note 19**). In case of isolating RNA from larger culture volumes, store RNA in several aliquots since repeated thawing and freezing may affect the quality of RNA. It is worth resuspending your total RNA in formamide even if you are planning to process the RNA straight away. It is essential to make sure that the RNA preparation is of good quality before performing further applications. Additionally, precipitation of RNA from formamide is a further purification step to get rid of salts carried over from parasite lysis.

3.5. Precipitation of RNA from Formamide

1. Dilute RNA-formamide solution with 2 vol of DEPC-treated dH_2O (if you are precipitating less than 10 µL of RNA, make volume up to 30 µL with DEPC-treated dH_2O). Add 1/10 vol of *RNase*-free 5 *M* ammonium acetate and 3 vol of ethanol, mix and incubate at –20°C for 1 h.
2. Spin at 15,000g at 4°C for 20 min. Take off the supernatant and wash the pellet with 500 µL of 70% ethanol (prepared with DEPC-treated H_2O).
3. Carefully decant the ethanol wash solution, let the pellet dry for 5 min and resuspend total RNA in an appropriate volume of DEPC-treated H_2O (*see* **Note 20**). Incubate at 55°C for 5 min and gently vortex the tube. Collect the liquid at the bottom of the tube by centrifugation for 5 s, place the tube on ice, and rapidly proceed with the planned RNA processing protocol.

3.6. Treatment of Total RNA with RNase-Free Dnase for RT-PCR Studies and Nuclease Protection Assays

Although the repeated acidic phenol:chloroform extraction of total RNA is a very effective means of removing genomic DNA from the RNA preparation, complete separation of DNA from RNA is never achieved. Minimal amounts of contaminating genomic

Extraction and Purification of P. falciparum Total RNA

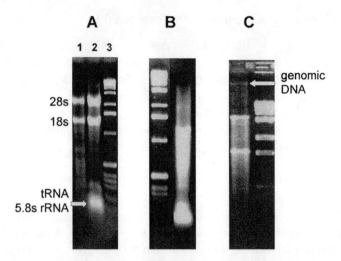

Fig. 1. *P. falciparum* total RNA separated on 1.2% agarose gels. **(A)** The sharp definition of the two 28s and 18s rRNA species, and the minimal smearing above, in between and below the rRNA signals in both RNA preparations are indicative for high quality total RNA. Total ring stage RNA (*lane 1*), total trophozoite stage RNA (*lane 2*), 1 kb DNA ladder (Gibco-BRL) (*lane 3*). **(B)** Partially degraded trophozoite stage total RNA. Note the intense smearing of degraded RNA below the 18s rRNA band, and the absence of the larger intact 28s rRNA molecules. **(C)** Slightly degraded total ring stage RNA with genomic contamination. Genomic DNA appears as a sharp high-molecular weight band on top of the lane.

DNA readily lead to false-positive products in nuclease protection assays and RT-PCR. It is therefore essential to completely remove genomic DNA templates from RNA preparations to obtain reliable results from these applications, especially when working with transcripts of low abundancy.

We routinely use RQ1 *RNase*-free *DNase* (Promega) and perform the *DNase* treatment according to the instructions of the supplier *(5)*, making a few alterations to the protocol:

1. Use 1 U of RQ1 *DNase* per 10 µg of RNA.
2. Add 1 µL (40 U) of *RNase* inhibitor rRNAsin (Promega).
3. For *DNase* treatment of less than 50 µg of RNA perform the reaction in a total volume of 50 µL. For larger amounts of RNA, do not exceed an RNA concentration of 1 µg/mL reaction volume
4. Extract the DNAse-treated RNA with 1 vol of acidic H_2O-saturated phenol and 1/3 vol of chloroform:isoamylalcohol (49:1). Precipitate the aqueous phase with 1/10 vol of *RNase*-free 5 *M* ammonium acetate and 3 vol of ethanol, and incubate at –20°C for 2 h. Spin the RNA at 15,000*g* at 4°C for 20 min in a microcentrifuge, wash the pellet once with 500 µL of 70% ethanol (prepared with DEPC-treated H_2O), and resuspend the pellet in an appropriate volume of DEPC-treated H_2O.
5. Proceed with your intended RNA processing without delay.

3.7. Other RNA Applications

3.7.1. Northern Analysis

RNA dissolved in formamide (*see* **Subheading 3.2., step 15**) can be directly used in Northern analysis.

3.7.2. Isolation of Poly(A+)-RNA from Total RNA

We have found that yield and quality of poly(A+)-RNA purified on oligo(dT)-cellulose from total RNA is indisputably better than poly(A+)-RNA directly extracted from intact parasites. We routinely use the Micro-FastTrack™ 2.0 Kit (Invitrogen) for purification of polyadenylated mRNA. However, kits of other suppliers may work equally well. Precipitate total RNA (*see* **Subheading 3.5.**) and proceed according to the manufacturer's instructions. Briefly, poly(A+)-mRNA hybridizes to the oligo(dT)-cellulose and is subsequently recovered after washing off poly(A–)-RNA species by low-salt elution.

4. Notes

1. Guanidinium salts are among the most effective protein denaturants (*6*).
2. We routinely use microcentrifuge tubes from unopened bags as provided by the supplier. Wear fresh gloves while removing the tubes from the bag, immediately close the lid of the tubes, and set these tubes aside for RNA extraction and analysis. It is not necessary to autoclave the microcentrifuge tubes.
3. Thawing of phenol at room temperature takes some time. Therefore, remove the phenol from –20°C at least 1 h before use.
4. DEPC reacts with histidine residues of proteins and will efficiently inactivate *RNases* (*7*). Compounds containing primary amine groups, such as Tris-HCl, will also react with DEPC and thus should not be treated with DEPC (*8*).
5. Alternatively, 10 vol of 0.025% saponin can be used to lyse the RBCs. With this procedure, the parasite pellet after centrifugation will be more visible.
6. Using less extraction buffer for parasite lysis will be sufficient to completely destroy the cells. However, larger lysate volumes will yield a clear aqueous supernatant after acid phenol:chloroform extraction, and will remove genomic DNA more efficiently.
7. The use of H_2O-saturated acidic phenol together with acidic 2 *M* sodium acetate (pH 4.0) added to the extraction buffer causes DNA and proteins to accumulate in the organic phase and interphase while RNA is retained in the aqueous supernatant.
8. Be careful not to touch the interphase when removing the aqueous supernatant to avoid contamination of the RNA fraction with genomic DNA. It is advisable to leave behind a small volume of supernatant, hence accepting a slightly lower yield of total RNA.
9. If isolating RNA from large culture volumes, RNA may also be precipitated and pelleted in 50 mL centrifuge tubes at 3,500 *g* for 40 min in a bench centrifuge in order to avoid the handling of dozens of microcentrifuge tubes.
10. When isolating RNA from cultures at low parasitemias or from blood samples, add 5 μL of an *RNase*-free glycogen (5 μg/μL). Glycogen acts as a carrier to aid efficient precipitation of RNA at low concentrations. Furthermore, after centrifugation the RNA pellet becomes visible.
11. Do not let dry the pellet completely or it will be difficult to resuspend the RNA.
12. **Step 9** introduces a second acidic phenol:chloroform extraction in order to efficiently remove genomic DNA from the RNA preparation. However, if minor DNA contamination is of no concern for your subsequent RNA application, the investigator may directly proceed to **step 14** of the protocol (without drying the RNA pellet in **step 8**).
13. At this step, always resuspend the RNA pellet in 450 μL of extraction buffer, irrespective of the initial culture volume.
14. Expect 5–10 μg of total RNA from a 10 mL culture (6–8% parasitemia) containing ring-stage parasites only. For asynchronous cultures or cultures containing mainly trophozoites and schizonts, expect 10–25 μg of total RNA.

15. Carefully open the tube, briefly, to get rid of residual ethanol vapors.
16. Under standard conditions, the concentration of total RNA can be roughly estimated by comparing the staining intensities of the RNA sample and the DNA size marker of known mass. However, keep in mind that ethidium bromide staining of double-stranded DNA is more efficient compared to single-stranded RNA. Alternatively, RNA size markers may be used for determination of RNA concentration.
17. The presence of two distinct bands (corresponding to the 28S and 18S rRNAs) on the agarose gel is indicative of total RNA of good quality. Furthermore, when the upper band is stained more intensively than the lower one, and when minimum smearing is observed above, in between and below the rRNA bands, the RNA is of high integrity. tRNAs and 5.8s rRNAs migrate at the leading edge of the gel and appear as a diffuse signal *(4)*. Contaminating genomic DNA will appear as a high molecular weight band on top of the gel.
18. Diluting isolated total RNA 1:50 usually is appropriate for determining RNA concentration. However, in order to obtain reliable results, the absorbance value measured by the spectrophotometer should not be smaller than 0.05.
19. Formamide effectively protects RNA from degradation *(9)*.
20. The volume and composition of RNA resuspension buffer used at this step depends on the subsequent RNA application intended to perform.

References

1. Chomczynski, P. and Sacchi, N. (1987) Single-step method of RNA isolation by acid guanidinium thiocyanate-phenol-chloroform extraction. *Anal. Biochem.* **162,** 156–159.
2. Cirgwin, J. M., Przybyla, A. E., MacDonald, R. J., and Rutter, W. J. (1979) Isolation of biologically active ribonucleic acid from sources enriched in ribonuclease. *Biochemistry* **18,** 5294–5298.
3. Ausubel, F., Brent, R., Kingston, R. E., Moore, D. D., Seidmann, J. G., Smith, J. A. and Struhl, K. (eds.) (1988) Guanidine Methods for total RNA preparation, in *Current Protocols in Molecular Biology*, Unit 4.2, Suppl. 36, Greene & Wiley, New York.
4. Farrell, R. E., Jr. (ed.) (1993) *RNA Methodologies: A Laboratory Guide for Isolation and Characterisation.* Academic Press, London.
5. Promega Technical Bulletin No. 518. *RQ1 RNAse-free DNAse.*
6. Cox, R. A.(1968) The use of guanidinium chloride in the isolation of nucleic acids, in *Methods in Enzymology,* vol. 12, (Grossman, L. and Moldave, K., eds.), Academic Press, Orlando, pp. 120–129.
7. Sambrook, J., Fritsch, E. F., and Maniatis, T. (1989*) Molecular Cloning: A Laboratory Manual.* Cold Spring Harbor Laboratory, Cold Spring Harbor, New York.
8. Ambion Technical Bulletin 159. *Working with RNA.*
9. Chomczynski, P. (1992) Solubilisation in formamide protects RNA from degradation. *Nucleic Acids Res.* **20,** 3791,3792.

15

Extraction and Purification of *Plasmodium* Parasite DNA

Hans-Peter Beck

1. Introduction

The isolation of high-quality genomic DNA is often the initial step for molecular genetic analyses and crucial to many applications within the molecular research of malaria. Criteria for quality of DNA are purity, stability, and integrity and size of the molecules. Quantity also often is an important issue, particularly for material with limited resources. There are probably as many DNA-isolation protocols and techniques available as malaria research groups. Nevertheless, only a few protocols fulfill the above-mentioned criteria, although generally they all work according to the same principle. Several commercial DNA isolation kits are also available, purifying DNA on affinity matrices. These kits are costly and are difficult to adapt to various amounts of starting material, but allow complete standardization, high throughput, and automation.

The protocol described in this chapter allows the isolation of DNA with extremely high quality and purity in a reasonable time and without use of expensive or rare materials *(1)*. The protocol can easily be adapted for large-scale applications and yields good quality DNA even with small amounts of sample material with little to no protein contaminants. **Figure 1** shows *P. falciparum* DNA isolated by this method and subsequently size fractionated on a 0.5% agarose gel.

Cells are initially washed and purified and membranes are lysed using detergents. This is followed by protein denaturation and separation making use of the properties of some organic solvents to denature and to precipitate proteins. After the separation, a solution containing nucleic acids, salt, and polysaccharides, as well as traces of protein remains. Eucaryotic nuclei also contain many strongly binding chromatin proteins, which often remain associated with nucleic acids, thus being carried over in the DNA preparation. This is avoided by the inclusion of a proteinase K digest prior to the organic protein denaturation.

If material from *P. falciparum* cultures is used, then there is no problem with contaminating DNA from human cells. However, if material from patients is used (blood samples) then DNA contamination with human DNA is unavoidable. For most applications, for example, PCR and DNA hybridization, this can be ignored, but if, for example, genomic libraries are to be produced from uncultured material, this poses a serious problem. The DNA content of a human cell is approximately 200 times that of a ring stage parasite and thus *Plasmodium* DNA might be highly underrepresented in a

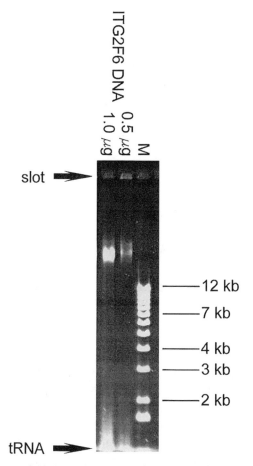

Fig. 1. Size separation of high molecular weight DNA from *Plasmodium falciparum* (ITG2F6) on a 0.5% agarose gel in TBE buffer. M: 0.05 µg 1-kb ladder (Gibco Life Technolgies). The DNA has been prepared according to the above protocol. The 1-µg sample has been vortex and pipetted without care and some breakage of DNA is visible. Also note the high concentration of RNA below 2.5 kb.

given sample. Separating lymphocytes from the blood sample by buffy-coat elimination or density-gradient centrifugation will substantially reduce the amount of human DNA but will never eliminate human DNA completely.

2. Materials
2.1. Equipment
1. Bench-top centrifuge.
2. Microfuge.
3. Water bath or heating block.
4. Pipetors.
5. Gel-electrophoresis equipment.

2.2. Reagents
1. Phosphate-buffered saline (PBS): 11.5 g Na_2HPO_4, 2.96 g $NaH_2PO_4 \cdot 2H_2O$, 5.84 NaCl per liter.

2. PBS with 0.05% (w/v) saponin (Sigma 8047-15-2).
3. Tris-EDTA (TE) buffer: 10 mM Tris-HCl, 1 mM EDTA, pH 7.6.
4. Tris-borate-EDTA (TBE)-buffer: 1 M Tris-HCl, 0.8 M boric acid, 2 mM EDTA, pH 8.0.
5. Proteinase K: 20 mg/mL in 10 mM Tris-HCl pH 8.0.
6. Sodium dodecyl sulfate (SDS): 20% (w/v) in H_2O.
7. TE-saturated phenol (pH 8.0) (*see* **Note 1**).
8. Chloroform.
9. 3 M sodium acetate, pH 4.5.
10. Ethanol, 99 and 75%.
11. DNA-grade agarose.
12. DNA-size-marker (e.g., 1-kb ladder; Gibco-Life Technologies).
13. Ethidium bromide (*see* **Note 2**).
14. Blood samples, culture material (*see* **Note 3**).

3. Methods

1. Transfer cultured material to a sterile tube and sediment erythrocytes by centrifugation at 800g for 5 min.
2. Discard the supernatant and add 5 pellet-volumes of ice-cold PBS with 0.05% (w/v) saponin. Mix and leave for 10 min on ice.
3. Pellet parasites for 5 min at 4000g.
4. Discard the supernatant and wash pellet twice with PBS.
5. Resuspend the pellet after the last wash in 1/100 of original culture volume TE.
6. Add proteinase K to a final concentration of 20 µg/mL.
7. Add 20% SDS to a final concentration of 0.5 %.
8. Carefully mix by inverting the tube slowly (*see* **Note 4**).
9. Incubate mixture for 2–12 h at 56–65°C (*see* **Note 5**).
10. Add 1 vol of TE-saturated phenol and 1 vol of chloroform (*see* **Note 6**).
11. Invert tubes for approx 10 min until a complete mixture of the organic and aqueous phase occurs. Do not vortex!
12. Spin at 15,000g for 10 min.
13. Transfer the upper aqueous phase into a new tube (*see* **Note 7**) and repeat the phenol–chloroform step (*see* **step 10**).
14. Spin at 15,000g for 10 min.
15. Transfer the upper aqueous phase into a new tube, add 1 vol of chloroform, and mix both phases carefully for 5 min.
16. Spin at 15,000g for 5 min.
17. Transfer aqueous phase to new tube (*see* **Notes 8** and **9**) and add 1/10 vol of 3 M Na acetate, pH 4.5 and 2.5 vol of 99% ethanol.
18. Invert tube several times to mix solution completely.
19. Freeze at –20°C for at least 1 h (*see* **Note 10**).
20. Spin at ~15,000g for 10 min.
21. Discard the supernatant.
22. Add 500 –1000 µL of 70% ice-cold ethanol.
23. Quickly vortex the tube.
24. Spin at ~15,000g for 5 min.
25. Discard the supernatant and air-dry tube (*see* **Note 11**).
26. Dissolve dried pellet in TE (*see* **Note 12**) to an approximate concentration of 0.5–1 mg/mL DNA (*see* **Notes 13** and **14**).
27. Check the quality of DNA by running 0.5–1 µg of the preparation on a 0.5% agarose gel. The DNA should give a broad but distinct band at approx 20 kb (**Fig. 1**). The smear below the band should not extend further down than 10–15 kb.
28. Store DNA at –20°C (*see* **Note 15**).

4. Notes

1. Prepare TE-saturated phenol by thawing an aliquot of crystallized phenol, add an equal volume of 1 M Tris-HCl, pH 8.0, and stir for 15 min. After mixing, allow the phases to separate and discard upper aqueous phase. Repeat until pH of the phenol is approximately at pH 8.0. Check pH of the phenol phase using pH-indicator paper. Add approx 1 vol of TE buffer, stir and allow phases to separate again. Repeat this step twice. Add sufficient TE to overlay the phenol phase completely. Store in a light-tight container at 4°C or at –20°C if not used frequently. Extreme care must be taken when handling phenol. Phenol is a strong denaturing agent and any contact with skin or eyes will be extremely dangerous. Wear gloves and goggles, and, if possible, work in a chemical hood. Spills on skin or eyes must be immediately rinsed extensively with warm water; do not use alcohol!
2. Ethidium bromide is potentially carcinogen and mutagen. Always wear gloves when handling ethidium bromide. We routinely stain our gels after the gel separation in a dedicated container close to the transilluminator to reduce carryover and spillage.
3. For better purification, it is advisable to perform a lysis of erythrocytes prior to the DNA isolation by adding 5 vol of ice-cold 5 mM Na-phosphate buffer, pH 7.2. Mix gently and spin briefly at 4000 rpm. Discard the supernatant and repeat the washing until supernatant remains clear. Above the parasite and nuclei pellet, erythrocyte ghosts can be seen (creamy white sediment); these can be discarded carefully.
4. From this step onward, extreme care must be taken to avoid breakage of the DNA through shearing. Never mix vigorously or vortex the sample. Each pipetting step should be done with pipet tips cut off at the tip to avoid shearing during pipetting. If using Pasteur pipets cut off the thin end of the pipet.
5. This step is flexible since no nucleases will be able to work at these SDS concentrations and will be digested by proteinase K.
6. Some protocols advise to add phenol first and to add chloroform just prior to the centrifugation. Other protocols use a mixture of phenol–chloroform–isoamyl alcohol (50:49:1). However, we have not observed any differences in the quality of DNA using either protocol. Phenol should be stored frozen in small aliquots under nitrogen to avoid oxidization. Oxidation products of phenol nick DNA and can complete destroy DNA. Some protocols add antioxidants to reduce oxidation.
7. Sometimes it is difficult to suck off the viscous aqueous phase. At this stage, carryover from the interphase is not problematic.
8. At this stage, make sure not to transfer any material from the interface, rather waste some material from the aqueous phase. Any carryover of interphase or organic phase will result in lower purity of DNA and might even inhibit subsequent enzymatic manipulations.
9. Some protocols recommend a further step using dimethylether in order to remove all phenol traces. Phenol completely dissolves in dimethylether building an aceotropic mixture which can be evaporated. However, phenol also dissolves in small concentrations in ethanol and with the subsequent ethanol precipitation there is no need for the dimethylether step.
10. With large DNA quantities, the precipitate will be visible immediately and the sample can be spun without prior freezing.
11. Do not overdry the pellets, since otherwise the DNA pellets will be hard to dissolve. In our laboratory we place the open tube for 3–5 min in a 65°C heating block.
12. If good yields have been obtained during the preparation it might take several hours for the DNA to dissolve. Do not speed up the process by pipetting or vortexing. Place closed tube into a heating block set at 65°C and incubate over night. The DNA is completely safe at these temperatures and dissolves readily. Dissolved high molecular weight DNA sometimes is difficult to manipulate even at lower concentrations and must be diluted further. Depending on the application, sometimes high concentrations are needed, then use pipets with cut-off tips for transferring material.

13. The DNA solution still contains substantial amounts of RNA. If RNA contamination is a problem, add DNase-free RNase A to a final concentration of 1 mg/mL (in 10 mM Tris-HCl, pH 8.0). Incubate for 1–2 h at 37°C. Repeat the complete phenol extraction step as before. To obtain highest quality of DNA, some protocols subsequently dialyze the DNA solution extensively against TE. However, in none of our applications this has occurred to be necessary and loss of material accompanied with dialysis outweighs the additional purity.
14. Purity and concentration of DNA can be estimated photometrically: dilute 1 µL of DNA in 500 µL of TE and measure the absorption at 260 nm in a UV spectrophotometer against TE as standard. In this dilution a DNA concentration of 250 µg/mL will give an $OD_{260\ nm} = 0.01$. In other words, an $OD_{260\ nm} = 1$ corresponds to 50 µg/mL DNA undiluted. Purity can be estimated by calculating the factor $OD_{260\ nm}/OD_{280\ nm}$, which should be ≥2.
15. Avoid frequent freeze thaw cycles which also causes DNA nicks and breakage, aliquot DNA solution or store under alcohol.

References

1. Kavenoff, R. and Zimm, B. H. (1973) Chromosome-sized DNA molecules from *Drosophila*. *Chromosoma* **41,** 1–27.

16

Southern Blotting of Parasite DNA

Tobias Spielmann

1. Introduction

Southern blotting, first described by E. M. Southern in 1975 *(1)*, is one of the cornerstones in molecular biology. The idea to immobilize DNA on a solid support was first proposed by Denhardt *(2)*. Based on this, methods were developed for the identification of specific sequences in dot blots and recombinant clones. The procedure subsequently described by Southern allowed the transfer of size-fractionated DNA molecules to a membrane. Hybridization of a specific probe to the immobilized DNA made it possible to study the genomic distribution of single and multicopy sequences.

Although many aspects of Southern hybridization have been improved since its first description, the basic concept remains the same. The DNA to be analyzed is separated by gel-electrophoresis, denatured *in situ*, and transferred to a nylon or nitrocellulose membrane. The relative position of the size-fractionated DNA fragments is retained on the membrane. In the following hybridization step, a labeled DNA probe binds to complementary sequences immobilized on the filter. Probe that has bound nonspecifically to the membrane is removed by subsequent washing. Detection of the remaining probe reveals the position of the probe on the membrane and therefore the size of the sequence complementary to the probe.

Many related procedures make use of hybridization to DNA immobilized on membranes. They differ mainly in the type of DNA that is blotted. Whereas dot and slot blots, colony or plaque lifts are techniques to immobilize unfractionated DNA, Southern blotting in a strict sense means the transfer of DNA separated by gel electrophoresis. This chapter will focus on the transfer of digested genomic DNA and the subsequent hybridization of a labeled probe. The protocol can easily be adapted to any kind of DNA size fractionated on an agarose gel (for pulsed field gels, *see* **Note 1**). However, the blotting procedure does not work for DNA separated on polyacrylamide gels, whose pore sizes are too small to allow an efficient transfer by diffusion *(3,4)*. In that case, the DNA should be blotted using an electroblotting apparatus according to the manufacturer's instructions.

There are many possible variations to carry out Southern analysis, including the way of DNA transfer, transfer buffer, type of membrane, hybridization solution, and method of probe labeling. The procedure outlined in this chapter has been found to be reliable, sensitive and reasonably fast. The genomic DNA is digested by restriction endonucleases, precipitated and run on an agarose minigel. The DNA contained in the gel is subsequently depurinated and transferred to a positively charged nylon membrane using an alkaline transfer buffer. The transfer itself is accomplished with a vacuum blotter. Vacuum transfer can result in a two- to threefold increase of hybridization signal compared to capillary transfer *(5,6)*. However, a device for vacuum transfer might not be available in every laboratory. Therefore, an alternative protocol describing upward capillary transfer is also provided. After transfer, the filter is prehybridized in modified Church and Gilbert solution *(4)*, a phosphate-based buffer. For hybridization, a labeled and denatured probe is added. The filter is then washed to remove nonspecifically bound probe and autoradiographed. A protocol for labeling of DNA with radionucleotides and random primers is provided. Nylon membranes can be reprobed several times. Therefore, **Subheading 3.5.** covers a method to strip filters for reuse. For a time-scale of the whole procedure, *see* **Note 2**.

2. Materials

2.1. General Materials

1. Tabletop microcentrifuge.
2. Water bath.
3. Heating block.
4. Tray shaker.

2.2. Restriction Digest of Genomic DNA

1. Restriction endonucleases.
2. 10X restriction buffers.
3. 3 *M* sodium acetate, pH 5.2.
4. Ethanol, 100 and 75%.

2.3. Size Separation of DNA by Agarose Gel Electrophoresis

1. Electrophoresis grade agarose (e.g., Gibco-BRL, Basle, CH).
2. 1X Tris-borate- EDTA (TBE) made up from 5X TBE (5 *M* Tris-HCl, 4 *M* boric acid, 10 m*M* EDTA, pH 8.0).
3. 1 µg/mL ethidium bromide in 1X TBE (*see* **Note 3**).
4. Gel loading buffer: 30% glycerol, 10 m*M* Tris-HCl, pH 7.0, 10 m*M* EDTA, pH 8.0, 0.2% bromophenol blue, 0.2% xylene cyanol.
5. DNA size marker, e.g., 1 kb ladder (Gibco-BRL).
6. Agarose gel unit (e.g., cat. no. HE33, Hoefer, San Francisco, CA).
7. Power supply.
8. UV-transilluminator and gel documentation system (*see* **Note 4**).
9. Microwave oven or magnetic stirrer/heater.

2.4. Transfer of DNA from Agarose Gels to Nylon Membranes

1. Depurination solution: 0.25 *M* HCl.
2. Transfer buffer: 0.4 *M* NaOH (*see* **Note 5**).
3. Filter paper (Whatman 3MM or equivalent).
4. Parafilm.
5. Saran wrap.
6. Positively charged nylon membrane (Hybond XL, Hybond N$^+$, Amersham [Pharmacia Biotech, Dübendorf, Switzerland], or equivalent).

2.5. Hybridization of Radioactive Probes to Immobilized DNA

1. Klenow polymerase (e.g., Roche Diagnostics, Rotkreuz, Switzerland).
2. 10X Klenow reaction buffer (e.g., Roche Diagnostics) (*see* **Note 6**).
3. dATP, dTTP, dGTP, 0.5 *M* each.
4. Bovine serum albumin (BSA), acetylated, 10 ng/µL (e.g., Promega, Wallisellen, CH).
5. Random hexanucleotides (*see* **Note 6**), 3 µg/µL (e.g., Gibco-BRL).
6. α^{32}-P-dCTP (3000 Ci/mmol, 10 µg/µL, Amersham) (*see* **Note 7**).
7. Stop solution: 0.02 *M* EDTA, pH 8.0.
8. Hybridization solution (*see* **Note 8**): modified Church and Gilbert buffer (0.5 *M* sodium-phosphate buffer, pH 7.2, 7% sodium dodecyl sulfate [SDS], 1 m*M* EDTA), shelf life is approx 1 mo, made up from stock solutions that are stable for several months (0.5 *M* Na_2HPO_4, 0.5 *M* NaH_2PO4, 20% SDS, 0.5 *M* EDTA, pH 8.0).
9. Washing solutions: low stringency: 2X sodium chloride/sodium citrate (SSC), 0.1% SDS, moderate stringency: 1X SSC, 0.1% SDS, high stringency: 0.1X SSC, 0.1% SDS. Made up from 20X SSC (3 *M* NaCl, 0.5 *M* Na_3citrate, pH 7.0) and 20% SDS.
10. Hybridization oven (e.g., cat. no. 400 HY, Bachofer, Reutlingen, Germany) or incubator up to 70°C.
11. X-ray film (Kodak BioMax MS or equivalent).
12. Film cassette.
13. Intensifying screen (Kodak Biomax Trans Screen-HE or equivalent).

2.6. Stripping of Membranes

1. Stripping solution: 0.1% SDS.

3. Methods
3.1. Restriction Digest of Genomic DNA

In order to choose the appropriate restriction enzymes for a genomic digest, it is important to consider the base composition of the DNA. If the DNA to be digested exhibits an A/T-content of 50%, all four bases occur at the same frequency. Therefore, a restriction enzyme recognizing a site consisting of four bases will cut more frequently than a six-cutter enzyme, because a 4-bp match is more likely to occur than a 6-bp match. However, the genome of *Plasmodium falciparum* exhibits a very high A/T-content (approx 82%). On an average, an enzyme recognizing a site consisting of four G/C bases cuts *P. falciparum* DNA less frequently than a six-cutter recognizing only A/T. Thus, in this example, the four-cutter will generate larger fragments than the six-cutter.

1. In a 1.5-mL reaction tube, on ice, combine 0.5–2 µg of parasite DNA (*see* **Notes 9 and 10**), 10 µL of 10X restriction buffer, and 10 units of restriction enzyme (*see* **Note 11**), and add dH_2O to 100 µL.
2. Incubate the reaction for 1.5 h at the temperature required for the restriction enzyme used (usually 37°C).
3. Briefly spin the tube in a microcentrifuge. Add another 10 U of restriction enzyme and incubate for 1 h at the same temperature (*see* **Note 12**).
4. Precipitate the digested DNA (*see* **Note 13**) by adding 10 µL (1/10 vol) of 3 *M* sodium acetate pH 5.2 and 250 µL (2.5 vol) of 100% ethanol. Mix thoroughly by inverting the tube several times.
5. Incubate on ice for 10 min.

6. Spin the tube in a microcentrifuge at >12,000g for 30 min.
7. Discard the supernatant. Wash the DNA pellet by adding 1 mL of 75% ethanol and spin for 5 min at >12,000g.
8. Discard the supernatant and dry the pellet in a speed vacuum or by inverting the open tube on a Kimwipe until all remaining ethanol has evaporated.
9. Resuspend the DNA pellet in 8 µL of dH$_2$O. Add 2.5 µL of gel loading buffer and store at 4°C until electrophoresis.

3.2. Size Separation of DNA by Agarose Gel Electrophoresis

To save time and money, it is convenient to use small gels to separate digested genomic DNA for Southern blotting. We have received good results with gels of 30 mL vol (gel size: 7 × 10 cm). If a high resolution is crucial, a bigger gel might be more appropriate.

In the protocol outlined below, a 1% agarose gel is made. The agarose concentration should be adjusted to meet the specifications of the particular experiment performed (*see* **Note 14**). Moreover, agarose concentration affects transfer efficiency. Therefore, transfer time should be increased for higher concentrated gels. An additional factor affecting transfer efficiency is the thickness of the gel, which should be kept minimal (<1 cm).

1. Boil 0.3 g of agarose in 30 mL of 1X TBE (in a microwave oven or on a magnetic stirrer/heater). Make sure the agarose is completely dissolved.
2. Let the solution cool at room temperature to approx 60°C (*see* **Note 15**).
3. Pour the molten agarose into the gel casting tray, use combs with wide but thin teeth. Allow the gel to set completely.
4. Place the gel in the electrophoresis tank, add 1X TBE running buffer until the gel is just covered with buffer and load the digested genomic DNA (in gel loading buffer) into the slots. Load a molecular weight marker into the slots on both the right and the left side of the gel.
5. Run the gel at 5–6 V/cm until the bromophenol blue has migrated three-quarters to the bottom of the gel.
6. Stain the gel in 1X TBE with 1 µg/mL ethidium bromide for 15 min (*see* **Note 3**)
7. Photograph the gel using a UV-transilluminator (*see* **Note 4**). Place a transparent ruler alongside for size reference (*see* **Note 16**).

3.3. Transfer of DNA from Agarose Gels to Nylon Membranes

Two blotting protocols based on alkaline transfer (*see* **Note 5**) are provided. One makes use of a vacuum blotter for transfer (*see* **Subheading 3.3.2.** and **Fig. 1C**), whereas the other relies on passive capillary action (*see* **Subheading 3.3.3.** and **Fig. 1A,B**). **Figure 1** can be used as a guide to set up the transfer.

3.3.1. Preparing the Gel for Transfer

1. Incubate agarose gel in 0.25 M HCl for 10 min with gentle agitation (*see* **Note 17**).
2. Briefly rinse the gel in dH$_2$O.
3. Incubate gel in 0.4 M NaOH for 20 min with gentle agitation.

3.3.2. Vacuum Transfer

1. Cut out 3 pieces of Whatman 3MM paper that should be larger than the gel in both dimensions, soak them in 0.4 M NaOH, and stack them on the porous screen of the vacuum chamber. Position the 3MM papers in such a way, that the window for the gel in the sealing mat will fit on top of them (*see* **Note 18**).

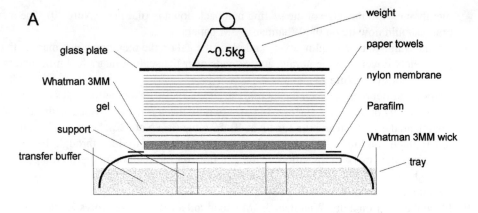
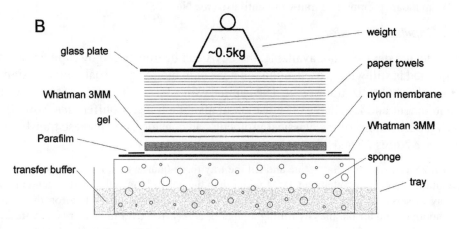
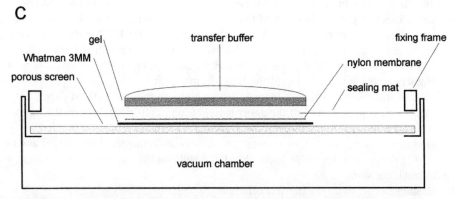

Fig. 1. Different setups for Southern transfer. (**A**) Upward capillary transfer using a Whatman 3MM wick. (**B**) Upward capillary transfer using a sponge. (**C**) Vacuum transfer.

2. If air bubbles are trapped under the Whatman 3MM papers, remove them by rolling a pipet or glass rod over the papers.
3. Cut out a piece of positively charged nylon membrane (Hybond XL, Amersham, or equivalent) about 1 cm larger than the gel in both dimensions (*see* **Note 19**). Prewet the nylon membrane in 0.4 *M* NaOH. Place the sealing mat on the porous screen. Fold back one edge of the sealing mat, exposing its underside and cover the exposed window with the

membrane. Roll the edge of the sealing mat back down in place (*see* **Note 20**). The membrane should now lie on the Whatman 3MM papers.
4. Fix the frame to the vacuum device and place the gel on the membrane in a manner that it completely covers the membrane and extends over the sealing mat a few millimeters on all four sides (*see* **Note 21**).
5. Remove air bubbles between the membrane and the gel with a pipettor a glass rod and add 0.4 M NaOH onto the gel surface. Make sure only the surface is covered with fluid.
6. Switch on the vacuum and run for 1 h at 40–50 mbar (*see* **Note 22**). Sequentially add more 0.4 M NaOH on top of the gel to replace the liquid that has drained through the gel.
7. After transfer, mark the position of the gel slots on the membrane by piercing a pencil through the slots.
8. Restain the gel to assess transfer efficiency.
9. Place the filter on a dry Whatman 3MM paper and air dry it (*see* **Notes 23** and **24**).
10. Wrap the membrane in Saran wrap until use (*see* **Note 25**).

3.3.3. Capillary Transfer

If a vacuum device is not available, the DNA can be transferred by capillary action. This method is still widely used, although a vacuum blotted gel usually gives a significantly higher hybridization signal. The capillary transfer method does not require special equipment and is reliable. The use of an alkaline transfer buffer (*see* **Note 5**), as well as modifications of the classical protocol, can considerably increase transfer efficiency (*see* **Notes 26** and **27**).

1. Bridge a tray with a glass plate and fill the tray three-quarters with 0.4 M NaOH.
2. Cut three papers of Whatman 3MM wider than the gel and long enough to extend into the tray to serve as wicks. Wet the papers in 0.4 M NaOH and place them on the bridge, smooth out all air bubbles by rolling a pipet or a glass rod over the paper (*see* **Note 27**).
3. Place the gel upside-down on the wick platform and surround it with Parafilm (*see* **Note 28**). Remove air bubbles trapped beneath the gel with a pipet or a glass rod.
4. Cut a sheet of positively charged nylon membrane (Hybond XL, Amersham or equivalent) slightly larger than the gel. Prewet the membrane in 0.4 M NaOH and place it on top of the gel. Make sure there are no air bubbles trapped between the filter and the gel.
5. Cut three sheets of 3MM paper to the size of the membrane. Wet the sheets with 0.4 M NaOH and stack them on top of the filter. Remove air bubbles.
6. Place a stack of paper towels (approx 10 cm high) on top of the 3MM paper.
7. Place a glass plate on top of the paper towels and put a weight (approx 0.5 kg) on top.
8. Allow the transfer to proceed for 12–16 h.
9. Dismantle the transfer setup, leaving the membrane on the gel. Turn the gel and the filter upside down on a dry 3MM paper. Mark the slots on the membrane by piercing a pencil through the slots.
10. Restain the gel to assess transfer efficiency.
11. Place the filter on a Whatman 3MM paper and air-dry it (*see* **Notes 23** and **24**).
12. Wrap the membrane in Saran wrap until used (*see* **Note 25**).

3.4. Hybridization of Radioactive Probes to the Immobilized DNA

3.4.1. Labeling of the Probe

We routinely use the High Prime Labeling Kit (Roche Diagnostics) to generate radioactive labeled probes with consistent results (*see* **Notes 7**, **29**, and **30**). The protocol outlined below is based on the same principle but the reagents are not premixed and can be purchased separately.

Southern Blotting of Parasite DNA

1. Dilute the DNA to be labeled to a concentration of 6 ng/µL (*see* **Note 31**).
2. Denature the DNA by heating it to 95–100°C for 5 min in a heat block or in a water bath and snap cool on ice or ice and ethanol.
3. While the DNA is denaturing, combine the following components in a 0.5-mL tube on ice:
   a. 2 µL of 10X Klenow buffer.
   b. 3 µL of dNTP mix (0.5 M dATP, 0.5 M dTTP, 0.5 M dGTP mixed 1:1:1).
   c. 1 µL of random hexanucleotides (3 µg/mL).
   d. 1 µL of BSA (10 ng/µL).
   e. 2 µL of dH$_2$O (*see* **Notes 6** and **32**).
4. Briefly spin the tube containing the denatured DNA and add 5 µL (approx 30 ng) to the mix prepared in **step 3**.
5. Add 5 µL of α^{32}-P-dCTP (3000 Ci/mmol, 10 µCi/µL) and 1 µL of Klenow enzyme (2 U/µL) (*see* **Notes 33** and **34**).
6. Incubate at 37°C for 30 min.
7. Stop the reaction by adding the labeling reaction into a screw cap tube containing 30 µL of 0.02 M EDTA, pH 8.0 (*see* **Notes 35** and **36**).

3.4.2. Hybridization and Washing

1. Place the nylon membrane in a hybridization tube and add at least 2 mL of preheated hybridization solution (modified Church and Gilbert buffer; *see* **Note 8**) per 10 cm^2 of membrane (*see* **Notes 37** and **38**).
2. Prehybridize the filter at 64°C for at least 30 min with constant agitation in a hybridization oven.
3. Denature the probe by heating to 95–100°C for 5 min and add it directly to an appropriate volume of hybridization solution preheated to 64°C (*see* **Notes 39** and **40**). Immediately proceed to **step 4**.
4. Discard the prehybridization solution and add the preheated solution containing the probe.
5. Hybridize overnight in a hybridization oven at 64°C with constant agitation (*see* **Note 41**).
6. Remove hybridization solution and store at 4°C (*see* **Note 42**).
7. Add an excess (a volume that occupies approx 30% of the tube) of preheated low stringency wash solution and incubate for 20 min in the hybridization oven at 64°C with constant agitation.
8. Discard the wash solution and repeat **step 7** with fresh low-stringency wash solution.
9. Discard and add an excess of preheated high-stringency wash solution. Repeat incubation for 10 min (*see* **Note 43**).
10. Discard the high-stringency wash solution and wrap the membrane in Saran wrap. Avoid trapping liquid in the Saran wrap, but do not allow the filter to dry out (*see* **Note 44**).
11. Autoradiograph the filter with an x-ray film and an intensifying screen at –70°C for 2–16 h (*see* **Note 45**).

3.5. Stripping of Membranes

Removing previously hybridized probes from nylon membranes works well, if the membranes were not allowed to dry out after hybridization (*see* **Note 45**). Nylon filters can be stripped several times with little loss of signal.

1. Boil a solution of 0.1% SDS, add the moist membrane, and leave boiling for 5 min.
2. Switch off the heating source and leave for 10 min.
3. Check the removal of the probe using a Geiger counter or, if desired, by autoradiography.
4. **Steps 1** and **2** may be repeated if the probe is difficult to remove.
5. Wrap the moist filter in Saran wrap.

4. Notes

1. DNA separated by pulsed-field gel electrophoresis has to be exposed to UV for 5 min (use UV-transilluminator) prior depuration (*see* **Subheading 3.3.1., step 1**: this step fractionates DNA in addition to depuration).

2. First day: Restriction digest and precipitation (3.5 h), gel electrophoresis (1.5 h), preparing the gel for transfer (0.5 h), blotting (1.5 h for vacuum transfer, otherwise overnight), prehybridization and simultaneous labeling (1 h), hybridization (overnight). Second day: Washing procedure (1.5 h), autoradiography (2–16 h).
3. Ethidium bromide is a powerful mutagen and moderately toxic. Wear gloves when working with solutions or gels containing this dye.
4. UV radiation is harmful, particularly to the eyes. Make sure that the UV light source is adequately shielded and wear protective goggles or if possible a Plexiglas shield covering the face.
5. This transfer technique is not suitable for nitrocellulose membranes, as they do not retain DNA at a pH higher than 9.0 and fall apart after prolonged exposure to alkali. Additionally, blots generated by alkaline transfer are prone to high backgrounds when a chemiluminescent detection system is used. In the latter two cases, a neutral transfer buffer should be used such as 20X or 10X SSC (*see* **Subheading 2.5., item 9** for buffer composition). Details for Southern analysis using nitrocellulose can be retrieved from **refs. 7** and **8**.
6. A 10X Klenow reaction buffer including 62.5 A_{260}/mL random hexanucleotides is available from Roche Diagnostics. Similar reaction buffers are available from other sources. In this case, additional random hexanucleotides can be omitted.
7. Observe safety regulations for handling, use, storage, and disposal of radioactive materials.
8. We routinely use the modified Church-Gilbert hybridization buffer for Southern and Northern hybridizations. The phosphate ions contained in this solution compete for nonspecific binding sites on the filter with both free nucleotides and the phosphate backbone of the probe. If a high background is observed, 1% BSA fraction V may be added to the hybridization solution to increase its blocking capacity.
9. The suggested amount of starting material should be more than sufficient to detect a single copy sequence in malaria parasites by an overnight exposure. However, if starting material is limited, sensitivity can easily be increased by the following modifications: use Ultrahyb (Ambion, Lugano, Switzerland), a commercially available hybridization solution containing formamide, and hybridize at 42°C. We routinely use this solution in Northern hybridizations to detect genes expressed at low levels. It is also possible to supplement the Church-Gilbert hybridization solution (*see* **Note 8**) with 10% dextran sulfate *(9)* or 10% polyethylene glycol (PEG) *(10,11)*, although the increased viscosity makes handling of these solutions difficult. A marked increase in sensitivity is also achieved if single-stranded RNA is used as a probe (*see also* **Note 29**).
10. If a strong background smear is observed after autoradiography, it might be advisable to reduce the amount of DNA loaded per lane. This often improves the signal-to-noise ratio.
11. If DNA has to be digested with two or more enzymes that share no suitable reaction buffer, the digests should be carried out one after another with intermittent precipitation steps.
12. A complete digest is essential for Southern analysis. However, many restriction enzymes lose their activity after less than 2 h at incubation temperature; therefore, another 10 U are added after 1.5 h.
13. Precipitation bears two advantages:
    a. The digest can be performed in a large volume.
    b. Precipitation rules out between sample variations in band migration that arise because of the dissimilar salt composition required for the digest of different restriction enzymes.
14. Be careful when handling gels of low agarose concentration for blotting, they are prone to break. **Table 1** shows the optimal separation range for gels with different agarose concentrations in 1X TBE, according to Sambrook et al. *(7)*.
15. Do not cast the gel at temperatures higher than 70°C, since this reduces the life time of combs (distorted teeth result in diffuse bands) and casting equipment.
16. If unavoidable, the gel may be stored wrapped in Saran wrap for a maximum of 1 d at this stage. But the bands will get "fuzzy" with increasing storage time, because of diffusion of DNA.

Table 1
Optional Separation Range for Gels with Different Agarose Concentrations

Agarose (%)	0.3	0.6	0.7	0.9	1.2	1.5	2.0
Separation range (kb)	5–60	1–20	0.8–10	0.5–7	0.4–6	0.2–3	0.1–2

17. This step leads to partial depurination of the DNA fragments, which results in strand cleavage. This is of advantage because the efficiency of transfer increases with decreasing fragment sizes. Do not extend this step, to avoid fragments that are too small.
18. If the sealing mats provided with the vacuum blotter are already used up for gels unsuitable for the experiment planned (i.e., the window does not fit your gel), a large piece of Parafilm can be cut to the same size and used as sealing mat.
19. Wear gloves when handling the membrane and avoid scratches if using forceps.
20. Instead of directly placing the membrane on the 3MM paper, the membrane is arranged on the sealing mat. This makes sure the membrane is positioned properly, completely covering the window of the sealing mat.
21. If the slots of the gel are broken, remove the upper part (from the slots to the top) of the gel. Make sure that the window cut out of the sealing mat is still smaller than the gel.
22. If unadjusted, the suction will increase for the first 10 min or so. Take care that it never exceeds 50 mbar, or the gel will be compressed resulting in a reduced transfer efficiency.
23. DNA blotted by alkali becomes covalently linked to positively charged nylon membranes. This makes crosslinking unnecessary.
24. In most protocols, it is recommended to rinse the membrane in 2X SSC at this stage. However, this step is not necessary and may even result in loss of signal strength with some brands of membrane if the transfer was performed in a neutral high salt buffer.
25. Filters wrapped in Saran wrap can be stored for at least 1 yr at 4°C.
26. A faster transfer can be achieved by downward capillary transfer *(12,13)*.
27. To increase transfer speed it is possible to replace the bridge and the wick by a sponge (larger than the gel) carrying three sheets of 3MM cut to the size of the sponge (*see* **Fig. 1B**).
28. The Parafilm prevents a direct contact between the Whatman papers serving as wick and the stack of absorbing paper towels. This would short-circuit the system, resulting in inefficient transfer.
29. The procedure makes use of random hexanucleotides to prime DNA synthesis at various positions on the template, generating many overlapping, labeled fragments from both strands *(14,15)*. Although there are procedures that generate probes with a higher specific activity than random priming, detection of a single copy gene in organisms with genomes of comparably low complexity (such as malaria parasites), poses no problem if sufficient starting material is available (*see* **Note 9**). The benefit of the random priming method lies in the possibility to label undefined mixtures of sequences, such as DNA libraries for instance, which can then be used for screening purposes. If a higher sensitivity is required, a RNA probe can be generated by in vitro transcription *(16)*. RNA probes are up to 10-times more sensitive than DNA probes generated by random priming. However, they are virtually impossible to remove from filters, a disadvantage that has been overcome by commercial strip and labeling kits (e.g., Strip-EZ™ RNA, Ambion). A third possibility is the use of an asymmetric PCR to generate a single-stranded DNA probe. The sensitivity achieved with a probe generated that way is about 3 to 5 times higher compared to probes obtained by random primer labeling.
30. Because of the health hazard associated with radioactivity, a variety of nonradioactive labeling and detection systems have been developed. Apart from the advantage to elimi-

nate hazardous isotopes, these probes have a much longer shelf life than radioactive probes. However, all the nonisotopic techniques we have used so far showed lower sensitivity and tended to give high background signals. Detection also adds several steps to the protocol. Therefore we recommend the use of ^{32}P for applications that require a high sensitivity like genomic Southern blots.

31. It is advisable to remove vector or adapter sequences from the template DNA prior to the labeling reaction. This step is indispensable for probes used to screen libraries or similar applications where cross-hybridization to vector or bacterial sequences might occur.
32. The labeling reaction is designed for a reaction volume of 20 µL, containing 50 µCi of [α^{32}-P]dCTP and 30 ng of template DNA, which is usually recommended. However, we routinely use half of the assay volume (containing 25 µCi of [α^{32}-P]dCTP and 15 ng of template DNA), with consistent results. As indicated in **Subheading 3.4.1., step 3**, in order to avoid inaccuracies resulting from pipetting small quantities, it is advisable to premix a larger volume of the components to obtain a stock solution for several reactions. The stock solution can be stored at –20°C. However, too many freeze–thaw cycles should be avoided.
33. Do not use [α^{32}-P]dCTP older than 2 wk, as this may result in spotty filters.
34. Because of the high A–T content of *P. falciparum* DNA, the use of radioactive dATPs or dTTPs would lead to heavier labeling of the probe. However, since the labeled nucleotides are present at much lower concentrations than the other precursors, the label concentration is the factor which limits the reaction rate and short products are frequently produced if the supply of dATP or dTTP is limiting. For this reason, we prefer to use radioactive dCTP for labeling.
35. For safety reasons, it is important to use screw-cap tubes (with seals), if the labeling reaction is denatured in boiling water, other tubes might burst.
36. Purification of the probe is usually not necessary. However, if a problem with the labeling reaction is suspected, it might be helpful to precipitate the DNA from the reaction and to check the incorporation of the radioactive nucleotides by comparing radiation of pellet and supernatant with a Geiger counter. Alternatively, a Sephadex G-50 column can be used to purify the probe.
37. If several blots are hybridized in the same tube, the use of a hybridization mesh (e.g., Labnet) should be considered to prevent the filters from sticking together.
38. The minimum volume of prehybridization solution indicated might be too small for large tubes containing small filters. Use enough; an excess does no harm.
39. If desired, the probe can be denatured after addition to the hybridization solution.
40. Because of the faster reassociation kinetics, the volume of the hybridization solution should be kept small. However, there should be enough liquid to cover the filter properly.
41. We get satisfactory results using a hybridization temperature of 64°C, even with very AT-rich probes that detect noncoding regions of the *P. falciparum* genome. If there is background, increase the temperature to 68°C. Alternatively, the temperature might be decreased, if no signal is detected.
42. Usually, there is still enough labeled probe present in the hybridization solution for reuse: denature the solution at 95–100°C for 5 min, cool it down to 64°C and add it to the prehybridized filter. Since ^{32}P decays rapidly, the solution should not be reused more than 1–2 wk after the original labeling reaction.
43. Stringency washes depend on the nature of the probe and the target to be hybridized. The lower the salt concentration and the higher the washing temperature, the higher the stringency. The washing procedure described is appropriate for perfectly matching probes. If a heterologous probe is used, or if different related target sequences exist, use the moderate stringency solution instead of the high stringency solution in the final washing step. Moreover, a reduction of the hybridization and washing temperature should be considered.

44. Drying of the membrane "fixes" the probe-template hybrid, making stripping of the probe virtually impossible.
45. Exposure time depends on the amount of starting material and the nature of the target DNA. If 0.5–2 µg of DNA were loaded per lane and the target sequence is a single copy gene, autoradiography for 5 h to overnight at –70°C with an intensifying screen is appropriate. If Ultrahyb (*see* **Note 9**) was used as hybridization solution with the same amount of target DNA, exposure time should be decreased to 2 h.

References

1. Southern, E. M. (1975) Detection of specific sequences among DNA fragments separated by gel electrophoresis. *J. Mol. Biol.* **98,** 503–517.
2. Denhardt D. T. (1966) A membrane-filter technique for the detection of complementary DNA. *Biochem. Biophys. Res. Commun.* **23,** 641–646.
3. Stellwag E. J. and Dahlberg A. E. (1980) Electrophoretic transfer of DNA, RNA and protein onto diazobenzyloxymethyl (DBM)-paper. *Nucleic Acids Res.* **8,** 299–317.
4. Church G. M. and Gilbert W. (1984) Genomic sequencing. *Proc. Natl. Acad. Sci. USA* **81,** 1991–1995.
5. Medveczky, P. Chang, C.-W. Oste, C., and Mulder C. (1987) Rapid vacuum driven transfer of DNA and RNA from gels to solid supports. *BioTechniques* **5,** 242.
6. Olszewska, E. and Jones, K.(1988) Vacuum blotting enhances nucleic acid transfer. *Trends Genet.* **4,** 92–94.
7. Sambrook, J. Fritsch, G. F. and Maniatis, T. (1989) *Molecular Cloning: A Laboratory Manual.* 2nd ed. Cold Spring Harbor Laboratory, Cold Spring Harbor, New York.
8. Ausubel, F. Brent, R. Kingston, R. E. Moore, D. D. Seidman, J. G. Smith, J. A., and Struhl, K. (1988) *Current Protocols in Molecular Biology,* Greene and Wiley, New York.
9. Wahl, G. M. Stern, M., and Stark, G. R. (1979) Efficient transfer of large DNA fragments from agarose gels to diazobenzyloxymethyl-paper and rapid hybridization by using dextran sulfate. *Proc. Natl. Acad. Sci. USA* **76,** 3683–3687.
10. Renz, M. and Kurz, C. (1984) A colorimetric method for DNA hybridization. *Nucleic Acids Res.* **12,** 3435–3344.
11. Amasino R. M. (1986) Acceleration of nucleic acid hybridization rate by polyethylene glycol. *Anal. Biochem.* **152,** 304–307.
12. Chomczynski, P. (1992) One-hr downward alkaline capillary transfer for blotting of DNA and RNA. *Anal. Biochem.* **201,** 134–139.
13. Lichtenstein, A. V. Moiseev, V., and Zaboikin, M. M. (1990) A procedure for DNA and RNA transfer to membrane filters avoiding weight-induced gel flattening. *Anal. Biochem.* **191,** 187–191.
14. Feinber A. P. and Vogelstein, B (1983) A technique for radiolabeling DNA restriciton endonuclease fragments to high specific activity. *Anal. Biochem.* **132,** 6–13.
15. Feinber A. P. and Vogelstein, B (1984) A technique for radiolabeling DNA restriciton endonuclease fragments to high specific activity. Addendum. *Anal. Biochem.* **137,** 266–267.
16. Melton D. A. and Krieg P. A. (1984) Efficient in vitro synthesis of biologically active RNA and RNA hybridization probes from plasmids containing a bacteriophage SP6 promoter. *Nucleic Acids Res.* **12,** 7035–7056.

17

SDS-PAGE and Western Blotting of *Plasmodium falciparum* Proteins

Roland A. Cooper

1. Introduction

Sodium dodecyl sulfate-polyacrylamide gel electrophoresis (SDS-PAGE) and Western blotting are complementary methods for separating and detecting the presence of a specific protein from a complex mixture. Proteins, from a cell extract, for example, are separated electrophoretically through a polyacrylamide gel. Next, these are transferred onto a nitrocellulose membrane by electrical current, preserving the original banding pattern from the gel. The membrane is probed with an antibody specific for the protein of interest, forming an antibody–antigen complex that can be visualized by a variety of techniques. This chapter outlines a procedure for the analysis of a specific protein(s) from cultured, asexual stage *Plasmodium falciparum* by SDS-PAGE and Western blotting, using a minigel system. The entire procedure can be completed in one day. Most of the methods outlined below are applicable to proteins from a wide variety of sources.

SDS-PAGE and Western blotting are indispensable procedures for basic studies of parasite biology and malaria immunology. For instance, they are utilized in the screening of antisera when producing poly- and monoclonal antibodies against *Plasmodium* antigens. Western blotting is useful in tracking a particular protein through purification procedures or physicochemical studies. Other questions addressed with these techniques include when a particular protein is expressed in the parasite life cycle, assessing relative expression levels and size polymorphisms of a protein from different strains of parasite *(1)*, or how exposure to a drug or biochemical intervention may alter the expression characteristics of a protein *(2)*. Sera from humans exposed to *Plasmodium* can be used to screen blots of parasite extracts, characterizing antigens recognized by antibodies from individuals or populations *(3,4)*. Western blotting will become particularly useful for monitoring protein expression as techniques for the genetic transfection of *P. falciparum* continue to develop *(5)*.

1.1. SDS-PAGE

SDS-PAGE, described by Laemmli *(6)*, is the most popular technique for the qualitative separation of protein mixtures. Formation of crosslinked polyacrylamide gels is a result of the copolymerization of acrylamide and bisacrylamide (*N,N'*-methylenebisacrylamide). Polymerization is catalyzed by free radical formation from persulfate and accelerated by *N,N,N',N'*-tetramethylethylenediamine (TEMED). Protein samples are prepared for SDS-PAGE by heating in a buffer containing the anionic detergent SDS and β-mercaptoethanol as a reducing agent. SDS extensively binds and denatures the proteins, forming roughly linear, negatively charged molecules that can be separated based on size by electrophoresis through an acrylamide gel. An electrical current passed through the gel causes the proteins to migrate toward the anode. Smaller proteins progress more rapidly through the matrix, while larger proteins are slowed due to frictional resistance of the gel.

1.2. Electroblotting

Electrophoretic transfer of proteins from polyacrylamide gels onto an adsorbent membrane was first described by Towbin et al. *(7)*, and is commonly referred to as Western blotting. Typically, the gel is placed vertically into a holder (inner core) between platinum electrodes submerged in a tank of transfer buffer. Application of an electric field causes proteins to migrate from the gel toward the anode, where they are immobilized on a membrane, usually composed of nitrocellulose. Once proteins have been transferred, all nonspecific binding sites on the membrane must be blocked with a solution containing inert protein. The membrane can then be exposed to a variety of probes to characterize or identify antigens of interest. Specific antibodies are the most commonly used tools to identify such proteins.

1.3. Detection of Protein by Enhanced Chemiluminescence

Chemiluminescence has become the method of choice for the detection of antigens on blotted membranes, replacing cumbersome ^{125}I labeling techniques and less sensitive colorimetric methods. Following transfer and blocking, the membrane is incubated with an antibody specific for the protein of interest. Next, the membrane is incubated with a horseradish peroxidase-conjugated secondary antibody specific for the primary antibody. After washing, the membrane is exposed to substrates that produce a light-emitting reaction catalyzed by horseradish peroxidase. The light signal is captured on X-ray film, indicating the presence and relative position of the antigen of interest. In the light-emitting reaction, peroxidase is first oxidized by hydrogen peroxide. The enzyme, together with hydrogen peroxide, catalyze the production of light by oxidizing the diacylhydrazide, luminol. The intensity and duration of the light signal is prolonged by the addition of phenolic enhancers to the reaction *(8)*. This method is highly sensitive, detecting proteins at the femtogram (10^{-15}) level in some cases.

2. Materials
2.1. Preparation of Protein Samples from P. falciparum
2.1.1. Equipment
1. Large centrifuge tubes, such as 50-mL blue-capped Falcon tubes.
2. Microcentrifuge tubes, with locking tops or screw caps.

2.1.2. Reagents
1. *Plasmodium falciparum*, maintained in culture *(9)*.
2. 10% Saponin solution.
3. SDS-PAGE reducing sample buffer (2X): to prepare 50 mL of 2X buffer, combine 9.6 mL

of 0.5 M Tris-HCl, pH 6.8, 10 mL of glycerol, 2 g (or 20 mL of a 10% solution) of sodium dodecyl sulfate (SDS), 0.65 mL of 0.5 M EDTA, 10 mg of bromphenol blue, and 1 mL of β-mercaptoethanol. Add H_2O to 50 mL and mix. Store tightly covered, in the dark. For prolonged storage, add β-mercaptoethanol just prior to use.
4. Phosphate-buffered saline (PBS): to prepare 1 L of 10X PBS, combine 80 g of NaCl, 2 g of KCl, 11.5 g of $Na_2HPO_4 \cdot 7H_2O$, and 2 g of KH_2PO_4. For a working solution, add 100 mL of 10X stock to 900 mL distilled H_2O (~pH 7.3).

2.2. SDS-PAGE Electrophoresis

2.2.1. Equipment

1. 25-mL side-arm flask with stopper.
2. Pipet tips for loading gels (*see* **Note 1**).
3. Syringe and needle (e.g., 6-mL syringe and a 23*G*, 1-in. needle).
4. Electrophoresis apparatus including tank and lid, inner core, glass plates, spacers (0.75 mm), backing plates, casting stand and combs (*see* **Note 2**).
5. Constant voltage power supply.

2.2.2. Reagents

1. TEMED. Store at 4°C.
2. 10% Ammonium persulfate (*see* **Note 3**).
3. 1.5 M Tris-HCl, pH 8.8: add 18.15 g of Tris-base to 70 mL of distilled water, and adjust pH to 8.8 with 10 N HCl. Adjust volume to 100 mL.
4. 0.5 M Tris-HCl, pH 6.8: add 6 g of Tris base to 70 mL of distilled water, and adjust pH with 10 N HCl to 6.8. Adjust volume to 100 mL.
5. 30% Acrylamide stock solution: combine 29.2 g of acrylamide and 0.8 g of *N,N'*-methylene-bisacrylamide. Adjust volume to 100 mL with distilled H_2O. Filter and store at 4°C in the dark for up to 2 mo (*see* **Note 4**).
6. 10% SDS solution. Store at room temperature (*see* **Note 5**).
7. Tris-glycine electrophoresis buffer. To prepare 1 L of 10X buffer (*see* **Note 6**), combine 30 g of Tris-base, 145 g of glycine, and 10 g of SDS. To prepare a working solution, dilute 100 mL of 10X stock with 900 mL of distilled H_2O. No pH adjustment is necessary.
8. Prestained SDS-PAGE protein standards (*see* **Note 7**).

2.3. Electroblotting

2.3.1. Equipment

1. Nitrocellulose or polyvinylidene fluoride (PVDF) transfer membrane (*see* **Note 8**).
2. Whatman 3MM filter paper, cut to a dimension slightly larger than transfer membrane. Two pieces are needed for each gel.
3. Transfer tank with lid, cassettes, cassette holder (inner core), and two blotting pads for each membrane (*see* **Note 9**).
4. Constant voltage power supply.

2.3.2. Reagents

1. Tris-glycine transfer buffer. To prepare 1 L of 10X buffer, combine 30 g of Tris-base and 145 g of glycine. To prepare a working solution, add 100 mL of 10X stock to 700 mL of distilled H_2O and 200 mL of methanol (*see* **Note 10**). No pH adjustment is necessary. Store at 4°C.
2. Ponceau S protein stain (optional): 0.5% Ponceau S (w/v) in 1% acetic acid (v/v).

2.4. Immunodetection of Proteins (Using a Chemiluminescent System)

2.4.1. Equipment

1. Film, such as Kodak X-Omat AR (cat. no. 165 1454) or Biomax ML (cat. no. 876 1520).
2. Film cassette (e.g., Amersham Hypercassette).
3. Film markers (e.g., Stratagene Glogos II Autorad Markers).
4. Plastic (Saran) wrap.
5. Dark room, film developing equipment, or automated developing machine.

2.4.2. Reagents

1. PBS-Tween: PBS with Tween-20, 0.1% (v/v): add 1 mL of Tween-20 to 1 L of PBS. Mix well.
2. Blocking buffer: PBS-Tween with 5% nonfat dry milk (*see* **Note 11**).
3. Poly- or monoclonal antibody specific for the protein of interest (*see* **Note 12**).
4. Horseradish peroxidase-conjugated secondary antibody, specific for primary antibody (*see* **Note 13**).
5. Reagents for enhanced chemiluminescent detection, such as a proprietary kits such as SuperSignal (Pierce), ECL (Amersham Pharmacia), and ImmunStar (Bio-Rad) give excellent results.

3. Methods

3.1. Preparation of Plasmodium falciparum for SDS-PAGE

This section describes the preparation of a 5-mL culture of asexual blood stage parasites for SDS-PAGE electrophoresis. It has been used effectively for culture volumes as large as 50 mL maintained at 5% hematocrit. Volumes can be scaled up or down as necessary. Buffer, centrifuges, rotors, tubes, and so on, should be kept at 4°C to reduce degradation of proteins by protease activity. Gloves should be worn during this procedure. Note that saponin, used to lyse red blood cells, will separate erythrocyte cytosol from parasites, but erythrocyte membranes remain associated with parasites *(10)*. Therefore, it is important to prepare a control sample of uninfected erythrocytes. To examine timing of protein expression, parasites can be maintained under sorbitol synchronization and harvested at specific stages *(11)*. Alternatively, mixed stage cultures may be separated by stage using a Percoll gradient *(12)*. Additionally, other life-cycle stages of the parasite may be processed for SDS-PAGE and Western blotting. The culturing and preparation of gametocytes and gametes for SDS-PAGE has been presented in detail *(13,14)*. Although more difficult to generate samples for analysis, sporozoites of *P. falciparum* have been examined by SDS-PAGE and Western blotting, both by analyzing whole infected mosquito salivary glands *(15)* and by isolation of pure sporozoites *(3)*. Oocysts of *P. falciparum* have been similarly studied by analysis of entire mosquito midguts *(15)*. A collegenase digestion procedure may be used on infected mosquito midguts to obtain pure oocysts *(16)*.

1. Transfer 5 mL of parasite culture, typically at 5% hematocrit and 5% parasitemia, to a large centrifuge tube (*see* **Note 14**). Add PBS to 50 mL. Pellet erythrocytes by spinning at 800*g* for 5 min in a benchtop centrifuge.
2. Aspirate the supernatant, resuspend the pellet in 50 mL PBS and centrifuge at 800*g* for 5 min. Remove the supernatant.
3. Resuspend pellet in 50 mL of PBS. Lyse red blood cells by adding 50 µL of 10% saponin solution. Rock tube until turbidity noticeably decreases, about 1–2 min. Centrifuge at 1800*g* for 10 min. Aspirate the supernatant (into flask containing appropriate disinfectant), which should be red as a result of the erythrocyte cytosol.

4. Resuspend pellet in 50 mL of PBS and spin at 1800g for 10 min. Aspirate the supernatant. The pellet will be much reduced at this point and will be brown in color due to hemozoin. Bring pellet volume to 200 µL in 1X SDS sample buffer (*see* **Note 15**). Mix the sample well to solubilize protein (*see* **Note 16**).
 5. Heat sample at 90°C for 5 min in a locking-top or screw-cap microfuge tube. Samples should be frozen at –20°C until use.

3.2. SDS Polyacrylamide Gel Electrophoresis

This outline describes the preparation of polyacrylamide minigels and the subsequent electrophoretic separation of a mixed protein sample through a Tris-glycine-buffered discontinuous system. Most SDS-PAGE run nowadays are in the ~7 × 8 cm minigel format, providing greatly improved separation times over larger systems. The procedure is applicable to many types of protein samples in addition to those from *P. falciparum*. A short 4% acrylamide stacking gel is poured over the separating gel to help concentrate proteins. Due to the large pore size of the stacking gel, mixed proteins will initially "stack" in a tight band at the beginning of electrophoresis, ensuring the sample will enter the separating gel at the same time. A gel thickness of 0.75 mm is suitable for running a 10 µL parasite sample. Thicker gels may be used to concentrate more protein sample, but this is partially offset by decreased transfer efficiency. Prepoured gels from various vendors are available and produce excellent results. However, they are expensive, and have limited shelf lives. Commercially poured gradient gels are particularly convenient, as their preparation is difficult and requires specialized equipment. Gradient gels are useful for separating a wide range of protein sizes, but are not as efficient at resolving proteins of similar size as a constant percentage gel.

We routinely use 5% acrylamide minigels to separate proteins as large as 330 kDa. However, the *P. falciparum* genome encodes some exceptionally large proteins, such as Pf332, with a molecular weight of ~750 kDa *(17)*. Wiesner et al. *(17)* have devised an electrophoretic system using polyacrylamide-agarose composite gels capable of handling these very large proteins. Of course, it is difficult to estimate the molecular weight of such proteins due to the lack of calibrated standards of this size.

 1. Clean the glass plates with detergent and water, rinse well, and dry.
 2. Assemble the glass plates, spacers, and clamp/backing device (collectively known as a sandwich-clamp assembly). Insert assembly into casting stand. This step will vary according to the individual manufacturer of the electrophoresis unit.
 3. In a 25-mL side-arm flask, prepare an appropriate (*see* **Note 17**) separating gel according to **Table 1**.
 4. After degassing, add ammonium persulfate and TEMED to the gel monomer solution and gently swirl. Pour the gel to the level marked (*see* **Note 19**). A 1-mL pipetor is useful to pour the gel. Pipet the solution smoothly against the larger glass plate, over the opening. The solution will flow between the plates.
 5. Add 100–200 µL of water-saturated *sec*-butanol on top of the separating gel. This will overlie the gel monomer, allowing it to polymerize with a level, smooth surface. Allow gel to polymerize 45 min to 1 h.
 6. While the separating gel is polymerizing, mix the stacking gel. To prepare 10 mL of stacking gel (sufficient for at least four minigels of 7 × 8 cm × 0.75 mm), combine 2.5 mL of 0.5 M Tris-HCl (pH 6.8), 6.0 mL of H_2O, 1.2 mL of stock acrylamide (37.5:1), 250 µL of 10% SDS, 50 µL of 10% ammonium persulfate, and 5 µL of TEMED. Degas 1 min before adding ammonium persulfate and TEMED.

Table 1
Preparation of Separating Gel[a]

Acrylamide gel (%)	1.5 M Tris-HCL, pH 8.8 (mL)	H_2O (mL)	Stock acrylamide (mL)
5.0	2.5	5.65	1.67
7.5	2.5	4.85	2.5
10.0	2.5	4.0	3.33
12.0	2.5	3.35	4.0
15.0	2.5	2.35	5.0

[a]To each, add: 100 µL of 10% SDS, 50 µL of 10% ammonium persulfate, and 5 µL of TEMED. Before adding ammonium persulfate and TEMED, the solution should be degassed for 5 min by connecting the stoppered side-arm flask to a vacuum pump (*see* **Note 18**).

7. After the separating gel has polymerized, pour off *sec*-butanol and unpolymerized acrylamide solution. Rinse the top of the gel with 0.5 M Tris-HCl, pH 6.8.
8. Add ammonium persulfate and TEMED to the stacking gel solution, swirl gently to mix, and pour to top of glass plates.
9. Insert combs. Avoid trapping bubbles (*see* **Note 20**). Some gel monomer may spill over the edge of the small glass plate. Add more gel solution if necessary. Allow gel to polymerize for 30–45 min.
10. Remove combs, attach the sandwich plates to the inner core according to manufacturer's instructions. Put assembly into tank.
11. Fill the inner chamber with electrophoresis buffer. The sample wells of the stacking gel should be filled with buffer. Carefully inspect for leaks. It may be necessary to reassemble the unit if it is leaking.
12. Add sufficient electrophoresis buffer into the tank to cover the lower electrode. The tank should be tilted to remove bubbles caught under the gel plates.
13. Remove excess gel from the glass plate above the stacking gel with the edge of a small metal spatula. Unpolymerized gel should be removed from the loading wells with a syringe containing transfer buffer. Insert the needle tip into each well and rinse gently with buffer.
14. Load the samples. Usually 10 µL per lane is sufficient. In the first lane, add 5–7 µL of prestained protein standard. Gently pipet the sample, inserting the pipet tip near the bottom of the well. The sample will stay in place due to the high density of the glycerol compared to the buffer (*see* **Note 21**).
15. Electrophorese at constant voltage according to manufacturer's recommendations. Bio-Rad MiniProtean II is run at 200V, Novex XCell II at 125V. Continue until blue dye front passes through the bottom of the gel. This will take approx 45 min (Bio-Rad) to 1.5 h (Novex).
16. When the run is finished, turn off power supply. Disassemble electrophoresis unit and remove the glass plates containing the gels.
17. Carefully pry the glass plates apart. The gel will remain on one of the plates. Remove and discard the stacking gel using the edge of the free plate. Loosen the gel from the glass by holding the plate, gel side down, in a small plastic container with transfer buffer. Gently shake the glass, and the gel should fall away.
18. Allow gel to equilibrate for 5 min (maximum) in transfer buffer before proceeding.

3.3. Electroblotting

1. Cut nitrocellulose membrane to a size slightly larger than the gel (*see* **Note 22**). Additionally, two pieces of 3MM filter paper are needed for each gel.
2. Wet nitrocellulose membrane in distilled water for 15 s, then place in transfer buffer for at least 10 min. Handle the membrane with gloved hands.

SDS-PAGE and Western Blotting

3. Submerge blotting pads in transfer buffer; two pads are required per gel. Press out air bubbles, as they inhibit protein transfer.
4. In a tray containing transfer buffer (about 2 cm deep), place the transfer cassette with the cathode side (the negative, or black side) on the tray bottom. Open cassette and place a soaked blotting pad on the cathode side, followed by a wetted filter paper, the gel, the membrane, a second piece of filter paper, and finally the second soaked pad (*see* **Notes 23 and 24**). Avoid introducing bubbles at all times.
5. Clamp the cassette sandwich together. The assembly must fit tightly. If not, the pads may be worn and should be replaced. Make sure membrane is on the anode (positive, red) side relative to the gel! Place cassette(s) in the holder, facing in the proper direction. Put the whole assembly into the tank, again in the correct orientation.
6. Fill tank with cold (4°C) transfer buffer.
7. Add ice to the cooling unit, and put in tank. Place a stir bar on the bottom of the tank.
8. Transfer at 100 V for 1 h, or overnight, in a cold room at 15 V (Bio-Rad Trans-Blot) (*see* **Note 25**). The apparatus should be run on a stir plate for even temperature and ion distribution. Novex XCell II does not require a cooling unit, and should be run at 25 V for 1.5–2 h.
9. Stop transfer by turning off power supply. Disassemble gel cassette, and remove membrane with flat-tipped forceps. The stained protein markers should be clearly visible on the membrane.
10. The membrane should either be blocked as in the following section, or first may be stained with Ponceau S (*see* **Note 26**).

3.4. Immunodetection with Chemiluminescence

Because protocols for detection of individual antigens can vary considerably, this procedure is intended to be a starting point. Although we get excellent results for many proteins following the steps outlined below, the protocol should be optimized as necessary. Blots should be kept on an orbital shaker/rocker in **steps 1–7**. Reserve a lane of the blot to probe with negative control antisera. This should be an antibody specific for a *Plasmodium* protein other than the one of interest. If using a polyclonal antibody, prebleed sera should be tested as well. Membranes should always be handled carefully while wearing gloves or using flat-tipped forceps.

1. Following electroblotting of the proteins, the membrane is transferred to a small plastic container, such as the lid of a pipet tip box. Gently rock the membrane in blocking buffer for 1 h at room temperature (*see* **Note 27**); 25 mL of blocking buffer per blot is sufficient.
2. Briefly rinse in PBS-Tween, then continue to wash in fresh PBS-Tween for 5 min.
3. Incubate membrane with primary antibody at desired titer, in a small volume of blocking buffer (*see* **Notes 28 and 29**).
4. Wash as in **step 2**.
5. Wash membrane in three changes of blocking buffer, 5–10 min each.
6. Incubate with secondary antibody, at a dilution recommended by the supplier, 30 min to 1 h, at room temperature (*see* **Note 13**).
7. Wash in 3 changes of PBS-Tween, 10 min each.
8. Prepare the reagents for chemiluminescent detection according to manufacturer's instructions. One milliliter per blot is sufficient.
9. Drain membrane well by holding a corner with forceps and blotting opposite edge on a paper towel. Place face-up on plastic wrap (*see* **Note 30**). Pipet 1 mL of developing reagent over the surface of each membrane. Allow to incubate for 1 min.
10. Drain membrane in the same manner as in **step 9**.
11. Place membrane face down on plastic wrap. Fold the plastic over the membrane to form an "envelope." Put envelope face-up into the developing cassette, using tape to secure it in

the cassette. Use a glow in the dark film marker in the cassette to align membrane with film after development.
12. In the darkroom, expose membrane to film. Initial film should be exposed about 10 min. If the film is too dark (overexposed), there will be plenty of signal remaining for shorter exposures.
13. After developing, place film over membrane in cassette. Align with film marker to record position of membrane and protein standards.
14. Several remedies can be tried if background is too high (*see* **Note 31**), or if the signal is too weak (*see* **Note 32**).
15. The blot can be stored frozen, or it can be stripped and reprobed with other antibodies (*see* **Note 33**) (*see* **Fig. 1**).

4. Notes

1. Specially elongated pipet tips are ideal for loading the gels. We use the MULTI Mini-Flex 10-µL microcapillary tip (cat. no. 53550-428, VWR).
2. We use the Mini-Protean II (Bio-Rad) when pouring our own gels, and the Xcell II (Invitrogen) when using prepoured gels. Both systems produce excellent results.
3. Ammonium persulfate solution should be prepared fresh prior to use. In practice, we use it for up to 1 wk.
4. Because acrylamide monomer is neurotoxic, great care should be exercised during handling, especially weighing of the powder. For this reason, we use commercially prepared solutions of 30% acrylamide:bisacrylamide (37.5:1) (cat. no. 161-0158, Bio-Rad). Gloves should be worn when working with the solution.
5. SDS may precipitate at cool temperatures. Warming the bottle in a water bath will redissolve the crystals.
6. 10X SDS-PAGE electrophoresis buffer can be purchased from many companies. It is much more economical to prepare it in the laboratory, however. The same holds true for 10X transfer buffer, 2X sample buffer and 10X PBS.
7. For general use, prestained, broad-range molecular weight standards are most useful (cat. no. LC5625, Invitrogen; cat. no. 161-0318, Bio-Rad). The calibrated standards allow for a fairly accurate estimate of the molecular weight of the separated proteins. While most proteins migrate based on the size (due to the constant size:charge ratio), certain proteins do not migrate as expected due to excessive charge, hydrophobicity, or glycosylation. Because the protein standards are stained, they also serve as a visual indicator of the quality of transfer following electroblotting.
8. Many brands of nitrocellulose sheets are very fragile. Trans-Blot nitrocellulose sheets from Bio-Rad (cat. no. 162-0116) are durable and effective. For most proteins, 0.45-µm pore size nitrocellulose works well, although if working with polypeptides smaller than 15 kDa, 0.22-µm pore size may help prevent loss of product through the membrane. For immunoblotting of most proteins, nitrocellulose is suitable, although PVDF is more resilient. Proteins on nitrocellulose membranes are more easily stained with Ponceau S than those blotted to PVDF. Because PVDF membranes are hydrophobic, they must be prewetted in methanol, then placed in transfer buffer and allowed to equilibrate at least 15 min. Certain proteins may bind to one type of membrane better than another. The ideal transfer conditions must be determined empirically.
9. The transfer apparatus is usually supplied as a complete kit, such as the Bio-Rad Mini Trans-Blot Module, or Novex Xcell II Blot Module.
10. It is believed that methanol decreases efficiency of electrophoretic transfer of large proteins (>100 kDa). Methanol may be omitted from the buffer if poor transfer appears to be a problem with large proteins.

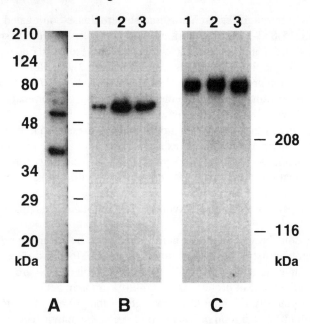

Fig. 1. Example of Western blots of *P. falciparum* proteins separated by SDS-PAGE. All blots were subject to high-salt PBS-Tween-20 and Triton X-100 washes as described (*see* **Notes 31** and **32**). Location of molecular weight standards are indicated by horizontal lines. (**A**) Plasmepsin II from the FCB strain of *P. falciparum*. Sample consisted of whole trophozoites prepared as described in **Subheading 3**. The larger band corresponds to the precursor enzyme, proplasmepsin II (51 kDa). The smaller band (37 kDa) is the active form of the enzyme. Samples were run on a 12% SDS polyacrylamide gel. Membrane was incubated with polyclonal antiplasmepsin II, 1:2500, overnight at 4°C in PBS-Tween-20 with 5% nonfat dry milk. Secondary antibody was horseradish peroxidase anti-rabbit IgG, incubated at 1:2500, 20 min at room temp. Antiplasmepsin II antibody was a generous gift from Ritu Banerjee and Daniel Goldberg (*2*). (**B**) HRP II from various fractions of *P. falciparum* trophozoites from the FCB strain (*1*). Lane 1: food vacuoles; lane 2: whole trophozoites; lane 3: 100,000g soluble fraction of trophozoite lysate. This protein is known to vary from 63 to 104 kDa, depending on parasite strain. In FCB, the molecular weight is between 55 and 65 kDa. This was the same gel and membrane as in panel A. Membrane was incubated with anti-HRP II monoclonal antibody, 1:1000, 1 h in PBS-Tween-20 with 5% nonfat dry milk, room temp. Secondary antibody was anti-mouse IgG at 1:25,000, 1 h at room temperature. (**C**) CG2 from 3 strains of *P. falciparum* (*18*). Lane 1: Dd2 strain; lane 2: HB3 strain; lane 3: FCB strain. A small size polymorphism is apparent in the ~330,000 kDa proteins. Samples were run on a 5% SDS-polyacrylamide gel. Membrane was incubated with affinity purified anti-CG2 monoclonal antibody, 1:1000, 1 h room temp in PBS-Tween-20 with 5% nonfat dry milk. Secondary antibody conditions were the same as in panel B.

11. Any nonfat powdered milk formulation will work. We use a generic brand. Mix buffer on a stir plate for several minutes before using. This solution will keep for a few days at 4°C.
12. Proteins blotted from SDS polyacrylamide gels will be essentially denatured on the membrane. Usually polyclonal antisera contain antibodies that will recognize the antigen in a denatured state. However, this may not be the case with monoclonal antibodies. If satisfactory results are not obtained with a monoclonal antibody, it may be necessary to separate the proteins under nondenaturing conditions.

13. We use Jackson ImmunoResearch horseradish peroxidase-conjugated donkey anti-mouse IgG (cat. no. 715-035-151) at a dilution of 1:25,000 (1 µL antibody solution in 25 mL of PBS-Tween with 5% nonfat dry milk) with excellent results. In our experience, following reconstitution, the antibody performs well for at least 2 yr if stored at 4°C.
14. Some *P. falciparum* proteins, such as HRP II and EBA-175 *(1,19)* are shed into the culture supernatant. A portion of the culture media can be reserved for analysis on Western blot. Mix media 1:1 with 2X sample buffer and heat as described.
15. Based on a 5-mL culture, at 5% hematocrit and 5% parasitemia, loading a 10-µL sample per well will yield approx 6×10^6 parasites per lane. If quantification of parasites is important, confirm hematocrit with a hemocytometer or by a capillary tube method. It is generally accepted that there are $\sim 10^{10}$ cells/mL packed erythrocytes.
16. Pipetting the sample up and down through the narrow opening of a 200-µL pipet tip is effective at dispersing the sample. Avoid excessive foaming, as this may promote protein breakdown.
17. The gel selection shown in **Table 2** guide is meant to be a starting point. Some proteins, based on their chemistry, have migration rates in SDS-PAGE different than predicted by their molecular weight. Commercially available gradient gels are especially useful for separating a large range of protein sizes. A 4–20% gradient gel is suitable for initial sample screening, particularly if the molecular weight of the protein of interest is uncertain. For more in-depth information on polyacrylamide gels, the author recommends Bio-Rad US/EG Bulletin 1156, Acrylamide Polymerization—A Practical Approach.
18. Oxygen inhibits the polymerization of monomer acrylamide solutions. This effect becomes more pronounced with low percentage acrylamide gels. For reproducibility, gel monomer should be degassed.
19. Before pouring gel, insert comb between the glass plates, into the proper orientation. Using a permanent marker, place a mark on the glass plate or the backing plate 1 cm below the level of the bottom of the comb. Fill the separating gel to this mark to allow ample room for the stacking gel. Additionally, because the gel wells can be difficult to distinguish, mark the position of the bottom of the wells. This will assist in accurate sample loading.
20. Introducing the comb at a slight angle will help prevent bubbles from getting trapped in the stacking gel.
21. It may help to briefly centrifuge samples prior to loading to help pellet viscous DNA. If the sample is too viscous to load, it should be diluted with 1X sample buffer.
22. For cutting filter paper and nitrocellulose, it is helpful to make a cardboard template slightly larger than the gel.
23. With some membranes, it is difficult to distinguish onto which side the stained protein standards have been blotted. Place a small pencil mark or number on the side of the membrane in contact with the gel. This will allow for proper orientation of the membrane during film exposure.
24. Gels of 10% acrylamide or higher are relatively easy to handle, and can be carefully laid directly onto the wet filter paper. Low-acrylamide gels should be "floated" onto a piece of filter paper submerged in ~1 cm deep transfer buffer. The gel can be manipulated with gloved hands and the rounded end of a small spatula. The gel must sit flat and unfolded on the filter paper. Some practice is required.
25. The cooling unit is not needed when the transfer is conducted in a cold room. However, the buffer should be stirred for optimal transfer.
26. Staining the membrane is useful to assess the quality of transfer, and it permits the blot to be cut into strips accurately if it will be screened against multiple antibodies. Stain 2 min with sufficient Ponceau S solution to cover the blot. Destain with several changes of distilled H_2O. Blots can then be blocked and processed as usual. The stain may be reused.
27. Nonfat dry milk may contain biotin, which causes excessive background when using a

Table 2
Gel Selection Guide

Acrylamide gel (%)	Approximate protein range (kDa)
5	90–350
7.5	50–200
10	30–150
12	15–100
15	5–70

biotin-streptavidin detection system. The milk can be omitted from the solution, as Tween-20 is a satisfactory blocking agent. Commercial blocking solutions are also available. Blocking overnight will sometimes help to decrease background, although 1 h is sufficient in most cases.

28. It will be necessary to test a range of titers if there is no previous information regarding the use of a particular antibody. A dilution range of 1/100–1/2000 can be utilized the first time. Time of incubation should also be optimized. This typically ranges from 1 h at room temperature to overnight at 4°C. The primary antibody solution may be reused two or three times. Sodium azide should be added (0.02% w/v) and the solution stored at 4°C.
29. Membrane clamping devices such as the multiscreen (Bio-Rad) allows a single blot to be probed with many different antisera, without cutting the membrane into strips. Such an apparatus is useful in that very small quantities of primary antibody are needed for screening.
30. A piece of plastic wrap may be stretched over a glass plate. It will adhere on its own, and can be smoothed with a damp paper towel.
31. Suggestions to reduce high background:
    a. The membrane may be washed in high-salt (0.5 M NaCl) PBS-Tween buffer (add 1 g NaCl per 50 mL PBS-Tween). Three washes at 5–10 min per wash, after both primary and secondary antibody incubations can reduce background.
    b. Following high-salt PBS-Tween washes, the membrane can be washed in three changes of PBS-Tween containing Triton-X100 (0.5% v/v), 5 min per wash. We use Triton X-100 washes following the secondary antibody incubation only.
    c. Increasing the dilution of the primary or secondary antibody, or incubating for a shorter time, may decrease background.
    d. Increase wash times.
    e. Block membrane overnight, or try another blocking agent, such as PBS-Tween with 1% (w/v) bovine serum albumin.
    f. The use of an affinity-purified primary antibody, or immunoprecipitating the protein of interest prior to SDS-PAGE, can increase quality of immunoblots *(20)*.
    g. Use PVDF membrane instead of nitrocellulose.
32. Suggestions to enhance a weak signal (keep in mind that such methods can increase background as well):
    a. Increase titer and/or time of incubation with the primary antibody. We frequently incubate overnight at 4°C with gentle rocking. Increasing titer and/or time of incubation of secondary antibody often leads to high background. Follow guidelines provided by the supplier.
    b. Use a more sensitive chemiluminescent reagent kit.
    c. Use buffers without Tween-20. Use PBS with 5% nonfat dry milk for blocking, incubating with antibodies and where required for washing, followed by washes with PBS (without milk).
    d. Load a more concentrated parasite sample.
33. After exposing to film, membranes may be stored for long periods by freezing at –20°C.

Leave them in the plastic wrap envelope when freezing. Prior to reprobing, the blot must be stripped. To prepare stripping buffer, combine 10 mL of 10% SDS, 6.25 mL of 0.5 M Tris-HCl (pH 6.8), and 0.35 mL of β-mercaptoethanol. Add distilled H_2O to 50 mL final volume. Incubate blot at 50°C for 30 min on a rocking platform or in a hybridization oven roller in the stripping buffer. Wash the membrane with three changes of PBS-Tween, 10 min each, at room temperature. To reprobe, begin with the blocking step and continue as described in **Subheading 3.4.** Signals will likely be weaker when probing a second or third time. Stripping time can be reduced or increased as necessary.

References

1. Rock, E. P., Marsh, K., Saul, A .J., Wellems, T. E., Taylor, D. W., Maloy, W .L., and Howard, R. J. (1987) Comparative analysis of the *Plasmodium falciparum* histidine-rich proteins HRP-I, HRP-II and HRP-III in malaria parasites of diverse origins. *Parasitology* **95,** 209–227.
2. Francis, S. E., Banerjee, R., and Goldberg, D. E. (1997) Biosynthesis and maturation of the malaria aspartic hemoglobinases plasmepsins I and II. *J. Biol. Chem.* **272,** 14,961–14,968.
3. Nardin, E. H., Nussenzweig, V., Nussenzweig, R. S., Collins, W. E., Tranakchit Harinasuta, K., Tapchaisri, P., and Chomcharn, Y. (1982) Circumsporozoite proteins of human malaria parasites *Plasmodium falciparum* and *Plasmodium vivax. J. Exp. Med.* **156,** 20–30.
4. Contreras, C. E., Pance, A., Marcano, N., Gonzalez, N., and Bianco, N. (1999) Detection of specific antibodies to *Plasmodium falciparum* in blood bank donors from malaria-endemic areas of Venezuela. *Am. J. Trop. Med. Hyg.* **60,** 948–953.
5. Waters, A. P., Thomas, A. W., van Dijk, M. R., and Janse C. J. (1997) Transfection of malaria parasites. *Methods* **13,** 134–147.
6. Laemmli, U. K. (1970) Cleavage of structural proteins during the assembly of the head of bacteriophage T4. *Nature* **227,** 680–685.
7. Towbin, H., Staehelin, T., and Gordon, J. (1979) Electrophoretic transfer of proteins from polyacrylamide gels to nitrocellulose sheets: Procedure and some applications. *Proc. Natl. Acad. Sci. USA* **76,** 4350–4354.
8. Thorpe, G. H. G., Kricka, L. J., Moseley, M. R., and Whitehead, T. P. (1985) Phenols as enhancers of the chemiluminescent horse radish peroxidase-luminol-hydrogen peroxide reaction: application in luminescence-monitored enzyme immunoassays. *Clin. Chem.* **31,** 1335–1341.
9. Trager, W. and Jensen, J. B. (1976) Human malaria parasites in continuous culture. *Science* **193,** 673–675.
10. Beaumelle, B. D., Vial, H. J., and Philippot, J. R. (1987) Reevaluation, using marker enzymes, of the ability of saponin and ammonium chloride to free *Plasmodium* from infected cells. *J. Parasitol.* **73,** 743–748.
11. Lambros, C. and Vanderberg, J. P. (1979) Synchronization of *Plasmodium falciparum* erythrocytic stages in culture. *J. Parasitol.* **65,** 418–420.
12. Kramer, K. J., Chow Kan, S., and Siddiqui, W. A. (1982) Concentration of *Plasmodium falciparum*-infected erythrocytes by density gradient centrifugation in Percoll. *J. Parasitol.* **68,** 336,337.
13. Carter, R., Ranford-Cartwright, L., and Alano, P. (1993) The culture and preparation of gametocytes of *Plasmodium falciparum* for immunochemical, molecular and mosquito infectivity studies, in *Protocols in Molecular Parasitology* (Hyde, J. E., ed.), Humana, Totowa, NJ, pp. 67–88.
14. Petmitr, P., Pongvilairat, G., and Wilairat, P. (1997) Large scale culture technique for pure *Plasmodium falciparum* gametocytes. *Southeast Asian J. Trop. Med. Public Health* **28,** 18–21.
15. Boulanger, N., Matile H., and Betschart, B. (1988) Formation of the circumsporozoite protein of *Plasmodium falciparum* in *Anopheles stephensi. Acta Trop.* **45,** 55–65.
16. Rosenberg, R. and Rungsiwongse, J. (1991) The number of sporozoites produced by individual malaria oocysts. *Am. J. Trop. Med. Hyg.* **45,** 574–577.
17. Wiesner, J., Mattei, D., Scherf, A., and Lanzer, M. (1998) Biology of giant proteins of *Plasmodium*: resolution on polyacrylamide-agarose composite gels. *Parasitol. Today* **14,** 38–40.
18. Su, X.-S., Kirkman, L. A., Fujioka, H., and Wellems, T. E. (1997) Complex polymorphisms in an ~330 kDa protein are linked to chloroquine-resistant *P. falciparum* in Southeast Asia and Africa. *Cell,* **91,** 593–603.
19. Camus, D. and Hadley, T. J. (1985) A *Plasmodium falciparum* antigen that binds to host erythrocytes and merozoites. *Science* **230,** 553–556.
20. Harlow, E. and Lane, D. (1999) Immunoblotting, in *Using Antibodies: A Laboratory Manual.* Cold Spring Harbor Laboratory Press, Cold Spring Harbor, NY, pp. 267–310.

18

Nested PCR Analysis of *Plasmodium* Parasites

Georges Snounou and Balbir Singh

1. Introduction

In the natural and untreated intermediate hosts, *Plasmodium* infections are maintained for extended and often lifelong periods. Parasitemia generally reaches its highest level in the first peak after inoculation, and only rises thereafter during increasingly brief and interspersed episodes. As the host acquires tolerance to higher numbers of multiplying parasites, the severity and duration of the clinical episodes that coincide with the parasite peaks diminish.

Microscopic examination of blood has until recently been the sole rapid and practical method to detect and identify *Plasmodium* parasites unequivocally. The main limitation of this method is the difficulty in detecting very low levels of parasites. In human infections, the probability of detecting the parasite diminishes rapidly as the parasitemia falls below 0.0005% or 25 parasites/microliter of blood (2.5 parasites per 800 white blood cells counted, assuming 8×10^3 white blood cell and 5×10^6 red blood cell per microliter of blood). Medically, this relative insensitivity is of little consequence because clinical episodes are only rarely associated with such low parasitemias. However, the fact that circulating parasites are undetectable throughout the major part of the infection is of some consequence to the epidemiological and biological perceptions of malaria *(1–8)*.

Amplification of DNA by the polymerase chain reaction (PCR) has provided the opportunity to devise highly sensitive methods of parasite detection *(9–16)*, and the specificity inherent to this method allows the unequivocal identification of the parasite species. The efficiency of the assay is markedly improved when a nested PCR strategy is adopted *(13,17–20)*. In this strategy, two rounds of amplification are carried out, with the product of the first reaction serving as the template for a second reaction where the oligonucleotide primers used hybridize to sequences contained within that product. In this manner, a single parasite genome can be detected routinely and reproducibly, and the sensitivity then depends solely on the quantity and nature of the initial DNA template. A single parasite can be reproducibly detected in the DNA template directly purified from 10 µL of blood (0.000002% parasitemia). When substantially higher volumes of blood are used, the host's genomic DNA might adversely affect the efficacy of amplification. Removal of the white blood cells before DNA purification would overcome this problem, and thus substantially lower parasitemias would become amenable to PCR analysis. Amplification assays are also suited to the detection and species identification of *Plasmodium* parasites in the insect vector *(21–27)*. This provides a definite improvement over microscopic examination of oocysts or sporozoites that differ little morphologically between the human parasite species.

Ultimately, any DNA sequence which proves unique to a particular parasite species, and which is not found in the genomes of the vertebrate or insect hosts, can be used as a target for PCR amplification. Although many such sequences are known for *P. falciparum*, a smaller choice is available for the other human malaria parasites and the *Plasmodium* species that infect simian, rodent, avian, and saurian hosts. The genes for the small subunit ribosomal RNA (ssrRNA) are often used for phylogenetic analysis of organisms and have been characterized from a wide range of *Plasmodium* species *(28–37)*. Sequence comparison has revealed the presence of DNA stretches unique to each of the parasite species or shared by all members of the *Plasmodium* genus, but not found in other organisms. These differences have been initially exploited to detect and differentiate the four human *Plasmodium* species, through hybridization to blotted total rRNA obtained from infected blood *(38,39)*. This strategy has been adapted to a nested PCR protocol that can be used for the detection of any malaria parasite and for the identification of the four *Plasmodium* species that infect humans *(13,40–43)*. Quantification methods with a single PCR amplification reaction using primers targeting *Plasmodium* genus-conserved ssrRNA gene sequences are described elsewhere *(41,44–46)*.

2. Materials

2.1. Designated Areas

Ideally, the following procedures should be carried out in physically separated rooms (*see* **Note 1**):

1. DNA template preparation from samples and their storage.
2. Storage, preparation, and aliquoting of PCR reagents.
3. Addition of DNA template to amplification reaction tubes.
4. Addition of template from the Nest 1 to the Nest 2 reactions.
5. Analysis and storage of PCR product (*see* **Note 2**).

2.2. Equipment

See **Note 3**.

1. Micropipetors and tips.
2. Thermal cycler.
3. Apparatus for agarose gel electrophoresis.
4. Photographic equipment and UV transilluminator.
5. Refrigerator (4°C) and freezer (–20°C).
6. Microcentrifuge.

2.3. Reagents

See **Note 4**.

1. PCR buffer (10X stock): 500 mM KCl, 100 mM Tris-HCl, pH 8.3, 20 mM MgCl$_2$ and 1 mg/mL gelatin. Store at 4°C or –20°C (*see* **Note 5**).
2. Deoxyribonucleoside triphosphate (dNTP) stock solution: 5 mM of each of the four dNTPs: dATP, dCTP, dGTP, and dTTP. Store working stocks at –20°C, and back-up stocks at –70°C.
3. Oligonucleotide primers: a 2.5 µM stock of each (*see* **Note 6**). Store working stocks at –20°C, and back-up stocks at –70°C.

4. AmpliTaq polymerase (Cetus) (*see* **Note 5**). Store at –20°C.
5. Mineral oil. Store at room temperature.
6. Loading buffer (5X stock): 50 m*M* Tris-HCl, pH 8.0, 75 m*M* EDTA, pH 8.0, 0.5% SDS, 30% w/v sucrose, 10% w/v Ficoll-Hapaque (average molecular weight, 400,000), and 0.25% w/v approximately of Orange G dye. If Ficoll is not available, use 40% sucrose. Store at room temperature.
7. Tris-borate-EDTA (TBE) buffer (10X stock): 1 *M* Tris-HCl, 1 *M* boric acid, 50 m*M* EDTA. A pH of approx 8.3 should be obtained without any adjustment. Store at room temperature.
8. High-resolution agarose such as NuSieve™ agarose or MetaPhor™ agarose (FMC Bioproduct, Inc.).
9. Ethidium bromide solution: 10 mg/mL in water. Extreme care should be taken when handling this solution as this chemical is highly carcinogenic. Store in the dark at 4°C.
10. Water (sterile).

3. Methods

3.1. Sample Collection and Template Preparation

Samples to be analyzed by PCR can be obtained in different forms either from the vertebrate or the insect hosts. It is beyond the scope of this chapter to provide an exhaustive review of the collection procedure and the various methods used to obtain DNA template suitable for PCR amplification protocols. Nonetheless, the following points should be considered.

1. Correct, consistent, and clear labeling is of the utmost importance. Care must be taken that a unique code or number is assigned to each sample, and that this code or number is legibly written on the label. Labeling of the sample tubes or filter paper spots must be done with waterproof indelible ink. Tubes are best labeled using sticky labels and marker pens suitable for cold storage. Keeping the tubes dry before and after freezing prolongs the life of the label. A further label on the top of the tube, though desirable, is not always practical.
2. The value of the highly sensitive PCR protocols lies in their ability to detect very low numbers of parasites. The potential advantage of PCR over microscopic examination is therefore best realized when a substantial volume of blood (≥5 µL) is screened in each PCR assay. Depending on the method of DNA isolation and the nature of the sample, the final volume of the DNA template might substantially exceed the original volume of the blood sample from which it was purified. Concentrating the DNA template, for example, by ethanol precipitation, should be considered in these cases.
3. For epidemiological studies where sampling is performed in remote areas, over lengthy periods of time, or by untrained staff, the simplest method of collection is strongly recommended. Finger-prick blood spotted and dried on filter paper offers the most practical and robust method of sample collection, storage, and transport. Commercially available kits for this type of collection, although relatively costly, provide a standardized sample volume per spot from which the DNA template can be rapidly obtained (for example Isocode™ Stix PCR preparation dipsticks from Schleicher and Schüll, Dassel, Germany).
4. The presence of PCR inhibitors in the DNA template can seriously reduce the sensitivity of the amplification assay. The anticoagulants heparin or EDTA when used in high concentration might be carried over in the template solution. The presence of large quantities of host DNA can also reduce the efficiency of the reaction, a problem particularly encountered if parasites are sought in tissue samples when fixative, such as formalin, can also cause problems. The nature of the support used for dried blood spot collection, the storage, and processing of such samples, can influence the efficiency of the assay *(47,48)*.

Finally, PCR inhibitors are often found in templates purified from mosquito samples *(21,49)*. The choice of DNA template preparation method and/or dissection of the midguts or the salivary glands before template preparation minimizes such inhibition *(50)*.

5. It is advisable to test the efficacy of the PCR protocol for each given combination of samples and template purification method. This is achieved by spiking uninfected samples with a known quantity of parasite DNA or serially diluted blood containing a known quantity of parasites, then subjecting these samples to routine storage before preparing the template and establishing the level of PCR detection sensitivity, and whether inhibition of the PCR assay occurs.

3.2. The Amplification Reactions

The amplification strategy used for the detection and identification of malaria parasites is the nested PCR. Two advantages of this method are a substantially increased detection sensitivity and a decreased susceptibility to minor variations in the amplification conditions and DNA template quality. The disadvantages are the additional time and materials required, and the substantially increased risk of contamination. The latter can be minimized provided the precautions and methodology described later are assiduously followed.

In the first amplification reaction (Nest 1), a pair of oligonucleotide primers (genus-specific rPLU1 and rPLU5), which will hybridize to sequences in the ssrRNA genes of any *Plasmodium* parasite, are used to amplify a 1.6–1.7 kb fragment of these genes (the size varies with the species). The product of this first reaction is then used as the DNA template for a second amplification reaction (Nest 2), in which the oligonucleotide primers used recognize sequences contained within the DNA fragment amplified in the first reaction (**Fig. 1**). Two types of Nest 2 reactions, distinguished by the nature of the oligonucleotide primers used, can then be performed.

1. Genus-specific: When the *Plasmodium* genus-specific primers rPLU3 and rPLU4 are used (**Fig. 1**), detection of a PCR product (product of approx 235 bp) will indicate the presence of malaria parasites in the sample. These primers have been successfully used to detect all the *Plasmodium* species and subspecies that infect humans and rodents, *P. gonderi*, *P. pitheci*, and *P. silvaticum* from nonhuman primates, and *P. gallinaceum* from birds. Moreover the sequences recognized by the four primers are found in the published sequences of the ssrRNA genes of another four species of parasites that infect monkeys or apes, *P. lophurae* of birds and *P. floridense* of saurians. These primers are therefore considered to be effective at detecting the presence of any *Plasmodium* species.
2. Species-specific: Oligonucleotide primers (**Fig. 1**) specific to the four *Plasmodium* species which infect humans are available for use in SEPARATE Nest 2 reactions (*see* **Note 7**): rFAL1 and rFAL2 for *P. falciparum* (product of 206 bp), rMAL1 and rMAL2 for *P. malariae* (product of 145 bp), rVIV1 and rVIV2 for *P. vivax* (product of 121 bp). Previously the primers rOVA1 and rOVA2 have been used for *P. ovale* *(13)*; however, observations in Southeast Asia have revealed that some lines of this parasite harbor variant ssrRNA sequences *(51,52)*. Experiments using such parasites show that the rOVA2 primer does not hybridize to the variant genes. Since the complete sequence of this gene is not currently known, *P. ovale* parasites are detected using rOVA1 and rPLU2, a genus-specific oligonucleotide primer (product of 226 bp). In all cases, the specific product is only obtained when DNA from the corresponding parasite species is present.

The *Plasmodium* genus-specific detection protocol makes this sensitive PCR method available to researchers wishing to detect any malaria parasite species without the prior

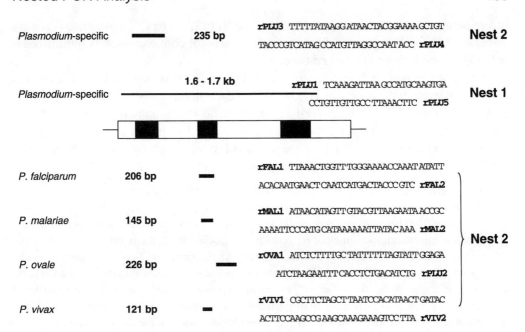

Fig. 1. Schematic representation of a plasmodial ssrRNA gene with the sequences and the approximate positions of the different oligonucleotide primer pairs. Thin or thick lines indicate the products of Nest 1 and Nest 2 reactions, respectively, with the expected sizes indicated. The name of each oligonucleotide primer is in bold, and is placed before the sequence (presented 5' to 3') for sense primers, and after the sequence for antisense primers. The sequences of the ssrRNA genes to which the *Plasmodium*-specific primers correspond, are highly conserved amongst all published sequences, although in some cases a few mismatches at the 5'-end of the primers are noted. The base pair position of these primers is given with respect to the ssrRNA A-type gene of *P. falciparum* (accession no. M19172): rPLU1 37–60 & rPLU5 1653–1673; rPLU3 132–161 & rPLU4 353–364. The sequences recognized by the human malaria species-specific primers are found in the following ssrRNA genes, and the base pair positions are as follows: *P. falciparum* C-type gene (accession no. M19173): rFAL1 664–693 & rFAL2 840–869; *P. vivax* A-type gene (accession no. U03079): rVIV1 666–695 & rVIV2 757–786; *P. malariae* A-type gene (accession no. M54897): rMAL1 686–715 & rMAL2 801–830; *P. ovale* A-type gene (accession no. L48987): rOVA1 770–799 & rPLU2 969–995 (rPLU2 is a *Plasmodium*-specific oligonucleotide primer).

necessity of obtaining specific sequence data for the species of interest. It can also provide a substantial saving in resources for epidemiological surveys, especially in areas with known low parasite prevalence, since large numbers of field samples can be rapidly prescreened for the presence of the human malaria parasites using two PCR reactions, and only those found positive will then be subjected to a further four Nest 2 reactions (where all four species are present) to determine the species compositions.

3.2.1. Setting Up the Amplification Reaction

For all the PCR amplifications, the volume used for each reaction is 20 µL (*see* **Note 8**). A master mix containing all the reagents, except for the DNA, is prepared and aliquoted into the reaction tubes (*see* **Notes 9–11**), and overlaid with mineral oil (*see* **Note 12**). It is worth making sure that the reagents are fully thawed out and vortexed before being used to prepare the master mix (*see* **Note 13**). The DNA template is always added last. The risk of cross-contamination and aerosol contamination is minimized if only ONE tube is open at any one time.

In order to minimize contamination, template addition is made after the aliquoted reaction mixtures have been overlaid with oil. Reaction mixtures for Nest 1 and Nest 2, as well as template addition for Nest 1, can be made and aliquoted in the same room. However, template addition for the Nest 2 reaction MUST be performed in a separate room. Product analysis MUST be done in a third room, in which none of the reagents used for PCR should ever be opened or, preferably, stored.

1. Remove the PCR buffer and oligonucleotide primers from the freezer and allow to thaw. The PCR buffer is stable indefinitely at room temperature. The oligonucleotide primers will not suffer from being placed repeatedly at ambient temperature for a few minutes at a time.
2. Add the appropriate volumes of reagents to the labeled master mix tube in the following order: water, PCR buffer, oligonucleotide primers. The final $MgCl_2$ concentration is 2 mM, variations in the $MgCl_2$ (1–4 mM) have not been found to affect the efficiency and specificity of the amplification. Each oligonucleotide primer is used at a final concentration of 250 nM. Decreased efficiency of amplification was observed with oligonucleotide primers at lower concentrations. The oligonucleotide pairs used in the Nest 1 and Nest 2 reactions are given in **Fig. 1**.
3. Remove the dNTP aliquot from the freezer, thaw, add appropriate amount to the master mix tube, and replace immediately in the freezer. The final concentration of each dNTP is 125 µM.
4. Remove the *Taq* polymerase from the freezer, add the appropriate amount to the master mix tube, and replace immediately in the freezer. A total of 0.4 units of enzyme is used for each 20 µL reaction (2 U/100 µL).
5. Mix the contents of the master mix tube thoroughly by vortexing (*see* **Note 14**).
6. Aliquot 20 µL of the master mix to each tube. Remember to open only one experimental tube at a time, closing it as soon as the master mix is aliquoted. This will minimize the possibility of contamination. The same tip can be used for all the tubes.
7. Add 50 µL of mineral oil to each tube, taking care to avoid splashes by adding the oil on the side of the wall at the top of the tube and allowing gravity to do the rest. It is very important that the same volume of oil is added to each tube (*see* **Note 15**). The same tip can be used for all the tubes.
8. DNA template addition must be made last and only after overlaying the master mix with mineral oil, so as to ensure a minimum risk of contamination. The appropriate amount of DNA template (usually 1 µL) is added by immersing the micropipet tip INSIDE the oil, preferably in the middle of the oil layer. The droplet will sink in the oil and merge with the master mix. This procedure will prevent aerosols escaping and contaminating other tubes (*see* **Note 16**).
9. When aliquoting the product of the Nest 1 reaction as a DNA template for the Nest 2 reaction, the following procedure must be used. Remove the DNA template aliquot from the Nest 1 tube from UNDER the oil, taking care that it is the product and not the oil that is removed! This is achieved by placing the tip of the micropipet in the middle of the aqueous layer, just below the oil and without touching the tube walls. Check that it is indeed the DNA template that is added to the Nest 2 tube, by observing the droplet as it is added to the oil layer in the Nest 2 tube. This procedure will prevent aerosols escaping and avoid contaminating other tubes.
10. Do NOT mix the contents by vortexing the tubes as this will result in an increased risk of contamination, and will necessitate a centrifugation step.

3.2.2. The Amplification Reactions

The cycling parameters for the PCR amplifications are as follows (*see* **Notes 15**, **17**, and **18**).

Step 1: 95°C for 5 min.   Initial denaturation
Step 2: X°C for 2 min.   Annealing
Step 3: 72°C for 2 min.   Extension
Step 4: 94°C for 1 min.   Denaturation
Step 5: Repeat **steps 2–4** for a total of 25 cycles (Nest 1) or 30 cycles (Nest 2)
Step 6: X°C for 2 min.   Final annealing
Step 7: 72°C for 5 min.   Final extension
Step 8: The reaction is completed by reducing the temperature to 25°C (*see* **Note 19**).

X = 58°C for Nest 1 (rPLU1/rPLU5) and all species-specific Nest 2 reactions (rFAL1/rFAL2; rMAL1/rMAL2; rOVA1/rPLU2; rVIV1/rVIV2).

X = 64°C for *Plasmodium*-specific Nest 2 reactions (rPLU3/rPLU4).

These conditions have been determined using a PTC-100 thermal cycler (MJ Research Inc.).

3.3. PCR Product Analysis

3.3.1. Electrophoresis

As a result of the very high sensitivity of the Nested PCR methodology, the amount of amplification product obtained from one parasite is sufficient for visualization by ethidium bromide staining following electrophoresis. The DNA products expected from all the different amplification reactions range between 121 bp to 235 bp (**Fig. 1**). The NuSieve™ agarose:agarose mixture has proved to be an ideal medium because of the ease of utilization, and the fact that a gel can be reused for a large number of times (*see* **Notes 20** and **21**). This is a particularly useful property since both types of agarose are relatively expensive. Although the suppliers of NuSieve agarose (FMC Bioproduct Inc.) recommend a 3% w/v gel with a 3:1 ratio of NuSieve agarose:agarose, equally successful resolution of product can be obtained with gels in which a cheaper ratio is used, namely, a 2% w/v gel with a 1:3 ratio of NuSieve agarose to agarose. MetaPhor™ gels are also suitable at 3% w/v, and can also be reused as above. Gels are made and electrophoresed in TBE buffer. The dissolved agarose mixture should be allowed to cool to approximately 55°C before pouring into the gel cast, and then allowed to set for at least 30 min (with an additional 30 min in the fridge for MetaPhor™ gels) before removing the combs and using.

In order to prepare the PCR product for electrophoresis, add 5 µL of loading buffer to the tubes at the end of the Nest 2 amplification. A brief centrifugation will bring the Loading Buffer under the oil layer. It is very important to mix the PCR reaction with the loading buffer before loading onto the gel. This is best done by repeated up-and-down pipetting just prior to loading onto the gel. The positive controls for the PCR amplification will be sufficient as a size marker. It is advisable to use 12.5 µL from the total of 25 µL when analyzing the PCR product, and to store the remainder, until the results have been interpreted and recorded correctly. The tracking dye, Orange G, migrates below the level of the smallest PCR product expected, thus electrophoresis is carried out until the Orange G reaches the end of the gel.

DNA is visualized under ultraviolet (UV) illumination, following staining by ethidium bromide. Ethidium bromide is a very carcinogenic chemical. Consequently, ethidium bromide is neither added to the gel nor to the TBE buffer. The gel is better stained following electrophoresis, by immersion in TBE buffer containing ethidium bromide (0.1–0.5 µg/mL) for a period of 30 min. It is then destained by placing in TBE buffer for a minimum of 10 min. Destaining times of up to 3 h are unlikely to result in appreciable loss of the DNA product bands.

3.3.2. Interpretation

A DNA product of approx 1.6–1.7 kb in size, which represents the product from the Nest 1 reaction and irrespective of the primers used in the Nest 2 reaction, is sometimes observed, particularly when the number of parasites in the samples is high, or when the oligonucleotide concentration or the number of cycles in the Nest 1 reaction are high.

The specificity of the genus- and the species-specific oligonucleotide primer pairs is very high, and the amplification products from the Nest 2 reactions have diagnostic sizes (**Fig. 1**). The efficiency of amplification by nested PCR is such that the amount of product obtained does not alter with a large range of parasite DNA in the original sample *(2,13)*. The observation of the specific PCR product of varying intensity is an indication that the copy number of the target is close to the detection limit, or that the amplification reaction might not be functioning at full efficiency, a common observation when PCR inhibitors are still present in the DNA template. When the amount of parasite material present in the sample under analysis is very low, inconsistency in the amplification might be observed. In other words, repeated amplification from the same sample could result in the detection of product in some cases, but not others, the "all-or-none" effect *(13)*. This is due to the very high sensitivity of the PCR analysis presented here, in which only a few parasite genomes are required to give a positive result. Thus, if the concentration of the target DNA in the aliquot used for the analysis is equivalent to less than one parasite, the result of the amplification will depend on the probability of picking this target DNA in an aliquot from the DNA template solution.

Two types of ssrRNA genes, differing in their sequence and size, are known to be present in *Plasmodium*, one expressed in the asexual stages and the other in the sexual stages *(53)*. Although, the oligonucleotide primers used for the PCR analysis are designed to recognize only one of the two ssrRNA gene types, amplification from the other gene type might occur when large quantities of parasite DNA are present in the sample. This probably accounts for the specific band of slightly higher molecular size that is frequently observed for samples that originally contain a large number of parasites.

4. Notes

1. There are two major problems encountered with the use of PCR. The first is the reduction of the amplification efficiency, which is easily detectable through the use of appropriate controls and which is mainly due to human error, problems with the equipment, or the quality of the reagents or the DNA template. The second is contamination, is invariably the result of human error. Two types of contamination are possible. In the first type, parasite material is transferred between samples (cross-contamination), or between the samples and the PCR reagents. In the second type, a PCR product from either of the two nested amplification reactions, comes into contact with the samples, the reagents, or the equipment. The second type of contamination is by far the more serious, and unfortunately the easiest to achieve. The extent of the precautions to be taken ultimately depends on the particular situation of each laboratory, and consequently a compromise will almost always

have to be reached. Paramount in the mind of the researcher must be is the adoption of the following procedures.
   a. The probability of contact between the PCR template and the PCR reagents and equipment is minimized.
   b. The PCR product NEVER comes into contact with either of the samples, the DNA template, the PCR reagents, or the equipment used for their handling.

   Thus, physically separated areas for the equipment and reagents must be assigned to the different procedures.

2. When space is at a premium, the first three areas could be combined, provided that care is taken to prevent cross-contamination between tubes. The addition of template for the Nest 2 reaction, however, must be performed in a separate room, and handling of the PCR product requires yet another room. It must be stressed that the setting up of PCR amplifications does not need to be performed under sterile conditions. Ultimately, only one room is required to be dedicated for PCR analysis, namely, the room in which the PCR product is handled. Only access to two other rooms, one for **steps 1–3** and one for **step 4**, will be needed. Since these procedures are relatively short in duration, this should pose little problem even in overcrowded laboratories. The rooms in which **steps 1–3** are carried out are considered as PCR clean, whereas those in which **steps 4** and **5** are performed are PCR dirty. The risk of contamination in the room assigned to **step 5** is a certainty, whereas in the room where **step 4** is carried out, the risk of contamination is minimal provided, good technique is used (*see* **Note 4**). It should be noted that any room can be considered as an absolutely PCR clean room for malaria work, provided that the template of the amplification reaction, namely any genetic material from *Plasmodium* species (tissue or blood samples, cultured parasites and lysates, bacterial clones or PCR product), have never been handled within this room.

3. All the equipment (tubes, pipet tips, and containers) to be used for the PCR analysis should be purchased new and dedicated to a specific procedure, and therefore, the appropriate room. Ideally three sets of micropipettes should be purchased, one for sample handling and DNA template preparation, one for PCR reagents handling, and, finally, one for PCR product handling. When using nested PCR, an additional micropipet must be obtained and dedicated for DNA template addition from the Nest 1 to the Nest 2 reaction. Positive displacement micropipets, which reduce the risk of contamination, are only advisable for handling solutions that contain DNA template. Filtered tips which are expensive, might, for example, be considered for the DNA template addition from the Nest 1 to the Nest 2 reaction; they are not essential to obtain contamination-free amplification. The most efficient method to decontaminate equipment is to soak it for a few hours in a $0.1–0.2\ M$ HCl, should this be possible. The acid depurinates DNA, and thus renders it refractive to PCR amplification. UV exposure of equipment and surfaces provides an efficient way to neutralize contaminating DNA, whereas autoclaving, although of some use, is not as efficient at degrading DNA.

4. All the chemicals to be used for the procedure must be obtained before initiating the project, and dedicated to PCR work. They must only be opened in an absolutely PCR clean room. Aliquoting, weighing, pH adjustments, and storage must be done with absolutely PCR clean materials. During storage, a Parafilm strip should be used as an extra seal around the closed lid of the bottles. Aliquoting from the stocks to obtain working solutions must be done under sterile and PCR clean conditions. The hierarchy of stocks and their backup should be considered carefully. It is strongly advisable to divide all chemicals and solutions into at least two, or preferably three, aliquots. The first aliquot is used as the working stock, and the second aliquot should be considered as the backup stock, and only used when the working stock runs out, or becomes "suspect." Finally, a reference stock should be considered. It is hoped that this stock should never have to be

used, but it could provide a backup should a disaster befall the laboratory, and consequently should be stored in a separate laboratory or building. Contamination of any one of the chemicals or stocks will render the results of the analysis totally invalid. Once a stock is made, it should be divided up immediately, and the working stock aliquot tested for efficiency and lack of contamination.

5. The composition of the PCR buffer depends on the *Taq* polymerase used. The buffer used should be the one provided with the *Taq* polymerase purchased. These enzymes are currently available from a variety of sources (bacterial strains and companies). The reaction conditions described in this chapter have been optimized using the AmpliTaq polymerase. It is advisable to establish the optimal conditions if other enzyme/buffer combinations are used *(43,54)*. This is best achieved using standard templates known to contain a defined number of parasite genomes (0.1, 1, 10, 100, 1000, and 10,000 genomes per aliquot). Optimal $MgCl_2$ concentrations should be first established, as this component usually has the most impact on the efficiency of the amplification reaction. The effective final concentration of Mg^{2+} might be inadvertently reduced by:
    a. The addition of EDTA, which is often used in the buffer used to resuspend the DNA template.
    b. The presence of large concentrations of dNTPs, since these molecules chelate Mg^{2+} ions in equimolar proportions.

6. The calculations used to obtain the concentration of the oligonucleotides described here have been performed assuming that:
    a. 330 g is the average molecular weight of each base.
    b. An optical density $OD_{260\,nm} = 1.0$ for a 1-cm path in a quartz cuvette, is equivalent to a solution containing 33 µg of oligonucleotide primer per milliliter of water.

    However, the above quantity is only a crude estimate. More accurate calculations can made by taking the base composition into account, and thus an extinction coefficient is obtained for each individual primer. This is the method employed by some companies who supply oligonucleotides. The sequences and the names of the oligonucleotide primers are given in **Fig. 1**.

7. Since the PCR products from the species-specific Nest 2 reactions differ in size, there is a temptation to perform all these reactions simultaneously in a single tube (multiplex PCR), thus saving time and money. This strategy has been tested with the oligonucleotide primers presented here, and was found to result in a substantial loss of sensitivity: when parasites were artificially mixed in widely differing ratios, the minor parasite species (100–10,000 times less than the other species depending on the total number of parasites in the aliquot tested) could not be detected in multiplex PCR Nest 2 reactions, but was invariably detected when individual Nest 2 reactions were performed. This phenomenon could be inherent to the primers used, since it was not apparently noted by other researchers who independently developed a nested PCR strategy for human *Plasmodium* species detection using the ssrRNA genes *(20)*. Nonetheless some sensitivity will invariably be lost as a result of competition between the different amplified fragments for the limited materials present in the reaction. "Multiplexing" is nonetheless an excellent cost-saving strategy, provided that the loss of sensitivity, whose extent can be established experimentally, falls within the tolerance of each particular investigation.

8. This small reaction volume has been found sufficient for the purposes of highly sensitive detection. Larger volumes will not improve the efficiency of detection, but will cost more in materials, particularly the expensive *Taq* polymerase. Nonetheless, when the volume of the template aliquot to be analyzed is large, or when dilution of possible PCR inhibitors in this template is sought, larger reaction volumes become necessary but usually only for the Nest 1 reaction.

9. PCR reactions do not need to be set up under sterile conditions. The tubes and pipet tips do

not require autoclaving, although they must be PCR clean. It should be noted that autoclaving does not destroy DNA totally, and cannot therefore, be used to "clean up" reagents and materials. Since the reagents are stored at $-20°C$, sterility is unnecessary. Moreover, once the reaction is initiated, the temperature does not fall below $58°C$ and is often at $94°C$, conditions hardly suitable for bacterial growth or endonuclease activity.

10. Despite what is written in most manuals, it is strongly recommended to AVOID the use of gloves when setting up PCR. Most types of gloves generate a large amount of static electricity, which will trap any aerosols that might be formed around the fingers, and promote the dissemination of these possible contaminants. In addition, gloves are not only uncomfortable but also expensive. Washing of hands before setting up the reactions ensures minimal contamination.
11. It is definitely preferable to have PCR tubes with a flat lid, on which the label can be written. It is also very wise to place the tubes in racks, in the same order as that of the samples, with this order being kept throughout the PCR analysis.
12. The use of mineral oil is crucial when performing nested PCR, since it provides an excellent barrier against cross-contamination. The addition of DNA template into the oil layer drastically reduces the risks of aerosol formation, and therefore, contamination with DNA template or the Nest 1 PCR product. The development of thermal cyclers with heated lids has obviated the need of an oil barrier against evaporation. For the reasons above, it is still strongly advisable to add mineral oil to these reactions. This is all the more important to minimize contamination when arrays of tube or 96-well plates are used, since multiple samples are simultaneously open to the atmosphere.
13. If the freezer where the reagents are stored is close by, and the procedure described here is followed, the requirement for ice during the setting up of the PCR can be avoided.
14. The master mix is stable at room temperature for several hours without loss of efficiency.
16. When a large number of samples are analyzed, the risks of misaliquoting, for example mistakenly adding two different templates to one tube, are relatively high. A brief loss of concentration can therefore invalidate the results, without this being necessarily noticeable by the researcher. A simple way to avoid this problem is to move each tube along the rack (or to a back or front row) immediately after template addition, remembering to close the lid and check that the number of the template sample corresponds to that on the PCR assay tube.
17. The optimization of the parameters has been performed using a PTC-100 thermal cycler (MJ Research Inc.), and with the AmpliTaq enzyme/PCR buffer combination described. It is quite possible that changes to cycle number, step times, annealing temperature, and even oligonucleotide and dNTP concentrations, could be made without altering the sensitivity and the resolution of the technique. If a different thermal cycler, or another enzyme + buffer combination are used, the reaction conditions described might have to be altered. In my experience, the amplification reactions using the oligos described here are sensitive to increased annealing temperatures. Optimization of the reaction conditions, and confirmation that the sensitivity and specificity of the assay are adequate, are best performed using a series of templates containing DNA derived from known numbers of parasite genomes (1, 10, 100, 1000, etc.) for each of the four *Plasmodium* species.
15. The heating and cooling rates and the accuracy of the temperature calibration are two parameters that can affect the efficiency of the amplification. The fit of the tube into the machine is extremely important as it affects the rate of heat transfer. The use of a drop of oil in the "holes" of the heating block in the thermal cycler can be of help unless the manufacturers advise against such a procedure. Tube thickness varies between different suppliers, and will also affect the rate of heat transfer, thus the indiscriminate use of different types of tubes should be avoided. The use of tubes specifically designed for a particular thermal cycler is therefore an advantage. The tube content volume will also alter

the rate of heat transfer, thus it is important to keep the volume of the mineral oil overlay constant.
18. The efficiency of amplification can be monitored by including a positive control, to which the template added represents the minimum quantity of DNA required for the detection of the PCR product, namely that obtained from 1 to 10 parasites. A further positive control to which 1000-fold more template is added will monitor the quality of the amplification reagents. A minimum of one negative control (no DNA template or human DNA) must be performed each time a reaction is performed. In order to confirm that insignificant levels of contamination arise when performing the PCR analysis, a run of alternate positive and negative samples should be subjected to the nested PCR protocol on a regular basis.
19. Once the reaction is complete, the DNA product is quite stable, since any potential contaminating nucleases will have been destroyed by the lengthy incubation at high temperatures. Consequently, the use of a soak temperature lower than 20°C at **Subheading 3.2.2., step 8** is of no value, and actually will only result in the thermal cycler working unnecessarily. In fact, in tropical countries where the humidity might be high, temperatures lower than 25°C will result in considerable water condensation on and around the heating block.
20. Following use, gels can be stored indefinitely at room temperature, provided they are submerged in TBE buffer. It is advisable to change the TBE buffer once or twice in the first few days of storage, as ethidium bromide and the tracker dye diffuse out of the gels.
21. In order to reuse a gel, break it into small pieces and reboil it making sure that all the agarose has dissolved. The use of a microwave is by far the most rapid and practical way to remelt a gel. Despite frequent reuse, up to 30 times in our experience, the resolving power needed for the present purpose and the physical integrity of the gel are retained. Two problems are likely to be encountered. The reduction in gel volume through the occasional loss of gel pieces, or, more commonly, through evaporation due to frequent reboiling. The latter will alter the agarose concentration but is easily remedied by the addition of distilled water. The accumulation of dust and other dirt particles in the gel will eventually become troublesome. The commonest source is the powder from the gloves that must be worn when handling the gels. This is minimized by prewashing the gloved hands and the use of lint-free towels. The amount of money saved when large numbers of gels are required is substantial. An important point to bear in mind is that reused gels will always contain small amounts of ethidium bromide and should therefore be handled with care.

References

1. Brown, A. E., Kain, K. C., Pipithkul, J., and Webster, H. K. (1992) Demonstration by the polymerase chain reaction of mixed *Plasmodium falciparum* and *P. vivax* infections undetected by conventional microscopy. *Trans. R. Soc. Trop. Med. Hyg.* **86,** 609–612.
2. Snounou, G., Pinheiro, L., Gonçalves, A., Fonseca, L., Dias, F., Brown, K. N., et al. (1993) The importance of sensitive detection of malaria parasites in the human and insect hosts in epidemiological studies, as shown by the analysis of field samples from Guinea Bissau. *Trans. R. Soc. Trop. Med. Hyg.* **87,** 649–653.
3. Black, J., Hommel, M., Snounou, G., and Pinder, M. (1994) Mixed infections with *Plasmodium falciparum* and *P. malariae* and fever in malaria. *Lancet* **343,** 1095.
4. Bottius, E., Guanzirolli, A., Trape, J.-F., Rogier, C., Konate, L., and Druilhe, P. (1996) Malaria: Even more chronic in nature than previously thought; evidence for subpatent parasitaemia detectable by the polymerase chain reaction. *Trans. R. Soc. Trop. Med. Hyg.* **90,** 15–19.
5. Roper, C., Elhassan, I. M., Hviid, L., Giha, H., Richardson, W. A., Babiker, H. A., et al. (1996) Detection of very low level *Plasmodium falciparum* infections using the nested polymerase chain reaction and a reassessment of the epidemiology of unstable malaria in Sudan. *Am. J. Trop. Med. Hyg.* **54,** 325–331.
6. Wagner, G., Koram, K., McGuinness, D., Bennett, S., Nkrumah, F., and Riley, E. (1998) High incidence of asymptomatic malaria infections in a birth cohort of children less than one year of age in Ghana, detected by multicopy gene polymerase chain reaction. *Am. J. Trop. Med. Hyg.* **59,** 115–123.
7. Snounou, G., Pinheiro, L., Antunes, A. M., Ferreira, C., and do Rosario, V. E. (1998) Non-immune

patients in the Democratic Republic of São Tomé e Principe reveal a high level of transmission of *P. ovale* and *P. vivax* despite low frequency in immune patients. *Acta. Trop.* **70,** 197–203.
8. Arez, A. P., Snounou, G., Pinto, J., Sousa, C. A., Modiano, D., Ribeiro, H., et al. (1999) A clonal *Plasmodium falciparum* population in an isolated outbreak of malaria in the Republic of Cabo Verde. *Parasitol.* **118,** 347–55.
9. Jaureguiberry, G., Hatin, I., d' Auriol, L., and Galibert, G. (1990) PCR detection of *Plasmodium falciparum* by oligonucleotide probes. *Mol. Cell. Prob.* **4,** 409–414.
10. Sethabutr, O., Brown, A. E., Panyim, S., Kain, K. C., Webster, H. K., and Echeverria, P. (1992) Detection of *Plasmodium falciparum* by polymerase chain reaction in a field study. *J. Inf. Dis.* **166,** 145–148.
11. Barker, R. H. J., Banchongaksorn, T., Courval, J. M., Suwonkerd, W., Rimwungtragoon, K., and Wirth, D. F. (1992) A simple method to detect *Plasmodium falciparum* directly from blood samples using the polymerase chain reaction. *Am. J. Trop. Med. Hyg.* **46,** 416–426.
12. Kain, K. C., Brown, A. E., Mirabelli, L., and Webster, H. K. (1993) Detection of *Plasmodium vivax* by polymerase chain reaction in a field study. *J. Infect. Dis.* **168,** 1323–1326.
13. Snounou, G., Viriyakosol, S., Zhu, X. P., Jarra, W., Pinheiro, L., Do Rosario, V. E., et al. (1993) High sensitivity of detection of human malaria parasites by the use of nested polymerase chain reaction. *Mol. Biochem. Parasitol.* **61,** 315–320.
14. Tirasophon, W. and Panyim, S. (1993) PCR for low-level detection of malaria parasites in blood. *Meth. Mol. Biol.* **21,** 205–211.
15. Seesod, N., Lundeberg, J., Hedrum, A., Åslund, L., Holder, A. A., Thaithong, S., et al. (1993) Immunomagnetic purification to facilitate DNA diagnosis of *Plasmodium falciparum*. *J. Clin. Microbiol.* **31,** 2715–2719.
16. Tirasophon, W., Rajkulchai, P., Ponglikitmongkol, M., Wilairat, P., Boonsaeng, V., and Panyim, S. (1994) A highly sensitive, rapid, and simple polymerase chain reaction-based method to detect human malaria (*Plasmodium falciparum* and *Plasmodium vivax*) in blood samples. *Am. J. Trop. Med. Hyg.* **51,** 308–313.
17. Wataya, Y., Arai, M., Kubochi, F., Mizukoshi, C., Kakutani, T., Ohta, N., et al. (1993) DNA diagnosis of falciparum malaria using a double PCR technique: A field trial in the Solomon Islands. *Mol. Biochem. Parasitol.* **58,** 165–167.
18. Arai, M., Mizukoshi, C., Kubochi, F., Kakutani, T., and Wataya, Y. (1994) Detection of *Plasmodium falciparum* in human blood by a nested polymerase chain reaction. *Am. J. Trop. Med. Hyg.* **51,** 617–626.
19. Khoo, A., Furata, T., Abdullah, N. R., Bah, N. A., Kojima, S., and Wah, M. J. (1996) Nested polymerase chain reaction for detection of *Plasmodium falciparum* infection in Malaysia. *Trans. R. Soc. Trop. Med. Hyg.* **90,** 40,41.
20. Rubio, J. M., Benito, A., Roche, J., Berzosa, P. J., Garcia, M. L., Mico, M., Edu, M., and Alvar, J. (1999) Semi-nested, multiplex polymerase chain reaction for detection of human malaria parasites and evidence of *Plasmodium vivax* infection in Equatorial Guinea. *Am. J. Trop. Med. Hyg.* **60,** 183–187.
21. Schriefer, M. E., Sacci, J. B., Jr., Wirtz, R. A., and Azad, A. F. (1991) Detection of polymerase chain reaction-amplified malarial DNA in infected blood and individual mosquitoes. *Exp. Parasitol.* **73,** 311–316.
22. Tassanakajon, A., Boonsaeng, V., Wilairat, P., and Panyim, S. (1993) Polymerase chain reaction detection of *Plasmodium falciparum* in mosquitoes. *Trans. R. Soc. Trop. Med. Hyg.* **87,** 273–275.
23. Stoeffels, J. A. W., Docters van Leeuwen, W. M., and Post, R. J. (1995) Detection of *Plasmodium* sporozoites in mosquitoes by polymerase chain reaction and oligonucleotide rDNA probe, without dissection of the salivary glands. *Med. Vet. Entomol.* **9,** 433–437.
24. Bouare, M., Sangare, D., Bagayoko, M., Toure, A., Toure, Y. T., and Vernick, K. D. (1996) Simultaneous detection by polymerase chain reaction of mosquito species and *Plasmodium falciparum* infection in *Anopheles gambiae* sensu lato. *Am. J. Trop. Med. Hyg.* **54,** 629–631.
25. Arez, A. P., Palsson, K., Pinto, J., Franco, A. S., Dinis, J., Jaenson, T. G. T., Snounou, G., and Do Rosario, V. E. (1997) Transmission of mixed malaria species and strains by mosquitoes, as detected by PCR, in a study area in Guinea-Bissau. *Parassitol.* **39,** 65–70.
26. Pinto, J., Arez, A. P., Franco, A. S., Do Rosario, V. E., Palsson, K., Jaenson, T. G. T., et al. (1997) Simplified methodology for PCR investigation of midguts from mosquitoes of the *Anopheles gambiae* complex, in which the vector and *Plasmodium* species can both be identified. *Ann. Trop. Med. Parasitol.* **91,** 217–219.
27. Wilson, M. D., -Okyere, A., Okoli, A. U., McCall, P. J., and Snounou, G. (1998) Direct comparison of microscopy and polymerase chain reaction for the detection of *Plasmodium* sporozoites in salivary glands of mosquitoes. *Trans. R. Soc. Trop. Med. Hyg.* **92,** 482,483.
28. Unnasch, T. R. and Wirth, D. F. (1983) The cloned rRNA genes of *P. lophurae*: A novel rDNA structure. *Nucleic Acids Res.* **11,** 8461–8472.
29. Gunderson, J. H., McCutchan, T. F., and Sogin, M. L. (1986) Sequence of the small subunit riboso-

mal RNA gene expressed in the bloodstream stages of *Plasmodium berghei*: evolutionary implications. *J. Protozool.* **33,** 525–529.
30. McCutchan, T. F., de la Cruz, V. F., Lal, A. A., Gunderson, J. H., Elwood, H. J., and Sogin, M. L. (1988) Primary sequences of two small subunit ribosomal RNA genes from *Plasmodium falciparum*. *Mol. Biochem. Parasitol.* **28,** 63–68.
31. Waters, A. P., Unnasch, T. R., Wirth, D. F., and McCutchan, T. F. (1989) Sequence of a small ribosomal RNA gene from *Plasmodium lophurae*. *Nucleic Acids Res.* **17,** 1763.
32. Waters, A. P. and McCutchan, T. F. (1989) Partial sequence of the asexually expressed SU rRNA gene of *Plasmodium vivax*. *Nucleic Acids Res.* **17,** 2135.
33. Goman, M., Mons, B., and Scaife, J. G. (1991) The complete sequence of a *Plasmodium malariae* SSUrRNA gene and its comparison to other plasmodial SSUrRNA genes. *Mol. Biochem. Parasitol.* **45,** 281–288.
34. Corredor, V. and Enea, V. (1994) The small ribosomal subunit RNA isoforms in *Plasmodium cynomolgi*. *Genetics* **136,** 857–865.
35. Qari, S. H., Goldman, I. F., Pieniazek, N. J., Collins, W. E., and Lal, A. A. (1994) Blood and sporozoite stage-specific small subunit ribosomal RNA- encoding genes of the human malaria parasite *Plasmodium vivax*. *Gene* **150,** 43–49.
36. Waters, A. P., Higgins, D. G., and McCutchan, T. F. (1993) Evolutionary relatedness of some primate models of *Plasmodium*. *Mol. Biol. Evol.* **10,** 914–923.
37. Escalante, A. A., Goldman, I. F., De Rijk, P., De Wachter, R., Collins, W. E., Qari, S. H., and Lal, A. A. (1997) Phylogenetic study of the genus *Plasmodium* based on the secondary structure-based alignment of the small subunit ribosomal RNA. *Mol. Biochem. Parasitol.* **90,** 317–321.
38. Waters, A. P. and McCutchan, T. F. (1989) Rapid, sensitive diagnosis of malaria based on ribosomal RNA. *Lancet* **1,** 1343–1346.
39. Lal, A. A., Changkasiri, S., Hollingdale, M. R., and McCutchan, T. F. (1989) Ribosomal RNA-based diagnosis of *Plasmodium falciparum* malaria. *Mol. Biochem. Parasitol.* **36,** 67–72.
40. Snounou, G., Viriyakosol, S., Jarra, W., Thaithong, S., and Brown, K. N. (1993) Identification of the four human malaria parasite species in field samples by the polymerase chain reaction and detection of a high prevalence of mixed infections. *Mol. Biochem. Parasitol.* **58,** 283–292.
41. Hulier, E., Pétour, P., Snounou, G., Nivez, M.-P., Miltgen, F., Mazier, D., and Rénia, L. (1996) A method for the quantitative assessment of malaria parasite development in organs of the mammalian host. *Mol. Biochem. Parasitol.* **77,** 127–135.
42. Jarra, W. and Snounou, G. (1998) Only viable parasites are detected by PCR following clearance of rodent malarial infections by drug treatment or immune responses. *Infect. Immun.* **66,** 3783–3787.
43. Singh, B., Bobogare, A., Cox-Singh, J., Snounou, G., Abdullah, M. S., and Rahman, H. A. (1999) A genus- and species-specific nested polymerase chain reaction malaria detection assay for epidemiologic studies. *Am. J. Trop. Med. Hyg.* **60,** 687–692.
44. Vernick, K. D., Barreau, C., and Seeley, D. C. J. (1995) *Plasmodium*: A quantitative molecular assay for the detection of sporogonic-stage malaria parasites. *Exp. Parasitol.* **81,** 436–444.
45. Vernick, K. D., Keister, D. B., Toure, A., and Toure, Y. T. (1996) Quantification of *Plasmodium falciparum* sporozoites by ribosomal RNA detection. *Am. J. Trop. Med. Hyg.* **54,** 430–438.
46. Briones, M. R. S., Tsuji, M., and Nussenzweig, V. (1996) The large difference in infectivity for mice of *Plasmodium berghei* and *Plasmodium yoelii* sporozoites cannot be correlated with their ability to enter into hepatocytes. *Mol. Biochem. Parasitol.* **77,** 7–17.
47. Cox-Singh, J., Mahayet, S., Abdullah, M. S., and Singh, B. (1997) Increased sensitivity of malaria detection by nested polymerase chain reaction using simple sampling and DNA extraction. *Int. J. Parasitol.* **27,** 1575–1577.
48. Färnert, A., Arez, A. P., Correia, A. T., Björkman, A., Snounou, G., and Do Rosário, V. (1999) Sampling and storage of blood and the detection of malaria parasites by polymerase chain reaction. *Trans. R. Soc. Trop. Med. Hyg.* **93,** 50–53.
49. Siridewa, K., Karunanayake, E. H., and Chandrasekharan, N. V. (1996) Polymerase chain reaction-based technique for the detection of *Wuchereria bancrofti* in human blood samples, hydrocele fluid, and mosquito vectors. *Am. J. Trop. Med. Hyg.* **54,** 72–76.
50. Arez, A. P., Lopes, D., Pinto, J., Franco, A. S., Snounou, G., and Do Rosário, V. E. (2000) *Plasmodium* sp.: optimal protocols for PCR amplification of low parasite numbers from mosquito (*Anopheles* sp.) samples. *Exp. Parasitol.* **94,** 269–272.
51. Kawamoto, F., Miyake, H., Kaneko, O., Kimura, M., Nguyen Thi Dung, Nguyen The Dung, et al. (1996) Sequence variation in the 18S rRNA gene, a target for PCR-based malaria diagnosis, in *Plasmodium ovale* from southern Vietnam. *J. Clin. Microbiol.* **34,** 2287–2289.
52. Zhou, M., Liu, Q., Wongsrichanalai, C., Suwonkerd, W., Panart, K., Prajakwong, S., et al. (1998) High prevalence of *Plasmodium malariae* and *Plasmodium ovale* in malaria patients along the Thai-Myanmar border, as revealed by acridine orange staining and PCR-based diagnoses. *Trop. Med. Int. Health* **3,** 304–312.

53. Rogers, M. J., Li, J., and McCutchan, T. F., (1998) The *Plasmodium* rRNA genes: developmental regulation and drug target in *Malaria: Parasite Biology, Pathogenesis, and Protection* (Sherman, I. W., ed.), ASM Press, Washington, DC, pp. 203–217.
54. Singh, B., Cox-Singh, J., Miller, A. O., Abdullah, M. S., Snounou, G., and Rahman, H. A. (1996) Detection of malaria in Malaysia by nested polymerase chain reaction amplification of dried blood spots on filter papers. *Trans. R. Soc. Trop. Med. Hyg.* **90,** 519–521.

19

RFLP Analysis

Hans-Peter Beck

1. Introduction

The genetic identification of malaria parasites has become an important task for epidemiological studies and for laboratory-based studies *(1–7)*. A large number of adapted field isolates have been cultured and widely circulated during the past years. However, only few of these have been clearly characterized by molecular means and in some instances it has been shown that parasites previously believed to be derived from different isolates show a similar to identical molecular makeup *(8)*. Similarly, parasite cultures believed to be identical show significant differences on a genetic level *(9)*. Such unreliability of sources of research might cause havoc to the studies and might contribute to discrepant findings.

Many studies rely on the ability to unequivocally identify parasite strains or broods, such as recombination studies with parasite crossing that require the detailed analysis of parents and progeny *(10)*. Antimalarial treatment studies also need the analysis of molecular markers for the assessment of subsequent infections *(3,4,7)*.

During the past years, several methods have been developed for the characterization of *Plasmodium* parasites. Such methods include isoenzyme analysis *(11)*, polymerase chain reaction (PCR) amplification of polymorphic loci and subsequent analysis *(5,6)*, random amplified polymorphic DNA (RAPD)-analysis *(12)*, and restriction fragment length polymorphism (RFLP)-analysis or molecular fingerprinting *(13)*. Depending on the needs of the study, these different methods will provide the necessary discrimination power and resolution. The major issue in the decision as to which method to use will be the availability of sample material and required resolution.

The RFLP technique (or molecular fingerprinting) allows a rapid assessment of large numbers of samples with high to very high discrimination power. However, RFLP techniques require large amounts of sample material and usually radioactive-labeled nucleotides. With the advent of PCR technology, RFLP genotyping has been less frequently used but remains a powerful tool for genotyping of laboratory-cultured parasites with high discrimination and for use in genetic linkage studies.

Essentially, RFLP analysis detects subtle differences within a given DNA sequence or DNA sequences, the closely related fingerprinting techniques detects differences within the whole genome by making use of multiple and randomly distributed sequences *(14)*. Total DNA is cut into specific fragments with restriction endonucleases recognizing short sequences. Differences between DNA sequences might be seen as

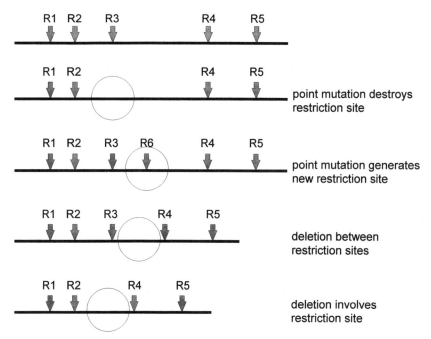

Fig. 1. Schematic representation of events changing the migration pattern of defined restriction fragments. R1–R6 represent the distribution of identical restriction sites along a given sequence. Radioactive probes corresponding to each of the resulting fragments will identify all mutational events and thus provide discrimination with high resolution.

the addition of new restriction sites, loss of restriction sites, or altered sizes of intervening sequences between two restriction sites (**Fig. 1**). These changes in the DNA will result in differences in the migration pattern of the fragments. However, it would be impossible to observe those changes within the complex smear of DNA fragments, and therefore a specific detection of the sequences to be analyzed is required. Usually, after size separation on agarose gels, the DNA is blotted onto a nylon membrane and hybridized with a radioactive labeled probe to visualize the specific banding pattern. The choice of probe to be hybridized is crucial and will determine resolution, discrimination power, and sensitivity. If differences within polymorphic sequences are of interest, this particular sequence is used for hybridization. In order to obtain a much higher discrimination, repetitive sequences, such as microsatellite sequences *(15)*, repeated sequences *(13)*, or sequences of gene families *(16)* have been used.

2. Materials
2.1. Equipment

1. Tabletop centrifuge.
2. Microfuge.
3. 37, 65, and 95°C heating blocks (or water baths).
4. Gel electrophoresis equipment.
5. UV transilluminator (320 nm).
6. Adjustable pipetors (P20, P200, P1000).
7. Trays.
8. Hybridization oven.

Genotyping: RFLP Analysis

9. Ice-maker.
10. Paper towels.
11. Whatman 3MM paper.
12. Charged nylon membranes.
13. X-ray-films.
14. ParaFilm™.
15. Darkroom and film development facilities.

2.2. Reagents

1. Restriction enzymes and buffers.
2. Proteinase K: 20 mg/mL in 10 mM Tris-HCl, pH 8.0.
3. Klenow polymerase (2 U/µL).
4. 10X Klenow buffer: 400 mM Tris-HCl, pH 7.5, 100 mM MgCl$_2$, 500 mM NaCl.
5. Deoxyribonuclease triphosphate (dNTP) mix: 0.5 mM dATP, 0.5 mM dTTP, 0.5 mM dGTP.
6. Random hexanucleotides (3 µg/mL).
7. α-^{32}P-dCTP (3000 Ci/mmol, 10 µCi/µL).
8. Tris-EDTA (TE) buffer: 10 mM Tris-HCl, 1 mM EDTA, pH 7.6.
9. Tris-borate-EDTA (TBE) buffer: 5 M Tris-HCl, 4 M boric acid, 10 mM EDTA, pH 8.0.
10. Phosphate-buffered saline (PBS): 11.5 g Na$_2$HPO$_4$, 2.96 g NaH$_2$PO$_4$·2H$_2$O, 5.84 NaCl per liter.
11. 1 M sodium-phosphate buffer, pH 7.2: prepare by adding 1 M NaH$_2$PO$_4$ to 1 M Na$_2$HPO$_4$ until pH 7.2 is reached.
12. 20X Sodium chloride-sodium citrate (SSC): 3 M NaCl, 0.5 M Na citrate, pH 7.0. Prepare 2X SSC working stock.
13. Hybridization solution (modified Church and Gilbert buffer) *(17)*: 0.5 M sodium phosphate buffer pH 7.2, 7% sodium dodecyl sulfate (SDS), 10 mM EDTA. Prepare 1 L of hybridization buffer by adding 350 mL of 20%SDS, 20 mL of 0.5 M EDTA and 130 mL of H$_2$O to 500 mL of 1 M sodium-phosphate buffer. Do not use the mixed buffer for more than 1 mo.
14. Low-stringency wash solution: 2X SSC and 0.1% SDS.
15. High-stringency wash solution: 0.1X SSC, 0.1% SDS.
16. SDS: 20% (w/v) in H$_2$O.
17. PBS with 0.05% (w/v) saponin.
18. Blue juice: TE containing 30% glycerol, 0.2% bromphenol blue, 0.2% xylene cyanol.
19. TBE buffer containing 0.01 mg/mL ethidium bromide.
20. Bovine serum albumin (BSA): 10 ng/µL.
21. Ethanol: 70%, 99%.
22. 3 M Na acetate (pH 4.5).
23. 0.25 N HCl.
24. 0.4 N NaOH.
25. 20 mM EDTA, pH 8.0.
26. TE-saturated phenol.
27. Chloroform.
28. DNA-grade agarose.
29. DNA-size marker.
30. High Prime Labeling Kit (Roche Diagnostics) (optional).

3. Methods

3.1. DNA Isolation

1. Grow parasite cultures to a density between 5 and 10% parasitemia (*see* **Note 1**).
2. Spin at 3500*g*.

3. Discard supernatant and add 5 pellet-volumes of ice-cold PBS with 0.05% (w/v) saponin. Mix and leave for 10 min on ice.
4. Pellet parasites for 5 min at 3500g.
5. Discard supernatant and wash pellet twice with PBS.
6. Resuspend the pellet in 1/100 of original culture volume TE.
7. Add 20% SDS (w/v in H_2O) to a final concentration of 0.5%.
8. Add proteinase K to a final concentration of 20 µg/mL (*see* **Note 2**).
9. Incubate at 65°C for 2–12 h.
10. Add 1 vol of TE-saturated phenol (pH 8.0) and 1 vol of chloroform, and mix for 5 min without vortexing (*see* **Note 3**).
11. Spin at 15,000g and remove upper aqueous phase.
12. Repeat phenol/chloroform extraction once followed by chloroform only extraction.
13. Add 1/10 vol of 3 M Na acetate (pH 4.5) and 2.5 vol of 99% ethanol, mix and freeze at –20°C for at least 1 h.
14. Spin 15 min at 15,000g.
15. Wash pellet with 70% ethanol.
16. Spin and air dry.
17. Dissolve pellet in TE (*see* **Note 4**).

3.2. Restriction Digest

1. Digest 10 µg of DNA with 10 U of appropriate restriction enzyme using the appropriate buffer (*see* **Note 5**).
2. Incubate for 2 h at the specified temperature (*see* **Note 6**).
3. Add 3 µL of blue juice (*see* **Note 7**).
4. Incubate for 5 min at 65°C.
5. Load sample onto gel (*see* **Fig. 2**).

3.3. Agarose Gel Electrophoresis

1. Prepare a 0.7% agarose gel in TBE buffer (*see* **Note 8**).
2. Load samples into slots; include a DNA size marker in parallel slots (*see* **Note 9**).
3. Run at 70–120 V (~6 V/cm), depending on gel size, until the bromphenol blue band reaches the end of the gel.
4. Incubate in TBE buffer containing 0.01 mg/mL ethidium bromide (*see* **Note 10**).
5. Visualize on UV transilluminator (312 nm).

3.4. Southern Blotting

1. Pretreat gel with 0.25 N HCl for 10 min (*see* **Note 11**).
2. Treat gel in 0.4 N NaOH for 20 min to denature DNA. Change NaOH once.
3. Prepare capillary blot as shown in **Fig. 3**:
    a. Place a glass plate on a support into a tray with 0.4 N NaOH. Make sure glass plate is higher than the level of liquid.
    b. Cut Whatman 3MM paper into two sheets wide as the gel, but longer to reach into the liquid in the tray.
    c. Soak sheets in 0.4 N NaOH and place on the glass plates.
    d. Place treated gel on the Whatman paper; avoid trapping air bubbles in between.
    e. Cut a sheet of positively charged nylon membrane (Hybond XL, Amersham, or equivalent) slightly larger than the gel (*see* **Note 12**).
    f. Prewet the membrane in 0.4 N NaOH and place it on top of the gel. Make sure not to trap air bubbles between filter and gel.
    g. Use ParaFilm™ to seal the gel in order to avoid short circuit between the Whatman paper at the bottom and filter paper above the gel.

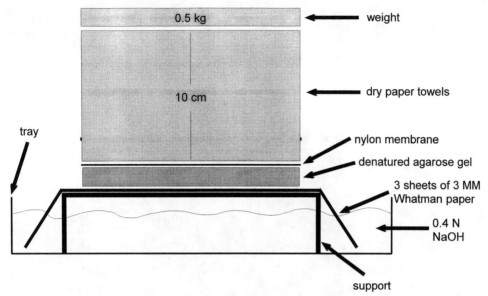

Fig. 2. Schematic assembly for Southern capillary blots. Assure that no close circuit can occur between the lower Whatman paper and the paper stack. Seal around the gel with ParaFilm™.

    h. Cut three sheets of 3MM paper to the size of the membrane.
    i. Soak the sheets in 0.4 N NaOH and place them on top of the filter. Remove trapped air bubbles by rolling a glass pipette of the sheets.
    j. Place a stack of dry paper towels (approx 10 cm high) on top of the 3MM paper.
    k. Place a glass plate on top of the paper towels and put a weight on top (approx 0.5 kg).
4. Allow the transfer to proceed for 12–16 h.
5. Dismantle the transfer setup, leaving the membrane on the gel. Turn gel and filter upside down on a dry 3MM paper. Mark the slots on the membrane using a pencil to punch through the slots.
6. Carefully remove the gel and restain the gel to assess transfer efficiency.
7. Rinse membrane quickly in 2X SSC, air dry, and immobilize DNA on the filter by placing the filter DNA side up in an UV-linker (Stratalinker, Stratagene) (*see* **Note 13**).

3.5. Probe Labeling

We routinely use the High Prime Labeling Kit (Roche Diagnostics) to generate radioactively labeled probes with consistent results. The protocol outlined below is based on the same principle but the reagents are not premixed, and can be purchased separately.

1. Dilute the DNA to be labeled to a concentration of 6 ng/µL (*see* **Note 14**).
2. Denature the DNA by incubating at 95–100°C for 5 min in a heating block or water bath.
3. Snap cool on ice or ice–methanol mixture.
4. While DNA is denaturing, add the following components to a screw-cap Eppendorf cup on ice: 2 µL of 10X Klenow buffer, 3 µL of dNTP mix, 1 µL of 3 µg/mL random hexanucleotides (*see* **Note 15**), 1 µL of 10 ng/µL BSA, and 2 µL of dH$_2$O.
5. Briefly spin the tube containing the denatured DNA and add 5 µL (approx 30 ng) to the labeling mix.
6. Add 5 µL of α-^{32}P-dCTP (3000 Ci/mmol, 10 µCi/µL) (*see* **Note 16**).
7. Add 1 µL of Klenow enzyme (2 U/µL).

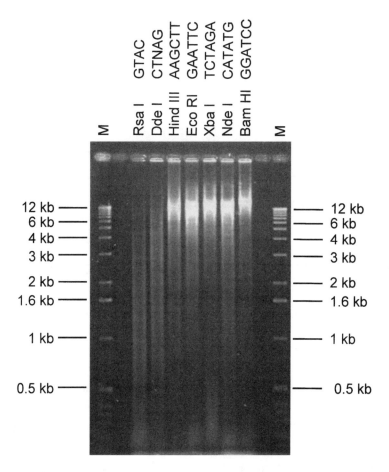

Fig. 3. *Plasmodium falciparum* DNA digested with various restriction enzymes and separated on a 0.7% agarose gel run in TBE buffer. The recognition sequence of the respective enzyme is given above the name. M: 1 kb-ladder (Gibco–LifeTechnologies). The high molecular weight band in the right 5 lanes emphasizes the fact that complete digests with uniformly distributed fragments can only be achieved with *Plasmodium falciparum* DNA by using restriction enzymes recognizing 4 bases only.

8. Incubate at 37°C for 30 min.
9. Stop the reaction by adding 30 µL of 20 mM EDTA, pH 8.0.

3.6. Hybridization

1. Place the nylon membrane in a hybridization tube and add 2 mL/10 cm^2 of prewarmed hybridization solution.
2. Prehybridize the filter at 64°C for 30 min in a hybridization oven.
3. Denature the labeled probe at 95–100°C for 5 min.
4. Discard prehybridization solution. Add fresh hybridization solution, prewarmed to 64°C.
5. Immediately add the denatured probe to the tube.
6. Hybridize overnight at 64°C with constant agitation.
7. Remove hybridization solution and store at 4°C.
8. Add an excess of preheated low-stringency wash solution. Incubate for 20 min in hybridization oven at 64°C with constant agitation.

9. Discard the wash solution and repeat **step 8** with fresh low-stringency wash solution.
10. Discard and add an excess of preheated high-stringency wash solution. Incubate for 10 min.
11. Discard the high-stringency wash solution. Place filter quickly on a dry sheet of Whatman 3MM paper to drain excess fluid and wrap the membrane in Saran wrap. Avoid trapping liquid in the Saran wrap. Do not allow the filter to dry out.
12. Autoradiograph the filter with a X-ray film (e.g., Kodak X-omat) and intensifying screen at −70°C for 2–16 h.

4. Notes

1. There is no need for synchronization. However, cultures in late trophozoite or schizont stage yield higher amounts of DNA than ring-stage cultures.
2. Both proteinase K and SDS prevent all action of nucleases. Incubation can vary between 2 h and overnight. Prior to adding SDS, the pellet must be well resuspended by pipetting TE up and down. Avoid vigorous pipetting after adding SDS because of shearing of the DNA.
3. Phenol must be water- or TE-saturated, not oxidized, and adjusted to a pH of 7.6. Oxidized phenol nicks DNA and the quality of DNA is low. Some protocols advise the researcher to add phenol first and to add chloroform just prior to the centrifugation, other protocols add a mixture of phenol:chloroform:isoamylalcohol (50:49:1). We have not observed any difference between these protocols. Protect skin and eyes from contact with phenol. Work when possible in a safety cabinet.
4. We usually dissolve the pellet in 1/500 or 1/1000 of the original culture volume.
5. The sequence to be digested should be screened for sites to achieve a high discrimination, that is, the digest should generate many fragments uniformly sized between 100 bp and 5 kb. If the sequence is unknown, usually a 4-bp cutter recognizing AT-rich sequences will provide sufficient discrimination. However, it is sometimes advisable to use two different restriction enzymes, which will result in increased discrimination power due to more uniformly cut fragments. Depending on the enzyme, this can be done simultaneously (same reaction) or must be done subsequently. If the reaction conditions differ significantly, a precipitation step between is recommended.
6. For some restriction endonucleases, it might be advisable to add another 10 units after the first incubation. Make sure that the volume of the digesting mixture is large enough that the glycerol concentration does not exceed 5% (restriction enzymes usually contain 50 % glycerol).
7. If the volume of the digest is larger than 10 µL, it is advisable to precipitate the digested DNA prior to loading. Subsequently, dissolve the DNA in 10 µL of TE. Similarly, high-salt restriction buffers and high glycerol concentrations render the migration of DNA fragments in gels and precipitation is also recommended.
8. Other buffer systems work equally well, although TBE gives the best resolution in our laboratory.
9. For the subsequent hybridization, markers that contain plasmid or phage fragments, also contained in the hybridization probe, might be useful. Such size markers can be commercially obtained, are batch-controlled with defined fragments, and are usually cheaper than those that are self-produced. Markers of the size 0.2–1.0 µg are usually clearly visible and can diluted more if hybridized. For a precise analysis of minor length differences, markers should be loaded after every second or third sample lane.
10. This staining step can be omitted if an aliquot of the digest is run separately to test whether the digest is complete. In completely digested samples, few distinct bands should be visible.
11. This step is necessary to depurinate long DNA molecules for improved transfer, but can be omitted if the restriction digest yields predominantly fragments below 5 kb.
12. Never touch the membrane with fingers; always wear gloves, that have been washed to rinse off talcum powder.

13. If no UV crosslinker is available, this step can either be omitted, because the alkali-transfer already binds single-stranded DNA covalently, or the additional crosslinking can be done by placing the membrane on Saran wrap DNA face-down on a UV transilluminator. Expose to UV for approx 30 s.
14. Denature more probe than used in the labeling reaction because small volumes evaporate despite closed lids.
15. Random hexanucleotides are cheaply commercially available (i.e., BRL LifeTechnologies).
16. For most labeling purposes, 25 µCi dCTP will be sufficient, but adjust the volume to a final volume of 20 µL.

References

1. Färnert, A., Rooth, I., Svensson, A., Snounou, G., and Björkman, A. (1999) Complexity of *Plasmodium falciparum* infections is consistent over time and protects against clinical disease in Tanzanian children. *J. Infect. Dis.* **179**, 989–995.
2. Färnert, A., Snounou, G., Rooth, I., and Björkman, A. (1997) Daily dynamics of *Plasmodium falciparum* subpopulations in asymptomatic children in a holoendemic area. *Am. J. Trop. Med. Hyg.* **56**, 538–547.
3. Brockman, A., Paul, R. E., Anderson, T. J., Hackford, I., Phaiphun, L., Looareesuwan, S., Nosten, F., and Day, K. P. (1999) Application of genetic markers to the identification of recrudescent *Plasmodium falciparum* infections on the northwestern border of Thailand. *Am. J. Trop. Med. Hyg.* **60**, 14–21.
4. Ohrt, C., Mirabelli-Primdahl, L., Karnasuta, C., Chantakulkij, S., and Kain, K. C. (1997) Distinguishing *Plasmodium falciparum* treatment failures from reinfections by restrictions fragment length polymorphism and polymerase chain reaction genotyping. *Am. J. Trop. Med. Hyg.* **57**, 430–437.
5. Viriyakosol, S., Siripoon, N., Petcharapirat, C., Petcharapirat, P., Jarra, W., Thaithong, S., et al. (1995) Genotyping of *Plasmodium falciparum* isolates by the polymerase chain reaction and potential uses in epidemiological studies. *Bull. World Health Organ.* **73**, 85–95.
6. Felger, I., Tavul, L., Kabintik, S., Marshall, V., Genton, B., Alpers, M., and Beck, H. P. (1994) Plasmodium falciparum: extensive polymorphism in merozoite surface antigen 2 alleles in an area with endemic malaria in Papua New Guinea. *Exp. Parasitol.* **79**, 106–116.
7. Irion, A., Felger, I., Abdulla, S., Smith, T., Mull, R., Tanner, M., et al. (1998) Distinction of recrudescences from new infections by PCR-RFLP analysis in a comparative trial of CGP 56 697 and chloroquine in Tanzanian children. *Trop. Med. Int. Health.* **3**, 490–497.
8. Kahane, B., Sibilli, L., Scherf, A., Jaureguiberry, G., Langsley, G., Ozaki, L.S., et al. (1987) The polymorphic 11.1 locus of Plasmodium falciparum. *Mol. Biochem. Parasitol.* **26**, 77–85.
9. Fandeur, T., Bonnefoy, S., and Mercereau-Puijalon, O. (1991) In vivo and in vitro derived Palo Alto lines of *Plasmodium falciparum* are genetically unrelated. *Mol. Biochem. Parasitol.* **47**, 167–178.
10. Walliker, D., Quakyi, I. A., Wellems, T. E., McCutchan, T. F., Szarfman, A., et al. (1987) Genetic analysis of the human malaria parasite *Plasmodium falciparum*. *Science* **236**, 1661–1666.
11. Sanderson, A., Walliker, D., and Molez, J. F. (1981) Enzyme typing of Plasmodium falciparum from African and some other Old World countries. *Trans. R. Soc. Trop. Med. Hyg.* **75**, 263–267.
12. Howard, J., Carlton, J. M., Walliker, D., and Jensen, J. B. (1996) Use of random amplified polymorphic DNA (RAPD) technique in inheritance studies of *Plasmodium falciparum*. *J. Parasitol.* **82**, 941–946.
13. Dolan, S. A., Herrfeldt, J.A., and Wellems, T. E. (1993) Restriction polymorphisms and fingerprint patterns from an interspersed repetitive element of *Plasmodium falciparum* DNA. *Mol. Biochem. Parasitol.* **61**, 137–142.
14. Jeffreys, A. J., Brookfield, J. F., and Semeonoff, R. (1985) Positive identification of an immigration test-case using human DNA fingerprints. *Nature* 6, 317, 818,819.
15. Su, X. Z., Carucci, D. J., and Wellems, T. E. (1998) *Plasmodium falciparum*: parasite typing by using a multicopy microsatellite marker, PfRRM. *Exp. Parasitol.* **89**, 262–265.
16. Carcy, B., Bonnefoy, S., Schrevel, J., and Mercereau-Puijalon, O. (1995) *Plasmodium falciparum*: typing of malaria parasites based on polymorphism of a novel multigene family. *Exp. Parasitol.* **80**, 463–472.
17. Church, G. M. and Gilbert, W. (1984) Genomic sequencing. *Proc. Natl. Acad. Sci. USA* **81**, 1991–1995.

20

Analysis of Gene Expression by RT-PCR

Peter Preiser

1. Introduction

To achieve a more complete understanding of malaria parasite biology and its interaction with its host, it is essential to be able to study the transcription and expression of parasite genes. Northern blot analysis and *RNase* protection assays are commonly used to study gene expression *(1)*. However, both procedures require a significant amount of input RNA. In the case of the human malaria species, only *Plasmodium falciparum* can be grown in culture *(2)*. This allows the production of a sufficient number of parasites to obtain the necessary amounts of RNA to use in these procedures. Unfortunately, when working with samples obtained from either animals or humans, the amount of RNA obtained is often insufficient to perform even one such experiment.

The polymerase chain reaction (PCR) *(3)* has been shown to be highly sensitive in the detection of malaria parasites *(4)*. Detection of a single parasite in a sample can be achieved by this method *(5,6)*. Furthermore, PCR allows the rapid and efficient screening of a large number of samples. Using RNA instead of DNA and including a reverse transcription (RT) step before the subsequent PCR has made it possible to study gene expression from extremely small amounts of RNA. RT-PCR has been used extensively to study the transcription of malaria parasite genes where the starting material is limited *(7–9)*. Recent work in *P. falciparum* and *P. yoelii* has shown that this approach can be used to study transcription in a single parasite *(10,11)*.

Like PCR, one of the advantages of RT-PCR is that it is relatively easy and quick, therefore allowing the analysis of a large number of samples in a very short time. The procedure for obtaining reliable and consistent RT-PCR results can be divided into three independent steps: RNA extraction, reverse transcription to make the cDNA, and finally the PCR step. Malaria parasites contain a significant amount of *RNases* and for that reason good quality RNA is difficult to obtain. To avoid any problems, it is critical that the cells are lysed rapidly in the presence of a strong protein denaturant like guanidine isothiocyanate *(12,13)*. Guanidine isothiocyanate denatures all proteins extremely quickly, leading to the inactivation of any *RNases* present in the sample. The denatured proteins are subsequently removed by a phenol/chloroform extraction step. The use of acidic phenol in the extraction step leads to the precipitation of DNA at the interface with the phenol/chloroform, leading to an enrichment of RNA. Still minor amounts of DNA are carried through the extraction procedure and need to be removed by treating the RNA with *DNase*. To produce cDNA, reverse transcription of the RNA can be primed with either random or gene-specific primers. The cDNA produced can then be used for either single or nested PCR. The efficiency of PCR varies from primer to primer and can even be effected by the thermal cycler used. It is therefore essential that the optimal PCR conditions for each primer pair are determined empirically. Depending on the amount of starting material and/or on the level of transcription of the gene studied, a single PCR step may not be sufficient, and it may be necessary to include a second PCR step using nested primers. This will increase the sensitivity by at least one order of magnitude. The greatest advantage of RT-PCR is that it is quick and extremely sensitive, but it should be remembered that it is normally not quantitative. In cases where it is important to establish differences in the levels of transcription, it may be necessary to resort to other approaches.

2. Materials

2.1. Equipment

1. Micropipets and tips (*see* **Note 1**).
2. Thermal cycler.

2.2. Reagents

The following solutions are required for the extraction of RNA and RT-PCR analysis (*see* **Note 2**).

1. Phosphate buffered saline (PBS). Store at room temperature.
2. Phenol:chloroform:isoamyl alcohol (25:24:1), acid equilibrated pH 4.7. Store at 4°C in the dark.
3. Phenol:chloroform:isoamyl alcohol (25:24:1), pH 6.7. Store at 4°C in the dark.
4. Chloroform:isoamyl alcohol (24:1). Store at room temperature.
5. *RNase*-free water (*see* **Note 3**). Store at room temperature.
6. Polyinosinic acid (PolyI): Prepare a 2 mg/mL stock solution in water. Store at –20°C.
7. Lysis buffer: 4 M guanidine isothiocyanate, 0.02 M sodium citrate, 0.5% sarcosyl, 0.1 M β-mercaptoethanol. Store at room temperature (*see* **Note 4**).
8. 2 M sodium acetate, adjusted to around pH 4.0–4.2 with glacial acetic acid. Store at room temperature.
9. Isopropanol. Store at room temperature.
10. 80% ethanol. Store at –20°C.
11. Random primers (0.75 µg/µL). Store at –20°C.
12. 5X reverse transcriptase buffer: 250 mM Tris-HCl (pH 8.3), 375 mM KCl, 15 mM MgCl$_2$. Store at –20°C (*see* **Note 5**).
13. 0.1 M dithiothreitol (DTT). Store at –20°C.
14. *DNase I*, *RNase*-free (10–50 U/µL). Store at –20°C.
15. RNasin ribonuclease inhibitor (20–40 U/µL). Store at –20°C.
16. Reverse transcriptase (200 U/µL) (*see* **Note 6**). Store at –20°C.
17. Deoxyribonucleotide triphosphate (dNTP) stock solution: 10 mM dATP, 10 mM dCTP, 10 mM dGTP, 10 mM dTTP in water. Store at –20°C.
18. 10X PCR buffer: 500 mM KCl, 100 mM Tris-HCl, pH 8.3, 0.01% (w/v) gelatin (*see* **Note 7**). Store at –20°C.
19. 25 mM MgCl2. Store at –20°C.
20. Primer stock solutions (*see* **Note 8**). Store at –20°C.
21. *AmpliTaq DNA* polymerase (5 U/µL) (Perkin Elmer). Store at –20°C.

Gene Expression by RT-PCR

22. Mineral oil (*DNase-, RNase-,* and protease-free). Store at room temperature.
23. 10X Tris-borate-EDTA (TBE): 1 M Tris-HCl, 1 M boric acid, 50 mM EDTA. A pH approximating 8.3 is obtained without any further adjustment. Store at room temperature.
24. 6X gel-loading buffer: 0.25% bromophenol blue, 0.25% xylene cyanol FF, 40% (w/v) sucrose in 1X TBE. Store at 4°C.
25. Agarose (*see* **Note 9**).

3. Methods

Depending on the amount of starting material available, a number of different approaches can be used to get good RT-PCR results. The procedure can be divided into three main sections. The RNA extraction step, the reverse transcriptase step to produce the cDNA template for the PCR reaction, and finally the PCR step and subsequent analysis of the product. **Figure 1** shows a schematic representation of the main steps involved in the protocol. While it is not essential to work in a sterile environment, great care should be taken at all steps to avoid any contamination of the samples. It is advantageous to use 1.5-mL Eppendorf tubes throughout the procedure, except when otherwise indicated. For all the steps involving RNA, diethyl pyrocarbonate (DEPC)-treated water is used.

3.1. Preparation of Cells

1. Spin infected RBC in centrifuge at 5000g for 10 min (*see* **Note 10**).
2. Remove the supernatant.
3. For larger samples >0.5 mL, wash the pellet in 5X vol of PBS (at 4°C). Do not resuspend cells by vortexing. Pipet slowly up and down. Spin as in **step 1** and remove the supernatant.
4. Samples can be stored at –70°C for a number of weeks without significantly affecting RNA integrity (*see* **Note 11**).

3.2. RNA Extraction

The efficient and reliable extraction of RNA as well as the subsequent removal of any DNA contamination is the essential step in this procedure. Two procedures are described here. The first (extraction I) can be used for the extraction of RNA from any amount of cells, while the second (extraction II) is only suitable for up to 1000 cells.

3.2.1. Extraction I

1. To the cell pellet, add 500 µL of lysis buffer (*see* **Note 12**).
2. If less than 1×10^6 cells are used, add 10 µL of PolyI (2 mg/mL).
3. Add 50 µL of 2 M sodium acetate (pH 4.0).
4. Add 500 µL of phenol–chloroform (pH 4.7) solution.
5. Vortex for 15 s.
6. Place tubes on ice for 10 min.
7. Centrifuge at 10,000g for 10 min at 4°C.
8. Carefully remove the aqueous upper layer into a new 1.5-mL tube (*see* **Note 13**).
9. Add an equal volume of isopropanol to the aqueous phase.
10. Vortex briefly.
11. Place tubes at –20°C for at least 30 min (*see* **Note 14**).
12. Centrifuge tubes at 10,000g for 15 min at 4°C (*see* **Note 15**).
13. Carefully remove all the supernatant with a pipet tip. Avoid dislodging the pellet. If the pellet does become dislodged, spin the tube for another 5 min.
14. To the pellet, add 100 µL of 75% ethanol. Do not vortex.

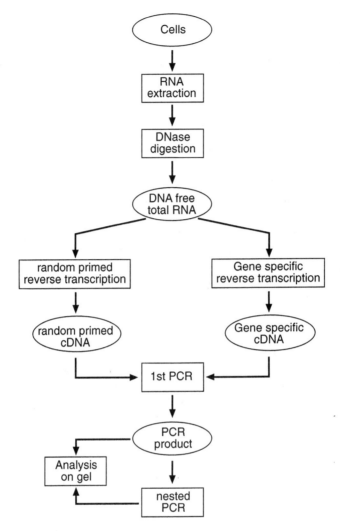

Fig. 1. Flow chart indicating the steps for RNA preparation and RT-PCR. Boxes indicate procedures, and circles indicate starting materials or products of a procedure.

15. Centrifuge tube at 10,000g for 5 min.
16. Remove supernatant as in **step 13**.
17. Air-dry the pellet for 5 min (*see* **Note 16**).
18. Resuspend the pellet in 20 µL of water (*see* **Note 17**). If >10⁶ cells were used, resuspend in 20 µL/10⁶ cells.
19. To each 20 µL of RNA, add 8 µL of 5X reverse transcriptase buffer, 2 µL of *DNase*, 1 µL of RNasin, and 9 µL of water (*see* **Note 18**). Mix by pipetting up and down.
20. Incubate for 1–2 h at 37°C.
21. Add 110 µL of lysis buffer.
22. Add 15 µL of 2 *M* sodium acetate.
23. Add 150 µL of phenol–chloroform (pH 6.7).
24. Vortex for 15 s.
25. Centrifuge at 10,000g for 5 min.
26. Transfer the aqueous top layer to a new 1.5-mL tube.

Gene Expression by RT-PCR

27. Add an equal amount of isopropanol.
28. Vortex.
29. Incubate for 30 min at –20°C (see **Note 14**).
30. Centrifuge at 10,000g for 15 min at 4°C.
31. Repeat **steps 13–17**.

3.2.2. Extraction II (<1000 Cells)

It is important that the starting volume of the cells in buffer is small (5 µL) (see **Note 19**).

1. Add 4 µL of reverse transcriptase buffer.
2. Incubate at 93°C for 3 min.
3. Place tubes on ice.
4. Add 31 µL of buffer containing 10 µL of PolyI, 2 µL of DTT, 4 µL of reverse transcriptase buffer, 1 µL of RNasin, 1 µL of *DNase*, and 13 µL of water (see **Note 20**).
5. Incubate at 37°C for 1 h.
6. Add 110 µL of lysis buffer.
7. Add 15 µL of 2 M sodium acetate.
8. Add 150 µL of phenol–chloroform.
9. Vortex for 15 s
10. Centrifuge at 10,000g for 5 min.
11. Transfer aqueous top layer to a new 1.5-mL tube.
12. Add equal amount of isopropanol.
13. Vortex.
14. Incubate for 30 min at –20°C (see **Note 14**).
15. Centrifuge at 10,000g for 15 min at 4°C. Carefully remove all of the supernatant with a pipet tip. Avoid dislodging the pellet. If the pellet does become dislodged, spin the tube for another 5 min.
16. To the pellet, add 100 µL of 75% ethanol. Do not vortex.
17. Centrifuge tube at 10,000g for 5 min.
18. Remove the supernatant as in **step 13**.
19. Air-dry the pellet for 5 min (see **Note 16**).

3.3. Preparation of cDNA

Efficient reverse transcription (RT) of the template RNA is crucial for the subsequent PCR step especially when very small amounts of RNA are available. Using random primers for the initiation of reverse transcription is in most cases the method of choice. In cases where a more sensitive technique is required, RT using gene-specific primers needs to be used.

3.3.1. Reverse Transcription (Using Random Primers)

1. To the RNA, add 2 µL of 10X PCR buffer, 2.4 µL of 25 mM MgCl$_2$, 2 µL of dNTP stock, 1 µL of RNasin, and water to a final volume of 16 µL (see **Notes 21 and 22**).
2. Mix by pipetting up and down.
3. For each sample analyzed, label two tubes (+ RT, -RT) (see **Note 23**).
4. Add 8 µL of the RNA mix to each tube (from **step 2**).
5. Add 1 µL of random primers to both tubes.
6. Add 1 µL of reverse transcriptase to +RT tube and 1 µL of water to the –RT tube, mix contents with pipet tip.
7. Incubate tubes for 10 min at 25°C followed by 1 h at 42°C.
8. Inactivate reverse transcriptase by heating the tube for 3 min at 93°C (see **Note 24**).
9. Proceed straight to the PCR step.

3.3.2. Reverse Transcription (Using Gene-Specific Primers)

1. Resuspend the RNA in 23 µL of water.
2. Add 2 µL of gene-specific primer stock (*see* **Note 25**).
3. Vortex briefly.
4. Spin down the liquid.
5. Incubate the sample for 10 min at 70°C.
6. In the meantime, label two tubes per sample (+RT, –RT).
7. Into each tube, add 4 µL of 5X RT buffer, 2 µL of DTT, 1 µL of dNTP stock, 0.5 µL of RNasin (*see* **Note 26**).
8. Briefly spin tubes from **step 5** and place on ice.
9. Add 11.5 µL from the sample into the +RT and –RT tube.
10. Add 1 µL of reverse transcriptase to +RT tube; add 1 µL of water to –RT tube. Mix with pipet tip.
11. Incubate for 1 h at 42°C.
12. Inactivate enzyme for 3 min at 93°C.
13. Proceed to the PCR step.

3.4. PCR

Although it is not the intention of this chapter to discuss the different approaches to optimizing PCR reactions, it is important to note that optimal PCR conditions for each primer pair needs to be established. Each primer pair combination uses conditions that are different from each other, and may even be influenced by the type of PCR machine used. It is therefore recommended that optimal conditions for your primers and PCR machine be established empirically. Generally, the two parameters that have the greatest impact on the efficiency of a PCR reaction are hybridization temperature and Mg^{2+} concentration. It is therefore advisable to establish these parameters using, for example, genomic DNA.

3.4.1. First PCR

3.4.1.1. USING cDNA PREPARED WITH RANDOM PRIMERS

1. To cDNA generated by random priming (10 µL), add 1.5 µL of 10X PCR buffer, 1 µL of each primer (*see* **Note 27**), x µL of $MgCl_2$, 0.5 µL of *Taq* polymerase, and water (final vol, 25 µL) (*see* **Note 28**).
2. Mix and spin briefly in centrifuge.
3. Overlay with one drop of mineral oil (*see* **Note 29**).
4. Place the tube in the PCR machine and run the program identified as most efficient for your primer pair (*see* **Note 30**).
5. The PCR product can be stored at 4°C until further use (*see* **Note 31**).

3.4.1.2. USING cDNA PREPARED WITH GENE-SPECIFIC PRIMERS

1. To cDNA generated with gene specific primers (20 µL), add 3 µL of 10X PCR buffer, 1 µL of 3' end primer and 2 µL of 5' end primer (*see* **Notes 27, 28,** and **32**) 1 µL of dNTPs, x µL of $MgCl_2$, 0.5 µL of *Taq* polymerase, and water (final volume 50 µL).
2. Mix and spin briefly in centrifuge.
3. Overlay with one drop of mineral oil (*see* **Note 29**).
4. Place the tube in the PCR machine and run the program identified as most efficient for your primer pair (*see* **Note 30**).
5. The PCR product can be stored at 4°C until further use (*see* **Note 31**).

Gene Expression by RT-PCR

3.4.2. Nested PCR

In a number of circumstances it is advantageous to use nested PCR. This assay has the advantage of giving the highest level of sensitivity (single-cell RT-PCR), as well as greater specificity. On the downside, it does require two primer pairs, and even more care must be taken to avoid contamination because of its extreme sensitivity.

1. The nested PCR reaction is carried out in a final volume of 25 µL. For each reaction, prepare a tube containing 2.5 µL of 10X PCR buffer, 1 µL of dNTPs, 1 µL of each nested primer, x µL of $MgCl_2$, 0.5 µL of *Taq* polymerase, and water to 23 µL (*see* **Note 33**).
2. Overlay with one drop of mineral oil (*see* **Note 29**).
3. Add 2 µL of the product of the first PCR reaction to the corresponding tube (*see* **Note 34**).
4. Place the tube into the PCR machine and run the program (*see* **Note 30**).
5. The PCR product can be stored at 4°C until further use (*see* **Note 31**).

3.5. Analysis of PCR Product

3.5.1. Gel Electrophoresis

PCR is an extremely sensitive technique and products normally can be detected by ethidium bromide staining and visualization under an UV light source. In most cases, a 1% agarose gel (Multipurpose agarose) can be used to separate the PCR product efficiently. Only in the cases when the product is small (100–500 bp) is it advantageous to use high resolution agarose (e.g., Metaphore agarose).

1. Prepare appropriate agarose gel (*see* **Note 35**).
2. Take 5 µL of each PCR reaction and mix with 1 µL of loading buffer.
3. Load and run gel.
4. Photograph gel on UV light-box.

3.5.2. Interpretation of Results

In most cases of RT-PCR, the results are very easy to analyze. For each RNA sample analyzed, two PCR products need to be run on a gel (± RT). In the case of the +RT tube, you should see a product of the expected size, while the –RT lane should be blank. The presence of a band in the appropriate lane indicates that the gene analyzed is transcribed in this sample. If there is any product detected in the –RT lane, it is most likely due to genomic DNA contamination (*see* **Note 36**). Alternatively, it could be due to contamination of solutions especially when doing nested PCR where aerosol contamination is a problem. If a product is seen in the –RT lane, no valid interpretation about the transcription of the gene can be made (*see* **Note 37**). **Figure 2, panel I**, gives an example of an RT-PCR reaction across the intron of the Exp-1 gene of *Plasmodium falciparum* *(14)* using random primed reverse transcription. Clearly highlighted is the difference between the PCR product produced from genomic DNA versus cDNA. **Figure 2, panel II**, shows an example of single-cell RT-PCR and PCR of the 235-kDa rhoptry protein multigene family of *P. yoelii* *(10)* using gene-specific reverse transcription and nested PCR. Note that in this circumstance no difference in size is seen between the PCR product obtained from cDNA and genomic DNA.

4. Notes

1. Since RT-PCR is an extremely sensitive procedure, extreme care should be taken to avoid contamination. Even the smallest amount of aerosol can contaminate the pipet and other

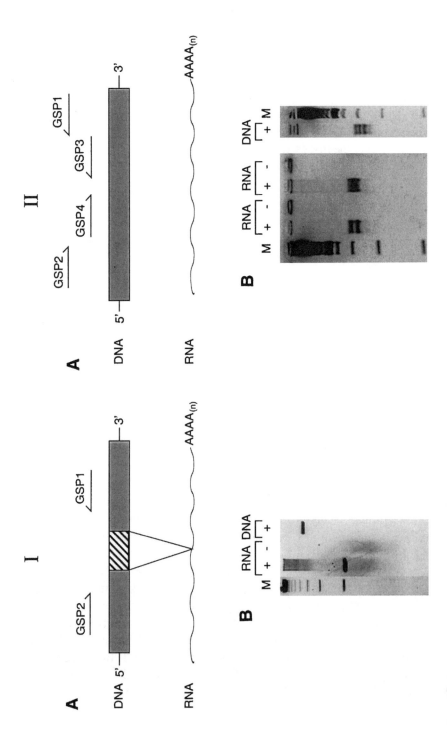

Fig. 2. RT-PCR using random primed reverse transcription across an intron-containing gene and a potential set of primers used for the RT-PCR protocol. GSP1 and GSP2 are gene specific primers spanning the intron (shaded box). (B) Typical example of results obtained with primers across the intron of the exp-1 gene of *P. falciparum* (*14*) using random primed reverse transcription followed by a single PCR step. The expected RT-PCR product is just under 100 bp, and the PCR product that forms genomic DNA is around 320 bp (**RNA+**: reverse transcriptase added; **RNA−**: no reverse transcriptase added; **DNA+**: genomic DNA added; **M**: 100-bp marker). **II.** Gene-specific

pieces of equipment. It is therefore advisable to use filter tips to avoid some of these problems.

2. Working with RNA requires special care especially in avoiding the introduction of *RNases*. *RNases* are found everywhere; it is therefore essential that gloves are worn at all times when working with RNA. Furthermore, all solutions should be ensured to be RNase free. This includes making up all solutions with *RNase*-free water. A number of buffers used in the procedure are available commercially and the suppliers are mentioned where appropriate.

3. *RNase*-free water can very easily be prepared in any laboratory. Use distilled water and add DEPC. It is recommended to prepare about 5–10 L at one time. Add a stirring bar and boil for a number of hours until the smell of DEPC has nearly completely disappeared. This is normally achieved with the loss of about 20% of the starting material, due to evaporation. Aliquot the water into 500-mL bottles and autoclave. Store at room temperature until needed. Be aware of any cloudiness (very small bits floating in the water when bottle is shaken) that may appear in an opened bottle, this normally indicates contamination and, in that case, it is recommended to use a new bottle.

4. Lysis buffer is commercially available from a number of companies that sell RNA isolation kits and there is very little advantage in using one over the other. It is easily prepared in the lab. Use the highest quality of reagents available and make up with *RNase*-free water. Mercaptoethanol is not added to the buffer until needed. Unlike with commercially available lysis buffer, it is important to test your own solutions on a test sample as there is evidence that some batches of the lysis solution do not work well. The lysis buffer can be stored at room temperature indefinitely.

5. Reverse transcriptase buffer is usually supplied with the enzyme. This buffer can be used for the reverse transcriptase step. It is only important to adjust the PCR buffer used in the subsequent step to account for potential differences between different RT-buffers.

6. All the experiments described here have been done using Superscript™ II RNaseH⁻ reverse transcriptase (Life Science Technologies), which in our hands consistently gives the best results. However, other reverse transcriptase enzymes can be used, especially if the RNA comes from >1000 cells.

7. PCR buffer is usually supplied with the enzyme.

8. Make primer stock solutions to a final concentration of 100 ng/µL. Since primers are precious, it is advised to aliquot stock solutions into smaller volumes (e.g., 100 µL) and store at –20°C. Use only one tube at a time. This avoids potentially losing the whole primer stock due to a contamination problem.

9. Depending on the expected size of the RT-PCR product, different types of agarose should be used. For small fragments <500 bp, MetaPhor agarose (FMC) is very good, while for larger products, normal multipurpose agarose is sufficient.

10. When the sample consists of a very small number of cells (e.g., a micromanipulated single cell), it is best to go straight into the extraction procedure reducing the amount of lysis buffer added accordingly. However, it is important that the volume of lysis buffer is at least 80% of the total volume.

Fig. 2. *(continued)* reverse transcription followed by nested PCR. **(A)** Schematic of a gene without intron and a potential set of primers used for the gene-specific RT-PCR protocol. GSP1 and GSP3 are gene-specific 3' end primers, while GSP2 and GSP4 are specific 5' end primers. **(B)** RT-PCR analysis of the 235 kDa rhoptry protein multigene family of *P. yoelii* *(10)*. Analysis of RNA from two individual schizonts by nested gene specific RT-PCR and the DNA from one schizont by nested PCR (**RNA+**: reverse transcriptase added; **RNA–**: no reverse transcriptase added; **DNA+**: genomic DNA added; **M**: 100-bp marker).

11. For long-term storage of cell samples without completing the RNA extraction procedure, it is best to add the appropriate amount of lysis buffer to the cells before freezing. RNA in a cell–lysis buffer mix is stable for some time even at room temperature, but can be stored for 1 mo at –20°C.
12. The protocol described here is designed for relatively small amounts of starting material. If the cell pellet is >100 µL, the lysis buffer should be doubled and the RNA should be extracted in two tubes. Similarly, if the sample after the addition of the lysis buffer is very ropy, it may be best to increase the amount of lysis buffer used.
13. After the centrifugation step, it is usual to see a whitish interface between the aqueous and phenol phase. This interface consists predominately of genomic DNA and some protein. Care should be taken to avoid removing this material with the aqueous phase, as this will make complete removal of DNA later on very difficult. If some of the material is picked up by accident, spin tube again for 5 min. In the case when the starting amount of material was very small, 1–1000 cells, the interface is often not visible and it may even be advisable to pick it up with the aqueous phase to ensure that no RNA is lost. In this case the *DNase* treatment used to remove contaminating DNA in the later stage of the procedure should be sufficient.
14. Do not place tubes at –70°C, as this will lead to the precipitation of salts as well as the RNA. The precipitated salts can lead to problems in the subsequent steps.
15. When precipitating very small amounts of RNA, it is often very hard to see a pellet at the bottom of the tube. It is therefore helpful to ensure that all tubes face the same way during the spin. This should allow removal of the supernatant without disrupting the "invisible" pellet.
16. Before drying the RNA pellet, make sure to remove as much of the supernatant as possible. Do not dry the pellet under vacuum, as it becomes extremely difficult to resuspend afterward. In most circumstances it is possible to resuspend the RNA straight after the removal of any residual supernatant.
17. If RNA does not dissolve straight away, heat the sample to 70°C and vortex vigorously.
18. Depending on the sensitivity of the primers and the abundance of the target RNA, RNA from 10^6 cells should give enough material to do between 10 and 50 RT-PCR reactions. It may therefore be convenient to only use a small proportion of your RNA for this step. The minimum amount of RNA needed to detect the gene of interest needs to be determined in control experiments. If less RNA is used as starting material, adjust volume to 20 µL with water.
19. If the starting volume of cells is significantly greater than 5 µL, extraction procedure I should be used. Volumes up to 18 µL can be accommodated by increasing the amount of 5X reverse transcriptase buffer added in **step 1** to 8 µL and eliminating the water and 5X reverse transcriptase buffer added in **step 4**. It is not recommended to try to pellet cells with a high-speed spin because this leads to cell lysis and subsequent loss of RNA.
20. When more than one sample is analyzed, it is recommended to prepare a master mix of all the reagents. Make a little more than is actually required to allow for pipetting errors. When preparing a master mix, always add the enzyme(s) last. Mix contents by pipetting up and down, before adding the appropriate amount to each sample.
21. To ensure that sufficient sample is available at the subsequent step, it is recommended to slightly increase the volume of the reaction to around 18 µL by increasing the volumes of all reagents accordingly.
22. In some circumstances, it is possible to combine *DNase* digestion and reverse transcription without the time-consuming phenol–chloroform extraction and precipitation of the RNA (extraction I: **steps 19–31**). To the RNA, add 2 µL of 10X PCR buffer, 2.4 µL of 25 mM $MgCl_2$, 2 µL of dNTP stock, 1 µL of RNasin, 1 µL of *DNase*, and water to a final volume

of 16 μL (*see* **Note 21**). Mix the sample and incubate for 1–2 h at 37°C. Heat-inactivate the *DNase* by incubating the sample for 5 min at 75°C *(15)*. Proceed to **step 3** of the protocol.
23. Use tubes that fit into the thermal cycler available in your laboratory.
24. **Steps 7** and **8** can be performed in a thermal cycler with a heated lid. Do not use a thermal cycler without a heated lid for these steps as evaporation can lead to less efficient cDNA generation.
25. It is advantageous to have at least two gene specific primers for the 3' end (GSP1 and GSP3 in **Fig. 2, panel IIa**). Use the most 3' primer (GSP1) for the RT step and subsequently use the internal one (GSP3) for the PCR step.
26. To ensure uniformity throughout all the samples, prepare a master mix of all the reagents. Prepare a little more master mix then needed to allow for pipetting errors, Add the RNasin last; add this master mix to each tube, changing pipet tips if there is any risk of contamination.
27. Best RT-PCR results are obtained with primers that give products of an expected size of 200–1000 bp. It is advantageous if the primers used span an intron-containing region, since this can serve as an additional control for DNA contamination (*see* **Fig. 2, panel II**).
28. Make up a master mix containing all the reagents, adding the *Taq* polymerase last. The amount of $MgCl_2$ to add to the reaction depends on the optimum conditions for the primer pair used. Remember to account for the 3 m*M* $MgCl_2$ that is present in the cDNA sample when calculating the final $MgCl_2$ concentration. Add 30 μL of master mix to each tube, changing pipet tips between each tube.
29. While PCR reactions using thermal cyclers with heated lids do not normally require a mineral oil overlay, it is nevertheless advisable to use mineral oil for this procedure. Mineral oil significantly reduces the risk of aerosol contamination. This is of greatest importance when PCR reactions are set up in the same room as where the product is analyzed.
30. Usually a standard PCR protocol can be used:
    Step 1: 93°C for 3 min.
    Step 2: 93°C for 1 min.
    Step 3: n°C for 1 min (n = hybridization temperature for primer pair).
    Step 4: 72°C for 1.5 min.
    Step 5: repeat 2–4 35X.
    Step 6: end.
31. PCR products in the PCR reaction buffer are stable at 4°C; for long-term storage, place samples at –20°C.
32. When using the same 3' end primer in the PCR that has been used to prime the reverse transcriptase reaction, it is not uncommon to obtained a lot of background. This is most likely due to mispriming of the 3' end primer during the RT step. The background signal will be eliminated if the product produced in the first PCR reaction is subsequently used as template for a nested PCR reaction. Alternatively, higher specificity is obtained when an additional primer internal to the one used at the RT step is used in the PCR reaction (*see* **Note 25**). In this case, use equal amounts (2 μL) of both primers in the PCR reaction. For example, use primer GSP1 to prime the RT reaction followed by GSP3 and GSP2 for the PCR reaction (**Fig. 2, panel II**).
33. Make up a master mix containing all the reagents, adding the *Taq* polymerase last. The amount of $MgCl_2$ to add to the reaction depends on the optimum conditions for the primer pair used. Add 23 μL of master mix to each tube.
34. To avoid contamination, it is important to place the pipet tip into the mineral oil layer before dispensing the PCR product from the first PCR reaction. No further mixing is necessary. A positive displacement pipet while not absolutely essential, is very useful at this step.
35. In most circumstances, running the gel in 1X TBE is appropriate. The use of Tris-acetate

(TAE) is only necessary if the PCR product is purified directly from the gel using certain DNA clean-up kits. It is usually possible to add the ethidium bromide to the agarose before the gel is poured. This eliminates the necessity to stain the gel at the end of the electrophoresis run. Ethidium bromide is toxic, so great care should be taken when handling it.

36. If a band is seen in the –RT control it is most often due to incomplete *DNase* digestion. Extending the digestion time or increasing the amount of added *DNase* usually is sufficient to eliminate this problem. If the problem persists even after extensive *DNase* digestion, contamination of the stock solutions may be the problem. The quickest way to solve this problem is to use new solutions (*see* **Note 8**); a more extensive way is to check all the solutions by PCR for contamination. If the problem still persists, try setting up the RT-PCR reactions, including *DNase* digestion, in a different room. At this time it may also be necessary to evaluate the whole procedure to detect where possible contamination can occur.
37. In the circumstances where the RT-PCR product spans an intron, some information can still be obtained even when there is DNA contamination. Since the RT-PCR product is smaller than the product produced with the same primers from genomic DNA, a smaller product should be seen in the +RT sample in addition to the larger band from the genomic DNA.

References

1. Sambrook, J., Fritsch, E. F., and Maniatis, T. (1989) *Molecular Cloning: A Laboratory Manual,* 2nd ed., Cold Spring Harbor Laboratory Press, Plainview, New York.
2. Trager, W. and Jensen, J. B. (1976) Human malaria parasites in continuous culture. *Science* **193,** 673–675.
3. Saiki, R. K., Gelfand, D. H., Stoffel, S., Scharf, S. J., Higuchi, R., Horn, G. T., et al. (1988) Primer-directed enzymatic amplification of DNA with a thermostable DNA polymerase. *Science* **239,** 487–491.
4. Snounou, G., Viriyakosol, S., Jarra, W., Thaithong, S., and Brown, K. N. (1993) Identification of the four human malaria parasite species in field samples by the polymerase chain reaction and detection of a high prevalence of mixed infections. *Mol. Biochem. Parasitol.* **58,** 283–292.
5. Snounou, G., Viriyakosol, S., Zhu, X. P., Jarra, W., Pinheiro, L., do Rosario, V. E., et al. (1993) High sensitivity of detection of human malaria parasites by the use of nested polymerase chain reaction. *Mol. Biochem. Parasitol.* **61,** 315–320.
6. Tirasophon, W., Ponglikitmongkol, M., Wilairat, P., Boonsaeng, V., and Panyim, S. (1991) A novel detection of a single *Plasmodium falciparum* in infected blood. *Biochem. Biophys. Res. Commun.* **175,** 179–184.
7. Smith, J. D., Chitnis, C. E., Craig, A. G., Roberts, D. J., Hudson-Taylor, D. E., Peterson, D. S., et al. (1995) Switches in expression of *Plasmodium falciparum* var genes correlate with changes in antigenic and cytoadherent phenotypes of infected erythrocytes. *Cell* **82,** 101–110.
8. Preiser, P. R. and Jarra, W. (1998) *Plasmodium yoelii*: differences in the transcription of the 235-kDa rhoptry protein multigene family in lethal and nonlethal lines. *Exp. Parasitol.* **89,** 50–57.
9. Wilson, R. J., Denny, P. W., Preiser, P. R., Rangachari, K., Roberts, K., Roy, A., et al. (1996) Complete gene map of the plastid-like DNA of the malaria parasite *Plasmodium falciparum*. *J. Mol. Biol.* **261,** 155–172.
10. Preiser, P. R., Jarra, W., Capiod, T., and Snounou, G. (1999) A rhoptry-protein-associated mechanism of clonal phenotypic variation in rodent malaria. *Nature* **398,** 618–622.
11. Chen, Q., Fernandez, V., Sundstrom, A., Schlichtherle, M., Datta, S., Hagblom, P., et al. (1998) Developmental selection of var gene expression in *Plasmodium falciparum*. *Nature* **394,** 392–395.
12. Chomczynski, P. and Sacchi, N. (1987) Single-step method of RNA isolation by acid guanidinium thiocyanate- phenol-chloroform extraction. *Anal. Biochem.* **162,** 156–159.
13. Chirgwin, J. M., Przybyla, A. E., MacDonald, R. J., and Rutter, W. J. (1979) Isolation of biologically active ribonucleic acid from sources enriched in ribonuclease. *Biochemistry* **18,** 5294–5299.
14. Simmons, D., Woollett, G., Bergin-Cartwright, M., Kay, D., and Scaife, J. (1987) A malaria protein exported into a new compartment within the host erythrocyte. *EMBO J.* **6,** 485–491.
15. Huang, Z., Fasco, M. J., and Kaminsky, L. S. (1996) Optimization of *DNase* I removal of contaminating DNA from RNA for use in quantitative RNA-PCR. *Biotechniques* **20,** 1012–1014, 1016, 1018–1020.

21

In Situ Detection of RNA in Blood- and Mosquito-Stage Malaria Parasites

Joanne Thompson

1. Introduction

The technique of *in situ* RNA hybridization provides a means of examining RNA expression within individual parasites at many stages of development in both the vertebrate and mosquito hosts. The protocols described in this chapter have been developed with the aim of preserving the morphology of the parasites within the infected host cells or tissues, so that many examples of each parasite stage can be identified and observed in a single experiment. The subcellular morphology of the parasite is also preserved, and the localization of abundant RNA transcripts can be directly visualized (*1,2;* **Fig. 1**). The parasite samples are extremely simple to prepare and do not require specialized purification procedures; *in situ* hybridization has, for example, been used successfully to demonstrate the expression patterns of a number of sexual-stage specific genes within gametocytes present in smears of a few microliters of infected blood (*1,3–5*).

In situ hybridization procedures have particular application in the detection of transcription products during the mosquito stages of the parasite's development, where analysis of gene expression using other detection procedures (Northern blotting and reverse-transcription-polymerase chain reaction [RT-PCR]) are inevitably complicated by the presence of very large numbers of blood-stage parasites within the mosquito bloodmeal during the first few days after transmission, and by the very large mass ratio of mosquito cellular material to parasites. The fact that ookinetes, oocysts, and sporozoites are supported within cellular tissues during their development in the mosquito has made these stages the most favorable for the development of standardized and reproducible *in situ* hybridization protocols. By contrast, there are a number of difficulties associated with the detection of mRNA in blood-stage parasites, most notably cell-to-cell variability in the robustness of the parasite and infected red blood cell membranes and the strong autofluorescence of red blood cell hemoglobin, all of which can give rise to poor parasite morphology and low detection sensitivity. The exceptionally high adenine-thymine (AT) content of the parasite genome and the presence of repetitive and low-complexity sequences, particularly in many blood-stage parasite surface proteins, can also create particular problems in the selection of specific probes.

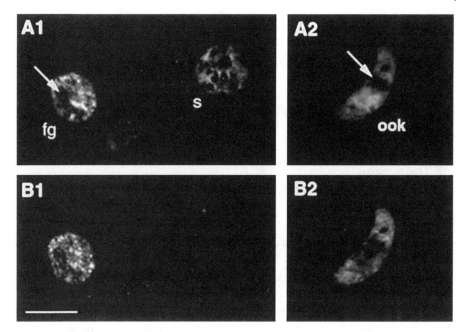

Fig. 1. *In situ* detection of Pbs21 and ribosomal RNA in *P. berghei* parasites. Female gametocytes (fg), schizonts (s) and ookinetes (ook) present within smears of *P. berghei*-infected blood *(1)* or ookinete cultures *(2)* were cohybridized with biotin-labeled rRNA **(A)** and digoxigenin-labeled Pbs21 **(B)** antisense RNA probes. Signals were detected with rhodamine-conjugated antidigoxigenin and fluorescein-conjugated antibiotin antibodies. Ribosomal RNA is detected throughout the cytoplasm of all parasite stages, but is not detected in nuclei (arrowed); the morphology of the parasites is, therefore, clearly visible. Pbs21 mRNA, by contrast, is detected in a stage-specific manner, only within female gametocytes and ookinetes, and is visible as punctate staining.

Described here are two tried and tested *in situ* hybridization protocols that have been adapted successfully to detect a variety or mRNA and rRNA transcripts in smears of red blood cells infected with *Plasmodium berghei* and *P. falciparum*, and in ookinetes, oocysts, and sporozoites present within infected *Anopheles stephensi* mosquitoes. In all cases, hybridization conditions are standardized by the use of antisense RNA probes that are labeled by in vitro transcription in the presence of the nonradioactively labeled hapten moieties; digoxigenin, biotin, or fluorescein. Particular emphasis has been placed throughout on strategies aimed at reducing the problems inherent in the detection of RNA expression in parasite material.

2. Materials
2.1. Equipment

1. Glass staining troughs. Rinse in distilled water and bake at 180°C.
2. Hybridization oven, ideally with a shaking platform (for *in situ* detection of RNA in blood-stage parasites; *see* **Subheading 3.2.**) or a fixture that can be adapted for rotation of Eppendorf tubes (for whole-mount *in situ* hybridization; *see* **Subheading 3.3.**).
3. Slides and cover slips: Wash slides and cover slips (22 × 22 mm) in 1 *M* HCl for 30 min, rinse in H₂O, wash in 95% ethanol for 30 min and air-dry. Coat slides in 1X Denhardt's solution (*see* **Subheading 2.2.**) in 3X SSPE overnight at 65°C. Rinse in water and fix in

In Situ RNA Detection

ethanol:acetic acid (3:1 v/v) for 20 min. Air-dry, rinse in H_2O, air-dry, and store at room temperature. Bake cover slips at 180°C for 2 h.

2.2. General Buffers

Prepare all solutions using autoclaved Milli-Q H_2O or equivalent high purity H_2O. Stock solutions should be either treated with diethyl pyrocarbonate (DEPC) or prepared from DEPC-treated H_2O wherever possible to destroy RNAse activity. Chemicals used in hybridization solutions should be molecular biology grade where possible.

1. Denhardt's solution (50X stock): Dissolve 1 g of Ficoll-Hypaque, 1 g of polyvinyl pyrrolidine and 1 g of bovine serum albumin (BSA) in 100 mL of DEPC-treated H_2O. Store at –20°C. The Denhardt's solution used in the hybridization buffer must be prepared from BSA that is certified to be RNAse-free; preprepared molecular biology-grade Denhardt's solution may be purchased for this purpose.
2. Phosphate-buffered saline (PBS) (10X stock): Combine 80 g of NaCl, 2 g of KCl, 11.5 g of $Na_2HPO_4 \cdot 7H_2O$, and 2 g KH_2PO_4, and make up to 1 L final volume. Sterilize by autoclaving or filtering, and store at 4°C. Working solution should be pH 7.2.
3. PBS-T: PBS plus 0.1% Tween-20: Prepare by adding 0.1 mL of Tween-20 per 100 mL PBS X1.
4. 4% paraformaldehyde/PBS solution: Dissolve 4 g of paraformaldehyde in 50 mL of H_2O and heat to 65°C. Add 10 µL of 10 M NaOH (the solution should clear immediately). Filter, add 10 mL of 10X PBS, and make up to 100 mL final volume. Prepare solutions fresh each time, and cool to 4°C before use.
5. SSPE (20X stock): Combine 174 g of NaCl, 27.6 g of $Na_2HPO_4 \cdot H_2O$, and 7.4 g of EDTA, and make up to 1 L final volume. Adjust to pH 7.4 with NaOH. Treat with DEPC, autoclave, and store at room temperature.
6. Sodium chloride-sodium citrate buffer (SSC) (20X stock): Combine 175 g of NaCl and 88 g of Na_3citrate$\cdot 2H_2O$, and make up to 1 L final volume. Adjust to pH 7.3 with NaOH. Treat with DEPC, autoclave, and store at room temperature.
7. Proteinase K (20 mg/mL stock): Prepare a 20 mg/mL stock in 0.2 mM EDTA, pH 7.6. Aliquot and store at –20°C; do not reuse aliquots. Alternatively, purchase proteinase K solutions that are stabilized by the addition of 1 mM calcium acetate and may be stored at 4°C (Boehringer Mannheim).
8. 0.1 M Triethanolamine (TEA): Dissolve 1.34 mL of triethanolamine in 100 mL of H_2O, and adjust to pH 8.0 with HCl.
9. EDTA, 0.2 M, 0.01 M.
10. 2 M LiCl (optional).
11. 0.2 M HCl.
12. 0.1 M Tris-HCl, 0.01 M EDTA, pH 8.0.
13. Ethanol, 50%, 70%, 90%, 100%.
14. Methanol, 25%, 50%, 75%, 100%.

2.3. Probe Preparation and Hybridization

1. 2.5X nucleoside triphosphate (NTP) mix: 2.5 mM ATP, 2.5 mM CTP, 2.5 mM GTP, 1.65 mM UTP, 0.85 mM Digoxigenin-, biotin-, or fluorescein-11-UTP.
2. T7, T3, or SP6 RNA polymerase, supplied with manufacturer's transcription buffer.
3. RNAse Inhibitor.
4. Formamide: Purchase molecular-grade, deionized solution. For long-term storage, aliquot and store at –20°C.
5. 50% dextran sulfate: Dissolve 2.5 g of dextran sulfate in 4 mL of DEPC-treated H_2O with rotation (several hours to overnight). Make volume to 5 mL with H_2O. Dextran sulfate

solution is extremely viscous; to reduce pipetting errors, prepare 200 µL aliquots to which other reagents can be added directly. Store at –20°C.

6. 10% Sodium dodecyl sulfate (SDS): Dissolve 5 g in 50 mL of DEPC-treated H_2O. Store at room temperature.
7. 10% Polyoxyethylenesorbitan monolaurate (Tween-20): Dissolve 5 mL in 50 mL of DEPC-treated H_2O. Store at room temperature.
8. 100 mg/mL heparin: dissolve porcine grade 1A heparin (or equivalent) at 100 mg/mL in 4X SSC. Store at 4°C.

2.4. Hybridization Buffer for In Situ Hybridization of Parasites in Blood Smears

Combine 200 µL of 50% dextran sulfate, 500 µL of formamide, 100 µL of 20X SSPE, 100 µL of 50X Denhardt's solution, 20 µL of 200 mg/ mL tRNA, and 50 µL of 10% SDS. Add appropriate volume of labeled probe, and make up to 1 mL final volume with DEPC-treated H_2O.

2.5. Whole Mount In Situ Hybridization

1. Prehybridization buffer: Combine 0.5 mL of formamide, 200 µL of 20X SSC, 10 µL of 100 mg/ mL heparin, 100 µL of 10% Tween-20 and 8-mL of DEPC-treated H_2O. Adjust to pH 4.5 with 1 M citric acid (~0.7 mL). Make up to 10 mL with DEPC-treated H_2O.
2. Hybridization buffer: Add 10 µL of 10 mg/mL ssDNA and 20 µL of 10 mg/mL yeast tRNA to 1 mL of prehybridization buffer.
3. 2 mg/mL glycine in PBS-T.

2.6. Signal Detection

1. Maleic acid buffer: 0.1 M maleic acid, and 0.15 M NaCl. Adjust to pH 7.5 with 10 M NaOH or solid NaOH. Autoclave and store at room temperature.
2. Blocking buffer (10X stock): Dissolve blocking reagent (Boehringer Mannheim) in maleic acid buffer to a final concentration of 10% (w/v). Blocking reagent takes several hours to dissolve, with stirring and heating to 50°C. Store at 4°C for a few days, or at –20°C for long-term storage. Prepare 1X working stock by diluting in maleic acid buffer.
3. Alkaline phosphatase (AP) buffer: 0.1 M Tris-HCl, 0.1 M NaCl, and 0.05 M $MgCl_2$. Adjust to pH 9.5 with NaOH. Store at room temperature. Do not use if precipitation has occurred.
4. Nitroblue tetrazolium/5-bromo-4-chloro-3-indoyl phosphate (NBT/BCIP) solution: Add 4.5 µL of NBT and 3.5 µL of BCIP to 1 mL of AP buffer. Prepare immediately before use and keep in the dark. High-quality NBT and BCIP solutions should be purchased commercially (Boehringer Mannheim), aliquoted, and stored at –20°C.

3. Methods
3.1. Preparation of Probe
3.1.1. Choice of Probe

The specificity and sensitivity of *in situ* RNA hybridization is greatly influenced by the complexity of the probe and the choice of labeling and detection systems. Many plasmodial genes contain regions of low complexity, exceptionally high AT content (common within untranslated regions), and extensive regions of repetitive sequence. Probes containing low-complexity sequences are likely to crossreact in a nonspecific manner with other gene transcripts leading, at least, to higher background and consequent reduction in detection sensitivity or to uninterpretable results. It is important, therefore, to avoid including low-complexity sequences within the probe wherever possible. In the case of *P. falciparum*, potential probe sequences may be compared with sequences in the *P. falciparum* genome databases as an additional precaution.

We have used antisense RNA probes of 100–400 nucleotides in the *in situ* hybridization protocols described below. They offer a number of advantages: RNA/RNA duplexes are relatively stable compared to RNA/DNA duplexes, permitting more stringent and standardized hybridization conditions, and the sense strand does not compete for probe during hybridization as it would with random-primed DNA probes. Antisense RNA probes can be synthesized and labeled at high efficiency with a variety of hapten-labeled UTPs by in vitro transcription in the presence of RNA polymerases.

3.1.2. Choice of Labeling System

Antisense RNA probes may be labeled with digoxigenin, biotin, or fluorescein moieties. Probes labeled directly with fluorescein are useful only for the detection of very abundant RNA transcripts. A range of enzyme-linked and fluorochrome-conjugated antidigoxigenin antibodies are available, however, which greatly enhance the sensitivity of indirect detection of digoxigenin-labeled probes (Boehringer Mannheim). Biotin-labeled probes can similarly be detected using antibiotin antibodies or conjugated avidin-streptavidin reagents. Two complementary-labeled probes may be used in a single hybridization experiment to detect the coexpression of two different RNA transcripts (*see* **Notes 1** and **2**).

Complete kits may be purchased from Boehringer Mannheim for in vitro transcription of digoxigenin or biotin-labeled RNA probes. Alternatively, hapten- or fluorescein-labeled nucleotides or nucleotide mixes can be purchased individually, and are more economical if a number of probes are synthesized. A method for transcribing labeled RNA is given below, which results in the incorporation of labeled UTP every 20–25 nucleotides (provided the RNA sequence is not excessively UTP-rich).

3.1.3. In Vitro Transcription of Hapten- or Fluorescein-Labeled Antisense RNA Probes

1. Subclone DNA encoding the desired probe sequence (100–400 bp) into appropriate transcription vectors containing initiation sites for T_7, T_3, or SP_6 RNA polymerases. Linearize the plasmids with restriction enzymes that give 5' overhangs. Alternatively RNA polymerase initiation sites may be attached to suitable DNA fragments during PCR amplification. Extract PCR fragments with phenol/chloroform and ethanol-precipitate in the presence of 2 M ammonium acetate to remove unincorporated deoxynucleoside triphosphates (dNTPs). Remember that transcription must take place in the appropriate orientation to produce antisense RNA transcripts.
2. Mix at room temperature 1–3 µg of linearized plasmid DNA or 200–400 ng of PCR product, 20 µL of 2.5 × NTP mix, 5 µL of 10X transcription buffer, 50 U of RNAse inhibitor, and 50 U of T_7, T_3, or SP_6 RNA polymerase in DEPC-treated H_2O, to a final volume of 50 µL. Incubate at 37°C for 2 h.
3. Stop reaction by the addition of 5 µL of 200 mM EDTA (pH 8.0). Remove unincorporated nucleotides (optional) by ethanol precipitation in the presence of 0.8-M LiCl. Resuspend the RNA transcript to a volume of 100 µL with DEPC-treated H_2O. Store labeled probe in H_2O or under ethanol at –20°C or –70°C.
4. The quality and quantity of transcribed RNA can be determined by gel electrophoresis and ethidium bromide staining in a 1.5% agarose gel (note that under these nondenaturing

conditions, the labeled RNA will appear at an anomalous molecular weight). The quantity of transcribed RNA should be much greater than that of the input DNA template.

3.2. In Situ *Detection of RNAs in Blood-Stage* Plasmodium

In this procedure, parasites or blood cells containing parasites are immobilized onto coated slides (*see* **Note 3**). Many of the subsequent steps are carried out by immersing the slides in 50–100 mL solutions within glass staining troughs. Solutions can be changed simply by pouring away one solution and rapidly pouring in the next. All steps are carried out with gentle shaking at room temperature unless stated.

1. Smear parasitized blood samples onto Denhardt's-coated slides. Using a diamond knife, mark ~1 cm^2 squares suitable for hybridization (2–3 hybridizations per slide are possible). Rapidly air-dry and fix immediately in ice-cold 100% ethanol for 2 min. Hydrate through a series of 90%, 70%, and 50% EtOH for 2 min each. Fix again in ice-cold 4% paraformaldehyde–PBS solution. Wash in 2X SSPE for 5 min.
2. Incubate in 0.1 M Tris-HCl, 0.01 M EDTA (pH 8.0) with 5 µg/mL proteinase K for 15–30 min at 37°C. This treatment is intended to permeabilize the cells and expose the RNAs. However, it may be necessary to titrate for optimal results, since the activity of the proteinase K and the conditions required to produce optimal morphology of the parasites may vary. Wash in 2X SSPE for 5 min.
3. Immerse in 0.2 M HCl for 15 min to strip basic proteins from the sample. Wash in 2X SSPE for 5 min.
4. Transfer slides to a rack that will fit into a trough containing a magnetic stirrer (a Petri dish with holes cut out can be used to prevent the stir-bar from crashing into the slides) and immerse in 0.1 M triethanolamine solution. Add acetic anhydride with stirring to a final concentration of 0.25% for 5 min (note that acetic anhydride is toxic and volatile so this step should be carried out in a fume hood). This treatment is reported to acetylate exposed amino acids and is important in reducing background signal. Wash in 2X SSPE and hold at room temperature until hybridization mix is ready.
5. Denature the antisense RNA probe(s) by heating to 65°C for 5 min, then chilling on ice. Add to the hybridization mix at a concentration of ~20 ng/mL.
6. Remove the slides from SSPE and allow to air-dry briefly, drying around the square marked for hybridization with a tissue.
7. Add 50–100 µL of hybridization solution containing the labeled probe(s) and cover with a cover slip taking care to exclude any bubbles.
8. Place slides in an airtight box containing filter paper soaked in 50% formamide/2X SSPE. Formamide is toxic and volatile, and therefore should be handled in a fume hood.
9. Hybridize for at least 16 h at 50°C.
10. Transfer slides into 2X SSPE and allow the cover slips to float off (forceps may be used gently to assist).
11. Transfer slides into a fresh staining trough containing 2X SSPE and wash for 1 h at room temperature with agitation.
12. Wash slides in 0.2X SSPE for 1 h at 50°C with agitation.
13. Wash slides in 0.2X SSPE for 30 min at room temperature.

3.2.1. Detection with Alkaline Phosphatase-Conjugated Antibodies

1. Incubate slides in maleic acid buffer for 10 min, and then in 1X blocking buffer for 30 min at room temperature.
2. Dry around the square marked for hybridization with a tissue, taking care that the sample does not dry. Place slides in an airtight box containing wet filter paper, and apply 100 µL of 1X blocking buffer containing a 1:500 dilution of alkaline phosphatase-conjugated antidigoxigenin antibody to the sample. Incubate for 3 h at room temperature.

In Situ RNA Detection

3. Transfer slides to a staining trough and wash in three changes of maleic acid buffer for 20 min each.
4. Incubate slides in alkaline phosphatase buffer for 5 min.
5. Dry around the square marked for hybridization with a tissue, taking care that the sample does not dry. Place slides in an airtight box containing wet filter paper, and apply 100 µL of NBT/BCIP solution. Incubate overnight in complete darkness (e.g., cover with aluminium foil).
6. Stop the reaction by incubating the slides in 10 mM Tris-HCl, 1 mM EDTA solution (pH 8.0) for 5 min at room temperature. DNA may be counterstained with 1 µg/mL 4,6-diamidino-2-phenylindole (DAPI) at this stage, although strong precipitate may mask the fluorescent signal.
7. Dry the slides, mount the samples in an aqueous mounting medium, and view immediately (precipitated substrate may disperse during storage). The signal is detected as a purple precipitate.

3.2.2. Detection with Fluorochrome-Conjugated Antibodies

1. Incubate slides in PBS-T for 30 min at room temperature.
2. Dry around the square marked for hybridization with a tissue, taking care that the sample does not dry. Place slides in an airtight box containing wet filter paper, and apply 100 µL of PBS-T containing appropriate fluorochrome-conjugated antibodies, diluted at the concentration recommended by the manufacturer; it is advisable to try a series of antibody concentrations to determine empirically the concentration giving the optimal signal:background noise ratio. Incubate for 1 h at room temperature in the dark.
3. Transfer slides to a staining trough and wash in three changes of PBS-T for 20 min each in the dark.
4. Dry the slides, mount the samples in an aqueous mounting medium containing antifadent reagents, and view immediately.

3.3. Whole-Mount In Situ Detection of RNAs in Mosquito-Stage Plasmodium

After dissection, all steps can be carried out in 1.5-mL Eppendorf tubes. Unless stated, all steps and washes are carried out with gentle rocking or rotation at room temperature. To change solutions, allow the guts to sink to the bottom of the tube and remove the liquid, leaving a little above the guts to reduce loss or damage.

1. Dissect infected mosquito midguts in PBS at 4°C and transfer to ice-cold 4% paraformaldehyde/PBS solution for 1–2 h for fixation (*see* **Note 4**).
2. Wash guts in 3 changes of PBS for 5 min each. Dehydrate the guts in a methanol series (25, 50, 75, and 100%) for 5 min each. Dissected guts store well at this stage in 100% methanol at –20°C.
3. Rehydrate guts through a reverse methanol series (25, 50, 75, and 100%) for 5 min each. Wash in three changes of PBS-T for 5 min each.
4. Permeabilize with 50 µg/mL proteinase K for 5 min.
5. Stop digestion by washing in 2 mg/mL glycine in PBS-T for 2 min. Wash in three changes of PBS-T for 5 min each.
6. Refix in cold 4% paraformaldehyde/PBS solution for 20 min. Wash in three changes of PBS-T for 5 min each.
7. Prehybridize guts in prehybridization buffer for at least 2 h at 70°C.
8. Replace prehybridization buffer with 100–200 µL hybridization buffer containing ~20 ng/mL denatured antisense RNA probe. Hybridize for at least 16 h at 70°C.
9. Add 800 µL of prehybridization buffer, preheated to 70°C. Wash twice in prehybridization buffer at 70°C for 10 min each. Wash twice in 50% hybridization buffer/PBS-T at 70°C for 10 min each. Wash five times in PBS-T at room temperature for 10 min each. Guts

tend to float in hybridization buffer, so take great care to avoid loss or damage to the guts during the initial wash steps.
10. Incubate guts in the appropriate fluorescent antihapten antibodies, used at a concentration of 2–20 μg/mL (or according to the manufacturer's instruction) for 1 h at room temperature.
11. Wash with five changes of PBS-T for 20 min each.
12. Transfer guts to aqueous mounting media containing antifadent reagents in microwell slides. Alternatively, attach strips of cover slip to the slides to produce shallow wells that will prevent the guts being excessively squashed when the cover slip is applied. View by confocal laser scanning microscopy (CLSM) to visualize the morphology of the parasites within the gut. Flat portions of the guts containing oocysts may be dissected for viewing by conventional microscoscopy, but it will not be possible to resolve the subcellular structures.

3.4. In Situ *Detection of RNA in Salivary Gland Sporozoites*

In situ detection of RNA within sporozoites can be performed using either of the two protocols described above. Salivary glands are, however, smaller and more difficult to handle than mosquito guts in the whole-mount procedure, so it is more convenient to immobilize them onto coated slides for processing (*see* **Subheading 3.2.**). In this case, infected salivary glands are dissected from the mosquito into PBS, transferred (with forceps) onto Denhardt's coated slides and allowed to dry completely with the use of a fan. In a single experiment, one of a pair of salivary glands may be left intact, while the other is disrupted so that some sporozoites are released onto the slide; in this way, sporozoites may be viewed both individually, at high resolution, by conventional fluorescence microscopy, and as bundles within the glands by CLSM.

3.5. Controls

1. The conventional control for *in situ* hybridization experiments is the sense strand RNA probe.
2. A good indication for the specificity of the *in situ* hybridization reaction is to view (and record) a variety of different parasite stages; differential expression in particular stages is good evidence that the probe is specifically hybridizing to a stage-specific gene transcript. Uninfected red blood cells, reticulocytes, and leukocytes, present in blood smears, and mosquito tissues also act as controls for hybridization specificity.
3. Where fluorochrome-conjugated detection systems can be used, it is particularly recommended to include a control RNA probe labeled with a different hapten that is detected with a complementary antibody. We have, for example, used a fluorescein-labeled antisense RNA probe corresponding to a region of *P. berghei* rRNA genes which is unique to *Plasmodium* spp and expressed at all stages of parasite development. When used in combination with a specific digoxigenin-labeled antisense probe, which is detected with a rhodamine-conjugated antidigoxigenin antibody, this probe serves not only as an internal control of the success and specificity of the *in situ* hybridization but also permits fluorescent visualization of the parasite cytoplasm and subcellular structures and, therefore, identification of the parasite stage. Unfortunately, NBT/BCIP precipitates mask fluorescent signals, preventing the use of control probes when the hybridization signal is detected using enzyme-linked antibodies.

4. Notes

1. Enzyme-linked antibodies have proved to be the most sensitive and specific method tested to detect *in situ* hybridization of digoxigenin-labeled probes. In this procedure, high affinity antidigoxigenin antibodies, conjugated to alkaline phosphatase, enzymatically precipi-

tate colorimetric substrates (NBT and BCIP) in reactions that can be allowed to proceed for several hours. The major drawbacks to the use of this system is that only one probe can be used in each experiment (the purple substrate masks fluorescent signals) and it is difficult to differentiate subcellularly localized signals from the crystals of hemazoin pigment present in many parasite stages.

2. The detection of hapten-labeled probes using fluorochrome-labeled antibodies or avidin-streptavidin systems is generally less sensitive than enzyme-linked detection methods. In addition, the high autofluorescence of the hemoglobin present in red blood cells can mask specific hybridization signals (especially within cells in a thick blood smear). Despite these limitations, the use of fluorescent detection systems is usually the method of choice as it permits the visualization of RNA subcellular localization and the codetection of two transcripts within the same cell.

3. There is great cell-to-cell variability in the robustness of the red blood cell and parasite membranes. As most of this variability appears to arise from differences in the thickness of the smear, and variations between different infected blood samples, it is difficult to establish hybridization conditions that give good parasite morphology with high sensitivity every time. The parasite membrane is more robust than the infected red blood cell membrane, so the morphology of mature parasites, which occupy almost the entire host cell, usually appears better than that of ring stages. The larger parasites of *P. falciparum* more reproducibly give good results than rodent parasites in blood stages.

   Hybridizations may be carried out on a number of samples at the same time, however, so different smear preparations and prehybridization procedures can be tested during a single experiment; for example, a series of slides can be incubated in the proteinase K solution for different periods of time (particularly when titrating the optimal digestion conditions for a new batch of proteinase K). Experimental cell-to cell variability makes it especially important to monitor and record the efficacy of the hybridization reaction within localized areas.

4. The presence of a large bloodmeal within the gut during the first 1–2 d after feeding can present great difficulties in preparing the tissue for visualization of ookinetes and early oocysts in the midgut wall. Sections of the gut can be dissected from the bloodmeal after hybridization, but the morphology of the gut is preserved better if the bloodmeal is removed during the initial dissection. One way to do this is to fix the dissected guts for ~30 s in 4% paraformaldehyde-PBS solution, return to PBS and make a tear along the length of the gut. Gently dip the gut into PBS a few times until the surface tension causes the bloodmeal to be released. This method is suitable for preparing infected *A. stephensi* midguts, and is particularly effective if mosquitoes are fed with an ookinete culture and the gut is dissected during the few hours before the peritrophic membrane has formed.

References

1. Thompson, J. and Sinden, R. E. (1994) *In situ* detection of Pbs21 mRNA during sexual development of *Plasmodum berghei*. *Mol. Biochem. Parasitol.* **68,** 189–196.
2. Shaw, M. K. Thompson, J., and Sinden, R. E. (1996) Localization of ribosomal RNA and Pbs21 mRNA in the sexual stages of *Plasmodium berghei* using electron microscope *in situ* hybridization. *Eur. J. Cell Biol.* **71,** 270–276.
3. Baker, D. A., Thompson, J., Daramola, O. O., Carlton, J. M., and Targett, G. A. (1995) Sexual-stage-specific RNA expression of a new *Plasmodium falciparum* gene detected by *in situ* hybridization. *Mol. Biochem. Parasitol.* **72,** 193–201.
4. Dechering, K. J., Thompson, J., Dodemont, H. J., Eling, W., and Konings, R. N. (1997) Developmentally regulated expression of Pfs16, a marker for sexual differentiation of the human malaria parasite *Plasmodium falciparum*. *Mol. Biochem. Parasitol.* **89,** 235–244.
5. Thompson, J., van Spaendonk, R. M. L., Choudhuri, R., Sinden, R. E., Janse, C. J., and Waters, A. P. (1999) Heterogeneous ribosome populations are present in *Plasmodium berghei* during development in its vector. *Mol. Microbiol.* **31,** 253–260.

22

Purification of Chromosomes from *Plasmodium falciparum*

Daniel J. Carucci, Paul Horrocks, and Malcolm J. Gardner

1. Introduction

Sequencing of the entire genome from the human malaria parasite, *Plasmodium falciparum*, began in earnest in 1996 with the formation of an international consortium of scientists and funding agencies *(1)*. Due to the instability of the highly adenine (A) and thymine (T)-rich *P. falciparum* DNA in large insert bacterial vectors, and to distribute the enormous task of completing the 30 Mb genome among several genome centers, the decision was made to sequence the *P. falciparum* genome by chromosome. Obtaining pure chromosome material, free of contaminating DNA and in sufficient yield, is of paramount importance in order to produce sequencing libraries that are as unbiased and as representative of the genome as possible. Minimizing the amount of cross-contaminating chromosomal DNA is critical to reduce costs and ensure accurate assembly of the sequence. To date, the sequences of two chromosomes from *P. falciparum* have been published, chromosome 2 *(2)* and chromosome 3 *(3)*, and it is anticipated that the remainder will be completed by the end of 2002.

This chapter summarizes the methods used to purify individual chromosomes from *P. falciparum* (clone 3D7) and is intended to provide guidance for the purification of chromosomes from additional species of *Plasmodium* as well as from other small genomes.

2. Materials

2.1. Equipment

1. Chef DR-III pulse field gel electrophoresis apparatus (Bio-Rad).
2. Benchtop centrifuge, capable of 4°C (Forma Scientific).
3. Microscope slides, glass cover slips.
4. Tubes: 1.5-mL Eppendorf, 50-mL conical.
5. Culture flasks, 150-mL.
6. Glass Pasteur pipets.
7. Glass plate (15 cm × 15 cm) cleaned with 70% isopropanol.

2.2. Reagents

1. Parasite wash buffer: 10 m*M* Tris-HCl (pH 8.0), 0.85% NaCl. Autoclave and store at 4°C.
2. Red blood cell lysis solution: 1% acetic acid in ddH$_2$O, prepared fresh immediately prior to use from concentrated glacial acetic acid.

3. 10% bleach solution (Clorox).
4. Proteinase K solution: 2 mg/mL proteinase K in 1% Sarkosyl, 0.5 M EDTA.
5. InCert® agarose (FMC).
6. 50 mM ethylenediaminetetracetic acid (EDTA), pH 8.0.
7. Agarose (SeaPlaque, GTG; chromosomal grade agarose, Bio-Rad).
8. 10X Tris-acetate (TAE) buffer (Quality Biologicals).
9. 10X Tris-borate-EDTA (TBE) buffer (Quality Biologicals).
10. Molecular weight markers: *Saccharomyces cerevisiae* and/or *Hansenula wingei* Chef size markers (Bio-Rad).
11. Ethidium bromide solution (5 μg/mL) in ddH$_2$O.

3. Methods

3.1. Parasite Preparation

Cultivate *Plasmodium falciparum* (clone 3D7) parasites to mature schizonts to maximize the nucleic acid concentration, using standard techniques *(4)* (*see* Chapter 46). In order to minimize possible alterations of the genome that can occur in continuous culture, keep parasite aliquots frozen until needed and then cultivate only as long as necessary to produce sufficient material for chromosome purification.

1. Remove a 100-μL aliquot of the parasite culture to a sterile Eppendorf tube for cell count determination. Prepare two thin films to assess parasite morphology and percentage parasitemia (*see* **Notes 1** and **2**).
2. Collect the culture from four large (150-mL) flasks in 50-mL conical tubes. Centrifuge the tubes at 500g for 10 min at 4°C in a benchtop centrifuge. Remove the supernatant using a 25-mL pipet and discard it into 10% bleach solution.
3. Add approx 10 mL of ice-cold parasite wash buffer to the tube, and resuspend the red cell pellet using a large bore pipet. If necessary, combine the contents of multiple tubes into one tube (no more than the contents of two flasks per one 50-mL tube). Fill the tube with ice-cold buffer and invert several times.
4. Centrifuge as above.
5. Remove the supernatant as above and thoroughly resuspend the parasite pellet in approx 25 mL of ice-cold wash buffer.
6. Add an equal volume of red cell lysis solution, gently mix by inversion, and place on ice for 5 min. The acetic acid treatment causes the red blood cells to lyse releasing the parasites. After several minutes, the contents of the tube will darken.
7. Centrifuge the freed parasites in a benchtop centrifuge at 2000g for 10 min at 4°C. Remove the supernatant carefully so as not to disturb the parasite pellet.
8. Wash the freed parasites by gently and thoroughly resuspending the parasite pellet in 10 mL of ice-cold parasite wash buffer, filling the tube with buffer and centrifuging the tube at 2000g as above (*see* **Note 3**). Repeat the wash process several times until the supernatant is essentially colorless (~3–4 washes) (*see* **Note 4**).
9. Determine cell counts using a hemocytometer, and determine parasite concentrations by Giemsa staining (*see* **Notes 1** and **2**).
10. Using the determination of the cell count and the percentage parasitemia, add sufficient buffer to resuspend the freed parasites to 1×10^9 parasites/milliliter.
11. Place the tube containing the parasites in a 50°C waterbath for approx 10 min.
12. Add an equal volume of molten 1% InCert agarose (FMC) in buffer prewarmed to 50°C, and mix gently using a wide bore pipet. Using the same pipet, transfer the parasite–agarose solution to a 1 cm × 1 cm × 10 cm Perspect gel mold plugged at one end with solidified Incert agarose (*see* **Note 5**).

Chromosome Purification

13. Allow the molds to cool to 4°C for approx 2 h.
14. Push the agarose-embedded parasites out of the mold into a 50-mL conical tube containing proteinase K solution (2 mg/mL proteinase K in 1% Sarkosyl, 0.5 M EDTA).
15. Incubate at 50°C for 48–72 h with one change of proteinase K solution.
16. Store the parasite "noodle" in 50 mM EDTA at 4°C until needed. Parasites stored in this manner have been used successfully for up to 1 yr with no observable chromosomal degradation.

3.2. Pulse Field Gel Electrophoretic Separation of Chromosomes

A Chef DRIII apparatus (Bio-Rad) is used for all chromosome separations. All gels are maintained at 14°C with a recirculating chilling unit at a setting of 7 (approx 1 L/min) on the variable speed pump. The conditions vary depending on the size of the chromosome and the stage of purification, the agarose type, and the buffer used.

1. Prepare 2 L of electrophoresis buffer (either 1X TAE or 0.5X TBE).
2. Add the correct amount of agarose to a 250-mL screw-top bottle, add 100 mL of buffer and melt in a microwave oven. Use heat protective gloves, long-sleeve clothing, and low settings on the microwave. Be sure the screw top is loose, otherwise the bottle could explode. Carefully, swirl the bottle to ensure the agarose is melted completely (take care to not allow the molten agarose to boil over). Place the bottle with a tightened top in a 60°C waterbath for at least 15 min.
3. Place the remaining buffer (~2 L) into the PFG apparatus, switch on the pump and then the recirculating cooler (*see* **Note 6**).
4. Assemble the gel-pouring apparatus as recommended by the manufacturer. Use a leveling platform to ensure the poured gel is level.
5. To prepare a gel for purification of large amounts of parasite material, tape the gel comb in an inverted position onto the comb holder so that the entire flat surface of the comb just comes into contact with the bottom of the gel mold. Flip the comb 90° so that the bottom edge of the comb is pointed to you. This horizontal surface will be used to arrange the parasite blocks prior to pouring the gel.
6. Take uniform parasite slices from the "noodle" using a glass cover slip "knife" and two offset microscope slides as thickness guides. It is convenient to use a glass plate (15 cm × 15 cm) cleaned with 70% isopropanol as a cutting surface. A glass pipet can be formed with a Bunsen burner to create a "loop" to retrieve the "noodle" from the Falcon tube.
7. Cut the thin slices in half. Parasite slices approximately one microscope slide thick (1–2 mm) and one-half to one-quarter of a single slice are sufficient per lane. The final parasite blocks are 1 cm × 0.25 mm–0.5 cm × 1–2 mm in dimension.
8. Using the glass cover slip and the glass loop to manipulate the slices, arrange the parasite slices side by side on the flat side of the gel comb with the long orientation of the blocks parallel with the long axis of the comb.
9. If necessary, include molecular weight standards (*Saccharomyces cerevisiae* and/or *Hansenula wingei* Chef size markers [Bio-Rad]) on the extreme ends of the comb.
10. Right the comb onto the gel mold and fix the parasite blocks by pipetting a small bead of molten (60°C) agarose at the junction where the parasites meet the bottom of the gel mold.
11. Allow the agarose to cool for 10 min and then pour the remaining agarose into the gel mold, just covering the parasite blocks. There should be several milliliters of molten agarose remaining in the 250-mL bottle.
12. Place the bottle containing the remaining agarose back in the 60°C waterbath. This agarose will be used to fill the space left by the comb.
13. Allow the agarose in the gel mold to cool to room temperature (~1 h), carefully remove the comb and working quickly fill the space with molten agarose using a pipet.

14. Allow the gel to cool to room temperature (approx 10 min).
15. Position the gel guide in the buffer tank.
16. Disassemble the gel mold and carefully lift the gel and the gel support from the mold and place in gel guide in the pulsed field gel electrophoresis apparatus. Close the lid.
17. Allow the gel to come to the temperature of the buffer (approx 10 min) and then switch on the power supply.

3.3. Chromosome Isolation

In general, initial chromosome purification is carried out using either chromosomal grade (Bio-Rad) or low melting temperature agarose at 0.8–1.0% in either 1X TAE or 0.5X TBE (*see* **Table 1**). However, if necessary, a second round of purification may be performed to concentrate the chromosome material in the agarose gel, to remove potentially contaminating DNA from other chromosomes and to isolate the chromosome in low-melting-point agarose. The second round of purification is performed by placing the excised chromosome material from the initial gel against an electrophoresis comb in a second gel mold and pouring molten low melting point agarose SeaPlaque (FMC) around the slice.

The chromosomes are excised from the gel using minimal exposure to ultraviolet (UV) light so as to minimize UV crosslinking and possible damage to the highly adenine-thymine *P. falciparum* DNA.

1. Cut vertical gel slices from the ends of the gel parallel to the migration of the DNA and sufficient from the edges to ensure that approx 0.5 cm of the chromosomes are included in the slice.
2. Stain the slices in ethidium bromide solution (5 µg/mL) for 20 min.
3. Visualize the gel using a UV fluorescence (320 nm) light box.
4. Cut notches corresponding to the desired individual chromosomes in the agarose gel, and use these notches as guides to cut the desired chromosome band from the original gel.
5. Place the unstained gel on a sheet of clear plastic wrap on the bench surface.
6. Place the notched slices in their original positions against the unstained gel.
7. Using the notched edges as guides, cut across the gel using a straightedge to excise the desired chromosome.
8. Place the gel slice either in a second gel mold apparatus for second round purification or in a Falcon tube in 50 mM EDTA at 4°C until needed.
9. Stain the gel with ethidium bromide and photograph with UV transillumination to verify the chromosome excision.

4. Notes

1. The cell count and percentage parasitemias are needed to determine the final concentration of parasites in the agarose blocks. To determine the cell count, note the total volume of the culture. Set up four Eppendorf tubes, marked 10, 10^2, 10^3, 10^4, containing 90 µL of PBS and add 10 µL of the culture to the first tube, gently mix the tube contents by flicking, and transfer 10 µL to the second tube. Repeat this for the remainder tubes. Using a hemocytometer and a cover slip, add 10 µL of the 10^3 and 10^4 tubes on each side of the hemocytometers under the cover slip. Count the number of cells from each side of the major grid. Use the dilution that that results in between 100–1000 cells (use other dilutions if needed). Determine the total cell count in the culture using the formula: Cell count × 10^4 × dilution factor (generally 10^3) × total volume.
2. To prepare a thin blood film, remove 100 µL of the culture into an Eppendorf tube and centrifuge for 2 min at 3000g in a benchtop microfuge. Remove 50 µL of the supernatant

Table 1
Gel Conditions for Separation of Chromosomes

Chromosomes	Purification	Agarose type (conc: %)	Buffer type	Block	Time (h)	Voltage (V/cm)	Field angle (degrees)	Ramp time (s)
1–4	Initial	SeaPlaque 1.2	0.5X TBE	1	90	3.7	120	180–250
1–4	Final	SeaPlaque 1.2	0.5X TBE	1	24	6	120	60–120
5–9 (5)	Initial/final	Gibco-BRL LMP 0.9	0.5X TBE	1	155	3.1	120	360
10–12	Initial	SeaPlaque LMP 1	0.5X TBE	1	60	3	120	200–400
10–12	Initial	SeaPlaque LMP 1	0.5X TBE	2	32	3	106	600–1000
10–12	Final	SeaPlaque	0.5X TBE	1	90	3	106	180–900
13–14	Initial	Chromo grade 0.8	1X TAE	1	48	3	106	500
13–14	Final	SeaPlaque LMP 1	1X TAE	1	48	3	106	500

and thoroughly resuspend the pellet by flicking the tube. Place 5 µL of the resuspended pellet on the near side of a glass microscope slide and without delay, using a second slide placed at a 45° angle to the first, draw the second side backward into the blood spot, and with a very rapid stroke, spread the blood spot across the first slide. Allow the slide to dry for 10 min and then fix the slide by flooding it with methanol. Tip off the methanol, allow the slide to dry for a few minutes and place in a 5% Giemsa solution in ddH$_2$O for 30 min. After rinsing the slide with distilled water, and allowing it to dry, determine the parasitemia using a ×100 oil objective, by averaging the counts of the number of infected cells per total number of red blood cells in 10 fields of view.

3. The released parasites are less dense than the infected red blood cells and thus require greater centrifugation.
4. It may be necessary to leave a significant amount of the hemolyzed blood supernatant in the first wash as the brown parasite pellet may be difficult to visualize.
5. Prepare both the Incert agarose and the "noodle" mold early in the process and allow it to cool in a 50°C waterbath. Use ~500 µL of the molten Insert agarose to plug the end of the Perspect mold, sealed with Parafilm and allow to cool at room temperature for about 30 min.
6. Be sure to switch the pump on first before the chiller unit otherwise the chiller unit may cause the buffer inside the chilling unit to freeze.

Acknowledgments

We thank Mr. James Pedersen and Ms. Sarah Lee for their technical assistance.

The opinions and assertions herein are those of the authors and are not to be construed as official or as reflecting the views of the US Navy or naval service at large. The work was supported by the Office for Research on Minority Health of the National Institutes of Health and by the Naval Medical Research and Development Command work units STO F 6.3a63002AA0101HFX, STO F 6.161102AA0101BFX, STO F 6.262787A00101EFX, and STEP C611102A0101BCX.

References

1. Hoffman, S. L., Bancroft, W. H., Gottlieb, M., James, S. L., Burroughs, E. C., Stephenson, J. R., and Morgan, M. J. (1997) Funding for malaria genome sequencing. *Nature* **387,** 647.
2. Gardner, M. J,. Tettelin, H., Carucci, D. J., Cummings, L. M., Aravind, L., Koonin, E. V., Shallom, S., et al. (1998) Chromosome 2 sequence of the human malaria parasite *Plasmodium falciparum*. *Science* **282,** 1126–1132.
3. Bowman, S., Lawson, D., Basham, D., Brown, D., Chillingworth, T., Churcher, C. M., et al. (1999) The complete nucleotide sequence of chromosome 3 of *Plasmodium falciparum*. *Nature* **400,** 532–538.
4. Trager, W. and Jensen, J. B. (1976) Human malaria parasites in continuous culture. *Science* **193,** 673–675.
5. Holloway, S. P. Gerousis, M., Delves, C. J., Sims, P. F., Scaife, J. G., and Hyde, J. E. (1990) The tubulin genes of the human malaria parasite Plasmodium Falciparum, their chromosomal location and sequence analysis of the alpha-tubulin 11 gene. *Mol. Biochem. Parasitol.* **43(2),** 257–270.

23

Construction of Genomic Libraries from the DNA of *Plasmodium* Species

Leda M. Cummings, Dharmendar Rathore, and Thomas F. McCutchan

1. Introduction

Genomic DNA libraries represent the total complement of the genetic information of an organism's DNA, as opposed to cDNA libraries, which contain only the protein encoding sequences expressed at a particular stage of the life cycle. Ideally, a genomic library contains the coding and control elements from all stages of the life cycle as well as a surfeit of sequence whose role in the storage and utilization of the organism's heritable information is unknown. Comparison of the arrangement of sequences in the two types of libraries has been used as an approach in understanding the process of dispensing informational content to the cell. For example, the comparison has provided technique for identifying and understanding control elements for gene expression. Further, the genomic conformation of a segment of genetic material and its counterpart in the messenger RNA pool, from which cDNA libraries are constructed, may be different for a number of reasons. Features of the processing of RNA such as the excision of introns, RNA processing and exon splicing have been discovered because of the power of comparing genomic and cDNA libraries.

Historically, the problem faced during the process of preparing genomic libraries related to making them as nearly representative of the genome as possible. Ideally, one wanted to be able to produce random cleavage and control the average size of the resultant fragments. Given this idyllic situation, one could calculate the number of DNA clones that would need to be produced to include every possible sequence. Realistically, however, some sequences either consistently underwent rearrangement in the host microbe carrying the recombinant DNA, or so retarded the growth of the host cell that its representation was no longer representative of its copy number in the genome. Initially, partial restriction nuclease digestion was used to produce DNA fragmentation. This often yielded good libraries, but they were not representative of the genome because the restriction nuclease cleavage sites themselves were not randomly distributed. This may still offer the best hope of producing large recombinant fragments in viral vectors or in cosmids. Vectors containing large DNA fragments, however, are now less frequently used for the general survey of genome sequences. With the advent of entire genome projects, advances in polymerase chain reaction (PCR) technology and methods for chromosome separation, the use of restriction nuclease-produced libraries are often reserved for more specialized purposes. Random cleavage by either DNase I or by physical shearing of DNA provides a more uniform cleavage than the restriction nuclease to produce controlled fragment sizes.

The most use for random libraries has shifted from identifying genes by sequence similarity to their use for both large-scale sequencing projects and projects designed to screen for biological function or reactivity to particular antisera. The automation of sequencing and screening along with improved methods for computer analysis of the output has made the screening of a large number of clones with smaller insert sizes less daunting.

The example used here for DNA preparation is a procedure developed for automated shotgun sequences, random shearing. The general usefulness of the procedure for preparing fragments for any type of library is apparent. The advantages in preparing DNA in this fashion are many. They include the fact that the procedures for shearing DNA are both reproducible and controllable. The resultant fragments have been exhaustively analyzed, and measures of their representation of the genome have been established in other systems. The resultant fragments can as easily be put into vectors suitable for immunoscreening, complementation analysis, and functional identification.

2. Materials

2.1. Isolation of DNA from Parasites

1. TSE buffer: 20 mM Tris-HCl, pH 8.0, 100 mM NaCl, 50 mM ehtylenediaminetetraacetic acid (EDTA).
2. Extraction buffer: 10 mM Tris-HCl, pH 8.0, 100 mM EDTA.
3. Tris-EDTA (TE): 10 mM Tris-HCl, pH 8.0, 1 mM EDTA.
4. 1X Tris-borate-EDTA (TBE) buffer: 90 mM Tris-borate, 2 mM EDTA.
5. 10% sodium dodecyl sulfate (SDS) (20X stock).
6. 500 µg/mL DNase free RNase A (25X stock).
7. 10 mg/mL proteinase K solution (100X stock).
8. Tris-buffered phenol, pH 8.0.
9. 10 M ammonium acetate solution.
10. 100% ethanol.
11. Saturated solution of cesium chloride.
12. Hoechst dye 33258.
13. Isopropanol.

2.2. Shearing of DNA and Preparation of a Library

1. Glycerol (Life Technologies, cat. no. 15514-011).
2. 3 M sodium acetate (Life Technologies).
3. Autoclaved milli Q® H$_2$O (Millipore Corporation, Chicago, IL).
4. Nebulizer chamber complete aeromist set with 7 ft. tube (cat. no. 4207DN, IPI Medical Products, Chicago, IL).
5. LM agarose (LM-MP) (cat. no. 1.441.345, Boehringer Mannheim).
6. 10 mg/mL ethidium bromide (cat. no. 15585-011, Life Technologies),
7. AgarAce (cat. no. M1743, Promega).
8. Phenol:chloroform:isoamyl alcohol (cat. no. 15593-031, Life Technologies).
9. Buffer-saturated phenol (cat. no. 15513-039, Life Technologies).
10. 1-butanol (cat. no. BT-105, Sigma).
11. Dimethylformamide (Life Technologies).
12. 1 M Tris-HCl, pH 8.0 (cat. no. 15568-025, Life Technologies).

13. 0.5 M EDTA, pH 8.0 (cat. no. 155575-020, Life Technologies).
14. 10X TBE buffer (cat. no. 15581-044, Life Technologies).
15. BAL 31 (cat. no. 213L, New England BioLabs).
16. T4 DNA polymerase (cat. no. 203S, New England BioLabs).
17. 10 mM deoxynucleoside triphosphate (dNTP) (cat. no. 1.969.064, Boehringer Mannheim).
18. 10 mg/mL bovine serum albumin (BSA) (Life Technologies).
19. BstXI adapters (cat. no. N408-18, Invitrogen).
20. T4 DNA ligase (cat. no. 716.359, Boehringer Mannheim).
21. ATP-dependent DNase (cat. no. E3101K, Epicenter Technology).
22. SOB plates (per liter): 20 g of bacto-tryptone, 5 g of Bacto-yeast extract, 0.5 g of NaCl, 2.5 mM KCl. Adjust to pH 7.0 with NaOH. Sterilize by autoclaving. Before use, add $MgCl_2$ to 10 mM.
23. 250 µg/mL isopropyl thiogalactopyranoside (IPTG) (cat. no. 15529-019, Life Technologies) (*see* **Note 1**).
24. 100 mg/mL X-gal (cat. no. 15520-034, Life Technologies) (*see* **Note 2**).
25. Electro-max DH10β competent cells (cat. no. 18290-015, Life Technologies).
26. Ultraviolet lamp UVL-18 (cat. no. 36553-124, VWR).
27. Water bath.
28. Refrigerated centrifuge.

3. Methods
3.1. Isolation of Plasmodium DNA

1. Collect parasitized red blood cells by spinning for 10 min at 1000g at room temperature.
2. Resuspend packed infected red blood cells (RBC) in 5–10 vol of cold TSE buffer.
3. Spin resuspended cells at 1000g for 10 min, discard supernatant, and wash cell pellet again.
4. Resuspend pellet in 15 mL of extraction buffer and add 750 µL of 10% SDS to obtain a final concentration of 0.5%.
5. Add 160 µL of proteinase K to obtain a final concentration of 100 µg/mL.
6. Add 640 µL of DNase free RNaseA to obtain a final concentration of 20 µg/mL.
7. Incubate the suspension at 50°C for 2 h with periodic gentle swirling.
8. Cool the suspension to room temperature and add equal volume of Tris-HCl-saturated phenol (pH 8.0).
9. Gently mix the two phases on an end-to-end shaker.
10. Separate the two phases by centrifugation at 5000g for 15 min at room temperature.
11. With a wide-bore pipet, transfer the viscous aqueous phase to a fresh tube and twice repeat the extraction with phenol.
12. After the third extraction, precipitate the DNA with 0.2 vol of 10 M ammonium acetate and 2 vol of room temperature ethanol.

3.2. Separation of Parasite and Host DNA by Equilibrium Centrifugation in CsCl-Hoechst Gradient

The contamination of parasite DNA with that of the host or other microbes is always a problem. The most difficult situation is found with avian and reptilian parasites, which are located in nucleated red blood cells. The *Plasmodium* species found in mammals infects nonnucleated red cells but purification schemes almost always yield a percentage of nucleated cells as well. Even a 1% contamination of nucleated host cells leads to a 50% contamination of the DNA preparation, due to the difference in the genome size of *Plasmodium* and higher eukaryotes. Most nucleated cells can be removed from the blood prior to its use for culturing *P. falciparum*. This then produces the purest DNA.

It should be remembered, however, that bacteria or other microbes could also be a contaminant regardless of the preparation used. Hoechst dye–CsCl centrifugation is a very efficient procedure for separating contaminating host DNA from parasite DNA for a large number of *Plasmodium* species *(1,2)*. Many of the *Plasmodium* species have a strong bias toward dA.dT in their genome with *P. berghei*, *P. falciparum*, *P. lophurae*, and *P. gallinaceum* having a dA.dT content of more than 80% *(2)*. The dA.dT content of their hosts are much lower. The method is based on the preferential binding of the bis-benzimide dye Hoechst 33258 to dA.dT-rich DNA sequences. Binding of the dye reduces the buoyant density of dA.dT-rich DNA to the extent that separation of mixtures can be achieved by isopycnic ultracentrifugation in cesium chloride gradients.

1. Dilute parasite DNA with TE and add to a saturated cesium chloride solution. The considerations relating to the extent of dilution involve a balance between minimizing the total volume and the need to obtain a final refractive index of 1.3950.
2. Dissolve Hoechst dye in water to final concentration of approx 1 mg/10 mL of water. Add a weight of Hoechst dye 33258 equal to the estimated amount of parasite DNA to the cesium chloride mixture. We have shown that separation occurs when the weight of the dye is within a 10-fold range of the parasite DNA. A yellow precipitate often forms initially. One should try to get the precipitate into solution, but if the DNA is protein-free, it will not disturb the gradient formation.
3. Centrifuge the solution in a vertical rotor (VTi50, Beckman) in an ultracentrifuge at 196,409g for 18 h at 25°C. Parasite DNA will separate into a tight band on the top on account of its lower density in comparison to the high-density host DNA (*see* **Note 3**).
4. Remove the parasite DNA by first puncturing the top of the tube with a needle to prevent vacuum formation when the sample is withdrawn. Then illuminate the tube in the dark using a hand-held long-wave UV light; protective eye gear must be used during this process. Remove parasite DNA from the side of the tube using a syringe fitted with a 16-gage needle.
5. Extract DNA four times with an equal volume of isopropanol (equilibrated with cesium chloride solution) to remove the dye.
6. Remove cesium chloride by dialyzing against TE (pH 8.0) (*see* **Note 4**).
7. Extract DNA with an equal volume of phenol, followed by phenol:chloroform:isoamyl alcohol (25:24:1), and chloroform.
8. Precipitate DNA with 0.1 vol of 3 *M* sodium acetate (pH 5.2) and 2.0 vol of chilled ethanol.

3.3. Preparation of Sheared Plasmodium DNA in a Controlled Fashion

The successful completion of a large-throughput library relies on the ability to establish the most complete set of random DNA fragments *(3)*. Ideally, such a library would have clones spanning every DNA segment with close to equal frequencies. In addition, the total number of independent clones in the library should be high enough to guarantee the representation of every genomic segment. Libraries are most likely to fit the random model if prepared from small inserts and from a narrow size range (which helps minimize differences in growth rate and DNA rearrangements). A random library is best constructed from mechanically sheared fragments, since any enzymatic cleavage is generally nonrandom. Uneven representation of restriction enzyme sites in some genomic regions may generate fragments that are not size-selectable and thus unclonable. *See* **Fig. 1**.

Simulations are performed to calculate the number of random clones necessary to obtain a complete genome sequence. This is performed according to the model pro-

Fig. 1. Size-selection of mechanically sheared *Plasmodium yoelii* total genomic DNA. *Lanes 1* and *6*, 1 kb DNA ladder; *Lanes 2–5*, total genomic DNA is mechanically sheared and size-fractionated by gel electrophoresis. Fractions contain fragments with average sizes of 1.0 kb (*lane 5*), 1.5 kb (*lane 4*), 2.0 kb (*lane 3*), and 2.5 kb (*lane 2*) are, respectively, loaded on a 1% agarose gel.

posed for shotgun sequencing *(4)* and applies also to genomic libraries made for reasons other than sequencing. Procedures for producing this type of library have been developed to facilitate sequencing, and data from such work provides the most conclusive evidence that the procedures are working to yield randomly cleaved fragments. Here, we describe shotgun-sequencing libraries as an example for the procedure to prepare DNA fragments for any inclusive library. **Table 1** shows the computer simulation of random sequencing as applied to the large-scale sequencing of the 1 Mb *Plasmodium falciparum* chromosome 2 *(5)*. In order to obtain an eightfold genome representation, 18,000 sequences from chromosome 2 shotgun library are needed. Theoretically, five double-strand gaps would require closure by other methods, assuming a representative and fully random library, based on an average sequence read length of 450 bp. A considerable saving of efforts can be achieved if the inserts are at least twice the average sequence read length so that each template can be sequenced from both ends without redundancy *(4)*. The Institute of Genome Research (TIGR) has significant experience in preparing shotgun libraries from bacterial genomic DNA, and modifications have been developed to further improve the construction of shotgun libraries since its first use in the large-scale sequencing of *Haemophilus influenza* *(6)*.

Table 1
Random Sequencing Prediction for *P. falciparum* Chromosome 2 (L) of 1,000,000 bp and Read Length (w) of 450 bp

	Unsequenced (%)	Unsequenced (bp)	DS gaps	Average gap length	Coverage
5000	10.54	105,399	527	200	2.3
10,000	1.11	11,109	111	100	4.5
15,000	0.12	1,171	18	67	6.8
18,000	0.03	304	5	56	8.1
20,000	0.01	123	2	50	9.0

Here, we describe in detail the construction of small-insert libraries from total genomic DNA and pulse-field gel-purified chromosomal DNA. In this method, randomly sheared, end-repaired DNA fragments are ligated to oligonucleotide adapters creating a 4-base noncomplementary overhang. This methodology is also applicable to BAC or YAC shotgun libraries. DNA inserts are then ligated to a plasmid vector with compatible cohesive ends.

In the *Plasmodium* genome sequencing project at TIGR, several chromosome-specific *Plasmodium falciparum* small-insert libraries have been constructed that are being used for shotgun sequencing. Due to its high AT content, *Plasmodium falciparum* DNA is minimally exposed to the UV light to prevent nicking. When ethidium bromide staining and visualization of DNA is necessary, a long-wave hand-held UV lamp (365 nm) is used.

3.3.1. Preparation of Vector DNA

Many types of vector can be used depending on the eventual use of the library. We have been regularly using a pUC19-derived vector. Vector DNA is obtained by large-scale plasmid extraction followed by CsCl gradient purification. Vector DNA is digested with *Bst*XI/*Not*I and subsequently dephosphorylated with bacterial alkaline phosphatase, followed by gel purification.

3.3.2. DNA Shearing

3.3.2.1. Total Genomic DNA

1. Dilute a 100-µL aliquot (100 µg/mL) of total genomic DNA to 1 mL final volume with 450 µL glycerol, 200 µL of 3 *M* sodium acetate, and 250 µL of H$_2$O. Chill to 0°C in a nebulizer chamber.
2. Connect the chamber to a nitrogen source at a pressure of approximately 4 psi for 60 s.
3. Ethanol precipitate the DNA (*see* **Note 5**).
4. Visualize a small sample on a 1% agarose gel. DNA fragments are usually in the size range 1–4 kb.
5. Select fractions containing 1–2 kb and 2–3 kb fragments by fractionation in 1% low-melting agarose gels.
6. Purify DNA fragments by digesting the agarose with AgarAce, according to the procedure described in **Subheading 3.3.2.2.**

3.3.2.2. Pulsed-Field Gel Electrophoresis-Purified DNA

Individual chromosomes or YACs, obtained by preparative pulsed-field gel electrophoresis (PFGE), can be used to prepare chromosome-specific libraries. High molecular weight DNA is fractionated in low-melting-point agarose.

1. Cut agarose blocks in 1-cm pieces. If DNA blocks have been stored in 50 mM EDTA, dialyze in several changes of 10 mM Tris-HCl, pH 8.0, 1 mM EDTA (TE) at 4°C for at least 16–18 h.
2. Equilibrate DNA blocks with 0.3 M sodium acetate prepared in TE.
3. Melt agarose blocks at 69°C for 15–20 min and then transfer to 40°C.
4. Digest agarose overnight with 10 µL of AgarAce per 100 mg of agarose (*see* **Note 6**).
5. Extract the mixture three times with phenol and subsequently with 1-butanol until the volume is reduced to approx 600 µL.
6. Ethanol-precipitate the DNA, wash with 75% ethanol, and resuspend in 100 µL of TE (pH 8.0) (*see* **Note 7**).
7. Visualize a small sample (2.5 µL) on a 1% agarose gel for size verification.

3.4. Generation of Blunt Ends

Shearing produces staggered ends with both 5' and 3' overlaps. These are removed by BAL-31 nuclease treatment, which yields about 70% blunt ends. DNA fragments are then end-polished by T4 DNA polymerase. This enzyme efficiently removes short 3' overhangs and fills in short 3' recesses.

1. To 97.5 µL of DNA, add 100 µL of 2X BAL-31 buffer, 1 µL of enzyme (1 U/µL), and H_2O to 200 µL. Incubate at 30°C for 3 min.
2. Stop the reaction by adding 10 µL of 0.5 M EDTA (pH 8.0).
3. Extract DNA with phenol:chloroform, precipitate with 2.5 vol of ethanol and resuspend in 100 µL of water.
4. To 25 µL of DNA (one-fourth of the total sheared DNA) add 5 µL of 10 mg/mL BSA, 50 µL of 10X T4 DNA polymerase buffer, 5 µL of dNTPs (100 µM each), 20 µL of T4 DNA polymerase (3 U/µL), and bring to 500 µL final volume with distilled water.
5. Incubate the reaction at 11°C for 15 min.
6. Stop the reaction with 20 µL of 0.5 M EDTA.
7. Add 50 µL of 3 M sodium acetate and extract once with phenol:chloroform.
8. Ethanol precipitate the DNA and resuspend in 25 µL.

3.5. Size Selection

DNA fragments are size-selected by electrophoresis on a 1% low-melting point agarose in 1X Tris-acetate (TAE). Fractions containing ~1.5 kb- and ~2.0 kb-fragments are typically selected for library construction.

1. Cut a block of gel containing DNA fragments of required size and add volume of 3 M sodium acetate corresponding to 1/10 of the block.
2. Melt agarose block at 69°C for 5–10 min.
3. Digest agarose as previously described.
4. Extract DNA twice with phenol:chloroform.
5. Precipitate DNA and resuspend in 20 µL of TE.
6. Visualize precipitated DNA by running a 1-µL sample on a 1% agarose gel.

3.6. Ligation of 3' CACA Adapters

The use of synthetic oligonucleotide as cloning adapters have been shown to increase the cloning efficiency by minimizing the formation of chimeric inserts and reducing the circularized vector background. End-polished inserts are ligated to oligonucleotides creating a 4-base noncomplementary overhang (CACA). The ligation to non-self-annealing pUC19/*Bst*XI/*Not*I ends (TGTG) produces less than 5% level of background colonies.

1. Resuspend the oligonucleotides in 18 µL of water (to 1 µg/µL).
2. To 10 µL of DNA (from previous step), add 2 µL of 10X ligase buffer, 2 µL of *Bst*XI adapters, and 5 µL of H$_2$O. Double-stranded adapters are prepared by incubation at 4°C for 30 min.
3. Add 1 µL of DNA ligase and incubate the reaction overnight at 4°C.
4. Remove the adapters from the insert–adapter complexes by fractionation in 1% low-melting-point agarose.
5. Remove the DNA from the agarose with AgarAce as previously described.
6. Ethanol precipitate the DNA and resuspend to 10 ng/µL.

3.7. Insert–Vector Ligation

The library is generated by ligating a molar excess (about 4:1) of the sheared DNA fragments with pUC19 DNA that has been digested with *Bst*XI/*Not*I, gel-purified, and treated with bacterial alkaline phosphatase.

1. To 25 ng of *Bst*XI/*Not*I/BAP pUC19 DNA, add 100 ng of 3' CACA-tailed inserts (*see* **Note 8**).
2. Add 5 µL of 10X T4 ligation buffer, 1 µL of T4 DNA ligase (400 U/µL), and water to 50 µL.
3. Incubate overnight at 16°C and heat-inactivate the enzyme at 70°C for 10 min.

3.8. Removal of Linear DNA with ATP-Dependent DNase

The shotgun libraries prepared for the majority of the TIGR microbial sequencing programs make use of the vector + insert fraction (v + i). The gel-purification of the (v + i) fraction has proven to be instrumental in obtaining low level of non-insert-bearing colonies. Most recently, Dr. Hamilton Smith (previously at TIGR; now at Celera Genomics) introduced an essential improvement in the library construction protocol. The ligated insert + vector fraction is treated with ATP-dependent DNase, which effectively digests open circles, leaving intact the closed circular molecules. A scheme of this cloning step is shown in **Fig. 2**. **Figure 3** shows the activity of the ATP-dependent DNase on pUC19 DNA to illustrate the validity of the principle.

This step eliminates the need for the (v + i) purification step by gel electrophoresis (which requires larger amounts of starting DNA). In addition, the UV-irradiation of DNA fragments to be cloned can be totally eliminated. These two factors are of crucial importance when cloning *Plasmodium* DNA.

1. Add 1 µL of ATP-dependent DNase and incubate the reaction at 37°C for 15 min.
2. Heat-inactivate the enzyme at 70°C for 10 min.
3. Dilute the sample 1:2 and electroporate 25 µL of electrocompetent cells with 1 µL of diluted sample. The electroporation is performed at 1.75 V, 200 Ω, 25 f.
4. Add immediately 500 µL of SOC medium prepared with 20% glycerol, and transfer cells to ice.

3.9. Library Plating

Amplification of the library is avoided to prevent preferential selection of rapidly growing clones. A serial dilution of the electroporation mixture is spread on SOB plates containing a 5-mL bottom layer of SOB (supplemented with 4X ampicillin (i.e., 200 µg/mL final) and a 15-mL top layer of SOB agar poured just prior to plating. X-gal incorporation into the top layer permits blue-white selection. Diluted aliquots are stored at –70°C and are withdrawn from the freezer as needed for plating.

Genomic DNA Libraries

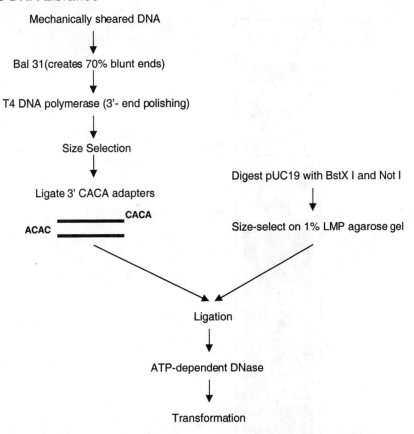

Fig. 2. General shotgun cloning scheme. Mechanically sheared DNA fragments are end-polished, ligated to 3'-CACA overhang oligonucleotide adaptors and subsequently ligated to pUC19 with compatible cohesive ends. The ligation mixture is treated with ATP-dependent DNase, which removes linear and open circle molecules.

Figure 4 shows a size range distribution of inserts obtained from a 1.5-kb gel-purified DNA fraction. DNA was obtained from mechanically sheared PFGE-purified chromosome 11 from *Plasmodium falciparum*. DNA prepared from these clones is digested with *Eco*RI and *Bam*HI to release the insert. The furthest right lane shows the migration of a 100-bp DNA ladder standard. The faint band migrating above the 2-kb marker represents pUC19 vector DNA. Note that insert size ranges from ~1 to ~1.8 kb. A tight insert size distribution is highly desirable to prevent unequal replication time.

A shotgun library prepared with *Plasmodium yoelii* total genomic DNA had 2.5×10^7 recombinants. A PFGE-purified chromosome 11 library from *Plasmodium falciparum* had 1.3×10^8 recombinants. Sequence analysis from clones derived from such libraries shows that they are random, with 5% contamination with *Escherichia coli* chromosomal DNA and vector sequences, which make them a very cost-effective and suitable reagent for large-scale sequencing programs.

3.10. Screening of Random Shear Libraries

The procedure described above for producing DNA fragments for sequencing can be used to produce inserts for almost any type of library. Vectors of choice change

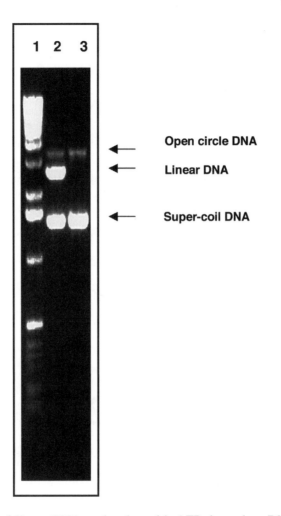

Fig. 3. Removal of linear DNA molecules with ATP-dependent DNase. pUC19 DNA untreated (*lane 2*) and treated with ATP-dependent DNase (*lane 3*) are separated on a 1% agarose gel. *Lane 1* contains a 1-kb DNA ladder.

yearly and new procedures for identifying clones by the biological activity of the DNA fragment that they carry are constantly being developed.

4. Notes
1. Prepare in water, sterilize by filtration through a 0.22-μm filter and store at –20°C.
2. Prepare X-gal solution with dimethylformamide, filtrate, and store at –20°C in a light-protected tube.
3. Sometimes the 35-kb plastid of all *Plasmodium* species will make a third band but, with most parasites DNA, it remains with the genomic DNA.
4. It is also possible to dilute the cesium chloride three-fold and then precipitate the DNA with 2 vol of ethanol. In this case, the pellet is washed twice with 70% ethanol to remove remaining cesium chloride. We prefer the dialysis approach.
5. To prevent DNA depurination, the sodium acetate solution is not adjusted to pH 5.0 (pH should be around 8.0).

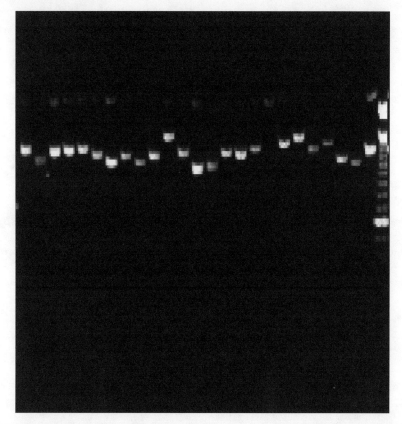

Fig. 4. Recombinant clones from a *Plasmodium falciparum* chromosome 11-specific shotgun library. DNA is digested with *Eco*RI and *Bam*HI for verification of insert size. The furthest right lane shows a 100-bp DNA ladder.

6. Incubation with AgarAce for small agarose blocks is performed at 40°C for 2–4 h. Additional amount of AgarAce can be added prior to the overnight incubation.
7. Precipitate the DNA in a 2-mL Eppendorf tube; samples can be pooled after DNA precipitation in case of very large volume of melted agarose.
8. For PFGE-purified DNA use half or all of the 3' CACA-tailed inserts, since DNA concentration might be difficult to estimate.

References

1. Dame, J. B. and McCutchan, T. F. (1987) *Plasmodium falciparum*: Hoechst dye 33258-CsCl ultracentrifugation for separating parasite and host DNAs. *Exp. Parasitol.* **64**, 264–266.
2. McCutchan, T. F., Dame, J. B., Miller, L. H., and Barnwell, J. (1984) Evolutionary relatedness of Plasmodium species as determined by the structure of DNA. *Science* **225**, 4664–4808.
3. Lander, E. S. and Waterman, M. S. (1988) Genomic mapping by fingerprinting random clones: A mathematical analysis. *Genomics* **2**, 231–239.
4. Edwards, A., Voss, H., Rice, P., Civitello, A., Stegemann, J., Schwager, C., et al. (1990) Automated DNA sequencing of the human HPRT locus. *Genomics* **6**, 593–608.
5. Gardner, M. J., Tettelin, H., Carucci, D. J., Cummings, L. M., Aravind, L., Koonin, E. V., et al. (1999) Chromosome 2 sequence of the human malaria parasite *Plasmodium falciparum:* Plasticity of a eukaryotic chromosome. *Science* **282**, 1126–1132.
6. Fleischmann, R. D., Adams, M. D., White, O., Clayton, R. A., Kirkness, E. F., Kerlavage, A. R., et al. (1995) Whole-genome random shotgun sequencing and assembly of *Haemophilus influenzae* Rd. *Science* **269**, 496–512.

24

Construction of a Gene Library with Mung Bean Nuclease-Treated Genomic DNA

Dharmendar Rathore and Thomas F. McCutchan

1. Introduction

Mung bean nuclease (MBN) has been used typically for its single-stranded nuclease activity *(1)*. Under defined reaction conditions, however, which include altered solvation and elevated temperature, hypersensitive sites surrounding coding regions become sensitive to nuclease cleavage *(2)*, while sequences within coding regions remain insensitive (**Fig. 1**). Mung bean cleavage of genomic DNA from *Plasmodium* species results in a solution of gene-containing fragments (**Fig. 2**) *(2)*. A significant proportion of the noncoding regions is reduced to shards, but it is not known what percentage of the genome this includes. The cleavage sites are not simply single-stranded bubbles in the DNA but stable nucleic acid structures that contain specific arrangements of paired and unpaired nucleotides *(3)*.

A study has investigated the relationship between the length of coding regions in *P. falciparum* and the length of the corresponding mung bean-digested fragment (**Fig. 1**). Using oligonucleotides representing the extreme 5' and 3' end of the genes, Vernick et al. *(4)* screened mung bean-digested DNA for ten different proteins and found that DNA coding for each protein was present in a single mung bean fragment. The size of the mung bean fragment vs the size of the coding region was graphed for each gene, and the relationship between them was analyzed by linear regression, which yielded the equation $y = 1.2x - 338$. The linear equation predicts that, under the described conditions, the mung bean fragments that contain genes should, on average, be about 1.2 times the size of the protein-coding region in the genomic DNA. DNA from both *Drosophila* and humans have been cleaved under slightly different conditions with similar success, but the number of analyses preformed to date is limited *(3)*. The size differential between fragment and coding region may be different with DNA from other organisms, but cleavage is precise.

This method makes it possible to identify intact genes in long stretches of DNA or produce gene libraries that can be screened with procedures like DNA hybridization *(4)*, antibody detection *(5)*, or functional complementation *(6)*. Although fragments are generated by the cleavage of structures rather than a specific primary sequence, they appear to be as sharply defined as restriction fragments by Southern blot analysis. A library prepared with fragments generated by mung bean digest represents an enrich-

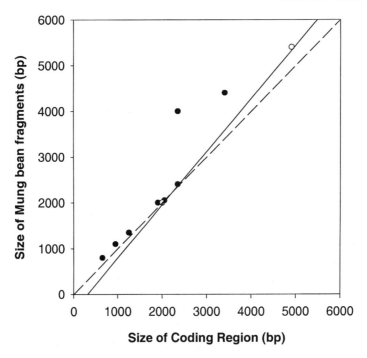

Fig. 1. Size of mung bean vs protein coding regions. Data points from mung bean fragments are shown as dotted circles. Data points from published accounts of other laboratories are shown as open circles. The solid line represents the linear equation, $y = 1.2x - 338$ derived from the data. The shaded line represents $y = x$, the expected result if the mung bean fragment sizes were the same size as the coding region.

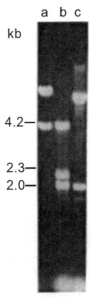

Fig. 2. Mung bean nuclease treated of DNA yields products that run as unit bands upon agarose electrophoresis. A plasmid, pPbSL7.8 (2) was treated with *Eco*RI (*lane a*), mung bean nuclease and *Eco*RI (*lane b*) or mung bean nuclease alone (*lane c*). The plasmid contains a DNA fragment from *P. berghei*. The fragment is 7.8 kb and contains the external transcribed spacer, the small subunit rRNA, ITS1, 5.8S RNA, ITS2, and 3 kb of the LSU rRNA.

ment of coding sequences and a very significant increase in the proportion of complete and intact coding regions *(7)*. Fragment representation is a function of their genomic copy number. In contrast to a cDNA library, the probability of obtaining a given sequence does not depend on the level of expression of the gene. Neither does representation depend upon the time of expression or tissue type. Mung bean-based genomic libraries have been made with numerous other protozoans, including trypanosomes *(8)*, *Giardia (9)*, *Toxoplasma (10)*, *Leishmenia (11)*, and *Babesia (12)*.

The process of preparing a mung bean nuclease digest library can be conveniently divided into the sections presented under **Subheading 2.**

2. Materials
2.1. Purification of Parasite DNA

1. TSE buffer: 20 mM Tris-HCl, pH 8.0, 100 mM NaCl, 50 mM EDTA.
2. Extraction buffer: 10 mM Tris-HCl, pH 8.0, 100 mM EDTA.
3. Tris-EDTA (TE) buffer: 10 mM Tris-HCl, pH 8.0, 1 mM EDTA.
4. 1X Tris-borate-EDTA (TBE) buffer: 90 mM Tris-borate, 2 mM EDTA.
5. 15% saponin solution (100X stock).
6. 10% sodium dodecyl sulfate (SDS) (20X stock).
7. 500 µg/mL DNase-free RNase A (25X stock).
8. 10 mg/mL proteinase K solution (100X stock) (Life Technologies, Rockville, MD).
9. Tris-HCl-buffered phenol, pH 8.0 (Life Technologies).
10. 10 M ammonium acetate solution.
11. 100% ethanol.
12. Saturated solution of cesium chloride.
13. Hoechst dye 33258 (Sigma, St. Louis, MO).
14. Isopropanol.

2.2. Mung Bean Digestion of DNA

1. 10X Mung bean nuclease buffer: 2 M NaCl, 300 mM sodium acetate, pH 4.6, 10 mM ZnSO$_4$.
2. Mung bean nuclease (40 U/µL) (Life Technologies).
3. Ultrapure formamide (*see* **Note 1**) (Life Technologies).
4. Tris-HCl-buffered phenol, pH 8.0 (Life Technologies).
5. Phenol:chloroform:isoamyl alcohol (25:24:1).
6. Chloroform.
7. 3 M sodium acetate.

2.3. Purification of DNA Fragments

1. Agarose.
2. QIAEXII DNA purification kit (Qiagen, CA).
3. Tissue culture medium (TCM) solution: 50 mM Tris-HCl, pH 7.4, 50 mM CaCl$_2$, 50 mM MgCl$_2$.

2.4. Preparing Blunt-Ended Fragments

1. 5X T4 DNA polymerase blunt-ending buffer: 165 mM Tris-acetate, pH 7.9, 50 mM magnesium acetate, 330 mM sodium acetate, 500 µg/mL bovine serum albumin (BSA) (Life Technologies).
2. T4 DNA polymerase (2 U/µL) (Life Technologies).

2.5. Attachment of Adapters and Phosphorylation of Ends

1. *Eco*RI adapters (1 mg/mL) (Life Technologies).
2. 5X Adapter buffer: 330 mM Tris-HCl, pH 7.6, 50 mM MgCl$_2$, 5 mM ATP (Life Technologies).
3. T4 DNA ligase (1 U/μL) (Life Technologies).
4. T4 polynucleotide kinase (10 U/μL) (Life Technologies).

2.6. Ligation of DNA Fragments

1. Plasmid vector pBluescript II SK(+) (Stratagene Cloning Systems, CA).
2. *Eco*RI (Life Technologies).
3. Calf intestinal phosphatase (Stratagene Cloning Systems).
4. 5X DNA ligase buffer: 250 mM Tris-HCl, pH 7.6, 50 mM MgCl$_2$, 5 mM ATP, 5 mM DTT, 25% polyethylene glycol 8000 (Life Technologies).
5. T4 DNA ligase (1 U/μL) (Life Technologies).

2.7. Transformation of Cloned Fragments

1. LB medium, pH 7.0 (per liter): 10 g of Bacto-tryptone, 5 g of Bacto-yeast extract, 10 g of NaCl.
2. LB plates containing 100 μg/mL of ampicillin.

2.8. Screening of Mung Bean Library

1. Screening of a library depends upon the approach. Essentially any procedure that is used with cDNA can be used with the mung bean nuclease library: complementation *(6)*, antibody *(5)*, hybridization *(4)*.

2.9. Amplification and Storage of Library

1. Ampicillin.
2. Glycerol.
3. LB medium and plates.

3. Methods

3.1. Separation of Parasite DNA Free From any Contaminating Host DNA

3.1.1. Isolation of Total DNA

1. Collect parasitized red cells by spinning for 10 min at 1000g at room temperature.
2. Resuspend packed infected red blood cells (RBC) with 5–10 vol of cold TSE buffer.
3. Spin resuspended cells at 1000g for 10 min, discard supernatant, and wash cell pellet again.
4. Rescue parasites from the red blood cells by lysing red blood cells with detergent, as described by Trager *(13)*. This is an optional step but often results in purer parasite DNA.
5. Resuspend cells in equal volume of TSE and lyse cells by adding 0.01 vol of 15% saponin solution to achieve a final concentration of 0.15%. Incubate at room temperature for 2 min.
6. To stop the reaction, add 10 vol of cold TSE and spin cells at 15,000g for 10 min in a refrigerated centrifuge.
7. Discard supernatant and wash pellet with 25 mL of TSE (*see* **Note 2**).
8. Resuspend pellet in 15 mL of extraction buffer and add 750 μL of 10% SDS to get a final concentration of 0.5%.
9. Add 160 μL of proteinase K to obtain a final concentration of 100 μg/mL.
10. Add 640 μL of DNase-free RNaseA to get a final concentration of 20 μg/mL.
11. Incubate the suspension at 50°C for 2 h with periodic gentle swirling.
12. Cool the suspension to room temperature and add equal volume of Tris-HCl-saturated phenol, pH 8.0.

P. falciparum Mung Bean Libraries

13. Gently mix the two phases with an end-to-end shaker (*see* **Note 3**).
14. Separate the two phases by centrifugation at 5000*g* for 15 min at room temperature.
15. With a wide-bore pipet, transfer the viscous aqueous phase to a fresh tube and twice repeat the extraction with phenol.
16. After the third extraction, precipitate the DNA with 0.2 vol of 10 *M* ammonium acetate and 2 vol of room temperature ethanol.

3.1.2. Purification of Parasite DNA by Equilibrium Centrifugation in CsCl–Hoechst Gradient

Unlike *P. falciparum* that is being cultured in vitro in nonnucleated RBCs, isolation of parasite DNA from the RBC of an infected animal presents the unique problem of host DNA contamination. Even a 1% contamination of nucleated host cells leads to a 50% contamination of the DNA preparation, due to the difference in the genome size of *Plasmodium* and higher eukaryotes *(14)*. Parasite DNA from nucleated avian and reptilian RBC present an even greater challenge.

Hoechst dye–CsCl centrifugation is a very efficient procedure for separating contaminating host DNA from parasite DNA for a large number of *Plasmodium* species *(14)*. Many of the *Plasmodium* species have a strong bias toward dA.dT in their genome with *P. berghei*, *P. falciparum*, *P. lophurae*, and *P. gallinaceum* having a dA.dT content of more than 80% *(15)*. The dA.dT content of their hosts are much lower. The method is based on the preferential binding of the bis-benzimide dye Hoechst 33258 to dA.dT-rich DNA sequences. Binding of the dye reduces the buoyant density of dA.dT-rich DNA to the extent that separation of mixtures can be achieved by isopycnic ultracentrifugation in cesium chloride gradients (**Fig. 3**).

1. Dilute parasite DNA with TE and add to a saturated cesium chloride solution. The considerations relating to the extent of dilution involve a balance between minimizing the total volume and the need to obtain a final refractive index of 1.3950. Dissolve Hoechst dye in water to a concentration of approx 1 µg/10 µL of water. Add a weight of Hoechst dye 33258 equal to the estimated amount of parasite DNA to the cesium chloride mixture. We have shown that separation occurs when the weight of the dye is within a tenfold range of the parasite DNA. A yellow precipitate often forms initially. One should try to get the precipitate into solution, but if the DNA is protein-free, it will not disturb the gradient formation.
2. Centrifuge the solution in a vertical rotor (VTi50, Beckman) in an ultracentrifuge at 196,409*g* for 18 h at 25°C *(14,15)*. Parasite DNA separates into a tight band on the top on account of its lower density in comparison to the high-density host DNA *(14,15)* (**Fig. 3**). Sometimes the 35-kb plastid of all *Plasmodium* species will make a third band but, with most parasite DNAs, it remains with the genomic DNA.
3. Remove the parasite DNA is by first puncturing the top of the tube a needle to prevent vacuum formation when the sample is withdrawn. Then illuminate the tube in the dark using a hand held long wave UV light; protective eye gear must be used during this process. Remove the parasite DNA from the side of the tube using a syringe fitted with a 16-gage needle.
4. Extract the DNA four times with equal volumes of isopropanol (that has been equilibrated with a cesium chloride solution) to remove the dye.
5. Remove the cesium chloride by dialyzing against TE. It is also possible to dilute the cesium chloride threefold and then precipitate the DNA with two volumes of ethanol. In this case the pellet is washed twice with 70% ethanol to remove remaining cesium chloride. We prefer the dialysis approach.

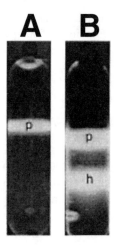

Fig. 3. Density bands formed by DNA after Hoechst dye CsCl centrifugation. (**A**) *Plasmodium falciparum* DNA isolated from culture. (**B**) *Plasmodium berghei* DNA isolated from infected erythrocytes of mice. p, parasite; h, host.

6. Extract the DNA with an equal volume of phenol followed by phenol:chloroform:isoamyl alcohol (25:24:1) and chloroform.
7. Precipitate the DNA with 0.1 vol of 3 M sodium acetate pH 5.2 and 2.0 vol of chilled ethanol.

3.2. Digestion of Genomic DNA with Mung Bean Nuclease

Genomic DNA is digested in the presence of formamide. Conditions for the digestion of genomic DNA with mung bean nuclease vary from one species to another (*2*) (**Table 1**). It is best to do several digestions as described below and test their quality by Southern blot analysis. The probes for testing the quality of the digestion products can be for any known gene in that the genes of a single genome all appear to be cleaved under similar conditions.

1. Ideally, an analytical digestion with 2 µg of genomic DNA is performed in different concentrations of formamide (*see* **Note 4**).
2. A typical set of reaction is performed as shown in **Table 2**.
3. Preincubate the reaction at 50°C for 5 min before adding the enzyme.
4. Mix the enzyme by gentle pipetting, and then incubate at 50°C for 20 min.
5. Stop the reaction by adding 200 µL of cold TE, pH 8.0.
6. Extract the digested DNA with an equal volume of phenol, followed by phenol-chloroform, and chloroform only extraction.
7. Precipitate the DNA with 0.1 vol of 3 M sodium acetate (pH 5.2) and 2.5 vol of chilled 100% ethanol.
8. Test an aliquot of each reaction by Southern blot analysis, and select the optimal digestion for preparation of the library.

3.3. Isolation of DNA Fragments from the Gel

Often the entire digest can be put into a vector. If size selection is desired, the DNA is run on an agarose gel to separate the digested products. Alternatively DNA size ranges can be excised from the gel. Ideally fragments can be divided into three catego-

Table 1
Optimal Concentration of Formamide for Different Species of *Plasmodium*

Parasite species	Host	% Formamide
P. falciparum	Primate	35
P. berghei	Rodent	45
P. yoelii	Rodent	45
P. lophurae	Avian	35
P. knowlesi	Primate	45

Table 2
Typical Reactions in Digestion of Genomic DNA

Contents	A (µL)	B (µL)	C (µL)
10X Mung bean nuclease buffer	10	10	10
Genomic DNA (100 µg/mL)	20	20	20
Formamide	35	40	45
BSA (1 mg/mL)	1	1	1
Water	33	28	23
Enzyme (4 U/µL)	1	1	1
Total	100	100	100

ries 100–500 bp, 0.5–2 kb, 2–8 kb for extraction and subsequent cloning. DNA can be recovered from the gel using any of the standard procedures.

3.3.1. Isolation with QIAEX II Kit

Most of the commercially available kits require different amounts of reagents depending upon the size of DNA fragments. We have used QIAEX II kit from Qiagen for isolation of DNA fragments from agarose gels. Depending upon the size of the fragments different amounts of QX1 (solubilization buffer) is required. *See* **Table 3**.

1. Add 30 µL of QIAEX II and incubate at 50°C for 10 min to solubilize the agarose and bind the DNA. Vortex vigorously every 2 min.
2. Centrifuge at 10,000*g* for 30 s and carefully remove the supernatant.
3. Wash pellet once with 500 µL of buffer QX1 followed by two washings in PE buffer.
4. Air dry the pellet for 15 min or until it turns white.
5. Resuspend pellet in 20 µL of water and incubate at 50°C for 5 min to elute the DNA.
6. Spin for 30 s and collect the supernatant containing DNA.

3.3.2. Freeze Thaw Technique

1. This method requires that the gel is made up of low melting point agarose *(16)*.
2. Melt the gel slice at 65°C in a heat block and mix with an equal volume of complete T-cell medium (TCM) solution, with vigorous vortexing.
3. Freeze the mixture at –70°C for 60 min, and then thaw at 65°C and vortex.
4. After three freeze–thaw cycles, extract DNA sequentially with equal volume of phenol, phenol-chloroform, and chloroform.
5. Precipitate DNA with 0.1 vol of sodium acetate and 2 vol of chilled ethanol at –70°C for 2 h.
6. Wash DNA pellet with 70% ethanol and resuspend in TE.

Table 3
Amount of Buffer Required for Various Sizes of Fragments

Fragment size	Volume of buffer QX1	Volume of buffer QX1 + 2 vol of H$_2$O
<100 bp	6	
100 bp to 4 kb	3	
> 4 kb		3

Table 4
Reaction of Blunt-Ending

Reagent	Volume (µL)
5X T4 DNA polymerase blunt-ending buffer	20
0.1 M Dithiothreitol	1
0.5 mM dNTP mix	20
DNA (2 µg)	x

3.4. End Filling

Mung bean nuclease fragments do not have blunt ends. Repairing the ends so that they can be ligated into a vector is the most critical step in preparing mung bean libraries and is almost always the problem when subsequent failure in transformation efficiency occurs. This is curious because mung bean nuclease is often used to produce blunt-ended fragments for cloning when the standard conditions are used (*1*). Several approaches have worked to produce highly representative libraries (*2,5,6*). We generate blunt ends on the digested DNA using T4 DNA polymerase.

T4 DNA polymerase can be used for generating blunt ends from 3' recessed ends, from 3' protruding ends, or from a population containing both (*16*). Both activities are probably essential for treatment of mung bean fragments. In the presence of all four deoxyribonucleoside triphosphates (dNTPs), the polymerase reaction proceeds much more rapidly than the exonuclease reaction (*see* **Note 5**). Thus a molecule with a 3' recessed end will be rendered blunt-ended when the polymerase activity of the enzyme extends the recessed strand in the 3' direction using the 5' overhang of the other strand as template. A molecule with a 3' protruding end will be rendered blunt-ended when the exonuclease activity of the enzyme digests the overhang from the 3' end until it reaches a double-stranded portion of the DNA. Once a blunt end is created it is maintained as an equilibrium state because as the exonuclease activity removes nucleotides from the 3' ends they are immediately replaced by the much more rapid polymerase activity. Creation of blunt end is necessary prior to adding adapters.

1. The reaction is mixed on ice as shown in **Table 4**.
2. With autoclaved distilled H$_2$O bring volume to 95 µL and add 5 µL of T4 DNA polymerase (2 U/µL).
3. Incubate the reaction at 11°C for 15 min followed by heat inactivation at 65°C for 15 min.
4. Subject the reaction to standard phenol, phenol-chloroform, and chloroform extractions and precipitate the DNA.
5. Wash DNA pellet with 70% ethanol, dry, and resuspend blunt-ended DNA in 10 mM Tris-HCl, pH 7.4.

6. Though DNA at this stage can be ligated directly into a blunt-ended vector, blunt-end ligations are not very efficient.

3.5. Attachment of Adapters and Phosphorylation

Adapters are short pieces of double-stranded DNA with one blunt end and one cohesive end, which encodes for a restriction site. Adapters are formed by annealing two oligonucleotides together, forming a phosphorylated blunt end and a nonphosphorylated cohesive end. The addition of an adapter to blunt ended fragments results in the creation of cohesive ends. A 20:1 molar ratio (adapter: insert) is required to obtain the maximal efficiency of the reaction.

1. The reaction is performed as shown in **Table 5**.
2. With autoclaved distilled H_2O bring volume to 45 µL and add 5 µL of T4 DNA ligase (1 U/µL).
3. Incubate the reaction at 16°C for a minimum of 20 h.
4. Heat-inactivate the enzyme and place the reaction on ice.
5. Add 3 µL of T4 polynucleotide kinase (10 U/µL) and incubate the reaction at 37°C for 30 min to phosphorylate the sticky ends.
6. Heat the reaction at 70°C for 10 min and place the reaction on ice.
7. Removed the unattached adapters from the genomic DNA fragments on an agarose gel and purify as described above. Spin columns can also be used to separate adapters from genomic fragments, but confirmation of the removal of excess adapters by gel electrophoresis is required. Their molar excess in the DNA would otherwise create a problem for subsequent ligation reactions.
8. Resuspend the DNA in 20 µL of water

3.6. Ligation of DNA Fragments

T4 DNA ligase catalyzes the ATP-dependent formation of a phosphodiester bond between the 3' hydroxyl end of a double-stranded DNA fragment and the 5' phosphate end of the same or another DNA fragment. A number of ligation reaction parameters are important, including DNA concentration, the molar ratio of insert to vector, temperature, buffer composition, and enzyme composition.

Purified DNA fragments can now be put into almost any plasmid or viral vector. We prefer to put mung bean fragments into phage and use up to 15% polyethelene glycol in the ligation mix. Often, however, it is useful to put fragments directly into plasmids. We use plasmid pBluescript II SK (+) vector as an example. It's a 2.96 kb high-copy number ColE1-based plasmid with ampicillin resistance.

When ligation reaction products are used to transform bacteria, a significant background of colonies containing no *Plasmodium* inserts may result from the religation of the ends of the vector molecules. Treating the linear vector DNA with alkaline phosphatase prior to the ligation can reduce this background. The phosphatase removes the 5' phosphate groups from each strand of the vector molecule, preventing T4 DNA ligase from forming phosphodiester bonds between the two ends of the vector.

1. Digest 2 µg of vector DNA with 10 U of *Eco*RI for 3 h at 37°C. The vector possesses a single *Eco*RI site in the multiple cloning region.
2. Add 10 U of calf intestinal alkaline phosphatase to the reaction and incubate for 1 h to dephosphorylate vector DNA.
3. Heat inactivate the enzymes at 65°C for 10 min.
4. Extract the DNA with phenol, phenol-chloroform, and chloroform, and precipitate as described previously.

**Table 5
Addition of Adapters to Blunt-Ended Fragments**

Adapter	Volume (μL)
2 μg blunt-ended DNA	x
5X Adapter buffer	10
*Eco*RI adapters (1 mg/mL)	10
0.1 *M* Dithiothreitol	7

**Table 6
Ligation of Fragments**

Fragment	Volume (μL)
Purified mung bean fragments	3
*Eco*RI-digested pBluescript DNA	1
5X Ligase buffer	2

5. Resuspend the plasmid DNA in 20 μL of water, and use for ligation (*see* **Note 6** and **Table 6**).
6. Adjust volume to 9 μL with autoclaved distilled H$_2$O and add 1 μL of T4 DNA ligase (1 U/μL) (*see* **Note 7**).
7. Incubate the reaction at 14°C for 16 h, and then transform.

3.7. Transformation

1. Thaw one vial of *Epicurian coli* XL 10-Gold ultracompetent cells on ice.
2. Aliquot 100 μL of the cells into a prechilled 15-mL polypropylene tube and add 4 μL of XL 10-Gold β-mercaptoethanol mix to the cells.
3. Incubate on ice for 10 min, swirling every 2 min
4. Add 5 μL of ligation reaction and incubate the tube on ice for 30 min.
5. Heat shock the tube in a water bath set at 42°C for 30 s followed by incubation on ice for 2 min.
6. Add 900 μL of preheated (42°C) NYZ$^+$ broth and incubate at 37°C for 1 h with shaking at 250 rpm.
7. Plate the bacterial suspension directly on the surface of 20 LB plates containing 100 μg/mL of ampicillin and incubate at 37°C for 16 h.

3.8. Screening of Mung Bean Nuclease-Digested DNA Library

There is nothing unique about screening a mung bean library. The library can be screened in any of the same ways that cDNA libraries are screened. Complementation of auxotrophic mutants with *Plasmodium* genes has been used successfully *(6)*. Screening using DNA probes *(4)* or antibody binding has also been useful *(5)*.

3.9. Amplification and Storage of Mung Bean Library

After preparation of fragments with termini that can be inserted into the vector of choice, the mung bean library is like any other *Plasmodium* DNA recombinant library. Amplification procedures depend upon vector and host. Influences depending upon the preparation of the inserted DNA probably make little difference. We do believe, on the basis of purely anecdotal information, that the mung bean fragments are less likely to rearrange in microbial hosts because the fragments have a more normal distribution of nucleotides and most A-T-rich regions have been eliminated.

4. Notes

1. Obtain deionized formamide and store in small aliquots (0.5 mL). Discard the solution if it does not freeze at −20°C. Thaw an aliquot for use and discard the unused portion.
2. Pellet at this stage can be stored at −70°C.
3. Complete mixing takes 30 min to 1 h.
4. Mung bean nuclease is used at a concentration of 2 U/μg of DNA in the reaction. Depending upon the source, enzyme is generally available at a concentration of 20–40 U/μL. To achieve the required concentration, enzyme is diluted in cold 1X enzyme reaction buffer.
5. The K_m of T4 DNA polymerase for dNTPs is approx 20 μM. Therefore the reaction mixture should contain at least 20 μM of each dNTPs.
6. DNA is not suspended in TE as EDTA inhibits the activity of T4 DNA ligase.
7. As the insert size is very heterogeneous, different concentrations of inserts should be tried for ligation, keeping the concentration of vector constant.

References

1. Johnson, P. H. and Laskowski, M. Sr. (1968) Sugar-unspecific mung bean nuclease I. *J. Biol. Chem.* **243,** 3421–3424.
2. McCutchan, T. F., Hansen, J. L., Dame, J. B., and Mullins, J. A. (1984) Mung bean nuclease cleaves Plasmodium genomic DNA at sites before and after genes. *Science* **225,** 625–628.
3. Vernick, K. D. and McCutchan, T. F. (1998) A novel class of supercoil-independent nuclease hypersensitive site is comprised of alternative DNA structures that flank eukaryotic genes. *J. Mol. Biol.* **279,** 737–751.
4. Vernick, K. D., Imberski, R. B., and McCutchan, T. F. (1988) Mung bean nuclease exhibits a generalized gene-excision activity upon purified *Plasmodium falciparum* genomic DNA. *Nucleic Acids Res.* **16,** 6883–6896.
5. Dame, J. B., Williams, J. L., McCutchan, T. F., Weber, J. L., Wirtz, R. A., Hockmeyer, W. T., et al. (1984) Structure of the gene encoding the immunodominant surface antigen on the sporozoite of the human malaria parasite *Plasmodium falciparum*. *Science* **225,** 593–599.
6. Kaslow, D. C. and Hill, S. (1990) Cloning metabolic pathway genes by complementation in *Escherichia coli*. Isolation and expression of *Plasmodium falciparum* glucose phosphate isomerase. *J. Biol. Chem.* **265,** 12,337–12,341.
7. Reddy, G. R., Chakrabarti, D., Schuster, S. M., Ferl, R. J, Almira, E. C., and Dame, J. B. (1993) Gene sequence tags from *Plasmodium falciparum* genomic DNA fragments prepared by the "genease" activity of mung bean nuclease. *Proc. Natl. Acad. Sci. USA* **90,** 9867–9871.
8. Brown, K. H., Brentano, S. T., and Donelson, J. E. (1986) Mung bean nuclease cleaves preferentially at the boundaries of variant surface glycoprotein gene transpositions in trypanosome DNA. *J. Biol. Chem.* **261,** 10,352–10,358.
9. Adam, R. D., Aggarwal, A., Lal, A. A, de La Cruz, V. F., McCutchan, T., and Nash, T. E. (1988) Antigenic variation of a cysteine-rich protein in *Giardia lamblia*. *J. Exp. Med.* **167,** 109–118.
10. Johnson, A. M., Illana, S., Dubey, J. P., and Dame, J. B. (1987) *Toxoplasma gondii* and *Hammondia hammondi*: DNA comparison using cloned rRNA gene probes. *Exp. Parasitol.* **63,** 272–278.
11. Muhich, M. L. and Simpson, L. (1986) Specific cleavage of kinetoplast minicircle DNA from *Leishmania tarentolae* by mung bean nuclease and identification of several additional minicircle sequence classes. *Nucleic Acids Res.* **14,** 5531–5556.
12. Tripp, C. A., Wagner, G. G., and Rice-Ficht, A. C. (1989) *Babesia bovis*: gene isolation and characterization using a mung bean nuclease-derived expression library. *Exp. Parasitol.* **69,** 211–225.
13. Trager, W. (1971) Malaria parasites (*Plasmodium lophurae*) developing extracellularly in vitro: incorporation of labeled precursors. *J. Protozool.* **18,** 392–399.
14. Dame, J. B. and McCutchan, T. F. (1987) *Plasmodium falciparum*: Hoechst dye 33258-CsCl ultracentrifugation for separating parasite and host DNAs. *Exp. Parasitol.* **64,** 264–266.
15. McCutchan, T. F., Dame, J. B., Miller, L. H., and Barnwell, J. (1984) Evolutionary relatedness of *Plasmodium* species as determined by the structure of DNA. *Science* **225,** 808–811.
16. Sambrook, J., Fritsch, E. F., and Maniatis, T. (1989) *Molecular Cloning: A Laboratory Manual.* Cold Spring Harbor Laboratory, Cold Spring Harbor, N.Y.

25

Construction of *Plasmodium falciparum* λ cDNA Libraries

David A. Fidock, Dharmendar Rathore, and Thomas F. McCutchan

1. Introduction

cDNA libraries represent the genetic information encoded in the messenger RNA (mRNA) of a particular tissue or organism. Construction of a cDNA library begins with the isolation of purified and full-length RNA. This isolation is made difficult by the fact that RNA molecules are exceptionally liable to ribonucleases that are very stable and active and require no cofactors to function (*see* **Note 1**). The first step in all RNA isolation protocols, therefore, involves lysing the cells in a chemical environment that results in the denaturation of ribonucleases.

The information encoded by the RNA is converted into a stable DNA duplex, which is then inserted into a self-replicating plasmid or phage vector. Relative abundance of the clone of interest can vary widely. Highly abundant messages can represent 10% or more of total mRNA, whereas rare messages can be present as single copies. The size of the library necessary to include the clone of interest is a direct reflection of the relative abundance of the mRNA of interest. In general, this abundance is not precisely known. If the mRNA of interest is relatively abundant, then the efficiency of generating clones is not so important and the choice of cloning strategy and vector should be based on the desired use for the clone. On the other hand, if the mRNA of interest is rare, then high cloning efficiency is of central importance.

The number of clones required to achieve a given probability that a low abundance mRNA will be present in the cDNA library is

$$N = \frac{\ln(1-P)}{\ln(1-1/n)}$$

where N = number of clones required, P = the probability desired (usually 0.99), and $1/n$ = the fractional proportion of total mRNA represented by a single type of rare mRNA.

Ideally, a representative cDNA library should contain at least one version of each full-length sequence of the total mRNA population. The construction of cDNA libraries is fundamental for discovering new genes and assigning gene function.

2. Materials

2.1. Isolation of Total RNA

1. Trizol reagent (Life Technologies, Gaithersburg, MD).
2. Chloroform.
3. Isopropanol.
4. Ethanol.
5. Diethyl pyrocarbonate (DEPC)-treated water (*see* **Note 2**).
6. Formaldehyde.
7. 10X MOPS buffer: 200 mM 3-[N-morpholino]propane-sulfonic acid (MOPS), pH 7.0, 50 mM sodium acetate, 10 mM ethylenediaminetetraacetic acid (EDTA).
8. RNA sample buffer: 50% formamide, 1X MOPS buffer, 7% formaldehyde, 0.04% bromophenol blue.
9. TE: 10 mM Tris-HCl, 1 mM EDTA, pH 7.0.

2.2. Separation of mRNA

1. Oligotex (Qiagen, Santa Clara, CA): 10% (w/v) suspension of resin in 10 mM Tris-HCl, pH 7.5, 500 mM NaCl, 1 mM EDTA, 0.1% SDS, 0.1% sodium azide.
2. 2X Binding buffer: 20 mM Tris-HCl, pH 7.5, 1 M NaCl, 2 mM EDTA, 0.2% sodium dodecyl sulfate (SDS).
3. Wash buffer OW2: 10 mM Tris-HCl, pH 7.5, 150 mM NaCl, 1 mM EDTA.
4. Elution buffer: 5 mM Tris-HCl, pH 7.5.

2.3. cDNA Synthesis

1. Moloney murine leukemia virus-reverse transcriptase (MMLV-RT).
2. RNase block ribonuclease inhibitor (40 U/µL).
3. First-strand methyl nucleotide mixture: 10 mM dATP, dGTP, and dTTP, plus 5 mM 5-methyl dCTP.
4. 10X First-strand buffer (Stratagene, La Jolla, CA).
5. 10X Second-strand buffer (Stratagene).
6. Second-strand dNTP mixture: 10 mM dATP, dGTP, and dTTP, plus 26 mM dCTP.
7. *Escherichia coli* RNase H (1.5 U/µL).
8. *E. coli* DNA polymerase I (9 U/µL).
9. 3 M sodium acetate.

2.4. Blunting of cDNA Termini

1. Blunting dNTP mixture: 2.5 mM dATP, dGTP, dTTP, and dCTP.
2. Pfu (proofreading) DNA polymerase (2.5 U/µL) (Stratagene).

2.5. Attachment of EcoRI Adapters

1. *Eco*RI adapters (0.4 µg/µL).
2. 10X ligase buffer: 500 mM Tris-HCl, pH 7.5, 70 mM MgCl$_2$, 10 mM dithiothreitol.
3. 10 mM rATP.
4. T4 DNA ligase (4 U/µL).

2.6. Phosphorylation of EcoRI Ends

1. T4 polynucleotide kinase (10 U/mL).

2.7. Digestion with XhoI

1. *Xho*I (40 U/µL).

2. *Xho*I buffer supplement (Stratagene).
3. 10X STE buffer: 1 M NaCl, 200 mM Tris-HCl, pH 7.5, 100 mM EDTA.

2.8. Size Fractionation

1. Chroma Spin 400 column (Clonetech, Palo Alto, CA).

2.9. Ligation of cDNA

1. Uni-ZAP XR vector (Stratagene).
2. 10X ligase buffer: 500 mM Tris-HCl, pH 7.5, 70 mM MgCl$_2$, 10 mM dithiothreitol.
3. 10 mM rATP.
4. T4 DNA ligase (4 U/µL).
5. Gigapack III gold packaging extract (Stratagene).
6. XL1-blue MRF'-competent cells.
7. SM buffer (per liter): 5.8 g of NaCl, 2 g of MgSO$_4$·7H$_2$O, 50 mL of 1 M Tris-HCl, pH 7.5, 5 mL of 2% (w/v) gelatin.
8. NZY broth (per liter): 5 g of NaCl, 2 g of MgSO$_4$·7H$_2$O, 5 g of yeast extract, 10 g of NZ amine, pH 7.5.
9. NZY agar (per liter): 1 L of NZY broth, 15 g of agar.
10. NZY top agar (per liter): 1 L of NZY broth, 0.7% (w/v) agarose.

3. Methods

Harvest 50–100 mL of *Plasmodium falciparum* infected erythrocyte culture at 4% hematocrit and 5% to 10% parasitemia. This will yield approx 50–200 µg of total RNA. Isolate parasites by gentle lysis of red blood cells with saponin without disrupting the parasites. The protocol for isolation of parasites by saponin treatment are described in Chapters 23 and 24 on preparation of genomic libraries.

3.1. Isolation of Total RNA

Total RNA is isolated by a single-step RNA isolation method using Trizol reagent (Life Technologies), a premixed monophasic solution of phenol and guanidine isothiocyanate used in an improved single-step RNA isolation method developed by Chomczynski and Sacchi *(1)*.

1. Resuspend parasite pellet in 5 mL of Trizol reagent and incubate at room temperature for at least 5 min to ensure the complete dissociation of nucleoprotein complexes. Trizol disrupts cells and dissolves cell components without damaging the RNA.
2. Add 1 mL of chloroform, cover, and shake vigorously for 30 s. Centrifuge at 10,000g for 15 min at 4°C. Addition of chloroform followed by centrifugation separates the solution into a colorless upper aqueous phase and a lower red, phenol:chloroform organic phase. RNA remains exclusively in the upper aqueous phase, whereas DNA and proteins are in the interphase and organic phase.
3. Avoiding the interphase, transfer the upper aqueous phase to a sterile centrifuge tube.
4. Add an equal volume of ice-cold isopropanol, mix well, and store at room temperature for 10 min.
5. Centrifuge at 10,000g for 15 min at 4°C to pellet the RNA.
6. Decant the supernatant and drain the sample on several layers of sterile Kimwipes for approx 5 min.
7. To remove isopropanol and salts, add 5 mL of ice-cold 75% ethanol to the RNA pellet, and mix the sample by vortexing.
8. Centrifuge at 10,000g for 10 min at 4°C to pellet the RNA.

9. Decant the supernatant; drain the tube by inversion on several layers of sterile Kimwipes for 5 min.
10. Resuspend the RNA pellet in 250 µL of DEPC-treated water and proceed on to quantitation of total RNA.

3.2. Quantitation of Total RNA

1. Remove a 5-µL aliquot of total RNA and dilute with 95 µL of TE buffer.
2. Read the A260 blanked against TE buffer, and calculate the amount of RNA recovered.
3. The RNA recovered may be determined by the formula:

$$\text{Total RNA} = A260 \times 40 \text{ µg/mL} \times 20 \times 0.25 \text{ mL}$$

where A260 is the absorbance of the solution at 260 nm, 40 µg/mL is a fixed conversion factor relating absorbance to concentration for RNA, 20 is the dilution factor, and 0.25 mL is the total volume.

3.3. Evaluation of RNA Preparation

1. Prepare agarose solution by melting 1.5 g of agarose in 85 mL of DEPC-treated water. Allow to cool to 50–60°C.
2. Add 10 mL of 10X MOPS buffer and 5.4 mL of 37% formaldehyde (12.3 M) solution and mix well by swirling. The final concentration of agarose is approx 1.5% and formaldehyde is 0.66 M.
3. Pour into RNAse-free gel cast.
4. Add 10 µL of RNA sample buffer to each sample.
5. Pipet repeatedly to resuspend. Centrifuge briefly and place on ice.
6. Add 1 µL of 1 µg/mL ethidium bromide solution to each sample. Heat all of the RNA samples and the molecular weight marker at 65–70°C for 5 min.
7. Chill on ice. Spin (5 s) to clear the microfuge tube walls of droplets.
8. Electrophorese at 75 V until bromophenol dye front is two-thirds of the distance to the end of the gel.

3.4. Separation of mRNA

The vast majority of cellular RNA molecules are tRNAs and rRNAs. Only 1% to 5% of total cellular RNA is mRNA. The actual amount of mRNA depends on the cell type and physiological state. Separation of mRNA from the total RNA pool is essential for the construction of cDNA libraries. Most mRNAs contain a poly(A) tail, while the structural RNAs do not. Poly(A) selection therefore enriches the mRNA. Aviv and Leder *(2)* first used oligo(dT) cellulose to bind poly(A)$^+$ message and thus achieved fractionation of mRNA. Enrichment for poly A$^+$ containing RNAs enhances the efficiency of the reverse transcriptase reaction to produce maximal amounts of cDNA. At least 100 µg of total RNA is required to obtain the 1 µg of (poly)A$^+$ RNA necessary for library construction.

We have routinely used the Oligotex mRNA purification kit (Qiagen) for purifying mRNA from the total RNA pool. The kit utilizes an affinity resin to isolate (poly)A$^+$ mRNA from the total mRNA preparation in less than 30 min.

1. Take 240 µL of total RNA at ≥ 0.4 mg/mL concentration, and add an equal volume of preheated 2X binding buffer.
2. Add 30 µL of Oligotex suspension (*see* **Note 3**) and mix the contents by gentle flicking.
3. Incubate the suspension at 65°C for 3 min to disrupt the RNA secondary structure.

P. falciparum cDNA Libraries

4. Incubate the suspension for 10 min at room temperature to allow hybridization between the (poly)A tails of mRNA and the dT_{30} oligos attached to the latex beads.
5. Spin at 15,000g for 2 min to pellet the Oligotex resin containing the mRNA.
6. Carefully remove the supernatant without disturbing the pellet (*see* **Note 4**).
7. Wash resin with 400 µL of OW2 wash buffer and transfer onto a spin column.
8. Centrifuge the spin column at 15,000g for 30 s in a microfuge and discard the flow through.
9. Repeat the washing step once again and transfer the spin column to a fresh 1.5-mL microcentrifuge tube.
10. Add 25 µL of elution buffer to the column and mix by gentle pipetting (*see* **Note 5**).
11. Elute the mRNA by centrifugation at 15,000g for 30 s. The mRNA is now ready for cDNA synthesis.

3.5. cDNA Synthesis

Conversion of mRNA into a double-stranded insert DNA is a two-step process. In the first step, Moloney murine leukemia virus-reverse transcriptase enzyme (MMLV-RT) synthesizes the first strand using mRNA as template and a poly dT-based oligo as primer. For synthesizing the first strand, a 50-nucleotide-long oligo dT-based oligo with a linker containing the *Xho*I restriction site is used as primer:

5' GAGAGAGAGAGAGAGAGAGAGAACTAGT**CTCGAG**TTTTTTTTTTTTTTTTTT 3'

The oligo introduces a *Xho*I enzyme site into the cDNA that is later used for cloning inserts in a sense orientation into the vector. To prevent the recognition of other *Xho*I sites in the cDNA, all cytosines introduced during first-strand synthesis are methylated by using 5-methyl dCTP instead of normal dCTP. in the second step, RNase H nicks the original mRNA bound to the first-strand cDNA. Nicked mRNAs serve as primers for DNA polymerase I to synthesize the second-strand cDNA.

After ligation of *Eco*RI adapters to the 5' end and digestion by *Xho*I to release the 3' *Xho*I sites that were incorporated into the cDNA by the primer linker, the resulting cDNAs have unique ends that can be directionally cloned into *Eco*RI/*Xho*I-digested vectors.

A large number of kits are commercially available for cDNA synthesis. We have regularly used the cDNA synthesis kit from Stratagene. One interesting alternative system that can preferentially amplify full-length cDNAs, the CapFinder System, is briefly discussed in **Note 6**.

3.6. Synthesis of First Strand

1. Add 5 µL of 10X first-strand buffer, 3 µL of first-strand methyl nucleotide mixture, 2 µL of linker primer, 12.5 µL of DEPC-treated water, and 1 µL of ribonuclease inhibitor. Mix gently.
2. Add 25 µL of purified mRNA obtained from the previous step and mix gently.
3. Incubate the reaction at room temperature for 10 min to allow the primer to anneal to the template.
4. Add 1.5 µL of mMLV-RT, mix, and incubate the reaction at 37°C for 1 h. After 1 h, store tube on ice.

3.7. Synthesis of Second Strand

1. To the tube containing the DNA-RNA hybrid after first-strand synthesis, add 20 µL of second-strand buffer, 6 µL of second-strand dNTP mixture, 111 µL of sterile water, 2 µL of RNase H, and 11 µL of DNA polymerase I (*see* **Note 7**).
2. Gently mix the contents of the tube and incubate the reaction at 16°C for 2.5 h.

3.8. Blunting of cDNA Termini

Once the cDNA synthesis is complete, the uneven termini of the double-stranded cDNA are repaired by the exonuclease and polymerase activities of Pfu DNA polymerase. This results in blunt-ended DNA fragments ready to be used for attachment of *Eco*RI adapters.

1. Add 23 µL of blunting dNTP mix and 2 µL of Pfu DNA polymerase. Incubate the reaction at 72°C for 30 min.
2. Remove the reaction and add 200 µL of phenol-chloroform and vortex. Centrifuge at 15,000g for 2 min, transfer the upper aqueous layer to a new tube, and extract with an equal volume of chloroform.
3. Collect the aqueous layer and precipitate cDNA with 0.1 vol of 3 M sodium acetate and 2 vol of chilled 100% ethanol.
4. Precipitate at –70°C for at least 4 h.
5. Spin at 15,000g for 1 h at 4°C.
6. Remove supernatant and wash pellet with 500 µL of chilled 70% ethanol. Spin at 15,000g for 2 min at room temperature.
7. Remove the ethanol; wash and dry the pellet in a speed vacuum.
8. Resuspend pellet in 7 µL of *Eco*RI adapters, and incubate at 4°C for at least 30 min to allow the cDNA to resuspend.

3.9. Attachment of EcoRI Adapters

Adapters are ligated to blunt ends to produce cohesive termini that can be used for cloning. Adapters resulting in *Eco*RI-based sticky ends are used. These adapters are composed of 9-mer and 13-mer oligonucleotides, which are complementary to each other, with an *Eco*RI cohesive end. The adapters have the following sequence:

5' AATTCGGCACGAG 3' 3' GCCGTGCTC 5'

The shorter oligonucleotide is kinased, which facilitates its ligation to the blunt end of the cDNA. The longer oligonucleotide is kept dephosphorylated to prevent it from ligating to other cohesive ends.

1. Add 1 µL of 10X ligase buffer, 1 µL of 10 mM rATP, and 1 µL of T4 DNA ligase to the tube containing 7 µL of blunted cDNA.
2. Incubate the tube at 8°C for 16 h.
3. Heat inactivate the ligase by incubating at 70°C for 30 min.

3.10. Phosphorylation of EcoRI Ends

To facilitate the ligation of adapter-attached cDNA into the vector, the unphosphorylated 5' ends of the adapter are phosphorylated using T4 polynucleotide kinase.

1. To the cDNA tube, add 1 µL of 10X ligase buffer, 2 µL of 10 mM rATP, 6 µL of sterile water, and 1 µL of T4 polynucleotide kinase.
2. Incubate the reaction at 37°C for 30 min.
3. Inactivate the kinase by heating at 70°C for 30 min in a heat block.
4. Spin down the condensation, and allow the reaction to cool down to room temperature for 5 min.

3.11. Digestion with XhoI

1. To the phosphorylated cDNA, add 28 µL of *Xho*I buffer supplement and 3 µL of *Xho*I enzyme.

P. falciparum cDNA Libraries

2. Incubate at 37°C for 2 h.
3. Add 5 µL of 10X STE buffer and 125 µL of 100% ethanol. Precipitate the reaction overnight at –20°C.
4. The next day, spin at 15,000g for 60 min at 4°C.
5. Resuspend the pellet in 60 µL of 1X STE buffer. The sample is now ready for size fractionation.

3.12. Size Fractionation

Size fractionation of cDNA is preferable to size fractionation of the initial mRNA, primarily because DNA is considerably less susceptible to degradation than is RNA. In addition, the small fragments that are generated during cDNA synthesis due to contaminating nucleases are also removed. These would otherwise be preferentially inserted and decrease the yield of long cDNA clones in the library.

We have used Chroma Spin 400 columns (Clontech Laboratories, Palo Alto, CA) to separate cDNA molecules. The resin comprises uniform microscopic beads of a hydrophilic porous material. Molecules larger than the pore size are excluded from the resin. These molecules quickly move through the gel bed when the column is centrifuged briefly, while molecules smaller than the pore size are held back. Thus, DNA is eluted from the column in order of decreasing molecule size. Chroma Spin 400 columns effectively separate fragments of less than 600 bp in length from the cDNA pool, resulting in an enrichment of complete cDNA sequences.

1. Snap the break-away end off a spin column, place it in a 2-mL collection tube, and spin at 700g for 5 min at room temperature.
2. Discard the collection tube, and place the column in a fresh tube.
3. Slowly apply the sample (60 µL) in the center of the gel bed surface and centrifuge at 700g for 5 min.
4. Discard the spin column and collect the purified sample at the bottom of the collection tube.
5. Extract and precipitate DNA with 0.1 vol of 3 M sodium acetate and 2 vol of chilled 100% ethanol.
6. Incubate at –70°C for 4 h. Pellet the DNA by spinning at 15,000g for 60 min at 4°C.
7. Wash the pellet with 100 µL of 70% ethanol and resuspend in 6 µL of sterile water.

3.13. Ligation of cDNA Into λ and Packaging

Processed cDNA can either be cloned in a λ vector, giving rise to a phage library or, alternatively, fragments can be ligated into a plasmid vector. A protocol for cloning into vectors is provided below. A protocol for ligation of cDNA into a plasmid vector and subsequent *E. coli* transformation is provided in Chapter 24.

The Uni-ZAP XR vector system (Stratagene) is a suitable choice of λ vectors for ligation. This system combines the high efficiency of λ library construction with the convenience of a plasmid system with blue-white color selection. Uni-ZAP XR vector is doubly digested with *Xho*I and *Eco*RI and can accommodate DNA inserts from 0–10 kb in length.

1. Take 2.5 µL of resuspended cDNA, add 0.5 µL of 10X T4 DNA ligase buffer, 0.5 µL of 10 mM rATP, and 1 µL of Uni-ZAP XR vector.
2. Add 0.5 µL of T4 DNA ligase and incubate overnight at 12°C.

3.14. Packaging of Ligated Arms

Packaging extracts are used to package recombinant λ phage with high efficiency. The single pack format of Gigapack III Gold packaging extracts (Stratagene) simplifies the packaging procedure and increases the efficiency and representation of libraries constructed from highly methylated DNA. Each packaging extract is restriction minus (HsdR⁻ McrA⁻ McrBC⁻ McrF⁻ Mrr⁻) to optimize packaging efficiency and library representation. Optimal packaging efficiencies are obtained with λ DNAs that are concatameric.

1. Remove a vial of packaging extract and thaw it slowly between your fingers until the extract just begins to thaw.
2. Add 1 μL of ligated DNA, mix gently, and incubate the tube at room temperature for 2 h.
3. Add 500 μL of SM buffer followed by addition of 20 μL of chloroform. Gently mix the contents.
4. Spin the tube briefly to sediment the debris, and then transfer the supernatant to a fresh tube. The supernatant containing the phage is now ready to be titered.

3.15. Titration of Packaged Library

E. coli cells used for titration of the packaged library and subsequent amplification should always be prepared fresh. XL1-Blue MRF' cells work well and are available from a number of biotechnology companies including Stratagene.

1. To prepare XL1-Blue MRF' cells, streak a bacterial glycerol stock onto a LB plate with 12.5 μg/mL tetracycline. Incubate overnight at 37°C.
2. Inoculate a single colony into NZY or LB medium, supplemented with 10 mM MgSO$_4$ and 0.2% (w/v) maltose.
3. Grow shaking at 37°C for 4–6 h, until bacteria attain an OD$_{600}$ of 0.6–1.0. Pellet at 500g for 10 min.
4. Gently resuspend the cells in half the original volume with sterile 10 mM MgSO$_4$. Dilute cells to an OD$_{600}$ of 0.5 with sterile 10 mM MgSO$_4$.
5. Mix 1 μL and 0.1 μL of packaged DNA with 200 μL of freshly prepared XL1-Blue MRF' cells at OD$_{600}$ = 0.5 (total of two tubes).
6. Incubate at 37°C for 15 min to allow the phage to attach to the cells.
7. Add 3 mL of NZY top agar maintained at 48°C, 15 μL of 0.5 M IPTG solution, and 50 μL of 250 mg/mL X-gal (in dimethylformamide).
8. Plate immediately onto 82 mm² NZY agar plates and allow plates to set for 10 min. Invert the plates and incubate overnight at 37°C.
9. Calculate the titer of the unamplified library stock as number of pfu per microgram of λ vector arms. Also calculate the percentage of recombinant (white) plaques, which should exceed 90%. If the results of the sample cDNA ligation, the test insert ligation, and the packaging positive control give desired results, package the remaining 4 μL of sample cDNA in four separate reactions (*see* **Note 8**).

3.16. Amplification of cDNA Library

The primary library should be stored at 4°C and be amplified immediately, or at worst no later than 7–10 d after construction. An amplified library is stable for many years when stored at –80°C in 7% dimethyl sulfoxide and may be screened thousands of times.

P. falciparum cDNA Libraries

1. Prepare *E. coli* host cell strain as described in **Subheading 3.15.** (XL1-Blue MRF', XL1-Blue, and LE392 are all appropriate choices). Dilute cells in 10 mM MgSO$_4$ to a final OD$_{600}$ of 0.5. For each 150 mm^2 plate, 600 µL of cells will be needed.
2. Mix aliquots of the packaged mixture containing about 50,000 pfu with 600 µL of cells in Falcon 2059 polypropylene tubes (*see* **Notes 9** and **10**).
3. Incubate the tubes containing phage and cells for 15 min at 37°C.
4. Add 6.5 mL of NZY top agar (maintained molten at 48°C) to each aliquot of infected bacteria and immediately spread evenly onto a freshly poured 150 mm^2 NYZ agar plate. Incubate at 37°C until the plaques attain no more than 1–2 mm (generally 10–11 h).
5. Overlay the plates with 8 mL of SM buffer. Store overnight at 4°C with gentle circular rocking. This allows diffusion of the phage into the SM buffer.
6. Tilt the plates and recover the bacteriophage suspension into a sterile polypropylene container (using 10-mL sterile pipets). Add an extra 2 mL of SM buffer to each plate. Using a 1-mL pipetor, squirt the solution over the entire plate surface to dislodge remaining phage from the agar. Tilt the plates and recover the remaining solution. Pool, measure the volume, and add chloroform to 5% final concentration. Mix well and incubate for 15 min at room temperature.
7. Remove the cell debris by centrifugation for 10 min at 500g.
8. Recover the supernatant and repeat centrifugation to remove remaining cell debris.
9. Transfer the supernatant to a sterile polypropylene or glass container and add chloroform to 0.3% final and DMSO to 7% final concentrations. These should be aliquoted into labeled cryotubes (1-2 mL/tube) and stored at –80°C.
10. Check the titer of the amplified library (should be 10^9–10^{10} pfu/mL). It is also useful to quality control the library and record its specifications (*see* **Notes 11** and **12**). The library is now ready for screening using readily available protocols.

4. Notes

1. It is critical that appropriate precautions be taken to protect the sample from ribonucleases normally present on human skin and in the environment. At minimum, latex gloves should be worn at all times and the reaction mixtures kept at 4°C and protected from the environment.
2. Add 0.1% DEPC to water, incubate overnight, and autoclave.
3. Incubate Oligotex suspension and 2X binding buffer at 37°C and mix immediately before use.
4. Save supernatant until certain that satisfactory binding and elution of (poly)A$^+$ mRNA has occurred.
5. Incubate elution buffer at 70°C in a water bath or heat block before use.
6. The CapFinder System (Clontech Laboratories) is a PCR-based method for the construction of cDNA libraries. This system utilizes a unique CapSwitch oligonucleotide along with a modified oligo(dT) primer to prime the first-strand reaction. The CapSwitch oligonucleotide serves as a short, extended template at the 5' mRNA end for reverse transcriptase. When the reverse transcriptase reaches the 5' end of the mRNA, the enzyme switches templates and continues replicating to the end of the oligonucleotide. This switching mechanism, in most cases, is dependent on the 7-methylguanosine cap structure present on the 5' end of all eukaryotic mRNAs. The resulting full-length single-stranded cDNA contains the complete 5' end of the mRNA as well as the sequence complementary to the CapSwitch oligonucleotide, which then serves as a universal PCR priming site (CapSwitch anchor) in the subsequent amplification.

   The CapSwitch-anchored single-stranded cDNA undergoes a long-distance PCR amplification to generate high yields of full-length, double-stranded cDNA. Only those oligo(dT)-primed single-stranded cDNAs having a CapSwitch anchor sequence at the 5' end can serve as templates and be exponentially amplified using the 3' and 5' PCR primers provided with the kit. In most cases, incomplete cDNAs and cDNA transcribed from (poly)A⁻ RNA will lack the CapSwitch anchor sequence and therefore cannot be amplified. This virtually eliminates library contamination by genomic DNA and (poly)A⁻ RNA.

In commonly used cDNA synthesis methods, 5' ends of genes tend to be underrepresented. This is particularly true for long mRNAs that have a persistent secondary structure. Furthermore, use of T4 DNA polymerase to generate blunt cDNA ends after second-strand synthesis commonly results in heterogeneous 5' ends that are 5–30 nucleotides shorter than the original mRNA. The selective amplification obtained with the CapFinder system results in a cDNA library with a higher percentage of full-length clones than libraries constructed using conventional second-strand methods. Clones isolated from CapFinder cDNA libraries are more likely to contain sequences corresponding to the complete 5' untranslated region and facilitate preliminary mapping of transcription start sites. On the other hand, CapFinder cDNA libraries may not be suitable for immunoscreening for certain proteins, because in some cases, 5'-untranslated regions contain stop codons in-frame with the initiating methionine in the expression vector.

7. The second-strand nucleotide mixture contains dCTP to reduce the probability of 5-methyl dCTP becoming incorporated in the second strand. This ensures that the restriction sites in the linker primer will be susceptible to restriction enzyme digestion.
8. A good *P. falciparum* cDNA library should comprise over 1 million pfu, with approx 90% recombinants as calculated using blue/white screening.
9. Do not exceed 300 µL of phage volume per 600 µL of cells.
10. Ideally, library amplification should begin with no fewer than 1 million primary phage, to ensure reasonable representation of rarer cDNA sequences in the amplified library. We prefer to amplify 1.5 million pfu, which corresponds to 30 large (150 mm²) plates. Titering of the amplified library should demonstrate approximately 1 million-fold amplification, with an amplified library titer of approx 10^9 pfu/mL.
11. As part of the quality control, we routinely PCR amplify 20 randomly chosen plaques using primers flanking the cloning sites and check for insert size. A good *P. falciparum* library should have an average insert size of about 1.5 kb, and inserts as long as 3.5–4.0 kb should be detectable. The use of primers flanking known introns from expressed genes is also recommended to check for lack of genomic DNA contamination. Another method of assessing for genomic DNA contamination is to use primers from genes known to be expressed at other stages of the parasite life cycle. Finally, it is important to note that *P. falciparum* libraries can easily be contaminated with abundant rRNA sequences, present as a result of their A-T richness and abundance in the starting total RNA population. This explains the critical need for preparing poly(A)⁺ mRNA by oligo-d(T) selection. It is useful to screen 1000–2000 pfu from the amplified cDNA library by plaque hybridization with a probe specific for the rRNA SSU sequence, which in our experience is the most abundant rRNA species in *P. falciparum* cDNA libraries. A good library should contain no more than 3% of SSU inserts. Note that the SSU sequence differs between parasite stages *(3)*.
12. Among the useful points to include when making a written record of a cDNA library are the organism, parasite stage, strain, directional or not, amplified or not, method of preparation of total and poly(A)⁺ RNA, vector and cloning sites, primers used for universal amplification, possibility of in vivo excision, number of pfu in the unamplified library and percentage of recombinant phage, titer of amplified library, method of storage, *E. coli* strain used for propagation, researcher and date of construction, and additional comments.

Acknowledgments

We thank Brenda Rae Marshall for editorial assistance.

References

1. Chomczynski, P. and Sacchi, N. (1987) Single step method of RNA isolation by acid guanidinium thiocyanate-phenol-chloroform extraction. *Anal. Biochem.* **162,** 156–161.
2. Aviv, H. and Leder, P. (1972) Purification of biologically active globin messenger RNA by chromatography on oligothymidylic acid-cellulose. *Proc. Natl. Acad. Sci. USA* **69,** 1408–1412.
3. Gunderson, J. H., Sogin, M. L., Wollett, G., Hollingdale, M., de la Cruz, V. F., Waters, A. P., and McCutchan, T. F. (1987) Structurally distinct, stage-specific ribosomes occur in Plasmodium. *Science* **238,** 933–937.

26

Production of Stage-Specific *Plasmodium falciparum* cDNA Libraries Using Subtractive Hybridization

David A. Fidock, Thanh V. Nguyen, Brenda T. Beerntsen, and Anthony A. James

1. Introduction

Subtractive hybridization techniques are designed to deplete one cDNA population (the "target") of sequences common to a second group of cDNAs (the "driver"), thereby effectively enriching the target population for unique sequences. The applications of this powerful methodology are widespread, and have included identification of genes expressed during embryonic development as well as genes expressed solely in specialized regions of the brain *(1–5)*.

There are a number of reasons why cDNA library-based subtractive hybridization can be useful to the study of Apicomplexan parasites such as *Plasmodium falciparum*. These parasites have complex life cycles involving vertebrate and insect hosts, with infection occurring sequentially in a series of tissues or cells during the cycle. These developmental stages are accompanied by changes enabling the parasite to recognize, invade, and develop within the next host-cell type. *Plasmodium falciparum* also undergoes profound changes during its postfertilization sexual stages (zygotes and ookinetes), as well as during sporogony and intracellular schizogony. Subtractive hybridization allows an assessment of the role of differential gene expression in controlling or accompanying these developmental changes. These procedures also are useful when coupled with gene amplification techniques because many of the developmental stages of the parasite do not lend themselves to large-scale production, making it difficult to obtain enough mRNA for cDNA library construction. Once a library is made, it can be amplified and used in a potentially unlimited number of subtractive hybridization procedures.

The protocol detailed in this chapter describes the entire subtractive hybridization procedure applied to cDNA libraries from two different *P. falciparum* stages. The success of this technique depends on the specific annealing of common target and driver nucleotide sequences achieved by incubating target antisense cDNA with an excess of driver cRNA (cRNA = complementary RNA generated from cDNA template). Complementary double-stranded sequences are then removed from unique single-stranded target cDNA by hydroxyapatite chromatography. Excess driver cRNA is removed by alkaline hydrolysis. The cDNA is amplified using the polymerase chain reaction (PCR)

From: *Methods in Molecular Medicine, Vol. 72: Malaria Methods and Protocols*
Edited by: Denise L. Doolan © Humana Press, Inc., Totowa, NJ

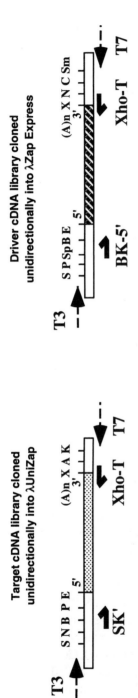
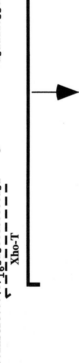

Fig. 1. Flowchart of subtractive hybridization procedures applied to *Plasmodium falciparum* cDNA libraries. This hybridization protocol relies on the use of a hydroxyapatite column to remove complementary target cDNA–driver cRNA duplexes. Excess driver cRNA is hydrolyzed and target-specific oligonucleotide primers are used to amplify subtracted target cDNA sequences. These subtracted sequences can be cloned into a plasmid vector to generate an easily manipulable subtracted library. (A)n, polyadenosine tail; A, *ApaI*; B, *BamHI*; C, *ClaI*; E, *EcoRI*; K, *KpnI*; N, *NotI*; P, *PstI*; S, *SalI*; Sm, *SmaI*; Sp, *SpeI*; X, *XhoI*.

and oligonucleotide primers specific for the target library. These gene amplification products are cloned into a convenient plasmid or λ-phage vector for subsequent analysis. This protocol is schematically represented in **Fig. 1**.

We have developed this protocol starting from the directional tag PCR subtraction method developed by J. G. Sutcliffe and colleagues *(5)*, which was optimized for plasmid cDNA libraries derived from higher eukaryotic organisms whose DNA is high in G-C content. Our protocol has modified this procedure in order to perform the subtraction using cDNA libraries cloned into λ-phage Zap vectors (Stratagene, La Jolla, CA). In our experience, *P. falciparum* sequences are far less subject to sequence "scrambling" and internal deletions in phage than in plasmid vectors. In addition, phage can accommodate much larger cDNA inserts than plasmids. Accordingly, the protocol has been adjusted to account for the differences between our phage vectors and the plasmid vectors used by Sutcliffe. Also, we have modified the cDNA and cRNA synthesis and purification and optimized the gene amplification conditions to accommodate the relatively high A-T content, about 76% of the *P. falciparum* genome.

This procedure requires that the target and driver libraries be made in λ Uni-Zap and Zap Express vectors (Stratagene), respectively. These vectors can be purchased with the phage arms already prepared for unidirectional cloning (5' *Eco*RI – 3' *Xho*I coupled to poly-dT). The parasite libraries can be efficiently made using the cDNA synthesis, cloning and phage manipulation protocols supplied by Stratagene. Note that λ Uni-Zap is a phagemid vector from which a plasmid can be excised leading to the recovery of pBluescript SK plasmids. Excision of the λ Zap Express phagemid results in the recovery of the pBK-CMV plasmid.

We have applied this protocol to generate a subtracted *P. falciparum* sporozoite cDNA library. This library was derived from a *P. falciparum* sporozoite-infected *Anopheles gambiae* salivary gland cDNA library (the target library cloned into λ Uni-Zap XR) from which we depleted uninfected *A. gambiae* salivary gland cDNAs (cloned into λ ZapExpress; driver library no. 1) as well as parasite sequences present in a *P. falciparum* asexual blood stage cDNA library (cloned into λ ZapExpress; driver library no. 2). Diagnostic PCR with parasite or mosquito genes of defined stage or tissue specificity demonstrated efficient subtraction of sequences common to the target and driver populations *(6)*.

2. Materials
2.1. Production of cDNA Libraries

1. TRIzol (Life Technologies, Gaithersburg, MD).
2. Oligotex (Qiagen, Valencia, CA).
3. RNase-free quartz cuvette (Amersham Pharmacia Biotech, Piscataway NJ).
4. ZAP Express cDNA/Gigapack Cloning Kit (Stratagene, La Jolla CA).
5. Uni-ZAP XR vector (Stratagene, the cDNA synthesis and Gigapack kits required for library construction with this vector are the same as for the ZAP Express vector).
6. *Escherichia coli* LE392 and XL1-blue cells.
7. SM buffer: 5.8 g of NaCl, 2 g of $MgSO_4 \cdot 7H_2O$, 50 mL of 1 *M* Tris-HCl, pH 7.5, 5 mL of 2% gelatin, H_2O to 1 L. Autoclave to sterilize and store at 4°C.
8. Chloroform ($CHCl_3$).
9. Dimethyl sulfoxide (DMSO).

2.2. Purification of Phage DNA

1. NZY/agar plates: 5 g of NaCl, 2 g of $MgSO_4 \cdot 7H_2O$, 5 g of yeast extract, 10 g of NZ amine (casein hydrolysate), adjust to pH 7.5 with NaOH. Make volume to 1 L. Add 15 g of Difco agar. Autoclave to sterilize. Allow about 80 mL per 150 mm² plate.
2. *E. coli* LE392 cells.
3. Lambda Maxi Kit (Qiagen).
4. 20 mg/mL Proteinase K.
5. 10% sodium dodecyl sulfate (SDS).
6. 3 *M* sodium acetate, pH 5.2.
7. 100% ethanol.
8. 1X Tris-EDTA (TE), pH 8.0.
9. Buffer-saturated phenol.

2.3. Preparation of Antisense-Strand cDNA

1. *Not* I restriction enzyme and OPA⁺ buffer (Amersham Pharmacia).
2. 1 *M* NaCl.
3. 1% Triton X-100.
4. 0.5 *M* EDTA, pH 8.5.
5. Diethyl pyrocarbonate (DEPC)-treated distilled H_2O.
6. *Xho*-T primer: 5'-AGA-TCC-CCC-TCG-AGT-TTT-TTT-TTT-TTT-TTT-3'.
7. SK' primer: 5'-CGC-TCT-AGA-ACT-AGT-GGA-TCC-C-3'.
8. *Taq* DNA polymerase.
9. 2 m*M* dNTP stocks.
10. Thermal cycler with heated bonnet.
11. Streptavidin Magnesphere® paramagnetic particles (cat. no. Z5481, Promega, Madison, WI).
12. Magnesphere technology magnetic separation stand (cat. no. Z5332, Promega).
13. Biotin- 16-dUTP (cat. no. 1093070, Roche Molecular Biochemicals, Indianapolis IN).
14. 0.5X sodium chloride-sodium citrate (SSC) buffer: 4.4 g of NaCl, 0.7 g of $NaH_2PO_4 \cdot H_2O$, 0.2 g of EDTA in final volume of 1 L. Adjust pH to 7.4 with NaOH.
15. 0.2 *N* NaOH, 0.05% Tween-20 solution.
16. 1 *M* Tris-HCl, pH 7.5.

2.4. Preparation of Sense-Strand cRNA

1. T3 Megascript kit (cat. no. 1388; Ambion, Austin TX).
2. [^{32}P]-UTP.
3. DNase I ("FPLC pure," Amersham Pharmacia).
4. Isopropanol.

2.5. Subtractive Hybridization

1. 2X Hyb solution: 1 *M* NaCl, 0.1 *M* HEPES pH 7.5, 4 m*M* EDTA, 0.2% SDS. To make, combine 10 mL of 5 *M* NaCl, 10 mL of 0.5 *M* HEPES pH 7.5, 0.4 mL of 0.5 *M* EDTA, 1 mL of 10% SDS, 28.6 mL of H_2O. Treat with DEPC and autoclave.
2. 50 m*M* phosphate buffer, pH 6.6, 0.2% SDS.
3. 120 m*M* phosphate buffer, 0.2% SDS.
4. 400 m*M* phosphate buffer, 0.2% SDS.
5. Bio-Gel HTP hydroxyapatite, DNA grade (cat. no. 130-0520, Bio-Rad, Hercules CA).
6. Open-ended water-jacketed chromatography column, 1 X 30 cm (cat. no. 737-6201, Bio-Rad).
7. Geiger counter.
8. Peristaltic pump, connected to a water bath maintained at 60°C.
9. 2-Butanol.

10. Centricon YM-30 columns (Amicon Bioseparations, Millipore, Bedford MA).
11. Glycogen (Roche Molecular Biochemicals).

2.6. Making the Subtractive cDNA Library

1. SizeSep 400 column (Amersham Pharmacia).
2. *Pst*I, *Xho*I, and *Sal*I restriction enzymes.
3. pGEM-3Zf(–) plasmid vector (Promega).
4. Electrocompetent *E. coli* DH5α cells (cat. no. 11319019, Life Technologies).
5. LB agar plates + ampicillin (50–100 µg/mL).
6. Microspin S-200 columns (Amersham Pharmacia).

3. Methods

3.1. Production of cDNA Libraries

Generation of a good cDNA library is dependent on preparing non-degraded, DNA-free RNA. Methods for RNA isolation and construction of λ-phage cDNA libraries can be found in an accompanying chapter of this volume (*see* Chapter 25). Our preferred method is to use the Trizol reagent on pelleted, washed parasites, with particular attention being given to the following procedures:

1. To produce poly(A)$^+$ mRNA, we have found the oligo (dT) resin (Oligotex) produced by Qiagen to be superior to other commercially available preparations. Ribosomal RNA contamination of *P. falciparum* cDNA libraries is a frequent problem, necessitating considerable effort and attention to reducing this to minimum levels. Our solution to this problem was to select poly(A)$^+$-RNA enriched by two rounds of binding and elution from separate Oligotex columns.
2. Quantification of poly(A)$^+$ mRNA is essential for optimizing cDNA synthesis. Accurate determination of mRNA yields can be a problem with *P. falciparum* because recovered amounts can be low (≤0.2 µg). A simple method for determining yields is to measure the UV absorbances at 260 nm and 280 nm (to calculate yield and purity) in a RNAse-free quartz cuvette (*see* **Note 1**).
3. Production of cDNA libraries works well using the cDNA synthesis and cloning kit developed by Stratagene. Our only modification of their well-designed protocol was to reduce the oligonucleotide primer, adaptor, and enzyme concentrations when using <1 µg of mRNA as the starting material (*see* **Note 2**). The prepared cDNA was size-fractionated over a Sephacryl S-500 column.
4. It is critical to amplify the primary cDNA library within 1–2 wk of preparation in order to prevent loss of phage. Again, Stratagene has a good protocol detailed in their literature supplied with the kit. If possible, amplify $1–3 \times 10^6$ pfu (plaque forming units). We do this by plating phage and freshly prepared *E. coli* XL1-blue MRF' or LE392 cells onto large (150 mm^2) NZY/agar plates. We generally apply 5×10^4 pfu per plate and incubate at 37°C until the plaques are confluent (about 10–11 h). Good amplification should give about a 10^6-fold increase in the titer. To elute the phage, add 5 mL of SM buffer to each plate and agitate gently on a rocker device at 4°C for 2 h. Tilt the plates and recover the eluate. Add another 2 mL of SM buffer and wash the solution across the top of the plate using a 1-mL pipetor. Tilt the plate again and recover the remaining solution. Pool the phage eluate into 250-mL CHCl$_3$-resistant centrifuge tubes. Add CHCl$_3$ to a final volume of 5%, mix by inverting, and spin at 6000*g* for 10 min at 4°C. This procedure will pellet the bacterial and agar debris, leaving the phage in the aqueous phase. Add CHCl$_3$ to a final volume of 0.3% and leave rocking overnight at 4°C. Centrifuge to remove remaining debris (twice if necessary). To the final supernatant, add CHCl$_3$ to 0.3% final volume, DMSO to 7% final

volume, aliquot and store at –80°C. Libraries prepared and stored in this manner should show no appreciable reduction in phage titer over a number of years.

3.2. Purification of Phage DNA

Once the target and driver cDNA libraries are constructed and their quality is assured, it is important to make pure, nondegraded phage DNA. This material serves for the subsequent steps of generating target antisense cDNA and driver sense-strand cRNA.

1. Plate out $1.5–2.0 \times 10^6$ plaque-forming units (pfu) from each library, at $5–6 \times 10^4$ pfu per large 150 mm^2 NZY/agar plate. LE392 or XL1-blue MRF' are suitable *E. coli* host strains. Protocols for amplifying the phage are provided with the Stratagene Gigapack cloning kits.
2. Elute the phage into SM buffer as described in **Subheading 3.1., step 4.**
3. Titer the amplified library using fresh LE392 cells. If the titer is $>> 2 \times 10^9$ pfu/mL and the total volume of the phage solution is over 200 mL, proceed to making phage DNA (*see* **Note 3**). Good quality phage DNA can be prepared using the Qiagen Lambda Maxi kit. Apply no more than the equivalent of 5×10^{12} pfu per Maxi column, otherwise the column will become saturated and yields will be less.
4. Purified phage DNA needs to be treated with proteinase K in order to get good yields of cRNA. Add 20 µL of proteinase K (20 mg/mL) and 20 µL of 10% SDS to 50-µg phage DNA. Increase reaction mix up to a final volume of 400 µL with dH$_2$O. Incubate at 37°C for 1 h.
5. Extract subsequently with an equal volume of phenol, phenol/CHCl$_3$, and finally CHCl$_3$. These organic solvent extractions are standard for nucleic acid preparations. An equal volume of solvent is added to the aqueous reaction mixture and the solution is gently agitated. A brief centrifugation separates the aqueous solution from the solvent. Use a clean pipet to remove the aqueous (top) layer to another tube. Note that phenol is highly corrosive and can be toxic at high doses. CHCl$_3$ is volatile. Please use and dispose of these reagents in a careful manner.
6. Precipitate the DNA in ethanol by addition of 1/9 vol of 3 *M* sodium acetate, pH 5.2 and 2.5 vol of 100% ethanol. Spin down the DNA in a microcentrifuge and wash the DNA pellet with cold 70% ethanol using standard procedures.
7. Resuspend the dried DNA pellet in 100 µL of 1X Tris-EDTA (TE), pH 8.0.
8. Quantify the yield of DNA using a spectrophotometer and run 1 µg on a 0.8% agarose gel. The ethidium bromide-stained DNA should migrate as a high molecular weight species of over 23 Kb.

3.3. Preparation of Antisense-Strand cDNA from the Target Library

The basic steps of preparing antisense-strand cDNA from the target library and sense-strand cRNA from the driver library are summarized in **Fig. 2**, including detail at the nucleotide level.

1. Linearize the phage DNA from the target library with *Not*I (GC|GGCCGC). This enzyme will cleave in the polylinker at the 5'-end of the *Eco*RI cloning site. Digest 4 µg of target phage DNA in a final volume of 40 µL in 1X OPA$^+$ buffer (Pharmacia) supplemented with 0.1 *M* NaCl and 0.1% Triton X-100 (final concentrations), plus 24 U *Not*I (Pharmacia). After a 90 min incubation at 37°C, add an additional 12 U of the enzyme and continue the incubation for 40 min. To stop the reaction, add 2 µL of 0.5 *M* EDTA pH 8.0, 4.4 µL of 3 *M* sodium acetate, pH 5.2, and 100 µL of chilled 100% ethanol. Precipitate the DNA, pellet by centrifugation, wash the pellet with 100 µL of 70% ethanol, and then dry the DNA pellet using standard procedures. Resuspend the DNA in 100 µL of DEPC-treated H$_2$O to give a final concentration of 0.04 µg/µL. Run 10 µL on a 0.8% agarose gel to check for linearization and lack of DNA degradation.

cDNA Libraries Using Subtractive Hybridization

Target library (in Lambda Uni-Zap, excision produces pBluescript SK):

```
      T3' primer ->                                    SK' primer ->
                          BstXI        EagI              SpeI
                    SacI       SacII    NotI    XbaI           BamHI
gaaattaaccctcactaaaggg aacaaaagctggagctccaccgcggtggcggc cgctctagaactagtgga-

              PstI
    SmaI      EcoRI adapter              XhoI     ApaI    KpnI
tcc cccgggctgcagg aattcggcacgag..Insert..(A)n.ctcgagggg ggcccggtacccaattc-
                                         <- Xho-T primer
gccctatagtgagtcgtattac aattc...
<- T7' primer
```

Linearize with *Not*I. PCR amplify biotinylated single strand antisense cDNA using Xho-T primer. Recover biotinylated single strand antisense cDNA on paramagnetic beads. Gives:

```
         XhoI                                        adaptor EcoRI
5' agatcccctcgag(t)16..Insert (negative strand)..ctcgtgccgaattc-

        SmaI       SpeI       EagI (cleaved)
PstI       BamHI      XbaI   NotI (cleaved)
ctgcagcccggggatccactagttctagagcggcc
```

Driver library (in Lambda ZapExpress, excision produces pBK-CMV):

```
      T3' primer ->                  PstI
                +1            SacI  BssHII    SalI  SpeI  BamHI
gaaattaaccctcactaaaggg aacaaaagctggagctcgcgcctgcaggtcgacactagtggatcc-

    EcoRI adapter              XhoI ScaI XbaI  NotI   ApaI  ClaI
aaag aattcggcacgag..Insert..(A)n.ctcgagagtacttctagagcggccgcgggcccatcgat-

     SmaI     KpnI
tttccacccgggtggggtaccaggtaagtgtacccaattc gccctatagtgagtcgtattac aattcactgg-
                                         <- T7' primer
```

Linearize with *Not*I. Make cRNA transcripts using T3 RNA polymerase. Gives:

```
                     PstI
          SacI BssHII       SalI  SpeI  BamHI    EcoRI adapter
5' gggaacaaaagctggagctcgcgcctgcaggtcgacactagtggatccaaag aattcggcacgag-

         XhoI ScaI XbaI  NotI (cleaved)
..Insert..(A)n.ctcgagagtacttctagagc
```

Fig. 2. Schematic representation of the production of single-strand antisense cDNA from the target library and sense-strand cRNA from the driver library. The lack of homology between the polylinker sequences flanking the inserts in the target and driver libraries is a critical factor for the successful outcome of this protocol and is depicted at the nucleotide level.

2. Single-strand antisense cDNA is generated from the linearized phage target DNA using the *Xho*-T primer (*see* **Subheading 2.3.**) and biotin-labeled dUTP in a PCR reaction (**Note**

4). Set up the reaction as follows: 1 µL (40 ng) of phage target DNA, 4 µL of 10X PCR buffer (Pharmacia; already contains $MgCl_2$), 4 µL of 2 mM dATP, 4 µL of 2 mM dGTP, 4 µL of 2 mM dCTP, 3 µL of 2 mM dTTP, 2 µL of 1 mM biotin-16-dUTP (gives a 3:1 ratio of cold dTTP to biotin-dUTP), 8 µL of *Xho*-T primer (5 mM), 2 µL of *Taq* DNA polymerase, and 8 µL of H_2O, to a final volume of 40 µL. This reaction is performed in thin-walled tubes in a thermal cycler using the following program: 1X (94°C 50 s, 48°C 45 s, 72°C 1min 30 s); 19X (93°C 30 s, 48°C 45 s, 72°C 1 min 30); 10X (92°C 30 s, 54°C 45 s, 72°C 1 min 30 s); 1X (72°C 5 min); hold at 4°C.

3. To precipitate the amplified cDNA products, add 4.4 µL of 3 M sodium acetate pH 5.2, 100 µL of ethanol. Precipitate and wash as described before (*see* **Subheading 3.2., step 6**). Resuspend in 50 µL of 0.5X sodium chloride-sodium citrate (SSC).
4. To recover single-strand biotinylated antisense cDNA, use streptavidin-coated paramagnetic beads (Promega; *see* **Note 5**) in a 1.5-mL microfuge tube as follows:
   a. Take 1 aliquot (600 µL) of beads and prewash 3 times with 0.5X SSC.
   b. Add single-strand target cDNA (resuspended in 0.5X SSC) and pipet to mix.
   c. Incubate 30 min at room temperature with gentle agitation every 5–10 min.
   d. Recover bound material by placing the microfuge tube into a magnetic stand (Promega; *see* **Note 6**).
   e. Remove the supernatant (while tube is in stand) and add 50 µL of 0.5X SSC solution to the bead mixture.
   f. Remove the tube from stand, resuspend beads by gentle pipetting of the solution over the beads.
   g. Place the tube back on stand and remove the supernatant.
   h. Repeat so as to have washed the bound material three times.
   i. To recover the bound cDNA: add 50 µL of solution containing 0.2 N NaOH, 0.05% Tween-20.
   j. Pipet vigorously to mix, then place tube back on stand and transfer supernatant (now with eluted cDNA) to a new tube. Quickly add 25 µL of 1 M Tris-HCl, pH 7.5, to neutralize the solution.
   k. To precipitate the DNA, add 8.33 µL of 3 M sodium acetate, pH 5.2, 187.5 µL of chilled 100% ethanol.
   l. Centrifuge, wash, dry and resuspend the purified biotinylated cDNA in 60 µL of DEPC-treated H_2O.
   m. Quantify the DNA (expect about 200–300 µg) and then extract once with phenol/$CHCl_3$ and once again with $CHCl_3$. Precipitate by adding 1/9 vol 3 M sodium acetate, pH 5.2, 2.5 vol ethanol (*see* **Subheading 3.2., step 6**). Resuspend in 5 µL of DEPC-H_2O.

3.4. Preparation of Driver Library, Sense-Strand cRNA

1. Linearize phage DNA (4.5 µg) from the driver library with *Not*I as described in **Subheading 3.3., step 1**. Extract the reaction once with phenol/$CHCl_3$ and once again with $CHCl_3$ to remove enzyme, then precipitate the DNA with ethanol (*see* **Subheading 3.2., step 6**). Assume that 0.5 µg of the phage DNA is lost, leaving 4 µg behind. Resuspend the DNA pellet in 10 µL of DEPC-treated H_2O.
2. Make sense-strand cRNA using T3 polymerase (Megascript kit; Ambion). As control, make cRNA from 1 µg test template (from the kit). Include [^{32}P]-UTP as a tracer molecule. Set up the reaction as follows (*see* **Note 8**): 5 µL of RNase-free H_2O, 4 µL of 10X transcription buffer, 4 µL of 75 mM ATP solution, 4 µL of 75 mM CTP solution, 4 µL of 75 mM GTP solution, 4 µL of 75 mM UTP solution, 1 µL of [^{32}P]UTP, 10 µL (4 µg) of linearized phage DNA, and 4 µL of T3 RNA Polymerase mix. Set up the reaction in 0.5-mL thin-walled PCR tubes (DEPC-treated). Mix gently and incubate for 6 h at 37°C in a thermal cycler equipped with a heated bonnet (preheated to prevent condensation).

cDNA Libraries Using Subtractive Hybridization

3. Add 2 µL of RNase-free DNase I and incubate at 37°C for 30 min. To precipitate, add 95 µL DEPC-treated H$_2$O, 15 µL of NH$_4$Ac and mix well. Then extract once with phenol/CHCL$_3$ and once again with CHCl$_3$. Precipitate cRNA with isopropanol. (This protocol is listed as Method 1 in Ambion's Megascript Kit). Chill the precipitate for at least 15 min at –20°C. Centrifuge for 15 min at maximum speed to pellet the RNA. Carefully remove the supernatant solution and resuspend the RNA in 60 µL of DEPC-treated H$_2$O.
4. Treat the driver cRNA a second time with DNase I by adding 1/9 vol of any RNase-free salt buffer (e.g., use AMV or RT buffer from any reverse transcriptase cDNA kit). Then use same conditions as for first DNase I treatment (*see* **Subheading 3.4., step 3**). Extract with phenol/CHCl$_3$ and once again with CHCl$_3$. Add 1/9 vol of 5 mM NH$_4$OAC/0.1 M EDTA to the cRNA solution and 1 vol of isopropanol. Precipitate the cRNA as before (*see* **Subheading 3.4., step 3**).
5. Resuspend the cRNA in 60 µL of DEPC-treated H$_2$O and quantify (use HCl/methanol-treated 50 µL quartz cuvets reserved exclusively for RNA work; *see* **Note 1**). Note that at least 50 µg of cRNA is required to perform the subtraction (*see* **Note 9**). Precipitate 50 µg cRNA in isopropanol as before (*see* **Subheading 3.4., step 4**). Store at –20°C until ready for use.

3.5. Subtractive Hybridization

1. Centrifuge the cRNA precipitated in isopropanol (*see* **Subheading 3.4., step 5**), wash the cRNA pellet in 70% ethanol and dry using standard procedures.
2. Dissolve the antisense-strand cDNA pellet (*see* **Subheading 3.3., step 4**) in 5 µL of DEPC-treated H$_2$O. Transfer the cDNA solution (containing 200-500 ng) to the dried driver cRNA pellet (containing about 50 µg driver cRNA). Resuspend the cRNA pellet by vortexing (keep the sample on ice as much as possible).
3. Add 5 µL of 2X Hyb solution to the cDNA tube to recover any remaining material (*see* **Subheading 2.5., item 1**). Transfer this solution to the cRNA solution tube for final volume of 10 µL. Transfer the cDNA/cRNA mixture to a 0.5-mL PCR tube (RNase-free) and heat in a preheated PCR block at 96°C for 90 s.
4. Incubate at 68°C for 24 h to allow hybridization of common target and driver sequences. This can be conveniently performed in a thermal cycler using the following conditions: 96°C for 90 s; ramp down to 68°C at –0.1°C/s; 68°C for 24 h.
5. After hybridization remove the reaction from the thermal cycler, add 90 µL of DEPC-treated H$_2$O and mix. Extract the aqueous phase once with phenol/CHCl$_3$, then once with CHCl$_3$. Add 1/9 vol of 3 M sodium acetate, pH 5.2, and 2.5 vol of chilled 100% ethanol. Precipitate under standard conditions, wash and dry pellet.
6. Dissolve the pellet in 0.5 mL of 50 mM sodium phosphate buffer (PB), pH 6.5, 0.2% SDS by repeated pipetting.
7. Add 10 mL of 50 mM PB, 0.2% SDS to 1 g of Bio-Gel HTP hydroxyapatite powder (DNA grade). Boil for 30 min (this treatment reduces nonspecific binding). For the following **steps 8–13**, be careful to always maintain some buffer above the gel. This will prevent the gel from running "dry."
8. Allow the hydroxyapatite suspension to settle for 10 min at room temperature and decant the buffer over the settled material. Resuspend the hydroxyapatite in 5–10 mL of 50 mM PB, 0.2% SDS, mix, let settle, and decant the buffer. Resuspend the settled hydroxyapatite in 5 mL of 50 mM PB, 0.2% SDS. Apply the resuspended material to a water-jacketed 1 × 30 cm chromatography column.
9. Connect the water-jacket to a circulating water bath maintained at 60°C. Add 4–5 mL of 50 mM PB, 0.2% SDS to the column. Using a pipet, load the hydroxyapatite as a slurry to the column. After the slurry has settled, connect the column to a peristaltic pump.

10. Wash the column with 6 mL of preheated (60°C) 50 mM PB, 0.2% SDS at a flow rate of 0.5 mL/min.
11. Preheat the resuspended cDNA/cRNA sample at 60°C and add the 0.5 mL sample to the column. Begin immediately to collect 0.5 mL fractions in Eppendorf or other collection tubes.
12. Once the sample has entered the hydroxyapatite, apply 6 mL of preheated (60°C) 50 mM PB, 0.2% SDS and continue collecting 0.5 mL fractions.
13. After the 6 mL of the 50 mM PB buffer has entered the hydroxyapatite, apply 6 mL of preheated (60°C) 120 mM PB/0.2% SDS and continue to collect samples.
14. Once the 6 mL of 120 mM buffer has entered the hydroxyapatite, apply 6 mL of preheated (60°C) 400 mM PB/0.2% SDS and continue sample collection.
15. Monitor the column output constantly with a Geiger counter to identify those fractions that contain radiolabeled, single-strand cRNA. This monitoring enables the identification of those samples containing the subtracted, single-strand target cDNA. The radiolabeled cRNA / subtracted cDNA should elute in or close to fractions 5–8 after application of the 120 mM phosphate buffer. Radioactive cRNA will also elute after application of the 400 mM phosphate buffer. This latter sample will correspond to duplexed cDNA/cRNA.
16. Collect the desired fractions from the 120 mM phosphate eluate (radioactive fractions plus 1–2 on either side). Pool the fractions that contain the subtracted cDNA (radioactive fractions plus one on each side), and reduce the volume by concentrating the DNA with 2-butanol *(7)*. Extract with $CHCl_3$ to remove the butanol. Then extract once with phenol/$CHCl_3$ and once again with $CHCl_3$.
17. Desalt and concentrate the subtracted cDNA sample by centrifugation in a Centricon YM-30 column (Amicon), according to manufacturer's instructions.
18. Measure the volume after column centrifugation (normally about 35 µL) and add H_2O to a final volume of 100 µL.
19. Add 4 µL of 0.5 M EDTA, pH 8.0, 2 µL of 20% SDS, 18 µL of 2 N NaOH. Incubate at 68°C for 30 min (this step hydrolyzes the driver cRNA).
20. Cool the sample mixture to room temperature and neutralize by adding 40 µL of Tris-HCl, pH 7.5, 12 µL of 2 N HCl, 7 µL of 5 M NaCl, 2.25 µL of glycogen (1 mg/mL), and 500 µL of ethanol.
21. Precipitate the subtracted cDNA at –70°C for 30 min. Centrifuge, discard supernatant, wash and dry pellet.
22. Resuspend the DNA pellet in 10 µL of H_2O. This is the subtracted cDNA sample (*see* **Note 10**).

3.6. Making and Screening the Subtracted cDNA Library

3.6.1. Making the Subtracted cDNA Library

The subtracted, single-strand target cDNA can now be amplified by PCR to produce double-strand inserts for a subtracted library.

1. Using the subtracted cDNA sample, set up a PCR reaction as follows: 1 µL of template (1/10 dilution of the 120 mM fraction or diluted control), 10 µL of 10X PCR buffer, 5 µL of DMSO, 4 µL of regular dNTP labeling mix (2 mM each dNTP), 2.5 µL of SK' primer (5 µM), 2.5 µL of *Xho*-T primer (5 µM), 74.5 µL of H_2O, and 0.5 µL of *Taq* DNA polymerase.
2. Run in 94°C preheated thermal cycler, program: 25 cycles at 94°C for 15 s; 56°C for 15 s; 64°C for 1 min.
3. Extract the PCR reaction once with phenol/$CHCl_3$ and once with $CHCl_3$.
4. Prepare a SizeSep 400 column and equilibrate it in ligation buffer (follow guidelines by the manufacturer, Amersham Pharmacia).

5. Load the phenolized PCR reaction on the column. Spin at 400g for 2 min and recover the flowthrough.
6. Digest the sample to completion with *Pst*I (cuts at the 5'-end) and *Xho*I (cuts at the 3' end). Heat inactivate the restriction enzymes at 68°C. Extract once with phenol and once with phenol/CHCl$_3$. Ethanol precipitate using the cDNA standard procedures (*see* **Subheading 3.2., step 6**). Resuspend in 20 µL of TE. Quantify 2 µL on an agarose gel.
7. Prepare pGEM-3Zf(–) vector by sequential digestion with *Sal*I and then *Pst*I (*Pst*I–digest the sample overnight). Note that *Sal*I and *Xho*I share compatible ends and that the ligated ends can be recleaved by *Taq*I. Heat inactivate the restriction enzymes at 68°C. Run on an agarose gel and purify the cleaved vector band. Quantify by running 5–10% of the recovered plasmid on an agarose gel. Ligate the plasmid vector and subtracted cDNA inserts overnight at 14°C (*see* **Note 12**).
8. Ethanol precipitate the ligation reaction. Spin, wash and dry the DNA pellet and resuspend in 10 µL of dH$_2$O. Electroporate 2 µL aliquots into *E. coli* DH5α electrocompetent cells. Plate out on LB plates + ampicillin (50–100 µg/mL; *see* **Note 13**).

3.6.2. Making a Subtracted cDNA Probe

Subsequent analysis of the library will depend on individual interests and may involve DNA or antibody hybridization with probes specific for individual genes. Another approach is to screen the subtracted library with the subtracted cDNA used for ligation. This will result in identification of a larger number of the abundantly expressed, target-specific cDNA clones. These clones should then be screened against the original target and driver libraries to verify specificity with the target population.

1. To make the subtracted cDNA probe, set up the following PCR reaction: 1 µL of template (1/10 dilution of the 120 mM fraction or diluted control), 10 µL of 10X PCR buffer, 5 µL of DMSO, 4 µL of modified dNTP labeling mix (2 mM d [G,T,C] TP, 0.2 mM dATP), 10 µL of [^{32}P] αdATP, 2.5 µL of S' primer (5 µM), 2.5 µL of *Xho*-T primer (5 µM), 63.5 µL of H$_2$O, and 0.5 µL of *Taq* DNA polymerase. Run in preheated thermal cycler, program: 25 cycles at 94°C for 15 s; 56°C for 15 s; 64°C for 1 min.
2. Purify radioactive probe using a MicroSpin S-200 column.
3. Use at least 5×10^5 cpm/mL to hybridize against subtracted cDNA plasmid library filters, using standard techniques (*see* **Note 14**).

4. Notes

1. To remove any RNAse activity from quartz cuvets, place the cuvet in a sterile 50-mL tube (Falcon) containing 40 mL of a 1:1 solution of concentrated HCl and methanol. Incubate with shaking at 37°C for at least 1 h. Remove the solvent by 3 or 4 rinses in DEPC-H$_2$O. Standard RNA procedures can be adopted to measure the RNA spectrophotometrically and recover the sample without degradation *(8)*.
2. We have been successful in generating *P. falciparum* cDNA libraries from as little as 0.1 µg poly(A)+ mRNA, resulting in a final library of 1.0×10^6 pfu, 93% recombinant and with no evidence of genomic DNA contamination by diagnostic PCR with primers spanning intron sequences. Production of cDNA libraries from such little amounts requires that primer, adaptor, and enzyme concentrations be decreased during the steps of cDNA synthesis relative to the amounts recommended in the Stratagene protocol. We scaled all reactions downward by 10-fold to construct the cDNA libraries. As a benchmark, when using 0.1 µg of mRNA, dilute all primer, adaptor and enzyme amounts 10-fold. Gene representation can be assessed by PCR screening of the cDNA library (using whole phage or purified phage DNA as the template) with a panel of diagnostic primers specific for

genes whose stage specificity is known. These primer pairs should include sequences flanking introns to check for the presence of contaminating genomic DNA.
3. A good phage amplification on plates should yield about $2.5–25 \times 10^{12}$ pfu total. No more than 5×10^{12} pfu should be applied per Qiagen λ Maxi column. We generally recovered about 100 µg pure phage DNA per 2.5×10^{12} pfu.
4. To control the representative production of biotinylated target cDNA, one can perform the PCR / streptavidin steps in duplicate and take 10% of the final volume for control PCR. This control PCR can use the primers *Xho*-T and SK' (*see* **Subheading 2.**) that are specific to the vector and can amplify all inserts in the target cDNA library. This PCR should be performed in parallel with purified target library phage DNA. These PCR samples should be electrophoresed on a 1% agarose gel and visualized after ethidium bromide staining. The smear patterns with the biotinylated antisense target cDNA and the target phage DNA should be similar.
5. We calculate that the amount of biotinylated antisense cDNA that we apply to one aliquot of streptavidin-coated paramagnetic beads represents about 1% of total binding capacity of the beads. Therefore, this procedure does not run into problems of saturation of the biotin binding sites on the bead surfaces.
6. We found the yield of biotinylated target antisense cDNA following bead purification to be about 200–350 ng.
7. In our protocol, we have incorporated a radiolabeled nucleotide into the cRNA reaction. This allows tracking of the location of the cRNA during hydroxyapatite chromatography. We assume that the single-strand driver cRNA and subtracted target cDNA coelute from the hydroxyapatite column and therefore we retain the radioactive fractions to recover the subtracted target cDNA. Alternatively, one could incorporate [^{32}P]-dATP into the PCR reaction for generation of the biotinylated target cDNA, to ensure during the subsequent hydroxyapatite column step that the single-stranded, nonassociated target cDNA was precisely recovered and separated away from the later peak of associated target cDNA/driver cRNA.
8. Ambion recommends using 1 µg of plasmid DNA per reaction. We use 2 µg of phage DNA, which is equivalent to a much lower molar concentration of DNA. Therefore we run 6-h reactions and pool two reactions at 2 µg each to generate enough cRNA. Also, we recommend two successive DNase I treatments to remove as much driver DNA as possible.
9. During initial familiarization with the technique of cRNA production, it is advisable to run samples (include a control) on agarose or polyacrylamide RNA gels, then overlay with Saran Wrap and expose to X-ray film in a light-sealed environment. Check for the presence of a 1.8-kb band in the control lane and compare this with the spread of sizes in driver cRNA lanes. The latter should show a smear with a range of sizes corresponding to that of the original cDNA inserts.
10. The subtracted cDNA precipitate may include what appears to be a SDS pellet. This will not interfere with subsequent PCR. Be sure to thoroughly mix by pipetting prior to subsequent steps.
11. The subtracted cDNA can be used in 1–2 rounds of additional subtraction to optimize the removal of sequences common to the target and driver cDNA/cRNA populations.
12. For an efficient ligation, we recommend adding 150 ng of vector plus 150 ng of insert in a total volume of 10 µL with 1 µL of ligase (1000 U/mL). This provides for a favorable kinetic reaction (*8*) and should result in a highly efficient ligation when beginning with well-digested and purified DNA templates.
13. Once the subtracted cDNA plasmid library has been transformed into *E. coli*, it is important to determine the total number of transformed colonies. Increasing the number of ligations and transformations will increase the overall size of the library. Using the above protocol, we generated a subtracted cDNA library of 1×10^6 colony-forming units. These

colonies can be pooled from the various LB + ampicillin plates and glycerol added to a final volume of 15%. The library can then be stored indefinitely as aliquots kept at –70°C. Scraping the surface of a partially thawed aliquot will recover enough cells to establish the library titer and plate the library out for subsequent analysis.
14. Given the high A-T content of *P. falciparum*, we recommend hybridizing at 54–56°C overnight, with the most stringent wash being 0.3X SSC, 0.5% SDS at 56°C (2 × 15 min).

Acknowledgments

This work was supported by a grant from the Burroughs Wellcome Fund to A. A. J. D. A. F. thanks the Pasteur Institute for their kind support of his postdoctoral training in the laboratory of A. A. J. We thank Lynn Olson for help in preparing the manuscript.

References

1. Casal, J. and Leptin, M. (1996) Identification of novel genes in *Drosophila* reveals the complex regulation of early gene activity in the mesoderm. *Proc. Natl. Acad. Sci. USA* **93,** 10,327–10,332.
2. Gautvik, K. M., de Lecea, L., Gautvik, V. T., Danielson, P. E., Tranque, P., Dopazo, A., Bloom, F. E., and Sutcliffe, J. G. (1996) Overview of the most prevalent hypothalamus-specific mRNAs, as identified by directional tag PCR subtraction. *Proc. Natl. Acad. Sci. USA* **93,** 8733–8738.
3. Saga, Y., Hata, N., Kobayashi, S., Magnuson, T., Seldin, M. F., and Taketo, M. M. (1996) MesP1: a novel basic helix-loop-helix protein expressed in the nascent mesodermal cells during mouse gastrulation. *Development* **122,** 2769–2778.
4. Tranque, P., Crossin, K. L., Cirelli, C., Edelman, G. M., and Mauro, V. P. (1996) Identification and characterization of a RING zinc finger gene (C-RZF) expressed in chicken embryo cells. *Proc. Natl. Acad. Sci. USA* **93,** 3105–3109.
5. Usui, H., Falk, J. D., Dopazo, A., de Lecea, L., Erlander, M. G., and Sutcliffe, J. G. (1994) Isolation of clones of rat striatum-specific mRNAs by directional tag PCR subtraction. *J. Neurosci.* **14,** 4915–4926.
6. Fidock, D. A., Nguyen, T. V., Ribeiro, J.M., Valenzuela, J. G., and James, A. A. (2000) Subtractive hybridization of *Plasmodium falciparum*: construction and validation of a cDNA library enriched in sporozoite-specific transcripts. *Exp. Parasitol.* **95,** 220–225.
7. Sambrook, J., Fritsch, E. F., and Maniatis, T. (1989) *Molecular Cloning: A Laboratory Manual,* 2nd Ed., Cold Spring Harbor Laboratory Press, Cold Spring Harbor, NY.
8. Dugaiczyk, A., Boyer, H. W., and Goodman, H. M. (1975) Ligation of *Eco*RI endonuclease-generated DNA fragments into linear and circular structures. *J. Mol. Biol.* **96,** 171–184.

27

Construction and Screening of YAC Libraries

Cecilia P. Sanchez, Martin Preuss, and Michael Lanzer

1. Introduction

The discovery by Burke et al. in 1987 *(1)* that yeast can accept large pieces of heterologous DNA as yeast artificial chromosomes (YAC) marked a milestone in the analysis of complex genomes. While standard prokaryotic cloning systems, such as plasmids and cosmids, are limited by the size of the DNA fragments they are able to incorporate, YACs allow for the stable integration of exogenous DNA fragments ranging from between 30 kb and several megabases *(2)*. Access to large fragments of cloned DNA has facilitated the physical mapping of large loci of interest, chromosomes, and entire genomes and provided a positional cloning approach to genes of interest. The basis of a YAC are the vector arms that contain all functions necessary for mitotic propagation in yeast, including a centromeric sequence (*CEN*), an autonomous replication sequence (*ARS*), telomere sequences (*TEL*), selectable markers (*TRP1* and *URA3*, mediating tryptophan and uracil autotrophy in suitable yeast hosts) and an interruptible marker containing the *Eco*RI cloning site (*SUP4-o*, mediating a red/white color selection for DNA insertion events). The most frequently used YAC vector is pYAC4 *(1)* which, besides the elements mentioned above, contains the *ColE1* replication origin and the ampicillin-selectable marker for propagation in *Escherichia coli* (**Fig. 1**).

YAC libraries have been generated for numerous organisms *(3)*, including the human malarial parasites *Plasmodium falciparum* and *Plasmodium vivax*, and several clone-specific YAC libraries are available for these two pathogens *(4–7)* (**Table 1**). In the case of *P. falciparum*, YAC technology has provided for the first time a stable source of DNA from this organism. *P. falciparum* DNA exhibits an unusually high A+T content that averages 82% and approaches 90–95% in intergenic regions. As a consequence of its high A+T content *P. falciparum* DNA is unstable in prokaryotic cloning systems and subjected to recombination and deletion events. The application of YAC technology to *P. falciparum* has substantially aided in the analysis of this parasite's chromosome organization as well as its biology, pathogenicity, and virulence. One prominent example is the characterization of multicopy gene families, such as the *var* gene family that encodes clonally variant adhesions, located on the surface of infected erythrocytes, and which mediates cytoadhesion of the parasitized erythrocyte to the endothelial lining of the microvasculature and uninfected erythrocytes *(8–11)*. The physical mapping of entire chromosomes by overlapping YAC clones has revealed that *var*

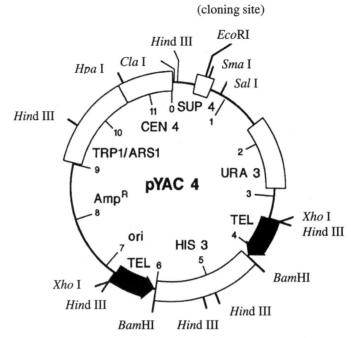

Fig. 1. Schematic drawing of the YAC cloning vector pYAC4. Functional elements as well as important restriction sites are indicated. *CEN4*, centromeric sequence; *ARS1*, autonomous replication sequence; *TEL*, telomere repeat sequences; *TRP1*, *URA3*, *HIS3*, tryptophan, uracil, and histidine selectable markers; *SUP4*, red/white color selection for DNA insertion events; *ori*, bacterial replication origin and *Amp^R*, ampicillin resistance gene.

Table 1
Representative YAC Libraries Derived from the Malarial Parasites, *Plasmodium falciparum* and *Plasmodium vivax*[a]

YAC libraries	Ref.	No. of YACs	Average insert size (kb)	Redundancy (fold)
P. falciparum FCR3	4	1056	100	3–4
P. falciparum 3D7	5	3552	75	9–10
P. falciparum Dd2	6	1440	150	7–8
P. vivax	7	560	180	3–4

[a]In addition, small YAC libraries have been made from DNA of the *P. falciparum* clones HB3, FCR3/B8, and ITO4/A4 *(5)*.

genes are predominantly located at subtelomeric regions *(12–15)* adjacent to *stevor* and *rif* genes, two other large gene families that encode variant antigens *(16,17)*. *P. falciparum* YACs have further aided in the ongoing genome sequencing project of this parasite.

Since its first description in 1987, YAC technology has been improved and simplified *(18,19)*, rendering it a routine technique for laboratories with a background in molecular biology. The construction of a YAC library consists of the following steps: isolation of high-quality genomic DNA; partial digestion of the genomic DNA; preparation of the YAC vector arms; ligation of the vector arms with the partially digested DNA; yeast transformation; and library organization and maintenance. YAC libraries

YAC Libraries

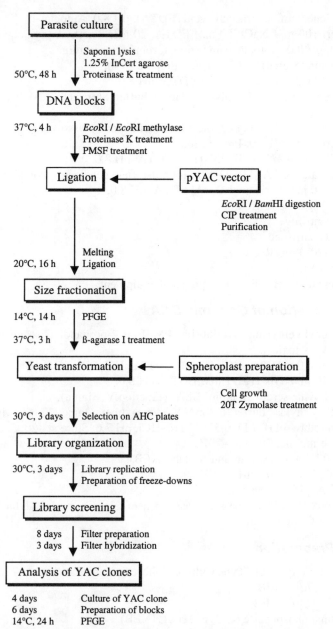

Fig. 2. Flow chart outlining the various steps involved in the generation and analysis of a YAC library.

are analyzed by either hybridization or polymerase chain reaction (PCR)-screening strategies (**Fig. 2**).

2. Materials
2.1. Preparation of Genomic P. falciparum DNA

1. 10X Trager buffer: 140 mM glucose, 580 mM KCl, 70 mM K$_2$HPO$_4$, 570 mM NaCl, 110 mM NaH$_2$CO$_3$, 10 mM NaH$_2$PO$_4$. Sterilize by filtration using a 0.22-μm filter unit (Nalgene). Store at 4°C.

2. 0.5 M ethylenediaminetetraacetic acid (EDTA), pH 8.0. Autoclave.
3. TSE buffer: 100 mM NaCl, 50 mM EDTA, 20 mM Tris(hydroxymethyl)aminomethane (Tris)-HCl, pH 8.0. Autoclave and store at room temperature.
4. 0.15% saponin (Sigma) in TSE buffer. Prepare fresh.
5. TE buffer: 10 mM Tris-HCl, 1 mM EDTA, pH 8.0.
6. 1.25% InCert agarose (FMC) dissolved in TE buffer. Store at 4°C. Melt at 68°C and equilibrate at 40°C prior to use.
7. 2 mg/mL proteinase K (Roche Molecular Biochemicals) dissolved in 0.5 M EDTA, 1% N-lauryl-sarcosinate, pH 8.0. Prepare fresh.
8. $T_{10}E_{50}$ buffer: 10 mM Tris-HCl, 50 mM EDTA, pH 8.0.
9. 10X Tris-borate-EDTA (TBE) buffer for pulse-field gel electrophoresis (PFGE): 890 mM Tris-base, 890 mM boric acid, 2 mM EDTA, pH 8.0.
10. SeaKem LE agarose (FMC).
11. Benchtop centrifuge.
12. High-speed centrifuge.
13. Castings molds (Bio-Rad).
14. 50°C water bath.
15. Chef III apparatus (Bio-Rad) or equivalent equipment.

2.2. Partial Digestion of Genomic DNA

1. 100 mM phenylmethylsulfonyl fluoride (PMSF) in 2-propanol. Dissolve at 68°C and store at –20°C. (Very toxic!).
2. $T_{10}E_{50}$ buffer: 10 mM Tris-HCl, 50 mM EDTA, pH 8.0.
3. TE buffer: 10 mM Tris-HCl, 1 mM EDTA, pH 8.0.
4. 10 mg/mL bovine serum albumin (BSA) (Fraction V, Sigma).
5. 10X *Eco*RI methylase buffer: 0.8 mM S-adenosylmethionine, 20 mM MgCl$_2$, 1 M NaCl, 10 mM dithiothreitol (DTT), 0.5 mM Tris-HCl, pH 7.6. Store at –20°C.
6. 100 mM spermidine. Store at –20°C.
7. *Eco*RI at 20 U/μL (New England Biolabs, NEB).
8. *Eco*RI methylase at 40 U/μL (NEB).
9. 0.5 M EDTA, pH 8.0. Autoclave.
10. 10 mg/mL proteinase K (Roche Molecular Biochemicals) dissolved in TE buffer. Prepare fresh.
11. 37°C and 50°C water baths.

2.3. Vector Preparation

1. pYAC4 vector (American Type Culture Collection, registration no. 67379).
2. *Eco*RI at 20 U/μL (NEB).
3. *Bam*HI at 20 U/μL (NEB).
4. Calf intestinal phosphatase (CIP) at 1U/μL (NEB).
5. 10X CIP buffer (NEB).
6. TE-saturated phenol/chloroform (1:1 v/v). Prepare fresh.
7. 100% Ethanol.
8. TE buffer: 10 mM Tris-HCl, 1 mM EDTA, pH 8.0.
9. T4 polynucleotide kinase at 10 U/μL (NEB).
10. 37°C water bath.
11. 68°C heating block.
12. Gel electrophoresis equipment.

2.4. Ligation

1. 1X ligation buffer: 30 mM NaCl, 10 mM MgCl$_2$, 0.75 mM spermidine, 0.3 mM spermine, 50 mM Tris-HCl, pH 7.6. Store at –20°C.

2. T4 polynucleotide kinase at 10 U/μL (NEB).
3. 100 mM adenosine triphosphate (ATP), pH 7.5. Store at –20°C.
4. 1 M DTT.
5. T4 DNA ligase at 400 U/μL (NEB).
6. 20°C and 37°C water baths.
7. 68°C heating block.
8. Gel electrophoresis equipment.

2.5. Size Fractionation

1. SeaKem low gelling agarose, molecular biology grade (FMC).
2. 10X TBE buffer for pulse-field gel electrophoresis (PFGE): 890 mM Tris-base, 890 mM boric acid, 2 mM EDTA, pH 8.0.
3. Yeast chromosome size marker (NEB).
4. 10 mg/mL ethidium bromide solution (Mutagenic!).
5. ß-agarase I at 1 U/μL (NEB).
6. Chef III apparatus (Bio-Rad) or equivalent equipment.
7. UV transilluminator and gel documentation system.
8. 68°C heating block.
9. 40°C water bath.

2.6. Yeast Spheroplast Preparation and Transformation

1. *Saccharomyces cerevisiae* strain AB1380 (American Type Culture Collection, reference no. 20843).
2. YPD medium: 1% Bacto-yeast extract (Difco Laboratories), 2% Bacto-peptone (Difco Laboratories), 2% glucose, pH to 5.8. After autoclaving, add the appropriate amount of filter-sterilized 40% glucose.
3. YPD agar: YPD medium, 2% Bacto-agar (Difco Laboratories).
4. 0.5% adenine hemisulfate (Sigma). Sterile filter.
5. 1 M sorbitol: Autoclave and store at 4°C.
6. Na-phosphate buffer, pH 7.6: 84.5 mL 1 M Na$_2$HPO$_4$ and 15.5 mL 1 M NaH$_2$PO$_4$ in 100 mL H$_2$O.
7. SPEM buffer: 1 M sorbitol, 0.1 M Na-phosphate buffer, pH 7.6, 10 mM EDTA, pH 8.0. Sterile filter. Add 30 mM of ß-mercaptoethanol prior to use.
8. 10 mg/mL of 20T Zymolase (ICN Inmunochemicals) dissolved in 10 mM NaH$_2$PO$_4$, pH 7.5. Store in 100-μL aliquots at –70°C. Use a fresh aliquot for each transformation.
9. STC solution: 1 M sorbitol, 10 mM CaCl$_2$, 10 mM Tris-HCl, pH 7.5. Sterile filter and store at 4°C.
10. PEG solution: 20% polyethylene glycol mol wt 8000, 10 mM CaCl$_2$, 10 mM Tris-HCl, pH 7.5. After autoclaving, the appropriate amount of a sterile 1 M CaCl$_2$ solution is added. A white precipitate may form in the solution on the addition CaCl$_2$. However, this does not seem to interfere with the transformation efficiency.
11. SOS medium: 1 M sorbitol, 6.5 mM CaCl$_2$, 0.25% Bacto-yeast extract (Difco Laboratories), 0.5% Bacto-peptone (Difco Laboratories). Filter sterilize and store at room temperature. Before use, add 20 μg/mL adenine, 20 μg/mL histidine, 30 μg/mL leucine, 30 μg/mL lysine, 20 μg/mL tryptophan, and 20 μg/mL uracil.
12. AHC medium: 1 M sorbitol, 0.17% Bacto-yeast nitrogen base without amino acids (Difco Laboratories), 1% casamino acids (Difco Laboratories), 20 μg/mL adenine hemisulfate. Add filter-sterilized glucose to a final concentration of 2% after autoclaving.
13. AHC-Top agar: AHC medium, 5% Bacto-agar (Difco Laboratories).
14. AHC agar: AHC medium, 2% Bacto-agar (Difco Laboratories).
15. 96- or 384-Well microtiter plates (NUNC).
16. 30°C shaking water bath.

17. Benchtop centrifuge.
18. 20°C, 37°C, and 50°C water baths.
19. 30°C incubator.

2.7. Replication and Long-Term Storage of YAC Library

1. YPD medium: 1% Bacto-yeast extract (Difco Laboratories), 2% Bacto-peptone (Difco Laboratories), 2% glucose. Adjust pH to 5.8. After autoclaving, add the appropriate amount of filter-sterilized 40% glucose.
2. YPD medium/glycerol (2:3).
3. 96 or 384 replicator (Sigma).
4. 30°C incubator.

2.8. YAC Filter Preparation and Analysis

1. YPD medium: 1% Bacto-yeast extract (Difco Laboratories), 2% Bacto-peptone (Difco Laboratories), 2% glucose. Adjust pH to 5.8. After autoclaving, add the appropriate amount of filter-sterilized 40% glucose.
2. YPD agar: YPD medium, 2% Bacto-agar (Difco Laboratories).
3. 20% Ca-propionate solution (Sigma). Filter-sterilize.
4. 4 mg/mL Novozyme (Sigma) dissolved in 1 M sorbitol, 0.1 M Na-citrate, 10 mM DTT, 10 mM EDTA, pH 8.0. Prepare fresh.
5. Denaturing solution: 0.5 M NaOH, 1.5 M NaCl.
6. Neutralizing solution: 1 M Tris-HCl, pH 7.4, 1.5 M NaCl.
7. 0.25 mg/mL proteinase K dissolved in 0.15 M NaCl, 0.1 M Tris-HCl, pH 7.4. Prepare fresh.
8. 23 × 23 cm culture plates (Nunc).
9. 3MM Whatman paper.
10. 22 × 22 cm Nylon transfer membrane Hybond-N+ (Amersham).
11. DNA UV crosslinker.
12. Replicator.
13. 30°C and 37°C incubators.

2.9. YAC Filter Hybridization

1. 1 M sodium phosphate buffer, pH 7.2: 68.4 mL of 1 M Na$_2$HPO$_4$ and 31.6 mL of 1 M NaH$_2$PO$_4$ in 100 mL H$_2$O.
2. Church buffer: 7% sodium dodecyl sulfate (SDS), 1 mM EDTA, 0.5 M Na-phosphate buffer, pH 7.2.
3. 10 mg/mL of yeast tRNA (Sigma) dissolved in TE buffer, 1% SDS.
4. DNA labeling kit (e.g., random priming DNA labeling kit from Roche Molecular Biochemicals).
5. Alpha-[^{32}P]dATP, 3000 Ci/mmol (Amersham).
6. Washing buffer: 0.1% SDS, 40 mM Na-phosphate, pH 7.2.
7. Stripping buffer: 0.1% SDS, 5 mM Na-phosphate, pH 7.2.
8. 94°C heating block.
9. Hybridization oven.
10. 90°C water bath.
11. Radiographic film, O-Max film (Kodak).

2.10. Analysis of YAC Clones by PFGE

1. AHC medium: 1 M sorbitol, 0.17% Bacto-yeast nitrogen base without amino acids (DIFCO Laboratories), 1% casamino acids (DIFCO Laboratories), 20 µg/mL adenine hemisulfate. Add filter-sterilized glucose to a final concentration of 2% after autoclaving.
2. AHC agar: AHC medium, 2% Bacto-agar (Difco Laboratories).

3. 1 M sorbitol. Autoclave and store at 4°C.
4. SPEM buffer: 1 M sorbitol, 0.1 M Na-phosphate, 10 mM EDTA, pH 8.0. Sterile filter. Add 30 mM of ß-mercaptoethanol prior to use.
5. 30 mg/mL of 20T Zymolase (ICN Inmunochemicals) dissolved in 10 mM phosphate buffer, pH 7.5. Prepare fresh.
6. 1.25% InCert agarose (FMC) dissolved in TE buffer.
7. 2 mg/mL of proteinase K (Roche Molecular Biochemicals) dissolved in 1% N-lauryl sarcosinate, 0.5 M EDTA, pH 8.0. Prepare fresh.
8. $T_{10}E_{50}$ buffer: 10 mM Tris-HCl, 50 mM EDTA, pH 8.0.
9. 10X TBE buffer for pulse-field gel electrophoresis (PFGE): 890 mM Tris-base, 890 mM boric acid, 2 mM EDTA, pH 8.0.
10. SeaKem LE agarose (FMC).
11. Yeast chromosome size marker (NEB).
12. 10 mg/mL ethidium bromide solution (mutagenic!).
13. 30°C incubator.
14. 30°C rotating shaker.
15. Casting molds (Bio-Rad).
16. Benchtop centrifuge.
17. Chef III apparatus (BioRad) or equivalent equipment.

3. Methods

3.1. Preparation of High Quality P. falciparum DNA

DNA is a fragile polymer that easily breaks when subjected to mechanical stress. In order to generate high molecular weight DNA, cells are embedded in agarose blocks and all subsequent manipulation steps are carried out using DNA embedded in agarose. *P. falciparum* is maintained in tissue culture as described by Trager and Jensen *(20)*.

1. Harvest cells (~5 × 10^{10} erythrocytes) at a parasitemia of approx 5%.
2. Wash the cells in 1X Trager buffer.
3. Measure the volume of the cell pellet and add 2 vol of a 0.15% saponin solution in order to lyse the erythrocytes. Incubate at room temperature for 5 min.
4. Add 10 vol of TSE buffer and collect the parasites by centrifugation (12,000g at 4°C for 10 min).
5. Wash parasite pellet in TSE buffer and collect by centrifugation as described above.
6. Resuspend parasites in TSE buffer at a concentration of ~5 × 10^8 parasites/mL.
7. Add an equal volume of melted 1.25% InCert agarose and mix gently.
8. Fill each casting mold with 100 µL of cell/agarose suspension. Keep the molds on ice for 5 min.
9. Place agarose blocks in proteinase K solution and incubate at 50°C for 48 h. Use 0.5 mL of solution for each block.
10. Wash blocks twice in $T_{10}E_{50}$ buffer for 15 min while gently shaking.
11. Store blocks in $T_{10}E_{50}$ buffer at 4°C until use.
12. Check the quality of the DNA by PFGE analysis (0.4% SeaKem LE agarose gel; 0.5X TBE buffer; voltage, 2.7 V/cm; running time, 36 h; initial switching time, 120 s; final switching time, 720 s; temperature, 18°C). Each block should contain between 5–10 µg of genomic DNA.

3.2. Partial Digestion of Genomic DNA

Prior to digestion with restriction endonuclease, the blocks containing the DNA of interest need to be treated with PMSF in order to inactivate any residual proteinase K from the DNA preparation. Partially digested genomic DNA for YAC cloning is then generated using a mixture of the restriction endonuclease *Eco*RI and *Eco*RI methylase. The proper conditions need to be determined experimentally for each DNA preparation by titrating both enzymes against each other. The ratio of both enzymes determines the average DNA size, which, for YAC cloning, should range from between 200 and 500 kb. It is recommended to start initially with 1 U of *Eco*RI and 30 U of *Eco*RI methylase per agarose block (5–10 µg of DNA) and then stepwise increase the concentration of the methylase until a ratio of 1:320 is reached. In some cases it may be also necessary to titrate the concentration of *Eco*RI in order to generate DNA fragments of the desired average size.

1. Dialyze blocks twice in $T_{10}E_{50}$ buffer containing 1 mM of PMSF (1 mL per block) for 30 min. Shake gently.
2. Wash blocks three times in $T_{10}E_{50}$ buffer (1 mL per block) for 30 min. Shake gently.
3. Wash blocks three times in $T_{10}E_{1}$ buffer (1 mL per block) for 30 min. Shake gently.
4. Transfer block into a sterile Eppendorf tube and add:
    a. 50 µL of BSA (10 mg/mL).
    b. 50 µL of 10X *Eco*RI methylase buffer.
    c. 13 µL of spermidine (100 mM).
    d. *Eco*RI restriction endonuclease.
    e. *Eco*RI methylase.
    f. Sterile H_2O to a final volume of 500 µL.
5. Incubate at 37°C for 4 h.
6. Stop the reaction by adding:
    a. 55 µL of 0.5 M EDTA, pH 8.0.
    b. 62 µL of proteinase K solution (10 mg/mL).
7. Incubate at 50°C for 90 min.
8. Wash the blocks once in TE buffer at 50°C for 30 min. Shake gently.
9. Incubate the blocks in TE buffer containing 1 mM PMSF at 50°C for 30 min, to inactive the proteinase K (1 mL of solution per block). Shake gently.
10. Wash the block twice in TE buffer at 50°C for 30 min.

The genomic DNA is now ready for ligation.

3.3. YAC Vector Preparation

The left and right YAC cloning vector arms are liberated from the plasmid pYAC4 by digestion with the restriction endonucleases *Eco*RI and *Bam*HI. pYAC4 can be propagated in suitable *E. coli* strains, such as DH5α.

1. Cleave purified pYAC4 plasmid DNA with *Eco*RI and *Bam*HI to completion.
2. Denature restriction endonucleases by heat treatment at 68°C for 10 min.
3. Dephosphorylate free 5' ends using calf intestinal phosphatase CIP (0.05 U/µg DNA) in 1X CIP buffer. Incubate at 37°C for 30 min.
4. Add EDTA, pH 8.0, to a final concentration of 5 mM and incubate the reaction at 68°C for 10 min to inactivate the CIP.
5. Purify the DNA by phenol/chloroform extraction followed by ethanol precipitation.
6. Resuspend the DNA in TE buffer at a final concentration of 2 µg/µL.

Digestion of the vector arms is monitored electrophoretically. Dephosphorylation is verified in ligation assays with and without the addition of T4 polynucleotide kinase.

3.4. Ligation

1. Wash agarose blocks three times in 1X ligation buffer at room temperature for 30 min.

2. Remove all liquid and place six agarose blocks into a 1.5-mL Eppendorf tube.
3. Add an equal amount of premade vector arm DNA. Assuming that each of the six agarose blocks contains 5–10 μg of genomic DNA, approx 60 μg of premade vector arm DNA are required. The mass ratio of 1:1 between genomic and vector arm DNA translates approximately into a 500-fold molar excess of vector arms versus genomic DNA.
4. Melt agarose blocks at 68°C for 10 min.
5. Allow genomic and vector arm DNA to equilibrate at 37°C for 2 h.
6. Add 1/10 vol of premade ligation mix (22 U/μL T4 DNA ligase, 5 mM ATP, 100 mM DTT in 1X ligation buffer).
7. Remove 10 μL of the sample and add 10 U of T4 polynucleotide kinase to it in order to verify ligation conditions. Self-ligation of the vector arms will occur and is monitored electrophoretically.
8. Incubate ligation reaction at 20°C for 16 h. The agarose will solidify during this time.

3.5. Size Fractionation

A size fractionation of the ligation reaction is highly recommended to remove the excess vector arms as well as small YACs that would otherwise be overrepresented in the YAC library due to their higher transformation efficiency. Size selection is carried out using PFGE.

1. Prepare a 1% agarose gel in 0.5X TBE, using a low gelling agarose such as SeaKem molecular grade. Tape up several teeth of the gel comb in order to generate a well large enough for the ligation sample. Load yeast chromosome size markers in the outside wells and seal all wells with molten agarose.
2. Run PFGE (voltage, 4.5 V/cm; running time, 18 h; initial switching time, 5 s; final switching time, 25 s; temperature, 14°C; field angle 120°; running buffer, 0.5X TBE)
3. After electrophoresis, cut off the gel lanes containing the size markers and a very small part of the lane containing the sample. Stain these pieces in 500 mL of running buffer containing 80 μL of a 10 μg/mL ethidium bromide solution for 30 min and examine them under UV light.
4. Mark the area of the gel that holds DNA fragments ranging from between 100 and 500 kb.
5. Reassemble the gel pieces with the rest of the gel, which was kept in running buffer during this time to avoid drying out.
6. Excise the region containing the high molecular weight DNA, using a sterile glass cover slip.
7. Place this gel slice into a 15-mL Falcon tube. Store at 4°C until use.
8. Stain the entire gel as described in **step 3**, examine by UV light and take a picture for documentation.
9. Melt the gel slice containing the size-fractionated DNA at 68°C for 10 min.
10. Allow the sample to cool to 40°C, add ß-agarase I to a final concentration of 50 U/mL and incubate at 40°C for 3 h. The DNA is now ready for transformation into yeast spheroplasts.

3.6. Yeast Spheroplast Preparation and Transformation

Saccharomyces cereviseae strain AB1380 is used as the host in YAC (*1*). As a selective medium we routinely use AHC-medium. The AHC medium is as selective as synthetic minimal media, allows for red/white color selection of positive transformants, but is much easier to prepare (*see* **Note 2**).

1. For high transformation efficiency, plate AB1380 yeast cells from a frozen glycerol stock onto a YPD plate and incubate the plate at 30°C for 48 h.
2. Take a single red colony and prepare an overnight culture in YPD medium supplemented with 0.002% adenine hemisulfate.
3. Inoculate 50 mL of YPD medium supplemented with 0.002% adenine hemisulfate with

the fresh overnight culture to an OD_{600} of 1.0. Incubate the culture at 30°C until an OD_{600} of 4.0 is reached. This corresponds to approx 3×10^7 cells/mL. The cell density should double every 90 min.
4. Collect the cells by centrifugation at 400–600g for 10 min at room temperature.
5. Wash the cells once with 20 mL sterile water, centrifuge as above.
6. Wash once with 20 mL of 1 M sorbitol, centrifuge as above.
7. Resuspend the cells in 20 mL of 30°C prewarmed SPEM buffer containing fresh ß-mercaptoethanol. The cell density should be close to 7.5×10^7 cells/mL. Remove sample, dilute it 10-fold in water and determine the OD_{600}. The OD_{600}-value obtained serves as the prespheroplast reference.
8. Add 45 µL of 20T Zymolase solution. Incubate cells by slowly shaking in a 30°C water bath for about 20 min. The extent of spheroplast formation is determined spectrophotometrically. Every 5 min a 50 µL aliquot is removed, 10-fold diluted in water and the OD_{600} value determined. As spheroplasts lyse in water due to the lack of a cell wall, the reduction in the OD_{600}-value is directly proportional to the amount of spheroplasts formed. 80–90% of the cells must be spheroplasts within 20 min to achieve optimal transformation efficiency. If not, it is strongly recommended to repeat the preparation.
9. Collect the spheroplasts by centrifugation at 200–300g for 4 min at room temperature.
10. Gently resuspend the cells in 20 mL of 1 M sorbitol, and centrifuge as described in **step 9**.
11. Wash the spheroplasts once in 20 mL of STC, centrifuge as above and finally resuspend them in 2 mL of STC. At this point the spheroplasts are stable at room temperature for at least an hour. The cell density is now determined microscopically using a hematocytometer. Adjust the volume to a final cell concentration of 6.5×10^8 cells/mL.
12. To 150 µL of spheroplasts, add 50 µL of ß-agarase I treated size-fractionated DNA (approx 80 ng) in a 15-mL Falcon tube. Incubate at 20°C for 10 min. Scaling up the reaction significantly reduces transformation efficiency (*see* **Note 3**).
13. Add 1.5 mL of PEG solution and mix by gently inverting the tube. Incubate at 20°C for 10 min.
14. Immediately centrifuge at 200–300g for 4 min at room temperature.
15. Carefully decant the supernatant and resuspend cells in 225 µL of SOS solution. Incubate at 30°C for 30 min.
16. Gently resuspend the settled spheroplasts. Add 8 mL of molten AHC-Top agar (prewarmed to 50°C) and gently invert to mix. Quickly pour onto an AHC plate that has been prewarmed to 37°C. For yeast transformation, AHC-Top agar and AHC plates need to contain 1 M sorbitol to osmotically buffer the spheroplasts. Allow plates to sit for 10 min at room temperature.
17. Incubate plates at 30°C for 7 d. Yeast clones are visible after 2–3 d. Those yeast clones that contain artificial chromosomes turn red on AHC-plates, while revertants and clones harboring pYAC4 remain white.
18. Red colonies are picked and transferred to 96- or 384-well microtiter plates containing 150 µL or 80 µL of YPD medium, respectively. Cells are grown for 2–3 d at 30°C.

3.7. Replication and Long-Term Storage of YAC Library

1. Using a 96 or 384 replicator, transfer YAC clones to a new set of microtiter plates containing an appropriate amount of YPD media in each well. Use a sterile replicator for each microtiter plate to avoid cross-contamination of YAC clones. Metal replicators can be sterilized as follows: Dip the pins of the replicator in 80% ethanol, flame, and then cool for several sec on a sterile YPD agar plate.
2. Grow YAC clones for 2–3 d at 30°C. During incubation approximately one-third of the media will evaporate. The YAC library can be stored at 4°C for up to 3 mo or be used to make frozen glycerol stocks.
3. To freeze down the YAC clones, add an appropriate volume of YPD media/glycerol (2:3

YAC Libraries

v/v) to each well to yield a final glycerol concentration of 20%. In the case of 96-well plates approx 50 µL YPD media/glycerol is usually added. Resuspend the cells in the medium, using a multichannel pipet.
4. Cells can now be stored at –80°C.

3.8. Analysis of YAC Libraries by Filter Hybridization

YAC libraries can be analyzed by PCR *(22)* or filter hybridization *(23)*. Although both techniques have been successfully applied to YAC libraries derived from malarial parasites *(5,12,15,24,25)*, filter hybridization is the method of choice because of convenience and speed. The analysis of a YAC library by filter hybridization involves the following steps: transfer of the yeast clones onto a nylon filter; processing of the filters to prepare the DNA; and subsequently screening of the YAC filter by hybridization.

1. Prepare a fresh replicate of the YAC library. Grow cells in YPD medium for 2 d at 30°C.
2. Prepare YPD agar plates. We usually use large square plates (23 × 23 cm) because they can hold the clones of up to six microtiter plates.
3. For each 22 × 22 cm plate, prepare four 3MM papers and one nylon filter of 22 × 22 cm in size. While the nylon filters can be bought sterile, 3MM paper needs to be sterilized prior to usage by either autoclaving or exposure to UV light. Most commercially available crosslinkers have preset conditions for sterilization.
4. Mark the top left corner of the 22 × 22 cm nylon filter using a pencil.
5. Soak two 3MM papers for each nylon filter in 200 mL of YPD media supplemented with 2.5 mL of a 20% Ca-propionate solution.
6. Place the two layers of wet 3MM paper into the plate cover and lie the nylon filter on top. Wait until the nylon filter is moist.
7. Now stamp the YAC clones onto the nylon filter, using the appropriate replicator (*see* **Note 4**).
8. Remove the filter and place it, colonies facing up, onto the YPD agar plate.
9. Incubate the plates inverted at 30°C for 2–3 d.
10. Place the filters YAC colony side up onto two layers of 3MM paper soaked in 50 mL of Novozyme solution. Novozyme digests the cell wall. Use the plate cover as an incubation chamber and the plate bottom as the cover after removal of the YPD agar.
11. Seal the plate with Saran wrap and incubate at 37°C overnight.
12. Place the nylon filter onto two layers of 3 MM paper soaked in denaturing solution. Incubate at room temperature for 15 min.
13. Let the nylon filters dry on fresh 3 MM paper for 10 min at room temperature.
14. Transfer the nylon filters to two layers of 3 MM paper soaked in neutralizing solution. Incubate at room temperature for 10–30 min.
15. Place the nylon filter in a 22 × 22 cm plate containing 250 mL of the proteinase K solution. Incubation at 37°C for 2–3 h. During this time the YAC clones will lose their reddish color.
16. Dry the membranes for 36 h on 3MM paper at room temperature.
17. UV-crosslink the YAC DNA to the nylon filter in a UV crosslinking chamber, according to the manufacturer's recommendations.

3.9. YAC Filter Hybridization

1. Preincubate the YAC filters at 50°C overnight in Church buffer supplemented with 0.2 mg/mL of yeast tRNA. Use approx 0.2 mL of church buffer per cm^2 of nylon filter.
2. Prepare a radiolabeled probe and denature the probe by heat treatment at 94°C for 10 min.
3. Add the radiolabeled probe to the prehybridized nylon filter. For best signal-to-noise ratio use 750,000–1,000,000 cpm of probe per milliliter of Church buffer. Incubate at 50°C overnight.
4. Wash the nylon filters in washing buffer at room temperature for 5 min.

5. Wash the nylon filters in prewarmed washing buffer at 65°C for 15 min.
6. Wrap the damp filters in Saran wrap and expose them to X-ray imaging film.
7. Filters can be washed more stringently if necessary and can be reused several times. Remove bound radioactive probes by incubating the filters in stripping buffer at 90°C for 30 min. Do not allow the filters to dry.

3.10. Analysis of Individual YAC Clones by PFGE

1. Streak the YAC clone of interest out on an AHC agar plate. Incubate at 30°C for 3 d.
2. Inoculate 20 mL of AHC media with the YAC clone and incubate at 30°C for 2 d while shaking at 175 rpm.
3. Collect cells by centrifugation at 400–600g for 10 min at 4°C.
4. Wash the cell pellet in 5 mL of water and centrifuge as described above.
5. Wash the cells in 5 mL of 1 M sorbitol and centrifuge as described above.
6. Resuspend the cells in 250 µL of SPEM, add 30 µL of 20T Zymolase solution.
7. Add immediately 400 µL of 37°C preequilibrated 1.25% InCert agarose, mix and add 100 µL to each casting mold.
8. Leave the blocks on ice for 5 min to solidify.
9. Remove the hardened agarose blocks from the casting molds and incubate them in 5 mL of SPEM supplemented with 50 µL of 20T Zymolase solution at 37°C for 7–8 h.
10. Replace the SPEM/Zymolase solution with 5 mL of proteinase K solution and digest the blocks at 50°C for 48 h.
11. Place the blocks in $T_{10}E_{50}$ for storage at 4°C. If embedded DNA is to be used for restriction analysis, follow **Subheading 3.2., steps 1–3**.
12. For PFGE analysis, prepare a 1.0% SeaKem LE agarose gel in 0.5X TBE buffer.
13. Carefully place the agarose blocks in the wells using glass cover slips, load a DNA size standard and seal all wells (including the empty ones) with LE agarose.
14. Run the PFGE using the following conditions: running buffer, 0.5X TBE buffer; voltage, 5.5 V/cm; running time, 24 h; initial switching time, 5 s; final switching time, 25 s; temperature, 14°C; field angle 120°.
15. Stain the gel in 500 mL of running buffer containing 80 µL of 10 µg/mL ethidium bromide for 30 min.

4. Notes

1. The success of a YAC library construction rises and falls with the quality and purity of the starting DNA material. DNA that is partially degraded, enzymatically or chemically modified will never produce a YAC library. Therefore, exposure of the DNA to UV light or X-rays is to be avoided as both conditions introduce strand breakages, depurination, and crosslinking events into the DNA. Exposure to X-rays becomes a problem when DNA is shipped by air freight as all packages, not only hand luggage and carry-ons, are routinely X-rayed for safety inspections. In some cases it may be necessary to collect parasite material from patients, for instance when an *in vitro* culture system is not available. An example is the malarial parasite *P. vivax*, which propagates within reticulocytes. As immature erythrocytes cannot be readily obtained in large quantities, in vitro culture conditions for *P. vivax* are difficult to establish. Access to parasite DNA, therefore, relies on either material obtained from *P. vivax* infected patients or monkeys. If parasite material for library construction is collected from host organisms, great care needs to be taken in order to remove any contaminating nucleated cells from the host prior to DNA preparation. For the construction of a *P. vivax* YAC library, we devised a two-step purification protocol *(6)*. Host leukocytes were initially removed from the erythrocytes using Plasmodipur filters (Organon Teknika). Erythrocytes infected with *P. vivax* were then concentrated using a single

step 16% Nycodenz gradient. Nycodenz (Sigma) is resuspended in PBS and centrifugation is carried out at 900g for 30 min at 15°C. The resulting material was found to be free of contaminating human DNA, as verified by both Southern and PCR analyses *(6)*.
2. It has been observed that YAC transformants often do not grow on synthetic minimal medium selective for both tryptophan and uracil autotrophy. This phenomenon has been attributed to the weak *TRP* promoter which limits expression of the *TRP* gene product. To overcome this problem selective pressure can be applied sequentially, first for uracil and subsequently for tryptophan autotrophy. Alternatively, transformants can be selected on AHC-medium as described herein. AHC medium is initially rich in amino acids. However, autoclaving breaks down most of the tryptophan. The residual amounts of tryptophan do not support permissive growth of trypothan auxotrophic clones, yet they are sufficient to allow the cells to recuperate after transformation and express the *TRP* gene encoded by the YAC vector arm.
3. It is tempting to scale up the yeast transformation reaction. However, there is ample evidence that scaling up the reaction significantly reduces transfromation efficiency. While processing and handling many reaction tubes at the same time is cumbersome and time consuming, it is still faster than repeating the experiment.
4. YAC clones can be transferred to nylon filters manually, using a replicator as described herein. This technique produces filters of high quality without the need of sophisticated and expensive equipment. However, if a larger number of filters are needed, as may be necessary for a large scale mapping effort, automated devices may be used for filter production. Besides speed, automated systems offer the advantage of arranging YAC clones at a higher density and in duplicates on the filter. This saves material and reagents and simplifies the acquisition and interpretation of the data, which can now proceed using an automated reader.

References

1. Burke, D. T., Carle, G. F., and Olso, M. V. (1987) Cloning of large fragments of exogenous DNA into yeast by means of artificial chromosomes vectors. *Science* **236,** 806–812.
2. Dausset, J., Ougen, P., Abderrahim, H., Billault, A., Sambucy, J. L., Cohen, D., and Le Paslier, D. (1992) The CEPH YAC library. *Behring Inst. Mitt.* **91,** 13–20.
3. Larin, Z., Monaco, A. P., and Lehrach, H. (1991) Yeast artificial chromosome libraries containing large inserts from mouse and human DNA. *Proc. Natl. Acad. Sci. USA* **88,** 4123–4127.
4. de Bruin, D., Lanzer, M., and Ravetch, J. V. (1992) Characterization of yeast artificial chromosomes from *Plasmodium falciparum*: construction of a stable, representative library and cloning of telomeric DNA fragments. *Genomics* **14,** 332–339.
5. Foster, J. and Thompson, J. (1995) The *Plasmodium falciparum* genome project: a resource for researchers. *Parasitol. Today* **11,** 1–4.
6. Camargo, A. A., Fischer, K., and Lanzer, M. (1997) Construction and rapid screening of a representative yeast artificial chromosome library from the *Plasmodium falciparum* stain Dd2. *Parasitol. Res.* **83,** 87–89.
7. Camargo, A. A., Fischer, K., Lanzer, M., and del Portillo, H. A. (1997) Construction and characterization of a *Plasmodium vivax* genomic library in yeast artificial chromosomes. *Genomics* **42,** 467–473.
8. Su, X., Heatwole, V. M., Wertheimer, S. P., Guinet, F., Herrfeldt, D. S., Peterson, D. S., et al. (1995) The large diverse gene family var encodes proteins involved in cytoadherence and antigenic variation of *Plasmodium falciparum*-infected erythrocytes. *Cell* **82,** 89–100.
9. Smith, J. D., Chitnis, C. E., Craig, A. G., Roberts, D. J., Hudson-Taylor, D. E., Peterson, D. S., et al. (1995) Switches in expression of *P. falciparum var* genes correlate with changes in antigenic and cytoadherent phenotypes of infected erythrocytes. *Cell* **82,** 101–110.
10. Baruch, D. I., Gormley, J. A., Ma, C., Howard, R. J., and Pasloske, B. L. (1996) *Plasmodium falciparum* erythrocyte membrane protein 1 is a parasitized erythrocyte receptor for adherence to CD36, thrombospondin, and intracellular adhesion molecule 1. *Proc. Natl. Acad. Sci. USA* **93,** 3497–3502.
11. Rowe, J. A., Moulds, J. M., Newbold, C. I., and Miller, L. H. (1997) *P. falciparum* rosetting mediated by a parasite-variant erythrocyte membrane protein and complement-receptor 1. *Nature* **388,** 292–295.
12. Rubio, J. P., Thompson, J. K., and Cowman, A. F. (1996) The *var* genes of *Plasmodium falciparum* are located in the subtelomeric region of most chromosomes. *EMBO J.* **15,** 4069–4077.

13. Thompson, J. K., Rubio, J. P., Caruana, S., Brockman, A., Wickham, M. E., and Cowman, A. F. (1997) The chromosomal organization of the *Plasmodium falciparum var* gene family is conserved. *Mol. Biochem. Parasitol.* **87,** 49–60.
14. Hernandez-Rivas, R., Mattei, D., Sterkers, Y., Peterson, D. S., Wellems, T. E., and Scherf, A. (1997) Expressed *var* genes are found in *Plasmodium falciparum* subtelomeric regions. *Mol. Cell. Biol.* **17,** 604–611.
15. Fischer, K, Horrocks, P., Preuss, M., Wiesner, J., Wunsch, S., Camargo, A. A., and Lanzer, M. (1997) Expression of *var* genes located within polymorphic subtelomeric domains of *Plasmodium falciparum* chromosomes. *Mol. Cell Biol.* **17,** 3679–3686.
16. Cheng, Q., Cloonan, N., Fischer, K., Thompson, J., Waine, G., Lanzer, M., and Saul, A. (1998). *Stevor* and *rif* are *Plasmodium falciparum* multicopy gene families which potentially encode variant antigens. *Mol. Biochem. Parasitol.* **97,** 161–176.
17. Gardner, M. J., Tettelin, H., Carucci, D. J., Cummings, L. M., Aravind, L., Koonin, E. V., et al. (1998) Chromosome 2 sequence of the human malaria parasite *Plasmodium falciparum. Science* **282,** 1126–1132.
18. Burke, D. T. and Olson, M. V. (1991) Preparation of clone libraries in yeast artificial-chromosome vectors. *Methods Enzymol.* **194,** 251–270.
19. Monaco, A. P. (1992) Pulsed-field gel electrophoresis, in *Methods in Molecular Biology,* vol. 12, (Burmeister, M. and Ulanovsky, L., eds.), Humana Press, Totowa, NJ, pp. 225–234.
20. Trager, W. and Jensen, J. B. (1976) Human malaria parasites in continuous culture. *Science* **193,** 673–675.
21. Green, E. D. and Olson, M. V. (1990) Systematic screening of yeast artificial-chromosome libraries by use of the polymerase chain reaction. *Proc. Natl. Acad. Sci USA* **87,** 1213–1217.
22. Brownstein, B. H., Silverman, G. A., Little, R. D., Burke, D. T., Korsmeyer, S. J., Schlessinger, D., and Olson, M. V. (1989) Isolation of single-copy human genes from a library of yeast artificial chromosome clones. *Science* **244,** 1348–1351.
23. Ross, M. T., Hoheisel, J. D., Monaco, A. P., Larin, Z., Zehetner, G., and Lehrach, H. (1992) High density gridded YAC filters: their potential as genome mapping tools, in *Techniques for the Analysis of Complex Genomes* (Anand, R., ed.), Academic Press, London, pp. 137–153.
24. Lanzer, M., de Bruin, D., and Ravetch, J. V. (1993) Transcriptional differences in polymorphic and conserved domains of a complete cloned *P. falciparum* chromosome. *Nature* **361,** 654–657.
25. Rubio, J. P., Triglia, T., Kemp, D. J., de Bruin, D., Ravetch, J. V., and Cowman, A. F. (1995) A YAC contig map of *Plasmodium falciparum* chromosome 4: characterization of a DNA amplification between two recently separated isolates. *Genomics* **26,** 192–198.

28

Episomal Transformation of *Plasmodium berghei*

Chris J. Janse and Andrew P. Waters

1. Introduction

Genetic transformation of an organism is a fundamental investigational tool that allows the researcher to gain an insight into the basic cellular biology. Numerous areas of parasite biology can be addressed simply by introducing plasmids that are maintained as episomes into the cell of choice and modifying phenotype through the expression of plasmid-borne transgenes. The methodologies to achieve this in a stable drug selectable fashion were developed for malaria parasites 7 yr ago *(1–4)*. These have already yielded insights into gene transcription promoter structure *(3,5)*, reporter gene expression *(2,3,6)*, the requirements for stage-specific localization of malaria proteins *(7,8)*, and organelle targeting *(9)*. Attempts have been made to complement genetic mutants through episomal transgene expression of a normal copy of the defective gene making use of a second selectable maker that is now available *(10,11)*. Little experimentation has yet been performed investigating what is the most obvious application that is the over expression of transgene-encoded proteins creating novel phenotypes such as drug resistance and dominant negative phenotypes. All transfection procedures take 2–3 wk in the *P. berghei* model and the integration of intact episomes under these conditions into the genome of *P. berghei* has not yet been demonstrated. In the human parasite, *P. falciparum* episome integration is a commonly observed mechanism, although a protracted selection procedure after transfection is necessary to obtain integrants *(12–14)*. In *P. berghei*, site-directed integration of linear pieces of DNA is relatively efficient (*see* Chapter 29), whereas integration of intact plasmids might not occur at an observable frequency under normal transfection procedures. Consequently, episomal transfection of *P. berghei* will probably not generate parasites, that might have confusing phenotypes as a result of unexpected integration into the parasite genome and the full range of strategic use of episomal transfection can be considered.

In this chapter, methods are described that are used in our standard protocol for the stable transfection of *P. berghei*. This is based on electroporation of purified mature schizonts with constructs that contain a selectable marker that can be selected in vivo. Transformed parasites are selected in vivo by treating rats or mice with the antimalarial drug pyrimethamine or WR99210 (*see* next page). With these methods both stable episomal transformation and targeted integration of DNA into the genome can be obtained. We will conclude with a brief discussion of recent observations about episome replication and inheritance that might influence parasite handling and experimental design.

Based on the growth kinetics of parasites during the selection of transfected parasites, we believe that the transformation efficiency is low (1000–10^4 transformed parasites in 6×10^8 parasites, assuming a daily 3–5 times multiplication rate in rats). This is 10- to 100-fold higher than the combined frequency of transformation and integration. We have not yet thoroughly tested the many variables in the transfection protocol to increase the efficiency. The protocol described in this chapter will reproducibly generate recombinant parasites harboring the plasmid introduced into the experiment. We have found the methodology to be completely reliable.

Mature schizonts containing fully developed merozoites are suitable target cells for introduction of DNA. Since these stages are no longer dependent for growth on their host cells, damage to the erythrocyte membranes caused by the procedures that introduce DNA into the parasites should not be a problem. Mature schizonts of *P. berghei* can be collected in large numbers by relatively simple culture and purification methods. In addition, the viability of merozoites of *P. berghei* does not rapidly decrease during the manipulation procedures of the schizonts.

Introduction of DNA into schizonts is in *P. berghei* more successful than transfection of the other intraerythrocytic stages, such as ring forms and trophozoites. Stable, episomally transformed parasites can be selected after transfection of these stages. However, at least up to now, we have not been able to achieve stable integration of foreign DNA by means of transfection of intraerythrocytic ring forms and trophozoites.

To date only three selectable markers (and only two selectable activities) exist for the transformation of *P. berghei*: the pyrimethamine resistant form of the *dhfr/ts* gene of *Plasmodium* and of *Toxoplasma gondii* and the pyrimethamine resistant *dhfr* gene of humans *(1,2)*. The latter gene confers not only resistance to pyrimethamine but also to the antimalarial drug, WR99210 *(10,11)*. Resistance to WR99210 can be selected for on a pyrimethamine-resistant background *(11)* and so in theory, episomal-based complementation of mutants created by site directed modification of the *P. berghei* genome is possible. Introduction of all three selectable markers into pyrimethamine sensitive *P. berghei* parasites results in a strong increase (approx 100–1000X) in pyrimethamine resistance which allows for a relatively simple *in vivo* selection procedure in order to obtain transformed parasites *(11)*. The *T. gondii dhfr/ts* gene is preferred to the *Plasmodium* gene as a selectable marker because it reduces (but does not eliminate due to homology in the flanking promoter and downstream regions) the chance of unwanted recombination with the endogenous *Plasmodium* gene and because it may confer resistance to higher levels of pyrimethamine. Pyrimethamine selection *in vivo* is preferred above WR99210 selection, since the latter drug has more toxic side effects in rodents than pyrimethamine. All vectors used to express foreign DNA in *Plasmodium* rely upon *Plasmodium* promoters. No heterologous promoter from species other than *Plasmodium* has been reported to be active in *Plasmodium*. Three heterologous promoters of *P. falciparum* have been tested in *P. berghei* and only the CAM promoter has been active (T. F. de Koning-Ward et al., unpublished results).

We describe here the tools and distinct methods used to achieve the stable transfection with the different selectable markers that are available with subsequent characterization of *P. berghei* parasites and transgene expression. A separate chapter deals with the specific disruption of genes in *P. berghei* (*see* Chapter 29).

2. Materials

2.1. Equipment

1. Low-speed swing-out centrifuge capable of generating 800–1200g and holding 50-mL Falcon tubes. Values for rpm are given for a Beckman GP benchtop centrifuge and should be converted.
2. Standard culture equipment: usually disposable plastics or sterilized 500-mL Ehrlenmeyer flasks.
3. A climate-controlled incubator capable of holding and shaking the flasks. Ideally this should contain a gas inlet and be capable of maintaining humidity.
4. Gas supply (5% CO_2, 10% O_2, 85% N_2) (5% CO_2, 5% O_2, 90% N_2 is also good).
5. Eppendorf benchtop microcentrifuge or equivalent.
6. Light microscope capable of ×1000 magnification (×100 objective, immersion oil) for parasite examination.
7. Glass slides, frosted at one end.
8. Electroporation apparatus capable of delivering 1.1 kV, 25 mF (e.g., Bio-Rad Gene Pulser; BTX ECM-399).
9. Electroporation cuvets (0.4cm gap between electrodes, e.g., Gene Pulser Cuvette, Bio-Rad, cat. no. 165-2088).
10. Disposable hypodermic syringes with 20- and 13-gage needles.
11. Wistar rats, ~200–250 g, female.
12. Swiss mice, female, 20–25 g.
13. Anesthesia equipment capable of delivering isofluorane (e.g., Vapex® 3 calibrated vaporizer, International Market Design, UK).

2.2. Reagents

1. Culture medium: RPMI 1640, supplemented with L-glutamine and 25 mM HEPES, without $NaHCO_3$ (if no HEPES is present, add 4.95 g of HEPES per liter culture medium). To prepare medium, dissolve 10.41 g of RPMI 1640 medium in 1 L of water (add powder slowly under continuous stirring, using a magnetic stirrer); add 1.75 or 2 g of $NaHCO_3$ (and HEPES if necessary); add 50.000 IU of Neomycin (stock-solution of 10,000 IU/mL; Gibco); sterilize by filtration through a 0.2-μm sieve; store frozen at –20°C in 100–200 mL bottle. Just before use of the medium for culture, add fetal calf serum at a concentration of 20% (v/v) to give complete culture medium.
2. Pyrimethamine (Sigma P-7771). For rats: 20 mg in 10 mL of H_2O. Since pyrimethamine is not very soluble in water, add 1 drop of Tween-20 and sonicate in bath for 10 min. Mix suspension well before injection. Inject 1 mL per 200 g rat = 10 mg/kg. For mice: 25 mg in 10 mL DMSO. Inject 0.1 mL per 25 g mice = 10 mg/kg.
3. WR99210 (Gift of David Jacobus, Jacobus Pharmaceuticals, Princeton, NJ): prepare as a 3 mg/mL or 4 mg/mL solution in 70% DMSO/30% H_2O.
4. Nycodenz: NycoPrep 1.150 (Nycomed PharmaAS, Oslo, Norway) 1.150 g/mL density, 290 Osm. To prepare from powder, dissolve 27.6 g of solid Nycodenz in 60 mL of buffered medium (5 mM Tris-HCl, pH 7.5 containing 3 mM KCl and 0.3 mM Na_2EDTA) and made up to 100 mL with the same medium (density at 20°C, 1.15 g/mL). Autoclave at 120°C for 20 min and store at 4°C.
5. Giemsa stain (Merck): Prepare two drops Giemsa per 1 mL of H_2O and use 4 mL per slide; fix 1 s in methanol, dry under fan, and stain for 25 min, Alternatively, for faster staining, prepare 4 drops Giemsa/mL H_2O and stain for 12 min. Rinse with tap water, and dry under fan (*see* **Note 1**).
6. Cytomix: 120 mM KCl, 0.15 mM $CaCl_2$, 10 mM K_2HPO_4/KH_2PO_4 pH7.6, 25 mM HEPES pH 7.6, 2 mM EGTA pH 7.6, 5 mM $MgCl_2$, adjust pH with KOH.

7. Appropriate plasmid DNA, 50 mg: prepared by column purification using standard methodologies.

3. Methods
3.1. Time Schedule Overview
1. *Day 1:* infect donor rat with *P. berghei*.
2. *Day 2:* treat recipient rat with phenylhydrazine.
3. *Day 4 or 5:* schizont isolation, electroporation and injection in recipient rat.
4. *Day 6–9:* pyrimethamine treatment.
5. *Day 7–13:* analysis for developing parasitemia and transfer to mice.

3.2. Handling of Animals

In general, personnel who hold the appropriate, nationally approved license for such work must carry out the handling and injection of animals. We routinely anesthetize the animals to ensure that the procedures are carried out correctly with minimal stress inflicted upon the animal. Check and conform to your national guidelines for the care of laboratory animals.

3.3. Infection of Rats (Day 1)

A rat is infected with *P. berghei* to serve as a source of blood-stage parasites for the culture and purification of schizonts. The schizont is the developmental stage that is used for introduction of foreign DNA by electroporation.

Infect a Wistar rat (200–250 g) via intraperitoneal inoculation with $1–2 \times 10^7$ infected erythrocytes on day 1 (in general, d 1 is a Thursday). Obtain infected erythrocytes from an infected mouse with a parasitemia between 5 and 15%. Dilute approx 8 drops of blood (20–30 µL) from either the tail or from the heart of the mouse in 1 mL of PBS (room temperature). Directly inject 0.5 mL of the suspension intraperitoneally (ip) (two injections, one on each side of the peritoneal cavity) using a 2-mL syringe. In general, the rat will have a parasitemia of 1–3% on d 4 or 5.

3.4. Phenylhydrazine Treatment of the Recipient Rat (Day 2)

The phenylhydrazine treatment initiates reticulocytosis in the recipient animal and provides an enriched source of target cells for the reticulocyte preferent *P. berghei* parasites (*see* **Note 2**).

Treat a Wistar rat (200–250 g) on d 2 with one ip injection of 0.30–0.35 mL of phenylhydrazine-HCl stock-solution (30–40 mg/kg bodyweight), using a 1-mL syringe.

3.5. In Vitro Culture of Schizonts (Day 4 or 5)

The conditions described here are developed for short-term cultures to allow for the development of ring forms into mature schizonts. In these cultures, reinvasion of merozoites is limited to a minimum due to the inability of mature schizonts of *P. berghei* to burst spontaneously and the low number of reticulocytes present in the cultures (*see* **Note 3**).

1. Collect 5–10 mL of infected heart blood with a parasitemia of 1–3% between 10:00 and 16:00 h at d 4 or 5. The blood is collected by cardiac puncture under anesthesia (10-mL syringe) from the rat that was infected on d 1.
2. Dilute the blood in 5–8 mL of complete culture medium (*see* **Subheading 2.2.**) to which 0.3 mL of stock-solution heparin is added.

3. Centrifuge (wash) for 8 min at 250g and remove culture medium from the erythrocyte pellet.
4. Resuspend the erythrocytes in 120–150 mL of complete culture medium in a sterile 500-mL Ehrlenmeyer flask (*see* **Note 4**).
5. Place the flask on a shaker (37°C, using a shaking water bath, oven or climate room) and connect it to the gas-system (gas mixture: 5% CO_2, 10% O_2, 85% N_2) (*see* **Note 5**).
6. Switch the shaker on after 5 min at a minimal speed to keep the cells in suspension.
7. Maintain the parasites in culture until the next morning (~09:00 h) at 37°C (*see* **Note 6**).
8. Take a small sample (~0.5 mL) from the cultures at 09:00 h for determination of the maturity of the parasites in an Eppendorf tube.
9. Pellet the erythrocytes at maximum speed in Eppendorf centrifuge for 5 s. Remove the supernatant.
10. Make a thin blood smear (film) of the erythrocyte pellet, fix by dipping briefly (~2 s in fresh methanol) and stain with Giemsa (*see* **Subheading 2.2.**).
11. Examine schizont development with a light-microscope at a ×1000 magnification (×100 objective, immersion oil).
12. If the culture has proceeded as normal, mature schizonts (8–16 nuclei) should comprise the predominant form visible in the smear. If this is not the case, continue the incubation if the culture and monitor every hour by parasite smear until the culture has >80% mature schizonts (*see* **Note 7**).

3.6. Purification of Mature Schizonts (Day 5 or 6)

Before electroporation of schizonts, the schizonts (1–3% of the total cell population) are separated from the uninfected erythrocytes that are present in the culture (*see* **Note 8**).

1. Prepare a 55% Nycodenz-phosphate-buffered saline (PBS) solution (v/v) (*see* **Subheading 2.3.**). In general, a total volume of 50 mL (27.5 mL Nycodenz stock-solution and 22.5 mL PBS) is used for a culture suspension of schizonts of 120–150 mL.
2. Divide the culture suspension of schizonts in four or five 50-mL tubes (~35 mL per tube)
3. Using a 10-mL pipet, pipet carefully and slowly 10 mL of Nycodenz solution *under* the culture suspension in each tube, so that a sharp partition is visible between the two suspensions.
4. Centrifuge for 20–30 min at 250g.
5. Collect carefully the "brown" (gray) layer at the interface between the two suspensions (in general, a total volume of ~30–40 mL is collected) (*see* **Note 9**).
6. Centrifuge for 8 min at 1500 rpm to pellet the schizonts; for this 'washing' step, add ~20 mL of culture medium (this can be obtained from the top of the tubes containing the density gradients) to the schizont-suspension collected from the interface.
7. Carefully remove the supernatant leaving as little as possible of the residual supernatant.
8. Resuspend the schizont pellet with Cytomix to a total volume of 100–120 mL.

3.7. Remarks on Preparation and Purification of Transfection Constructs

The construction of DNA vectors for genetic transformation has been described elsewhere *(3,15)*. Here, we will describe only the final steps of the preparation and purification of these constructs.

1. The transfection plasmid is generally isolated from *E. coli* Sure cells (Stratagene, La Jolla, CA) using Qiagene Maxiprep (Qiagen), starting from 500 mL of bacteria grown in standard Luria-Bertani (LB) medium. The DNA pellet is resuspended in TNE (10 mM Tris-1 mM EDTA, 100 mM NaCl (pH 8.0) at a concentration of 500 ng/µL to 1 µg/µL.
2. Store the DNA at –20°C before electroporation.

3. Use 25–40 µg vector DNA per transformation.
4. Just before electroporation, adjust to volume is adjusted to 300 µL with incomplete Cytomix.

3.8. Transfection of Schizonts by Electroporation (Day 5 or 6)

The electroporation conditions have been modified from those that were described in earlier publications and more closely mirror those described by Fidock and Wellens *(10)*.

1. Obtain purified schizonts from a culture of infected blood from one rat in Cytomix (100–120 µL) and mix with the DNA suspension (DNA constructs dissolved in 300 µL Cytomix) in an Eppendorf tube.
2. Transfer this 400 µL suspension to a prechilled (4°C) 0.4-cm electroporation cuvette.
3. Subject the suspension to an electric pulse using a Bio-Rad electroporation apparatus (1.1 kV, 25 mF, usually yielding a time-constant of 0.8–1.4 ms).
4. Immediately place the cuvet on ice for 3–5 min.

3.9. Injection of Transfected Schizonts into the Rat (Day 4 or 5)

We estimate that 50–80% of the schizonts (merozoites) is killed by the electroporation procedure. This estimation is based on comparison of the parasitemias of the injected rats 4 h after inoculation of electroporated and nonelectroporated purified schizonts

1. At 20–30 min before electroporation of the parasites, place a rat that has been treated with phenylhydrazine on d 2 (*see* **Note 10**) at 37°C. Holding a rat at this temperature for a period of 20–30 min simplifies the procedure of intravenous injection by inducing the swelling of the tail vein.
2. Inject the complete suspension (400 mL) intravenously into a tail-vein of the anesthetized rat, using a 1-mL syringe with a 12*G* needle. Generally, we inject the animal between 10:30 and 11:00 h.
3. Determine the parasitaemia of the rat at 4 h after infection, using a Giemsa-stained blood smear. In general, the parasitemia ranges between 0.1–1%.

3.10. Selection of Resistant Parasites by Pyrimethamine Treatment of the Rat (Day 5 or 6 to Day 9 or 10)

After inoculation of transfected parasites into a rat, the rat is treated on four consecutive days with a single ip dose of 10 mg/kg bodyweight pyrimethamine. The treatment commences 1 d after infection of the rat, to allow the parasites to complete one full developmental cycle without drug pressure. Since the transfected schizonts are inoculated between 10:30 and 11:00 and one cycle takes 22–24 h, the drug is administered to the infected animals in the afternoon. In our typical schedule, we treat the animals between 15:00 and 18:00 on the 4 d. Directly after the first two treatments, we observe a rapid drop in parasitemia to undetectable levels.

1. For each dose, inject ip 1 mL of pyrimethamine stock-solution into an anesthetized animal.
2. Prepare a Giemsa-stained blood smear on d 11 to check for the efficacy of the treatment. In a successful treatment, no parasites will be detected by examination of 20–40 microscope fields (×1000 magnification) in a blood smear.
3. Prepare Giemsa-stained blood smears from d 13–20 to determine parasitemias. In successful experiments, the parasitemias increase to levels between 0.3 and 1.0% on d 13–18. If one expect that the introduction of DNA may have an effect on the growth rate of the parasites, extension of the period of follow-up may be necessary.

P. berghei Transformation

4. At a parasitemia of approx 1%, transfer parasites from the *parent population* to naive mice (*see* **Subheading 3.12.**) and collect 1 mL of heart blood by heart puncture under anesthesia for cryopreservation in liquid nitrogen.

3.11. Cryopreservation of the Parent Population 1 (Days 13–18)

1. At a parasitemia of approx 1% after transfer of parasites from the parent population 1 to naive mice (*see* **Subheading 3.10.**), collect 1 mL of heart blood by heart puncture under anesthesia (from the rat or from a mouse).
2. Mix the blood with 1 mL of a 30% (v/v) glycerol/PBS solution containing 0.6 mL of heparin stock solution.
3. Divide suspension in 4 aliquots, transfer to cryotubes (0.5 mL per tube), keep for 5 min at 4°C and freeze directly in liquid nitrogen.

3.12. Transfer of Resistant Parasites of Parent Population 1 from the Rat (or Mouse) to Naive Mice (Days 13–18)

Parasites are transferred from the parent population 1 in the rat to naive mice to allow for a second round of pyrimethamine selection and to produce sufficient numbers for genotype and phenotype analysis.

1. At a parasitemia of approx 1%, using a 1-mL syringe, collect 10–20 drops of tail blood from the rat in 0.8 mL of PBS for ip infection of 4 mice.
2. Inject 4 mice with 0.2 mL of the erythrocyte suspension.
3. Treat these mice with pyrimethamine (0.1 mL of DMSO/pyrimethamine solution, 10 mg/kg body weight) on 3 consecutive days.
4. Prepare Giemsa-stained blood smears from d 3 after injection to check for the increasing parasitemia.
5. At a parasitemia between 1 and 15% these mice are used for the following:
   a. Cryopreservation of parasites (parent population 2; *see* **Subheading 3.13.**).
   b. Phenotypic analysis in Giemsa-stained blood smears (e.g., growth rate, gametocyte production).
   c. Collection of DNA for genotype analysis in particular plasmid rescue.

3.13. Cryopreservation of Parent Population 2 (Day 16–25)

1. At a parasitemia of approx 1–5%, collect 1 mL of heart blood by heart puncture under anesthesia from 1 of the 4 mice.
2. Mix the blood with 1 mL of a 30% (v/v) glycerol/PBS solution, containing 0.1 mL of heparin stock solution.
3. Divide suspension in 4 aliquots, transfer to cryotubes (0.5 mL per tube), keep for 5 min at 4°C, and freeze directly in liquid nitrogen.

3.14. Cloning of Parasites from Parent Population 2

We try to perform the cloning procedure in an as standardized a manner as possible. In most of our experiments, 30–50% of the mice will become positive. Therefore, we assume that we inject 0.3–0.5 viable parasites per mouse instead of the calculated 2 parasites. If only 1 parasite is injected per mouse, the number of positive mice drops markedly. The number of mice used in one cloning experiment is dependent on the number of different genotypes in the parent population, the estimated ratio of these different genotypes and the number of clones it may be necessary to analyze. If we expect that only one genotype is present, we inject 10 mice (*see* **Note 11**).

1. Inject 1 or 2 mice with infected erythrocytes (transfected parasites) from the parent population 2. Infected erythrocytes can be obtained either from a parasitemic mouse or from a frozen stock in liquid nitrogen (in the latter case, we thaw the frozen sample at room temperature and inject 0.01–0.2 mL cell suspension (blood in 30% PBS-glycerol) per mouse). These mice serve as a source of parasites for cloning.
2. At a parasitemia of 0.1–0.5%, collect 10–15 µL tail blood using a heparinized capillary and resuspend in 1 mL of complete culture medium at room temperature.
3. Place a small sample of this suspension into a Burker cell counter. At a ×400 magnification, determine the red blood cell density in the cell counter and calculate the number of infected cells in the cell suspension (based on the erythrocyte density in the cell counter and the parasitemia, determined using a Giemsa-stained blood smear).
4. Dilute the cell suspension with complete culture medium to give a suspension of 15 infected cells per milliliter.
5. Inject 10–30 mice with 0.2 mL iv of the suspension (2 parasites per mouse), under anesthesia.
6. At 8 days after injection, prepare Giemsa-stained blood smears from these mice. Typically, the parasitemia is between 0.05 and 0.1% in 30–50% of the mice. The other mice remain negative.
7. At a parasitemia of 1–3%, collect some tail blood for PCR analysis, transfer the parasites to a naive mouse if desired (e.g., if DNA is required for Southern analysis), and collect 1 mL of heart blood for cryopreservation.

3.15. Plasmid Rescue from Episomally Transformed P. berghei

Plasmid rescue is performed in order to confirm that the parasite population under analysis contains both the plasmid construct used in the original electroporation procedure and that the episome is unrearranged. This is necessary to confirm that observations upon the phenotype are a result of the plasmid construct. Plasmid cure is also used for this purpose (*see* **Subheading 3.16.**).

1. Thaw two tubes containing 200 µL of competent *E. coli* cells on ice.
2. Add to one tube 0.5–1.0 µg genomic DNA of transfected parasites. To the second tube, add 0.5–1.0 µg genomic DNA of wild-type parasites as a negative control.
3. Incubate on ice for 30 min.
4. Heat-pulse the cells at 42°C for 30 s without agitation.
5. Incubate on ice for 3 min.
6. Add 800 µL of SOC medium.
7. Transfer the cells to a 15-mL plastic tube and incubate for 1 h at 37°C with agitation.
8. Spread 100 µL of cells from each culture on different LB ampicillin plates.
9. Incubate the plates overnight at 37°C.
10. If bacterial colonies are present, pick 10 bacterial colonies from the plate with bacteria that are transformed with DNA of the transfected parasites.
11. Grow an overnight culture of each single bacterial colony in 4 mL of LB medium containing 50–100 µg/mL Ampicillin.
12. Isolate plasmid DNA from 1.5 mL of the suspension culture using the alkaline lysis method described by Qiagen.
13. Analyze the plasmid DNA using restriction enzyme digestion.

3.16. Curing Episomes from a P. berghei Population

Transformed parasites can be cured of plasmids by growth in the absence of selection pressure (i.e., drug). To date, a centromere has not been functionally identified in *Plasmodium*. Inclusion of a centromere on a transforming plasmid in the parasite would

P. berghei Transformation

allow equal segregation of plasmid copies into the daughter cells produced by parasite replication. In the absence of the centromeric element segregation is random and a parasite maintains a plasmid at some cost (*see* **Note 12**). Therefore, extended cultivation of parasites in the absence of selective pressure results in a parasite population that is effectively cured of the original transforming plasmid. Any phenotypic change that accrued from the presence of the plasmid should have disappeared. This test is performed to demonstrate the link between phenotype and presence of the plasmid. It is achieved by simple mechanical passage of infection on a weekly basis to a naive animal.

1. Place two drops of blood collected via a tail vein bleed from an infected mouse (with a parasitemia of approx 10%) into 8 mL of PBS.
2. Inject two naive mice with 500 µL of the suspension ip.
3. At 1 wk later, the parasitemia should be 10% in the newly infected animals. Transfer the parasites to two more animals as described above and sacrifice the donor animals for parasite DNA preparation.
4. After isolation of parasite genomic DNA, test for the presence of the plasmid by plasmid rescue (*see* **Subheading 3.15.**) and PCR.
5. Repeat the cycle of infection and analysis.
6. Confirm that parasite populations have been cured of the plasmid by drug challenge on an infected animal.
7. The entire procedure can be completed within 6 wk.

4. Notes

1. The quality of the staining pattern is dependent on the pH of the water. If it is too acid then large blue nonspecific precipitates will result. In this case, use phosphate-buffered saline (pH 7.2) to dilute the Giemsa.
2. A rat is treated with phenylhydrazine (PHZ) to induce reticulocytosis. This rat is later used for iv injection of the purified and transfected schizonts. Since *P. berghei* has a strong preference for reticulocytes, invasion of the parasites is more efficient in phenylhydrazine treated rats than in untreated rats. Moreover, in the phenylhydrazine-treated rat the number of multiply infected erythrocytes is reduced. Multiply infected erythrocytes containing more than two parasites will not support the growth of these parasites into healthy mature schizonts.
3. Blood stages of *P. berghei* are cultured in RPMI 1640 medium (pH 7.3) to which fetal calf serum has been added. In general, the parasites are maintained in vitro for only one developmental cycle: ring forms/young trophozoites are allowed to develop into mature schizonts during a period of 16–23 h. Blood is collected from the positive rat (infected at d 0) at a parasitemia of 1–3% at d 4 or 5 between 10:00 and 16:00 h (higher parasitemias are undesirable, since many red blood cells will become multiply infected). In animals that are kept under the normal day/night light regime, the development of the parasites is relatively synchronous. In these animals, bursting of the schizonts and invasion by merozoites occurs in the early morning between 04:00 and 06:00 h. Therefore most parasites are in the ring form/young trophozoite stage in the morning working hour. The infected blood is incubated at 37°C. The next day at 09:00 h, all parasites have developed into mature schizonts that do not burst under our in vitro conditions. Schizonts of *P. berghei* containing mature merozoites can survive for several hours and can be manipulated without bursting and without the loss of viability.
4. We use these flasks for our in house "automatic" continuous gassing system in which the cultures are continuously gassed throughout the complete culture period. Continuous gassing is, however, not necessary for these cultures. Cultures can therefore be maintained in closed, plastic 250-mL culture flasks, which are gassed once at the beginning of the culture period.

5. Gas mixtures used for the culture of *P. falciparum* are also suited for *P. berghei*. The most important factor for both parasite species in the gas mixture is the reduced concentration of O_2 as compared to air.
6. The temperature is critical since the rate of development is highly dependent on the temperature. At temperatures above 38.5°C, parasites will degenerate. At temperatures below 37°C, the parasites will develop into healthy parasites but the developmental time will be extended. Even at a temperature of 30°C, the parasites are able to reach the mature schizont stage, but the development of ring forms into schizonts will take more than 48 h.
7. Healthy, viable schizonts are recognized by the presence of 12–16 (24) "free" merozoites within one red blood cell and one cluster of pigment (hemozoin). Smearing the cells on the microscope slides often damages the red cell membrane and the merozoites are visible as more or less clustered free parasites. A purple (red) defined compact nucleus and a dot of blue cytoplasm is characteristic of viable merozoites. Approximately 15–25% of the parasites in these smears are single-nucleated (young) gametocytes. Degenerate schizonts often show a compact morphology in which the separate merozoites are difficult to recognize. Be careful not to mistake developing schizonts (which are still in the process of nuclear division prior to budding off of the merozoites) for degenerated schizonts. Illustrations of the parasite for as they appear in in vivo and in vitro cultures can be found at: http://www./lumc.nl/1040/research/malaria/malaria.html.
8. We start the purification procedure between 09:00 and 10:00 h. Starting later in the morning results in a higher percentage of degenerated schizonts. For the density gradients we use Nycodenz instead of Percoll. Although Percoll is used by many workers to separate parasite stages, our experience is that the viability of parasites is less affected (in fact, no effect) by Nycodenz.
9. The schizonts (and gametocytes and old trophozoites, if present) will collect at the interface of the two suspensions, while the uninfected cells will pellet on the bottom of the tubes. The method of collection is similar to using a vacuum cleaner to suck up the parasitized cells at the interface.
10. Local heat application on the tail using either hot water or, more effectively, a heat lamp can also be used to make the tail vein more prominent and ease injection. This is the single most difficult step in the procedure and the cause of most experimental failures. This is due to the rigidity of the rat dermis. Experienced and skilled personnel are needed to carry this out routinely.
11. Parasite clones are obtained by the method of "limiting dilution." We prefer to clone the parasites after two full rounds of pyrimethamine selection of the parasites. The first round is the treatment in the rat resulting in parent population 1. The second round is the treatment of parent population 1 in mice (on 3–4 consecutive days), resulting in parent population 2. (We do not normally clone parasites that are transfected with a single, circular construct. The demonstration of the presence of the unaltered construct by Southern analysis and plasmid rescue is usually sufficient for these parasite populations. In these populations, parasites that do not contain episomes as a result of the uneven segregation of the episomes during nuclear division are produced continuously. We only clone these parasites when more than one construct is present in the population.)
12. It is important to note here that transformed parasites that contain episomes grow slower than transgenic parasites containing DNA integrated into their genome. The slower growth rate is due to the unstable segregation during schizogony of the episomes, resulting in the production of merozoites in each cycle that do not contain the episomes. These parasites are therefore sensitive to pyrimethamine during their development into the next schizont generation and are killed by the treatment. We have indications that up to 50–60% of merozoites produced from episome containing parasites lack the episomes. The episome positive parasites may contain up to 20–40 copies of the episome per nucleus.

References

1. van Dijk, M. R., Waters, A. P., and Janse, C. J. (1995) Stable transfection of malaria parasite blood stages. *Science* **268,** 1358–1362.
2. Wu Y., Kirkman L. A., and Wellems, T. E. (1996) Transformation of *Plasmodium falciparum* malaria parasites by homologous integration of plasmids that confer resistance to pyrimethamine. *Proc. Natl. Acad. Sci. USA* **93,** 1130–1134.
3. Crabb, B. S., Triglia, T., Waterkeyn, J. G., and Cowman, A. F. (1997) Stable transgene expression in *Plasmodium falciparum*. *Mol. Biochem. Parasitol.* **90,** 131–144.
4. van der Wel, A. M., Tomas, A. M., Kocken, C. H. M., Malhotra, P., Janse, C. J., Waters, A. P., and Thomas, A. W. (1997) Transfection and *in vivo* selection of the primate malaria parasite, *Plasmodium knowlesi*. *J. Exp. Med.* **185,** 1499–1503.
5. de Koning-Ward, T. F., Sperança, M. A., Waters, A. P., and Janse, C. J. (1999) Analysis of stage specificity of promoters in *Plasmodium berghei* using luciferase as a reporter *Mol. Biochem. Parasitol.* **100,** 141–146.
6. de Koning-Ward, T. F., Thomas, A. W., Waters, A. P., and Janse, C. J. (1998) Stable expression of green fluorescent protein in blood and mosquito stages of *Plasmodium berghei*. *Mol. Biochem. Parasitol.* **97,** 247–252.
7. Margos, G., van Dijk, M. R., Ramesar,J., Janse, C. J., Waters, A. P., and Sinden, R. E. (1998) Transgenic expression of a mosquito stage malarial protein, Pbs21, in blood stages of transformed *Plasmodium berghei* and the induction of an immune response upon infection. *Infect. Immun.* **66,** 3884–3891.
8. Kocken, C. H. M., van der Wel, A. M., Dubbeld, M. A., Narum, D. L., van de Rijke, F. M., van Gemert, G.-J., et al. (1998) Precise timing of expression of a *Plasmodium falciparum* derived transgene in *P. berghei* is a critical determinant of subsequent subcellular location. *J. Biol. Chem.* **273,** 15,119–15,124.
9. Waller, R. F., Keeling, P. J., Donald, R. G., Striepen, B., Handman, E., Lang-Unnasch, N., et al. (1998) Nuclear-encoded proteins target to the plastid in *Toxoplasma gondii* and *Plasmodium falciparum*. *Proc. Natl. Acad. Sci. USA* **95,** 12,352–12,357.
10. Fidock, D. A. and Wellems, T. E. (1997) Transformation with human dihydrofolate reductase renders malaria parasites insensitive to WR99210 but does not affect the intrinsic activity of proguanil. *Proc. Natl, Acad, Sci. USA* **94,** 10,931–10,936.
11. de Koning-Ward, T. F., Fidock, D. A., Thathy, V., Menard, R., van Spaendonk, R. M. L., Waters, A. P., and Janse, C. J. (2000) The selectable marker human dihydrofolate reductase enables sequential genetic manipulation of the *Plasmodium berghei* genome. *Mol. Biochem. Parasitol.* **106,** 199–212.
12. Crabb, B. S. and Cowman, A. F. (1996) Characterization of promoters and stable transfection by homologous and non-homologous recombination in *Plasmodium falciparum*. *Proc. Natl. Acad. Sci. USA* **93,** 7289–7294.
13. Crabb, B. S., Cooke, B. M., Reeder, J. C., Waller, R. F., Caruana, S. R., Davern, K. M., et al. (1997) Targeted gene disruption shows that knobs enable malaria-infected red cells to cytoadhere under physiological shear stress. *Cell* **89,** 287–296.
14. Triglia T., Wang P., Sims P. F., Hyde J. E., and Cowman A. F. (1998) Allelic exchange at the endogenous genomic locus in *Plasmodium falciparum* proves the role of dihydropteroate synthase in sulfadoxine-resistant malaria. *EMBO J.* **17,** 3807–3815.
15. Tomas, A. M., van der Wel, A. M., Thomas, A. W., Janse, C. J., and Waters, A. P. (1998) Transfection systems for animal models of malaria. *Parasitol. Today* **14,** 245–249.

29

Gene Targeting in *Plasmodium berghei*

Vandana Thathy and Robert Ménard

1. Introduction

Gene targeting, the technology that permits inactivation or modification of a gene by homologous recombination, has been reproducibly established in two plasmodial species: *Plasmodium falciparum (1)*, the most important pathogen for humans, and *Plasmodium berghei (2)*, which infects rodents. In both species, the red blood cell (RBC) stages of the parasite are transformed and targeting constructs integrate into the genome exclusively via homologous recombination.

The *P. berghei* transformation system (**Fig. 1**) offers a double advantage over the *P. falciparum* system. At the genetic level, published reports *(3–8)* indicate that targeted *P. berghei* clones are obtained more reproducibly (they represent frequently more than 50% of the parasites in the first-generation resistant population), and more rapidly (typically in less than 1 mo following parasite transformation). In addition, gene modification strategies, which rely on targeting constructs that contain large fragments of untranslated sequences, are facilitated by the lower A/T-richness (and thus greater stability in *Escherichia coli*) of such sequences in *P. berghei* than in *P. falciparum*. At the phenotypic level, the *P. berghei* system permits the in vivo analysis of liver infection by sporozoites. The rodent system is therefore ideal for addressing the function of any protein expressed in preerythrocytic stages of the parasite. As with *P. falciparum*, however, the transformation of RBC stages of the parasite and the lack of inducible promoter technology in *Plasmodium* have so far hampered analysis of essential genes expressed in the RBC stages of the parasite.

Only one selectable marker has proved successful so far for generating *P. berghei* recombinants (a dihydrofolate reductase thymidylate synthase (*DHFR-TS*) variant gene that confers resistance to the drug pyrimethamine). Nonetheless, because the *Plasmodium* genome is haploid in the RBC stages of the parasite and because many proteins of interest are encoded by single-copy genes, one selectable marker is sufficient for undertaking a functional analysis of any malarial protein that is not essential for parasite replication in RBCs. We present here the methodology and protocols for gene inactivation or modification in *P. berghei*.

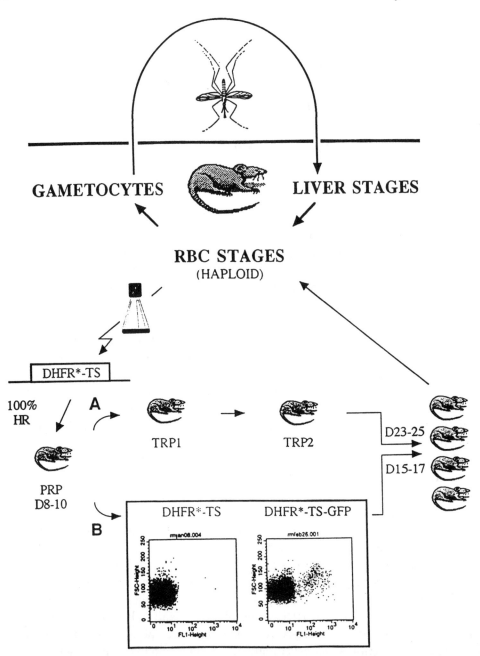

Fig. 1. Gene targeting using pyrimethamine- and FACS-based selection in *Plasmodium berghei*. *P. berghei* schizonts obtained after an overnight culture of infected blood are transformed with linear targeting constructs containing a mutant *DHFR-TS* (DHFR*-TS) gene that confers resistance to pyrimethamine. The incoming constructs integrate into the genome of the haploid parasites only by homologous recombination. A transformed parental resistant population (PRP) emerges by d 8–10 postelectroporation (PE). **(A)** The PRP is transferred to a new rat yielding a rapidly expanding transfer resistant population (TRP1) by d 10–12 PE. TRP1 is treated with pyrimethamine and is transferred to a third rat, yielding TRP2. During expansion of TRP2, the genomic DNA of the PRP and TRP1 is analyzed to evaluate the presence and

2. Materials

2.1. Construction of Targeting Plasmids

1. Bacterial vector: A high copy number plasmid that can be propagated and selected in *Escherichia coli,* such as pUC or pBluescript.
2. Positive selectable marker: The selectable marker most commonly used is the pyrimethamine-resistant variant of the *P. berghei DHFR-TS* bifunctional gene *(2,9)* constitutively expressed from its own 5'- (2.2 kb) and 3'- (1 kb or truncated to 0.4 kb) untranslated regions (UTR). A derivative of this cassette, PyrFlu, confers pyrimethamine resistance and directs a fluorescence signal via a *DHFR-TS-GFPmut2* fusion gene under the control of 2.2 kb of 5'- and 0.4 kb of 3'-UTR of *P. berghei DHFR-TS (10).*
3. A targeting sequence originating from the genomic DNA of the same *P. berghei* strain (NK65) as that used as the recipient in transformation experiments.
4. *E. coli* DH5α competent cells or *Epicurian coli* XL1-blue MRF' Kan Supercompetent Cells (Stratagene, La Jolla, CA, cat. no. 200248).

2.2. Transfection Protocol

2.2.1. Preparation of Targeting Construct

1. *E. coli* DH5α competent cells or *Epicurian coli* XL1-blue MRF' Kan Supercompetent Cells.
2. Standard Luria-Bertani (LB) medium.
3. Ampicillin, sodium salt (Sigma, St. Louis, MO, cat. no. A-9518): 100 mg/mL in sterile H_2O. Store in aliquots at –20°C.
4. DNA extraction: Qiagen Maxiprep (Qiagen, Valencia, CA).
5. TE buffer (10 mM Tris-HCl, pH 8.0, 1 mM EDTA) or Buffer EB (10 mM Tris-HCl, pH 8.5) (Qiagen).
6. Appropriate restriction enzymes (New England Biolabs, Beverly, MA).
7. Seakem LE agarose (FMC BioProducts, Rockland, ME).
8. Extraction of DNA fragments after agarose gel electrophoresis: Gene Clean (BIO101, Vista, CA).
9. Purification of DNA: QIAquick PCR Purification Kit (Qiagen), or phenol-chloroform extraction and ethanol precipitation.

2.2.2. Preparation of Recipient Parasites

1. Pyrimethamine-sensitive *P. berghei* strain NK65 (*see* **Note 1**).
2. *Anopheles stephensi*, maintained at 21°C and 80% relative humidity.
3. Syrian hamsters, 4–8 wk old.
4. Wistar rats, 100–150 g.
5. Large rat restrainer (Braintree Scientific, Braintree, MA, cat. no. RTV-180).
6. 1-mL tuberculin syringes with detachable needles; needle gage and length: 27-gage × 1/2 in.
7. 6-mL syringes with 21-gage needles.
8. Heparin solution (200 USP units (U)/mL): Dilute heparin sodium (10,000 U/mL) 1/50 in phosphate-buffered saline (PBS). Store at 4°C.

Fig. 1. *(continued)* proportion of correct targeting events in the resistant populations. Following genomicanalysis, the TRP2 is cloned by limiting dilution into rats, yielding clonal populations by d 23–25 PE. **(B)** Use of the *DHFR-TS-GFPmut2* (DHFR*-TS-GFP) selectable marker will confer a fluorescent signal on transformed, pyrimethamine-resistant parasites allowing for the enrichment of recombinant parasites by FACS selection from the total resistant population (panel). Fluorescent infected RBCs are sorted, and then cloned into rats to produce clonal recombinant populations by d 15–17 PE. Mosquitoes are fed on rats infected with recombinant clones when mature gametocytes are present, and the phenotypes of the subsequent parasite stages analyzed.

9. Complete culture medium: RPMI medium 1640 with 25 mM HEPES buffer and L-glutamine (Gibco-BRL, Life Technologies, Grand Island, NY, cat. no. 22400-089), supplemented with 20% fetal bovine serum (FBS) (Gibco-BRL, cat. no. 16000-036), and 50 IU/mL neomycin (Sigma, cat. no. N-1142). Filter using **item 10** below. Prepare fresh each time. Keep at 37°C.
10. Corning disposable 500-mL filter system with 0.22-µM pore size (Fisher Scientific, cat. no. 09-761-5).
11. 500-mL Erlenmeyer flasks with side arms, and matching one-hole rubber stoppers with 4 in. long glass tubes. Assemble flasks and stoppers, plug side arms and glass tubes with cotton, cover with aluminum foil, and autoclave.
12. Falcon 2098 50-mL polypropylene conical tubes (Fisher, cat. no. 14-959-49A).
13. Tabletop centrifuge with a swinging rotor and no-brake option (e.g., Beckman, Model TJ-6).
14. CO_2 and N_2 tanks.
15. CO_2/O_2 incubator, large enough to accommodate the platform shaker (*see* **item 16** below), and maintained at 10% O_2, 5% CO_2, 85% N_2; 37°C.
16. Innova 2000 portable platform shaker, 13 × 11 in. (Fisher, cat. no. 14-278-104).
17. Materials for preparing blood smears:
    a. Methanol, absolute.
    b. Giemsa stain, modified (Sigma, cat. no. GS-500). Dilute 1:20 in H_2O before use.
    c. Precleaned microscope slides.

2.2.3. Electroporation

1. PBS 1X without Ca^{2+} and Mg^{2+}, pH 7.2 (Gibco-BRL, cat. no. 20012-027).
2. Accudenz A.G. (Nycodenz) powder (Accurate Chemical & Scientific Corporation, Westbury, NY, cat. no. AN-7050).
3. Accudenz stock solution: Dissolve 27.6 g of Accudenz powder in 100 mL of stock solution buffer (5 mM Tris-HCl, pH 7.5, 3 mM potassium chloride, 0.3 mM EDTA). Sterilize by autoclaving. Cover with aluminum foil and store at 4°C.
4. Falcon 2098 50-mL polypropylene conical tubes (Fisher, cat. no. 14-959-49A).
5. Tabletop centrifuge with a swinging rotor and no-brake option (e.g., Beckman, Model TJ-6).
6. Disposable Pasteur pipets.
7. Electroporation cuvettes, 0.4-cm electrode gap (Bio-Rad Laboratories, Hercules, CA, cat. no. 165-2088). Chill on ice before use.
8. Gene Pulser (Bio-Rad).
9. 1-mL tuberculin syringes with detachable needles; needle gage and length: 27 × 1/2 in.
10. Phenylhydrazine hydrochloride stock solution (25 mg/mL): Dissolve 250 mg of phenylhydrazine (Sigma, cat. no. P6926) in 10 mL of 0.9% (w/v) NaCl. Store in aliquots at –20°C (*see* **Note 2**).
11. Sprague-Dawley rats, 55–60 g.
12. Small rat restrainer (Braintree Scientific Inc., cat. no. RTV-170).

2.2.4. Parasite Selection

1. Pyrimethamine solution (1.25 mg/mL): To 20 mL of PBS, add 25 mg of pyrimethamine (Sigma, cat. no. P-7771, stored in the dark), 300 µL of 1 N HCl and 70 µL of dimethyl sulfoxide (Sigma). Vortex and sonicate for 20 min at room temperature. Prepare fresh each time.
2. Materials for preparing blood smears (*see* **Subheading 2.2.2.**).
3. Glycerol (Fisher).
4. Alsever's solution: Dissolve 10.25 g of dextrose, 4 g of sodium citrate, 0.275 of citric acid, and 2.1 g of sodium chloride in 500 mL of H_2O, pH ~6.8 (or Sigma, cat. no. A-3551). Store at 4°C.
5. Freezing solution: Mix 1 part of glycerol with 9 parts of Alsever's solution. Prepare fresh each time.

Gene Targeting

6. Wheaton Cryule 2-mL cryogenic vials (Fisher, cat. no. 03-341-18c).
7. Liquid nitrogen tank(s).
8. Fluorescence microscope with fluorescein isothiocyanate (FITC) filter settings.
9. Becton Dickinson FACScan machine with an argon laser tuned at 488 nm.
10. Coulter Epics Elite sorter (Coulter, Miami, FL).

2.2.5. Parasite Cloning

1. RPMI medium 1640 supplemented with 20% FBS (*see* **Subheading 2.2.2.**).
2. Hemocytometer.
3. 1-mL 28-gage insulin syringes.
4. Sprague-Dawley rats, ~55–75 g.
5. Small rat restrainer (*see* **Subheading 2.2.3.**).

2.2.6. Preparation and Analysis of Parasite Genomic DNA

1. PBS (*see* **Subheading 2.2.3.**).
2. Plasmodipur filter units (Euro-diagnostica, Arnhem, The Netherlands, cat. no. 8011).
3. 12-mL luer-lok syringes.
4. Saponin (Fluka Chemical Corp., Milwaukee, WI, cat. no. 84510).
5. RNase A, 100 mg/mL (Qiagen) for use with the QIAamp Blood Kit.
6. QIAamp Blood Kit (Qiagen).
7. Southern blot analysis: DIG DNA Labeling Kit (Boehringer Mannheim, Roche Molecular Biochemicals, Indianapolis, IN, cat. no. 1 175 033); DIG Luminescent Detection Kit for Nucleic Acids (Boehringer Mannheim, cat. no. 1 363 514).
8. Appropriate restriction enzymes (New England Biolabs).

3. Methods

The following protocol is based on gene targeting experiments in a pyrimethamine-sensitive, gametocyte-producer clone of *P. berghei* strain NK65 *(3,5,7,10)*, and is derived from techniques that have previously been described *(11–13)*.

3.1. Construction of Targeting Plasmids

The components of targeting plasmids of both the replacement and the insertion types are (a) a plasmid backbone, (b) a positive selectable marker, and (c) a homologous targeting sequence of sufficient length, isolated from the genomic DNA of a *P. berghei* strain isogenic to the one being transformed. The vector backbone is derived from a high copy number plasmid that can be propagated in *Escherichia coli*, such as pUC or pBluescript, to ensure high yields of the targeting construct for electroporation. In our transformation experiments, we have utilized a selection cassette containing the pyrimethamine-resistant, *DHFR-TS* mutant gene from *P. berghei* expressed by its own flanking regions *(2,9)*, borne by plasmid pMD204 (~5 kb) and derivatives pMD205 and pPyrFlu. Plasmid pMD205 lacks only the *EcoRV-SmaI* fragment (~0.6 kb) of distal 3' UTR present in pMD204 *(5,7,10)*. Plasmid pPyrFlu (~5 kb) contains a *DHFR-TS-GFPmut2* fusion gene *(10,14)* and confers pyrimethamine-resistance and a concomitant GFPmut2-based fluorescence signal on transformed RBC stages of the parasite (*see* **Note 3**). The pyrimethamine resistant *DHFR-TS* gene from *Toxoplasma gondii* under the control of *P. berghei* expression sequences *(12)* may also be used, minimizing the frequency of gene conversion events that may occur at the endogenous *DHFR-TS* locus when using a homologous *DHFR-TS* resistance cassette.

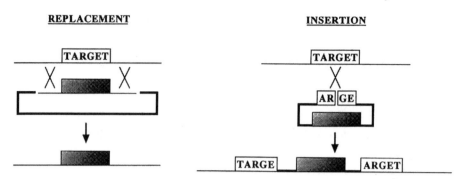

Fig. 2. Gene inactivation using replacement or insertion targeting plasmids. The coding sequence of the target gene (*open box*) is flanked by untranslated sequences (*thin lines*). The resistance cassette is symbolized by a *shaded box*, and the bacterial plasmid sequences by *thick lines*. The replacement construct contains two regions of homology on either side of the selectable marker and is cut to release a linear fragment with one region of homology at each end. The liberated insert of the replacement construct promotes a double crossover leading to the allelic exchange of the wild-type gene by the disrupted copy. The insertion construct contains an internal fragment of the target gene. Linearization of the construct within the targeting sequence (gap) promotes plasmid integration via a single crossover, generating two truncated copies of the target gene in the final recombinant locus.

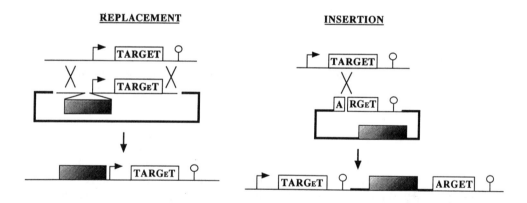

Fig. 3. Introduction of a subtle gene modification using replacement or insertion targeting plasmids. The coding sequence of the target gene (*open box*) is flanked by untranslated sequences (*thin lines*) bearing the gene promoter (*arrow*) and 3' signals (*circle*) necessary for normal gene expression. The subtle modification in the coding sequence is shown by the modification of the E to ε. The resistance cassette is symbolized by a *shaded box*, and the bacterial plasmid sequences by thick lines. The replacement construct contains an upstream region of homology consisting of the 5' UTR of the target gene, and a downstream region of homology starting upstream from the gene promoter and ending after the mutation in the coding sequence. The recombinant locus generated by a double crossover event contains one full-length copy of the gene bearing the mutation and flanked by the UTRs necessary for expression. The insertion construct contains the 3' part of the gene bearing the mutation followed by the 3' UTR. The plasmid is linearized (gap) within the region of homology upstream from the mutation. The recombinant locus generated by a single crossover event duplicates the region of homology and places the mutation in the first, full-length, and expressed copy of the gene.

Gene Targeting

The basic design of replacement or insertion targeting plasmids that can be used to promote gene inactivation *(13)* or gene modifications through single recombination events are shown in **Figs. 2** and **3**, respectively. The main feature of the plasmid transformation system for homologous recombination in *P. berghei*, but not in *P. falciparum* *(1,15)*, is that linearization of the targeting construct is essential for the rapid isolation of homologous recombinants.

3.1.1. Gene Inactivation

See **Fig. 2**.

3.1.1.1. USING REPLACEMENT PLASMIDS

The replacement construct contains two regions of homology separated by the selectable marker. To increase the likelihood of generating a null mutation, the selectable cassette should be designed to replace all or part of the coding sequence of the target gene. Prior to electroporation, the replacement construct should be cut at both ends of the homologous fragment, which generates a linear molecule with free homologous ends. Alternatively, the construct may be cut in the plasmid backbone. Integration occurs via a double crossover between homologous sequences in the construct and in the genome, leading to the allelic exchange of the target gene by the disrupted exogenous copy (*see* **Note 4**).

3.1.1.2. USING INSERTION PLASMIDS

The insertion construct should contain an internal fragment of the target gene cloned as a single continuous sequence upstream or downstream of the selectable marker. Linearization of the construct at a unique restriction site within the targeting sequence prior to electroporation promotes plasmid integration via a single crossover between homologous sequences in the construct and in the genome. This leads to integration of the entire plasmid at the target locus and duplication of the region of homology, generating two truncated copies of the gene in the final structure (for other outcomes of recombination events, *see* **Note 5**). The restriction site used for plasmid linearization should be located at least 250 bp away from the ends of the homology region to ensure that exonuclease activity that may progress from the site of linearization does not eliminate one homologous arm and prevent crossover formation *(5,7)*.

3.1.2. Gene Modification

Several strategies may be used for gene modification. First, a modified version of the gene can be expressed via autonomously replicating episomes in a parasite line bearing a null mutation in the gene, a strategy that requires two selectable markers. The modified gene can also be created by a single recombination event at the wild-type locus, promoted by a targeting plasmid of the replacement or the insertion type (**Fig. 3**).

3.1.2.1. USING REPLACEMENT PLASMIDS

The selectable marker should be inserted upstream (or downstream) from the promoter region of the target gene. The transformed linear fragment will therefore contain an upstream region of homology consisting of 5' UTR of the gene, and a downstream region of homology that starts upstream from the gene promoter and ends after the mutation in the coding sequence. However, although the recombinant locus generated by the expected double crossover should contain the modified gene flanked by the UTRs necessary for expression, the selectable marker ends up inserted upstream (or downstream) from the regulatory sequences of the modified gene, which may disrupt a closely linked locus or affect its expression. Therefore, it may be advisable to design the replacement construct with an upstream region of homology consisting of sequences lying immediately upstream from the target gene (0.5–1.0 kb), so that the recombinant locus contains unaltered sequences up to the target gene (*see* **Note 6**). Our experience of gene modification using replacement plasmids is that the frequency of recombinants that have undergone the expected double crossover is reduced (often less than 10% of the resistant parasite population compared to a proportion of frequently 50% or more of the expected integrants obtained using insertion plasmids, *see* next subheading). Moreover, mutations located less than 500 bp away from one end of the replacement fragment are frequently corrected during recombination.

3.1.2.2. USING INSERTION PLASMIDS

The targeting sequence in the insertion plasmid must lack the 5' end of the gene (to generate a truncated second copy) and end after its 3' regulatory elements (to allow expression of the first copy). Two crucial considerations in the design of these plasmids *(7)* are (a) the distance between the linearization site and one edge of the targeting sequence (the short arm of homology), which should be at least 250 bp to allow productive recombination, and (b) the distance between the linearization site and the subtle mutation, which should be at least 500 bp in order for the mutation to be maintained during plasmid integration (*see* **Note 7**). When the plasmid is linearized within the region of homology upstream from the mutation, the recombinant locus generated by a single crossover event should contain a first copy of the gene that is full-length, modified and expressed, and a second copy that is truncated (lacking the 5' part of the gene) and not expressed (lacking the promoter sequence). This plasmid design thus allows modifications to be introduced in the distal region of a gene. Conversely, to introduce modifications in the proximal region of a gene, the targeting sequence bearing the mutation must contain the 5' promoter sequence of the target gene and lack the 3' end of its coding sequence. Linearization of this type of plasmid downstream of the mutation would then place the mutation in the second, full-length, and expressed copy of the gene. Note that the final loci contain uninterrupted sequences located upstream from the target gene (up to the first duplicate) as well as downstream (starting from the second duplicate), therefore minimizing the risk that plasmid integration may affect neighboring genes. In our hands, the insertion strategy has proven the most effective in generating *P. berghei* subtle mutants *(7)*.

3.2. Transfection Protocol

3.2.1. Preparation of Targeting Construct

The targeting plasmid is isolated from *E. coli* DH5α or *Epicurian coli* XL1-blue supercompetent cells using Qiagen Maxiprep, starting from 500 mL of bacteria grown overnight in standard LB medium containing 0.1 mg/mL of ampicillin. The DNA pellet is resuspended in TE buffer or buffer EB (Qiagen) at a final concentration of 1–5 µg/µL and digested overnight with the appropriate restriction enzymes. When attempting to

Gene Targeting

integrate the construct into a target locus, complete digestion prevents transformation of circular plasmids. The linearized molecule, or the liberated insert of a replacement construct can be purified from an agarose gel after 0.7% agarose gel electrophoresis using Gene Clean. However, traces of uncut insertion plasmid, or presence of the replacement plasmid backbone after liberation of the insert in the transfected DNA preparation, do not impede the recovery of the correct integration events. Therefore the DNA may also be purified after an overnight restriction digestion by standard phenol-chloroform extraction and ethanol precipitation, or by using the QIAquick PCR purification Kit (Qiagen) according to manufacturer's instructions. The DNA is finally resuspended or eluted with sterile water, TE buffer or Buffer EB, at a concentration of 500 ng to 1 µg/µL and stored at –20°C before electroporation. Just prior to electroporation, 10–50 µg of the linear DNA is mixed with PBS to a final volume of 300 µL.

3.2.2. Preparation of Recipient Parasites

1. Syrian hamsters are infected with a pyrimethamine-sensitive, gametocyte producer clone of *P. berghei* strain NK65 by exposure to the bite of infected *Anopheles stephensi* mosquitoes at 16–18 d postfeeding, or by intraperitoneal injection of 4×10^5 sporozoites in RPMI 1640. Infective sporozoites can be obtained by dissecting and grinding the salivary glands of infected mosquitoes in RPMI 1640 at 16–18 d postfeeding, centrifuging at 1000*g* for 2 min to remove cellular debris, and counting in a hemocytometer. RBC stages of the parasite are collected from the hamster by cardiac puncture when the parasitemia is 5–10%, 6–7 d after infection, and transferred to Wistar rats (see below). RBC stages resulting from a sporozoite-induced rodent infection should not be passaged more than once before infection of Wistar rats (*see* **Note 1**).
2. For infection of Wistar rats (donors), the hamster blood is diluted to 2% in PBS and 0.45 mL is injected intravenously into the tail vein using a 1-mL tuberculin syringe with a 27-gage needle. For easier handling, the rats are placed in an animal restrainer for the tail vein injections. The parasites collected from one donor rat constitute the recipient parasites of one electroporation experiment.
3. In vitro culture of RBC stages: At ~3 d after infection, the parasitemia of the donor rats should be between 1 and 3%. To collect schizonts for one transfection experiment, the blood (~5 mL) from one donor rat is first cultured under standard sterile culture conditions (*see* below) in a gently shaken Erlenmeyer flask for 15–16 h:

    a. Collect the blood by cardiac puncture using a 6-mL heparinized syringe with a 21-gage needle.
    b. Mix the blood with 10 mL of complete culture medium (warmed to 37°C) containing 0.25 mL of heparin solution (200 U/mL) in a 50-mL falcon tube, and centrifuge at 200*g* for 8 min in a swinging rotor at room temperature.
    c. Discard the supernatant. The RBC pellet should be ~ 3 mL.
    d. Complete to 20 mL with culture medium, mix gently and transfer to a 500-mL Erlenmeyer flask containing 100 mL of culture medium at 37°C. Wash the 50-mL Falcon tube with another 30 mL of culture medium and add it to the flask so that the total volume in the flask is 150 mL.
    e. Place the Erlenmeyer flask in the incubator containing a gas mixture of 10% O_2, 5% CO_2, 85% N_2, at 37°C and shake gently at 77–80 rpm (~ 0.065*g*) overnight for 15–16 h. During this time, the majority of the parasites mature into schizonts that do not rupture to release merozoites.

The quality of the in vitro culture seems to be important for the success of the transformation experiment, and can vary considerably from one flask to another. Thus, we perform the experiments in duplicate and choose the cultures that have the highest schizont:gametocyte ratios at the end of the culture. A high ratio (>3) is usually associated with a high parasitemia (>0.2%) 4-h postelectroporation, which is our best predictive value of a successful gene targeting experiment.

3.2.3. Electroporation

After 15–16 h of in vitro growth, the quality of the culture is checked by a Giemsa-stained smear of a drop of the RBC pellet obtained from a 0.5-mL aliquot of culture suspension. A majority of mature schizonts should be visible at a frequency of 3 to 6 schizonts per field. The mature schizonts from each culture are then separated from uninfected RBCs on density gradients in the following manner:

1. Add 27.5 mL of Accudenz stock solution to 22.5 mL of PBS to make a 55% density gradient. For each culture flask, prepare four 50-mL Falcon tubes and pour ~35 mL of culture suspension in each. Carefully add 10 mL of Accudenz/PBS gradient at the bottom of each Falcon tube and centrifuge for 25 min at 200g in the swinging rotor (no brake). After centrifugation, a pellet of uninfected RBCs, the white gradient, a dark ring at the interface consisting mainly (>90%) of schizont-infected RBCs, and the orange clear medium, should be visible from the bottom to the top of the tube.
2. Collect the ring of schizonts (~10 mL) at the interface of each Falcon tube with a Pasteur pipet. Pool two rings (from two Falcon tubes) in a new Falcon tube. To wash and pellet the schizonts, complete the volume to 30 mL with culture medium and centrifuge for 10 min at 200g in the swinging rotor.
3. Carefully remove the supernatant leaving <1 mm above the schizont pellet and pool the two pellets to have a final preparation of 100–150 µL, that results from one culture flask and contains ~10^8 schizonts that liberate ~10^9 merozoites during electroporation.
4. Immediately, mix the 100 µL schizont suspension with 300 µL of PBS containing 10–50 µg of linearized DNA, prepared as described above. Transfer the mixture (400 µL) into a chilled electroporation cuvette and deliver a single electric pulse with a Bio-Rad apparatus (800 V, 25 µF). The time constant is usually 0.8–1.4 ms. Place the cuvette on ice and rapidly (within 5–10 min) inject 0.2 mL of the electroporated parasites (merozoites) into the tail vein of each of two young (55–60 g), phenylhydrazine-treated (optional; *see* **Note 2**) Sprague-Dawley (SD) rats using a 1-mL tuberculin syringe with a 27-gage needle.

3.2.4. Parasite Selection

The parasitemia of the recipient SD rats is checked by Giemsa-stained blood smears prepared from a drop of blood obtained from the tail of the animals, beginning 4-h postelectroporation (D 0) and daily thereafter. At 4-h postelectroporation, the parasitemia should be 0.2 to 1.0%, consisting mainly of ring forms and young trophozoites.

1. Allow the parasites to complete one cycle of asexual division in the absence of drug pressure. Starting 27–30 h after electroporation, treat the recipient rats with a single intraperitoneal (ip) injection of 25 mg/kg pyrimethamine (*see* **Note 8**) every 24 h until replicating forms of the parasite are undetectable in blood smears (gametocytes can persist in the blood for longer periods). It usually takes 3–5 d for the parasitemia to decline to undetectable levels.
2. A population of pyrimethamine-resistant parasites, the parental resistant population (PRP), usually emerges at 8–10 d postelectroporation (or at 6–7 d postelectroporation for parasites transfected with episomal constructs). Treat the rats harboring the PRP for 2–3 con-

Gene Targeting

secutive days with pyrimethamine as described above. During this treatment period, the parasitemia should increase ~10-fold per day.

3. Passage the parasites of the PRP on the day of their detection (parasitemia ~0.01%) to a naive SD rat by IV injection of ~50–100 µL of tail blood (collected in an Eppendorf tube containing a drop of heparin) diluted to 200 µL with PBS. This yields a rapidly expanding transfer-resistant population (TRP1). Treat the TRP1, like the PRP, with 2–3 consecutive doses of pyrimethamine, starting at the day of their detection. The passage of the PRP is routinely done because we frequently obtain pyrimethamine-resistant parasites that expand slowly in the original recipient rat or are even cleared from the bloodstream before the parasitemia reaches 1%, perhaps due to the development of an antiparasite immune response.

4. If the parasites were transfected with targeting constructs bearing the PyrFlu selectable marker *(10)*, fluorescence of tagged RBC stages can be used for detection and further selection of recombinant parasite populations following pyrimethamine selection, by fluorescence-activated cell sorting (FACS) (*see* **Note 3**).

   a. Recombinant parasites within the PRP and TRP1 can be detected by examination of a drop of blood collected from the tail of an infected rat and diluted in PBS, by fluorescence microscopy using a standard FITC filter setting. Fluorescence displayed by parasites bearing PyrFlu-based episomes is usually stronger than that displayed by parasites bearing integrated constructs, reflecting differences in copy number of the plasmid in the two types of recombinants (up to 20 per nucleus in the former *[16]*).
   b. Analytical FACS can also be used to distinguish RBCs infected with recombinant parasites generated with the PyrFlu cassette from those infected with nontargeted but pyrimethamine-resistant parasites. The FACS is performed with a dilution of whole blood obtained by tail bleeding with PBS to the concentration of 10^6 cells/mL, and analysis using a FACScan machine equipped with an argon laser tuned at 488 nm. This is best done shortly after tail bleeding of the rat since the log fluorescence intensity of the parasites will decay with time.
   c. Fluorescent RBCs can be sorted from non-fluorescent cells at the speed of 500 cells/s using a Coulter Epics Elite sorter. Both populations can initiate an infection in rats when injected IV soon after sorting. This step may be especially useful when the desired recombinant parasites constitute a small proportion (<10%) of the pyrimethamine-resistant parental population as determined by Southern blot. The sorted fluorescent population can be cloned directly by IV injection into SD rats of a dilution that contains 1 parasite in 0.2 mL of PBS.

5. When the parasitemia induced by the PRP and/or the TRP1 exceeds 2%, collect the blood from the rats by cardiac puncture. An aliquot (~1 mL) of the blood is used for cryopreservation of the resistant parental populations. The parasites are cryopreserved by mixing 0.3 mL of blood with 0.6 mL of freezing solution in a 2-mL cryogenic vial, and frozen directly in liquid nitrogen. Use the remainder of the blood for passage to 1 or 2 naive SD rats for parasite cloning and for extraction of parasite genomic DNA (*see* **Subheadings 3.2.5.** and **3.2.6.**). The parasites should be passaged by IV injection to induce a parasitemia in the recipient rat(s) of ~0.01% at 2 d after transfer (TRP2). The genomic DNA of the PRP/TRP1 is subjected to Southern blot analysis to assess the presence of correctly targeted clones and evaluate their proportion in the total resistant population.

3.2.5. Parasite Cloning

Recombinant parasites in the TRP2 are cloned by limiting dilution when the rat parasitemia is between 0.1–1.0%, if at least 10% of the parasites contain the desired modification (*see* **Note 9**).

1. Determine the parasitemia of the donor rat harboring the TRP2. Ensure that no infected RBCs contain more than one parasite by careful examination of the blood smear.
2. Collect ~200 µL of blood from the donor rat and count the number of RBCs/µL of blood in a hemocytometer using a 200-fold dilution of whole blood in PBS.
3. Dilute the blood in 20% FBS-RPMI 1640 to a solution containing < 0.1 parasite/µL.
4. Inject an appropriate number of young (~55–75 g) SD rats in the tail vein using 1-mL U-100 insulin syringes, with a blood dilution that contains 1 parasite in 0.2 mL of PBS.
5. Parasites will become detectable in blood smears 6–7 d later in a proportion of the rats injected. To maximize the likelihood of injecting no more than one parasite, we consider valid only those cloning experiments where not more than 40% of the rats injected become positive.
6. When the parasitemia reaches ~2% collect the blood from the rats by cardiac puncture. Prepare frozen stocks of the clonal parasite populations as described above, and use the remainder of the blood to extract parasite genomic DNA for analysis by Southern blotting.
7. The cryopreserved stock of correctly targeted clones can then be used to infect SD rats or hamsters for transfer to mosquitoes and phenotype analysis. Thaw the cryopreserved stock at 37°C, and initiate the rodent infection by ip injection. Allow sugar-starved mosquitoes to feed on the animals as soon as they contain mature gametocytes that readily exflagellate on a microscope slide.

3.2.6. Preparation and Analysis of Parasite Genomic DNA

1. When the parasitemia of an animal is ~1% or greater, collect the blood by cardiac puncture.
2. Filter the blood to remove leukocytes through a Plasmodipur filter: Mix the blood with an equal volume of PBS and pour into a 12-mL syringe barrel that is connected to a Plasmodipur filter. Gently insert the plunger into the syringe barrel and filter the blood dropwise into a 50-mL Falcon tube. Wash the filter with 10 mL of PBS and collect the wash in the same Falcon tube.
3. Pellet the RBCs by centrifugation (700g, 10 min). Discard the supernatant.
4. Resuspend the RBC pellet in a total volume of 50 mL of PBS. Add 0.1 g of saponin and mix the solution vigorously to lyse the RBCs. Lysis is almost always immediate; otherwise the solution can be left for 10 min at room temperature for lysis to proceed.
5. Pellet the freed parasites (2000g, 10 min) in the 50-mL Falcon tube. Discard the supernatant. To wash the parasites, resuspend the pellet in 1 mL of PBS, transfer to a 1.5-mL Eppendorf tube and pellet (2000g, 5 min). Discard the supernatant.
6. Resuspend the parasite pellet in 200 µL of PBS and add 4 µL of RNase A (100 mg/mL).
7. Parasite lysis and DNA extraction is then performed using the QIAamp Blood Kit (Qiagen) according to the *Blood and Body Fluid Protocol* (see manufacturer's handbook). The DNA is usually eluted with 50-100 µL of elution buffer (Buffer AE, Qiagen) and stored at –20°C.
8. Analyze the genomic DNA by PCR and Southern blot hybridization as previously described, to reveal the presence of the desired recombination events *(13)*. For Southern blot analysis, label probes with DIG-dUTP by random priming using the DIG DNA Labeling Kit (Boehringer Mannheim) and detect chemiluminescence with CSPD using the DIG Luminescent Detection Kit For Nucleic Acids (Boehringer Mannheim).

3.3. Phenotypic Analysis of Recombinant Clones

One important aspect of gene targeting experiments is to ensure that the defective or new phenotype of a modified parasite clone is due to the lack or modification of the protein of interest. One control is to verify that independent recombinant clones (originating from independent transformation experiments) display the same altered phenotype. Another is to take advantage of gene inactivation or modifications created by

Gene Targeting

insertion plasmids and the intrachromosomal recombination that may occur between the duplicated sequences in the recombinant locus. If the parasites that have reverted to a wild-type locus after plasmid excision have regained a wild-type phenotype, then the altered phenotype is likely due to the planned mutation. It must be pointed out that both these controls can only rule out the possibility that the altered phenotype may result from unrelated genome modifications, which may occur during recombinant selection (the *Plasmodium* genome is known to be prone to large rearrangements during repeated cycles of mitotic parasite multiplication *[17]*). They do not, however, distinguish between an altered phenotype is likely due to the lack or modification of the protein of interest or to a side effect of the recombinant locus on expression of other, closely linked or distant loci (polar effect). Only typical complementation demonstrates that the observed phenotype is due (directly or not) to the lack or modification of the protein of interest, when reintroduction of the wild-type gene in the modified parasite restores the wild-type phenotype. This strategy requires a second selectable marker. A human DHFR gene that confers resistance to the drug WR99210 has recently been used successfully for complementation studies in *P. berghei* recombinants that express the pyrimethamine-resistant *DHFR-TS* gene *(18–20)*. It must also be kept in mind that complementation methods that rely on expression of the wild-type gene from nonintegrated episomes are limited by the high variability in the copy number of the episomes, as well as by the need to apply selective pressure for maintaining the plasmid in replicating parasites. Ideally, the plasmid containing the wild-type gene should thus be linearized prior to electroporation and integrated at the inactivated/modified locus.

4. Notes

1. For gene targeting experiments involving genes expressed at posterythrocytic stages of the parasite, it is important to ensure that targeted clones are gametocyte producers. Gametocyte production can be rapidly lost with repeated erythrocytic cycles and parasite transfers to new animals, and non-gametocyte producer lines rapidly overgrow the gametocyte-producer population *(17)*. Thus parasites that are used as recipients for the electroporation experiments should have gone through a minimal number of schizogonic cycles, i.e., should originate from a recent sporozoite-induced infection.
2. Merozoites of *P. berghei* preferentially invade reticulocytes. Recipient rats can be treated with a single ip dose of phenylhydrazine hydrochloride (30–40 mg/kg body weight) 5 d prior to injection with the electroporated material, to induce reticulocytosis, which could enhance the invasion rate of merozoites. We have found that phenylhydrazine treatment is not necessary to obtain a high parasitemia at 1 d postelectroporation and does not significantly influence the outcome of gene targeting experiments.
3. Pyrimethamine selection alone often yields a proportion of resistant nontargeted mutants, which appear like wild-type by Southern blot analysis. These may arise from gene conversion of the endogenous *DHFR-TS* by the incoming mutated copy in the resistance cassette, or from spontaneous mutations in the endogenous *DHFR-TS*. The concomitant fluorescent signal from targeted clones that bear the PyrFlu selectable marker allows a further selection and enrichment of recombinants by flow cytometry.
4. Short linker sequences of the vector left at one or both ends of the liberated insert of a replacement plasmid do not impede the correct allelic exchange and are not incorporated in the final structure *(3,5)*. Because no sequence is duplicated in the recombinant locus created by a double crossover event promoted by this type of replacement construct, the structure is stable and cannot revert to the wild-type configuration. Undesired events may include double crossovers involving concatemers of linear molecules *(13)*.

5. Undesired events during integration of insertion plasmids usually consist of insertion of more than one copy of the plasmid arranged in tandem repeats *(13)*; however, if the targeting plasmid does not contain a full-length copy of the target gene, all of the intermediate copies are truncated. Because the region of homology is duplicated in the recombinant locus created by plasmid insertion, the two copies can recombine to excise the plasmid and regenerate the wild-type gene. We have found such spontaneous plasmid excision events in TRAP⁻ sporozoites generated by integration of an insertion plasmid *(5)*.

6. Such a replacement plasmid contains a duplicated sequence. It is unstably maintained in *E. coli*, and often undergoes intramolecular recombination that results in the appearance of a smaller plasmid in addition to the original construct, during propagation in *E. coli* (V. Thathy and A. Nunes, unpublished results). This could lead to difficulties in purifying high amounts of the construct for electroporation.

7. In *P. berghei*, insertion plasmids integrate via a double-strand gap repair mechanism *(7)*. When using the *TRAP* gene as a target, mutations introduced within the targeting sequence close (< 450 bp) to the linearization site were often repaired by double-strand gap enlargement or mismatch correction, whereas mutations located 600 bp and farther from the linearization site were frequently maintained in the recombinant loci.

8. Pyrimethamine given at a dose of 25 mg/kg body weight is the most effective dose for parasites of the NK65 strain of *P. berghei*. However this dose is toxic for rats less than 50 g, and rats of this weight should not be used as recipients of the electroporated parasites.

9. If <10% of the parasites in the parental resistant population consist of the correct recombinants upon Southern analysis, a subcloning step can be performed to enrich the recombinant population in the absence of the PyrFlu selectable marker. A number of rats are injected with pools of 5 or 10 parasites. The gDNA of each parasite population is then subjected to Southern analysis to evaluate the proportion of recombinants present. The population containing the highest proportion of integrants can then be chosen for parasite cloning by limiting dilution.

References

1. Wu, Y., Kirkman, L. A., and Wellems, T. E. (1996) Transformation of *Plasmodium falciparum* malaria parasites by homologous integration of plasmids that confer resistance to pyrimethamine. *Proc. Natl. Acad. Sci. USA* **93,** 1130–1134.
2. Van Dijk, M. R., Janse, C. J., and Waters, A. P. (1996) Expression of a *Plasmodium* gene introduced into sub-telomeric regions of *Plasmodium berghei* chromosomes. *Science* **271,** 662–665.
3. Ménard, R., Sultan, A. A., Cortes, C., Altszuler, R., van Dijk, M. R., Janse, C.J., Waters, A. P., Nussenzweig, R. S., and Nussenzweig, V. (1997) Circumsporozoite protein is required for development of malaria sporozoites in mosquitoes. *Nature* **385,** 336–340.
4. Crabb, B. S., Cooke, B. M., Reeder, J. C., Waller, R. F., Caruana, S. R., Davern, K. M., et al. (1997) Targeted gene disruption shows that knobs enable malaria-infected red cells to cytoadhere under physiological sheer stress. *Cell* **89,** 287–296.
5. Sultan, A. A., Thathy, V., Frevert, U., Robson, K. J. H., Crisanti, A., Nussenzweig, V., Nussenzweig, R. S., and Ménard, R. (1997) TRAP is necessary for gliding motility and infectivity of Plasmodium sporozoites. *Cell* **90,** 511–522.
6. Triglia, T., Wang, P., Sims, P. F. G., Hyde, J. E., and Cowman, A. F. (1998) Allelic exchange at the endogenous genomic locus in *Plasmodium falciparum* proves the role of dihydropterate synthase in sulfadoxine-resistant malaria. *EMBO J.* **17,** 3807–3815.
7. Nunes, A., Thathy, V., Bruderer, T., Sultan, A. A., Nussenzweig, R. S., and Ménard, R. (1999) Subtle mutagenesis by ends-in recombination in malaria parasites. *Mol. Cell. Biol.* **19,** 2895–2902.
8. Lobo, C. A., Fujioka, hr., Aikawa, M., and Kumar, N. (1999) Disruption of the Pfg27 locus by homologous recombination leads to loss of the sexual phenotype in *P. falciparum*. *Mol. Cell* **3,** 793–798.
9. Van Dijk, M. R., Waters, A. P., and Janse, C. J. (1995) Stable transfection of malaria parasite blood stages. *Science* **268,** 1358–1362.
10. Sultan, A. A., Thathy, V., Nussenzweig, V., and Ménard, R. (1999) Green fluorescent protein as a marker in *Plasmodium berghei* transformation. *Infect. Immun.* **67,** 2602-2606.
11. Janse, C. J. and Waters, A. P. (1995) *Plasmodium berghei*: The application of cultivation and purification techniques to molecular studies of malaria parasites. *Parasitol. Today* **11,** 138–143.

12. Waters, A. P., Thomas, A. W., van Dijk, M. R., and Janse C. J. (1997) Transfection of malaria parasites, in *Methods: A Companion to Methods in Enzymology-Analysis of Apicomplexan Parasites,* vol. 13, Academic Press, New York, NY, pp. 134–147.
13. Ménard, R. and Janse, C. (1997) Gene targeting in malaria parasites, in *Methods: A Companion to Methods in Enzymology-Analysis of Apicomplexan Parasites,* vol. 13, Academic Press, New York, NY, pp. 148–157.
14. Cormack, B. P., Valdivia, R. H., and Falkow, S. (1996) FACS-optimized mutants of the green fluorescent protein (GFP). *Gene* **173,** 33–38.
15. Crabb, B. S. and Cowman, A. F. (1996) Characterization of promoters and stable transfection by homologous and nonhomologous recombination in *Plasmodium falciparum. Proc. Natl. Acad. Sci. USA* **93,** 7289–7294.
16. Van Dijk, M. R., Vinkenoog, R., Ramesar, J., Vervenne, R. A. W., Waters, A. P., and Janse, C. J. (1997) Replication, expression, and segregation of plasmid-borne DNA in genetically transformed malaria parasites. *Mol. Biochem. Parasitol.* **86,** 155–162.
17. Janse, C. J. (1993) Chromosome size polymorphism and DNA rearrangements in *Plasmodium. Parasitol. Today* **9,** 19–22.
18. de Koning-Ward, T. F., Fidock, D. A., Thathy, V., Ménard, R., van Spaendonk, R. M., Waters, A. P., and Janse, C. J. (2000) The selectable marker human dihydrofolate reductase enables sequential genetic manipulation of the *Plasmodium berghei* genome. *Mol. Biochem. Parasitol.* **106,** 199–212.
19. Sultan, A. A., Thathy, V., deKoning-Ward, T. F., and Nussenzweig, V. (2001) Complementation of *Plasmodium berghei* TRAP knockout parasite using human dihydrofolate reductase gene as a seletable marker. *Mol. Biochem. Parasitol.* **113,** 151–156.
20. Thathy, V., Fujioka, H., Gantt, S., Nussenzweig, R., Nussenzweig, V., and Ménard, R. (2002) Levels of CS protein in the *Plasmodium* oocyst control sporozoite budding and morphology. *EMBO J.* (in press).

V

Immunological Techniques

30

Peptide Vaccination

Valentin Meraldi, Jackeline F. Romero, and Giampietro Corradin

1. Introduction

Vaccination is the most cost-effective measure to control the pathological effects of an infectious agent *(1,2)*. With the combined use of molecular cloning and of modern sequencing methods, the sequence of many protein components present in different infectious agents have been elucidated, opening the way to the design and development of various vaccination strategies which include peptides, proteins and DNA. Peptide vaccination is *per se* the most simple, straightforward, and, in our opinion, safest approach to vaccination of the world population. For peptide, we intend protein fragments of any length obtained from chemical synthesis *(3–5)*. One drawback of using short peptides is the limitation of the intervention to defined members of the population that carry the major histocompatibility complex (MHC) antigen(s) which the peptides bind to (MHC restriction). This limitation can, in principle, be overcome by using a physical or chemical mixture of short peptides or to lengthen the size of the peptide fragment in order to cover the MHC antigens of the entire population *(6,7)*.

Another disadvantage of synthetic peptides is the possible contamination with chemical derivatives generated during the synthesis or cleavage of peptides from the resin that may require lengthy purification procedures to obtain a pure product, especially if it is intended for human use. Notwithstanding, peptide synthesis allows certain flexibility in vaccine design, as for example, the modification of the N and C termini to prevent degradation *(8)*.

Peptides tend to be less immunogenic than the whole organism or individual protein *(9)*. Thus, powerful adjuvants need to be added to the formulation in order to elicit a full immunological response. Overcoming these limitations, protection from malaria infection has been obtained through peptide immunization *(10–15)*.

In the following sections, we describe in detail the purification of synthetic peptides with size exclusion chromatography and reverse phase-high performance liquid chromatography (RP-HPLC), and we discuss the dissolution of peptide antigen and preparation of vaccine formulations.

2. Materials

The catalog numbers in parenthesis indicate articles we used in our work.

2.1. Size Exclusion Chromatography

1. Sephadex G-10, for peptides with mol wt <1000 Da (Pharmacia, Uppsala, Sweden, cat. no. 17-0010-01). Sephadex G-25 Fine, for peptides with mol wt <2000 Da (Pharmacia, cat. no. 17-0032-01). Sephadex G-50 Superfine, for peptides with mol wt between 2000 and 10,000 Da (Pharmacia, cat. no. 17-0041-01).
2. Glass columns of various dimensions from Bio-Rad (Hercules, CA).
3. 50% (v/v) acetic acid (Fluka Chemika, Buchs, Switzerland, cat. no. 45731) solution.
4. Peristaltic pump.
5. UV detector.
6. Recorder.
7. Fraction collector.

2.2. HPLC Purification

1. Solution A: 100% water (ROM pure Chemistry, Romil Ltd., Cambridge, UK, cat. no. H950) and 0.1% trifluoroacetic acid (TFA) (Fluka Chemika, cat. no. 91699).
2. Solution B: 100% acetonitrile (Fluka Chemika, cat. no. 00692) and 0.1% TFA (Fluka chemika, cat. no. 91699).
3. Isopropanol (Fluka Chemika, cat. no. 59310).
4. Nitric acid (Fluka Chemika, cat. no. 17078).
5. 1-mL syringe, from Plastipak® (Becton Dickinson, NJ, cat. no. 308.400).
6. Needle "type 3" with end beveled at 90° (Hamilton, Reno, NE).
7. HPLC apparatus.
8. Reverse-Phase HPLC (RP-HPLC) columns (25 cm length × 10 mm diameter) from Vydac™ (Hesperia, CA) of the Separations Group (Hesperia, CA) or any other manufacturer:C4 (cat. no. 214TP10), C8 (cat. no. 208TP10), and C18 (cat. no. 218TP10).
9. Fraction collector.

2.3. Quality Control

2.3.1. Analytical HPLC

For analytical reverse phase HPLC, the same materials as in the purification procedure are used, except for the choice of the columns. They have a smaller diameter in order to avoid the phenomena of diffusion and to save peptide materials and solvents. RP-HPLC columns (25-cm length × 4.6-mm diameter) can be obtained from Vydac™ of the Separations Group or any other manufacturer: C4 (cat. no. 214TP54), C8 (cat. no. 208TP54), and C18 (cat. no. 218TP54).

2.3.2. Mass Spectrometry

1. MALDI-TOF Mass spectrometry, Voyager-DE™ RP from PerSeptive Biosystems or any other similar instrument.
2. Software BioSpectrometry™ Workstation.
3. Sinapinic acid (3,5-dimethoxy-4-hydroxycinnamic acid) (Sigma Chemical Co., St. Louis, MO, cat. no. D 7927) is used as matrix for peptides and proteins with mol wt >10,000 Da. For peptides and proteins with mol wt <10,000 Da, the matrix is composed of α-cyano-4-hydroxycinnamic acid (CHCA) (Sigma, cat. no. C 2020).

2.3.3. Absorption Spectroscopy

1. Varian UV-visible Spectrophotometer Cary™ 1E (Varian Australia Pty Ltd., VIC, Australia) or any other equivalent instrument.

2. Software Cary™ OS/2 system version 2.5.
3. Quartz cuvets of 1.0-mL and 0.5-mL vol.

2.4. Subcutaneous and Intraperitoneal Injection of Mice

1. Syringe of 1-mL and 2-mLcapacity from Plastipak® (Becton Dickinson, cat. no. 308.400 and cat. no. 300.185).
2. Needle of 25-gage 5/8 (0.5 mm diameter x 16 mm length) from Microlance 3® (Becton Dickinson, cat. no. 300600).
3. 1.5-mL microtubes with safe lock from Eppendorf® (cat. no. 0030120.086, Hamburg, Germany).
4. 0.9% NaCl (Fluka Chemika, cat. no. 71380) solution in sterile distilled water.
5. Dimethylsulfoxide (DMSO) (Sigma, cat. no. D-5879).
6. Incomplete Freund's Adjuvant (IFA) from Difco Laboratories® (cat. no. 0639-60-6, Detroit, MI, USA).
7. Complete Freund's Adjuvant® (CFA) from Difco Laboratories (cat. no. 3113-60-5 with *M. tuberculosis* and cat. no. 0638-60-7 with *M. butyricum*).
8. Montanide® ISA (Incomplete Seppic Adjuvant) 720 and ISA 51 adjuvants (Seppic, Paris, France).
9. Adjuvant OM-174®, from OM Pharma (1212 Meyrin 2, Switzerland).
10. Adjuvant QS-21 (Antigenics, Framingham, MA).
11. Adjuvant aluminum hydroxide, $Al(OH)_3$ in H_2O (Pasteur Merieux S&V, France).
12. Vortex apparatus.
13. Sonicator with horn tip (Branson Sonifier 250, Danbury, CT, cat. no. 100-413-016).
14. Mouse holder for subcutaneous injection.

3. Methods
3.1. Size Exclusion Chromatography

The gel filtration is used after the cleavage of the peptide from the resin as a prepurification step to remove all the synthesis reagents, side products and partially truncated sequences *(16)*. The gel filtration material is composed of particles with a network of pores that form a molecular sieve. Molecules of small size penetrate into these pores, are retained and eluted in the "total volume" of the column. Peptides and polypeptides are too voluminous to diffuse in the pores and are eluted more quickly, in the "void volume" of the column.

1. Follow manufacturer's instructions for swelling the resin (column material or stationary phase).
2. Resuspend the resin (the volume of the column plus 10%) in 50% (v/v) acetic acid (mobile phase) with a resin:mobile phase ratio of 1:2.
3. Stir delicately the resin to form a homogeneous suspension.
4. Pack the column by pouring delicately the resin down a glass rod onto the inside wall of the glass column with the appropriate dimensions. Avoid as much as possible the formation of air bubbles.
5. Let the resin settle down for 30 min to 1 h.
6. Connect the top of the column to a buffer reservoir and equilibrate the column with the mobile phase (50% (v/v) acetic acid) at a flow rate of 10–15 mL/h.
7. Dissolve the lyophilized peptide in the smallest possible volume of 50% (v/v) acetic acid. It should be less than 5% of the volume of the column.
8. Disconnect the top of the column from the buffer reservoir and allow the buffer to completely penetrate into the resin. The column should not run dry.

9. Add delicately the peptide solution with a Pasteur pipet on the top of the stationary phase. Avoid as much as possible the swirling of the resin.
10. Let the peptide solution penetrate into the stationary phase, then add delicately some solvent of the mobile phase to start the size exclusion chromatography.
11. Record the absorbance at 280 nm in order to obtain a precise chromatogram of the prepurification.
12. Collect the fractions and analyze them individually by ninhydrin reaction (Kaiser test) or mass spectrometry *(17,18)*.

3.2. HPLC Purification

RP-HPLC has proven to be an invaluable tool for the purification and analysis of peptides and polypeptides. The technique of RP-HPLC relies on the fact that polypeptides are adsorbed onto the hydrophobic, stationary phase composed of aliphatic chains attached to a silica surface (*see* **Note 1**). The polypeptide remains adsorbed until the organic component of the mobile phase reaches a critical concentration. The peptide is then quickly eluted with little interactions with the reversed phase, which accounts for the sharp peaks and high resolution of this technique *(17,18)*.

1. Dissolve the lyophilized peptide in buffer A. If the peptide is not soluble in buffer A, it can be dissolved in 50% (v/v) acetic acid (*see* **Note 2**). On the contrary, if the peptide is very hydrophobic, try to dissolve it in acetonitrile.
2. Program the gradient for the elution. For example: start with 100% buffer A and end with 100% buffer B with a slope of 1% per min. The optimal flow rate using a column of 10-mm diameter is generally 3 mL/min.
3. Inject manually or automatically the peptide solution onto the RP-HPLC (*see* **Note 3**).
4. Collect the fractions (*see* **Note 3**). The resolution of the separation is increased with small volume fractions.
5. Wash the column with 100% buffer B or optionally with a mixture of one part of nitric acid 0.1 N and 4 parts of isopropanol (*see* **Note 3**).

3.3. Quality Control

3.3.1. Analytical HPLC

For the analytical RP-HPLC, the same procedure as described for the purification is used, except for the flow rate that is maintained at 1 mL/min or even lower. The quantity of sample analyzed per HPLC run varies between 10 and 100 µg *(17,18)*.

The retention time or volume are determined according to the chromatogram and are characteristic for each component.

The width of the elution peak at half of its height gives a good appreciation of the purity of the polypeptide. A non-symmetrical elution peak reveals the presence of two or more components with the same elution time (*see* **Note 4**).

3.3.2. Mass Spectrometry

Mass spectrometry (MS) has become a powerful technique for determining the mass of macromolecules and in addition, primary structure of polypeptides and proteins can be obtained (*see* **Note 4**) *(19)*. This technique has the great advantage of sensitivity and accuracy. In fact, only picomoles of sample are required and they do not need to be homogeneous. The MS technique is so developed today that it allows the sequencing of small peptides in the range of 20 amino acids *(20)*. In complementation with other techniques, the MS can also give some information about the structure of the polypeptide *(21)*.

1. Clean the sample plate and handle it only by the edge. If you wear gloves, use only powder-free gloves.
2. Prepare samples at a concentration of 1 to 10 pmol/µL just before loading the sample plate (*see* **Notes 5–7**).
3. Mix samples and matrix in microtubes before applying them to the sample plate when you are working with concentrated peptide solutions. On the contrary, if your samples are diluted, mix them with the matrix directly on the sample plate. To prevent matrix from drying, apply samples first.
4. Load 1.0 to 2.0 µL of sample on the sample plate. Do not touch the sample well with the pipet tip, it may cause uneven crystallization.
5. Perform the MS measurements.

3.3.3. Absorption Spectroscopy

The aromatic amino acids phenylalanine, tyrosine, tryptophan, and also disulfide bonds have the capacity to absorb light in the near-ultraviolet range. This physical property is used in the laboratory to determine the concentration of peptide solution using the molar extinction coefficient ε [cm$^{-1}M^{-1}$], (Cysteine: $\varepsilon = 300$ with $\lambda_{max}=250$ nm; phenylalanine: $\varepsilon = 197$ and $\lambda_{max} = 257.4$ nm; tyrosine: $\varepsilon = 1420$ with $\lambda_{max} = 274.6$ nm; tryptophan: $\varepsilon = 5600$ with $\lambda_{max} = 279.8$ nm) *(22)*. The number of aromatic residues and disulfide bonds can be determined from the UV absorbance spectrum under denaturing conditions so that its spectrum is the sum of its constituent residues *(23)*. This can be used to determine the degree of purity of a peptide solution (*see* **Note 4**).

1. Dissolve the lyophilized peptide in 50% (v/v) acetic acid.
2. Scan the blank solution using a quartz cuvet in the UV range of 240 to 320 nm.
3. Scan the sample solution in the same range.
4. Calculate the concentration of the peptide solution by using Lambert's law:

$$\log I_0/I = OD = \varepsilon \, c \, D$$

I_0: intensity of the incident light; I: intensity of the transmitted light; where OD: optical density; c (M) = concentration; ε (m$^{-1}M^{-1}$) = molar extinction coefficient; and D (cm) = length of the cuvette *(18)*.

3.4. Storage

Peptides are stored at –20°C or at 4°C in powder form, or at –20°C when dissolve in DMSO or buffer. Do not store peptides for long periods in 50% (v/v) acetic acid or TFA (*see* **Note 8**).

3.5. Dissolution of Peptides and Preparation of Antigen for Immunization

Peptides are generally soluble in 50% (v/v) acetic acid, but if this is not the case, they can be dissolved in TFA. After peptides are fully dissolved in 50% (v/v) acetic acid or TFA, they may be further diluted with water. For the enzyme linked immunosorbent assay (ELISA) assay, if peptides are not soluble in the ELISA buffer, dissolve them in 50% (v/v) acetic acid and let the solution evaporate under a hood until dryness. Continue then as indicated in the ELISA protocol. For T-cell proliferation, CTL and enzyme-linked immunospot (ELISPOT) assays, peptides are generally dissolved at a high concentration (10–20 mg/mL) in DMSO and directly diluted in the assay medium (*see* **Notes 3** and **9**). Before dissolving a peptide in a salt-containing buffer, try first on a small scale to verify whether the peptide is soluble in the buffer.

The peptide solution for vaccination should be free of contaminants, in particular from bacteria or bacterial products, in order to prevent any sepsis or extensive inflammatory reactions. Therefore, it is recommended to filter the antigen solutions through a microporous filter (0.22-µm pore size) before adding the adjuvant, and to maintain the peptide solution under sterile conditions. Measures to avoid contamination of the antigen solution with other substances, such as endotoxin or organic components, must be taken at any stage of the peptide purification, preparation, and storage.

3.6. Preparation of IFA or CFA Emulsions

IFA and CFA have long been considered as the golden standards of adjuvants. CFA is a mixture of nonmetabolizable oil (mineral oil), a surfactant (Arlacel A) and mycobacteria (*M. tuberculosis* or *M. butyricum*). It is prepared as water-in-oil vaccine emulsions. However, the reactogenicity and toxicity that CFA induces has made its use unacceptable even in laboratory animals. Therefore, CFA should be used only for the first immunization (priming); the subsequent immunizations should be performed with IFA or other adjuvants. IFA contains the same oil/surfactant mixture but does not contain any mycobacteria. These two adjuvants are not accepted for use in human vaccines *(24)*.

1. Mix thoroughly the source bottle of CFA in order to ensure a homogenous distribution of mycobacteria.
2. Dissolve the desired amount of peptide antigen in water or buffered saline or, if not soluble, in DMSO (*see* **Subheading 3.5.**).
3. Add one part of antigen solution to one part of IFA or CFA in a 2-mL syringe with its end closed with paraffin (*see* **Note 9**).
4. Emulsify the solution by filling and emptying the syringe or by vortexing.
5. Sonicate this emulsion with a horn tip sonicator with an output power of 80 W for 2 to 3 s until a thick emulsion has been formed. In order to avoid heating, the syringe is kept on ice.
6. Once you have obtained a homogenous oil-in-water emulsion, transfer the content with the plunger into a 1-mL syringe for the injection.
7. Keep the syringe at a temperature of approx 4°C.

3.7. Preparation of Montanide® ISA Emulsions

3.7.1 Montanide® ISA 51

Montanide® ISA 51 is an adjuvant based on mannide oleate in mineral oil solution. It is a clear, slightly colored product of low viscosity, especially recommended by the manufacturer for formulations with a high content of antigen *(25,26)*.

1. Bring the adjuvant to room temperature (RT). Once opened, the ampoule of adjuvant Montanide® ISA 51 is stored at 4°C.
2. Dissolve the desired amount of peptide antigen in water or buffered saline or, if not soluble, in DMSO (*see* **Subheading 3.5.**).
3. Add one part of Montanide® ISA 51 to one part of antigen solution in a 2-mL syringe.
4. Emulsify the solution by filling and emptying the syringe or by vortexing.
5. Sonicate the emulsion with a horn tip sonicator at the power of 80 W for 2 to 3 s until a thick emulsion has been formed. In order to avoid heating, the syringe is kept on ice. This step is optional.

6. Transfer the obtained water-in-oil preparation with the plunger into a 1-mL syringe for the injection.

3.7.2 Montanide® ISA 720

Montanide® ISA 720 is an oil adjuvant containing natural metabolizable oil and a highly refined emulsifier from the mannide monooleate family. It is designed for the production of water-in-oil injectable preparations. It is a ready-to-use product that can be mixed directly with an aqueous solution to provide stable and very fluid vaccine preparations.

The recommended ratio is 3 parts of antigen solution to 7 parts of Montanide ISA 720 *(27,28)*.

The procedure for the preparation of the vaccine emulsion is the same as for Montanide ISA 51. Once opened, the ampoule of adjuvant Montanide ISA 720 is stored at 4°C.

3.8. Preparation of QS-21 Solution

The adjuvant QS-21 is a complex glycolipid derived from saponin rich bark extracts of the Quillaja saponaria Molina tree. QS-21 is easy to prepare; it is not an emulsion *(29,30)*.

1. Solubilize the QS-21 powder in water at pH greater than 5.0 or in buffered saline at pH between 6.0 and 7.4. Its solubility is highest between pH 6.0 and 7.0. At alkaline pH, the QS-21 can be hydrolyzed and can lose an essential fatty acid moiety.
2. The QS-21 solution can be filter sterilized through a 0.22-µm filter.
3. Dissolve the desired amount of peptide antigen in water or buffered saline, or if not soluble, in DMSO (*see* **Subheading 3.5.**).
4. To obtain the vaccination preparation, the QS-21 solution is simply mixed with the antigens. No emulsification is required. The manufacturer advises to use 20 µg of adjuvant QS-21 per mouse.
5. Optionally QS-21 can be added to antigen/aluminum hydroxide mixtures.
6. Store the QS-21 solution at 2–8°C (if sterile) or at –20°C. Freeze/thaw processes do not cause alteration of the adjuvant.

3.9. Preparation of OM-174® Solution

The adjuvant OM-174® is a novel glucosamine disaccharide related to lipid A of Escherichia Coli. OM-174 is a clear solution that permits the rapid and easy formulation of soluble vaccine preparations *(31,32)*.

1. Incubate the adjuvant at 37°C for 10 min or at RT for 1 h.
2. Vortex vigorously for 3 min.
3. Dissolve the desired amount of peptide antigen in water or buffered saline or, if not soluble, in DMSO (*see* **Subheading 3.5.**).
4. Mix the necessary amount of OM-174 with the aqueous antigen solution or in 0.9% NaCl (*see* **Note 11**). The manufacturer recommends using 50 µg of adjuvant OM-174 per mice.
5. Vortex the mixture again for 3 min and transfer into the syringe for immunization.
6. Keep the vaccine solutions between 2 to 8°C before use. The adjuvant OM-174 is stored at 4°C.

3.10. Preparation of Aluminum Hydroxide Salt

Aluminum hydroxide salt can be used to adsorb peptides and proteins in a ratio of 50 to 200 µg proteins/mg aluminum hydroxide. Adsorption of proteins is dependent on the pI (isoelectric point) of the protein and the pH of the medium. Proteins with lower pI adsorb to the positively charged aluminum ion more strongly than proteins with higher pI (*33–35*). Aluminum salts are generally weaker adjuvants than emulsion adjuvants and, because of their short-term depot effect, booster injections are often needed. Due to their mild inflammatory reactions they are the primary adjuvants utilized in humans.

1. Bring the adjuvant preparation to room temperature.
2. Dissolve the desired amount of peptide antigen in water or buffered saline or, if not soluble, in DMSO (*see* **Subheading 3.5.**).
3. Add the antigen to the aluminum hydroxide solution. Each mouse receives 1 mg of aluminum hydroxide.
4. The adsorption of the antigen on the aluminum hydroxide salt is performed at RT for 30 min.
5. Centrifuge the solution at $1000g$ for 10 min at 4°C.
6. Wash the precipitate two times with PBS.
7. The concentration of adsorbed peptide is established by measuring the concentration of peptide in the collected supernatants (*see* **Subheading 3.3.3.**).
8. Resuspend the precipitate at the desired concentration with PBS for the injection.
9. The aluminum hydroxide is stored at 4°C (*see* **Note 12**).

3.11. Subcutaneous Injection of Mice

1. Fill the 1-mL syringe with the solution to inject and remove the air bubbles.
2. Remove the mouse from the cage by gently grasping the tail at the base.
3. Place the mouse in the mouse holder and maintain its position by firmly pulling its tail.
4. Immobilize the tail by gentle rearward traction.
5. Clean the injection site to remove debris that may result in contamination or infection.
6. Bend the tail at its first third near its base between the thumb and forefinger.
7. Insert the needle under the skin at the bend until reaching the base of the tail.
8. Inject with moderate pressure and speed in order to avoid reflux of the injection solution (*see* **Note 13**).
9. Remove delicately and slowly the needle and place the immunized mouse in a separate cage.

3.12. Intraperitoneal Injection of Mice

1. Fill up the 1-mL syringe with the solution to inject and remove the air bubbles.
2. Remove the mouse from the cage by gently grasping the tail at the base.
3. Place the mouse on a wire-bar cage lid to permit grasping. Apply gentle rearward traction on the tail.
4. Approach the nape of the mouse from the rear with the free hand. Firmly grasp the skin behind the ears with the thumb and index finger.
5. Put the tail beneath the little finger of the hand that is holding the mouse.
6. Expose the animal's abdomen.
7. Clean the injection site to remove debris that may result in contamination or infection.
8. Insert the needle into the lower left or right quadrant of the abdomen, avoiding the abdominal midline.
9. Inject with moderate pressure and speed.
10. Remove delicately the needle and place the immunized mouse in a separate cage.

3.13. End Point Immunological Analysis

Immunological readouts are detailed elsewhere in Part V of this volume (Chapters 31–42), and include the following:

1. T-cell proliferation.
2. CTL response.
3. ELISPOT assay.
4. Antibody response (ELISA).
5. Protection.

4. Notes

1. The choice of the RP-HPLC columns for the purification and for the analytical purposes depends on the molecular weight and hydrophobicity of the peptide. C4 columns are recommended for the separation of polypeptides larger than 4000–5000 Da or for very hydrophobic peptides of any size. C18 columns are recommended for the separation of peptides smaller than 4000–5000 Da, enzymatically digested fragments and synthetic peptides.
2. The injection peak from the 50% (v/v) acetic acid used to dissolve the peptide can be separated from the rest of the chromatogram by programming a plateau of buffer A before the gradient.
3. Cross-contamination is a real risk when dealing simultaneously with a number of peptides. Precautions should be taken to avoid it. These precautions are (a) gel exclusion and HPLC columns have to be thoroughly washed; (b) physical barriers (for example, filters on each container) have to be installed during lyophilization; (c) extreme care and thorough cleaning of dispensable pipettes and bulbs for Pasteur pipets. This is especially important in the case of CTL peptides that can be active even at 10^{-12} M or lower concentrations.
4. Purity of the final product is a rather subjective appreciation and the level of purity required depends on its applications. For vaccine development and diagnostics, it should be of the highest possible level. That means a single peak in mass spectrometry and analytical HPLC analysis, with purity exceeding 95%. For large polypeptides this is difficult to obtain if a high yield is also required.
5. A higher concentration of TFA may enhance sample ionization and improve sensitivity in mass spectrometry analysis.
6. Many samples adhere strongly to plastic tubes and pipet tips. You can minimize sample loss by preparing samples in 30% acetonitrile with 5 to 10% TFA.
7. Do not dilute sample with phosphate-buffered saline (PBS) or any other buffer. A high salt concentration can interfere with sample ionization in mass spectrometry analysis.
8. During stocking, peptides may adsorb water and assume a gluey aspect. In this case, dissolve them in a minimum amount of 50% (v/v) acetic acid (this could take as long as a few hours, slight heating at about 40°C may also help), dilute with water and lyophilize again.
9. For fast screening, peptide preparations can be used directly after cleavage from the resin without further purification. Only occasionally, we could observe toxicity in in vitro culture.
10. The adjuvant CFA is a potential hazard for laboratory personnel. Avoid any contact with skin and splashing in the eyes.
11. The adjuvant OM-174 tends to be viscous at concentrations greater than 2.5 mg/mL when mixed with a 0.9% NaCl solution.
12. The aluminum hydroxide salt preparation should not be frozen or lyophilized in order to avoid the agglomeration and precipitation of the salt.
13. Small mice (5 to 6 wk of age) should be immunized subcutaneously with a maximum volume of 50 µL. Elder mice can be immunized subcutaneously with a volume of 100 µL.

References

1. Gander, B., Merkle, H. P., and Corradin, G. (eds.) (1997) *Antigen Delivery System*. Harwood Academic Publishers, Amsterdam, The Netherlands.
2. *State of the World's Vaccines and Immunization*. WHO/UNICEF document. (1996) WHO/GPV/96.04.

3. Merrifield, R. B. (1986) Solid phase synthesis. *Science* **232,** 341–374.
4. Atherton, E., Logan, C. J., and Shepard, R. C. (1988) Peptide synthesis. Part II. Procedures for solid phase synthesis using Na-fluorenylmethoxycarbamylamino-acids on polyamide supports: synthesis of substance P and of acyl carrier protein 65–74 decapeptides. *J. Chem. Soc.* (Lond.) **1,** 538.
5. Pennington, M. W. and Dunn, B. M. (eds.) (1994) *Peptide Synthesis Protocols. Methods in Molecular Biology,* vol. 35. Humana Press, Totowa, NJ.
6. Sinigaglia, F., Guttinger, M., Romagnoli, P., and Takacs, B. (1990) Malaria antigens and MHC restriction. *Immunol. Lett.* **25,** 265–270.
7. Hoffman, S. L., Berzofsky, J. A., Isenbarger, D., Zeltser, E., Majarian, W. R., Gross, M., and Ballou, W. R. (1989) Immune response gene regulation of immunity to *Plasmodium berghei* sporozoites and circumsporozoite protein vaccines. Overcoming genetic restriction with whole organism and subunit vaccines. *J. Immunol.* **142,** 3581–3584.
8. Heimbrook, D. C., Saari, W. S., Balishin, N. L., Fisher, T. W., Friedman, A., Kiefer, D. M., et al. (1991) Gastrin releasing peptide antagonists with improved potency and stability. *J. Med. Chem.* **34,** 2102–2107.
9. Valmori, D., Pessi, A., Bianchi, E., and Corradin, G. (1992) Use of human universal antigenic tetanus toxin T cell epitopes as carriers for human vaccination. *J. Immunol.* **149,** 717–721.
10. Renggli, J., Valmori, D., Romero, J. F., Eberl, G., Romero, P., Betschart, B., and Corradin, G. (1995) CD8$^+$ T-cell protective immunity induced by immunization with *Plasmodium berghei* CS protein-derived synthetic peptides: evidence that localization of peptide-specific CTL is crucial for protection against malaria. *Immunol. Lett.* **46,** 199–205.
11. Migliorini, P., Betschart, B., and Corradin, G. (1993) Malaria vaccine: immunization of mice with a synthetic T cell helper epitope leads to protective immunity. *Eur. J. Immunol.* **23,** 582–585.
12. Blum-Tirouvanziam, U., Beghdadi-Rais, C., Roggero, M. A., Valmori, D., Bertholet, S., Bron, C., Fasel, N., and Corradin, G. (1995) Elicitation of specific cytotoxic T cells by immunization with malaria soluble synthetic polypeptides. *J. Immunol.* **153,** 4134–4141.
13. Tam, J. P., Clavijo, P., Lu, Y., Nussenzweig, V., Nussenzweig, R., and Zavala, F. (1990) Incorporation of T and B epitopes of the circumsporozoite protein in a chemically defined synthetic vaccine against malaria. *J. Exp. Med.* **171,** 299–306.
14. Wang, R., Charoenvit, Y., Corradin, G., Porrozzi, R., Hunter, R. L., Glenn, G., et al. (1995) Induction of protective polyclonal antibodies by immunization with a *Plasmodium yoelii* circumsporozoite protein multiple antigen peptide vaccine. *J. Immunol.* **154,** 2784–2793.
15. Wang, R., Charoenvit, Y., Corradin, G., De La Vega, P., Franke, E. D., and Hoffman, S. L. (1996) Protection against malaria by Plasmodium yoelii sporozoite surface protein 2 linear peptide induction of CD4$^+$ T cell- and IFN-g-dependant elimination of infected hepatocytes. *J. Immunol.* **157,** 4061–4067.
16. Porath, J. and Flodin, P. (1959) Gel filtration: a method for desalting and group separation. *Nature* **183,** 1657–1659.
17. Walker, J. M. (ed.) (1994) *Basic Protein and Peptide Protocols. Methods in Molecular Biology,* vol. 32. Humana Press, Totowa, NJ.
18. Dunn, B. M. and Pennington, M. W. (eds.) (1994) *Peptide Analysis Protocols. Methods in Molecular Biology,* vol. 36. Humana Press, Totowa, NJ.
19. Smith, J. B., Thevenon-Emeric, G., Smith, D. L., and Green, B. (1991) Elucidation of the primary structures of proteins by mass spectrometry. *Anal. Biochem.* **193,** 118–124.
20. Stachowiak, K., et al. (1988) Rapid protein sequencing by the enzyme-thermospray LC/MS method. *J. Amer. Chem. Soc.* **110,** 1758–1765.
21. Stachowiak, K., Otlewski, J., Polanowski, A., and Dyckes, D. F. (1990) Monitoring protein cleavage and concurrent disulfide bond assignment using thermospray LC/MS. *Pept. Res.* **3,** 148–154.
22. Bailey, J. E. (1966) Ph.D. Thesis, *London University.*
23. Edelhoch, H. (1967) Spectroscopic determination of tryptophan and tyrosine in proteins. *Biochemistry* **6,** 1948–1954.
24. Migliorini, P., Boulanger, N., Betschard, B., and Corradin, G. (1990) Plasmodium berghei subunit vaccine: repeat synthetic peptide of circumsporozoite protein comprising T- and B-cell epitopes fail to confer immunity. *Scand. J. Immun.* **31,** 237–242.
25. Perlaza, B. L., Arevalo-Herrera, M., Brahimi, K., Quintero, G., Palomino, J. C., Gras-Masee, H., et al. (1998) Immunogenicity of four *Plasmodium falciparum* preerythrocytic antigens in *Aotus lemurinus* monkeys. *Infect. Immun.* **66,** 3423–3428.
26. Franke, E. D., Corradin, G., and Hoffman, S. L. (1997) Induction of protective CTL responses against the *Plasmodium yoelii* circumsporozoite protein by immunization with peptides. *J. Immunol.* **159,** 3424–3433.
27. Lawrence, G. W., Saul, A., Giddy, A. J., Kemp, R., and Pye, D. (1997) Phase I trial in humans of an oil-based adjuvant SEPPIC MONTANIDE ISA 720. *Vaccine* **15,** 176–178.
28. Pye, D., Vandenberg, K. L., Dyer, S. L., Irving, D. O., Goss, N. H., Woodrow, G. C., et al. (1997) Selection of an adjuvant for vaccination with the malaria antigen, MSA-2. *Vaccine* **15,** 1017–1023.

29. Jacobsen, N. E., Fairbrother, W. J., Kensil, C. R., Lim, A., Wheeler, D. A., and Powell, M. F. (1996) Structure of the saponin adjuvant QS-21 and its base-catalyzed isomerization product by ^{1}H and natural abundance ^{13}C NMR spectroscopy. *Carbohydr. Res.* **280**, 1–14.
30. Cleland, J. L., Kensil, C. R., Lim, A., Jacobsen, N. E., Basa, L., Spellman, M., et al. (1996) Isomerization and formulation stability of the vaccine adjuvant QS-21. *J. Pharm. Sci.* **85**, 22–28.
31. Hoffman, S. L., Edelman, R., Bryan, J. P., Schneider, I., Davis, J., Sedegah, M., et al. (1994) Safety, immunogenicity and efficacy of a malaria sporozoite vaccine administered with monophosphoryl lipid A, cell wall skeleton of mycobacteria and squalene as adjuvant. *Am. J. Trop. Med. Hyg.* **51**, 603–612.
32. Onier, N., Hilpert, S., Reveneau, S., Arnould, L., Saint-Giorgio, V., Exbrayat, J. M., et al. (1999) Expression of inducible nitric oxide synthase in tumors in relation with their regression induced by lipid A in rats. *Int. J. Cancer.* **81**, 755–760.
33. Chang, M. F., White, J. L., Nail, S. L., and Hem, S. L. (1997) Role of the electrostatic attractive force in the adsorption of proteins by aluminum hydroxide adjuvant. *PDA J. Pharm. Sci. Technol.* **51**, 25–29.
34. Al-Shakhshir, R., Regnier, F., White, J. L., and Hem, S. L. (1994) Effect of protein adsorption on the surface charge characteristic of aluminum-containing adjuvants. *Vaccine* **12**, 472–474.
35. Pellegrini, V., Fineschi, N., Matteucci, G., Marsili, I., Nencioni, L. Puddu, M., et al. (1993) Preparation and immunogenicity of an inactivated hepatitis A vaccine. *Vaccine* **11**, 383–387.

31

DNA Vaccination

Richard C. Hedstrom and Denise L. Doolan

1. Introduction

There is increasing recognition that the unique antigenicities of the different stages of the *Plasmodium* spp. life cycle, the requirement for distinct immune mechanisms targeting these different stages, the immense allelic variation of parasites in the field, and the genetic heterogeneity of the target population pose enormous obstacles to the development of a vaccine against malaria. It is likely that an efficacious vaccine will need to induce different immune responses against multiple targets expressed at distinct stages of the parasite's life cycle (i.e., a multivalent, multistage, multiimmune response vaccine) *(1)*. However, how vaccines will be designed to produce protective broad-ranging immune responses has not been obvious based on conventional vaccine delivery systems.

The recent technology of DNA vaccination *(2,3)* has facilitated a revolutionary approach to developing an effective vaccine against complex pathogens, such as the *Plasmodium* spp. parasites that cause malaria, for which vaccines are not yet available. A major advantage of the DNA vaccine technology lies with its ability to induce preferentially CD8[+] T-cell immune responses that have to date been difficult to induce by the more traditional vaccines. The capacity of DNA vaccines to induce the CD8[+] T-cell dependent cytotoxic T-lymphocyte (CTL) and interferon-γ (IFN-γ) responses has now been established in mice, monkeys and humans, in malaria as well as in other disease systems *(4–8)* (*see also* <http://www.DNAvaccine.com>). Since these immune responses are thought to be critical in mediating protection in a number of disease systems, DNA-based vaccines are gaining widespread acceptance as a potential vaccine delivery system of choice. Additional advantages include their simplicity of design, modification and large-scale production, ease of mixing, stability, and lack of requirement for a cold-chain.

One uniquely powerful aspect of the DNA technology lies in the potential to enhance or redirect the immune response to the target antigen. These manipulations can be accomplished at a number of levels, either individually or in combination. First, for example, the vaccinating vector could be designed to coexpress "adjuvant" molecules that may enhance and/or sustain the immune response, either simultaneously with the target gene (immunomodulatory cytokines, chemokines, or ligands for costimulatory molecules) or incorporated into the vector backbone (CpG immunostimulatory motifs). Second, the vaccinating gene itself could be altered (mammalian codon usage; *see* codon usage database <http://www.kazusa.or.jp/codon/>) for enhanced expression of the encoded gene. Third, the vaccine could be modified such that the DNA encoded antigen may be preferentially targeted into specialized processing or presentation pathways or cellular locations (ubiquitination, secretory sequences, cytosol, or endoplasmic reticulum targeting sequences). Fourth, specific components of the vector such as promoter elements or termination sequences could be manipulated so as to control the amount and duration of target gene expression (cytomegalovirus [CMV] and Rous sarcoma virus [RSV] promoters) and thereby optimize and maintain long-term protective immunity. Finally, the simplicity of the DNA approach implies that it should be possible to combine many DNA sequences, each encoding different antigens (multivalent) from one or more stages of the life cycle (multistage), and thereby broaden the immune response (multiimmune response).

In this chapter, we describe the procedure for large-scale plasmid preparation and double-banded cesium chloride gradient ultracentrifugation to produce plasmid DNA for in vitro transfections and in vivo immunizations, and describe techniques for intradermal and intramuscular immunization of mice with plasmid DNA vaccines for subsequent assessment of immunogenicity and/or protective efficacy.

2. Materials
2.1. Equipment
2.1.1. Production and Purification of Plasmid DNA

1. 37°C shaking incubator.
2. Microcentrifuge.
3. Refrigerated centrifuge (Sorval RC2B or equivalent).
4. Sorvall GSA rotor.
5. Ultracentrifuge (Beckman XL-90 or equivalent).
6. Ultracentrifuge rotors (Beckman Vti 70.1 and Vti 90; or Sorval TV865B and TV1665; or equivalents).
7. Ultraclear QuickSeal tubes (13 × 51 mm, Beckman, cat. no. 344075; 16 × 76 mm, Beckman, cat. no. 342413; 35 mL, Sorval, cat. no. 03141; 6 mL Sorval, cat. no. 03945).
8. Tube sealer for Quick Seal tubes.
9. pH meter.
10. 2-L glass flasks, sterile ($n = 6$–12).
11. 1.5-mL microcentrifuge tubes.
12. 50-mL polypropylene centrifuge tubes.
13. 250-mL polypropylene centrifuge bottles with O-ring caps ($n = 12$–24).
14. 5-mL, 10-mL, and 25-mL sterile disposable plastic pipets.
15. Pipets (20, 200, and 1000 µL) and sterile tips.
16. 5-mL or 10-mL syringes.
17. 18-gage needles.
18. Dialysis cassettes, 1-3 mL or 3-15 mL vol (e.g., Slide-A-Lyzer dialysis cassettes; Pierce, cat. no. 66425 or no. 66410).
19. Wooden applicator sticks (optional).
20. Miracloth (Calbiochem, cat. no. 475855).

2.1.2. Transient Transfection

1. T75 (75 cm^2) flasks.
2. 6-Well plates.

2.1.3. Western Blotting and Immunoblotting

1. Polyvinylidene fluoride (PVDF) membrane (Millipore Immobilon-P).
2. Whatman 3MM filter paper.
3. Sponges.
4. Sodium dodecyl sulfate-polyacrylamide gel electrophoresis (SDS-PAGE) and transfer apparatus (*see* Chapter 17).
5. Transfer apparatus.
6. Saran wrap.
7. HyperFilm EC (Kodak).

2.1.4. In Vivo Immunizations

1. Disposable sterile 0.3-mL insulin syringes with a 28-gage 1/2 in. needle (Becton-Dickinson, cat. no. 30901).
2. Disposable sterile 1.0-mL insulin syringes with a 28-gage 1/2 in. needle (Becton-Dickinson, cat. no. 330909).
3. Forceps and razor blades (sterile).

2.2. Reagents

2.2.1. Production and Purification of Plasmid DNA

1. Terrific broth (Gibco-BRL, cat. no. 22711-022; or *see* **Subheading 2.3.**).
2. Cesium chloride (Gibco-BRL, cat. no. 15542-020).
3. Ribonuclease A (RNase A) (Sigma, cat. no. R 5503).
4. Lysozyme (Sigma, cat. no. L 6876).
5. Ethidium bromide (10 mg/mL; Gibco-BRL, cat. no. 15585-011 or equivalent).
6. Glycerol.
7. 3 M sodium acetate, pH 5.2 (Quality Biologica,l cat. no. 351-035-060; or *see* **Subheading 2.3.**).
8. SDS (Bio-Rad, cat. no. 161-0301 or Quality Biological, cat. no. 351-066-101).
9. Sodium hydroxide (e.g., 5 N NaOH; J.T. Baker, cat. no. 5677102).
10. Sodium chloride.
11. Tris base (Gibco-BRL, cat. no. 15504-012).
12. 1 M Tris-HCl, pH 8.0 (Gibco-BRL, cat. no. 351-007-100; or *see* **Subheading 2.3.**).
13. 1 M Tris-HCl, pH 7.5 (Gibco-BRL, cat. no. 15567-027; or *see* **Subheading 2.3.**).
14. 0.5 M EDTA solution, pH 8.0 (Quality Biological, cat. no. 50-118-5; or *see* **Subheading 2.3.**).
15. 20X sodium chloride-sodium citrate (SSC), pH 7.0 (Digene, cat. no. 3400-1024).
16. 50X Tris acetate (TAE) buffer (Quality Biological, cat. no. 351-008-130; or *see* **Subheading 2.3.**).
17. Agarose, ultra-pure (Gibco-BRL, cat. no. 15510-027).
18. Glycine.
19. Hydrochloric acid.
20. Glacial acetic acid.
21. Ethanol.
22. Methanol.
23. Isopropanol.

2.2.2. Transient Transfection

1. RPMI 1640 medium with HEPES (25 mM) and L-glutamine (2 mM) (Gibco-BRL, cat. no. 22400-089, or equivalent).
2. Opti-MEM reduced serum medium (Gibco-BRL, cat. no. 31985) or alternative (RPMI, Dulbecco's modified Eagle's medium [DMEM]).

3. Fetal calf serum (FCS), heat-inactivated (Sigma, cat. no. F4135 or equivalent).
4. Penicillin-streptomycin (100X) (10,000 U/10,000 µg), liquid (Gibco-BRL, cat. no. 15140-015 or equivalent).
5. Nonessential amino acids, 10 mM (100X), liquid (Gibco-BRL, cat. no. 11140-050 or equivalent).
6. L-Glutamine, 200 mM (100X), liquid (Gibco-BRL, cat. no. 25030-081 or equivalent).
7. Calcium-free and magnesium-free PBS (BioSource, cat. no. P312-000), or equivalent.
8. 0.05% Trypsin: 0.53 mM EDTA (Gibco-BRL, cat. no. 25-300-062).
9. FuGENE-6 transfection reagent (Boehringer Mannheim, cat. no. 1-814-443).

2.2.3. Western Blotting and Immunoblotting

1. Acrylamide, bisacrylamide (Bio-Rad).
2. SDS protein gel loading solution (Quality Biological, cat. no. 351-082-030).
3. Coomassie brilliant blue R-250 (Bio-Rad, cat. no. 161-0400).
4. Polyvinyl alcohol (PVA), 1 µg/mL.
5. Western-light chemiluminescent immunoblot detection system (Tropix, Bedford MA); includes I-Block, 40, Nitro-Block, and CSPD®.
6. Primary and secondary antibodies for immunodetection.
7. MilliQ or distilled water.

2.3. Preparation of Stock Solutions

2.3.1. Production and Purification of Plasmid DNA Analysis of In Vitro Expression

1. Terrific broth:

    a. Dissolve 47.0 g of commercially available Terrific Broth powder in 1000 mL of ddH$_2$O and add 4 mL of glycerol according to manufacturer's specifications. Autoclave to sterilize. Store at room temperature.

    b. Alternatively, to 900 mL of ddH$_2$O, add 12 g of bactotryptone, 24 g of Bacto-yeast extract, and 4 mL of glycerol. Stir or shake until solutes have dissolved. Autoclave to sterilize. Allow to cool to approx 60°C, and add 100 mL of 0.17 M KH$_2$PO$_4$/0.72 M K$_2$HPO$_4$. Store at room temperature.

2. 0.17 M KH$_2$PO$_4$/0.72 M K$_2$HPO$_4$: Dissolve 2.31 g of KH$_2$PO$_4$ and 12.54 g of K$_2$HPO$_4$ in 90 mL of ddH$_2$O. Adjust final volume to 100 mL. Autoclave to sterilize.

3. 1.0 M Tris-HCl: Dissolve 121.1 g of Tris-HCl in 900 mL of ddH$_2$O. Adjust to desired pH with concentrated HCl: pH 7.4, 70 mL of HCl; pH 7.6, 60 mL of HCl; pH 8.0, 42 mL of HCl. Allow solution to cool to room temperature before making the final adjustments to the pH (*see* **Note 1**). Make volume up to 1 L. Autoclave to sterilize. Store at room temperature.

4. 0.5 M EDTA, pH 8.0: Add 186.1 g of Na$_2$(EDTA).H$_2$O) to 800 mL of H$_2$O. Stir vigorously on a magnetic stirrer. Adjust pH to 8.0 with NaOH (approx 20 g of NaOH pellets) (*see* **Note 2**). Autoclave to sterilize.

5. Cell resuspension solution (50 mM Tris pH 7.5, 10 mM EDTA, 10 mg/mL RNase A, 30 mg/mL lysozyme): Add 25 mL of 1.0 M Tris-HCl, pH 7.5 and 10 mL of 0.5 M EDTA to 900 mL of ddH$_2$0 and make volume up to 1 L. Alternatively, add 7.5 mL of 1.0 M Tris-HCl, pH 7.5 and 3 mL of 0.5 M EDTA to 289.5 mL of ddH$_2$O and make volume up to 1 L. Autoclave to sterilize and store at room temperature. For each cleared lysate to be made from 2 L of cell culture, prepare fresh cell resuspension solution by adding 300 µL of RNase A stock solution and 300 µL of lysozyme stock solution to 300 mL of 50 mM Tris-HCl/10 mM EDTA buffer.

6. Cell lysis solution (0.2 M NaOH, 1% [w/v] SDS):

DNA Vaccination

a. Add 6 mL of 10 N NaOH and 15 mL of 20% SDS to 279 mL ddH$_2$O. Autoclave to sterilize.
b. Alternatively, dissolve 4 g of NaOH in 400 mL of ddH$_2$O. Add 5 g of SDS and stir to dissolve. Use gentle heat to dissolve if necessary. Make volume up to 500 mL. Autoclave to sterilize.

7. Neutralization solution (1.32 M potassium acetate, pH 4.8): Dissolve 117 g of potassium acetate in 600 mL of ddH$_2$O, add 63 mL of glacial acetic acid, and make volume up to 1 L. Alternatively, dissolve 58.5 g of potassium acetate in 300 mL of ddH$_2$O, add 31.5 mL of glacial acetic acid, and make volume up to 500 mL. Filter sterilize. Store at room temperature.
8. RNase A stock solution (10 µg/mL): Dissolve 100 µg of RNase A in 10 mL of 10 mM Tris-HCl, pH 7.5/15 mM NaCl. Heat to 100°C for 15 minutes. Allow to cool slowly to room temperature. Prepare 300-µL aliquots in 1.5-mL microfuge tubes and freeze at –20°C.
9. Lysozyme stock solution (30 µg/mL): Dissolve 150 mg of lysozyme in 5 mL of ddH$_2$O. Prepare 300-mL aliquots in 1.5-mL microfuge tubes and freeze at –20°C.
10. TE buffer (10 mM Tris-HCl, pH 7.5, 1 mM EDTA).
    a. Add 10 mL of 1 M Tris-HCl, pH 7.5 and 2 mL of 0.5 M EDTA to 900 mL of ddH$_2$O. Make volume up to 1 L. Sterilize by autoclaving.
    b. For dialysis, add 40 mL of 1 M Tris-HCl, pH 7.5 and 8 mL of 0.5 M EDTA to 4 L of ddH$_2$O.
11. CsCl/TE solution: Dissolve 55 g of CsCl in 50 mL of TE buffer.
12. Isopropanol saturated with 20X SSC: Add 25 mL of isopropanol to 25 mL of 20X SSC. Mix well and allow phases to separate. For ethidium bromide extractions, use isopropanol top layer.
13. Phenol:chloroform: Mix equal volumes of phenol and chloroform. Equilibrate the mixture by extracting several times with 0.1 M Tris-HCl, pH 7.6. Store equilibrated mixture under an equal volume of 0.0$1$ M Tris-HCl, pH 7.6, at 4°C in dark glass bottles.
14. 3 M sodium acetate pH 5.2: Dissolve 40.81 g of sodium acetate in 70 mL of ddH$_2$O. Adjust to pH 5.2 using glacial acetic acid. Adjust volume up to 100 mL. Sterilize by autoclaving.
15. 5 M NaCl: Dissolve 29.2 g of NaCl in 80 mL of ddH$_2$O. Adjust volume up to 100 mL. Sterilize by autoclaving.
16. 50X TAE: Combine 242 g of Tris base, 57.1 mL of glacial acetic acid, and 100 mL of 0.5 M EDTA, pH 8.0. Adjust volume to 1 L with ddH$_2$O.
17. 5X agarose gel loading buffer (Bluejuice) (0.25% bromophenol blue, 0.25% xylene cynal, 15% ficoll): Combine 0.25 g of bromophenol blue, 0.25 g of xylene cynal, and 15 mL of ficoll. Adjust volume to 100 mL with ddH$_2$O.

2.3.2. Transient Transfection

1. UM449 growth medium: RPMI-1640, 2 mM glutamine, 10% FCS: to 435 mL RPMI 1640, add 5 mL of penicillin-streptomycin (10,000 U/10,000 µg), 5 mL of 200 mM L-glutamine, 5 mL of nonessential amino acids (10 mM), and 50 mL of FCS. Filter-sterilize through a 0.22-µm filter. Store at 4°C. Use within 1 mo.

2.3.3. Western Blotting and Immunoblotting

1. Acrylamide stock solution: 30% (w/v) acrylamide, 0.8% bisacrylamide.
2. SDS-PAGE electrode buffer: To prepare, combine the reagents in the specified volume of ddH$_2$O shown in **Table 1**.
3. Coomassie blue stain: To prepare, mix 400 mL of methanol, 2 g of Coomassie brilliant blue R-250, 120 mL of glacial acetic acid, and 480 mL of ddH$_2$O.

Table 1
Preparation of SDS-PAGE Electrode Buffer

Reagent	ddH2O		
	1 L (g)	4 L (g)	15 L (g)
Tris-HCl	6.0	24.0	90.0
SDS	1.0	4.0	15.0
Glycine	28.8	115.2	432.0

4. Destain solution: To prepare, mix 680 mL of ddH$_2$O, 200 mL of methanol, and 120 mL of glacial acetic acid.
5. Transfer buffer: To prepare, combine 9.7 mL of 20 mM Tris base, 18.0 mL of 60 mM glycine, and 800 mL of methanol, and make up to a final volume of 4 L.
6. Blocking buffer: To prepare, combine 0.8 g of 0.2% I-block, 40 mL of 10X phosphate-buffered saline (PBS), 300 mL of ddH$_2$O, and heat for 40 s. Add 0.4 mL of 0.1% Tween-20 and make up to final volume of 400 mL.
7. Assay buffer: To prepare, combine 200 mL of ddH$_2$O, 2.4 mL of 0.1 M diethanolamine, 5 drops (using a Pasteur pipet) of concentrated HCl, and 250 µL of 1 M MgCl$_2$. Make up to final volume of 250 mL with ddH$_2$O.

3. Methods

See **Note 3**.

3.1. Production and Purification of Plasmid DNA

3.1.1. Large-Scale Cell Culture

1. Inoculate 5–10 mL of Terrific broth (supplemented with the appropriate antibiotic; typically, 50–200 µg/mL of ampicillin or kanamycin) in a 50-mL conical tube with a single colony from a freshly streaked plate of bacteria.
2. Incubate at 37°C with shaking at 250 rpm for at least 4 h (4–12 h) to prepare a log-phase culture.
3. Add 200 mL of Terrific broth (supplemented with the appropriate antibiotic) to each of 6–12 sterile 2-L flasks, depending on desired yield. Prewarm to 37°C.
4. Inoculate each flask with 400–800 µL of the log-phase culture.
5. Incubate at 37°C with shaking at 250 rpm overnight.

3.1.2. Preparation of Cleared Lysate

This procedure is modified from that described in **ref. 9**.

1. Transfer cell culture from each of the flasks to a 250-mL centrifuge bottle. Balance.
2. Centrifuge in GSA rotor at 6000 rpm for 10 min at 10·C.
3. Pour off the supernatant, and allow bottle to drain briefly upside-down.
4. With a sterile wooden stick (or 5-mL pipet), stir the cell pellets into a smooth paste.
5. Add 50 mL of cell resuspension solution to each bottle and completely resuspend the cell pellets.
6. Incubate on ice for 15 min.
7. Add 50 mL of cell lysis solution to each bottle and mix gently by inversion. Do not vortex.
8. Incubate on ice for 15 min with occasional gentle mixing.
9. Add 50 mL of neutralization solution to each bottle and mix gently but thoroughly. Do not vortex.

10. Incubate on ice for at least 15 min.
11. Centrifuge in GSA rotor at 10,000 rpm for 15 min at 4°C.
12. Pour the cleared supernatant through Miracloth into a clean centrifuge bottle.
13. Add 90 mL (0.6 vol) of isopropanol.
14. Incubate on ice for 30–60 min (or overnight).
15. Centrifuge at 10,000 rpm for 30 min at 4°C, to pellet the nucleic acids.
16. Pour off the supernatant, being very careful not to lose the pellet sticking to the bottom and sides of the bottle.
17. Air-dry the pellets briefly, resuspend in a minimal volume of TE buffer, and combine.

3.1.3. Purification of Plasmid DNA by Cesium Chloride Gradient Ultracentrifugation

See **Table 2**.

1. Prepare 4, 6, or 8 gradients in 16 × 76 mm tubes according to the expected yield of plasmid DNA (i.e., use more tubes for high-copy number plasmids).
2. Measure the volume of DNA solution obtained from the cleared lysate in **Subheading 3.2**.
3. Add 1.1 g of cesium chloride for each 1 mL of DNA solution. Mix thoroughly to dissolve; gentle heating may be required.
4. Add ethidium bromide to a final concentration of 200 µg/mL. Mix.
5. Transfer the DNA/ethidium bromide solution into the ultracentrifuge tubes, using a needle and syringe. Fill the tubes to the shoulder, topping off with CsCl solution (1.1 g CsCl/1.0 mL TE) if required.
6. Balance the tubes carefully to within 10 mg, seal, and check for leaks.
7. Spin at 64,000 rpm overnight at 18°C, in a Vti70.1 rotor or other appropriate ultracentrifuge rotor.
8. Carefully remove the lower band containing the supercoiled plasmid DNA with a syringe and 18-gage needle inserted just below the band (see **Note 4**).
9. Pool the plasmid DNA obtained from all tubes, mix, and load into 4, 6, or 8 tubes (16 × 76 mm) depending on plasmid quantity. Top off with CsCl solution (1.1 g CsCl/1.0 mL TE) if required.
10. Balance the tubes carefully to within 10 mg, seal, and check for leaks.
11. Spin at 70,000 rpm for 4–6 h at 20°C, in a Vti90 rotor.
12. Remove the lower band containing supercoiled plasmid DNA as above.
13. Remove the ethidium bromide by extracting 7 or 8 times with an equal volume of isopropanol saturated with CsCl or 20X SSC. Extract a further 1 or 2 times after all color is removed to ensure complete removal of ethidium bromide (see **Note 5**).
14. Transfer the DNA solution to 3-mL or 15-mL dialysis cassettes (depending on the volume), preequilibrated in TE buffer for 1–2 min, using a 18-gage needle attached to an appropriate sized syringe. After injecting the DNA into the dialysis cassette through one syringe port, remove the air bubble within the cassette by pulling on the syringe plunger.
15. Dialyze against 3–4 × 4-L changes of TE at 4°C with gentle magnetic stirring, over a period of approx 2 d (exchange of sample to dialysis volume should be at least 1:300).
16. Remove sample from dialysis cassette: using a 3-mL or 10-mL syringe, fill dialysis cassette with air, rotate cassette so that the sample in the corner where the needle has been inserted, and remove sample from the cassette by pulling back the syringe plunger. Transfer sample to a sterile 50-mL tube (or 15-mL tube according to volume).
17. Precipitate the DNA by adding 1/10 vol of 3 M sodium acetate (pH 4.8) and 2 vol of absolute ethanol. Mix and store at –20°C for at least 30 min to overnight. A large white, stringy precipitate of plasmid DNA should form.

Table 2
Purification of Plasmid DNA

No. of tubes	DNA solution (mL)	CsCl (g)	Ethidium bromide (mL)	Final volume (mL)
4	30	33	1.0	52.4
6	50	55	1.6	78.6
8	70	77	2.1	104.8

18. Centrifuge at 5000 rpm for 15 min in a tabletop centrifuge, to pellet the DNA.
19. Wash with 20 mL of 70% ethanol, air-dry, and resuspend in TE or PBS. For in vivo immunizations, plasmid DNA should be gently dissolved into injectable PBS or saline, by allowing the PBS or saline to sit on the pellet for a number of hours to days at 4°C. It is important to both minimize pipetting of the DNA and to completely dissolve the DNA. In our experience, the absorbance continues to increase with time for a couple of days so we routinely allow the DNA to dissolve for a couple of days at 4°C.
20. Quantitate by measuring the OD at 260 and 280 nm (typically, dilute sample 1/200 in H_2O).

    1 OD_{260} = 50 mg/mL DNA; pure DNA should have a 260/280 ratio of 1.8

21. Digest an aliquot of the plasmid DNA preparation with selected restriction enzymes known to cut the DNA plasmid at two or three sites, and run uncut and cut samples (0.1-, 0.5-, and 1.0-µg loads) on a 0.8% agarose gel to confirm that the basic characteristics of the DNA plasmid preparation are as expected. If desired, the plasmid DNA may be sequenced (usually, the plasmid is sequenced when first constructed, prior to large-scale plasmid preparation).
22. Prepare a datasheet detailing the construction and characteristics of the plasmid DNA.

3.2. Analysis of In Vitro Expression

3.2.1. Cell Culture of UM449 Cells

1. Maintain UM449 cells at 37°C in an atmosphere of 5% CO_2 by splitting 1:20 every 3–5 d (or 1:10 2 d before or 1:4 one day before use) in growth medium (RPMI 1640, 2.0 m*M* L-glutamine, 10% FCS). To split cultures:
    a. Remove spent medium from the T75 culture flask.
    b. Replace with 4 mL of trypsin:EDTA solution.
    c. Allow cells to detach (~5 min), and then completely detach by gentle tapping. Disperse cells gently by pipetting 3–6X using a 5-mL pipet, avoiding aeration.
    d. Dispense 0.2 mL (1:20) into 15 mL of fresh growth medium in a T75 flask. Record passage number on flask.
    e. Replace working cell line every 20 passages with low passage isolate maintained in liquid nitrogen.
2. On d 1 (Monday AM), split cells at 1:4 (one or more flasks) for next day seeding and at 1:20 (one flask) to maintain the working cell line, as described in **step 1**.

3.2.2. Transient Transfection

1. On d 2 (Tuesday AM):
    a. Remove spent medium from the T75 culture flask, and rinse with 10 mL of calcium-free and magnesium-free PBS, to remove any nonadherent cells.
    b. Harvest cells as described above, except that the entire volume (rather than 0.2 mL) is added to an equal volume of growth medium in a centrifuge tube.
    c. Centrifuge at 1000 rpm for 2–3 min.
    d. Gently resuspend the cell pellet in growth medium in 1/5 of the original culture volume.

DNA Vaccination

e. Count (see **Note 6**). Adjust cell concentration to 200,000 cells per 2 mL, with growth medium.
f. Dispense 2 mL (200,000 cells) into each well of a 6-well plate. Do not swirl cells in well, rather tap to distribute cells over the entire well.
g. Incubate at 37°C in an atmosphere of 5%CO_2 for ~24 h, until ~60–90% confluent.

2. On d 3 (Wednesday AM), transfect the UM449 cells with plasmid DNA in vitro:
   a. Calculate the number of transfection samples (X), and add an additional sample to compensate for working losses.
   b. Prepare a master mix with serum-free medium by adding 9 µL of FuGENE-6 transfection reagent to 92 µL of serum-free medium (e.g., Opti-MEM, RPMI 1640, or DMEM) in a 12 × 75 mm tube (polypropylene or polystyrene), for each sample; i.e., for X samples, mix $(X + 1) \times 9$ µL of FuGENE-6 and $(X + 1) \times 92$ µL of serum-free medium). The order of addition is important; ensure that the undiluted FuGENE-6 reagent is added directly into the medium and does not come into contact with the side of the tube, to prevent adsorption to the plastic tube and to prevent solubilization of plasticizers from the surface. Transfection can be accomplished in either the presence or absence of serum. Since transfection efficiency may improve in the presence of serum, in most cases, the type of cell culture used for routine culture of the cells can also be used during transfection. However, the FuGENE-6 transfection reagent/DNA complex formation must be performed in medium that does not contain serum.
   c. Stand 5 min at room temperature. During this incubation period, label microfuge tubes for each DNA sample, and then add 2 µL (2 mg) of DNA to the respective tube.
   d. Dropwise, add the 98 µL diluted FuGENE-6 transfection reagent to the tube containing the 2 µL DNA sample, and mix.
   e. Stand 15–20 min at room temperature.
   f. Meanwhile, remove the growth medium from the UM449 cell cultures in the 6-well plates, and replace with 2 mL of fresh cell growth medium (with serum).
   g. Dropwise, add the 100 mL DNA/FuGENE-6 solution to the cells. Swirl the plate to ensure even dispersal. It is not necessary to remove the reagent–DNA complex from the cells prior to assay, particularly when testing for gene expression 24–48 h after transfection.
   h. Incubate overnight at 37°C in an atmosphere of 5% CO_2.
   i. The next day, change the medium, and continue incubation for a total of 48 h posttransfection.

3.2.3. Analysis of In Vitro Expression

1. On d 5 (Friday PM), harvest the culture supernatants and cells:
   a. Pipet off ~1 mL and centrifuge for at least 10 min at top speed to remove intact cells and cell debris.
   b. Meanwhile, remove all supernatant from the well, and add 250 µL of 2X SDS-PAGE sample buffer to each well (see **Note 7**), covering the entire surface by tapping.
   c. Scrape with a blue tip to ensure complete lysis of all cells.
   d. Remove cell lysate to an appropriately labeled microcentrifuge tube (e.g., label as "cells," "well #," and "date"). Mix the centrifuged supernatant 1:1 with 2X SDS-PAGE sample buffer, and label tube appropriately (e.g., label as "supernatant," "well #," and "date"). Store all tubes at –70°C.
2. Carry out SDS-PAGE according to standard procedures (see Chapter 17), with the specifications as given in **Table 3**.
3. Immunoblot according to standard procedures (see also Chapter 17). Briefly:
   a. Cut PVDF membrane to size (Millipore Immobilon-P).

Table 3
SDS-PAGE Gel Specifications[a]

Components (for 2 × 0.75 mm gels)	7.5% (mL)	9.0% (mL)	10% (mL)	12% (mL)
Running gel				
Tris-HCl buffer (1.5 M, pH 8.8)	8.75	8.75	8.75	8.75
Acrylamide stock	8.46	10.15	11.83	14.20
ddH2O	13.79	12.10	10.42	8.05
20% SDS	310	310	310	310
10% AP:TEMED	60:7	60:7	60:7	60:7
Stacking gel				
Tris-HCl buffer (0.5 M, pH 6.8)	3.75			
Acrylamide stock	3.25			
ddH2O	9.00			
20% SDS	80			
10% AP:TEMED	176:17.6			

[a]*See* **Subheading 2.** for details of the SDS-PAGE electrode buffer, Coomassie blue stain, and destain solutions.

  b. Soak in 100% methanol. While sloshing, slowly add 4 vol of ddH$_2$O until the methanol is 20% (v/v).

  c. Place membrane in transfer buffer to soak for at least 15 min.

  d. Assemble blot by submerging a sponge and placing on it two sheets of Whatman 3MM filter paper. Remove together onto the frame of the transfer apparatus, black-to-back. Center the membrane, then carefully lay on the gel (by hand). Wet another two sheets of Whatman 3MM filter paper and place on the top of the membrane, followed by another sponge. Insert the assembled blot into the transfer apparatus.

  e. Run overnight at 25 V in a cold room (start with warm fresh transfer buffer).

  f. Mark the leading edge of standard protein marker bands using a marker pen, cut the membrane to the same size as the gel, and remove the gel. Mark the leading edge points inside.

  g. Wash with PBS for at least 5 min. During this period, prepare the polyvinyl alcohol solution and the blocking buffer.

  h. Soak the membrane briefly in 1 μg/mL polyvinyl alcohol.

  i. Place in blocking buffer at 4°C all day.

  j. Place in primary antibody in blocking buffer at 4°C overnight.

  k. Wash twice with 50 mL of blocking buffer.

  l. Place in secondary antibody in blocking buffer (1:10,000, 5 μL/50 mL) for 1 h.

  m. Wash three times with 50 mL of blocking buffer. Prepare assay buffer.

  n. Sit in Nitro-Block (1:20) in assay buffer.

  o. Wash twice in 50 mL of assay buffer, 5 min per wash.

  p. Sit for 5 min in CSPD® 1:100 in assay buffer.

  q. Drain the membrane and mount between Saran wrap all bubbles squeezed out. Tape edges and mount on an old film.

  r. Expose to HyperFilm ECL (start with ~2 min exposure).

3.3. In Vivo Immunogenicity

3.3.1. Intramuscular Immunization with Plasmid DNA

The tibialis anterior muscles of mice are injected with 50 μL (100 μL total volume, using both legs) of DNA (usually 1 mg/ml DNA) in saline, using a disposable sterile

DNA Vaccination

0.30 mL insulin syringe with a 28-gage 1/2 in. needle fitted with a plastic collar cut from a micropipet tip (yellow tip). The collar length is adjusted to limit the needle tip penetration to a distance of about 2 mm into the central part of the muscle.

1. Using aseptic technique where possible, prepare syringes for intramuscular injection as follows:
    a. Use one of the insulin syringes as a "template" to measure the 2 mm distance.
    b. On a flat sterile surface (we use the top of a pipet box), using sterile forceps, slide a yellow tip, narrow end first, down the needle until it touches the base of the syringe, which is positioned with the needle bevel up so that the opening is visible.
    c. Notch the tip with a razor blade just below the bevel of the "template" needle (about 2 mm from the tip of the needle).
    d. Remove the syringe and hold on to the tip with sterile forceps while cutting through completely at the notch using a sterile razor blade. It is convenient to prepare many collars before proceeding to attach the collars to the receiving syringes (*see* **Note 8**).
    e. Dab some adhesive around the base of the receiving needle using another yellow tip as an applicator (standard bathroom and tile sealant works well, but silicone-type adhesives are not very satisfactory).
    f. Seat the base of the collar (wide end) into the adhesive.
    g. Stand the syringe on end (plunger down) and allow the adhesive to dry overnight.
2. Position the mouse in the restrainer (we use a slotted beaker) by grasping the protruding hind leg such that the animal faces you as it is attempting to crawl up the beaker. This positioning facilitates injection of the correct muscle by placing the leg in an anterior position, that is, with the "knee-cap" facing up.
3. To better visualize the correct muscle, and to sterilize the area, swab the muscle area with alcohol, using either a commercially available alcohol swab or a cotton-tipped applicator dipped in alcohol (the latter is preferable if a large number of mice are to be immunized, since the swaps dry out relatively quickly). The tibialis anterior muscle is the bandlike muscle that wraps along the outside of the calf. We recommend that you carry out a preliminary study to define the muscle that is being targeting, by injecting a mouse muscle with a marker dye such as India ink or trypan blue and then dissecting the leg to expose the muscle area. If correct technique has been followed, the marker dye will be restricted to the tibialis anterior muscle, which will be stained uniformly.
4. Equilibrate the plasmid DNA samples to room temperature prior to immunization. Draw up the desired volume into the collared 28-gage 1/2 in. syringe (100 mL/mouse; one 0.3-mL syringe is sufficient for 3 mice). Use a separate syringe for each DNA sample.
5. Inject 50 µL of plasmid DNA per muscle, over period of 1–2 s. Keep the needle in place for a further 1–2 s after injection to help prevent leakage.
6. Optimal dosage and regime should be determined for each antigen-encoding plasmid. We recommend starting with 3 immunizations of 100 µg/100 µL doses (1 mg/mL plasmid DNA) at 3-wk intervals.

3.3.2. Intradermal Immunization with Plasmid DNA

Mice are injected intradermally using a 28-gage 1/2 in. insulin syringe (without collar) 1–2 cm from the base of the tail with DNA (2 mg/mL) in saline. The injection should result in a bleb that lasts for at least 30 s. If no bleb results, the injection was probably subcutaneous. The area of the tail around the site should noticeably blanch. It is important to avoid leakage, which can be best accomplished by minimizing the volume (we recommend that the DNA for intradermal immunization be at twice the concentration of that used for intramuscular immunization, that is, at 2 mg/mL) and by injecting slowly. Delivery of 100-μg doses thus requires two injections of 25-μL vol. The second injection should be 1–2 cm distal to the first injection site.

1. Position the mouse in the restrainer such that the tail protrudes and the base of the tail is easily accessible.
2. Equilibrate the plasmid DNA samples to room temperature prior to immunization. Draw up the desired volume into a noncollared 28-gage 1/2 in. syringe (50 μL/mouse; one 0.3-mL syringe is sufficient for 6 mice). Use a separate syringe for each DNA sample.
3. Position the needle in the base of the tail, bevel up, such that approximately half the length of the needle is in the skin just below the surface (you should be able to see the outline of the needle under the skin).
4. Inject 25 μL of plasmid DNA over a period of 5–10 s. Keep the needle in place for a further 1–2 s after injection to help prevent leakage.
5. Withdraw the needle, reposition approx 1–2 cm distal to the first injection site, and repeat the injection.
6. Optimal dosage and regime should be determined for each antigen-encoding plasmid. We recommend starting with 3 immunizations of 100 μg doses at 3-wk intervals.

4. Notes

1. The pH of Tris-HCl solutions is temperature-dependent and decreases approx 0.03 pH units for each 1°C increase in temperature.
2. The disodium salt of EDTA will not go into solution until the pH of the solution is adjusted to approx pH 8.0 by the addition of NaOH.
3. All waste containing ethidium bromide must be collected and turned over to the HazMat and Chemical Hygiene Office for disposal. All acidic or basic waste (below pH 2.0 or above pH 10.0) will be collected and stored for disposal by the Chemical Hygiene Office.
4. The waste CsCl–ethidium bromide solution should be collected and stored as aqueous waste.
5. The isopropanol–ethidium bromide waste should be collected and stored as organic waste.
6. Mix 10 μL of cell suspension into 90 μL of 0.4% trypan blue in RPM1. Count viable cells using a hemocytometer. Calculate the concentration of viable cells as follows: number of cells per 4 quadrants (not less than 250) divided by 4×10 (trypan blue dilution) $\times 10^4$ = number of viable cells per milliliter.
7. Be careful using 2-mercaptoethanol because certain antibody determinants are not recognized after reduction.
8. Alternatively, a guillotine-type device can be constructed with a groove into which the tip can be placed, and with a razor blade mounted such that it will cut the tip at approx 5 mm from the end (when placed on the needle, the end of the tip collar must be 2 mm from the end of the needle, just below the bevel opening).

References

1. Doolan, D. L. and Hoffman, S. L. (1997) Multi-gene vaccination against malaria: a multi-stage, multi-immune response approach. *Parasitol. Today* **13,** 171–178.
2. Wolff, J. A., Malone, R. W., Williams, P., Chong, W., Acsadi, G., Jani, A., et al. (1990) Direct gene transfer into mouse muscle in vivo. *Science* **247,** 1465–1468.
3. Ulmer, J. B., Donnelly, J. J., Parker, S. E., Rhodes, G. H., Felgner, P. L., Dwarki, V. J., et al. (1993) Heterologous protection against influenza by injection of DNA encoding a viral protein. *Science* **259,** 1745–1749.
4. Schneider, J., Gilbert, S. C., Hannan, C. M., Degano, P., Prieur, E., Sheu, E. G., et al. (1999) Induction of CD8+ T cells using heterologous prime-boost immunisation strategies. *Immunol. Rev.* **170,** 29–38.
5. Dubensky, T. W, Jr., Liu, M, A., and Ulmer, J. B. (2000) Delivery systems for gene-based vaccines. *Mol. Med.*. **6,** 723–732.

6. Liu, M. A. and Ulmer, J. B. (2000) Gene-based vaccines. *Mol. Ther.* **1,** 497–500.
7. Seder, R. A. and Hill, A. V. (2000) Vaccines against intracellular infections requiring cellular immunity. *Nature* **406,** 793–798.
8. Doolan, D. L. and Hoffman, S. L. (2001) DNA-based vaccines against malaria: status and promise of the Multi Stage DNA-based Vaccine Operation (MuStDO). *Inter. J. Parasitol.* **31,** 753–762.
9. Sambrook, J., Fritsch, E. F., and Maniatis, T. (1989) *Molecular Cloning: A Laboratory Manual.* Cold Spring Harbor Laboratory Press, Cold Spring Harbor, NY.
10. Hedstrom, R. C., Doolan, D. L., Wang, R., Malik, A., and Hoffman S. L. (1998) Immunogenicity of *P. falciparum* DNA vaccines. *Int. J. Mol. Med.* **2,** 29–38.

32

Assessing Antigen-Specific CD8+ and CD4+ T-Cell Responses in Mice After Immunization with Recombinant Viruses

Moriya Tsuji

1. Introduction

To determine the immunogenicity of vaccine candidates, it is essential to establish an accurate and convenient methodology for assessing the level of antigen-specific T-cell responses, and several methodologies have been developed. These include the T-cell proliferative assay as determined by tritium-labeled thymidine (^{3}H-TdR) incorporation, cytotoxic assays such as the chromium release assay, cytokine enzyme-linked immunosorbent assays (ELISAs) or the T-cell enzyme-linked immunospot (ELISPOT) assay to determine the amounts of cytokines released by activated T cells, and most recently, the newly developed tetramer assay. All these methods, each of which have both advantages and disadvantages, have been used to measure the level of cell-mediated immune responses in animals and humans infected with microbes or immunized with microbial antigens or various vaccines.

The frequency of antigen-specific cytotoxic CD8+ T cells has been routinely assessed by limiting dilution analysis using the chromium release assay. Recently, however, it has been demonstrated that the ELISPOT assay is more sensitive than the standard chromium release assay, and provides quantitative data on T-cell precursors entirely comparable to those obtained by the limiting dilution analysis (1). In addition, when a peptide representing a CD8+ T cell epitope is used as an antigen, it has been shown that in vitro or in vivo treatment with anti-CD8+, but not with anti-CD4+ antibodies, can completely abrogate the response detected by the ELISPOT assay (1), indicating that the CD8+ ELISPOT assay detects only antigen-specific CD8+ T cells. As for CD4+ T cells, virtually no methodology to assess the frequency of antigen-specific CD4+ T cells existed until the ELISPOT assay was established. It is now established, however, that the ELISPOT assay can also quantify the frequency of CD4+ T-cell precursors that are specific for an antigen. Like the CD8+ ELISPOT, when a peptide representing a CD4+ T cell epitope is used as an antigen, it has been demonstrated that the CD4+ ELISPOT assay detects antigen-specific CD4+ T cells. Furthermore, the ELISPOT assay provides much more accurate information on the frequency of the antigen-specific, functional T cells than that of the tetramer assay, although the tetramer assay requires much easier manipulation with a shorter period of time. Thus, the ELISPOT assay is now widely used for assessing the frequency of epitope-specific T cells in various infectious diseases model *(2–5)*.

Currently, many approaches have been taken for developing malaria vaccines. Candidate vaccines are based on recombinant subunit proteins and peptides, naked DNA plasmids, and live attenuated viral and bacterial vectors *(6)*. Among these, the use of certain microbial organisms as vectors, engineered for the expression of foreign genes, offer many attractive features. Besides providing considerable choice of nonpathogenic or highly attenuated intracellular microbial agents, they have the potential for inducing not only humoral, but also cellular immunity *(7)*. Since it has been demonstrated that malaria-specific T cells, including both $CD8^+$ and $CD4^+$ T cell subsets, have a major role in the immune response against the preerythrocytic liver stages of this parasite, several attempts have been made to induce protective antiplasmodial T cells in vivo, using various recombinant viral and bacterial vectors expressing Plasmodial proteins as immunogens *(8–11)*.

In this chapter, I describe the methodology of characterizing the induction and persistence of T cell-mediated antiplasmodial immune responses elicited by a recombinant adenovirus expressing the *Plasmodium yoelii* circumsporozoite (CS) protein in mice. The recombinant adenovirus expressing the CS protein of *P. yoelii* was generated and screened for the expression of the CS protein by an immunoradiometric assay (two-site immunoradiometric assay [IRMA]) *(12)*, as described previously *(8)*, and named AdPyCS. The methodology described here includes the immunization of mice with the adenoviral vector, the isolation of lymphocytes from the spleens and livers of the immunized mice, and the determination of the relative frequency of malaria epitope-specific $CD4^+$ and $CD8^+$ T cells by ELISPOT assay *(8)*.

2. Materials

2.1. Cell Culture Medium

1. Dulbecco's modified Eagle's medium (DMEM) (Gibco-BRL, Grand Island, NY).
2. Sodium bicarbonate (2 g/L) (Gibco-BRL).
3. Nonessential amino acids (100X) (Gibco-BRL).
4. L-glutamine (100X) (Gibco-BRL).
5. 5×10^{-5} M 2-mercaptoethanol (2ME) (Gibco-BRL).
6. 5 mM HEPES (Gibco-BRL).
7. Gentamicin (100X) (Gibco-BRL).
8. Penicillin/streptomycin (100X) (Gibco-BRL).
9. 5–10% Fetal calf serum (FCS) (Hyclone, Logan, UT).

2.2. EL4 Cell Supernatant

1. EL4 cells (ATCC, Rockville, MD).
2. Phorbol myristate acetate (PMA) (Sigma, St. Louis, MO).
3. 0.22-µm cellulose acetate 500-mL filter system (Corning-Costar, Corning, NY).
4. DMEM cell culture medium, with 5% FCS.

2.3. Isolation and Purification of Lymphoid Cells

2.3.1. From Livers

1. Hank's balanced salt solution (HBSS) supplemented with calcium and magnesium (Sigma).

2. 50-mesh Stainless steel sieve (Bellco, NJ).
3. 300-mesh Screen (Bellco).
4. 20-mL Syringe pestle (Monoject, St. Louis, MO).
5. Percoll (Amersham Pharmacia, Uppsala, Sweden).
6. 10X phosphate-buffered saline (PBS) (Gibco-BRL).
7. Heparin (Elkins-Sinn, Cherry Hill, NJ).
8. Ammonium chloride (Fisher Scientific, Fair Lawn, NJ).
9. Tris-HCl, pH 7.2 (Sigma).
10. Erythrocyte lysis buffer: 0.1 M ammonium chloride/20 mM Tris-HCl, pH 7.2.
11. DMEM medium supplemented with 10% FCS.
12. DMEM cell culture medium supplemented with 10% FCS and 1–2% EL4 supernatant.

2.3.2. From Spleens

1. Nylon mesh (Tetko, Elmsford, NY).
2. Microscope slides (Fisher Scientific).
3. DMEM medium supplemented with 10% FCS.
4. DMEM cell culture medium supplemented with 10% FCS and 1–2% EL4 supernatant.

2.4. ELISPOT Assay

2.4.1. In General

1. 96-well Multiscreen HA nitrocellulose plates (MAHAS no. 4510; Millipore, Bedford, MA).
2. PBS.
3. Ionomycin (Sigma).
4. Tween-20 (Fisher Scientific).
5. Tris-HCl (Sigma).
6. Peroxidase-labeled streptavidin (Kirkegaard & Perry Laboratories, Gaithersburg, MD).
7. 3-3' diaminobenzidine tetrahydrochloride (DAB) (Sigma).
8. H_2O_2 (Sigma).
9. DMEM cell culture medium supplemented with 10% FCS and 1–2% EL4 supernatant.

2.4.2. IFN-γ ELISPOT

1. Anti-mouse interferon-γ monoclonal antibody (IFN-γ MAb), clone R4-6A2 (cat. no. 18181D; Pharmingen, San Diego, CA).
2. Biotinylated anti-mouse IFN-γ MAb, XMG1.2 (cat. no. 18112D; Pharmingen).

2.4.3. IL-5 ELISPOT

1. Anti-mouse interleukin-5 (IL-5) MAb, TRFK-5 (Pharmingen).
2. Biotinylated anti-mouse IL-5 MAb, TRFK-4 (Pharmingen).

2.4.4. CD8+ T-Cell ELISPOT

1. Peptide SYVPSAEQI, an H-2K^d restricted cytotoxic T lymphocyte (CTL) epitope of the *P. yoelii* CS protein (Bio-Synthesis, Lewiville, TX).
2. P815 plasmacytoma (ATCC).

2.4.5. CD4+ T-Cell ELISPOT

1. Peptide YNRNIVNRLLGDALNGKPEEK, containing a CD4+ T cell epitope of the CS protein (Bio-Synthesis).
2. A20 B-cell lymphoma (ATCC).

3. Methods

3.1. Preparation of EL4 Supernatant

1. Culture EL4 thymoma cells at a concentration of 1×10^6 cells/mL for 24 h in the presence of DMEM cell culture medium containing 5% FCS and 10 ng/mL of PMA.
2. Collect culture supernatant by centrifugation at 370g for 10 min.
3. Filter the supernatant through a 0.22-µM filter.
4. Aliquot and store at –20°C.
5. Typically, the supernatant should contain approx 2000–3000 U/mL of IL-2 activity, as determined by CTLL2 assay (13).

3.2. Immunization of Mice

1. Thaw aliquot(s) of the recombinant adenovirus (stored frozen at –70°C) gradually on ice (*see* **Note 1**).
2. Dilute the virus to the desired concentration in PBS containing 1% of heat-inactivated normal mouse serum, and sonicate briefly for 5 s (*see* **Note 2**).
3. Inject groups of at least 3 BALB/c (H-2^d) mice with optimal doses (10^{8-9} plaque-forming units [Pfu] per mouse depending of the batches) of AdPyCS by the subcutaneous or intramuscular route. Both routes have been shown previously to be the optimal routes for the induction of high level of CS-specific T cell responses following AdPyCS immunization (8). Subcutaneous injection is done at both sides of the tail base with a volume of 50 µL each. Intramuscular injection is done in both thighs with a volume of 50 µL each. Inject a control group of mice with the equivalent Pfu of a recombinant adenovirus expressing an irrelevant protein, such as *Escherichia coli* lacZ (14).
4. At 12–14 d after immunization, recover the lymphocytes from the spleen and liver of immunized mice and perform an ELISPOT assay as detailed in **Subheadings 3.3.** and **3.4.** to determine the number of activated IFN-γ/IL-5 producing $CD8^+/CD4^+$ T cells.
5. To determine the persistence of CS-specific $CD8^+$ and $CD4^+$ T cells in the spleen or liver of mice given a single dose of AdPyCS, immunize groups of at least 3 BALB/c mice subcutaneously with an optimal dose of AdPyCS, and recover splenocytes 2, 4, or 8 wk after immunization.

3.3. Isolation and Purification of Lymphoid Cells

Mice are sacrificed 12–14 d after immunization and activated T-cell populations, present in their liver and spleen, are characterized and quantified using the ELISPOT assay.

3.3.1. Isolation and Purification of Lymphocytes from the Liver

Lymphocytes are obtained from the liver according to the method described by Goossens et al. (15).

1. Harvest the liver from euthanized mice and wash in ice-cold HBSS supplemented with calcium and magnesium.
2. Disrupt the liver in a 50-mesh stainless steel sieve, using a sterile 20-mL syringe pestle.
3. Filter the resulting suspension through a 300-mesh screen; the cell-free stroma will remain on the screen.
4. Suspend the cells in cold HBSS, 2–3 livers per 40 mL of HBSS.
5. Centrifuge at 370g for 10 min at 4°C.
6. Thoroughly resuspend the dissociated cells from 2–3 livers in 45 mL of HBSS solution containing 35% Percoll, 3.5% of 10X PBS and 200 U of heparin.
7. Centrifuge at 500g for 10 min at 20°C to separate hepatocytes from the lymphocytes.

8. Resuspend the pellet (which consists mostly of lymphocytes) in 0.1 M ammonium chloride/20 mM Tris-HCl, pH 7.2, for 1 min on ice, to lyse the remaining erythrocytes.
9. Wash the remaining cells 3X in DMEM supplemented with 10% FCS, at 370g for 10 min at 4°C.
10. Resuspend the liver lymphocytes in culture medium (DMEM supplemented with 10% FCS and 1–2% EL4 supernatant), and determine the cell counts.

3.3.2. Isolation and Purification of Lymphocytes from the Spleen

1. Harvest the spleen from euthanized mice.
2. To harvest the splenocytes, gently tease the spleen by grinding it between two autoclaved microscopic slides in a sterile petri dish containing 5 mL of DMEM supplemented with 10% FCS.
3. Pass the cell suspension through a nylon mesh.
4. Wash the cells 3X in DMEM supplemented with 10% FCS, at 370g for 10 min at 4°C.
5. Resuspend the cell pellet in culture medium (DMEM supplemented with 10% FCS and 1–2% EL4 supernatant), and determine the cell counts.

3.3. Quantification of Epitope-Specific CD8+ T-Cells by the ELISPOT Assay

3.3.1. IFN-γ CD8+ ELISPOT Assay

1. Coat 96-well multiscreen HA nitrocellulose plates with 75 µL per well of anti-mouse IFN-γ MAb, R4, at a concentration of 10 µg/mL in sterile PBS.
2. Incubate overnight at room temperature.
3. Wash the wells 3X with DMEM containing 10% FCS, at 370g for 10 min at room temperature.
4. Incubate for 2–3 h at 37°C in the presence of DMEM containing 10% FCS.
5. Wash the wells once more (*see* **Note 3**).
6. Incubate the major histocompatibility complex (MHC)-compatible target cells, P815, at a concentration of 1×10^7 cells/mL in the presence of 1 µM of the synthetic peptide representing the CD8+ T-cell epitope of the CS protein (SYVPSAEQI) for 1 h. Use P815 cells, not pulsed with the peptide, as negative control target cells.
7. γ-Irradiate the P815 cells at 10,000 rads.
8. Wash 3 times with DMEM culture medium supplemented with 10% FCS and 1–2% EL4.
9. Resuspend the cells at a concentration of 1×10^6 cells/mL in DMEM culture supplemented with 10% FCS and 1–2% EL4 supernatant.
10. Add the target cells to the ELISPOT wells at a concentration of 1×10^5 cells per 100 µL per well (target cells) (*see* **Note 4**).
11. Add serially diluted lymphocytes, starting from 5×10^5 cells per well (effector cells), in a volume of 100 µL per well. Include as positive controls lymphocytes incubated with PMA (1 µg/mL) and ionomycin (3 µM).
12. Coculture target and effector cells for 26–28 h at 37°C and 5% CO_2 (*see* **Note 5**).
13. Wash wells extensively (4 or 5 times) with PBS-0.05% Tween-20 (PBS-T).
14. Add 100 µL/well of the biotinylated anti-mouse IFN-γ MAb, XMG1.2, at a concentration of 2 µg/mL in PBS-Tween.
15. Incubate overnight at 4°C or 3–4 h at room temperature.
16. Wash wells extensively (4 or 5 times) with PBS-0.05% Tween-20 (PBS-T).
17. Incubate with 100 µL/well of peroxidase-labeled streptavidin diluted in PBS-Tween for 45 min.
18. Wash wells 4 times with PBS–T, and then 3 times with PBS alone (*see* **Note 6**).
19. Develop spots by adding 50 mM Tris-HCl, pH7.5 containing 1 mg/mL of 3,3'-diaminobenzidine tetrahydrochloride dihydrate (DAB) plus 0.5 µL/mL of 30% H_2O_2 (*see* **Note 7**) for 10 to 15 min.
20. Determine the number of spots corresponding to IFN-γ secreting cells using a stereomicroscope.

3.3.2. IL-5 CD8⁺ ELISPOT Assay

To determine the number of epitope-specific IL-5 producing CD8⁺ Th2-type (helper) T cells, the ELISPOT assay is performed as detailed above for IFN-γ, except that:

1. Target and effector cells are cocultured for 48 h (not 26–28 h).
2. The capture antibody is purified rat anti-mouse IL-5 MAb TRFK-5 (10 µg/mL) (not anti-mouse IFN-g MAb, R4-6A2) *(16)*.
3. The secondary antibody is anti-mouse IL-5 MAb TRFK-4 (5 µg/mL) (not anti-mouse IFN-γ MAb, XMG1.2) *(16)*.

3.4. Quantification of Epitope-Specific CD4⁺ T Cells by the ELISPOT Assay

The procedure for the detection of epitope-specific CD4⁺ T cells that produce IFN-γ (Th-1 type) is basically the same as that used for detection of CD8⁺ T cell IFN-g responses, except that:

1. MHC-compatible target cells expressing both MHC class I and class II, namely, the A20 B cell lymphoma *(17)*, are used instead of P815 cells (which express MHC class I but not MHC class II).
2. 10 µM of the synthetic peptide YNRNIVNRLLGDALNGKPEEK, which contains an epitope recognized by CS-specific CD4⁺ T cells *(17)*, is used instead of 1 µM of the PyCS CD8⁺ T cell epitope SYVPSAEQI.

The procedure for the detection of plasmodium specific CD4⁺ T cells which produce IL-5 (Th-2 type) is basically the same as that used for detection of CD4⁺ T cell IFN-γ responses, except that the pair of anti-mouse IL-5 MAbs is used instead of the pair of IFN-γ MAbs.

4. Notes

1. It is important to thaw the recombinant adenovirus gradually, on ice, to retain infectivity.
2. Dilution of the recombinant adenovirus must be done in the presence of 1% normal mouse serum to avoid the virus binding to the tube. The virus suspension should be briefly sonicated for 5 s to prevent the aggregation of the virus.
3. After coating the plates with antilymphokine antibodies, wells should be washed extensively to remove free antibodies in the medium.
4. After pulsing the target cells with peptides followed by γ-irradiation, the cells should be washed 3 times to remove excess free peptides.
5. For optimal activation of the T cells, the final culture medium for cultivating the cells in the ELISPOT plates should be supplemented with 10% FCS and 1–2% EL4 supernatant, which contains the amounts of IL-2 equivalent to 30 U/mL.
6. After washing the wells 4 times with PBS–Tween-20, the plates should be further washed 3 times with PBS alone before adding peroxidase-labeled streptavidin, to remove the detergent.
7. When developing the plates, DAB should be dissolved in 50 mM Tris-HCl, pH 7.5.

References

1. Miyahira, Y., Murata, K., Rodriguez, D., Rodriguez, J. R., Esteban, M., Rodrigues, M. M., and Zavala, F. (1995) Quantification of antigen specific CD8+ T cells using an ELISPOT assay. *J. Immunol. Meth.* **181**, 45–54.
2. Busch, D. H. and Pamer, E.G. (1998) MHC Class I/peptide stability: Implications for immunodominance, in vitro proliferation, and diversity of responding CTL. *J. Immunol.* **160**, 4441–4448.
3. Butz, E. A. and Bevan, M.J. (1998) Massive expansion of antigen-specific CD8+ T cells during an acute virus infection. *Immunity* **8**, 167–175.

4. Lalvani, A., Brooks, R., Hambleton, S., Britton, W. J., Hill, A. V. S., and McMichael, A. J. (1997) Rapid effector function in CD8+ memory T cells. *J. Exp. Med.* **186,** 859–865.
5. Murali-Krishna, K., Altman, J. D., Suresh, M., Sourdive, D. J. D., Zajac, A. J., Miller, J. D., et al. (1998) Counting antigen-specific CD8+ T cells: A reevaluation of bystander activation during viral infection. *Immunity* **8,** 177–187.
6. Kwiatkowski, D. and Mash, K. (1997) Development of a malaria vaccine. *Lancet* **350,** 1696–1701.
7. Ertl, H. C. J. and Xiang, Z. (1996) Novel vaccine approaches. *J. Immunol.* **156,** 3579–3582.
8. Rodrigues, E. G., Zavala, F., Eichinger, D., Wilson, J. M., and Tsuji, M. (1997) Single immunizing dose of recombinant adenovirus efficiently induces CD8+ T cell-mediated protective immunity against malaria. *J. Immunol.* **158,** 1268–1274.
9. Kumar, S., Miller, L., Quakyi, I., Keister, D., Houghten, R., Maloy, W., et al. (1988) Cytotoxic T cells specific for the circumsporozoite protein of *Plasmodium falciparum*. *Nature* **334,** 258–260.
10. Aggarwal, A., Kumar, S., Jaffe, R., Hone, D., Gross, M., and Sadoff, J. (1990) Oral *Salmonella*: malaria circumsporozoite recombinants induce specific CD8+ cytotoxic T cells. *J. Exp. Med.* **172,** 1083–1090.
11. Li, S., Rodrigues, M. M., Rodriguez, D., Rodriguez, J. R., Esteban, M., Palese, P., et al. (1993) Priming with recombinant influenza virus followed by administration of recombinant vaccinia virus induces CD8+ T-cell-mediated protective immunity against malaria. *Proc. Natl. Acad. Sci. USA* **90,** 5214–5218.
12. Tsuji, M., Corradin, G., and Zavala, F. (1992) Monoclonal antibodies recognize a processing dependent epitope present in the mature CS protein of various plasmodial species. *Paras. Immunol.* **14,** 457–469.
13. Mossmann, T. R., Cherwinski, H., Bond, M. W., Giedlin, M. A., and Coffman, R. L. (1986) Two types of murine helper T cell clone. I: Definition according to profiles of lymphokine activities and secreted proteins. *J. Immunol.* **136,** 2348–2357.
14. Yang, Y., Nunes, F. A., Berencsi, K., Furth, E. E., Gonczol, E., and Wilson, J. M. (1994) Cellular immunity to viral antigens limits E1-deleted adenoviruses for gene therapy. *Proc. Natl. Acad. Sci. USA* **91,** 4407–4411.
15. Goossens, P. L., Jouin, H., Marchal, G., and Milon, G. (1990) Isolation and flow cytometric analysis of the free lymphomyeloid cells present in murine liver. *J. Immunol. Meth.* **132,** 137–144.
16. Yamamoto, M., Fujihashi, K., Beagly, K. W., McGhee, J. R., and Kiyono H. (1993) Cytokine synthesis by intestinal intraepithelial lymphocytes: both γ/δ T cell receptor-positive and α/β T cell receptor-positive T cells in the G_1 phase of cell cycle produce IFN-γ and IL-5. *J. Immunol.* **150,** 106–114.
17. Takita-Sonoda, Y., Tsuji, M., Kamboj, K., Nussenzweig, R. S., Clavijo, P., and Zavala, F. (1996) *Plasmodium yoelii*: Peptide immunization induces protective CD4+ T cells against a previously unrecognized cryptic epitope of the circumsporozoite protein. *Exp. Parasitol.* **84,** 223–230.

33

Assessing CD4+ Helper T-Lymphocyte Responses by Lymphoproliferation

Isabella A. Quakyi and Jeffrey D. Ahlers

1. Introduction

CD4+ helper T cells play a central role in the development of malaria immune responses and a large number of epitopes from the sporozoite, sexual, and asexual stage of malaria proteins have been tabulated in both mice and humans *(1–8)*. It is clear that the incorporation of antigenic determinants stimulating helper T cells is important for the induction of antibody responses *(9–11)* and cytotoxic T cell responses *(12–15)* stimulated by peptide or subunit vaccines. Helper T cells (Th) also may act as effectors themselves, by secreting cytokines that may significantly influence the clinical course following infection.

Proliferative T-cell responses and IL-2 production are generally associated with CD4+ Th1-type cells following primary infection or immunization. Th1 cells produce IL-2 and IFN-γ, whereas Th2 cells are characterized by the production of IL-4, IL-5, IL-10, and IL-13 following in vitro antigenic stimulation. Functionally, Th1 responses in the mouse initiate delayed hypersensitivity, activate macrophages, and regulate production of IgG2a, whereas Th2 responses mediate humoral immune responses providing cognate help for IgG1 and IgE antibody secretion *(16)* (*see* **Note 1**). In this chapter, we describe our approach using CD4+ T helper cell responses to identify unique T-cell epitopes in asexual *Plasmodium falciparum* protein that may induce malaria-specific parasiticidal T cells.

In contrast to antibodies which recognize protein antigens in their intact form, both helper and cytotoxic T lymphocytes (CTL) generally recognize protein antigens on the surface of an antigen-presenting cell (APC), after processing by that cell to unfold or cleave the protein into fragments that then associate with major histocompatibility molecules on the cell. The response has been found to be highly focused on a limited number of discrete sites on the protein molecule rather than broadly directed at all segments of the protein. This general feature of the T-cell response is known as immunodominance *(17–19)*. CD4+ helper T cells generally recognize peptide fragments of exogenous proteins that are taken up by specialized APC, such as dendritic cells, macrophages, and B cells, into endosomes where they are cleaved into short peptides that bind to class II major histocompatibilty complex (MHC) molecules that transport them to the cell surface *(20)*. In any case, short synthetic peptides can be cleaved by serum and cell surface proteases to lengths that can bind directly to class II MHC molecules on the cell surface without going through the respective protein-processing pathways. This finding has led to the empirical approach of defining T-cell epitopes by first immunizing animals with the whole-protein molecule or recombinant subunit fragments and then screening antigen-reactive cells for proliferation using overlapping synthetic peptides spanning the whole-protein molecule. Helper T-cell responses are generated using soluble protein antigen in adjuvant. The same principle of identifying epitopes applies to human studies in which antigen-reactive cells obtained from peripheral blood lymphocytes following infection are tested against overlapping peptides or recombinant protein fragments spanning the whole protein.

1.1. Identification of Antigenic Peptides That Interact with MHC II Molecules

Underlying the selection of particular peptide determinants by both class I and class II molecules is the sequence polymorphism of individual class I heavy chain and class II α/β chains. In general, differences between alleles map to the peptide binding-groove and lead to different specificities for peptide binding to different MHC molecules. Since different peptides are preferentially presented by different MHC molecules, one would need to have either multiple epitopes binding to most MHC molecules represented in the population or, preferably, a small number of highly immunogenic epitopes presented by multiple MHC, in a synthetic peptide vaccine. One solution to this problem has come from the observation that certain segments of the human immunodeficiency virus type 1 (HIV1) envelope protein sequence contained multiple overlapping epitopes presented by different MHC class II molecules in the mouse and humans *(21)*. Identifying such Th cell cluster peptide regions (multideterminant) in a protein thus appears to be a good approach to providing promiscuous helper activity in any vaccine.

Identification of epitopes by the overlapping peptide method is both cost- and labor-intensive. For example, to perform proliferative assays using 15 amino acid long peptides overlapping by 5 amino acids spanning a given protein of length n, one would need to synthesize and assay $(n/5) - 1$ peptides. For the merozoite surface protein (MSP-1), the lead malaria vaccine candidate, of 1744 amino acid residues this would require testing 347 peptides. In an effort to facilitate the identification of T-cell epitopes, a computer algorithm called AMPHI was developed based on the tendency of T-cell antigenic sequences to form amphipathic α-helices *(22–25)*. Although the exact mechanism for a preference for peptide helices in binding to MHC molecules is uncertain, this structural feature is highly correlated to antigenic activity *(26)*. We have applied this approach to selecting Th cell epitopes from *Plasmodium falciparum* asexual stage proteins (**Fig. 1**) *(27)*.

Another approach for selecting T-cell antigenic sites is based on the identification of peptide-binding motifs for several major MHC and HLA alleles *(28,29)*. These motifs have permitted the identification of epitopes (from protein antigens whose sequences are known) with a high likelihood of being immunogenic for both Th and CTL responses, although fewer motifs for both mouse and human MHC class II molecules have been defined. EpiMer, a novel new computer algorithm, has been shown to be highly predictive of clustered T cell sites, or motif-rich regions *(30,31)*. The regions selected by EpiMer may be more likely to act as multideterminant binding peptides than randomly chosen peptides from the same antigen, due to their concentration of MHC-binding motif matches.

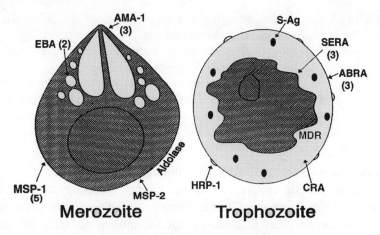

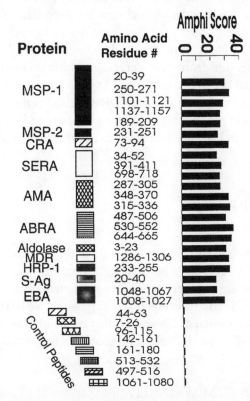

Fig. 1. Diagrammitc representation of merozoite and trophozoite asexula stage proteins. 22 putative Th-cell epitopes and 8 control peptides selected by AMPHI. Modified from Quakyi et al. *(27)* and Mshana et al. (manuscript in preparation).

There are several types of Th-cell epitopes measurable by testing proliferative responses following peptide immunization. Gammon et al. *(32)* have divided Th-cell epitopes into three convenient and testable categories.

1. Immunodominant Th-cell epitopes are the major epitopes in a protein that are processed by APC and efficiently presented by MHC molecules. Direct immunization with such peptides stimulates a strong proliferative response that is crossreactive with the native

protein. Often immunodominant epitopes of the parasite are found in regions of the protein that is highly variable due to immune pressure.

2. Subdominant Th-cell epitopes are determinants within a protein that recall in vitro proliferation to a more variable and lesser extent, and often show no measurable response from immunization with the whole protein. However, direct immunization with peptide stimulates a strong proliferative response in T cells that also respond to the native protein.

3. Cryptic Th-cell epitopes are peptides that are not effectively processed and presented by the host cell APC. Immunization with a cryptic peptide induces a response to itself but does not crossreact with the native protein.

Furthermore, in the rational design of synthetic peptide vaccines it is important to identify conserved peptide epitopes and avoid epitopes that may be deleterious (i.e., elicit suppressive Th-cell responses) to the development of protective immunity. Recent studies in malaria suggest that Th cells specific for antigen and their interactions with human leukocyte antigens (HLA) may affect the outcome of immunity to malaria (33). In a small study in Gambia, we found an association between HLA-DR13 and positive T lymphocyte proliferative responses to variant peptides of immunodominant Th-cell domains of the circumsporozoite (CS) protein (7,34). Recently, HLA-Bw53 was shown to be associated with protection from both cerebral malaria and severe malarial anemia. In addition, HLA-DRB1*1302, a subtype of DR13, was strongly associated with protection from severe malarial anemia. Indeed DRB1*1302 is the HLA class II allele with the highest frequency in individuals of African ancestry (35). Therefore identifying peptide epitopes that bind to this human class II molecule may be valuable for vaccine development.

2. Materials

2.1. Mouse CD4+ T-Cell Assays

2.1.1. Mice

F1 hybrid mice such as (C57BL/6 × C3H/He) F1 and (A.SW × BALB/c) F1 mice (H-2^{bxk} and H-2^{sxd}, respectively) may be used as a genetically defined model of an outbred population to study the immune response to predicted sequences. To map MHC restriction of responses, C57BL/10 congenic strains such as B10.A(5R), B10.BR, B10.S(9R), and B10.D2 may be used. These strains represent four different haplotypes and express H-2^b, H-2^k, H-2^s, and H-2^d class II molecules respectively.

2.1.2. Medium

Complete T-cell medium (CTM) consists of equal volumes of RPMI 1640 (with 2 mM L-glutamine and 25 mM HEPES) (Gibco) and Eagles Hank's amino acid medium (EHAA or Clicks) (Biosource), supplemented with 10% prescreened fetal bovine serum (FBS), 5×10^{-5} M 2-mercaptoethanol, 100 U/mL penicillin G, and 100 µg/mL streptomycin. Sterile filter through a 0.22-µm filter.

2.1.3. Immunization

We immunize 6–12 wk old mice with 10–30 µg of recombinant parasite protein antigen or synthetic peptide emulsified in 50 µL of complete Freund's adjuvant (Difco), subcutaneously at the base of the tail (see **Note 2**). Generally, we immunize three mice of each strain per experiment. Although cell yield is dependent on both the age and strain of mice immunized, three mice provide a minimum of 7×10^7 cells sufficient for 150 microtiter wells.

1. Suck equal volumes of antigen (in aqueous solution) and adjuvant into individual 1-mL glass syringes with Teflon plunger and Teflon Luer lock attachments (Hamilton).
2. Expel the air and connect the syringes with a teflon 26-gage connector 1 in. in length.
3. Inject the aqueous phase into the oil and pass the mixture back and forth between syringes 30–40 times until viscous.

Mice may also be immunized with whole-parasite antigen or parasitized red blood cells, which are taken up by the host APC and presented to T cells *(36)* (*see* **Note 3**).

2.2. Human CD4+ T-Cell Assays

2.2.1. Peptides

To identify CD4+ epitopes on blood-stage parasite proteins, we routinely prepare a panel of synthetic peptides representing conserved regions of parasite proteins not homologous to human sequences and predicted to be T-cell epitopes. Our strategy has been to define conserved epitopes that would induce T cells responsive to the whole parasite. However, our data suggest that most conserved regions within malaria parasites cannot induce parasite-reactive T cells, even though T cells specific for the relevant peptide can be induced *(37)* (*see* **Note 4**).

Obtain synthetic peptides prepared using solid-phase techniques as described in **ref. 38** with purity assessed by high pressure liquid chromatography (HPLC). Reconstitute lyophilized peptides in distilled water. When peptides are insoluble in water, reconstitute in either 10 mM NaOH or 10 mM HCl. Desalt all peptides prior to use and test for toxicity and nonspecific mitogenicity at 50 mg/mL in a human antitetanus toxoid or purified protein derivative (PPD) lymphocyte proliferation assay. Store peptide stock solutions at a concentration of 10 mg/mL at –20°C.

2.2.2. Antigens

1. Phytohemagglutinin (PHA) (Sigma Chemical Co., St. Louis, MO.): Dissolve in phosphate buffered saline at a concentration of 100 mg/mL, aliquot, and store frozen at –20°C. Thaw a fresh aliquot for use as needed.
2. PPD (Connaught Laboratories, Ontario, Canada): Reconstitute at a concentration of 1 mg/mL, aliquot, and store frozen at –20°C. Thaw a fresh aliquot for use as needed.
3. Crude total malaria antigen (MA): Prepare from very high parasitemic asexual stage in vitro cultures, aliquot and store at –70°C.
4. Parasite preparations: Prepare from live or Percoll-enriched, killed or sonicated and store in 1-mL aliquots of 1×10^8 parasites at –70°C.
5. Red blood cell (RBC) antigen controls: Aliquot at 1×10^8/mL and store at –70·C.
6. Tetanus toxoid (TT) (Commonwealth Biomedical Research Laboratories, Australia): 5 flocculation activity units per milliliter.

2.2.3. Medium

1. CTM: RPMI 1640/epidemic hepatitus-associated antigen (EHAA) (Biofluids), or minimal Eagle's medium (MEM) (Gibco-BRL), supplemented with 2.0 mM L-glutamine (Gibco-BRL), 25 mM HEPES (Bioscience), 0.5 M sodium pyruvate (Bioscience, Lenexa, KS), 0.01 mM nonessential amino acids (Gibco-BRL), 50 U penicillin/streptomycin (Gibco-BRL), and 10% normal human serum (NHS) from blood group AB types (Interstate Blood Bank, Memphis TN or C6 Diagnostics, Germantown, WI).

2. Incomplete RPMI 1640 (IRPMI): As above, but without serum.

2.2.3. Subjects

Our ongoing research collaborations with international malaria research laboratories, especially laboratories in malaria endemic populations, have made it possible to test these peptides in defined populations of whites who have
1. recovered from P. falciparum malaria;
2. been exposed, but never clinically infected;
3. never been exposed or infected;
4. been living in malaria endemic regions of Gambia, Ghana, Ivory Coast, and Cameroon.

2.2.4. Venipuncture

1. Obtain informed consent from human volunteers in test and control populations to participate in the project.
2. Using aseptic technique, collect peripheral blood by venipuncture, in heparinized vacutainer tubes. Where vacutainer tubes are not available, syringes containing preservative-free sodium heparin, 20–50 U/mL of blood can be used. Exercise extreme caution when drawing blood; the need for good sterile technique and biosafety practices cannot be overemphasized. Subsequent steps should be carried out in a laminar flow hood.

3. Methods

3.1. T-Cell Proliferation in Mice

3.1.1. Lymph Node Proliferation Assay

Helper T cells are required for optimal cytotoxic T-cell responses that kill pathogen-infected cells as well as for antibody responses to protein antigens. T-cell proliferation and interleukin 2 (IL-2) production represent the most convenient general methods for assessing helper T cells of the CD4$^+$ phenotype, which generally recognize antigen in association with class II MHC molecules on the APC or the B cell to be helped. The experimental methods described here have been used successfully in our laboratory to identify immunodominant T-cell epitopes within *P. falciparum* circumsporozoite and Pfs25 sexual stage and MSP1 proteins as well as in the HIV–1IIIB envelope protein.

1. At 10 d postimmunization, remove draining lymph nodes (inguinal and periaortic) from the mice using aseptic technique and place them in a small 60 × 15 mm plastic Petri dish containing 2 mL of CTM.
2. Make a single cell suspension by pressing the nodes in a circular fashion using the blunt end of a 3-mL disposable sterile syringe.
3. Wash cells 2X with CTM in 15-mL conical tubes, centrifuging at 1200 rpm for 10 min each time.
4. Resuspend lymph node cells at a concentration of 5–6 × 10^6/mL in CTM in 15-mL conical tubes.
5. Pipet 100 µL of cells using a repeat pipetor into triplicate or quadruplicate cultures of a 96-well flat bottom plate (Costar). We have found that the addition of 5 × 10^6 naive irradiated (3000 rad) syngeneic splenocytes usually increases the proliferative response and reduces medium-only background proliferation. It is convenient to prepare the plates by adding the various test antigens and peptides to the culture wells before adding lymph node cells. This will reduce risk of cross-contaminating cultures.
6. Add test peptides or recombinant proteins at appropriate concentration (e.g., 20 µ*M*, 6 µ*M*, and 2 µ*M,* covering a range of 1 log) in a volume of 100 µL CTM to triplicate or quadruplicate wells, to give a final concentration of 10, 3, and 1 µ*M* (*see* **Note 5**).

7. Include a medium control, a negative control peptide, and a PPD control in each plate.
8. Incubate cultures for 5 d at 37°C in an atmosphere of 5–6% CO_2.
9. Pulse the cultures with 1 µCi of ^{3}(H) thymidine (Amersham) in a 20 µL vol for 18–24 h.
10. Harvest cells onto glass fiber filter mats using a microtiter well plate harvester (Tomtec).
11. Dry filter mats in a microwave oven, seal in a plastic bag, and count in an LKB beta-plate scintillation counter.
12. Calculate geometric means, SEM and a stimulation index (SI), and student's *t* values comparing experimental and control replicates. An SI>2 or 3 indicates a positive response, provided that the experimental replicates are significantly different from the controls cultured with medium only ($p < 0.05$).

Although identification of helper T-cell epitopes are best identified in proliferation assays of draining lymph node cells following a single subcutaneous immunization, helper T-cell responses can also be measured in spleen-cell cultures following multiple immunizations. In that case, Th-cell responses are routinely measured after in vitro stimulation of spleen cell cultures obtained 13–14 d following immunization, and cells are pulsed with (^{3}H) thymidine on d 3 and harvested 18–24 h later.

3.2. T-Cell Proliferation in Humans

3.2.1. Introduction

Although we have found a good correlation between Th epitopes identified in several proteins between mice and humans *(2,39)*, ultimately the utility of identifying mouse epitopes may be limited for human vaccines. Studies and data from our research on malaria vaccine development, and that of others, support the central role of T lymphocytes, especially helper T cells. Numerous studies have shown that lymphocytes from immunized animals and humans proliferate in response to antigens to which they are sensitized. This in vitro response correlates with the existence of antigen-specific memory T cells. Lymphocytes also proliferate in response to mitogens, including PHA, poke-weed mitogen, and concanavalin; and also in response to different histocompatible antigens. During T-cell proliferation, quiescent cells undergo complex changes involving cell division, differentiation, and new DNA synthesis. Lymphocytes increase in size, cytoplasm becomes more extensive, nuclei are visible in the nucleus, and lymphocytes resemble blast cells. Such specific cellular T-cell activation allows these cells to exert regulatory and effector activity on B cells, T cells, and non-B and Non-T cells. The evaluation of mitogen- and/or antigen-induced lymphocyte proliferation can be quantitated by determination either of the percentage of lymphoblast or of increased DNA synthesis by addition of a radiolabeled precursor of DNA (usually tritiated thymidine). Thus have evolved the lymphocyte proliferation assay for human peripheral blood lymphocytes, pioneered by Nowell *(40)*, a simple and straightforward protocol widely used in the study of lymphocyte biology, specifically, as an efficient test for Th-cell function in malaria and other parasitic and infectious diseases. Thus the assay is a useful diagnostic tool for detecting immunological memory to a variety of antigens and pathogens, and can also be used to assess immunological recall to common antigens to which most individuals are exposed. The assay involves the in vitro culture of lymphocyte population in the presence or absence of selected antigen or mitogen. Changes induced in the stimulated groups are compared with changes in unstimulated population. Here, we outline the basic steps in the protocol and discuss advantages and limitations using this approach to assess helper cell functions.

3.2.2. Preparation of Peripheral Blood Mononuclear Cells

Separation of peripheral blood mononuclear cells (PBMC) is by density gradient centrifugation.

1. Bring all liquid reagents to room temperature before use.
2. Diluted whole blood 1:2 to 1:4 with calcium-free sterile saline, PBS and/or IRPMI.
3. Carefully layer diluted blood over Ficoll-Hypaque (Pharmacia) or on lymphocyte separating medium (calcium free balanced cell solutions minimize clumping of cells). Optimal results are obtained by carefully layering no more than three times the volume of diluted blood onto a given volume of Ficoll-Hypaque. Depending on the volume of blood, tubes can be of narrow diameter (15-mL conical tubes) or wider diameter (50-mL conical tubes). Diluted blood can be carefully layered onto gradient with the tube held at an angle, or separating medium can be carefully underlayered in the bottom of the tube.
4. Centrifuge the tubes at 400g (1800–2000 rpm) for 30 min at 22–25°C.
5. Collect the separated PBMC from the interface between the Ficoll and plasma, using a Pasteur pipet.
6. Wash two times using calcium-free incomplete culture medium, at 1600 rpm (approx 300g) for 6 min. This step is to wash out the Ficoll.
7. Resuspend cells after final wash in CTM, to ensure good viability.
8. Count cells and determine cell viability by trypan blue exclusion, using a hemocytometer.

3.2.3. Human T-Cell Proliferation Assay

Proliferation of PBMC in vitro to defined antigens is a simple method that is used to measure helper T-cell function. Over the years we have utilized this protocol to evaluate antigen recall responses to defined sporozoite-, sexual-, and asexual-stage antigens in diverse human population.

1. Calculate the working antigen concentration(s) with respect to the total volume of culture per well (200 µL).
2. Prepare control and test peptide solutions at 10X strength (100 µM working concentration; 10 µM per well final concentration peptide concentration determined previously in a peptide standardization assay to give optimum results in human proliferation assay; *see* **Note 6**), PHA and PPD at 20 µg per well, TT at 30 LF U/mL, and MA and RBC at 1×10^6 parasites per well.
3. Under sterile conditions, precharge 96-well round bottom plates (Costar) with 20 µL of medium, control antigens (PHA, PPD or TT, MA, and RBC) and n defined test peptides, in a predetermined pattern on the plate; test each control antigen and peptide in quadruplicate. Peptide is added at 10X concentration in a volume of 20 µL/well, to give a final peptide concentration of 10 µM when cells are added in 200 µL. Antigen-charged plates can be kept wrapped in foil and frozen at –20°C and used as needed.
4. Adjust purified PBMC (obtained from previous malaria-exposed donors and unexposed controls) to a final concentration of 1×10^6/mL in CTM (*see* **Note 7**).
5. Add 200 µL of PBMC suspension (2×10^5 cells) into each well of temperature calibrated thawed peptide charged sterile 96-well round-bottom plastic tissue culture plates. PBMCs from all subjects are thus tested for response to no antigen control, control antigens, PHA, PPD as positive recall antigen when screening populations in Africa, and TT if test population is white.
6. Incubate the 96-well cultures of PBMCs in a moist, 37°C incubator in an atmosphere of 5–7% CO_2 for 6 d.
7. Pulse cultures with 1 µCi/well of [^{3}H] thymidine (Amersham), for 18–24 h. Prior to pulsing

plates with radiolabeled thymidine, 20–40 µL of culture supernatant can be collected from each culture well and stored in separate sterile 96-well plates to test for cytokine production.
8. Harvest cells using a cell harvesting apparatus onto glass fiber filter mats. There are several harvesting models, but we find the Tomtec harvester is very efficient and user-friendly.
9. Assess incorporation of ^{3}H-thymidine on a beta plate counter. We currently use the Wallac-Pharmacia model.

3.2.4. Cryopreservation of Cells

The handling of cells prior to and following freezing and thawing is as important as the freezing process itself, since cell recovery, viability, and assay reproducibility can be affected from the use of poor protocol.

1. Prepare cryoprotective medium, 20% dimethyl sulfoxide (DMSO) in serum-free RPMI and cool to 4°C.
2. Isolate PBMC from whole blood using standard density gradient methods.
3. Suspend cells in medium containing 20% heat-inactivated FCS.
4. Examine cell suspension for purity, adjust cell concentration to twice that of the desired final concentration and place on ice. For PBMC, resuspend at 25×10^6 cells/mL.
5. Label the freezing vials with the name of the donor, the cell concentration and the date, and place vials on ice to cool at 4°C.
6. Add an equal volume of freezing medium dropwise (with mixing) to cell suspension. This gives a final freezing suspension of complete RPMI 1640 plus 10% FCS plus 10% DMSO.
7. Pipet 0.5 mL of cells into each chilled vial (12.5×10^6 cells/vial). A variety of containers have been demonstrated to be adequate for the freezing and thawing of lymphocyte suspension, including Nunc vials, Beckman vials, and glass vials.
8. Transfer to a controlled-rate-freezer. It is important that the cell remain in the freezing mixture for a minimum time before being frozen. Once frozen, transfer the ampules to the vapor phase of liquid nitrogen. In the absence of a controlled-rate freezing devise, place freezer vials into a styrofoam box at the bottom of an ultralow freezer, cells will remain viable for several months. Ideally cell should be stored long-term frozen in liquid nitrogen.

3.2.5. Thawing and Washing of Cryopreserved Cells

See **ref. 41**.

1. Warm tissue culture medium containing 10% human serum to 37°C.
2. Remove cells to be thawed from freezer and place on dry ice.
3. Thaw one vial at a time. Using forceps, remove one vial from dry ice and thaw rapidly with shaking in a 37°C water bath.
4. Just before the last crystal has melted, remove vial from water bath and transfer the contents of the vial (using a 1-mL disposable pipet) to a 15-mL round-bottom tube (Corning). Add cell suspension gently down the side of the tube (see **Note 8**).
5. Place the tube at room temperature, and gently dilute out the DMSO by adding RPMI with 10% human serum dropwise with gentle mixing until a DMSO concentration of 4% is achieved.
6. Make up volume to 15-mL with medium.
7. Centrifuge at speeds not exceeding 200g, approx 1000–1500 rpm, for 10 min.
8. Pour off or aspirate medium, and resuspend in fresh CTM.

3.2.6. Calculation of Results

1. Calculate mean values from the stimulated and unstimulated counts per minute, either by subtraction to get the stimulation difference or by ratio of mean counts per minute of

stimulated vs unstimulated to get the SI. Background counts in the absence of antigen and mitogen can range from 160 to 2000 counts per minute. Range of counts for mitogenic response to PHA are 50,000 to 150,000, for immunological recall to common antigens, for example, BCG/PPD and TT, are 5000 to 100,000 counts per minute, and for antigen-specific recall, are 5000 to 200,000 counts per minute.

2. Statistical analysis includes geometric means of quadruplicate samples (2×10^5 cells/well) and a T value (an index of assay consistency and significance of data). A response is typically considered positive if the SI is >2 and if the T value = 2.20 (95% confidence level).

3.2.7. Developing T-Cell Clones Reactive to Malaria Protein Antigens

We and others have demonstrated that population proliferative responses to epitopes on malaria proteins that are vaccine candidate may not necessarily be malaria specific and may be due to memory cells specific for crossreactive organisms *(1,8,42–45)*. Indeed, in a study where we analyzed population proliferative responses to conserved epitopes from MSP1, MSA2, AMA1, SERA, ABRA, and EBA, we found that peptides that were well recognized by both exposed and nonexposed groups contained sequence homology with other known pathogens and proteins (e.g., *Escherichia coli*, *Neisseria*, and adenovirus); one peptide shared 90% sequence homology with the CD37 leukocyte marker).

3.2.8. Generation of Epitope-Specific T-Cell Clones

The generation of epitope-specific T-cell clones and subsequent functional studies as to the MHC restriction element and biological activity, is one way to examine the antigenic specificities to population bulk PBMC responses to defined epitopes. There are a number of methods and variations that can be used to make such T-cell clones. We discuss here a method developed by Currier et al. *(45)*, and adapted for round-bottom wells that has worked consistently well in our hands *(8)*. The method is a limiting dilution cloning using mitogenic stimulation (**Fig. 2**).

1. Obtain PBMCs from whole blood by centrifugation over Ficoll-HyPaque, as detailed in **Subheading 3.2.2.**, and resuspend in CTM.
2. Plate the cells at 2×10^5 cells (200 µL) per well in 96 well round bottom plates with antigen whole cell, purified protein, recombinant protein, and or defined peptide.
3. Culture cells for 6 d in a 37°C water-jacketed incubator gassed with 5% CO_2 to maintain a pH of 7.2 to 7.3.
4. On d 6, gently resuspend all wells, remove 40 µL of cell suspension into a separate plate, and return the parent plate to the incubator.
5. Pulse the wells of 40 µL cell aliquots with 1 µCi of [^{3}H]thymidine, in 5 µL of RPMI, and incubate for 18 h.
6. Harvest assays to determine wells that are giving the maximum and significant DNA synthesis in response to the test antigen.
7. Pool all positive wells, collect the blasts, and layer over Ficoll-hypaque.
8. Centrifuge at 300*g* for 25 min.
9. Collect viable T cells at interface layer, wash 2X in fresh CTM and resuspend in CTM.
10. Initiate a second round of stimulation by incubating lymphoblastoid T cell lines (5×10^5) with peptide and irradiated autologous APC in a 48-well microtiter plate in CTM.
11. Culture for 6 d.
12. Pulse 40-µL aliquots with 1 µCi [^{3}H]thymidine for 18 h, harvest, count cells, pool all positive responder well from the parent plate, Ficoll, collect cells from interface, wash and resuspend all viable cells, as detailed above.

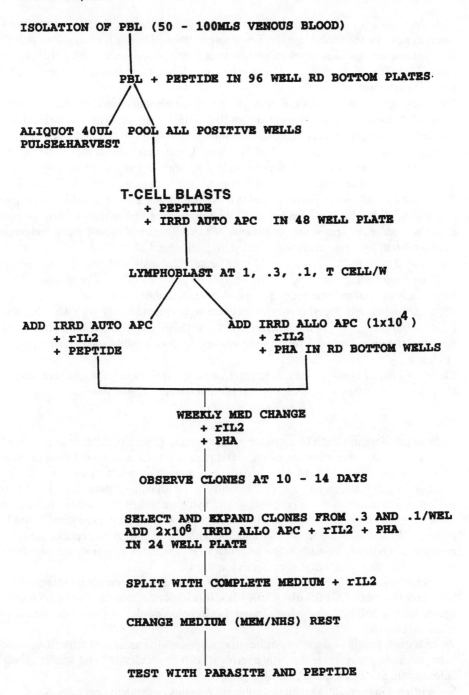

Fig. 2. T-cell cloning by limiting dilution and mitogenic stimulation.

13. Collect the T cell line about 7–14 d after an appropriate number of stimulation cycles.
14. Resuspend cells, layer over Ficoll-HyPaque, and centrifuge at 300g for 25 min.
15. Collect viable T cells from interface layer and wash two times in fresh medium.
16. Carry out limiting dilution cloning by plating out Ficolled antigen-specific blast cells in round bottom wells. A cell concentration of 3, 1, 0.3, and 0.1 gives a good range of dilu-

tions from which clones with high stringency (i.e., greater than 95% probability of clonality) can be selected. Plate the T-cell line in 96-well round bottom plates at 200 µL final volume per well and incubate with irrigated autologous feeders at 1×10^4 plus antigen or PHA. For the generation of high-stringency clones, there should be at least 120–180 wells of each dilution.
17. Incubate plates for at least 4 wk, during which time the medium should be changed once per week. Medium should be supplemented with 10% NHS, 0.5 µg/mL PHA, and 50 U/mL rIL-2.
18. Add fresh autologous feeder cells with medium change after the second week of culture.
19. Examine for positive wells after 3–4 wk in culture. Usually dilutions at which less than 10 or 15% of the wells are positive are best for scaling-up high-stringency clones. Poisson analysis can be used to determine the percentage probability of clonality for each dilution sample.
20. Select and expand clones from 0.3 and 0.1/well by adding 2×10^6 autologous or allogeneic PBMC (3000 Rad) to selected in 24-well plates. For restimulation in tissue culture flasks 10^6 feeder cells per square centimeter of surface area will be sufficient, and medium should contain final concentration of 0.5 µg/mL PHA and 25 U rIL-2.
21. After 5–7 d with PHA, clones should proliferate vigorously and can be split and given fresh medium containing 25 U/mL rIL-2. At this point T cells can be driven with IL-2 indefinitely or restimulated every 3–4 wk as described above.
22. Set aside aliquots of T-cell clones in medium supplemented only with 10% NHS until cells are fully rested (7–14 d). When rested, the cells are rounded and nonmitotic. This is the stage of cells used for all future assays to test for both peptide and parasite specificity of clones.
23. Clones of interest at rest phase can be frozen in aliquots to liquid nitrogen. For long-term clones, it is important to periodically test for mycoplasma contamination.

4. Notes

1. Th2 cell proliferation and Th2 cytokine production are generally difficult to measure following a primary immunization of mice and require clonal expansion by repeated antigenic stimulation every 10–14 d in vitro for detection. By contrast, in an endemic disease setting where humans undergoing chronic infection with malaria parasites and other concurrent parasitic infections, a preexisting Th2 cytokine milieu may favor Th2 responses allowing detection of proliferative CD4+ Th2 cell responses from peripheral blood lymphocytes (PBL) following in vitro stimulation with whole antigen or peptide. Since a Th1 response is generally considered to be protective in malaria infection, the proliferative assay is valuable for identifying protective epitopes.
2. In order to obtain a satisfactory adjuvant, we routinely dilute recombinant proteins solubilized in 8 M urea to 1X PBS using 10X PBS. It is important that the final salt concentration be approx 150 mM, as we have found that hypotonic solutions generally do not form stable emulsions.
3. When testing T cells for in vitro proliferative responses to parasitized red cell antigens, it is important to preincubate the cells overnight with host cell APC and wash thoroughly before adding to T-cell cultures (36).
4. Since in vivo immunization with recombinant proteins or synthetic peptides is not a practical approach to identify epitopes in humans, primary in vitro immunization with peptide-pulsed dendritic cells has been proposed as a means to unveil both dominant and subdominant epitopes in a given protein (37).
5. It is advisable to filter peptides in serum containing medium before use with low protein-binding filters (Millex-GV, Millipore). Peptides stock solutions that are not readily soluble in ddH$_2$O can be made soluble by dilute acid (10 mM HCl) or base (10 mM NaOH) and filtered at a high concentration. Dilutions are made from this stock in either serum-free medium or medium supplemented with 10% FBS and 100 µL added per well of a 96-well plate.

6. Positive peptide-specific responses should be subsequently titrated at 0.1, 1, 10, and 100 µM concentrations.
7. It is preferable to set up proliferation assays as soon as blood samples are drawn or at least within 24 h of collection of samples. Results from fresh cell assays are more likely to represent the immune status of the individual, and remaining cells can be cryopreserved. Also, cryopreservation allows for sequential samples of the same individual to be analyzed in the same assay. However it is not always possible to draw blood on day convenient to set up an assay; for example, the laboratory may be too far from the field where samples are obtained, and cells may have to be transported to established laboratories.
8. Cells that are frozen are initially sensitive, so pipetting and mixing should be kept at a minimum. All procedures of cryopreservation and thawing should be carried out under sterile conditions.

References

1. Good, M. F., Quakyi, I. A., Saul, A., Berzofsky, J. A., Carter, R., and Miller, L. H. (1987) Human T clones reactive to the sexual stages of Plasmodium falciparum malaria: High frequency of gamete-reactive T cells in peripheral blood of non-exposed donors. *J. Immunol.* **138,** 306–311.
2. Dontfraid, F., Cochran, M. A., Pombo, D., Knell, J. D., Quakyi, I., Kumar, S., et al. (1988) Human and murine CD4 T-cell epitopes map to the same region of the malaria circumsporozoite protein: Limited immunogenicity of sporozoites and circumsporozoite protein. *Mol. Biol. Med.* **5,** 185–196.
3. Good, M. F., Pombo, D., Quakyi, I. A., Riley, E. M., Houghten, R. A., Menon, A., et al. (1988) Human T cell recognition of the circumsporozoite protein of Plasmodium falciparum. Immunodominant T cell domains map to the polymorphic regions of the molecule. *Proc. Natl. Acad. Sci. USA* **85,** 1199–1203.
4. Good, M. F., Berzofsky, J. A., and Miller, L. H. (1988) The T-cell response to the malaria circumsporozoite protein: an immunological approach to vaccine development. *Annu. Rev. Immunol.* **6,** 663–688.
5. Good, M. F., Pombo, D., Maloy, W. L., De la Cruz, V. F., Miller, L. H., and Berzofsky, J. A. (1988) Parasite polymorphism present within minimal T-cell epitopes of Plasmodium falciparum circumsporozoite protein. *J. Immunol.* **140,** 1645–1650.
6. Good, M. F., Miller, L. H., Kumar, S., Quakyi, I. A., Keister, D., Adams, J., et al. (1988) Limited immunological recognition of critical malaria vaccine candidate antigens. *Science* **242,** 574–577.
7. DeGroot, A. S., Johnson, A. H., Maloy, W. L., Quakyi, I. A., Riley, E. M., Menon, A., et al. (1989) Human T cell recognition of polymorphic epitopes from malaria circumsporozoite protein. *J. Immunol.* **142,** 4000–4005.
8. Quakyi, I. A., Currier, J., Fell, A., Taylor, D. W., Roberts, T., Houghten, R. A., et al. (1994) Analysis of human T cell clones specific for conserved peptide sequences within malaria proteins: paucity of clones responsive to intact parasites. *J. Immunol.* **153,** 2082–2092.
9. Quakyi, I. A., Miller, L. H., Good, M. F., Ahlers, J. D., Isaacs, S. N., Nunberg, J. H., et al. (1995) Synthetic Peptides from P. falciparum sexual stage 25-kDa protein induce antibodies that react with the native protein: the role of IL-2 and conformational structure on immunogenicity of Pfs25. *Peptide Res.* **8,** 335-344.
10. Hirunpetcharat, C., Tian, J.-H., Kaslow, D. C., van Rooijen, N., Kumar, S., Berzofsky, J. A., et al. (1997) Complete protective immunity induced in mice by immunization with the 19kDa carboxy terminal fragment of the merozoite surface protein-1 (MSP1$_{19}$) of Plasmodium yoelii expressed in Saccharomyces cerevisiae: correlation of protection with antigen-specific antibody titer, but not effector CD4+ T cells. *J. Immunol.* **159,** 3400–3411.
11. Tian, J.-H., Miller, L. H., Kaslow, D. C., Ahlers, J., Good, M. F., Alling, D. W., et al. (1996) Genetic regulation of protective immune response in congenic strains of mice vaccinated with a subunit malaria vaccine. *J. Immunol.* **157,** 1176–1183.
12. Kumar, S., Miller, L. H., Quakyi, I. A., Keister, D. B., Houghten, R. A., Maloy, W. L., et al. (1988) Cytotoxic T cells specific for the circumsporozoite protein of Plasmodium falciparum. *Nature* **334,** 258–260.
13. Weiss, W. R., Sedegah, M., Beaudoin, R. L., Miller, L. H., and Good, M. F. (1988) CD8+ T cells (cytotoxic/suppressors) are required for protection in mice immunized with malaria sporozoites. *Proc. Natl. Acad. Sci. USA* **85,** 573–576.
14. Weiss, W. R., Mellouk, S., Houghten, R. A., Sedegah, M., Kumar, S., Good, M. F., et al. (1990) Cytotoxic T cells recognize a peptide from the circumsporozoite protein on malaria infected hepatocytes. *J. Exp. Med.* **171,** 763–773.
15. Doolan, D. L., Houghten, R. A., and Good, M. F. (1991) Assessment of human cytotoxic T cell activity using synthetic peptides: potential for field application. *Peptide Res.* **4,** 125–131.

16. Fargeas, C., Wu, C. Y., Nakajima, T., Cox, D., Nutman, T., and Delespesse, G. (1992) Differential effect of transforming growth factor beta on the synthesis of Th1- and Th2-like lymphokines by human T lymphocytes. *Eur. J. Immunol.* **22,** 2173–2176.
17. Berzofsky, J. A. (1991) Mechanisms of T cell recognition with application to vaccine design. *Molec. Immunol.* 28, 217–223.
18. Gammon, G., Klotz, J., Ando, D., and Sercarz, E. E. (1990) The T cell repertoire to a multi-determinant antigen: clonal heterogeneity of the T cell response, variation between syngeneic individuals, and in vitro selection of the T cell specificities. *J. Immunol.* **144,** 1571–1577.
19. Berzofsky, J. A. and Berkower, I. J. (1993) Immunogenicity and antigen structure, in *Fundamental Immunology* (Paul, W. E., ed.), Raven Press, NY, pp. 235–282.
20. Germain, R. N. and Margulies, D. H. (1993) The biochemistry and cell biology of antigen processing and presentation. *Annu. Rev. Immunol.* **11,** 403–450.
21. Hale, P. M., Cease, K. B., Houghten, R. A., Ouyang, C., Putney, S., Javaherian, K., et al. (1989) T cell multi-determinant regions in the human immunodeficiency virus envelope: toward overcoming the problem of major histocompatibility complex restriction. *Int. Immunol.* **1,** 409–415.
22. Berzofsky, J. A., Cease, K. B., Berkower, I. J., Margalit, H., Cornette, J., and DeLisi, C. (1986) Immunodominance of amphipathic peptides and their localization on the cell surface for antigen presentation to helper T cells, in *Progress in Immunology* (Cinader, V. B. and Miller, R. G., eds.), Academic Press, NY, pp. 255–265.
23. Margalit, H., Spouge, J. L., Cornette, J. L., Cease, K., DeLisi, C., and Berzofsky, J. A. (1987) Prediction of immunodominant helper T-cell antigenic sites from the primary sequence. *J. Immunol.* **138,** 2213–2229.
24. Margalit, H., DeLisi, C., and Berzofsky, J. A. (1988) Computer predictions of T-cell epitopes, in *The Molecular Approach to New and Improved Vaccines*. (Woodrow, G. C. and Levine, M. M., eds.), Marcel Dekker, NY.
25. Berzofsky, J. A. (1989) Structural features of T-cell recognition: spplications to vaccine design. *Phil. Trans. Roy. Soc. Lond.* **B 323,** 535–544.
26. Cornette, J. L., Margalit, H., Berzofsky, J. A., and DeLisi, C. (1995) Periodic variation in side-chain polarities of T-cell antigenic peptides correlates with their structure and activity. *Proc. Nat. Acad. Sci. USA* **92,** 8368–8372.
27. Quakyi, I. A., Taylor, D. W., Johnson, A. H., Allotey, J. B., Berzofsky, J. A., Miller, L. H., et al. (1992) Development of a malaria T-cell vaccine for blood atage immunity. *Scand. J Immunol.* **36,** 9–16.
28. Rammensee, H.-G., Friede, T., and Stevanovíc, S. (1995) MHC ligands and peptide motifs: first listing. *Immunogenetics* **41,** 178–228.
29. Sette, A., Vitiello, A., Reherman, B., Fowler, P., Nayersina, R., Kast, W. M., et al. (1994) The relationship between class I binding affinity and immunogenicity of potential cytotoxic T cell epitopes. *J. Immunol.* **153,** 5586–5592.
30. DeGroot, A. S., Jesdale, B. M., and Berzofsky, J. A. (1998) Prediction and determination of MHC ligands and T cell epitopes, in *Immunology of Infection*. (Kaufmann, S. H. E. and Kabelitz, D., eds.) Academic Press, London, pp. 79–108.
31. DeGroot, A. S., Meister, G. E., Cornette, J. L., Margalit, H., DeLisi, C., and Berzofsky, J. A. (1997) Computer prediction of T-cell epitopes, in *New Generation Vaccines*. (Levine, M. M., Woodrow, G. C., Kaper, J. B., and Cobon, G. S., eds.), Marcel Dekker, NY, pp. 127–138.
32. Gammon, G., Shastri, M., Logswell, J., Wilbur, S., Sadegh-Nasseri, S., Krzych, U., Miller, A., and Sercarz, E. (1987) The choice of T cell epitopes utilized on a protein antigen dependes on multiple factors distant from as well as at the determinant site. *Immunol. Rev.* **98,** 53–73.
33. Hill, A. V. S., Allsopp, C. E. M., Kwiatkowski, D., Anstey, N. M., Twumasi, P., Rowe, P. A., Bet al. (1991) Common West African HLA antigens are associated with protection from severe malaria. *Nature* **352,** 595–600.
34. Riley, E. M., Allen, S. J., Bennett, S., Thomas, P. J., O'Donnell, A., Lindsay, S. W., et al. (1990) Recognition of dominant T cell-stimulating epitopes from the circumsporozoite protein of Plasmodium falciparum and relationship to malaria morbidity in Gambian children. *Trans. R. Soc. Trop. Med. Hyg.* **84,** 648–657.
35. Johnson, A. H., Dunston, G., Rosen-Bronson, S., Hartzman, R. J., and Hurley, C. K. (1986) *Polymorphism of the HLA-D Region in American Blacks. HLA in Asia-Oceania 1986*. (Aizwa, M., ed.), Hokkaido University Press, Sapporo, Japan. p. 548.
36. Wasserman, G. M., Kumar, S., Ahlers, J., Ramsdell, F., Berzofsky, J. A., and Miller, L. H. (1993) An approach to development of specific T lymphocyte lines using preprocessed antigens in murine malaria Plasmodium vinckei vinckei. *Infect. Immun.* **61,** 1958–1963.
37. Celis, E., Tsai, V., Crimi, C., DeMars, R., Wentworth, P. A., Chesnut, R. W., et al. (1994) Induction of anti-tumor cytotoxic T lymphocytes in normal humans using primary cultures and synthetic peptide epitopes. *Proc. Natl. Acad. Sci. USA* **91,** 2105–2109.

38. Houghten, R. A.(1985) General method for the rapid solid-phase synthesis of large numbers of peptides: specificity of antigen-antibody interaction at the level of individual amino acids. *Proc. Natl. Acad. Sci. USA* **15,** 5131–5135.
39. Berzofsky, J. A., Pendleton, C. D., Clerici, M., Ahlers, J., Lucey, D. R., Putney, S. D., et al. (1991) Construction of peptides encompassing multi-determinant clusters of HIV envelope to induce in vitro T-cell responses in mice and humans of multiple MHC types. *J. Clin. Invest.* **88,** 876–884.
40. Nowell, P. L. (1960) Phytohemagglutinin: an indicator of mitosis in cultures of normal human leucocytes. *Cancer Res.* **20,** 462–466.
41. Maluish, A. E. and Strong, D. M. (1993) Lymphocyte proliferation, in *Manual of Clinical Immunology*. American Society for Microbiology, pp. 274–281.
42. Rzepczyk, C. M., Ramasamy, R., Mutch, D. A., Ho, P. C.-L., Battistutta, D., Anderson, K. L., et al. (1989) Analysis of human T cell response to two Plasmodium falciparum merozoite surface antigens. *Eur. J. Immunol.* **19,** 1797–1802.
43. Jones, K. R., Hickling, J. K., Targett, G. A. T., and Playfair, J. H. L. (1990) Polyclonal in vitro proliferative responses from non-immune donors to Plasmodium falciparum antigens require UCHL1+(memory) T cells. *Eur. J. Immunol.* **20,** 307–315.
44. Fern, J. and Good, M. F. (1992) Promiscuous malaria peptide epitope stimulates CD45Ra T cells from peripheral blood of nonexposed donors. *J. Immunol.* **148,** 907–913.
45. Currier, J., Sattabonkot, J., and Good, M. F. (1992) "Natural" T cells responsive to malaria: evidence implicating immunological cross-reactivity in the maintenance of TCR a/b+ malaria-specific responses from non-exposed donors. *Int. Immunol.* **4,** 985–994.

34

Limiting Dilution Analysis of Antigen-Specific CD4+ T-Cell Responses in Mice

Elsa Seixas, Jean Langhorne, and Stuart Quin

1. Introduction

In all rodent models of erythrocytic-stage malaria infections, the T-cell response to the parasite has been shown to be crucial for the development of protective immunity *(1–5)* and in some cases is thought to play a role in the pathology of malaria *(6,7)*. To determine which effector functions of T cells may be important in controlling parasites or in causing disease, it is necessary to determine the nature of the CD4+ T-cell response in a quantitative manner.

The magnitude of the response to different parasite proteins can be dependent on several factors, such as accessibility of the protein to different antigen processing cells, and to the antigen-processing pathway, as well as the major histocompatibility complex (MHC) of the host. An estimate of the frequency of CD4+ T cells responding to different parasite proteins, protein fragments, and individual peptides from infected or immune animals would indicate the relative immunodominance of particular parasite molecules or parts of molecules. A single assay in which the different parameters of the CD4+ T cell response to the parasite, such as proliferation, cytokine production, help for antibody and cytotoxicity, could be measured, would also be of value in determining the functional capacity of malaria-specific T cells.

There are several methods that allow the frequency of responding T cells to be estimated; limiting dilution analysis to determine precursor frequencies *(8,9)*, enzyme-linked immunospot (ELISPOT) assays *(10,11)* or intracellular staining with flow cytometric analysis to measure the numbers of T cells producing different cytokines *(12,13)*, and, more recently, MHC tetramers containing specific peptides to enumerate the numbers of specific T cells directly *(14,15)*. Although limiting dilution assays (LDA) are relatively labor-intensive when compared with the other methods, they still offer several advantages. First, measurement of the functional capacity of the CD4+ T cell (e.g., CD4+ T cell help for antibody production or cytotoxic activity) can be incorporated into the assay. Second, several parameters can be assayed from the same microcultures. Third, in contrast to the use of MHC tetramers, which is limited in each case to single peptides, the limiting dilution assay can be used to determine the response to infected erythrocytes, parasite proteins, or fragments thereof, as well as individual peptides. Finally the response of relatively rare specific T cells can be measured, unlike detection of cytokine-producing cells by flow cytometry where frequencies of less than 1 in 1000 responding cells would be difficult to enumerate.

In this chapter, we describe a method for estimating the precursor frequency of $CD4^+$ T cells from mice infected with *Plasmodium chabaudi chabaudi (AS)* specific for antigens expressed in infected erythrocytes. The assay can be readily modified to determine $CD4^+$ T-cell responses to other rodent malaria parasites and to defined malarial proteins or peptides. The assay is a two-step culture, which allows the measurement of $CD4^+$ helper T cell activity in the first part of the assay, and proliferation and cytokine production of responding clones of cells after a 2-d restimulation step.

The principle of the limiting dilution assay is to coculture many replicates of different dilutions of $CD4^+$ T cells with an excess of antigen-presenting cells and antigen. The number of nonresponding microcultures at each T-cell dilution is then used to estimate the precursor frequency using the Poisson distribution. It is important that only $CD4^+$ T cells are limiting and that each microculture should contain sufficient antigen, antigen-presenting cells, and B cells such that if there is a specific T cell present in the well, then a response can be measured. In the assays described here, the antigen-presenting cells in the initial culture period contain predominantly B cells from malaria-immune mice. A positive specific antibody response in a microculture indicates the presence of at least one precursor helper T cell. A subsequent second round of antigen stimulation for a further 2 d allows sufficient clonal expansion for the detection of cytokines and proliferation.

In the following sections, a detailed description is given of the preparation of $CD4^+$ T cells and antigen-presenting cells, measurement of $CD4^+$ T cell "help," and cytokine production, as well as how to set up and analyze an LDA.

2. Materials

2.1. Limiting Dilution Assay

2.1.1. Equipment

1. Flow cytometer with filters for 488 and 633 nm.
2. Cell harvester.
3. β-Counter.
4. CO_2 incubator (7% CO_2, 37°C).
5. Centrifuge for 15-mL, 50-mL, and flow cytometry tubes.
6. Cellular depletion magnet (Dynal).
7. Magnetic activated cell sorting (MACS) magnet (VarioMACS, MiniMACS, MidiMACS, or SuperMACS from Miltenyi Biotec).
8. MACS columns (MACS MS+/RS+, Miltenyi Biotec) (*see* **Note 1**).
9. Sterile scissors and forceps.
10. Sterile sieves.
11. 96-Well round-bottom plates with lids, tissue culture-grade, sterile.
12. Needles and syringes.
13. Plastic cell strainer (Falcon).
14. 96-Well round-bottom non-sterile plates.
15. 15-mL and 50-mL Tubes.
16. Hemocytometer.

2.1.2. Reagents

1. Hanks' washing buffer: Hanks' buffered salt solution (HBSS) with 5% fetal calf serum (FCS) and 12 mM HEPES buffer. Keep sterile at 4°C.
2. Complete Iscove's modified Dulbecco's medium (IMDM): with 10% FCS, 10 U/mL penicillin, 100 mg/mL streptomycin, 1 mM L-glutamine, 0.5 mM sodium pyruvate, 6 mM HEPES and 5×10^{-5} M β-mercaptoethanol. Store sterile at 4°C in the dark.
3. Erythrocyte lysing buffer: NH_4Cl-Tris-HCl, pH 7.2. Prepare stocks of 0.17 M Tris-HCl, pH 7.65 (solution 1) and 0.16 M NH_4Cl (solution 2). Make a dilution of 1-part solution 1 to 50-parts solution 2 and adjust the pH to 7.2. Sterile filter and keep at 4·C.
4. MACS buffer: Phosphate-buffered saline (PBS) with 0.5% bovine serum albumin (BSA) and 2 mM EDTA (*see* **Note 2**). Check that pH 7.2, sterile filter and keep at 4°C.
5. Flow cytometry buffer: PBS with 1% BSA, 5 mM EDTA, and 0.01% NaN3. Filter before use.
6. "Low-tox" rabbit complement (Cedar Lane). Store at –80°C (*see* **Note 3**).
7. DNAse (Sigma): Dilute 1 g in 100 mL of PBS. Store aliquots at –20°C.
8. Collagenase: Dilute to a concentration of 8 mg/mL in HBSS. Make aliquots and store at –20°C.
9. Streptavidin-coated microbeads (Miltenyi Biotec). Keep at 4°C.
10. Dynabeads M-450 (sheep anti-rat IgG) (Dynatech).
11. Monoclonal anti-mouse CD4 labeled with biotin.
12. Lympholyte (Cedar Lane).
13. Trypan blue solution.

2.2. ELISA

2.2.1. Equipment

1. Enzyme-linked immunosorbent assay (ELISA) reader with filters for 405 nm and 490 nm.
2. ELISA plate washer (*see* **Note 4**).
3. Box with lid for incubations in humid conditions (*see* **Note 5**).
4. Multichannel and/or multidispensing pipets.
5. Maxisorb 96 well plates (Nunclon, Denmark): cytokine ELISA.
6. Polysorb 96 well plates (Nunclon, Denmark): malaria-specific Ig ELISA.

2.2.2. General Reagents

1. Coat diluting buffer: PBS, 0.05% sodium azide.
2. Blocking buffer: PBS, 1% BSA, 0.3% Tween-20, 0.05% sodium azide.
3. Secondary Ig/conjugate-diluting buffer: PBS.
4. Washing buffer: 0.9% NaCl, 0.005 M K_2HPO_4, 0.005 M KH_2PO_4, 0.025% Tween-20.
5. Substrate (*p*-nitrophenylphosphate sodium salt (PNPP) (Sigma): Dilute to 1 mg/mL in diethanolamine buffer (*see* **Note 6**).
6. Diethanolamine buffer: 48.5 mL of diethanolamine, 400 mg of $MgCl_2$ $6H_2O$, 100 mg of sodium azide, and 450 mL of H_2O; adjust to pH 9.8 by adding conc. HCl (*see* **Note 7**).

2.2.3. Specific Reagents for Cytokine ELISA

1. Coating antibodies: For interleukin-4 (IL-4): 24.G2 (ATCC) and interferon (IFN): 0.3B2 *(16)* (*see* **Note 8**).
2. Biotinylated antibodies: For IL-4: 1.D.11 (ATCC) and IFN: 0.3B2 *(16)* (*see* **Note 8**).
3. Standard: A source of recombinant IL-4/IFN-γ (*see* **Notes 8** and **9**).
4. Streptavidin alkaline-phosphatase conjugate solution (Southern Biotechnology) (*see* **Note 8**).

2.2.4. Specific Reagents for Malaria-Specific Ig ELISA

1. Coating parasite preparation (*see* **Subheading 3.**): Dilute 100 µL of parasite prep in 400 µL of 50 mM Tris-HCl, pH 7.5, 1 mM EDTA, 0.5% sodium dodecyl sulfate (SDS), pH 8.0. Dilute in PBS, 0.05% sodium azide until OD280 nm = 0.05 (*see* **Note 8**).
2. Anti-mouse Ig alkaline phosphatase conjugated antibodies (Southern Biotechnology) at 200 µg/mL.

2.2.5. Specific Reagents for Preparation of Parasite Antigen

1. Parasite lysis buffer: 50 mM Tris-HCl, pH 7.5, 1 mM EDTA, 0.5% SDS.
2. PBS.
3. 10% saponin.
4. Heparin.
5. Apoproten (Sigma).
6. CF11 columns; CF11 (Whatman) and 5-mL syringes.

3. Methods
3.1. Introduction

The limiting dilution assay requires CD4+ T cells from the spleens of *P. chabaudi*-infected mice and from uninfected mice as controls, a source of antigen-presenting cells and, in the example given here, *P. chabaudi*-infected and uninfected erythrocytes as antigen. Antigen-presenting cells in the first part of the assay are a population of B cells, macrophages, and dendritic cells from the spleens of *P. chabaudi* immune mice (preferably at least 5 wk postinfection). The frequency of malaria-specific B cells in the spleens of *P. chabaudi* immune C57BL/6 or BALB/c mice has been found to be approx 1 in 10,000 (*17*). To ensure that all wells can give a positive response when a helper T cell is present, approx 30,000 immune B-cell-enriched spleen cells are used as antigen-presenting cells. After 7 to 10 d culture of many replicates of different CD4+ T-cell concentrations with the antigen-presenting cells and antigen, the supernatants are removed to determine the presence of malaria-specific antibodies (measured by an ELISA assay). Freshly prepared irradiated spleen cells and antigen are then added to the microcultures, which are incubated for a further 48 h. After this time, supernatants are removed to measure cytokines by ELISA or bioassay, and proliferation of the responding T cells is determined by the incorporation of [^{3}H]-thymidine.

3.2. CD4+ T-Cell Preparation
3.2.1. Collection of Spleens and Cell Preparation.

1. Use sterile forceps and scissors to remove the spleens necessary for the assay and collect in a 15-mL tube containing Hanks' washing buffer.
2. Place the spleens in a Petri dish, and carefully remove traces of fat without cutting the tissue.
3. Disrupt the spleen gently through the sieve with a sterile syringe plunger into a Petri dish with Hanks' washing buffer. Rinse the sieve and the plate and transfer to a 50-mL tube.
4. Centrifuge at 300g for 10 min at 4°C (all the centrifugation steps are carried out at 300g for 10 min at 4°C except in the case of Lympholyte (*see* **Subheading 3.3.2.**).
5. Pour off the supernatant, add 2 mL per spleen of buffer for lysing erythrocytes, and incubate 10 min at room temperature.
6. Add 20 mL of Hanks' washing buffer and centrifuge.

LDA Analysis

7. Resuspend the cells in 5 mL of Hanks' washing buffer and pass through a plastic cell strainer to obtain a better homogenization of the cell suspension.
8. Count live and dead cells using trypan blue on a hemocytometer.
9. Keep an aliquot of at least 5×10^5 cells for flow cytometry.

3.2.2. Selection of CD4+ T Cells

There are two ways to obtain CD4+ T cells: by positive or negative selection.

3.2.2.1. POSITIVE SELECTION OF CD4+ T CELLS

1. Take a volume of spleen cell suspension containing 10^8 cells and centrifuge.
2. Dilute the anti-mouse CD4 labeled with biotin in the MACS buffer in order to have 1 mL of the appropriate antibody dilution per 10^8 cells. Keep it on ice (see **Note 10**).
3. Discard the supernatant and resuspend the pellet in the MACS buffer with appropriate dilution of anti-CD4 antibody.
4. Incubate for 20 min on ice.
5. Add 10 mL of cold MACS buffer and centrifuge at $300g$ for 10 min at 4°C.
6. Dilute the streptavidin microbeads at 1:10 in ice-cold MACS buffer in order to have 1 mL of the mixture per 10^8 cells and keep it on ice.
7. Add the appropriate volume of beads to the pellet and incubate for 10 min on ice.
8. Add 10 mL of ice-cold MACS buffer and centrifuge at $300g$ for 10 min at 4°C.
9. Take column (MS+/RS+) from refrigerator immediately before use, place in a magnet (VarioMACS) and wash through with 500 µL of cold MACS buffer.
10. Change the collection tube and run 500 µL of cell suspension through. Wait until this passes through the column before adding another 500 µL of cells in a buffer because this will overload the column.
11. Wash the column three times with 500 µL of cold MACS buffer each time to remove unbound cells.
12. Collect this as the negative fraction, label the tube, and store on ice.
13. Remove the column from the magnet and place over a 15-mL tube.
14. Wash with 1-mL of cold MACS buffer using aplunger to gently push through.
15. Collect this as the positive fraction, label the tube, and store on ice.
16. Count the cells and aliquot a fraction for flow cytometry. Check the purity of the fraction by flow cytometry (see **Note 11**).

3.2.2.2. NEGATIVE SELECTION TO OBTAIN CD4+ T CELLS

For the negative selection of CD4+ T cells the following monoclonal antibodies can be used to remove non-CD4+ and monocytes: anti-MHC class II (M51114), anti-CD8 (53.6.72), and anti-Fc-receptor (2.4G2).

1. Add 5 mL of antibody solution containing approx 10 µg/mL of each antibody for 10^8 cells.
2. Incubate this for 30 min at 4°C, slowly rotating.
3. Resuspend in the appropriate amount of sheep anti-rat IgG magnetic beads (200 mL/10^8 cells) diluted in Hanks' washing buffer. Using the magnet to bind the B cells, wash three times in the washing buffer. After the last wash resuspend the beads in 2 mL of Hank's washing buffer per 200 ≤L of beads.
4. Wash the cell suspension three times with 50 mL of Hanks' washing buffer by centrifugation.
5. Resuspend the cell pellet in 3 mL of Hanks' washing buffer per 10^8 cells and add the appropriate amount of bead suspension.

6. Incubate for 15 min slowly rotating at 4°C.
7. Remove the labeled cells by incubating for 5 min in the magnet.
8. Take the supernatant and repeat **step 7**.
9. Wash the cells once by centrifugation and resuspend them in Iscove's medium.
10. Count the cells and take an aliquot for flow cytometry.

3.3. Antigen-Presenting Cells

3.3.1. Collection of Spleens and Cell Preparation

1. Collect the spleens from immune mice of the same strain and MHC haplotype as those used to isolate T cells.
2. Prepare an enzyme mix consisting of 1.5 mL of Hanks' washing buffer without FCS, 780 µL of collagenase and 250 µL of DNase per spleen.
3. Place the spleens in a Petri dish containing the enzyme mix, snip off the end, and gently scrape out cells by using two forceps (*see* **Note 12**).
4. Leave this to incubate 30 min at 37°C in CO_2 incubator.
5. Make a single-cell suspension by transferring the cells to a 50-mL tube using a 10-mL syringe and 19-gage needle.
6. Add Hanks' washing buffer with 5% FCS to rinse the Petri dish and centrifuge.
7. Add 2 mL per spleen of buffer for lysing erythrocytes and leave for 10 min at room temperature.
8. Add 20 mL of Hanks' washing buffer and centrifuge.

3.3.2. T-Cell Depletion

1. To remove all T cells, prepare an antibody mix solution using 2 mL per spleen of the following hybridoma supernatants containing approx 10 µg/mL of specific anti-CD4 (YTS 191), anti-CD8 (YTS 169) *(18)*, and anti-Thy-1 (J1j) *(19)* in the supernatant and add the rabbit complement diluted to a final concentration of 1:20 in the antibody mix (*see* **Note 13**).
2. Resuspend cells with half the above antibody mix, leave it in the water bath at 37°C for 30 min. Shake mixture gently through this period of time.
3. Fill up to 40 mL with Hanks' washing buffer and centrifuge.
4. Discard supernatant and resuspend cells in the other half of the antibody mix. Incubate again under the same conditions.
5. Fill up to 40 mL with Hanks' washing buffer, centrifuge, and resuspend the pellet in 2 mL of Hanks' washing buffer.
6. Prepare two 15-mL tubes with 3 mL of lympholyte per tube (lympholyte must be at room temperature before use).
7. Layer 1 mL of cell suspension gently into the tube containing lympholyte.
8. Centrifuge at 450g for 20 min at room temperature without the brake.
9. Collect the interphase between lympholyte and medium carefully from all the tubes in a 50-mL tube, add Hanks' washing buffer and centrifuge.
10. Wash three times by centrifugation at 300g using Hanks' washing buffer.
11. Resuspend in complete Iscove's medium, count the cells, and take an aliquot to assess the purity of the cells by flow cytometry as before.

3.4. Antigen

1. Collect 100 µL of blood from the tail of one mouse infected with a rodent malaria parasite and one uninfected mouse into tubes with 10 mL of complete Iscove's medium and 500 U/mL heparin (*see* **Note 14** and Chapter 5). Blood should be taken using a sterile, heparinized pipet. For a *P. chabaudi* infection, in which the parasites are synchronous, it is possible to

take mature trophozoites for maximal protein expression.
2. Centrifuge at 650g for 5 min at 4°C.
3. Pour off the supernatant carefully, and wash the erythrocytes by centrifugation three times with complete Iscove's medium.
4. Make a 0.1% suspension (v/v) in complete Iscove's medium of both infected and uninfected blood. Calculate the total volume needed to add 20 mL per well and take appropriate amount from the pellet, adding this to the required volume of complete Iscove's medium.

3.5. Final Setup of the Assay

1. Before distributing the different cell populations, dilute the antigen-presenting cells to a concentration of 3×10^5 cells/mL, vortex well and add 100 µL to all the wells.
2. Then dilute the CD4$^+$ T cells in serial two, five or tenfold dilutions using the vortex to mix immediately before adding 100 µL to the wells.
3. Finally, add the antigen or the normal red blood cells to the respective wells (*see* **Subheading 3.4.**).
4. Incubate for 7 to 10 d at 37°C in a humidified 7% CO_2 incubator.

For experimental cultures add:

1. 100 µL per well of antigen-presenting cells.
2. 100 µL perwell of CD4$^+$ T cells except on control wells.
3. 20 µL of a 0.1% suspension of infected erythrocytes or normal erythrocytes as antigen.

For control cultures add:

1. 100 µL of antigen-presenting cells per well.
2. 100 µL complete Iscove's medium.
3. 20 µL of a 0.1% suspension of infected erythrocytes or normal erythrocytes as antigen.

Note that a minimum of 24 microcultures per each CD4$^+$ T-cell dilution is necessary for accurate estimation of the precursor frequency. In addition, there should be the equivalent number of microcultures of antigen-presenting cells and antigen (infected and normal erythrocytes) in order to determine the background response in the absence of CD4$^+$ T cells. *See* **Fig. 1**.

3.6. Restimulation (d 7–10)

A restimulation step is necessary because after 7 d in culture the clonal expansion is insufficient to allow detection of cytokines.

1. Take spleens from uninfected mice of the same strain used in the original assay. Calculate the number of spleens required by the number of cells needed (3×10^5 cells per well).
2. Place spleens in a Petri dish containing the same enzyme mix as detailed in **Subheading 3.3.1.**
3. Snip off the end of the spleen, tease out the cells, and incubate the cell suspension in the enzyme solution (*see* **Subheading 3.3.1.**) at 37°C for 30 min in CO_2 incubator.
4. Transfer cells into a 50-mL tube using a 19-gage needle and 10-mL syringe, adding extra Hanks' washing buffer to wash the plate and centrifuge.
5. Pour off the supernatant and resuspend it in the buffer for lysing erythrocytes (2 mL per spleen).
6. Leave this for 10 min at room temperature.
7. Fill with 20 mL of Hanks' washing buffer and centrifuge.
8. Irradiate the cells at 30 Gy.

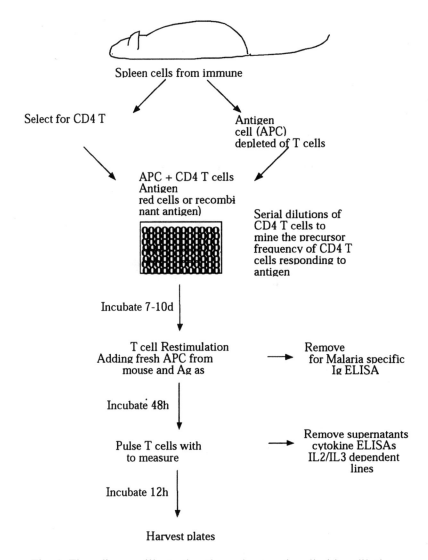

Fig. 1. Flow diagram illustrating the main steps in a limiting dilution.

9. Wash the cells by centrifugation with complete Iscove's medium.
10. Calculate the required volume of cell suspension to add 200 µL per well at a concentration of $1–2 \times 10^6$ cells/mL.

3.6.1. Antigen

Collect and prepare antigen as detailed in **Subheading 3.4.**

3.6.2. Setting up the Culture

1. Remove 200 µL of culture supernatant from the original assay plates and transfer them into new nonsterile plates. Supernatants are used to test for malaria-specific Ig (*see* **Subheading 3.10.3.**).
2. Add 200 µL per well of the cell suspension prepared above.
3. Add 20 µL of antigen or normal red blood cells per well.
4. Incubate for 48 h in 7% CO_2 incubator.

3.7. Flow Cytometry Analysis

3.7.1. Introduction

The purity of the CD4$^+$ T cells and antigen-presenting cells is determined by flow cytometry. For positively selected CD4$^+$ T cells, the antibody used to select the cells in the magnetic beads is directly conjugated to biotin. Therefore the cells can be directly visualized in the flow cytometer by using streptavidin-phycoerythrin (PE), -peridinin chlorophyll protein (PerCP) or other streptavidin-conjugated fluorochromes. For negatively selected CD4$^+$ T cells, an anti-CD4 antibody fluorescein isothiocyanate (FITC-), PE-, or biotin-labeled can be used. To determine what (if any) are the contaminating cells, the positively or negatively selected CD4$^+$ T cells can be additionally labeled with anti-CD3, anti-CD8, anti-Ig (for B cells), B220 (for B cells), and anti-γδ (for γδ cells).

3.7.2. Method

1. Dilute the labeled antibodies (FITC, PE, biotin, APC, or other) in flow cytometry buffer at the working concentration.
2. Centrifuge at 3000g for 10 min at 4°C.
3. Distribute a minimum of 5×10^5 cells per tube.
4. Add to the cell pellet 25 µL of the appropriate antibody dilution.
5. Incubate for 10–15 min on ice and in the dark.
6. Add 2 mL of flow cytometry buffer to each tube and centrifuge at 300g for 10 min at 4°C.
7. Add 25 µL of the second antibody and incubate as in **step 5**.
8. Wash the cells as in **step 6**.
9. Resuspend in 200 µL of flow cytometry buffer and acquire the cells. If not acquiring directly, cells can be fixed by adding 1% paraformaldehyde after the final washing step.

The steps of the method are the same for single-, two-, three-, or four-color staining. The number of colors used for the staining will determine the number of steps and washes in the method.

3.8. Pulsing with [^{3}H]Thymidine

1. Take out 200 µL of supernatant from each well and transfer to new nonsterile plates. This will be used to test for the presence of cytokines (*see* **Subheading 3.2.**). If testing for IL-2/IL-3, 100 µL should be transferred to sterile 96-round-bottomed plates (*see* **Subheading 3.8.**).
2. Add 100 µL per well of complete Iscove's medium.
3. Add 1.875 Bq of [^{3}H]thymidine solution (20 µL of stock dilution).
4. Incubate for a minimum of 8 h.
5. Harvest the cells using a cell harvester, and count [^{3}H]thymidine present on filter mats using a beta-counter.

3.9. Measurement of IL-2 and IL-3 Production by Limiting Dilution Cultures

Supernatants can be assayed for IL-2 and IL-3 content upon incubation with CTLL-2 (for IL-2) *(20)* and FDCP (for IL-3) *(21)* cell lines.

1. Incubate 5000 CTLL-2 or FDCP cells with 50 µL of supernatants for 48 h at 37°C, 7% CO_2 in complete Iscove's medium. A standard should also be set up using recombinant IL-2 or IL-3 and cells without cytokines as a background. Cells should be washed three times with 50 mL of IMDM complete to make sure no residual cytokines remain in the culture medium.
2. After 48 h, add 1.875 Bq of [^{3}H]thymidine solution (20 µL of stock dilution) for 12 h.

3. Harvest cells using a cell harvester. Actual amounts of cytokine can be determined from the standard curve.

3.10. ELISA for Measuring Cytokines and Malaria-Specific Ig in LDA Supernatants

3.10.1. Introduction

The ELISA method allows detection of a variety of cytokines released by activated $CD4^+$ T cells in response to antigenic stimulation in an LDA. In addition, $CD4^+$ T-cell "help" for B cells can also be measured indirectly by assaying levels of parasite specific antibody in the microcultures. Both techniques involve measuring cytokines or malaria-specific antibody by solid-phase ELISA.

For the cytokine ELISA, 96-well plates are coated with specific anticytokine monoclonal antibodies. The secondary antibody is also specific for the cytokine in question. This antibody is biotinylated, and, after the addition of streptavidin-alkaline phosphatase, the amount of antibody bound to the cytokine, and hence the amount of cytokine, can be measured after development of the reaction with an alkaline phosphate-specific substrate. For the malaria-specific antibodies, plates are coated with a preparation of malaria parasite lysate and then, after the antibody has bound, specific alkaline phosphatase-labeled anti-mouse antibody can be added and then developed as above to elucidate the proportion of malaria-specific antibody in the microcultures. Production of antibody of different isotypes can be determined by the specific antiisotype reagents. Alternatively total malaria-specific Ig can be measured using a whole anti-mouse Ig *(22)*.

3.10.2. Cytokine ELISA

1. Coat the plates with anti-IFN/IL-4 (1–10 µg/mL) (50 µL per well) for 1–2 h at 37°C, 3 h at room temperature, or overnight at 4°C. Plates should be kept in a dark, humid box and at 37°C unless specified. Plates should also be sealed to avoid evaporation.
2. Block for 1 h (200 µL per well) with a blocking buffer.
3. Wash the plates three times in a washing buffer.
4. Add test samples (50 µL per well) and standards appropriately diluted twofold for 10 dilutions in the same medium as the test sample. Leave 2 blank wells with just medium. Add supernatants from the microcultures with only APC and Ag and no T cells to calculate the background. Incubate for 1 h.
5. Wash the plates three times as before.
6. Add biotinylated anti-IFN/IL-4 at appropriate dilution in PBS only (50 µL per well). Incubate 1 h.
7. Wash extensively (at least 5 times).
8. Add streptavidin-alkaline phosphatase solution (50 µL per well) for 1 h. After incubation wash plates as before.
9. Add substrate (50 µL per well) at 1 mg/mL in diethanolamine buffer and allow color to develop at room temperature. This can take around 10 min for the IFN ELISA and around 15–20 min for the IL-4 and malaria-specific Ig ELISAs. Read OD_{405nm} (Ref 490 nm) on an ELISA reader.

3.10.3 Malaria-Specific Ig ELISA

3.10.3.1. Preparation of Parasite Antigen for Coating ELISA Plates

1. Take blood from approx 20–30 mice at the peak of infection while the majority of parasite are in the late trophozoite stage and collect in a heparinized tube.

LDA Analysis

2. Run the pooled blood over 2 × 5-mL sterile CF11 columns prewetted with PBS to remove leukocytes. Wash through with PBS and collect in 50-mL tube. Make up column effluent to 50 mL and centrifuge at 650g for 10 min. Remove supernatant carefully with a pipet.
3. Wash the pellet twice in PBS as above, and then add 1–2 drops of saponin with a Pasteur pipet. Incubate at room temperature for 5 min. The solution should be dark brown. Take care not to remove the red blood cell ghosts that are just above the pellet.
4. Wash at least three times in cold PBS until supernatant becomes colorless (*see* **Note 15**).
5. Discard supernatant (not ghosts) and resuspend in 1 mL of PBS with apoprotein as a protease inhibitor (see manufacturer's protocol for concentration). Store at –20°C.

3.10.3.2. ALTERNATE PROTOCOL

The protocol is as in **Subheading 3.10.2.** except for the following steps:

1. Parasite preparation is used to coat the plates.
2. Anti-mouse alkaline phosphatase-conjugated Ig is added at appropriate dilution (50 µL per well) and incubated for 1 h.
3. Eliminate **step 8**.

3.11. Calculating the Precursor Frequency of Responding $CD4^+$ T Cells

3.11.1. Introduction

This section deals with the calculations involved in determining the precursor frequency of $CD4^+$ T cells responding to a designated antigen in an LDA. Precursor frequencies can be calculated from the number of T cells proliferating or producing cytokines in response to antigen. An example of such a calculation is given below. A more detailed explanation of how to calculate the precursor frequencies is given, together with a computer program, in the second edition of **ref. *23***.

The number of $CD4^+$ T cells added to each microculture are serially diluted such that for each dilution of T cells there will be a fraction of nonresponding cultures. This means that initial assays have to be performed to determine the range of T-cell concentrations in which a fraction of positive responses can be obtained. For accurate estimates of precursor frequencies at least three dilutions of T cells containing a fraction of nonresponding cultures is necessary (more is better).

Precursor frequencies are determined by calculating the proportion of nonresponding wells at each dilution of T cells (F_0). A responding culture is defined as one whose response (proliferation, cytokines, antibodies) exceeds the mean background response of antigen-presenting cells and antigen without T cells by more than three standard deviations. Only the cultures that are negative are used in the calculation, that is, nonresponding and therefore contain no precursors.

Using the following zero-order term of the Poisson distribution the precursor frequency can be calculated:

$$\mu = -\ln F_0$$

where μ represents the average number of precursors per well and F_0 represents the fraction of nonresponding cultures.

This relationship holds only when the only component limiting in the cultures is the cell that is being titrated (*see* **Fig. 2**).

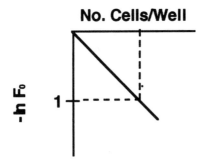

Fig. 2. Calculating the precursor frequency of CD4 T cells respondeing to antigen in an LDA by interpolating at the level of 37% ($-\ln F_0 = 1$) of nonresponding wells.

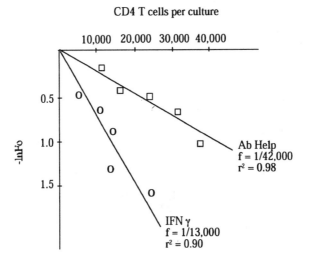

Fig. 3. Graph displaying the actual values plotted alongside the best-fitting line. The goodness-of-fit was determined by a regression analysis. By interpolating at the level of 37% nonresponding wells, i.e., when $-\ln F0 = 1.0$, then it is possible to estimate the number of CD4 T cells in the microcultures that responded to the designated antigen. In this case, 1 in every 13,000 CD4 T cells are producing IFN in response to malarial antigen, and 1 in every 42,000 are providing sufficient B cell "help" such that malaria-specific Ig is synthesized.

3.11.2. Estimation of Precursor Frequencies

The data are fitted to a straight line with a zero intercept by the method of least squares. This assumes that the kinetics of the response conform to a "single hit curve." The goodness-of-fit can either be determined by a chi-square or by regression analysis.

The precursor frequency is calculated from the concentration of input cells equivalent to 37% nonresponding cultures ($-\ln F_0 = 1$) or from the slope of the regression line (**Fig. 2**).

Figure 3 gives an example of the best-fitting line to the experimental points with an intercept of zero. The calculated precursor frequencies with the regression coefficient (goodness of fit) are given. **Figure 4** gives an example of how data may be best displayed to illustrate overall cytokine production on a well-to-well basis.

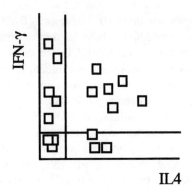

Fig. 4. Example diagram showing a method of displaying the different responses of individual microcultures in an LDA. The IL-4 and IFN responses have been determined for each microculture. Along the x-axis is the IL-4 response of each microculture and along the y-axis is the IFN response. This allows visualization of the overall cytokine profile in the wells. In this case there are some wells in which only one of the cytokines is being produced and some where both are being produced. The horizontal and vertical lines on the graph represent the mean background levels plus three standard deviations of the IL-4 and IFN repsonse in the assay.

4. Notes

1. The MACS columns should be in the refrigerator overnight before the day of use or at least a few hours before time of use.
2. It is important to keep the MACS buffer cold during the entire procedure. While the column is running keep the buffer on ice or in the refrigerator, taking it out only when needed. It is very important to degas the buffer before use because if air goes into the column, obstruction will occur and the selection will not work properly
3. Rabbit complement should only be taken out of the $-80°C$ just before use and be always kept on ice and not refrozen. Each batch should be pretested for the optimal dilution.
4. ELISA plate washing. Depending on the number of plates comprising the assay it may be convenient to use an Immunowash 12 (Nunc). This allows thorough washing of each well and washes 12 wells at a time. Alternatively, a wash bottle may be used and the buffer decanted between washes.
5. A plastic box with lid, which keeps out the light, should be used. Soaking paper towels in water and laying them in the bottom of the box will generate suitable humidity.
6. Substrate solution is light-sensitive and should not be stored for more than a few weeks (see manufacturer's instructions). If there is any discoloration, the substrate should not be used.
7. Diethanolamine buffer should be made up the day before use and stirred overnight at 4°C in the dark.
8. Antibodies must first be affinity-purified using protein G according to the manufacturer's protocols. The appropriate concentration of coating, biotinylated Ig and streptavidin alkaline-phosphatase to use should be predetermined in advance by carrying out serial dilutions and assessing sensitivity. In the case of the coating parasite preparation different OD must be tested before use *(16)*.
9. Starting concentration of recombinant cytokines is approx 250 U/mL but this should be established by first carrying out a serial titration of the standard such that a gradation from maximum to minimum is seen over about 10 dilution steps.
10. The anti-CD4 biotin labeled antibody has to be tested before to be able to determine which is the working concentration.

11. One can expect to obtain approximately 10^8 viable cells from an uninfected spleen. This number is approximately threefold higher in an immune animal. From an immune animal following positive selection approx 1.5×10^7 CD4$^+$ T cells and after complement lysis approx 4×10^7 splenic APCs can be isolated from spleen. The CD4$^+$ T cells resulting from positive selection should be >96% pure. The contaminating cells are likely to be B cells.
12. Cells are gently teased out of the spleen to avoid damaging fragile dendritic cells.
13. Hybridoma supernatants used in the antibody mix for the T-cell depletion need to be tested before to know at which concentration to use to obtain a good depletion. These specific antibodies are used here because they lyse in a complement-dependent manner.
14. For the parasite antigen preparation it is important that the infected erythrocytes contain as many parasites as possible in the late trophozoite stage. This allows extraction of a large amount of parasite protein. As *P. chabaudi chabaudi* is a synchronous infection it is possible to bleed mice when the majority of parasites are at this stage. It is also possible to use alternative antigens like recombinant malarial proteins and specific peptides.
15. Parasite preparation should be brown and not red in color. Red coloration indicates that the red cell lysis was incomplete, and this stage should be repeated.
16. It is better to compare frequencies within a single animal than to pool animals because there can be a great deal of variation in actual precursor frequencies from animal to animal. However, the nature of the T-cell response in each animal can be the same.

References

1. Meding, S. and Langhorne, J. (1991) CD4$^+$ T cells and B cells are necessary for the transfer of protective immunity to *Plasmodium chabaudi chabaudi*. *Eur. J. Immunol.* **21**, 1433–1438.
2. Kumar, S. and Miller, L. H. (1990) Cellular mechanisms in immunity to blood-stage infection. *Immunol. Lett.* **25**, 109.
3. Kumar, S., Good, M. F., Vinetz, J. M., and Miller, L. H. (1989) Interdependence of CD4+ T cells and malarial spleen in immunity to *Plasmodium vinckei vinckei*. *J. Immunol.* **143**, 2017–2023.
4. Brinkman, V., Kaufmann, S. H., and Simon, H. H. (1985) T cell mediated immune response in murine malaria: differential effects of antigen-specific Lyt cell subsets in recovery from *Plasmodium yoelii* in normal and T cell-deficient mice. *Infect. Immun.* **47**, 737–742.
5. Amante, F. H. and Good, M. F. (1997) Prolonged Th1-like response generated by a *Plasmodium yoelii*-specific T cell clone allows complete clearance of infection in reconstituted nude mice. *Parasite Immunol.* **19**, 111–126.
6. Hirunpetcharat, C., Finkelman, F., Clark, I. A., and Good, M. F. (1999) Malaria parasite-specific Th1-like T cells simultaneously reduce parasitemia and promote disease. *Parasite Immunol.* **21**, 319–329.
7. Grau, G. E., Piguet, P. F., and Engers, H. D. (1986) L3T4+ T lymphocytes play a major role in the pathogenesis of murine cerebral malaria. *J. Immunol.* **137**, 2348–2354.
8. Langhorne, J. and Simon, B. (1989) "Limiting dilution analysis of the T cell response to *Plasmodium chabaudi chabaudi* in mice". *Parasite Immunol.* **11**, 545–559.
9. Langhorne, J., Meding, S., Eichmann, K., and Gillard, S. S. (1989) The response of CD4+ T cells to *Plasmodium chabaudi chabaudi*. *Immunol. Rev.* **112**, 71–94.
10. Elghazali, G., Perlmann, H., Rutta, A. S. M., Perlmann, P., and Troye-Blomberg (1997) Elevated plasma levels of IgE in *Plasmodium falciparum*-primed individuals reflect an increase ratio of IL-4 to interferon-gamma (IFN-γ)-producing cells. *Clin. Exp. Immunol.* **109**, 84–89.
11. Hartemann, A. H., Richard, M.-F., and Boitard, C. (1999) Absence of significant Th2 response in diabetes-prone non-obese diabetic (NOD) mice. *Clin. Exp. Immunol.* **116**, 225–230.
12. Openshaw, P., Murphy, E. E., Hosken, A. N., Maino, V., Davis, K., Murphy, K., et al. (1995) Heterogeneity of intracellular cytokine synthesis at the single-cell level in polarized T helper 1 and T helper 2 populations. *J. Exp. Med.* **182**, 1357–1367.
13. Mascher, B., Schlenke, P., and Seyfarth, M. (1999) Expression and kinetics of cytokines determined by intracellular staining using flow cytometry. *J. Immunol. Meth.* **223**, 115–121.
14. Egan, M. A., Kuroda, M. J., Voss, G., Schmitz, J. E., Charini, W. A., Lord, C. I., et al. (1999) Use of major histocompatibility complex class I/Peptide/beta2M tetramers to quantitate CD8(+) cytotoxic T lymphocytes specific for dominant and nondominant viral epitopes in simian-human immunodeficiency virus-infected rhesus monkeys. *J. Virol.* **73**, 5466–5472.
15. Altman, J. D., Moss, P. A. H., Goulder, P. J. R., Barouch, D. H., McHeyzer-Williams, M. G., Bell, J. I., et al. (1996) Phenotypic analysis of antigen-specific T lymphocytes. *Science* **274**, 94–96.

16. Slade, S. and Langhorne, J. (1989) Production of interferon-gamma during infection with *Plasmodium chabaudi chabaudi*. *Immunobiol.* **179,** 353–365.
17. Langhorne, J., Gillard, S., Simon, B., Slade, S., and Eichmann, K. (1989) Frequencies of CD4+ T cells reactive with *Plasmodium chabaudi chabaudi*: distinct response kinetics for cells with Th1 and Th2 characteristics during infection. *Int. Immunol.* **1,** 416–424.
18. Cobbold, S. P., Jayasuriya, A., Nash, A., Prospero, T. D., and Waldmann, H. (1984) Therapy with monoclonal antibodies by elimination of T-cell subsets *in vivo*. *Nature* **312,** 548–551.
19. Bruce, J., Symington, F. W., McKearn, R. J., and Sprent, J. (1981) A monoclonal antibody discriminating between subsets of T and B cells. *J. Immunol.* **127,** 2496–2501.
20. Grabstein, K. Eisenmann, J. Mochizuki, D. Schanebeck, P. Gordon, P. Hopp, T, et al. (1986) Purification to homogeneity of B cell stimulating factor: a molecule that stimulates proliferation of multiple lymphokine-dependent cell lines. *J. Exp. Med.* **163,** 1405–1414.
21. Dexter, T. Garland, J. Scott, D. Scholnick, E., and Metcalf, D. (1980) Growth of factor-dependent haematopoietic precursor cell lines. *J. Exp. Med.* **152,** 1036–1047.
22. Langhorne, J., Evans, C., Asofsky, R., and Taylor, D. (1984) Immunoglobulin isotype distribution function of malaria-specific antibodies produced in infection with *Plasmodium chabaudi adami* and *Plasmodium yoelii*. *Cell. Immunol.* **87,** 452–461.
23. Lefkovits, I. snd Waldmann, H. (1999) *Limiting Dilution Analysis of Cells in the Immune System.* University Press, Oxford, UK.

35

Cell Trafficking

Malaria Blood-Stage Parasite-Specific CD4⁺ T Cells After Adoptive Transfer into Mice

Chakrit Hirunpetcharat and Michael F. Good

1. Introduction

CD4$^+$ T cells play an important role in immunity to blood-stage malaria parasites and in disease pathogenesis. The role of CD4 $^+$ T cells has been demonstrated by selective depletion in vitro *(1–3)*, by adoptive transfer of T cells to immunodeficient mice *(1,4–7)*, and by the ability of human T cells to inhibit parasite growth in vitro *(8)*. Both types of CD4$^+$ T cells (Th1 and Th2) have been shown to be effective in malaria infection *(6)*. Factors that affect cell activation, function, and life-span of parasite-specific T cells would be expected to have a significant impact on parasite survival, including the poor immunogenicity of parasite antigens *(9)*, antigenic polymorphism *(10–12)*, immunosuppression as a result of malaria infection *(13)*, and the need for continuous malaria exposure to maintain immunological memory *(14)*. Antigen-driven deletion or anergy of immunological responses is a major regulatory strategy to control potentially harmful responses *(15)* and may be used by infectious organisms to their advantage, as shown by the exhaustive deletion of lymphocytic choriomeningitis virus-specific CD8$^+$ T cells *(16)*, and the deletion of *Plasmodium berghei*-specific T cells after adoptive transfer into immunodeficient mice *(17)*.

Animal models provide a great tool for studying immunological T-cell responses, including cell tracing. This chapter describes a protocol for investigating and tracing T cells specific for malaria parasites, including *P. berghei* and other antigens, including ovalbumin, by labeling the cells with 5-(and 6-)-carboxyfluorescein diacetate succinimidyl ester (CFSE), a membrane-permeant reactive tracer. This amine-reactive tracer can passively diffuse into cells and bind to cytoplasmic components of cells. It is colorless and nonfluorescent until the acetate group is cleaved by intracellular esterases to yield highly fluorescent, amine-reactive fluorophores *(18)*. Cells labeled with CFSE can be detected for at least 8 wk after injection into mice *(19)*.

2. Materials

1. Mice: 6–8 wk-old female BALB/c and athymic BALB/c nude mice.
2. *P. berghei* ANKA.

3. Ovalbumin (Sigma).
4. Complete Freund's adjuvant (CFA; Sigma).
5. Eagle's minimum essential medium (EMEM): 14.02 g EMEM (Trace Scientific) supplemented with 0.1 mM nonessential amino acids, 292 mg/mL L-glutamine, 10 mM HEPES, 2 g/L sodium bicarbonate, 20 µg/mL gentamicin. Dissolve EMEM in 1000 mL of distilled water, filter through a 0.2-µm membrane and store at 4°C.
6. Fetal bovine serum (FBS).
7. Complete medium: 90 mL EMEM, 10 mL FBS, 5×10^{-5} M 2-mercaptoethanol (2-ME).
8. 5- (and 6-) carboxyfluorescein diacetate succinimidyl ester (CFSE; Molecular Probes).
9. Ficoll-Hypaque (Pharmacia, Biotech AB, Uppsala, Sweden).
10. Phosphate-buffered saline (PBS), pH 7.2.
11. Erythrocyte lysis buffer: 450 mL of 0.16 M (8.56 g/L) ammonium chloride, 50 mL of 0.17 M Tris-HCl, pH 7.65. Adjust pH to 7.2 with HCl, filter through a 0.2-µm membrane and store at 4°C.
12. 0.1% Trypan blue in PBS.
13. Stainless steel mess.
14. 1-mL Hamilton gas-tight syringes.
15. 26- and 18-gage needles.
16. 1-mL Syringes.
17. 6-, 24-, and 96-Well tissue culture plates.
18. 10- and 50-mL Centrifuge tubes.
19. 10×35 mm Petri dishes.
20. 15×60 mm Petri dishes.
21. Flow cytometer with 488-nm argon laser.

3. Methods

3.1. Generation of T-Cell Lines Specific for P. berghei and Ovalbumin

3.1.1. Immunization of Mice with Antigen

1. Freeze and thaw *P. berghei*-pRBC suspension (6×10^8 cells/mL) at −70°C and 37°C, respectively, for three rounds.
2. Sonicate the parasite on ice for 30 s and stop for 30 s, for four rounds.
3. Draw 0.5 mL of *P. berghei*–pRBC lysate in a 1-mL Hamilton gas-tight syringe and 0.5 mL of CFA in the other one, using 18-gage needles.
4. Connect the two syringes with a connecting tube (~ 2 mm diameter) via luer locks. Force the mixture repeatedly back and forth between the two syringes for 15 min (*see* **Note 1**).
5. Remove the connecting tube and attach the syringe with a 26-gage needle.
6. Inject 50 µL of the emulsion subcutaneously via each hind-leg footpad of each mouse (*see* **Note 2**). Four to six mice are typically used for a study.

3.1.2. Collection of Lymph Node Cells

1. At 7–9 d after immunization, sacrifice the mice and collect the draining inguinal and popliteal lymph nodes in a Petri dish containing 3 mL of EMEM medium.
2. Prepare a single cell suspension by gently pressing the lymph nodes with the flat-end side of a plunger of a 10-mL syringe.
3. Transfer the cell suspension into a 10-mL centrifuge tube and add EMEM medium to 10 mL final volume.
4. Stand the tube at room temperature for 5–10 min to let cell debris settle down to the bottom of the tube.
5. Collect the cell suspension into a new tube without disturbing the cell debris.

6. Centrifuge the tube at $400g$ (1400 rpm for Beckman GS-6 centrifuge) for 10 min.
7. Aspirate the supernatant and resuspend cells with 10 mL of EMEM.
8. Centrifuge the tube again at $400g$ for 10 min.
9. Aspirate the supernatant and resuspend cells with 2 or 3 mL of complete medium.
10. Count the number of cells using 0.1% trypan blue/PBS solution (*see* **Note 3**).
11. Adjust cells to 1×10^6 cells/mL in complete medium.

3.1.3. Growing and Maintenance of T-Cell Lines In Vitro

1. Add 2 mL of 1×10^6 lymph node cells/mL into each well of 24-well tissue culture plates.
2. Add 20 µL of 1×10^8 cells/mL of *P. berghei*–pRBC in each well, to give a final concentration of 1×10^6 pRBC/mL.
3. Incubate at 37°C, 5% CO_2 for 4 d.
4. Aspirate half of the supernatant from each well, resuspend cells and transfer into a 50-mL centrifuge tube.
5. Underlay the cell suspension with 10 mL of Ficoll-Hypaque.
6. Centrifuge at $400g$ for 20 min.
7. Collect cells at interface layer between Ficoll-Hypaque and supernatant and transfer into a new tube.
8. Centrifuge the tube at $400g$ for 10 min and discard supernatant.
9. Resuspend cells with EMEM medium, centrifuge again at $400g$ for 10 min and discard the supernatant.
10. Resuspend cells with complete medium and count cell number.
11. Adjust cells to a concentration of 1×10^6 cells/mL in complete medium.
12. Add 1 mL of the cell suspension into each well of 24-well tissue culture plates.
13. Add 1 mL of 2×10^6 cells/mL of irradiated normal spleen cells (*see* **Note 4**) to each well.
14. Incubate at 37°C, 5% CO_2 for 10–14 d to rest cells (*see* **Note 5**).
15. Aspirate 1 mL of supernatant from each well.
16. Add 1 mL of 2×10^6 cells/mL of new irradiated normal spleen cells and 20 µL of 1×10^8 cells/mL of *P. berghei*–pRBC in each well (*see* **Note 6**).
17. Incubate at 37°C, 5% CO_2 for 4 d.
18. For expanding and maintaining T cell lines, perform protocol of resting cells and stimulating cells as described in **steps 4–14** and in **steps 15–17**, repeatedly (*see* **Note 7**).

3.2. Labeling T Cells with CFSE

1. After resting cells for 10–14 d, collect cells into a 50-mL centrifuge tube.
2. Underlay the cell suspension with 10 mL of Ficoll-Hypaque.
3. Centrifuge at $400g$ for 20 min.
4. Collect cells at the interface layer and transfer into a new tube.
5. Wash cells twice with EMEM and once with PBS, by centrifugation at $400g$ for 10 min.
6. Resuspend cells to 1×10^7 cells/mL in PBS.
7. Add 2 µL of 5 mM CFSE to 1 mL of the cell suspension (i.e., 10 µM CFSE at final concentration) (*see* **Note 8**).
8. Incubate at 37°C for 30 min.
9. Wash cells twice with ice-cold 10% FCS/EMEM and once with EMEM by centrifugation at $400g$ for 10 min.
10. Resuspend cells to a concentration of 5×10^7 cells/mL in PBS and keep on ice (*see* **Note 9**). Cells are now ready for transfusion into mice.

3.3. Adoptive Transfer of CFSE-Labeled T Cells

1. Warm BALB/c nude mice under a 150-W lamp for 5 min, to enlarge the tail veins.

2. Fix a mouse in a mouse holder and secure the tail.
3. Inject 0.2 mL of 5×10^7 cells/mL intravenously into a lateral vein of recipient BALB/c nude mice using a 1-mL syringe with a 26-gage needle.

3.4. Challenge of Mice with Malaria Parasites

To determine the effect of malaria parasite infection on malaria parasite-specific T cells, recipient mice which have been passively transferred with CFSE-labeled *P. berghei*-specific T cells or ovalbumin-specific T cells are challenged with a fresh preparation of parasitized red blood cells within 4–24 h after receiving the CFSE-labeled T cells. Control mice receive CFSE-labeled *P. berghei*-specific CD4$^+$ T cells only and CFSE-labeled Ova-specific CD4$^+$ T cells with and without parasite challenge infection.

1. To prepare fresh live *P. berghei*–pRBC from a donor BALB/c mouse infected about one week previously, cut the tail tip to get 2–3 drops of blood in a sterile Eppendorf tube containing 1 mL heparin/PBS. Use one drop of blood to make a blood film.
2. Stain the blood film with Diff-Qiuk staining solutions, and determine percent parasitized red blood cells under a light microscope.
3. Wash the blood twice with PBS and count the cell number.
4. Calculate to correct the number of parasitized red blood cells.
5. Adjust parasitized red blood cells to a concentration of 5×10^6 cells/mL in PBS.
6. Inject 0.2 mL of the cell suspension intraperitoneally using a 1-mL syringe with a 26-gage needle.

3.5. Collection of Peripheral Blood and Organs

To investigate cell migration to different organs or tissues, peripheral blood and organs are collected and prepared for single cell suspension for assessment of adoptive cells by flow cytometry.

3.5.1. Peripheral Blood

1. Sacrifice mice using CO_2 gas and fix on a styrofoam board with 26-gage needles.
2. Drench mouse with 70% ethyl alcohol and insert a 26-gage needle coupled with 1-mL syringe from the sternum to heart.
3. Draw blood slowly (able to collect about 0.6–0.8 mL per mouse) and put into a 10-mL tube containing 5 µL of heparin.
4. Dilute the blood with equivalent volume of PBS.
5. Underlay the diluted blood with 2 mL of Ficoll-Hypaque.
6. Centrifuge at 400g for 20 min.
7. Collect cells at interface layer into a new tube.
8. Wash cells twice with PBS by centrifuging at 400g for 10 min.
9. Resuspend cells with 0.1% BSA/0.1% NaN$_3$/PBS, and count the number of cells using 0.1% trypan blue/PBS solution.
10. Adjust cell concentration to 1×10^6 cells/mL with 0.1% BSA/0.1% NaN$_3$/PBS.

3.5.2. Spleen, Lymph Nodes, and Liver

1. Collect the spleen, draining poplitial, and inguinal lymph nodes, and liver into ice-cold EMEM medium in 15×60-mm Petri dishes.
2. Press each organ with a blunt end plunger of a 10-mL syringe.
3. Collect cells into a 10-mL centrifuge tube and stand the tube for 10 min to settle cell debris.
4. Transfer cell suspensions into new tubes without disturbing the cell debris.
5. Underlay the cell suspension with 2 mL of Ficoll-Hypaque.

6. Centrifuge at 400g for 20 min.
7. Collect cells at the interface between Ficoll-Hypaque and medium into a new 10-mL tube, and add EMEM to 10 mL final volume.
8. Centrifuge tubes at 400g for 10 min.
9. Aspirate supernatant and resuspend the cell pellet with EMEM medium.
10. Centrifuge at 400g for 10 min.
11. Resuspend cells with 1 mL of 0.1% BSA/0.1% NaN$_3$/PBS, and count the number of cells using 0.1% trypan blue/PBS solution.
12. Adjust cell concentration to 1×10^6 cells/mL with 0.1% BSA/0.1% NaN$_3$/PBS.

3.5.3. Lung

Cells from lung are prepared according to the protocol used by Baumgarth and Kelso *(20)* as follows.

1. Flush *in situ* with 20 mL of PBS via cannulation of the heart, to remove the intravacular blood pool.
2. Remove the lung into ice-cold EMEM medium in 15 × 60-mm Petri dishes.
3. Cut the lung into small pieces.
4. Remove the medium, and add EMEM supplemented with 10% FCS, 216 mg of L-glutamine per liter, 5×10^{-5} M 2-mercaptoethanol, antibiotics, *DNase I* (50 U/mL; Boehringer Mannheim, Germany), and collagenase I (250 U/mL; type 4197, Worthington, Freehold, NJ).
5. Incubate for 90 min at 37°C on a rocker.
6. Pass the digested lung tissue through a stainless steel mesh and collect into a centrifuge tube, and add EMEM to 10 mL final volume.
7. Underlay the cell suspension with 2 mL of Ficoll-Hypaque.
8. Centrifuge at 400g for 20 min.
9. Collect cells at interface layer into a new tube.
10. Wash cells twice with PBS by centrifuging at 400g for 10 min.
11. Resuspend cells with 1 mL of 0.1% BSA/0.1% NaN$_3$/PBS, and count the number of cells using 0.1% trypan blue/PBS solution.
12. Adjust cell concentration to 1×10^6 cells/mL with 0.1% BSA/0.1% NaN$_3$/PBS.

3.5.4. Bone Marrow

1. Dissect the muscle around the two femurs and cut them into pieces, in a 10 × 35-mm Petri dish containing 2 mL EMEM medium.
2. Disperse bone marrow cells from the femurs by fixing a femur with a forcep and injecting medium into marrow matrix many times, using a 1 mL-syringe with a 26-gage needle.
3. Collect the cells into a 10-mL tube and add medium to 10 mL volume.
4. Centrifuge the tube at 400g for 10 min.
5. Aspirate supernatant, and resuspend cells with 3 mL of erythrocyte lysis buffer.
6. Incubate at 37°C for 7 min.
7. Add 7 mL of EMEM medium.
8. Centrifuge at 400g for 10 min.
9. Aspirate supernatant, and resuspend cells with 10 mL EMEM.
10. Centrifuge at 400g for 10 min.
11. Resuspend cells with 1 mL of 0.1% BSA/0.1% NaN$_3$/PBS, and count the number of cells using 0.1% trypan blue/PBS solution.
12. Adjust cell concentration to 1×10^6 cells/mL with 0.1% BSA/0.1% NaN$_3$/PBS.

3.6. Fluorescence-Activated Cell Sorter Analysis

The procedure of flow cytometry for determination and demonstration of the profile of CFSE-labeled T cells is based on the protocol described in Chapter 5 *(21)*, using CELLQuest software operating on the Macintosh computer platform, and the methodology of tracing CFSE-labeled cells *(22)*.

4. Notes

1. To test whether the emulsion is completely formed, release the connecting tube from one syringe, attach a 26-gage needle and place one drop of emulsion onto the surface of water in a beaker. The emulsion should not disperse. If the emulsion disperses, remove the 26-gage needle, replace the connecting tube and perform mixing again for another 5 min or more to get a complete emulsion. The formation of the emulsion may be quickened by placing the mixture in a 4°C refrigerator for 30 min before forcing back and forth repeatedly in the syringes. An alternative preparation of emulsion can be to mix in a microcentrifuge tube, using a 1-mL syringe with an 18-gage needle to draw up and expel the mixture repeatedly.
2. For ovalbumin immunization, emulsify the mixture of 0.5 mL of 1 mg/mL ovalbumin in PBS and 0.5 mL of CFA using 1-mL Hamilton syringes, and inject mice as described in **Subheading 3.1.**
3. Mix 10 μL of cell suspension with 90 μL of 0.1% trypan blue/PBS solution and fill in a hematocytometer chamber (approx 10 μL). Count four squares. The concentration of cells per milliliter is calculated by multiplying the number of cells counted in four squares with 25,000 (that is, average number per square $\times 10^4 \times$ dilution factor).
4. Irradiated spleen cells are used as feeder cells and antigen presenting cells in culture. Spleens are collected from naive syngenic mice and a single-cell suspension is prepared (one spleen yield about 1×10^8 lymphocytes). Cell debris is removed. Red blood cells are lysed by treatment with erythrocyte lysis buffer. Spleen cells are washed twice with EMEM medium, resuspended in complete medium, and irradiated with 2500 rad. Alternatively, spleen cells at the concentration of 5×10^7 cells/mL are treated with 50 μg/mL of mitomycin C at 37°C for 30 min. Cells are then washed four times with EMEM before use.
5. As observed under an inverted microscope, resting cells are smaller in size than activated cells.
6. For generating T-cell lines specific for ovalbumin, immunize each mouse with 100 μg of the antigen emulsified with CFA as described above, collect lymph node cells and culture the cells using the same protocol for generating *P. berghei*-specific T cells as described above but in the stimulation phase use ovalbumin at 400 μg/mL.
7. The reason for stimulating and resting antigen-specific T cells for at least three rounds is to obtain a highly specific response. The obtained cells are nearly all CD4$^+$ and are reactive to *P. berghei* but not to normal red blood cells nor PPD in case of *P. berghei*-specific T cells, and are reactive to ovalbumin but not to PPD in case of ovalbumin-specific T cells, as demonstrated in lymphoproliferation assays.
8. A 5-m*M* stock solution of CFSE (mol wt = 557.47) is prepared by dissolving the powder with dimethylsulfoxide and is kept at –20°C and protected from light.
9. To ensure that the cells were labeled with CFSE, check the cells under a fluorescent microscope or by flow cytometer with 488-nm argon laser and a 525-nm bandpass filter, by which all the cells will give bright green fluorescence or a high sharp peak of fluorescent intensity, respectively.

References

1. Kumar, S., Good, M. F., Dontfraid, F., Vinetz, J. M., and Miller, L. H. (1989) Independence of CD4$^+$ T cells and malarial spleen in immunity to *Plasmodium vinckei vinckei*. Relevance to vaccine development. *J. Immunol.* **143**, 2017–2023.

2. Suss, G., Eichmann, K., Kurry, E., Linke, A., and Langhorne, J. (1988) Roles of CD4- and CD8-bearing T lymphocytes in the immune response to the erythrocytic stages *of Plasmodium chabaudi*. *Infect. Immun.* **56,** 3081–3088.
3. Podoba, J. E. and Stevenson, M. M. (1991) $CD4^+$ and $CD8^+$ T lymphocytes both contribute to acquired immunity to blood stage *Plasmodium chabaudi AS*. *Infect. Immun.* **59,** 51–58.
4. Brake, D. A., Long, C. A., and Weidanz, W. P. (1988) Adoptive protection against *Plasmodium chabaudi aadami* malaria in athymic nude mice by a cloned T cell line. *J. Immunol.* **140,** 1989–1993.
5. Meding, S. J. and Longhorne, J. (1991) $CD4^+$ T cells and B cells are necessary in the transfer of protective immunity to *Plasmodium chabaudi chabaudi*. *Eur. J. Immunol.* **21,** 1433–1438.
6. Taylor-Robinson, A. W., Phillip, R. S., Severn, S., Moncada, S., and Liew, F. Y. (1993) The role of T_H1 and T_H2 cells in a rodent malaria infection. *Science* **260,** 1931–1934.
7. Taylor-Robinson, A. W. and Phillip, R. S. (1993) Protective $CD4^+$ T cell lines raised against *Plasmodium chabaudi* show characteristics of either Th1 and Th2 cells. *Parasite Immunol.* **15,** 301–310.
8. Fell, A. H., Silins, S. L., Baumgarth, N., and Good, M. G. (1996) *Plasmodium falciparum*-specific T cell clones from non-exposed and exposed donors are highly diverse in TCR bata chain V segment usage. *Int. Immunol.* **8,** 1877–1887.
9. Quakyi, I. A., Currier, J., Fell, A., Taylor, D. W., Roberts, T., Houghten, R. A., et al. (1994) Analysis of human T cell clones specific for conserved peptide sequences within malaria proteins. Paucity of clones responsive to intact parasites. *J. Immunol.* **153,** 2082–2092.
10. Brown, K. N. and Brown, I. N. (1965) Immunity to malaria: antigenic variation in chronic infections of *Plasmodium knowlesi*. *Nature (London)* **209,** 1286–1288.
11. Zevering, Y., Khamboonruang, C., and Good, M. F. (1994) Natural amino acid polymorphisms of the circumsporozoite protein of *Plasmodium falciparum* abrogate specific human $CD4^+$ T cell responsiveness. *Eur. J. Immunol.* **24,** 1418–1425.
12. Biggs, B. A., Gooze, L., Wycherley, K., Wollish, W., Southwell, B., Leech, J. H., et al. (1991) Antigenic variation in *Plasmodium falciparum*. *Proc. Natl. Acad. Sci. USA* **88,** 9171–9174.
13. Riley, E., Jobe, O., Blackman, M., Whittle, H. C., and Greenwood, B. M. (1989) *Plasmodium falciparum* schizont sonic extracts suppress lymphoproliferative responses to mitogens and antigens in malaria-immune individuals. *Infect. Immun.* **57,** 3181–3188.
14. Zevering, Y., Khamboonruang, C., Rungruengthanakit, K, Bathurst, I., Barr, P. Vibulahai, L., et al. (1994) Lifespans of human T cell responses to determinants from the circumsporozoite proteins of *Plasmodium falciparum* and *P. vivax*. *Proc. Natl. Acad. Sci. USA* **91,** 6118–6122.
15. Nossal, G. J. V. (1983) Cellular mechanisms of immunologic tolerance. *Ann. Rev. Immunol.* **1,** 33–62.
16. Moskophidis, D., Lechner, F., Pircher, H., and Zinkernagel, R. M. (1993) Virus persistance in acutely infected immunocompetent mice by exhausive of antiviral cytotoxic effector T cells. *Nature (London)* **362,** 758–761.
17. Hirunpetcharat, C. and Good, M. F. (1998) Depletion of *Plasmodium berghei*-specific $CD4^+$ T cells adoptively transferred into recipient mice after challenge with homologous parasite. *Proc. Natl. Acad. Sci. USA* **95,** 1715–1720.
18. Haugland, R. P. (1996) Membrane-permeant reactive tracers for long-term cell labeling, in *Handbook of Fluorescent Probes and Research Chemicals* (Spencer, M. T. Z., ed.), Molecular Probes, Eugene, OR, pp. 329–364.
19. Weston, S. A. and Parish, C. R. (1990) New fluorescent dyes for lymphocyte migration studies analysis by flow cytometry and fluorescence microscopy. *J. Immunol. Meth.* **133,** 87–97.
20. Baumgarth, N. and Kelso, A. (1996) In vitro blockade of gamma interferon affects the influenza virus-induced humoral and the local cellular immune response in lung tissue. *J. Virol.* **70,** 4411–4418.
21. Otten, G., Yokoyama, W.M., and Holms, K.L. (1995) Flow cytometry analysis using the Becton Dickinson FACScan, in *Current Protocols in Immunology* (Coligan, J. E., Kruisbeek, A. M., Margulies, D. H., et al., ed.) Greene Publishing and Wiley-Intersciences, pp. 5.4.1–5.4.19.
22. Lyons, A. B. and Parish, C. R. (1994) Determination of lymphocyte division by flow cytometry. *J. Immunol. Meth.* **171,** 131–137.

36

Assessing Antigen-Specific Proliferation and Cytokine Responses Using Flow Cytometry

Catherine E. M. Allsopp and Jean Langhorne

1. Introduction

There is evidence in rodent models of malaria and indirect evidence in human malaria of the role of T cells in protective immunity to the disease *(1–3)* and in the pathology of malaria infections *(4,5)*. Measurement of this T-cell response is important for two reasons: to determine which parasite proteins may be involved in acquisition of immunity, and to characterize the effector mechanisms of the responding cells.

There are several ways of determining cell function, some of which depend upon the cell type that is being examined. Some assays for $CD4^+$ T-cell function are an indirect readout, measuring the effect the T cell has on other cell types such as macrophages or B cells. A direct method of measuring $CD4^+$ T-cell responses is to measure proliferation in response to antigenic stimuli. Proliferation is often measured by incorporation of a radioactive marker, tritiated thymidine, or a fluorescent marker bromodeoxyuridine (BrdU) into the DNA of the dividing cells. Other ways of measuring proliferation include the use of fluorescent dyes that incorporate into the membrane or cytoplasm of cells. The dye is distributed equally between daughter cells as division occurs, giving a reduction in fluorescence intensity, which can be followed visually by flow cytometry. Three commonly used dyes include PKH26, which is compatible with rhodamine or phycoerythrin detection systems, PKH67 which is compatible with fluorescein detection systems *(6)*, and carboxylfluorescein diacetate succinimidyl ester (CFSE) which is also compatible with fluorescein detection systems *(7)*.

Use of dye assays is more informative than thymidine incorporation assays because they can be combined with labeling using monoclonal antibodies specific for cell surface proteins, to determine the phenotype of the responding cell. This is useful when measuring responses of one particular cell type in mixed populations of cells such as whole-spleen cell preparations or whole peripheral blood mononuclear cells. Removing the cell subtype from the mixed environment may have subtle effects on the cell activity that will be avoided by using this fluorometric assay system. The assay for cell division by labeling with CFSE or PKH26 and the use of cell surface markers can be extended to provide an extremely powerful assay that allows not only division and cell phenotype to be determined but also allows cytokines produced by the cells to be measured.

From: *Methods in Molecular Medicine, Vol. 72: Malaria Methods and Protocols*
Edited by: Denise L. Doolan © Humana Press, Inc., Totowa, NJ

Fluorometric assays have several advantages over thymidine incorporation assays. The cells are labeled with the fluorochrome before being placed in culture. Therefore, the assay provides a cumulative measure of the T-cell response during the whole-culture period, rather than the division over the short time window that is provided by thymidine pulsing at the end of the culture period. Further, flow cytometry allows multiparameter analysis of cellular responses within a single sample. The activity of different cells within a mixed population can be elucidated without the need for prior cell purification *(8)*. Finally, using a program called Modfit™, the kinetics of the individual responding cells can be charted *(9)*.

This chapter will describe methods used to isolate human peripheral blood mononuclear cells and their subsequent labeling with the dyes for proliferation studies. Although both CFSE and PKH can be used in this assay, examples will be restricted to PKH26. Combining PKH26 labeling with two other procedures, cell phenotype characterization using antibodies to surface markers and measurement of cytokine production at the single cell level by intracellular labeling with antibodies, is also discussed.

2. Materials

2.1. Equipment

1. Sterile 50-mL syringes and sterile needles.
2. Refrigerated benchtop centrifuge for 50-mL and 15-mL tubes and 96-well plates.
3. Analytical flow cytometer with filters suitable for the detection of common fluorochromes including fluorescein isothiocyanate (FITC), phycoerythrin (R-PE), peridinin chlorophyll protein (PerCP), and allophcocyanin (APC).
4. Refrigerator or 4°C incubator.
5. Humidified CO_2 incubator.
6. Sterile 96-well U-bottomed tissue culture plates with lid.
7. Sterile 50-, 15-mL tubes, and tubes suitable for flow cytometry.
8. Hemocytometer.
9. Irradiation source.

2.2. Reagents

2.2.1. Separation of Peripheral Blood Mononuclear Cells from Whole Blood

1. RPMI 1640 for washing. Store sterile at 4°C. Use at room temperature.
2. Complete RPMI medium: RPMI 1640 with 100 U/mL penicillin, 100 mg/mL streptomycin, 10 m*M* HEPES, 2 m*M* L-glutamine, 10% normal human serum. Store sterile at 4°C. Use at room temperature.
3. Ficoll-Hyaque (Pharmacia Biotech). Store sterile at room temperature. Use at room temperature.

2.2.2. Staining Cells with PKH26 Dye Before Culture

1. 1 *M* stock solution of PKH26 (Sigma). Store sterile at room temperature. Dilute to working stock of 3 μ*M* in CGL-DIL as required; 1 mL of a 3-μ*M* dye solution is required to stain 2×10^7 cells.
2. CGL-DIL diluent (Sigma). Store sterile at room temperature
3. Normal human serum. Heat-inactivate at 56°C for 1 h and store sterile, in aliquots, at –20°C.
4. Complete RPMI medium (*see* **Subheading 2.2.1.**).

2.2.3. Intracellular Staining For Cytokine Production

1. Phorbol 12-myristate (PMA) (Sigma): 2 mg/mL stock in dimethyl sulfoxide (DMSO). Store aliquots at –80°C. Prepare a 1-μg/mL working stock as required by diluting the 2 mg/mL

Assessing Antigen-Specific Proliferation and Cytokine Responses

stock at 1 in 2000 in complete RPMI medium. Use at a final concentration of 50 ng/mL PMA (add 10 µL of the 1 µg/mL working stock per 200-µL culture volume)

2. Ionomycin (Calbiochem): 2 mg/mL stock in DMSO. Store aliquots at −80°C. Prepare a 10 µg/mL working stock as required by diluting the 2 mg/mL stock at 1 in 200 in complete RPMI medium. Use at a final concentration of 500 ng/mL ionomycin (add 10 µL of the 10 µg/mL working stock per 200 µL culture volume).
3. Brefeldin A (Sigma): 1 mg/mL stock in 100% ethanol. Store aliquots at −80°C. Prepare a 0.1 mg/mL working stock as required by diluting the 1 mg/mL stock at 1 in 10 in complete RPMI medium. Use at a final concentration of 10 µg/mL Brefeldin A (add 20 µL of the 0.1 mg/mL working stock per 200 µL culture volume).
4. Permeabilization buffer: Phosphate-buffered saline (PBS), pH 7.2 with 0.5% saponin, 2% bovine serum albumin (BSA), 0.2% sodium azide. Filter and store sterile at 4°C.
5. Fluorescein-activated cell sorter (FACS) buffer: PBS, pH 7.2 with 2% BSA, 0.02% sodium azide. Store at 4°C. Filter sterilize before use.
6. Fixation buffer: PBS, pH 7.2 with 4.0% paraformaldehyde. Filter and store sterile at 4°C. Use at 4°C.
7. Fluorochrome labeled antibodies for cytokines: Commonly tested cytokines include interleukins, interferons, and tumor necrosis factors: IL-2, IL-4, IL-10, IFN-γ, IL-12, and TNF-α and -β. The choice of fluorochrome that is conjugated to the antibody is important and is discussed in detail in **Note 1**. Store all antibodies in the dark at 4°C. Dilute to working concentration in the appropriate buffer before use.
8. Control antibodies: These should include unconjugated mouse antibodies and fluorochrome-conjugated isotype control antibodies (*see* **Note 2**). Store all antibodies in the dark at 4°C. Dilute to working concentration in the appropriate buffer before use.

2.2.4. Surface Labeling of Cells for Phenotype Determination

1. FACS buffer (*see* **Subheading 2.2.3.**).
2. Fluorochrome-labeled antibodies to some common cell surface markers include CD3, CD4, CD8 (T cells), CD20 (B cells), CD14 (myelomonocytic cells), and CD16 (natural killer cells and neutrophils). Activation markers such as IL-2R-α and also CD45 isotype are commonly assessed. As mentioned in **Subheading 2.2.3.**, fluorochromes should be chosen with care and depend on the fluorochrome conjugate that has already been used for assessing cell division and intracellular cytokine production (*see* **Note 1**).
3. Control antibodies (*see* **Note 2**).
4. Fixation buffer (PBS, pH 7.2 with 1% paraformaldehyde). Store sterile at 4°C. Use at 4°C.

3. Methods
3.1. Introduction

To measure cell division by flow cytometry, PBMC must first be labeled with the appropriate dye (PKH or CFSE) and then cultured, before analysis on the flow cytometer at the end of the culture period. PBMC (fresh or cryopreserved) are required as a source of effector cells, and autologous cells (irradiated to prevent division) are required as antigen presenting cells; these can be Epstein-Barr virus (EBV)-transformed B cell lines, PBMC or monocyte-derived dendritic cells. Antigen-specific proliferation is measured by labeling the effector cells with PKH dye (unlabeled cells are included as a control) and incubating these with autologous antigen-presenting cells and the antigen of interest (titrated to ensure the optimal working concentration) for 3–7 d. The length of time that the cells are cultured depends upon factors such as the type of antigen used (mitogens will induce readily detectable and strong proliferation after just 3 d) and the precursor frequency of the effector cell function being determined. In some experiments, restimulation may be required after the first 3–7 d of culture to expand the effector cell population to detectable levels *(10)*. A negative control is included as a zero baseline against which to measure cell division in the presence of antigen. Nonspecific stimulation may be controlled for by the addition of an irrelevant antigen control. PKH can still be followed on the flow cytometer after 2 wk. In those cells that remain undivided, some bleaching of the dye is occasionally observed but divided and undivided cells are still easily distinguishable.

Effector PBMC are plated into 96-well U-bottomed tissue culture plates (minimum of 1×10^5 cells per well, but 2.5×10^5 cells per well is preferable). Antigen-presenting cells are irradiated to prevent them dividing and then plated at 2×10^4 cells per well with either antigen, control irrelevant antigen or no antigen. Triplicate wells must be included for each condition. The cells are incubated for the appropriate number of days in a 37°C humidified incubator with 5% CO_2. If the assay is designed to determine cell phenotype and cytokine production, enough replicate samples must be included such that the appropriate controls for the antibody labeling steps can be carried out. This is because the cells are labeled with both cytokine-specific and surface protein-specific antibodies *in situ* in the culture dish.

To determine the level of division of total cells, the cells may be harvested directly after culture and run on the flow cytometer. However, this will only provide a readout of relative total number of dividing cells. To determine the phenotype of responding cells and the cytokine production of individual cells within the culture, two further steps are required before the cells are analyzed on the flow cytometer.

Cytokine production is measured using intracellular labeling with monoclonal antibodies specific for the cytokines of interest *(11)*. Cytokine production by the cells is amplified by a brief incubation with PMA and ionomycin and then retarded in the Golgi by a short incubation with Brefeldin A. The cells are fixed in formaldehyde and permeabilized to allow cytokine-specific antibodies to enter the cytoplasm. Once sequential incubations with the cytokine-specific antibodies followed by washing are completed, the cells are resealed and kept in the dark until the next step of the assay. The next step involves the identification of surface molecules on cells to determine their phenotype. This relies on monoclonal antibodies specific for the surface molecules. The cells are incubated sequentially with each antibody, washed, and resuspended in fixation buffer. The cells are analyzed on the flow cytometer and the readout from the cells after the labeling steps will be more specific, showing the phenotype of the dividing cells and the cytokines produced.

3.2. Preparation of Cells for Culture

3.2.1. Separation of Lymphocytes from Peripheral Blood

1. Collect 10–50 mL of blood by venipuncture into a sterile tube with preservative-free heparin.
2. Dilute the blood 1 in 2 with RPMI 1640.
3. Overlay 20 mL of diluted blood onto 10 mL of Ficoll-Hypaque
4. Centrifuge cells in a bench top centrifuge at 400*g* for 20 min at room temperature, with no brake.
5. Collect the lymphocyte layer at the interface of the buffy coat and Ficoll-Hypaque.
6. Wash twice with RPMI 1640, at 400*g* for 5 min room temperature, with braking.
7. Resuspend the cell pellet in 5 mL of complete medium.
8. Count cells on a hemocytometer, using trypan blue to determine viability.

Assessing Antigen-Specific Proliferation and Cytokine Responses

3.2.2. Thawing of Cell Samples Stored in Liquid Nitrogen

1. Remove vial(s) of cells from the liquid nitrogen, and thaw at room temperature.
2. As soon as the ice melts, transfer the cells into a tube containing 5–10 mL of RPMI 1640.
3. Centrifuge immediately, at 400g for 5 min at room temperature.
4. Wash once, at 400g for 5 min at room temperature, to remove traces of the freezing medium containing DMSO.
5. Resuspend the cells in complete medium.
6. Count cells on a hemocytometer, using trypan blue to determine viability.

3.2.3. Preparation of Antigen-Presenting Cells

1. Resuspend autologous PBMC in complete medium and irradiate (300 Gy) to prevent division without destroying their ability to take up and present antigen (*see* **Note 3**).
2. Count the cells on a hemocytometer and determine viability with trypan blue.
3. Resuspend in complete medium at a concentration of 2×10^5 cells/mL.
4. Ensure that there are enough cells to add 2×10^4 APC (100 µL) to each well of 2×10^5 effector cells that are to be tested.

3.2.4. PKH26 Staining of PBMC

1. Obtain slightly more PBMC than the total number needed for the assay. Wash once with medium without serum (*see* **Note 4**) in a sterile 15-mL centrifuge tube, at 400g for 5 min.
2. Resuspend the cell pellet in the appropriate diluent for the dye; in this case, CGL-DIL (*see* **Note 5**). Use 1 mL diluent per 2×10^7 cells.
3. Prepare a 3 µM PKH26 dye solution by diluting the 1 M PKH26 dye stock in CGL-DIL. Use 1 mL of 3 µM dye solution per 2×10^7 cells.
4. Add the dye to the resuspended cells (1 mL of 3 µM dye per 1 mL cell suspension, 2×10^7 cells) and mix gently, but thoroughly, for 1 min only at room temperature (*see* **Note 5**).
5. Immediately add 2 vol of 100% normal human serum (4 mL serum per 2×10^7 cells) and mix gently, but thoroughly, for a further min.
6. Add complete medium to fill the 15-mL tube. Centrifuge at 400g for 5 min at room temperature. Discard the supernatant.
7. Resuspend the cell pellet in fresh complete medium. Count the cells on a hemocytometer using trypan blue to determine viability (*see* **Note 6**).
8. Check for correct dye incorporation by running a small aliquot of the stained cells on the flow cytometer.

3.2.5. Setting Up Cell Proliferation Cultures

1. Aliquot 2×10^4 antigen presenting cells (100 µL of a 2×10^5 solution of APC resuspended in complete medium) (*see* **Subheading 3.2.3.**) into each well of a U-bottomed 96-well sterile tissue culture plate with a lid.
2. Add 2×10^5 PKH-labeled PBMC (100 µL of a 2×10^6 solution of cells resuspended in complete medium) (*see* **Subheading 3.2.4.**) as effector cells to each well.
3. Add antigen at the appropriate concentration (usually 10–20 µL of antigen diluted in complete medium is added to each well, giving a final volume of just over 200 µL).
4. Each antigen condition should be assayed in triplicate. For each antigen, include:
   a. A no antigen control.
   b. A mitogen control.
   c. An irrelevant antigen control.

   For those assays where intracellular cytokine labeling and cell phenotype labeling will be carried out, each antigen condition must be plated out with enough replicates to allow

appropriate antibody labeling controls to be included, as all labeling steps will be carried out in the culture plate (*see* **Note 7**).
5. Place the cultures at 37°C in a humidified, 5% CO_2 incubator for 3–7 d, depending on the antigen.
6. If a restimulation step is required, after the first 3–7 d of culture, remove 100 µL of the original culture (avoiding the cell pellet) and add 100 µL of fresh antigen and antigen presenting cells diluted in complete RPMI medium. Return cell cultures to the incubator for a further week before analysis (*see* **Note 8**).

3.3. Monoclonal Antibody Labeling and Analysis by Flow Cytometry

In an experiment where both intracellular cytokines and cell surface molecules are to be determined, the intracellular staining is usually carried out first so as to avoid disturbing the surface marker labeling step. If the experiment is set up to determine cell phenotype only, then the following two sections (**Subheadings 3.3.1.** and **3.3.2.**) can be eliminated; proceed directly to **Subheading 3.3.3.**

3.3.1. Inhibition of Cytokine Secretion with Brefeldin A Treatment

1. Remove the cell culture plate from the incubator. Add PMA and ionomycin to the cells, to a final concentration of 50 ng/mL PMA and 500 ng/mL ionomycin. Culture volume is 200 µL per well; add 10 µL of 1 µg/mL PMA working stock, and add 10 µL of 10 µg/mL ionomycin working stock.
2. Incubate the cells for an additional 2 h at 37°C in a humidified 5% CO_2 incubator.
3. Remove the cell culture plate and centrifuge at 400g for 5 min at room temperature.
4. Remove 20 µL of medium from each well, and replace with 20 µL of Brefeldin A working stock (0.1 mg/mL) to give a final concentration of 10 µg/mL Brefeldin A.
5. Mix each well gently, and then return culture plate to the 37°C, 5% CO_2 incubator for an additional 4 h.
6. Remove plate and place on ice for 10 min.
7. Centrifuge the plate at 400g for 5 min at 4°C (all subsequent centrifugation steps are at 400g for 5 min at 4°C). Remove the supernatant.
8. Resuspend the cells in at least 150 µL of ice-cold FACS buffer, and wash once at 400g for 5 min at 4°C.
9. Resuspend the cells in a final volume of 100 µL of FACS buffer.
10. Add 100 µL of 4% paraformaldehyde solution to the resuspended cells and mix thoroughly.
11. Incubate at room temperature, in the dark, for at least 20 min (*see* **Note 9**).
12. Centrifuge the cells at 400g for 5 min at 4°C, remove the supernatant, and wash once with 150 µL of ice-cold FACS buffer (400g for 5 min at 4°C).
13. Resuspend the cells in a final volume of 200 µL of FACS buffer. The cells can be stored at this stage at 4°C, in the dark, for a maximum of 3 d before continuing to the next stage.

3.3.2. Intracellular Labeling of Cytokines with Monoclonal Antibodies

1. Centrifuge the cells (from **Subheading 3.3.1.**) at 400g for 5 min at 4°C. Remove the supernatant.
2. Resuspend the cell in 150 µL of permeabilization buffer/well, mixing thoroughly but gently.
3. Incubate the cells for 10 min at room temperature, in the dark.
4. To prevent nonspecific antibody binding, incubate the cells with a control mouse antibody diluted in permeabilization buffer (approx 300 µg/mL) for 10 min, on ice, in the dark (*see* **Note 2**).
5. Wash cells once with 150 µL of permeabilization buffer, at 400g for 5 min at 4°C.
6. Resuspend in cytokine-specific antibody, diluted to the correct concentration in permeabilization buffer.

7. Incubate the cells sequentially for 20 min with the relevant cytokine specific antibodies (*see* **Note 10**), washing once between each labeling step with permeabilization buffer (*see* **Note 11**). The order of antibody incubations will depend upon the isotypes of the antibodies used and whether they are directly conjugated to fluorochromes (*see* **Note 1**).
8. After the last incubation, wash the cells twice with at least 150 µL of permeabilization buffer each time, at 400g for 5 min at 4°C, to remove any unincorporated antibody.
9. Resuspend the cell pellet in 150–300 µL of FACS buffer to reseal the plasma membrane. Ensure that all assay wells are resuspended in the same final volume.
10. Store at 4°C in the dark until analysis.

3.3.3. Surface Labeling of Cells to Determine Phenotype

1. Centrifuge the cells (from **Subheading 3.3.2.**) at 400g for 5 min at 4°C. Remove the supernatant.
2. Resuspend the cells in control mouse antibody diluted in FACS buffer (approx 300 µg/mL), mixing thoroughly but gently. Incubate the cells for 10 min on ice, in the dark (*see* **Note 2**).
3. Wash cells once with 150 µL of FACS buffer at 400g for 5 min at 4°C.
4. Resuspend the cells sequentially in each of the surface molecule-specific antibodies (diluted to the correct working concentration in FACS buffer), incubating on ice and in the dark for a minimum of 15 min for each antibody, washing between each different antibody. The order of antibody incubations is critical; this is discussed in detail in **Note 1**.
5. After the last antibody incubation, wash the cells twice in FACS buffer and resuspend in 150–300 µL of ice cold fixation buffer. Ensure that all assay wells are resuspended in the same final volume.
6. Store the cells at 4°C in the dark until analysis.

3.3.4. Acquisition of Labeled Cells on the Flow Cytometer

1. Run a test sample run on the FACScan to determine the length of time required to collect enough events for meaningful analyses (this will vary according to the prevalence of the cell type to be examined). Define the viable lymphocyte population using forward and 90° light scatter (*see* **Note 12**).
2. Set the flow cytometer to collect events for a fixed length of time (usually between 80–100 s is sufficient), not a fixed number of events. This is critical so as to permit the relative number of viable cells recovered in each well to be determined because this is a measure of the magnitude of the cellular response.
3. Acquire events from all sample wells for the same fixed length of time, to allow comparison throughout the assay (*see* **Note 13**).

3.4. Analysis of Cell Division, Phenotype, and Function

3.4.1. Introduction

This section discusses the analysis of the flow cytometric data. The details of the analysis described here are based on the use of the Cell-Quest software used in association with the Becton Dickinson FACScan and FACScalibur machines. Measurement of the magnitude of the cellular response is described, and how this is combined with intracellular cytokine labeling and with cell phenotyping. Analysis of the data plots that are acquired on the flow cytometer is also described. The data are presented as relative total numbers of dividing cells, or as percentages of dividing cells. Three- and four-color labeling with different monoclonal antibodies makes it possible to plot diagrams of division (PKH fluorescence intensity) against cell phenotype markers and/or against cytokine markers to show exactly which cell responds and how it responds *(8)*.

It is critical that the flow cytometer be set to collect events for a fixed time for these analyses. If a fixed time is not used, then the numbers of cells in each well will not reflect the magnitude of the response (*see* **Note 13**). It is also critical that within a single experiment, the cells in each different well must be resuspended in exactly the same volume of liquid at the end of the antibody labeling and washing steps prior to acquisition.

3.4.2. Analysis of Total Cell Division Within a Mixed Cell Population

The first plots assessed after PKH labeling of cells are to determine whether the PKH label has been incorporated correctly. This is shown diagrammatically in **Fig. 1A**, which shows a histogram plot of labeled cells (high PKH fluorescence intensity) and unlabeled cells (low PKH fluorescence intensity). To determine the magnitude of the response after a period of in vitro culture, cell viability is assessed by drawing a density or dot plot of forward and 90° scatter of the events collected within a fixed time (**Fig. 1B**). A gate is drawn around the viable cells on this dot plot and these gated viable cells are then analyzed for PKH fluorescence intensity. Those cells that have divided will have a low fluorescence intensity (denoted by the shaded peak of cells), whereas those that have not divided will have a high fluorescence intensity (denoted by the clear peak of cells). Gating or marking those areas within the plot of low or high PKH fluorescence and then looking at the event counts within those regions will give the number of cells that are dividing or are not dividing in the culture.

The results of this assay will give values for total mixed cell populations, not cell subtype values. The data are presented as relative total numbers of cells dividing or percentage of cells dividing. Since the number of events are collected for a fixed time, the numbers of events in each of the regions specified will reflect the magnitude of the antigen-specific response. The triplicate wells for each sample condition allow for the mean and standard deviation of the data to be calculated, giving an indication of the reproducibility of the assay.

3.4.3. Analysis of Cytokine Production and Cell Phenotype

To determine whether cytokine production has occurred during culture and to determine the phenotype of the cytokine producing cells, those cells labeled with PKH and cultured with antigen for the prescribed time are prepared for intracellular labeling with cytokine specific antibodies (*see* **Subheading 3.3.2.**) and for surface labeling (*see* **Subheading 3.3.3.**). A cell viability gate is drawn and the level of cell division within this gate is analyzed as described previously (*see* **Subheading 3.4.2.**). The example in **Fig. 2** shows PKH-labeled cells that have been labeled after culture with a monoclonal antibody specific for IFN-γ and then with an antibody specific for CD4. The phenotype of the dividing cells is shown in the contour plot (**Fig. 2A**). PKH label (y-axis) is drawn against the antibody label for CD4 (x-axis). The dividing cells are those in the shaded area (lowered PKH fluorescence intensity), and the nondividing cells are shown in the unshaded area (these cells still have high PKH fluorescence intensity). The numbers of cells that are dividing or not dividing and that are CD4$^+$ or CD4$^-$ can be clearly determined by counting the numbers of cells in each quadrant (UR, LR, UL, LL). The cells that have CD4 on their surface and dividing are shown in the lower, right-hand quadrant of the contour plot.

Assessing Antigen-Specific Proliferation and Cytokine Responses

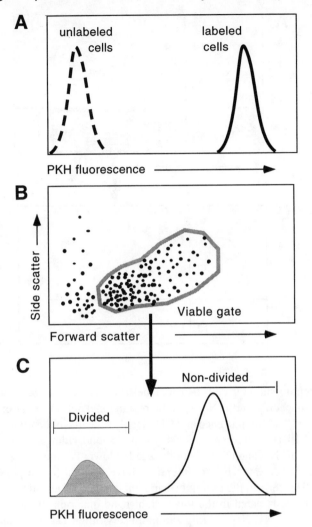

Fig. 1. Plots showing data analysis of PKH-stained cells. (**A**) PKH fluorescence of labeled and unlabeled cells. (**B**) After culture, the PKH-labeled cells are prepared for flow cytometry. The cell viability is assessed by forward and side scatter and a viability gate is drawn. (**C**) The levels of PKH fluorescence in this viable gate are assessed. Those cells with low PKH fluorescence have divided and those with high fluorescence have not divided. By gating on these cells, the numbers or percentages of cell division that has occurred can be determined.

To determine which cytokine is being produced by the cells which are both dividing and are CD4, a plot of dividing CD4$^+$ cells (*x*-axis) is drawn against IFN-γ production (*y*-axis). Those cells that are producing IFN-γ are in the top right-hand quadrant of the contour plot. The numbers of cells within this region of the plot can be counted and can be presented either as a percentage or as a relative total number of dividing CD4 cells producing IFN-γ.

The examples shown in the two figures give a comprehensive overview of how to determine the magnitude of a response, and how to specify the phenotype of the responding cells and the type of response that is induced using three different labeling steps. Depending upon the type of flow cytometer used and the number of different

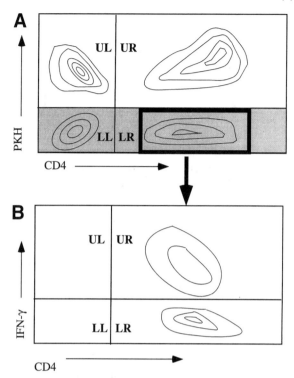

Fig. 2. Contour plots showing cells that are labeled with three different fluorochromes, PKH- and then with CD4-specific and IFN-γ-specific monoclonal antibodies after the culture period. (**A**) Contour plot of PKH fluorescence and CD4 label. All cells in the upper two quadrants (UL and UR) have high PKH fluorescence intensity and are nondividing cells (corresponding to the unshaded peak in Fig. 1C) and the cells in the shaded lower two quadrants are those that have lower PKH fluorescence intensity and therefore have divided (corresponding to the shaded peak in Fig. 1C). The cells in the two right-hand quadrants (UR an LR) are all CD4$^+$ cells. (**B**) Contour plot of the cells gated in the lower right-hand quadrant of panel A. These cells are analyzed for prescence of IFN-γ labeling within this dividing CD4$^+$ population. As all the cells gated from panel A that are plotted in this second contour plot are CD4$^+$, the two left-hand quadrants contain no cells. The cells in the top right-hand quadrant of the plot are CD4$^+$ and produce IFN-γ.

monoclonal antibodies used, a much wider variety of analyses can be carried out on a relatively limited sample size, comparing many more cytokines and also phenotyping many more cell subsets.

4. Notes

1. The fluorochrome conjugates for each antibody chosen should have different emission spectra from that of the dye used for proliferation measurement. PKH-26 emits at a peak of 567 nm so antibodies labeled with FITC (520 nm), biotin in conjunction with either streptavidin-PerCP (677 nm), or streptavidin-APC (660 nm) should be used. Antibodies conjugated to R-PE (576nm) cannot be used with PKH26. If more than one biotin-conjugated antibody is to be used to label cells in the same well, the first biotin-conjugated antibody label must be directly followed with the chosen second-step reagent before the next biotin-conjugated antibody is added to the well. The antibodies used in those experiments must be titrated before use to determine the optimal working concentration. Once

this is established, the antibodies can be diluted to the correct working concentration in the appropriate buffer. For intracellular labeling, the appropriate buffer is permeabilization buffer and for surface labeling, is the FACS buffer. The volume of diluted antibody solution used at each step is usually 20 µL.
2. Control antibodies must be included in flow cytometry experiments. An unconjugated mouse immunoglobulin with an irrelevant specificity should be included to act as a blocking antibody, blocking any nonspecific binding to cellular components. The antibody will also bind to the Fc receptor that may be present on the cells thus preventing nonspecific binding via Fc receptors. The second type of control antibodies required are those with irrelevant specificities, which do not bind specifically to the marker being labeled. These antibodies should have the same isotype as the antibodies used to label the cells and should be conjugated to fluorochromes to allow detection of any nonspecific binding that may occur to be observed.
3. Cell lines are more robust than PBMC, requiring higher doses of irradiation to prevent division. Autologous EBV-transformed B cell lines require 500–600 Gy. Dendritic cells, derived by culture of monocyte precursors with IL-4 and GM-CSF, do not divide so do not need irradiation before use as antigen-presenting cells. If you are unsure whether your antigen-presenting cells divide, control wells of antigen-presenting cells labeled with the PKH dye should be included in the culture plate, with and without antigen (but without added effector cells).
4. PKH26 is absorbed by lipids present in normal human serum. If a final serum-free wash is not carried out, the PKH dye will be absorbed into the lipid present in the serum and the efficiency of the cell labeling will decrease.
5. There are several different PKH dyes that can be obtained and these are used to label different cell types. Each of the dyes has its own specified diluent solution. When labeling with the dyes, care must be taken to ensure that the correct diluent is used.
6. The diluent for PKH and the dyes are slightly toxic to cells. For this reason the incubation steps with stain and diluent are kept relatively short. A few cells are lost with each staining and washing step; therefore, calculate final cell numbers required in the assay to include a loss of approx 20% of initial cell numbers.
7. When setting up the cultures in the 96-well U-bottomed plates for phenotype analysis and cytokine production analysis, controls for the different stages of the experiment must be included.
    a. The first type of controls are those needed to determine whether the cultures have been set up correctly, and whether the PKH staining has worked or has affected the ability of the cells to divide. Controls for this include: (a) PKH-unlabeled cells with and without antigen; (b) PKH-labeled cells with and without antigen; (c) PKH-labeled cells with a positive control antigen or a mitogen to determine cell viability; (d) If possible, it is advisable to also include an irrelevant antigen control of PKH-labeled cells. The PKH-labeled and the PKH-unlabeled cells also serve as controls for flow cytometry.
    b. The second type of controls are those required to control for the antibody labeling of cells. These are control mouse antibodies that are not conjugated to a fluorochrome; these act as blockers of nonspecific binding.
    c. The third type of controls are isotype controls. These are fluorochrome-conjugated antibodies of the same species (e.g., mouse) with the same isotype as those antibodies used for cell marker labeling. These isotype controls have an irrelevant specificity and will reveal any nonspecific binding that occurs.
    d. A fourth level of control must be also considered. The number of control wells will depend on the number of different antibodies used. For multiparameter analysis, the same cell is labeled with different antibodies conjugated to fluorochromes with differing emission spectra. As these spectra overlap to a greater or lesser degree, compensa-

tion values on the flow cytometer must be preset to account for this overlap. To do this, the cells of interest are labeled with each of the antibodies separately and the amplification and compensation settings adjusted for each fluorochrome. The compensation controls for a multiparameter assay are discussed further in **Note 12**.

Once several different antibodies are to be added to the same sample well, the order of antibody addition becomes relatively important. If the cells are not conjugated to a fluorochrome, an antiisotype-conjugated antibody should be used to reveal the binding of the primary antibody. Care must be taken to ensure that only the unconjugated antibody becomes labeled; therefore, as with the second step reagent that binds biotin, unconjugated antibodies should be labeled with their second step reagent, before addition of other antibodies.

8. If the antigen-specific cells being analyzed are only present at a low frequency, ensuring that a response is large enough to be detected may require restimulation or expansion steps of some kind. The cultures may either be restimulated by addition of further antigen-presenting cells and antigen after the initial culture period, or by the addition of IL-2, or both.

9. Culture of cells with PMA and ionomycin may change the levels of expression of some surface molecules. A decrease in CD4 and in CD8 expression has been noted in both mouse and human cells. Changes in surface expression may also result from the effects of Brefeldin A (CD14 expression may be reduced).

10. The cells are fixed at this stage of the experiment to prevent them from making more cytokine and from exporting new cytokine after the effect of Brefeldin A has worn off (Brefeldin A does not irreversibly stop export from the Golgi). This may cause some problems at later stages, since some antibodies are incompatible with fixation and may give peculiar and unexpected results on the flow cytometer. This can only be discovered by pretesting each of the antibodies to be used on fixed cells before commencing with this experiment. In some cases where fixation seriously interferes with the way in which the antibodies bind, surface labeling can be performed on the cells before they are fixed.

11. Extensive washing is required between each labeling reaction to remove any unincorporated antibody from the cells. Each wash at this stage is carried out in permeabilization buffer, to keep the cell cytoplasm open to allow entry of the monoclonal antibodies. If intracellular staining and surface staining are to be carried out on the same cells, intracellular staining is always done first to avoid any disruption of surface labeled molecules that might otherwise occur.

12. The position of viable cells on a forward by 90° scatter plot should be tested and confirmed using propidium iodide gating on unfixed cells before commencing this experiment. Since fixation allows the propidium iodide dye into the cells, fixed cells will all appear dead if propidium iodide is used. The position of the viable cells on the plot of forward versus 90° scatter will depend on how the flow cytometer has been set up to detect the light scatter (the levels of the detectors will be changed according to the physical attributes of the cells to be examined) and also on the type of cells that are analyzed.

13. The flow cytometer is set to act as a sophisticated Coulter counter. The cytometer is able to count the number of cells per well over a fixed time. If the initial cell number placed in each well before culture was the same, and if the volume that the cells were resuspended in prior to analysis was the same, then the number of cells collected within that fixed time will reflect the magnitude of the response that has occurred in that well. Not only can proportions of cells be determined but also the relative number of each different cell type can be determined. Combining the cell counting technique with the multiple labeling of cells means that the number of cells within a culture of a certain phenotype producing a certain cytokine can be identified. As the samples are collected for a fixed time, this number is a measure of the magnitude of the specific response.

References

1. Brake, D. A., Long, C. A., and Weidanz, W. P. (1988) Adoptive protection against *Plasmodium chabaudi adami* malaria in athymic Nude mice by a cloned T cell line. *J. Immunol.* **140,** 1989–1993.
2. Troye-Blomberg, M., Perlmann, H., Patarroyo, M. E., and Perlmann, P. (1983) Regulation of the immune response in *Plasmodium falciparum* malaria. II Antigen specific proliferative responses *in vitro. Clin. Exp. Immunol.* **53,** 345–353.
3. Malik, A., Egan, J. E., Huoghten, R. A., Sandoff, J. C., and Hoffman, S. L. (1991) Human cytotoxic T lymphocytes against the *Plasmodium falciparum* circumsporozoite protein. *Proc. Natl. Acad. Sci. USA* **88,** 3300–3304.
4. Grau, G. E., Piguet, P. F., Engers, H. D., Louis, J. A., Vassalli, P., and Lampert, P. H. (1996) L3T4+ lymphocytes play a major role in the pathogenesis of murine cerebral malaria. *J. Immunol.* **137,** 2348–2354.
5. McMurray, D. N. (1984) Cell mediated immunity in nutritional deficiency. *Proc. Food. Nutri. Sci.* **8,** 193–228.
6. Horan, P. K. and Slezak, S. E. (1989) Stable cell membrane labeling. *Nature* **430,** 167,168.
7. Lyons, A. B. and Parish, C. R. (1994) Determination of lymphocyte division by flow cytometry. *J. Immunol. Meth.* **171,** 131–137.
8. Allsopp, C. E., Nicholls, S. J., and Langhorne, J. (1998) A flow cytometric method to assess antigen specific proliferative responses of different subpopulations of fresh and cryopreserved human peripheral blood mononuclear cells. *J. Immunol. Meth.* **214,** 175–186.
9. Yamamura, Y., Rodriguez, N., Schwartz, A., Eylar, E. Bagwell, B., and Yano, N. (1995) A new flow cytometric method for quantitative assessment of lymphocyte mitogenic potentials. *Cell. Mol. Bio. Noisy le Grand.* **41, S121(Suppl. 1)**.
10. Kallas, E. G., Gibbons, D. C., Soucier, H., Fitzgerald, T., Treanor J. J., and Evans, T. G. (1999). Detection of intracellular antigen specific cytokines in human T cell populations. *J. Infect. Dis.* **179,** 1124–1131.
11. Openshaw P., Murphy, E. E., Hosken, N. A., Maino, V., Davis, K., Murphy, K., and O'Garra, A. (1995) Heterogenicity of intracellular cytokine synthesis at the single-cell level in polarized T helper 1 and T helper 2 populations. *J. Exp. Med.* **182,** 1357–1367.

37

Cytokine Analysis by Intracellular Staining

Aftab A. Ansari and Ann E. Mayne

1. Introduction

The advent of molecular techniques for the first time allowed for the specific biological characterization and a more clear understanding of the function of molecules synthesized by leukocytes termed cytokines. In addition, availability of the recombinant forms of these cytokines with defined sequences and in select cases molecular structure allowed for a rationale to group the growing list of cytokines and their cognate receptors. Cytokines in general are proteins or glycoproteins that are synthesized by a number of cell lineages, the best characterized are those synthesized by the hematopoietic cell lineages. Cytokines in general are either secreted, expressed on the cell membrane and/or held in a reservoir form within the cells that synthesize them. Several cytokines are synthesized in a biologically inactive form (procytokine) and need to be cleaved to release the active form, a notable example being tumor necrosis factor-α (TNF-α). Certain cytokines are held as storage depots within tissue areas within which their function is manifest. Thus cytokines that perform hematopoietic growth and differentiation appear to be stored as depots within stromal cell layers of the bone marrow, whereas those involved in tissue injury and repair are localized to the extracellular matrix, skin and bone and so on. Most, if not all, induce their biological effect by binding to their cognate receptors, which provide intracellular activation of signal transduction and second-messenger pathways. Most cytokines also appear to serve as growth factors or differentiation-inducing molecules for hematopoietic cell lineages. Ever since the cloning, sequencing, and characterization of the first wave of cytokines, there have been arguments and discussion as to the distinction between cytokines, chemokines and select molecules that are restricted in their function to specific organs such as the central nervous system (CNS). This issue is highlighted by the omission of insulin, which is not included as a cytokine, but is clearly a molecule with cytokine-like properties. Thus, at present the definition and inclusion of molecules termed cytokines is somewhat vague and appears to be largely dictated by the line of study that leads to the generation and identification of a new molecule.

Initially, cytokines were named by either the cell type that synthesized the cytokine or by the function of the molecule, for example, T-cell growth factor and tumor necrosis factor. It is now clear that utilization of such functions as the basis for nomenclature can pose potential problems, for example, tumor necrosis factor is no longer thought of as an anticancer agent, but, in fact, considered one of the proinflammatory cytokines. Such issues contributed to the use of sequential numbers for their nomenclature, such as interleukin-1 (IL-1). This too has posed problems because some cytokines given a sequential number, such as IL-8, have later been shown to belong in fact to the chemokine family both by structure and function. Initially cytokines were quantitated biologically, that is, by the function characterized for the molecule. These were most often carried out utilizing supernatant fluids from in vitro-activated hematopoietic cells. A classic example of this type of assay was the ability of T-cell growth factor—previously termed T-cell replacing factor (TCGF, TRF)—now known as IL-2, to induce the proliferation of lymphoid cells *(1–3)*. For certain research questions, such a bioassay is the most optimal, as it does define levels of functionally active molecules rather than a biochemically defined amount, which by definition contains both the biologically active and inactive forms of the molecule. For a number of cytokines such bioactivity assays have been standardized using a defined cell line (which expresses the cognate receptor and is dependent on the cytokine for growth and proliferation) and recombinant forms of the cytokine. A unit of activity is usually defined as the amount (by weight) of the cytokine that induces 50% maximal bioactivity. Enzyme-linked immunosorbent assay (ELISA)-type assays with a monoclonal antibody used for capture soon followed these bioassays, and an enzyme-conjugated heterologous antibody used to detect the captured cytokine. In some cases, a second enzyme-conjugated monoclonal antibody is used especially if directed against a distinct epitope of the cytokine or the same monoclonal antibody if the cytokine constituted multiple similar epitopes (epitopes here are defined operationally, as it is recognized that most antibodies recognize conformational sequences and not linear sequences). ELISA-type assays were soon followed by enzyme-linked immunospot (ELISPOT) assays *(4,5)*, in which the frequency alone and/or coupled with the phenotype of cells within a mixed population synthesizing a particular cytokine is determined, which is ideal for a number of research protocols. In addition to the standard ELISA and ELISPOT assays, since the cytokines are cloned and sequenced, a number of laboratories set up Northern blotting and polymerase chain reaction (PCR)-based semi-quantitative assays for defining the relative amounts of mRNA being synthesized by a given cell population. This was soon followed by a more precise quantitative assay for mRNA levels for the cytokines with the design and successful utilization of competitor sequences coding for the given cytokine (competitive reverse transcriptase [RT]-PCR) and real-time PCR assays. For select studies, a number of laboratories spent considerable effort to define *in situ* hybridization techniques for the quantitation of the number of cells synthesizing mRNA for a particular cytokine. This technique is clearly very useful in tissue specimens that can only be assayed as sections from fresh frozen tissues and for select cytokines in paraformaldehyde or formalin-fixed sections of tissues.

The fact that most cytokines can function in an autocrine and/or paracrine fashion led to the realization that interpretation of data on measurements of cytokines from cell populations ex vivo and/or in vitro cultured cells may be erroneous because cells synthesizing a given cytokine are most likely rapidly using up the cytokine in question. Thus, measurements of cytokine in the plasma or in vitro culture supernatant fluids is the net level of cytokine with an unknown amount being consumed and/or bound to its cognate receptor. In addition, such measurements clearly did not allow for defining the precise cell lineage(s) synthesizing the cytokine and/or utilizing the cytokine in question. In select studies, such knowledge is critical for the precise definition of the chain of events leading to a biological function being analyzed. This need was highlighted by the general acceptance of the Th1 and Th2 paradigm of immune regulation *(6,7)*. Thus, while a number of laboratories have provided data that assign specific cell-surface markers that distinguish Th1 from Th2 cells, most laboratories up to now use the cytokine profile as a more reliable marker to distinguish Th1 from Th2 prototype of cell subsets *(8–10)*. It is also to be noted that while the initial studies identified a clear dichotomy between Th1 and Th2 cells based on the synthesis of a defined set of cytokines by each subset, it has been clear for some time that such rigid dichotomy in fact does not exist, and overlapping cytokines are the general rule. The previous dichotomy was based on data derived primarily using murine tissues and detailed studies of cloned T-cell lines from other species, such as humans, led to the realization that the Th1/Th2 dichotomy needs to be reevaluated. It is now generally accepted that the synthesis of IFN-γ and, to a lesser extent, IL-2 by a given T-cell represents a prototype Th1 subset, whereas the synthesis of IL-4, IL-5, and, to a lesser extent, IL-10 by a given T-cell represents the Th2 subset. Intense studies on defining the role cytokines play in the regulation of immune response and the identity of the cell lineage origin of the cytokines in a wide variety of experimental and clinical studies is currently being undertaken. This has led to the development of the technique of combined cell surface marker and intracellular cytokine analysis using flow microfluorometry *(11)*, the subject of this chapter. It should be kept in mind that analysis of a given cell population for subsets that are synthesizing a given cytokine using the intracellular staining technique provides a single snap shot view of the event and as such the data needs to be interpreted within that context. It logically follows that optimum conditions need to be defined for the question being posed and the description provided below is primarily technical in nature.

For most immunological studies, there are generally three replicate sets of cell samples (if sufficient cell numbers are available) required to provide the needed data. The three sets of replicate samples include a set that is analyzed for the cell lineage specific constitutive synthesis of the cytokines being studied, the second set consists of cells cultured in vitro with mitogens (such as phytohemagglutinin protein [PHA-P] or lipopolysaccharide [LPS]) or cell activation agents such as PMA + Ionomycin (PMA is a protein kinase C activator and ionomycin is a Ca^{2+} ionophore, and together these two reagents lead to potent activation of T cells), respectively, that will allow for defining the maximal potential of a given cell lineage within the mixed population to synthesize the cytokines being studied. The third set includes cells cultured for a defined period of time with the specific antigen or peptide-pulsed antigen-presenting cell that will provide data on the antigen-specific induced cytokine(s) being synthesized by a given cell lineage among the mixed population. The three sets can be viewed as negative control, positive control, and the experimental sets, respectively, depending on the study being undertaken.

Combined cell surface phenotype and intracellular staining basically involves staining the cells with appropriate fluorochrome conjugated antibodies against cell surface markers, followed by fixation of the cells with paraformaldehyde (which crosslinks proteins and prevents their loss through leakage). The paraformaldehyde-fixed cell population is then treated with a cell-permeabilizing agent, usually a detergent that induces sufficient space between cell membrane proteins so as to allow agents such as antibodies to gain intracellular access. These cell surface-stained paraformaldehyde-fixed and permeabilized cells are then incubated with appropriate fluorochrome-conjugated antibodies against defined cytokines, washed, and then subjected to flow microfluorometric analysis. To increase the sensitivity for the intracellular detection of a given cytokine, prior to the entire staining and treatment protocol, the cells are cultured with a protein transport inhibitor such as monensin or Brefeldin-A that serves to trap and accumulate intracytoplasmic cytokines being synthesized facilitating an ease in their detection. Thus, culturing cells with such protein transport inhibitor followed by cell surface staining, fixing, membrane permeabilization, and intracellular cytokine staining allows for the identification of the precise cytokine synthesis pattern of phenotypically identified cells with the intensity of staining representing a relative quantitative estimate of the cytokine being synthesized by the phenotypically identified cell. The advantage of this technique is that compared to the other assays, such as ELISA-based assays and RT-PCR assays, which represent data obtained on entire mixed population of cells, the technique described in the following sections yields rapid identification of the cytokines being synthesized by individual cells within a mixed population.

Yet another approach for the analysis of cytokine-synthesizing cells is currently established in a number of laboratories. The technique utilizes the use of (at present) MHC class I tetramers (and in the near future MHC class II tetramers), which are loaded with specific peptides and thus serve to identify predominantly CD8+ T cells with specificity for a specific peptide-bound MHC molecule. Use of this technique requires knowledge of the MHC class I restricting molecule and the peptide of the antigen that specifically is presented by the MHC class I molecule. Thus, not only can the precise frequency of the peptide-specific CD8+ T cells be defined, but, following incubation and coculture with the cognate peptide bearing MHC class I expressing APCs, the number of tetramer-positive CD8+ T cells that synthesize a particular cytokine intracellularly can also be enumerated. The general protocol for determining these sorts of data is essentially similar to the one described above for defining the frequency of lineage-specific CD4+ or CD8+ T cells, for example, that are synthesizing a particular cytokine, except that the frequency of any given tetramer staining cells is normally quite low, except for highly skewed immune responses. Thus, in the use of tetramer technology in the case of malaria studies, there is a requirement for a knowledge of the dominant MHC alleles in the population being studied, knowledge of the malaria specific peptides that serve as targets, and a need for large numbers of cells for the analysis, especially if a battery of tetramers are to be utilized. As the genome of each of the malaria parasite species becomes unraveled, it will lead to the identification of the immunodominant epitopes and depend on the population expressing MHC alleles, which would then allow for the successful use of tetramer technology. For the time being, therefore, we describe below our general protocol for analysis of antigen specific intracellular cytokine synthesizing CD4 and CD8 cells.

2. Materials

See **Note 1**.

2.1. Media

1. Culture medium: RPMI 1640 supplemented with 10% heat inactivated fetal calf serum (FCS), 100 U/mL of penicillin, 100 µg/mL streptomycin and 2 mM L-glutamine.
2. Wash medium: PBS supplemented with 1% FCS, pH 7.4.
3. Staining medium: PBS supplemented with 1% FCS, 0.1% sodium azide, pH 7.4.
4. Fixing medium: prepare fresh each time by adding 400 mg of paraformaldehyde to 10 mL of warm PBS (4% w/v). Incubate the tube at 56°C for 1–2 h to dissolve the paraformaldehyde and then adjust to pH 7.4. Cover tube with aluminum foil.
5. Permeabilization buffer: Phosphate-buffered saline (PBS) supplemented with 1% FCS, 0.1% sodium azide and 0.1% saponin, pH 7.4.

2.2. Cell Activation Agents and Other Chemicals

1. Phorbol myristate acetate (PMA) in 100% ethanol.
2. Ionomycin (Ca^{2+} salt) in dimethyl sulfoxide (DMSO).
3. Anti-human CD3 monoclonal antibody.
4. PHA-HA16 (Burroughs-Wellcome).
5. Bacterial lipopolysaccharide *Escherichia coli* (LPS) in PBS.
6. Brefeldin-A in DMSO (or Monensin).
7. *DNAse I* (60,000 Dornase U/mL) in PBS.
8. 4% Paraformaldehyde.
9. Saponin.
10. Ficoll-Hypaque.

2.3. Monoclonal Antibodies

Depending on the type of flow cytometer available (i.e., FACScan or FACSCaliber), either 3- or 4-color analysis, respectively, can be performed. For simplicity, the 3-color protocol is described herein.

1. Fluorescein isothiocyanate (FITC), peridinin chlorophyll protein (PerCP), and phycoerythrin (PE)-conjugated isotype control antibodies.
2. FITC anti CD4.
3. PerCP anti CD8 (and if needed PerCP conjugated anti CD56 and PerCP conjugated anti CD14).
4. PE-conjugated anti-IL-2, anti-IL-4, anti-IL-6, anti-IL-10, anti-IL-12, anti-IFN-γ, and anti-TNF-α.

3. Methods

3.1. In Vitro Cell Culture

For purposes of this chapter, the technique being described is primarily focused on the analysis of lymphoid mononuclear cells (*see* **Note 2**) (**Fig. 1**). Clearly, analysis of other cell lineages may require adjustments and changes to the protocol. In addition, a protocol is described for the analysis of intracellular levels of the cytokines IL-2, IL-4, IL-6, IL-10, IL-12, IFN-γ, and TNF-α by $CD4^+$ T cells, $CD8^+$ T cells and non-CD4/$CD8^+$ T cells. Additional combinations of cell surface markers and cytokines can be analyzed as required for the research question being posed; for example, the inclusion of the analysis of the cytokines synthesized by CD16/$CD56^+$ natural killer (NK) cells, $CD14^+$ monocytes, activated $CD3^+$, $CD25^+$ or $CD3^+$, $CD69^+$ cells. Appropriate adjust-

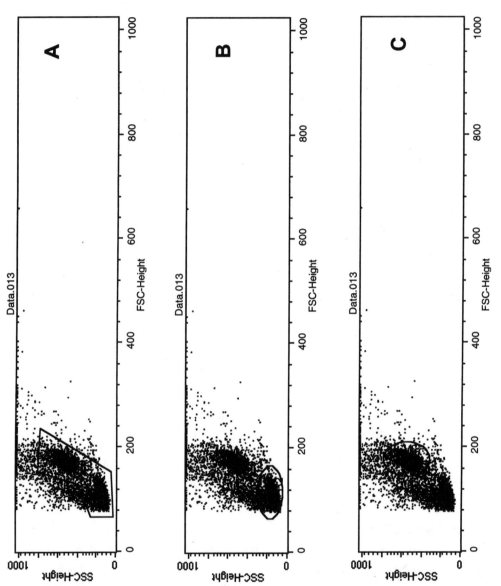

Fig. 1. Flow microfluorometric (FMF)-assisted forward light scatter (FSC) and side scatter (SSC) profile human PBMCs showing the approximate gating of lymphoid cells from monocytes. Data can thus be acquired on both lymphocytes and monocytes (**A**), only on lymphocytes (**B**), or only on monocytes (**C**).

ments in the protocol can thus easily be made. To keep the protocol simple, it is recommended that FITC- and PerCP-conjugated antibodies are used for the two cell surface markers and PE-conjugated anticytokine antibodies are used for the detection of intracellular cytokines.

3.2. Culture Media

Standard tissue culture medium can be used, such as RPMI 1640 supplemented with 2 mM L-glutamine, 1% penicillin-streptomycin, and 10% heat-inactivated (56°C for 1 h) FCS. Other tissue culture medium that is previously defined by a laboratory to be optimal for the source and species of lymphoid cells being studied can also be used.

3.3. Cell Activation Agents

In general, as described in **Subheading 1.**, analysis of three sets of replicate cell populations provides a good estimate of the cytokine profile. This includes analysis of the constitutive levels of cytokines synthesized by specific phenotypically identified cells, the maximal potential of the phenotypically identified cell to synthesize specific cytokines and the experimental condition being studied such as the cytokine profile of phenotypically marked cells upon exposure to a given antigen. Clearly such an analysis requires the availability of sufficient cell numbers but is described herein in case such numbers are indeed available. It is to be noted that there is normally very little cytokine and very few cells that synthesize a given cytokine "constitutively." Thus such an analysis of "constitutive" synthesis merely provides background and negative control values. However, in select conditions, such as acute exposure to antigens, for example, during acute infections, chronic infections, and in some autoimmune diseases, such analysis of "constitutive" synthesis of cytokines is deemed extremely important because interpretation of data derived from the analysis of the cells performed only following antigen- or mitogen-activation will be difficult as it will not allow a clear distinction between antigen-induced effects from the nonantigen-induced effects. Additionally, use of certain of the more commonly utilized mitogens or activation agents leads to down modulation of cell surface markers such as CD4, for example, which makes it difficult to definitively ascribe synthesis of particular cytokines to this lineage, although one can determine this by subtraction techniques described below (*see* **Note 3**). Finally, it is also clear that different cell-activation agents are required for the study of the expression of different cytokines. Thus, while anti-CD3, PHA-P, or PMA + ionomycin are each excellent for the detection of cytokines predominantly synthesized by T cells, other cytokines such as IL-6 and IL-12 require the activation of monocytes, which is better accomplished with the use of polyclonal mitogens, such as bacterial lipopolysaccharide (LPS). A general protocol thus follows and can be adjusted to include multiple cell activation agents depending on the cytokines of interest.

3.4. In Vitro Cell Activation

The activation of Ficoll-Hypaque density gradient purified human peripheral blood mononuclear cells (PBMC) is described herein, although other cell preparation such as whole-blood protocols, use of lymphoid cells from other tissues such as lymph nodes, spleen, and bone marrow can also be studied. Note that several of the reagents utilized below are nonpolar (PMA, ionomycin, etc.) and therefore they tend to coat the plastic surface of the tubes that are commonly used for tissue culture and cellular studies. Therefore it is highly recommended that fresh reagents be prepared with each staining protocol, otherwise the effective concentration will vary and lead to suboptimal results.

1. Label three sets of 11 sterile polystyrene round-bottomed tubes 1 through 11 for each cell sample to be analyzed with cell sample identifier and date. Label the first set 1C through 11C, the second set 1M through 11M, and the third set 1E through 11E (C for control/constitutive, M for mitogen/maximal or positive control, and E for experimental, such as antigen).
2. Adjust the PBMC to 1×10^6 cells/mL of media described above and dispense 1 mL to each of the 33 tubes (11 per set). This is overkill in terms of cell concentration. One can easily get by with 0.25×10^6 cells per tube which not only cuts down on the total number of cells needed from 33×10^6 cells to $8–9 \times 10^6$ cells but also cuts down on the amount of antibody reagents required to stain the cells. The higher concentration is just easier when analyzing by FACS because the samples (number of cells, usually analysis of a single sample involves 10,000 cells) runs faster.
3. Dispense 10 μL of 2.5 μg/mL of PMA and 10 μL of 100 μg/mL of ionomycin to tubes marked 1M through 8M. Dispense 10 μL of 1 μg/mL of LPS to tubes marked 9, 10, and 11M. One can utilize PHA-P (1 μg/ml) or immobilized anti-CD3 (precoat the tubes with 10 μg/mL of anti-CD3 mAb) + soluble anti-CD28 MAb (2 μg/mL) instead of the PMA + ionomycin for tubes 1M through 8M.
4. Dispense 10 μM of peptide antigen of interest or a previously defined optimal concentration of the antigen to be studied in a volume of 10 μL to the tubes marked 1E through 11E. The addition of anti-CD49d and CD28 MAb (2 μg/mL each) has been shown to increase the sensitivity of the detection of the antigen specific induction of cytokines and can be used at this stage.
5. Dispense 2 μL of 1 mg/mL Brefeldin-A to each of the 33 tubes (*see* **Note 4**). Note that monensin can be switched for Brefeldin-A. Also note that the concentration of Brefeldin-A has been reduced from the usual 10 μL of 1 mg/mL to 2 μL of 1 mg/mL because the cells are being incubated for 12–14 h instead of the usual 4–6 h (*see* **Note 5**). If shorter incubation time is chosen, one can increase the concentration of Brefeldin-A.
6. Mix the tubes well, cover with aluminum foil to keep the cells protected from light from hereon and incubate for a total of 12–14 h (for some cytokines 6 h is optimal but for a compromise so that the entire spectrum of cytokines can be analyzed simultaneously, 12–14 h of incubation appears to suffice) in a 37°C incubator with 7% CO_2 and a humidified atmosphere. Incubate the tubes at a slant of 5°, which also seems to optimize the analysis of intracellular cytokines.
7. Following incubation, add 10 μL of 10 mg/mL *DNAse* to each of the 33 tubes and mix well. This *DNAse* helps break cell clumps and is needed for the cultures containing PMA + ionomycin which induces potent cell activation but also leads to some cell death (which also occurs in some in vivo-activated cells that are being cultured from patients), which causes clumping and poses a problem during flow analysis. Other standard antigen-specific cultures may not require the addition of *DNase* and can be left out the tubes marked 1E through 11 E.

3.5. Blocking of Fc Receptors

In efforts to reduce background nonspecific staining by antibodies, FcR can be blocked by the addition of either anti-FcR antibodies (such as clone 2.4G2) purified immunoglobulin (Ig) from the same species and of the same isotype as the antibodies used for cell surface and anticytokine staining. Most of these antibodies are either of murine or rat origin and most are of the IgG1 and IgG2a isotype. Such isotype controls

Intracellular Cytokine Staining

at a concentration of 1 µg/mL are added to each tube, and the tubes are incubated for 15 min at 4°C. To each tube is added 1.0 mL of cold (4°C) media for washing, and the tubes are then centrifuged at 150g and the media discarded. Resuspend each cell population in 100 µL of RPMI 1640 with 20% FCS.

3.6. Cell Surface Staining

As described above, the optimum concentration of antibodies to be used to stain CD4 and CD8 needs to be established by each laboratory. Most commercial sources recommend 1.0 µg of antibody per 10^6 cells. Our laboratory has found that in fact <0.5 µg of antibody is sufficient to stain 10^6 cells and if 0.25×10^6 cells are used one can utilize 0.1 µg of the antibody. Add the following antibodies in a volume of 10 µL to each of the set of 11 tubes:

1. FITC IgG1 + PerCP IgG1
2. FITC IgG1 + PerCP IgG1
3. FITC IgG1 + PerCP IgG1
4. FITC anti-CD4 + PerCP anti-CD8
5. FITC anti-CD4 + PerCP anti-CD8
6. FITC anti-CD4 + PerCP anti-CD8
7. FITC anti-CD4 + PerCP anti-CD8
8. FITC anti-CD4 + PerCP anti-CD8
9. FITC IgG1 + PerCP IgG1
10. FITC anti-CD4 + PerCP anti-CD8
11. FITC anti-CD4 + PerCP anti-CD8

Vortex gently and incubate the three sets of 11 tubes at 4°C for 30 min covered with aluminum foil. Add 2 mL of cold RPMI 1640 + 20% FCS to each of the 33 tubes. Mix and centrifuge at 150g for 10 min at 4°C. Decant the supernatant fluid.

3.7. Fixation of Cells

1. Prepare fresh 4% paraformaldehyde in PBS, pH 7.4, in a fume hood. Make sure you check the pH prior to use.
2. Add 100 µL of the 4% paraformaldehyde to each of the 33 tubes.
3. Mix and incubate the tubes at 4°C for 20 min.
4. Add 2 mL of PBS, pH 7.4, containing 1% FCS and 0.1% (w/v) sodium azide to each of the 33 tubes.
5. Centrifuge at 150g for 10 min at 4°C.

3.8. Permeabilization of Cells

Resuspend cells in 50 µL of a freshly prepared solution of permeabilizing buffer (PBS containing 1% FCS, 0.1% sodium azide, 0.1% saponin, pH 7.4).

3.9. Staining of Cells

Add the following PE-conjugated antibodies (see the protocol below for tubes 1–11) in a volume of 5 µL. Again, most commercial sources recommend 1.0 µg/mL MAb for 10^6 cells. We routinely use 0.25×10^6 cells and add 0.1 µg of MAb in the permeabilization buffer (this buffer is needed because the effect of saponin is reversible and therefore requires its continuous presence for the MAb to gain intracellular access), which is more than sufficient. Incubate the cells at 4°C for 30 min. Wash the

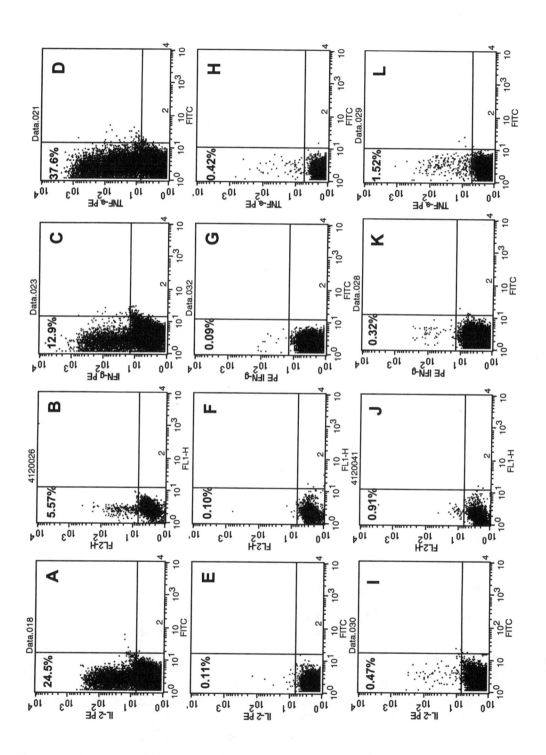

cells twice with 2 mL of cold permeabilization buffer and after the last wash, resuspend the cells in 0.5 mL of PBS containing 1% FCS and 0.1% sodium azide (w/v). Perform fluorescein-activated cell sorter (FACS) analysis on each tube.

1. FITC IgG1 + PerCP IgG1 + PE-conj. rat IgG2a
2. FITC IgG1 + PerCP IgG1+ PE conj. rat IgG1
3. FITC IgG1 + PerCP IgG1 + PE-conj. mouse IgG1
4. FITC anti-CD4 + PerCP anti-CD8 + PE-conj. anti-IL-2
5. FITC anti-CD4 + PerCP anti-CD8 + PE conj. anti-IL-4
6. FITC anti-CD4 + PerCP anti-CD8 + PE-conj. anti-IL-10
7. FITC anti CD4 + PerCP anti-CD8 + PE-conj. anti-TNFα
8. FITC anti-CD4 + PerCP anti-CD8 + PE-conj. anti-IFN-γ
9. FITC IgG1 + PerCP IgG1 + PE conj. rat IgG2a
10. FITC anti-CD4 + PerCP anti-CD8 + PE-conj. anti-IL-6
11. FITC anti-CD4 + PerCP anti-CD8 + PE-conj. anti-IL-12

Tubes 1, 2, 3, and 9 are isotype controls and need to be used to compare the profile of each of the appropriately stained cell preparation. Analysis is straightforward, a total of 10,000 cells are analyzed per sample. After forward and side scatter gates are set, one can gate on lymphocytes and obtain frequency of CD4 and CD8 lymphoid cells that express the relevant cytokine, and the mean channel ratio (MCR) provides the relative mean intensity of expression of the cytokine. In the case of select cytokines such as IL-6 and IL-12, one can gate on monocytes and obtain the frequency of monocytes that express such cytokines. Alternatively, one can use CD14 and gate on CD14$^+$ cells and obtain similar data. Finally, one can gate on all lymphoid and monocytes and obtain data on the relevant frequency of the appropriate subset synthesizing a particular cytokine (*see* **Note 6**) (**Fig. 2**).

4. Notes

1. Storage of reagents: Certain reagents such as PE- and PerCP-conjugated antibodies cannot be frozen and need to be stored at 4°C and in the dark. Most fluorochrome-conjugated proteins need to be stored in the dark, and cell samples stained with such reagents need to be stored in the dark. Certain reagents such as Brefeldin-A, and paraformaldehyde can be prepared as stock solutions, for example, at 20X and then diluted to the desired concentration, pH adjusted just prior to use and kept ice cold. As a rule of thumb, most antibodies do not perform well when repeatedly frozen and thawed. It is ideal to determine how many samples one will be staining per unit time and aliquot the antibodies in a volume that will be required for one staining protocol and freeze this volume in bullet tubes kept at –70°C.
2. Samples from patients with infectious diseases: Platelet abnormalities and/or clotting

---

Fig. 2 *(opposite page)*. PBMCs from a malaria-exposed individual were incubated overnight with either PMA + ionomycin (**A–D**) or an irrelevant OVA peptide (**E–H**) or 10 μM of a previously defined reactive malaria peptide (LSA13) (**I–L**) and stained for cell surface markers and for intracellular cytokines such as cytokines IL-2 (**A, E,** and **I**), IL-6 (**B, F,** and **J**), IFN-γ (**C, G,** and **K**) and TNF-α (**D, H,** and **L**), according to the protocol described in the text above. The numbers within each graph define the frequency of PBMCs that are synthesizing each of the cytokines in response to PMA + ionomycin, an irrelevant peptide, and the malaria peptide. A similar analysis can be performed by gating on specific subsets of the PBMCs such as CD3 (all T cells), CD4 (helper T cells), and CD8 (cytotoxic T cells), and so forth.

abnormalities are often encountered in samples from patients with select diseases. It is recommended that if the blood sample is either hemolyzed or contains microclots that this sample not be used for study, as this may result in skewing of the data. Also if the blood is collected in acid citrate dextrose (ACD), make sure that the blood is diluted in Ca^{2+}-, Mg^{2+}-free Hanks' balanced salt solution (HBSS), otherwise the whole sample will clot.

3. Use of agents that modulate cell surface molecules: It is known that use of agents such as PMA alone and/or in combination with ionophore leads to marked down regulation of the CD4 molecule which is sustained for considerable period of time. Thus, studies of intracellular cytokine expression by $CD4^+$ T cells using PMA + ionomycin-induced cell activation poses a problem. One can infer the frequency of $CD4^+$ T cells by deducting the values obtained of the $CD8^+$ T cells from those obtained with the $CD3^+$ (total) T cells with obvious limitations to the interpretation of the data. Similar caution is advised with the use of Brefeldin-A in the above protocol because it is known that Brefeldin-A downmodulates CD-14 expression and if used for long period of time leads to toxicity.

4. Use of Brefeldin-A: There are two issues with regards to Brefeldin A use. First of all, it is known that Brefeldin-A treatment of the cells will not only trap cytokines but will also inhibit presentation of protein antigens by major histocompatibility complex (MHC) molecules *(12–14)*. Thus, if one is studying intracellular cytokine profile of T cells activated with appropriate antigen and/or peptide pulsed APC, it would be only appropriate to use a cell line, B cells, or other non-T cell APC that has been previously pulsed with the antigen-peptide and add it to the PBMC cultures during the activation step with Brefeldin-A. Second, one needs to titrate the amount of Brefeldin-A based on the time interval selected for the detection of intracellular cytokines. The longer the culture period, the lower the dose of Brefeldin-A should be used otherwise it leads to cell death.

5. Optimal time for analysis: There is clearly a compromise in the interval of time that is required for optimal detection of the various cytokines. The extreme examples in our hands are the cytokines TNF-α, IFN-γ, and IL-4. TNF-α and IFN-γ signals are generated rather rapidly (within 4–6 h of cell activation), whereas the signals for IL-4 are optimally observed after 24–48 h of cell activation. Thus, one has to be careful in the analysis and interpretation of data. This is the reason our laboratory has chosen a compromise time interval of 12–14 h of cell activation. Again, controls are the key issue. For example, if samples can be obtained from a person before and after an experimental procedure, it is ideal as the person serves as his/her own control. If a lot of samples are going to be analyzed, such as a population study, it is recommended that a pool of PBMC be obtained (such as by leukopheresis) and aliquots cryopreserved. For each experimental analysis, a cryopreserved aliquot can thus be thawed out and run in parallel for quality control.

6. Cytokine inhibitors, receptor antagonists, and virally coded cytokine inhibitors and homologues: It is becoming increasingly clear that there are a number of issues that one needs to be aware of during the course of cytokine measurements. There is now clear evidence that besides the previously recognized soluble receptors that exist in the sera of some individuals (particularly those with certain diseases) that complex cytokines, there are plasma proteins that are known to interfere with cytokine synthesis and assays *(15,16)* including our personal observations for the presence of such inhibitory proteins in the assays for human IL-4, IFN-γ, and IL-18. In addition, depending on the sample source and the viruses inherent in the sample, one needs to be aware that there are proteins encoded by viruses that bind to select cytokines such as the IL-1β binding protein encoded by pox viruses (a problem when one uses vaccinia constructs for antigen presentation by APCs and performs IL-1 analysis of the cocultures), and viral homologues for cytokine and cytokine receptors such as IL-10 encoded by Epstein-Barr virus (EBV) and macrophage

inflammatory protein-1α (MIP-1α) encoded by cytomegalovirus (CMV) for example *(17–20)*.

References

1. Wecker, E., Schimpl, A., Hunig, T., and Kuhn, L. (1975) A T-cell produced mediator substance active in the humoral immune response. *Ann. N.Y. Acad. Sci.* **249,** 258–263.
2. Holbrook, N. J., Smith, K. A., Fornace, A. J., Jr., Comeau, C. M., Wiskocil, R. L., and Crabtree, G. R. (1984) T-cell growth factor: complete nucleotide sequence and organization of the gene in normal and malignant cells. *Proc. Nat. Acad. Sci. USA* **81,** 1634–1638.
3. Smith, K. A. (1984) Interleukin-2. *Ann. Rev. Immunol.* **2,** 319–333.
4. Czerkinsky, C., Andersson, G., Ekre, H. P., Nilsson, L. A., Klareskog, L., and Ouchterlony, O. (1988) Reverse ELISPOT assay for the clonal analysis of cytokine production I. Enumeration of gamma-interferon secreting cells. *J. Immun. Meth.* **110,** 29–36.
5. Czerkinsky, C., Moldoveanu, Z., Mestecky, J., Nilsson, L. A., and Ouchterlony, O. (1988) A novel two colour ELISPOT assay I. Simultaneous detection of distinct types of antibody secreting cells. *J Immunol. Meth.* **115,** 31–37.
6. Mosmann, T. R. and Coffman, R. L. (1989) Th1 and Th2 cells: different patterns of lymphokines secretions lead to different functional properties. *Ann. Rev. Immunol.* **7,** 145–173.
7. Cherwinski, H. M., Schumacher, J. H., Brown, K. D., and Mosmann, T. R. (1987) Two types of mouse helper T cell clones III. Further differences in lymphokines synthesis between Th1 and Th2 clones revealed by RNA hybridization, functionally monospecific bioassays and monoclonal antibodies. *J. Exp. Med.* **166,** 1229–1244.
8. Nagata, K., Tanaka, K., Ogawa, K., Kemmotsu, K., Imai, T., Yoshie, O., et al. (1999) Selective expression of a novel surface molecule by human Th2 cells in vivo. *J. Immunol.* **162,** 1278–1286.
9. Xu, D., Chan, W. L., Leung, B. P., Hunter, D., Schielz, K., Carter, R. W., et al. (1998) Selective expression and functions of interleukin-18 receptor on T helper cells (Th) type 1 but not Th2 cells. *J. Exp. Med.* **188,** 1485–1492.
10. Xu, D., Chan, W. L., Leung, B. P., Huang, F. P., Wheeler, R., Piedrafita, D., et al. (1998) Selective expression of a stable cell surface molecule on type 2 but not type 1 helper T cells. *J. Exp. Med.* **187,** 787–794.
11. Picker, L. J., Singh, M. K., Zdraveski, Z., Treer, J. R., Walsdrop, S. L., Bergstrom, P. R., et al. (1995) Direct demonstration of cytokine synthesis heterogeneity among human memory/effector T cells by flow cytometry. *Blood* **86,** 1408–1419.
12. Yewdell, J. W. and Bennink, J. (1989) Brefeldin A specifically inhibits presentation of protein antigens to cytotoxic T lymphocytes. *Science* **244,** 1072–1075.
13. Kakiuchi, T., Takatsuki, A., Watanabe, M., and Nariuchi. H. (1991) Inhibition by Brefeldin-A of the specific B cell antigen presentation to MHC-class II restricted T cell. *J. Immunol.* **147,** 3289–3295.
14. Nguyen, Q. V. and King, R. L. (1997) Brefeldin-A induced alterations in processing of MHC-class II-Ii complex depends upon microtubular function. *Am. J. Hematol.* **54,** 282–287.
15. Campbell, D. E., Georgiu, G. M., and Kemp, S. (1999) Pooled human immunoglobulin inhibits IL-4 but not IFN-γ or TNF-α secretion following in vitro stimulation of mononuclear cells with staphylococcal superantigens. *Cytokine* **11,** 359–365.
16. Zhang, J. G., Hilton, D. J., Wilson, T. A., McFarlane, C., Roberts, B. A., Moritz, R. L., et al. (1997) Identification, purification, and characterization of a soluble interleukin (IL)-13 binding protein: evidence that it is distinct from the cloned IL-13 receptor and IL-4 receptor alpha chains. *J. Biol. Chem.* **272,** 9474–9480.
17. Blomquist, M. C., Hunt, L. T., and Barker, W. C. (1984) Vaccinia virus 19-kilodalton protein: relationship to several mammalian proteins including 2 growth factors. *Proc. Nat. Acad. Sci. USA* **81,** 7363–7367.
18. Moore, K. W., O'Garra, A., de Waal Malefyt, R., Viera, P., and Mosmann, T. R. (1993) Interlekin-10. *Ann. Rev. Immunol.* **11,** 15–90.
19. Ahuja, S. K., Gao, J. L., and Murphy, P. M. (1994) Chemokine receptors and molecular mimicry. *Immunol. Today* **15,** 281–287.
20. Neote, K., DiGregorio, D., Mak, J. Y., Horuk, R., and Schall, T. J. (1993) Molecular cloning, functional expression and signaling characteristics of a C-C chemokine receptor. *Cell* **72,** 415–425.

38

Assessment of Antigen-Specific CTL- and CD8⁺-Dependent IFN-γ Responses in Mice

Katrin Peter, Régine Audran, Anilza Bonelo, Giampietro Corradin, and José Alejandro López

1. Introduction

Mosquito injection of *Plasmodium* sporozoites into the bloodstream is followed by a short transit period before invasion of the hepatocyte, the only host cell expressing class I major histocompatibility complex (MHC) molecules. Liver stages develop within the hepatocyte for periods that vary according to the species, and they express various stage-specific proteins. Finally, schizonts are released into the bloodstream and invade the red blood cells, the final host cell before the continuation of the parasite cycle upon a subsequent mosquito bite.

Protection to malaria could be efficiently obtained by immunization with irradiated sporozoites, a procedure that prevents parasite development upon repeated challenge with live sporozoites *(1,2)*. Studies in human volunteers and animal models have shown that protection is dependent on both antibodies and T-cell responses *(3–8)*. Among the targets for the immune response are the antigens expressed during the intrahepatic part of the life cycle, potentially presented in the context of MHC class I molecules to CD8⁺ lymphocytes. Moreover, CD8⁺ lymphocyte activity has been shown to be sufficient in protecting against disease in a mouse model *(9)*.

CD8⁺ lymphocyte activity is classically described as cytolytic, hence their designation of cytolytic T lymphocytes (CTL). Specific CTL encounter target cells presenting antigen (peptides) by a MHC class I molecule and lyse them. This process is either the result of the release of molecules such as perforin or the activation of death signals, such as those mediated by CD95/Fas and the interaction with its ligand *(10)*. This cytolytic function can be evaluated with the chromium release assay, an assay first developed in 1968 by Brunner et al. *(11)*, which continues to be the most specific detection and a sensitive system for the measurement of cytolytic activity.

Additionally to the lytic activity, CTL also release interleukins upon antigen recognition; moreover, in the case of mouse malaria, the protective effect of CD8⁺ lymphocytes appears to be mediated mainly through the release of interferon-γ (IFN-γ), and not through direct lysis of the infected cells. FAS and/or perforin-deficient mice are equally protected from *Plasmodium berghei* infection upon immunization with irradiated sporozoites *(12)*. For this reason, the evaluation of the protective effect of CD8⁺

lymphocytes in the course of malaria should take into account the production of interleukins, particularly IFN-γ upon antigen recognition. One sensitive assay currently available to fulfill this task is the enzyme-linked immunospot (ELISPOT), a double-sandwich enzyme-linked immunosorbent assay (ELISA)-based method that detects the production of interleukins upon antigen-specific activation *(13)*.

Despite the fact that an effective CTL response is elicited upon immunization with short synthetic peptides *(14)*, protection is achieved only when specific lymphocytes are passively transferred before the sporozoite challenge *(15)*. This finding suggests that a minimal number of specific CTL should be located in the liver at the time of sporozoite invasion in order to be able to block the development of the infection. This finding highlights the need for evaluating the presence of malaria specific lymphocytes in the liver. Detection of liver lymphocytes is now feasible *(16)*, and also the study of their function has become available *(17)*. Antigen-specific lymphocytes can efficiently reach the liver upon an infection with *P. berghei* or *P. yoelii* and their antigen-specific activity can be evaluated (V. Meraldi and G. Corradin, unpublished data).

Various proteins of the liver stages appear to be implicated in the protective effect of the immunization with irradiated sporozoites. CTL epitopes have been described in the circumsporozoite (CSP), thrombospondin-related anonymous protein/sporozoite surface antigen 2 (TRAP/SSP2), liver-stage antigen (LSA) proteins, and others. The function of CSP in the immune response has been particularly studied; partial or complete protection in the mouse model may be conferred by CSP specific $CD8^+$ or $CD4^+$ T cells *(9,15,18,19)*, and their role in controlling malaria infection is well documented (reviewed in **ref. 20**).

The study of the CTL lymphocyte function in mice is now facilitated with the existence of valuable reagents and standardized techniques. In this chapter, we describe in detail two methods for the study of the malaria-specific CTL function in mice: the chromium release assay and the IFN-γ ELISPOT. We also describe the cell preparations necessary for these assays.

2. Materials

Note that the catalog numbers in parentheses indicate items used by the authors.

2.1. Cell Isolation and Cell Culture

2.1.1. Lymph Node Cell Preparation and Culture

1. Culture medium (CM): Dulbecco's modified essential medium (DMEM) with glutamax-I and high glucose (cat. no. 31966-021, Gibco-BRL, Life Technologies, Paisley, UK) supplemented with 10% fetal calf serum (FCS) (cat. no. S0115, Seromed®, Biochrom KG, Berlin, Germany), 10 mM HEPES buffer (cat. no. 15630-056, Gibco-BRL), 50 µM ß-mercaptoethanol (cat. no. 805740, Merck-Schuchardt, Hohenbrunn, Germany), 100 IU/mL penicillin-streptomycin (cat. no. 15140-114, Gibco-BRL), and 30 U/mL interleukin-2 (IL-2) (supernatant from EL-4 mouse cell cultures, **ref. 21**).
2. Wash medium: DMEM with glutamax-I and high glucose (cat. no. 31966-021, Gibco-BRL) supplemented with 2% FCS (cat. no. S0115, Seromed®) and 10 mM HEPES buffer (cat. no. 15630-056, Gibco-BRL).
3. 24-Well tissue culture-treated polystyrene plates (cat. no. 3524, Costar®, Corning, NY).
4. Cell strainer, 70 µm (cat. no. 2350, Falcon®, Becton Dickinson, Franklin Lakes, NJ).
5. Tissue culture dish 60 × 15 mm (EASY GRIP™, Falcon®, cat. no. 3004).

CD8+ T-Cell Responses: Murine CTL Assays

6. 15-mL Disposable conical polypropylene tubes (cat. no. 366060, NUNC™, Nalge Nunc International Corp., Naperville, IL).
7. Syngeneic cells as antigen-presenting cells (APC): i.e., P815 (plasmacytoma cell line, H-2^d haplotype) in logarithmic growth phase.
8. γ-Irradiation source.

2.1.2. Spleen Cell Preparation and Culture

1. Culture medium: Use CM medium without IL-2.
2. Wash medium: Use same medium as for lymph nodes.
3. 6-Well tissue culture-treated polystyrene plates (cat. no. 3506, Costar®).
4. Cell strainer 70 μm (cat. no. 2350, Falcon®).
5. Tissue culture dish 60 × 15 mm (EASY GRIP™, cat. no. 3004, Falcon®).
6. 15-mL disposable conical polypropylene tubes (cat. no. 366060, NUNC™).

2.1.3. Liver Cell Preparation

1. PBS (10X) (cat. no. 14200-067, Gibco-BRL).
2. Percoll™ (cat. no. 17-0891-01, Amersham Pharmacia Biotech AB, Uppsala, Sweden).
3. DMEM with glutamax-I and high glucose (cat. no. 31966-021, Gibco-BRL).
4. FCS (cat. no. S0115, Seromed®).
5. Cell strainer, 70 μm (cat. no. 2350, Falcon®).
6. Culture dish, 94 × 16 mm (cat. no. 46185, Biocon, Paul Boettger, Bodenmais, Germany).
7. 50-mL disposable conical polypropylene tubes (cat. no. 2070, Falcon®).

2.2. Assessment of CTL Activity

2.2.1. Chromium Release Assay

1. Assay medium: DMEM with glutamax-I and high glucose (cat. no. 31966-021, Gibco-BRL) supplemented with 5% FCS (cat. no. S0115, Seromed®) and 10 mM HEPES buffer (cat. no. 15630-056, Gibco-BRL).
2. Wash medium: As described above for lymph node preparation.
3. 96-Well V-bottom microplate (cat. no. 651180, Cellstar®, Greiner Labortechnik, Frickenhausen, Germany).
4. Sodium chromate (^{51}Cr), 1 mCi/mL (cat. no. NEZ-030S, Steri-Pack, NEN™, Boston, MA).
5. Labeling buffer: 2 mg/mL bovine serum albumin (BSA) PBS.
6. Syngeneic target cells: i.e., P815 in logarithmic growth phase.
7. γ-Irradiation counter.

2.2.2. IFN-γ ELISPOT Assay

1. Assay medium: Use CM medium.
2. Saturation medium: DMEM with glutamax-I and high glucose (cat. no. 31966-021, Gibco-BRL) supplemented with 10% FCS (cat. no. S0115, Seromed®) 10 mM HEPES buffer (cat. no. 15630-056, Gibco-BRL).
3. 96-Well sterile filtration plate (cat. no. MAHAS4510, Multiscreen®-HA Millipore S.A. Molsheim, France).
4. PBS (10X) (cat. no. 14200-067, Gibco-BRL).
5. Polyoxyethylene-sorbitane Monolaurate (Tween 20) (cat. no. P-5927, Sigma, St. Louis, MO).
6. BSA (cat. no. 05490, Fluka Chemie AG Buchs, Switzerland).
7. Primary antibody anti-mouse IFN-γ (rat IgG1): produced from Hybridoma cell line R4-6A2 (American Type Culture Collection, Manassas, VA).
8. Secondary antibody anti-mouse IFN-γ AN-18.17.24 *(22)* biotinylated with sulfo-NHS-

LC-biotin (Pierce, Rockford, IL).
9. Streptavidin alkaline phosphatase-conjugate (cat. no. 1089 161, Boehringer Mannheim, Germany).
10. 5-Bromo-4-chloro-3-indolyl phosphate/nitro blue tetrazolium tablets (cat. no. B-5655, Sigma Fast™, Sigma).
11. Syngeneic antigen-presenting cells (APC).
12. Filter, 0.2 μm (cat. no. 16534, Sartorius, Sartorius AG, Goettingen, Germany).
13. γ-irradiation source.
14. Stereomicroscope.

3. Methods

Evaluation of the cytotoxic T-lymphocyte (CTL) response in mice involves a prior in vivo stimulation with antigen (e.g., synthetic peptide vaccination; *see* Chapter 30). With the ELISPOT assay, the CTL activity can often be evaluated directly ex vivo. However, the classical chromium release assay is less sensitive and the frequency of CTL is often too low to be directly detected. This problem can be overcome by in vitro stimulation of bulk cultures with either the antigen directly (spleen) or with a population of stimulating cells presenting the desired antigen (e.g., lymph nodes) in order to expand the specific CTL to amounts that are detectable. In some cases several rounds of in vitro stimulation may be necessary in order to detect a CTL activity. The following sections describe how to isolate cells from different organs and how to culture them where necessary for subsequent evaluation of CTL activity.

3.1. Preparation and Culture of Lymph Node Cells
3.1.1. Primary Lymph Node Cell Cultures

1. Sacrifice mice and remove inguinal and periaortic lymph nodes under sterile conditions.
2. Keep lymph nodes in a 5-mL vol of wash medium in a small tissue culture dish on ice until further processing (*see* **Note 1**).
3. Transfer lymph nodes into a cell strainer, break them on the strainer with the help of a sterile plastic syringe piston and collect them into 5 mL of wash medium in a small tissue culture dish.
4. Transfer cell suspension into a 15-mL conical tube. Rinse cell strainer and tissue culture dish once with 5 mL of wash medium and combine the two suspensions.
5. Let tissue debris sediment for 10–15 min and transfer supernatant to a fresh 15-mL conical tube.
6. Centrifuge at 450g for 5 min and wash cells twice more with 5 mL of wash medium (*see* **Note 2**).
7. Resuspend lymph node cells in culture medium at a concentration of 5×10^6 cells/mL in culture medium and aliquot 1 mL per well in a 24-well plate.
8. Pulse syngeneic APC with a final concentration of 1 μM peptide (antigen) for 1 h at 37°C. Use approx 1×10^6 cells/mL and gently resuspend them once or twice during incubation period.
9. Wash APC two to three times with 5 mL of wash medium, at 450g for 5 min.
10. Irradiate cells with 10,000 rad.
11. Centrifuge cells and resuspend APC in culture medium at a concentration of 2×10^5 cells/mL. Add 1 mL to each well containing the lymph node cells.
12. Incubate cultures at 37°C and 5% CO_2 for 7 d (*see* **Note 3**).

3.1.2. In Vitro Restimulation of Primary Lymph Node Cell Cultures

1. Transfer cells from primary lymph node cell culture (*see* **Subheading 3.1.1.**) into a 15-mL conical tube, centrifuge and wash twice more with 5 mL of wash medium (*see* **Note 2**).
2. Resuspend cells at a concentration of 5×10^5 cells/mL in culture medium and aliquot 1 mL

per well in a 24-well plate.
3. Pulse syngeneic APC with 1 µM peptide (antigen) for 1 h at 37°C. Use approx 1×10^6 cells/mL and gently resuspend them once or twice during incubation period.
4. Wash APC twice with 5 mL of wash medium and irradiate them with 10,000 rad.
5. Centrifuge cells and resuspend them at a concentration of 4×10^5 cells/mL in culture medium. Add 2×10^5 cells in a volume of 0.5 mL to each well containing the CTL.
6. Remove spleen from syngeneic mouse under sterile conditions and break organ on a cell strainer as described in **Subheading 3.2.1.** Wash cells twice with 5 mL of wash medium.
7. Irradiate spleen cells with 5000 rad, centrifuge and resuspend them at a concentration of 10×10^6 cells/mL in culture medium. Add 5×10^6 cells/well in a volume of 0.5 mL to wells containing CTL and APC.
8. Keep cultures at 37°C and 5% CO_2. After 2–3 d, cells will need to be diluted (*see* **Note 4**).

3.2. Preparation and Culture of Spleen Cells

3.2.1. Primary Spleen Cell Cultures

1. Sacrifice mice and remove spleen under sterile conditions. Keep spleen in 10 mL of wash medium in a 15-mL conical tube on ice until further processing (*see* **Note 1**).
2. Transfer spleen into a cell strainer, break the organ with the help of a sterile plastic syringe piston, and collect the cells into 5 mL of wash medium in a small tissue culture dish.
3. Transfer cell suspension into a 15-mL conical tube. Rinse cell strainer and tissue culture dish with 5 mL of wash medium and combine the two suspensions.
4. Let tissue debris sediment for 10–15 min and transfer supernatant to a fresh 15-mL conical tube.
5. Centrifuge at 450g for 5 min and wash two times more (*see* **Note 2**).
6. Resuspend cells at a concentration of 6×10^6 cells/mL in culture medium. Plate cells in a 6-well plate at a concentration of 30×10^6 cells per well in a volume of 5 mL culture medium.
7. Add peptide (antigen) to the culture (final concentration 1 µM).

3.2.2. In Vitro Restimulation of Primary Spleen Cell Cultures

The restimulation of primary spleen cell cultures is performed in a similar way as described above for restimulation of primary lymph node cell cultures, using CM medium (*see* **Subheading 3.1.2.**).

3.3. Preparation and Culture of Liver Cells

1. Sacrifice mice and cut portal vein to get rid of the blood.
2. Remove liver under sterile conditions and keep in a 50-mL conical tube containing 10 mL of PBS (4°C) until further processing.
3. Transfer liver into a cell strainer, break organ on the strainer with the help of a sterile plastic syringe piston, and collect cells in 20 mL of PBS in a tissue culture dish.
4. Transfer cell suspension into a 50-mL tube. Rinse cell strainer and tissue culture dish with 10 mL of PBS and combine the two suspensions.
5. Centrifuge for 5 min at 800g at 4°C and wash once with 50 mL of PBS.
6. Resuspend the pellet in 15 mL of 40% Percoll/DMEM, mix well, and layer carefully onto 15 mL of 80% Percoll/DMEM (Percoll 100% = 90 mL Percoll and 10 mL of 10X PBS). DMEM and Percoll should be brought to room temperature before use.
7. Centrifuge for 10 min at 580g at room temperature, without brake.
8. Eliminate the upper part of the Percoll, which contains the hepatocytes, and take the remaining part of the liquid.
9. Centrifuge and wash twice in 50 mL of PBS/5% FCS. Centrifuge at 800g at 4°C for 5 min.
10. Resuspend cells in 1 mL of assay medium for ELISPOT analysis. The cell recovery should be approx $2-4 \times 10^6$ cells per liver.

3.4. Assessment of CTL Activity

3.4.1. Chromium Release Assay

This assay is based on the principle that target cells highly susceptible to CTL mediated lysis are labeled with ^{51}chromium (^{51}Cr). They are plated at 1000–2000 cells per well in microtiter plates. To the same wells, specific effector cells (CTL) in an effector:target (E:T) ratio ranging from 0.1 to 200 are added. After a 4-h incubation period, the released ^{51}Cr is measured. Each effector cell is estimated to eliminate five target cells. To obtain a significant lysis (i.e., 50% of target cells) it is therefore necessary to have at least 100–200 specific CTL effectors. Unfortunately, the frequency of CTL precursors in cell suspensions directly isolated from organs is too low to be directly detected by this method. This problem can be overcome by stimulating the bulk cultures in vitro with antigen in order to expand the specific CTL to amounts to be detected by the chromium assay. However, the assay provides only a qualitative indication of the presence of specific CTL activity but no information about the frequency of specific CTL.

1. Resuspend target cells in 30 µL of labeling buffer (ideally approx 0.5×10^6 cells). Add 50 µL of fresh ^{51}Cr and incubate at 37°C for 1 h in the presence (1 µM) or absence of peptide (*see* **Notes 5–7**).
2. Transfer appropriate number of CTL to a conical 15-mL tube, centrifuge, wash twice with wash medium, and determine cell count.
3. Make different cell dilutions of the CTL in a Greiner 96-well plate in assay medium. Start with approx $2-3 \times 10^5$ cells and make 5–8 1:3 dilutions in final volumes of 100 µL per well.
4. Wash target cells three times with 5 mL of wash medium. Resuspend cells in assay medium at a concentration of 1×10^4 cells per mL. Add 100 µL of suspension (1000 cells) to each well containing effector cells.
5. For a control, use four wells with target cells and 100 µL medium (spontaneous release) and 4 wells with target cells plus 100 µL of 1 M HCl (total release).
6. Incubate the effector plus target cell culture at 37°C for 4 h.
7. Centrifuge the plate, remove 100 µL of the supernatant and count in a gamma counter.
8. Calculate the percent lysis as $100 \times$ [(experimental – spontaneous release)/(total – spontaneous release)].

The results of the chromium release assay can either be expressed as lytic units (LU) or as graphs of the percentage of lysis versus E:T ratios or effector cell numbers. LU provide a quantitative comparison of the relative activities of different effector cell populations, whereas the latter approach allows qualitative ranking of CTL activities. An LU is arbitrarily defined as the number of lymphocytes required to yield a selected lysis value. For this purpose the percent lysis versus the effector cell number is plotted and a lysis value (e.g., 30%) is selected through which most of the declining titration curves pass. Ideally, titration plots for each of the different sources of effector cells should be straight and parallel in this region. The LU per 10^6 cells is extrapolated from the value of the effector cell number. For example, if 5×10^4 cells from a particular sample give 30% lysis, then the sample has 20 LU per 10^6 cells.

3.5. IFN-γ ELISPOT Assay

The ELISPOT assay takes advantage of the local concentration of cytokine released around an activated T cell and allows the detection and enumeration of single cytokine-secreting cells. Cells plated into wells of a nitrocellulose-lined microtiter plate that have been preincubated with a cytokine-specific monoclonal antibody are stimulated with antigen-pulsed APC to trigger the specific secretion of cytokines. To detect cytokines as discrete spots, an alkaline phosphatase-labeled antibody is used. In general, each spot represents a single cytokine-producing cell. This assay is approximately 30–300 times more sensitive than the chromium release assay and a frequency of 1 cell in 20,000 to 40,000 can be detected. In addition, the assay is fast because in vitro stimulation may not be required.

Steps 1–7 are performed under sterile conditions.

1. Coat assay plate with 50 µL of primary antibody in sterile PBS at optimal concentration.
2. Incubate overnight at 4°C in humid environment.
3. Add 150 µL of saturation medium and incubate for 1–2 h at 37°C.
4. Wash three times with PBS.
5. Wash CTL twice with 5 mL of wash medium and add cells in a 100 µL volume to plates, making 5–8 1:5 dilutions. Start with approx 1×10^6 cells.
6. Pulse APC with 1 µM peptide for 1 h at 37°C. As a control, use nonpulsed APC. Irradiate APC with 10,000 rad and add to wells containing cells (1×10^5 cells per well).
7. Incubate for 24 h at 37°C and 5% CO_2.
8. Wash plate three times with PBS/0.01% Tween-20 (*see* **Note 8**).
9. Wash three times with PBS.
10. Add 50 µL of secondary antibody in PBS/1% BSA.
11. Incubate for 2 h at 37°C.
12. Wash plate three times with PBS/0.01% Tween-20.
13. Wash three times with PBS.
14. Add 50 µL of streptavidin alkaline phosphatase (1:1000 in PBS) and incubate for 30 min at 37°C.
15. Wash plate three times with PBS/ 0.01% Tween-20.
16. Wash three times with PBS.
17. Dissolve Sigma Fast™ tablets in distilled water (1 tablet per 10 mL), filter through 0.2 µm and add 50 µL to each well. (Tablets are more easily dissolved if kept at 37°C for 1 h.)
18. Incubate at least 5 min at room temperature until you see spots. Stop the reaction by rinsing the plate under tap water.
19. Let the plate dry and count spots using a stereomicroscope or with a computer assisted video image analysis, which allows a more objective enumeration of spots. Specific spots have well-defined dark blue centers, and blurred contours, whereas spots from cell debris are much smaller with sharper polygonal contours.

4. Notes

1. During the whole process of cell preparation, lymph node and spleen cells should be kept on ice as much as possible.
2. Cells are always centrifuged at 450g for 5 min, unless otherwise indicated.
3. For the first week of in vitro culture, it is normally not necessary to dilute the cells.
4. CTL are better cultured in relatively high density (0.7–1.0 $\times 10^6$ cells/mL). Dilutions of cultures should therefore be performed in small steps (1:2 or 1:3).
5. Sterile conditions are not necessary for the chromium release assay since only a 4-h incubation period is needed.
6. The ^{51}Cr labeling is done according to standard security measures concerning the use of radioelements.
7. The ^{51}Cr has a half-life of 28 d. If the chromium stock is kept for several weeks, the amount added for the target cell labeling should approximately be doubled each week. Alternatively, the counting time could be proportionally increased.
8. For the washing steps, plates can be directly dipped into a bath of PBS or PBS-Tween-20.

References

1. Clyde, D. F., Most, H., McCarthy, V. C., and Vanderberg, J. P. (1973) Immunization of man against sporozoite-induced falciparum malaria. *Am. J. Med. Sci.* **266,** 169–177.
2. Nussenzweig, R. S.,Vanderberg, J., Most, H., and Orton, C. (1967) Protective immunity produced by the injection of X-irradiated sporozoites of *Plasmodium berghei. Nature* **216,** 160–162.
3. Nardin, E. H. and Nussenzweig, R. S. (1993) T cell responses to pre-erythrocytic stages of malaria: role in protection and vaccine development. *Annu. Rev. Immunol.* **11,** 687–727.
4. Tsuji, M., Mombaerts, P., Lefrancois, L., Nussenzweig, R. S., Zavala, F., and Tonegawa, S. (1994) Gamma delta T cells contribute to immunity against the liver stages of malaria in alpha beta T-cell-deficient mice. *Proc. Natl. Acad. Sci. USA* **91,** 345–349.
5. Weiss, W. R., Mellouk, S., Houghten, R. A., Sedegah, M., Kumar, S., Good, M. F., et al. (1990) Cytotoxic T cells recognize a peptide from the circumsporozoite protein on malaria-infected hepatocytes. *J. Exp. Med.* **171,** 763–773.
6. Schofield, L., Villaquirán, J., Ferreira, J. A., Schellekens, H., Nussenzweig, R. S., and Nussenzweig, V. (1987) Gamma interferon CD8+ T cells and antibodies required for immunity to malaria sporozoites. *Nature* **330,** 664–666.
7. Malik, A., Egan, J. E., Houghten, R. A., Sadoff, J. C., and Hoffman, S. L. (1991) Human cytotoxic T lymphocytes against the *Plasmodium falciparum* circumsporozoite protein. *Proc. Natl. Acad. Sci. USA* **88,** 3300–3304.
8. Tsuji, M., Romero, P., Nussenzweig, R. S., and Zavala, F. (1990) CD4+ cytolytic T cell clone confers protection against murine malaria. *J. Exp. Med.* **172,** 1353–1357.
9. Romero, P., Maryanski, J. L., Corradin, G., Nussenzweig, R. S., Nussenzweig, V., and Zavala, F. (1989) Cloned cytotoxic T cells recognize an epitope in the circumsporozoite protein and protect against malaria. *Nature* **341,** 323–325.
10. Lowin, B., Hahne, M., Mattmann, C., and Tschopp, J. (1994) Cytolytic T-cell cytotoxicity is mediated through perforin and Fas lytic pathways. *Nature* **370,** 650–652.
11. Brunner, K. T., Mauel, J., Cerottini, J.- C., and Chapuis, B. (1968) Quantitative assay of the lytic action of immune lymphoid cells on 51-Cr-labelled allogeneic target cells in vitro; inhibition by isoantibody and by drugs. *Immunology* **14,** 181–196.
12. Renggli, J., Hahne, M., Matile, H., Betschart, B., Tschopp, J., and Corradin, G. (1997) Elimination of *P. berghei* liver stages is independent of Fas (CD95/Apo- I) or perforin-mediated cytotoxicity. *Parasite Immunol.* **19,** 145-148.
13. Czerkinsky, C., Andersson, G., Ekre, H. P., Nilsson, L. A., Klareskog, L., and Ouchterlony, O. (1988) Reverse ELISPOT assay for clonal analysis of cytokine production. Enumeration of gamma-interferon-secreting cells. *J. Immunol. Methods* **110,** 29-36.
14. Widmann, C., Romero, P., Maryanski, J. L., Corradin, G., Valmori, D. (1992) T helper epitopes enhance the cytotoxic response of mice immunized with MHC class-I-restricted malaria peptides. *J. Immunol. Methods* **155,** 95–99.
15. Renggli, J., Valmori, D., Romero, J. F., Eberl, G., Romero, P., Betschart, B., and Corradin, G. (1995) CD8+ T-cell protective immunity induced by immunization with *Plasmodium berghei* CS protein-derived synthetic peptides: evidence that localization of peptide-specific CTL is crucial for protection against malaria. *Immunol. Lett.* **46,** 199–205.
16. Ohteki, T., Seki, S., Abo, T., and Kumagai, K. (1990) Liver is a possible site for the proliferation of abnormal CD3+4-8- double-negative lymphocytes in autoimmune MRL-lpr/lpr mice. *J. Exp. Med.* **172,** 7–12.
17. Walker, P. R., Ohteki, T., Lopez, J. A., MacDonald, H. R., and Maryanski, J. L. (1995) Distinct phenotypes of antigen-selected CD8 T cells emerge at different stages of an in vivo immune response. *J. Immunol.* **155,** 3443–3452.
18. Migliorini, P., Betschart, B., and Corradin, G. (1993) Malaria vaccine: immunization of mice with a synthetic T cell helper epitope alone leads to protective immunity. *Eur. J. Immunol.* **23,** 582–585.
19. Roggero, M. A., Filippi, B., Church, P., Hoffman, S. L., Blum-Tirouvanziam, U., López, J. A., et al. (1995) Synthesis and immunological characterization of 104-mer and 102-mer peptides corresponding to the N- and C- termianl regions of the *Plasmodium falciparum* CS protein. *Mol. Immunol.* **32,** 1301–1309.
20. Hoffman, S. L., Franke, E. D., Hollingdale, M. R., and Druilhe, P. (1996) Attacking the infected hepatocyte, in *Malaria Vaccine Development. A Multi-Immune Response Approach.* (Hoffman, S. L., ed.), American Society for Microbiology, Washington, DC, pp. 35–75.
21. Farrar, J. J., Fuller-Farrar, J., Simon, P. L., Hilfiker, M. L., Stadler, B. M., and Farrar, W. L. (1980) Thymoma production of T cell growth factor (interleukin-2). *J. Immunol.* **125,** 2555–2558.
22. Prat, M., Gribaudo, G., Comoglio, M., Cavallo, G., and Landolfo, S. (1984) Monoclonal antibodies against murine g-interferon. *Proc. Natl. Acad. Sci. USA* **81,** 4515–4519.

39

Assessing Antigen-Specific CD8+ CTL Responses in Humans

Denise L. Doolan

1. Introduction

Townsend and colleagues (1) were the first to demonstrate, in 1984, that target cells that had been transfected with single viral RNA segments could be specifically recognized by cytotoxic T lymphocytes (CTL). Recombinant vaccinia viruses that expressed single-gene products were subsequently used to identify major target antigens for CTL (2,3). The demonstration that CTL could recognize transfectants that expressed only fragments of a given target antigen (4) led to the use of overlapping fragments and synthetic peptides to localize epitopes. It was subsequently established that target cells that had been incubated with short synthetic peptides corresponding to the sequence of the influenza virus nucleoprotein (NP) could be specifically recognized by NP-specific CTL in a major histocompatability complex (MHC) class I restricted manner (5,6). Bjorkman et al. (6,7) subsequently resolved the structure of the human class I molecule HLA-A2 by x-ray crystallography and provided evidence of a putative peptide binding site. These studies provided the foundation for characterizing CTL epitopes and CTL responses in many disease systems. Numerous CTL epitopes have been now identified, in all cases recognized in the context of a single human lymphocyte antigen (HLA) allele or family of closely related (HLA degenerate) alleles, and the use of synthetic peptides for defining the types of immune responses induced by experimental vaccination or natural exposure to the target pathogen has been now well documented.

In this chapter, I describe methods for the generation and assay of peptide-specific and antigen-specific CTL, and supporting protocols for obtaining effector cells from human peripheral blood and generation of short-term and long-term target cell lines.

2. Materials

2.1. Equipment

1. 37°C, 5% CO_2 incubator.
2. Benchtop centrifuge.
3. Microcentrifuge.
4. 50-mL Polypropylene centrifuge tubes.
5. 15-mL Polypropylene centrifuge tubes.
6. 10-mL Snap-top tubes (Nunc, cat. no. 2059).
7. 1.5-mL Microcentrifuge tubes.

8. 2-, 5-, 10-, and 25-mL Sterile disposable pipettes (individually wrapped).
9. 3-mL Sterile transfer pipets.
10. 96-Well round-bottomed plates.
11. 24-Well flat-bottomed plates.
12. 48-Well flat-bottomed plates (optional).
13. 25 cm^2 (T-25) Flasks.
14. 75 cm^2 (T-75) Flasks (optional).
15. Filter units, 500-mL, 0.22 µm (polycarbonate).
16. Nunc cyrotubes (1.8 mL) (preferably external thread; Nunc, cat. no. 375418).
17. Pipetman P-20, P-200, and P-1000 (P-2, P-10, and P-100, optional).
18. Pipet tips, 10–200 µL and 200–1000 µL, sterile (for Pipetman P-20, P-200, and P-1000).
19. Pipet aid, portable.
20. Hemocytometer (counting chamber) and hemocytometer cover slips.
21. Microscope with 10X or 20X ocular.
22. Counter, laboratory hand-held tally.
23. Tube racks, for 15-mL tubes, 50-mL tubes, and cryotubes.
24. Syringes, 20-mL and/or 50-mL or vacutainers (BD green top, with heparin) (for blood collection).
25. Cryofreezing containers (Mr. Freezy, Nalgene, cat. no. 5100-0001).
26. –70°C – –20°C freezers.
27. Liquid nitrogen storage.
28. Gamma counter.
29. Betaplate liquid scintillation counter (optional).

2.2. Reagents

All reagents are of tissue culture grade.

1. RPMI 1640 medium with HEPES (25 mM) and L-glutamine (2 mM) (Gibco cat. no. 22400-089 or equivalent). Store at 4°C. Supplement with fresh 2 mM L-glutamine every 7 d.
2. Pooled normal human AB serum (NHS), heat inactivated (ICN, cat. no. 29-309-49 or equivalent). Store 10-mL or 50-mL aliquots at –20°C.
3. Fetal calf serum (FCS), heat-inactivated (Sigma, cat. no. F4135 or equivalent). Store 50-mL aliquots at –20°C.
4. Penicillin-streptomycin (100X) (10,000 U/10,000 µg), liquid (Gibco, cat. no. 15140-015 or equivalent). Store 5-mL aliquots at –20°C (or at 4°C for up to 1 mo).
5. Nonessential amino acids, 10 mM (100X), liquid (Gibco, cat. no. 11140-050 or equivalent). Store 5-mL aliquots at –20°C (or at 4°C for up to 1 mo).
6. L-Glutamine, 200 mM (100X), liquid (Gibco, cat. no. 25030-081 or equivalent). Store 5-mL aliquots at –20°C (or at 4°C for no longer than 1 wk).
7. Sodium pyruvate, 100 mM (100X), liquid (Gibco, cat. no. 11360-070 or equivalent). Store 5-mL aliquots at –20°C (or at 4°C for up to 1 mo).
8. Ficoll-Hypaque (Pharmacia, cat. no. 17-1440-03) Store at room temperature before use.
9. Phytohemagluttin-P (PHA) (Sigma, cat. no. L9132) (*see* **Note 1**).
10. Recombinant human IL-2 (rIL-2) (e.g., Peprotech, Natick, MA).
11. Recombinant human IL-7 (rIL-7) (e.g., Peprotech)
12. 2% Trypan blue in phosphate-buffered saline (PBS). Dilute to 0.2% or 0.4% working concentration.
13. 2% Acetic acid or red cell lysing buffer (Sigma, cat. no. R7757) (optional).
14. Dimethyl sulfoxide (Sigma, cat. no. D-2650).
15. Isopropanol.
16. ^{51}Cr sodium chromate solution, 1 mCi/mL (Dupont New England Nuclear, Boston, MA) or equivalent.

17. Heparin (preservative-free).

2.3. Preparation of Stock Solutions

1. Complete culture medium (RPMI + 10%NHS): RPMI-1640, 2 m*M* glutamine, 10% normal human serum (NHS): To 430 mL of RPMI 1640, add 5 mL of penicillin-streptomycin (10,000 U/10,000 µg), 5 mL of 200 m*M* L-glutamine, 5 mL of 10 m*M* nonessential amino acids, 5 mL of 100 m*M* sodium pyruvate, and 50 mL of NHS. Filter-sterilize through a 0.22-µm filter (see **Note 2**). Store at 4°C. Use within 1 mo.
2. Wash medium (RPMI + 2% NHS): To 490 mL RPMI 1640, add 10 mL of NHS. Filter-sterilize through a 0.22-µm filter. Store at 4°C.
3. 2% Trypan blue in PBS: Dissolve 2 g of trypan blue in 100 mL of PBS (see **Note 3**). Store at room temperature.
4. Freezing solution A: RPMI medium supplemented with 80% FCS: add 80 mL of FCS per 20 mL of RPMI. Filter-sterilize through a 0.22-µm filter. Store at –20°C or prepare fresh as required.
5. Freezing solution B: 40% RPMI, 20% FCS, 20% glycerol, 20% dimethyl sulfoxide (DMSO): to 40 mL of RPMI 1640, add 20 mL of FCS, 20 mL of glycerol, and 20 mL of DMSO. Filter-sterilize through a 0.22-µm filter. Store at –20°C or prepare fresh as required. Note that it is important to add DMSO to FCS, not vice versa, since DMSO will react with the plastic.
6. Freezing solution C: 20% FCS, 20% glycerol, 10% DMSO in RPMI medium: add 20 mL of FCS, 20 mL of glycerol, and 10 mL of DMSO to 50 mL of RPMI. Filter sterilize through a 0.22-µm filter. Store at –20°C or prepare fresh as required.
7. Phytohemagglutinin-P (PHA) stock: Prepare a stock of 1 mg/mL PHA by reconstituting the sterile powder (5 mg/vial) with 5 mL of RPMI (without supplements). Filter-sterilize through a 0.22-µm filter, aliquot in 100–500 µL vol, and store at –20°C. Use at a dilution of 1 in 500 (20 µL/10 mL), to give a final concentration of 2 µg/mL PHA, according to manufacturer's specifications.
8. Mitomycin C: Prepare a working stock of 50 µg/mL mitomycin C by reconstituting the sterile powder (2 mg/vial) with 40 mL of RPMI. Filter-sterilize through a 0.22-µm filter, protect from light, and store at 4°C. Use solution at this 50 µg/mL concentration.

3. Methods

Carry out all cell culture work aseptically in a laminar flow cabinet, using sterile materials. Maintain cell cultures at 37°C in an atmosphere of 5% CO_2.

3.1. Isolation of Peripheral Blood Mononuclear Cells

1. Obtain informed consent from all blood donors and conduct all procedures in accordance with the ethical standards of local and federal committees on human experimentation.
2. Collect whole blood by venipuncture into a vacutainer or heparin-containing syringe (expel air in 20-mL or 50-mL syringe, and add heparin stock directly to the syringe, to a final concentration of 10 U heparin/mL blood) (see **Notes 4** and **5**).
3. Dilute whole blood 1 to 1.5 with RPMI medium (i.e., dilute 20 mL of blood to 30 mL final volume with 10 mL of RPMI).
4. Aliquot 12–15 mL of Ficoll-Hypaque (warmed to room temperature) into 50-mL tubes.
5. Carefully overlay Ficoll-Hypaque with 30 mL of diluted blood (ratio of Ficoll-Hypaque:blood should be approx 1:3).
6. Alternatively, aliquot 30 mL of diluted blood per 50-mL tube and slowly underlay diluted blood with 12–15 mL of Ficoll-Hypaque.
7. Centrifuge at 300*g* for 25 min at room temperature, with the brake off.
8. Collect plasma (diluted 1/1.5) from the top, aliquot, and store at –20°C or –70°C.

9. Carefully collect the unfractionated peripheral blood mononuclear cells (PBMCs) banded at the interface between the Ficoll-Hypaque and plasma, with a pipet using a circular motion. Transfer to a 50-mL tube.
10. Adjust volume to 50 mL with wash medium (wash no. 1).
11. Centrifuge at at $360g$ for 10 min at room temperature.
12. Decant or aspirate supernatant, leaving approx 1–2 mL per tube. Resuspend cell pellet by tapping side of tube. Adjust volume to 50 mL with cold wash medium (wash no. 2).
13. Centrifuge at $300g$ for 10 min.
14. Decant or aspirate supernatant, leaving approx 1–2 mL per tube. Resuspend cell pellet by tapping side of tube. Adjust volume to 50 mL with wash medium (wash no. 3).
15. Centrifuge at $300g$ for 5 min.
16. Completely decant or aspirate supernatant. Resuspend cell pellet in complete medium, in 1/10 vol of whole blood collected (i.e., resuspend cell pellet from 100 mL of whole blood in 10 mL of complete medium). Place cell suspension on ice, tightly capped.
17. For each sample, add 90 µL of trypan blue to one well of a round-bottomed 96-well plate. Add 10 µL of cell suspension to 90 µL of trypan blue (1 in 10 dilution).
18. Determine cell count with a hemocytometer, using trypan blue to differentiate viable cells. Optional: Count in the presence of 2% acetic acid to lyse the RBCs.
19. Resuspend cells at a concentration of 1×10^7 cells/mL in complete medium.
20. Freeze according to standard freezing procedures (*see* **Subheading 3.2.**) or set up assays using fresh PBMCs.

3.1.1. Option 1

If blood is collected by venipuncture into heparin-containing syringe and the plasma is to be stored, the following steps may be substituted for **Subheading 3.1., steps 2–5**:

1. Collect whole blood into heparinized syringe.
2. Aliquot 20–50 mL of heparinized blood per 50-mL tubes.
3. Centrifuge at $300g$ for 10 min at room temperature.
4. Carefully remove the upper plasma layer, aliquot, and store at –70°C.
5. Transfer the lower cell suspension to a 50-mL tube.
6. Dilute cells with 2X packed volume of wash medium (i.e., from 20 mL of whole blood, dilute 10 mL of packed cells with 20 mL of wash medium to final volume of 30 mL).
7. Layer 30 mL of diluted blood on 12–15 mL of Ficoll-Hypaque (warmed to room temperature).

Proceed as described in **Subheading 3.1., steps 7–20**.

3.1.2. Option 2

If blood is collected using a vacutainer and the plasma is to be stored, the following steps may be substituted for **Subheading 3.1., steps 2–5**:

1. Collect whole blood into a vacutainer contain sodium heparin.
2. Centrifuge at 1000 rpm for 10 min at room temperature.
3. Carefully remove the upper plasma layer, aliquot, and store at –70°C.
4. Transfer lower cell suspension to a 50-mL tube.
5. Dilute cells with 2X packed volume of wash medium (i.e., from 10 mL of whole blood, dilute 5 mL of packed cells with 10 mL of wash medium to final volume of 15 mL), using this step to completely rinse the vacutainer.
6. Layer 15 mL of diluted blood on 5–7.5 mL of Ficoll-Hypaque (warmed to room temperature).

Proceed as described in **Subheading 3.1., steps 7–20**, substituting 15-mL for 50-mL washes if blood volumes are small.

3.2. Freezing of PBMCs

All procedures should be carried out on ice.

1. Prepare freezing solutions A (80% FCS in RPMI) and B (20% FCS, 20% glycerol, and 10X DMSO in RPMI). Place on ice or at –20°C until required.
2. Centrifuge PBMC sample, remove supernatant, and resuspend cells at a concentration of 20×10^6 cells/mL in RPMI medium supplemented with 80% FCS (freezing solution A).
3. Place cell suspension on ice for at least 15 min (15–120 min).
4. Label cryotubes and place on ice to precool.
5. Prepare an ice-water bath for use in the biohazard hood. Transfer individual cell suspensions to tray as required.
6. On ice, dropwise with constant mixing, add an equal volume of ice-cold (freezing solution B) to the cell suspension (final concentration, 10×10^6 cells/mL in 50% FCS, 10% DMSO, 10% glycerol, 30% RPMI).
7. Immediately aliquot 1 mL of the cell suspension into precooled cryotubes, at a final concentration of 10×10^6 cells/1 mL per vial.
8. Place cryotubes on ice until all samples for that volunteer are completed.
9. To freeze, float cryovials in a beaker of isopropanol overnight at –70°C (*see* **Note 6**); place cryovials in a styrofoam container or commercially available Mr. Freezy container for at least 2 h but no longer than 24 h at –70°C; or place cryovials in a rate freezer (position D for 30 min, then position A for 15 min), before transfer to liquid nitrogen for long-term storage.

Alternatively, substitute the following for **Subheading 3.2., steps 1–6**:

1. Prepare freezing solution C of 20% FCS, 20% glycerol, 10% DMSO in RPMI medium. Place on ice or at –20°C until required.
2. Centrifuge PBMC sample, and remove supernatant.
3. Place cell suspension on ice for at least 15 min (15–120 min).
4. Label cryotubes and place on ice to precool.
5. Prepare an ice-water bath for use in the biohazard hood. Transfer individual cell suspensions to tray as required.
6. On ice, dropwise with constant mixing, add the ice-cold freezing solution to the cell suspension, to a final concentration of 10×10^6 cells/mL.

3.3. Thawing of PBMCs

All procedures should be carried out at room temperature.

1. Remove cryovials from liquid nitrogen and place on dry ice.
2. Preheat water bath to 37°C.
3. Prewarm complete medium to room temperature.
4. Thaw the cells at 37°C, with continuous shaking (do not leave cells stationary at 37°C or at room temperature for long periods).
5. As soon as the last ice disappears, transfer the contents of the vial to a 15-mL tube.
6. Immediately add complete medium to the cells dropwise (approx 1 drop/15 s initially, as specified below) with shaking to a final volume of approx 10 mL.
   a. 0–2 min 1 drop/15 s = 4 drops/min; 0.2 mL vol added
   b. 2–4 min 1 drop/10 s = 6 drops/min; 0.3 mL vol added to 0.5 mL total
   c. 4–6 min 1 drop/5 s = 12 drops/min; 0.6 mL vol added to 1.1 mL total
   d. 6–9 min 1 drop/1 s = 60 drops/min; 4.5 mL vol added to 5.6 mL total
   e. 9–10 min 5.5 mL vol added to 10.0 mL final volume

7. Centrifuge at 140g for 10 min, at room temperature.
8. Discard supernatant and resuspend the cell pellet in 10 mL of complete medium.
9. Centrifuge at 140g for 10 min. Decant or aspirate supernatant. Repeat wash.
10. Resuspend the cell pellet in 1 mL of complete medium.
11. Count the cells (1 in 10 dilution in 2% trypan blue).
12. Adjust cell concentration to a final concentration of 10×10^6 cells/mL with complete medium.

3.4. Generation of Phytohemagglutinin P-Activated Lymphoblasts

1. Generate PHA blasts by stimulating PBMC with 2 µg/mL PHA (20 µL PHA/10 mL) in vitro in 25-cm^2 flasks, at a cell concentration of approx $1-3 \times 10^6$ cells/mL.
2. Maintain cultures at 37°C in an atmosphere of 5% CO_2 by weekly or twice weekly medium changes.
3. Restimulate with 2 µg/mL PHA at approx weekly then two weekly intervals (for 4–6 wk).
4. Maintain long-term cultures without further PHA stimulation.

3.5. Generation of Lymphoblastoid Cell Lymphomas (LCLs)

3.5.1. Protocol 1

1. To approx $5-10 \times 10^6$ PBMC in a 10-mL snap-top tube, add 2 mL of EBV stock diluted at 1 in 2 with complete medium supplemented with 10% FCS.
2. Incubate overnight at 37°C in an atmosphere of 5% CO_2.
3. Remove 1 mL of the supernatant and replace with the same volume of medium containing 1 µg/mL cyclosporin A.
4. Maintain cells at 37°C in an atmosphere of 5% CO_2 until clumping is visible in the bottom of tube.
5. Transfer cultures into 24-well tissue culture plates in a final volume of 2 mL per well.
6. Maintain by weekly medium changes in the presence of cyclosporin A (1 µg/mL) for approx 3 wk.
7. When cell transformation is apparent, transfer cultures into upright 25-cm^2 flasks.
8. Maintain with weekly medium changes without cyclosporin A, and split when necessary. Do not oversplit, since lack of cell–cell contact will cause cultures to fail. Transfer to 75-cm^2 flasks when cell numbers are sufficient; maintain upright for 1–2 d and then culture long-term in 75-cm^2 flasks lying flat.

3.5.2. Protocol 2

1. To approx $5-10 \times 10^6$ PBMC in a 10-mL snap-top tube, add 0.25 mL of EBV stock. diluted at 1 in 2 with complete medium supplemented with 10% FCS.
2. Incubate 1 h at 37°C in an atmosphere of 5% CO_2.
3. Wash cells twice with RPMI supplemented with 2% FCS.
4. Add 3 mL of complete medium supplemented with PHA to 2 µg/mL final concentration.
5. Serially dilute the cell suspension twofold in the PHA-supplemented complete medium, in a final volume of 2 mL per well in a 24-well tissue culture plate.
6. Maintain cultures at 37°C in an atmosphere of 5% CO_2 by weekly medium changes (without PHA) for approx 3 wk.
7. When cell transformation is apparent, transfer cultures into upright 25-cm^2 Roux flasks, using the lowest dilution wells only.
8. Maintain cultures by weekly medium changes and split when necessary.

3.6. Mitomycin C Treatment of Cells

Mitomycin C treatment can be used to block division of the stimulator cells (APCs), if a γ-irradiator is not available. Blocking the division of stimulator cells is recommended to provide a clear distinction between responder and stimulator cell populations. A preferred alternative to mitomycin C treatment is γ-irradiation using 2000 rads (irradiation doses as high as 10,000 rads may be required for tumor cells).

1. Centrifuge the stimulator cells in a 10-mL snap-cap tube, and remove supernatant.
2. Add 0.5-2.0 mL of the 50 µg/mL mitomycin C stock to the cell pellet, depending on cell number.
3. Incubate for 30 min at 37°C in a humidified 37°C, 5% CO2 incubator (or at 37°C in a tightly capped tube, protected from the light).
4. Wash cells thoroughly three times with 14 mL medium at 300g for 5 min. Extensive washing is critical, since any carryover of mitomycin C may block the function of responder cells. If cells are irradiated rather than treated with mitomycin C, a single wash is sufficient.
5. Count using a hemocytometer. Recovery of cells after mitomycin C treatment may be as low as 50%.
6. Resuspend the mitomycin C-treated stimulator cells at the desired cell concentration.
7. If desired, the adequacy of mitomycin C treatment or irradiation can be determined by measuring radiolabeled thymidine incorporation.

3.7. Assessment of Peptide Toxicity

Synthetic peptides are tested for toxicity prior to use in vitro lymphoproliferation or CTL assays.

1. Plate PBMC in quadruplicate at 200,000 cells per well in a volume of 200 µL of complete medium in a round-bottomed 96-well plate in the presence of: (a) 0.3 µg/mL tetanus toxoid; (b) 0.3 µg/mL tetanus toxoid and 30 µg/mL peptide; and (c) no antigen control.
2. Incubate cultures for 6 d, at 37°C in an atmosphere of 5% CO_2.
3. Pulse wells with 0.5 µCi [^{3}H]thymidine for 16 h (overnight).
4. Harvest cultures and assess uptake by liquid scintillation spectroscopy (see Chapter 38).
5. A peptide is considered not toxic if the mean + 1 standard error of radioactive incorporation of the peptide tetanus-toxoid stimulated cells is not lower than the mean –1 standard error of radioactive incorporation of the cells stimulated with tetanus toxoid alone.

3.8. Generation of Peptide-Specific CTL Effectors

Generation of CTL involves culturing the responder cells, which contain a population of CTL precursors (pCTL), with a population of stimulator cells that express the desired antigen. The method of generating CTL detailed below is based on that described by Hogan et al. (10) for influenza A virus matrix peptide-specific CTL. Briefly, 3×10^6 PBMC are stimulated in primary in vitro culture, in the presence of 1–10 µg/mL peptide, in a final volume of 2 mL of complete medium in a 24-well tissue culture plate for 7 days (primary CTL assay) or for 8 d (secondary CTL). Recombinant human interleukin 2 (rIL-2) (50 U/mL) is added on d 2 of culture. For secondary CTL, on d 8, 10^6 of the primed responder cells are restimulated in secondary culture with 9×10^6 γ-irradiated (2000 rad) or mitomycin C-treated (50 µg/mL) autologous PBMC in 10 mL of complete medium in an upright 25-cm^2 flask, in the presence of 1–10 µg/mL peptide and 50 U/mL rIL-2, for an additional 5 d. Secondary cultures are tested for cytotoxicity on d 13. If desired, established autologous PHA blasts or LCL lines may substituted for fresh autologous PBMC as antigen presenting cells (APCs) in the secondary stimulation.

1. *Day 0:* Culture PBMC at a concentration of 3×10^6 cells/2 mL complete medium in 24-well plates (1–4 wells per peptide) in the presence of each HLA-matched peptide (1–10 µg/mL) or negative control peptide (e.g., HLA-A2.1 restricted HIV or HBV peptides) or positive control peptide (e.g., HLA-A2.1 restricted FLU peptides or CMV peptides) for 7 d, at 37°C. Prepare replicate wells to ensure that the required numbers of effector cells will be generated. Cell recovery after 7 d is usually 50–100% of the responder cells initially plated (*see* **Note 7**).
2. *Day 0:* Generate PHA blasts target cells by stimulating autologous PBMCs at a concentration of 3×10^6 cells/2 mL complete medium in a 24-well plate or $10–15 \times 10^6$ cells/10 mL complete medium an upright 25-cm² flask, in the presence of 0.2% (v/v) PHA. Note that this step can be omitted if a long-term autologous cell line has been previously established (*see* **Subheadings 3.4.** and **3.5.**).
3. *Day 2:* Add rIL-2 (50 U/mL) to effector cultures at 48 h after the initiation of culture. Do not add rIL-2 to PHA blast cultures.
4. *Day 6:* Prepare CTL targets, for the primary CTL assay as described in **Subheading 3.10.**
5. *Day 7:* Conduct CTL assay (primary cultures) as described in **Subheading 3.11.**
6. *Day 8:* Restimulate effector cultures for secondary assay as follows. Harvest primed responder cells from effector cultures and count. Aliquot 1×10^6 primed responder cells (approximately contents of one 24-well) per 25-cm² flask. Pulse autologous PHA blasts or PBMCs with 1–10 µg/mL specific peptide, for 30–90 min. Irradiate peptide-pulsed PHA blasts or fresh PBMCs at 2000 rad, and wash (300g for 5 min). Add 9×10^6 of those irradiated antigen presenting cells per 1×10^6 primed responder cells, in a final volume of 10 mL of complete medium in an upright 25-cm² flask. Add rIL-2 to a final concentration of 10 U/mL. Incubate cultures for an additional 5 d (13 d total), at 37°C in an atmosphere of 5% CO_2.
7. *Day 12:* Prepare CTL targets, for the secondary CTL assay, as described in **Subheading 3.10.**
8. *Day 13:* Conduct CTL assay (secondary cultures) as described in **Subheading 3.11.**

3.8. Generation of Virally-Stimulated CTL Effectors

Recombinant canary-pox (ALVAC) virus or other recombinant pox viruses expressing the antigen of interest can also be used for stimulation of CTL effector cells, or as target cells (*see* **Note 8**). Transient transfection systems can yield satisfactory results but these methods are not yet optimal for routine use. Intact recombinant protein which requires in vivo processing and presentation is not satisfactory.

1. To generate effector cells for primary CTL assays, infect 20% of total PBMCs (20% of 3×10^6 cells per well = 0.6×10^6 cells per well × number of wells) with recombinant virus at a multiplicity of infection (MOI) of 5 plaque-forming units (Pfu) per cell in a total volume of 100–200 µL, for 90 min at 37°C in 15-mL polypropylene tubes, with shaking every 15–20 min.
2. Wash infected cells twice at 300g for 10 min.
3. Add infected cells to the remaining PBMCs (80% of total PBMCs = 2.4×10^6 cells per well × number of wells).
4. Culture at a concentration of 3×10^6 cells/2 mL in 24-well plates for 7 d.
5. Add rIL-2 (20 U/mL) at 48 h after the initiation of the culture.
6. To generate effector cells for secondary CTL assays, add recombinant IL-7 (100 U/mL) at the initiation of culture, and add rIL-2 (20 U/mL) at 48 h.
7. On d 8, infect 2×10^6 autologous PHA blasts with recombinant virus at a MOI of 5 PFU/cell in a total volume of 100–200 µL, for 90 min at 37°C in 15-mL polypropylene tubes, with shaking. Wash infected cells twice at 300g for 10 min.
8. Add 2×10^6 of these virus-infected stimulator cells per 10×10^6 cultured responder cells in a 10 mL vol in a 25-cm² flask (final concentration = 1×10^6 responder cells/mL).

9. Add rIL-2 (20 U/mL) and rIL-7 (100 U/mL).
10. Culture for an additional 5 d.

3.10. Preparation of CTL Targets

On d 6 (primary assay) or d 12 (secondary assay), incubate PHA blast target cells overnight with or without 1–10 µg/mL peptide (0.5–1.0×10^6 cells per peptide) in a volume of 1 mL of complete medium, in a 15-mL conical tube with a loose cap or a 24-well plate. Alternatively, on the day of the assay, pulse PHA blast target cells with 1–10 µg/mL peptide in a volume of 50–100 µL in a 15-mL conical tube for 30–60 min before labelling with chromium. Target cells should have a high viability at the time of the assay, since targets of poor viability will leak more ^{51}Cr resulting in a high spontaneous release value.

3.11. CTL Assay

On d 7 (primary assay) or d 13 (secondary assay), cytolytic activity is assessed by conventional chromium release *(5)* using autologous or MHC-matched PHA blasts (or control MHC-mismatched PHA blasts) as target cells. Briefly, target cells are plated in triplicate at a concentration of 5000 cells/100 µL per well in a round-bottom 96-well plate in the presence of peptide-stimulated effector cells at various E:T ratios, with or without 1–10 µg/mL specific peptide. CTL activity is measured in a standard 6-h (peptide targets) or 5-h (if virally infected targets) chromium release assay. Labeled target cells expressing the same antigen are recognized and lysed by the effector CTL, releasing radioactivity into the supernatant. Follow standard radiation safety procedures when working with ^{51}Cr solution and ^{51}Cr labeled target cells.

1. Centrifuge target cells at 1200 rpm for 5 min to pellet the cells. Completely remove the supernatant, leaving not more than 50 µL residual medium.
2. Label target cells by adding 50–100 µCi of ^{51}Cr sodium chromate solution for 90 min at 37°C, with shaking every 20 min *(see* **Note 9**). During this labeling period, prepare effectors.
3. Harvest effector cells from 24-well plates into 15-mL (or 50-mL depending on number of wells harvested) conical tubes, making sure to wash wells thoroughly to recover all cells. Make volume up to 15 mL.
4. Centrifuge effectors at 1200 rpm for 5 min at room temperature.
5. Count using a hemocytometer and adjust effector cells to a concentration of 4×10^6 cells/mL (4×10^5 cells/100 µL = 80:1 effector:target ratio). Dilute an aliquot 1 in 2 to yield the desired effector:target ratios of 40:1, 20:1, and 10:1; aliquot 100 µL of each E:T ratio in triplicate wells *(see* **Note 10**). Alternatively, aliquot 100 µL per well for E:T = 80:1, 50 µL per well for E:T = 40:1, 25 µL per well for E:T = 20:1, 12.5 µL per well for E:T = 10:1, and make volume up to 100 µL with complete medium. Maintain cells at room temperature or on ice in a tightly capped tube until assay (for up to 4 h, provided the cells are resuspended occasionally). Once effector cells are plated, keep at 37°C in an atmosphere of 5% CO_2.
6. If other effector:target ratios are desired, adjust effector cell concentration as appropriate: for example, for an E:T ratio of 100:1, adjust effector cell concentration to 5×10^6 cells/mL. E:T ratios of 100:1, 30:1, 10:1, and 3:1 are also typical.
7. At the end of the 90-min labeling period, wash target cells three times at 1200 rpm for 5 min.
8. Resuspend target cells in 500–1000 µL of complete medium.
9. Count using a hemocytometer (1:2 dilution) and adjust target cells to a concentration of 5×10^4 cells/mL (5000 cells/100 µL).

10. Add peptide to the appropriate target at a concentration of 2–20 µg/mL (2X final concentration). Include one set of target cells without peptide (if there are five test peptides, prepare 5X vol of target without peptide).
11. Aliquot 100 µL per well peptide-labeled target cells to the respective effector cells (5000 cells per well, for each of the E:T ratios) (final volume, 200 µL per well).
12. Aliquot 100 µL per well target cells without peptide to a parallel set of effectors.
13. Aliquot 100 µL per well peptide-labeled target cells to six wells without effector cells. Add 100 µL of complete medium to three of these wells (medium control, for spontaneous release) and add 100 µL of 5% Triton-X100 or 10% SDS to the other three wells (maximum release).
14. Centrifuge the plate at 50g for 3 min to promote contact between the effector cells and the target cells.
15. Incubate at 37°C in an atmosphere of 5% CO_2 for 6 h (*see* **Note 11**).
16. Harvest using the Skatron harvesting system. Alternatively, centrifuge plate for 5 min at 210g rpm to pellet the cells, and harvest 100 µL of the supernatant for counting.
17. Count ^{51}Cr in a gamma scintillation counter, 1–2 min per sample.
18. Percent lysis is determined as (experimental release – spontaneous release)/(maximum release – spontaneous release) × 100. Percent specific lysis is determined by subtracting the percent lysis (control) from percent lysis (specific peptide). Maximum release is determined from supernatants of target cells lysed by the addition of 10% SDS or 5% Triton X-100. Spontaneous release (medium control release) is determined from the supernatants of target cells incubated without added effector cells. Results are expressed as the mean of triplicate determinations (*see* **Notes 12** and **13**). Results are presented as either percent lysis, or lytic units (*see* **Note 14**).

4. Notes

1. The source of PHA is critical, and must be tissue-culture grade. In our hands, the optimal reagent is Sigma, cat. no. L9132. We and others have experienced problems with other sources of PHA.
2. Medium supplemented with 10% NHS is difficult to filter through a 0.22-µm filter; use of a prefilter is recommended. Alternatively, prepare medium with all components except the NHS, filter sterilize through a 0.22-µm filter, and then add sterile NHS to 10% (v/v) (50 mL/500 mL medium). Commercially available NHS is sterile and can be purchased in volumes of 100 mL.
3. Do not use water to prepare the trypan blue stock, since this would result in lysis of the cells.
4. Use of heparin concentrations lower than 10 U/mL may result in clumping of cells and reduced cell yield.
5. On average, the PBMC yield from human blood is approx $1.0–2.0 \times 10^6$ PBMCs/mL. i.e., approx 2×10^7 PBMCs can be expected from 20 mL of blood.
6. Isopropanol cools at a rate of approx 1°C/min.
7. If desired, $CD8^+$ T cells may be used instead of PBMC populations as responders cells, with irradiated autologous peptide-pulsed PBMC or PHA blasts as stimulators, at a responder:stimulator ratio of 1:10 (1×10^5 $CD8^+$ enriched T cells and 1×10^6 APCs per 48 wells). Positively select $CD8^+$ T cells from PBMCs samples by magnetic cell sorting using the magnetic activated cell sorting (MACS) system (Miltneyi Biotec, Germany). Aliquot 1×10^5 $CD8^+$ cells per well, in 48-well plates. For APCs, γ-irradiate (2000 rad) autologous peptide pulsed PBMCs or PHA blasts and incubate in 48-well plates at a density of 1×10^6 cells/well, in serum-free medium in the presence of specific peptide (1–10 µg/mL) and 1 µg/mL of $β_2$ microglobulin at room temperature for 1–2 h. Then wash the stimulator cells and add back to 48-well plates at a concentration of 1×10^6 cells per well. Incubate the cultures at 37°C in an atmosphere of 5% CO_2 for 7–10 d. Add rec. IL-2 after 48 h at a concentration of 50 U/mL.

8. If recombinant viruses are to be used for infection of both stimulator cells for effector cultures, and target cells for the CTL assay, it is important to use different viral constructs where possible to reduce the specific virus–virus immune responses which may mask the response of interest.
9. ^{51}Cr has a half-life of 28 d. Therefore, the radioisotope should be used while it is fresh. Stocks of up to 1-mo old will result in satisfactory labeling, but the volume of isotope added per target should be increased.
10. Effector to target (E:T) ratios of 10:1, 20:1, 40:1, and 80:1 = 150:1 × triplicate wells at 5000 targets per well × 2 targets (test and control)/effector = 4.5×10^6 effector cells. Assume 50% loss in culture during primary stimulation. Assume at least twofold amplification during secondary restimulation. Adjust effector cultures accordingly. For example, for primary CTL assays, initiate culture with 9.0×10^6 cells = three wells at 3.0×10^6 cells per well. For secondary CTL assays, one flask of 10×10^6 cultured responder cells is sufficient for assay.
11. Effector cells will bind to target cells at room temperature, but cell lysis will not occur until the temperature in increased to 37°C. Shorter incubations (e.g., 4 h) may be adequate in some cases, but longer incubations are recommended for samples with little cytotoxic activity.
12. Results are expressed as the mean percent specific lysis of triplicate determinations. The mean standard error should be less than 10% of the mean value. Specific lysis may be defined as significant if it exceeds the 95% confidence interval for all values, or if greater than 10 or 15%, as arbitrarily specified.
13. Spontaneous (medium control) release should always be less than 20% of the maximum release; ideally, less than 5–10%. If greater than 20%, disregard the results of the assay.
14. One lytic unit is arbitrarily defined as the number of lymphocytes required to yield 30% lysis. It is determined by plotting percent specific lysis values versus the log of the effector cell number for each effector cell preparation, and then determining the lysis value (30%) through which the titration curves pass. Using this value, the number of lytic units (LU) per 10^6 effector cells is determined. If the titration curve of an effector cell population fails to reach the selected lysis value, refer to the activity as less than or greater than xLU, where x is the calculated minimum or maximum level. Often the data are not appropriate for LU calculations because the lysis values fail to titrate through the selection point or because the titration plots are not parallel.

References

1. Townsend, A. R. M., McMichael, A. J, Carter, N. P., Huddleston, J. A., and Brownlee, G. G. (1984) Cytotoxic T cell recognition of the influenza virus nucleoprotein and hemagglutinin expressed in transfected mouse L cells. *Cell* **39,** 13–25.
2. Gotch, F. M., McMichael, A. J., Smith, G. L., and Moss, B. (1986) Identification of the viral molecules recognized by influenza specific human cytotoxic T lymphocytes. *J. Exp. Med.* **165,** 408–416.
3. Yewdell, J. W., Bennink, J. R., Smith, G. L., and Moss, B. (1985) Influenza A virus nucleoprotein is a major target antigen for cross reactive anti-influenza A virus cytotoxic T lymphocytes. *Proc. Natl. Acad. Sci. USA* **82,** 1785–1789.
4. Townsend, A. R. M., Gotch, F. M., and Davey, J. (1985) Cytotoxic T cells recognize fragments of influenza nucleoprotein. *Cell* **42,** 457–467.
5. Townsend, A. R. M., Rothbard, J., Gotch, F. M., Bahadur, G., Wraith, D., and McMichael, A. J. (1986) The epitopes of influenza nucleoprotein recognized by cytotoxic T lymphocytes can be defined with short synthetic peptides. *Cell* **44,** 959–968.
6. Gotch, F., Rothbard, J., Howl, K., Townsend, A., and McMichael, A. (1987) Cytotoxic T lymphocytes recognize a fragment of influenza virus matrix protein in association with HLA-A2. *Nature* (London) **326,** 881,882.
7. Gotch, F. M., McMichael, A. J., Smith, G. L., and Moss, B. (1986) *J. Exp. Med.* **165,** 408–416.
8. Bjorkman, P. J., Saper, M. A., Samraoui, B., Bennett, W. S., Strominger, J. L., and Wiley, D. C. (1987a) Structure of the human class I histoincompatibility antigen, HLA-A2. *Nature* **329,** 506–512.

9. Bjorkman, P. J., Saper, M. A., Samraoui, B., Bennett, W. S., Strominger, J. L., and Wiley, D. C. (1987b) The foreign antigen-binding site T cell recognition regions of class I histoincompatibility antigens. *Nature* **329,** 512–518.
10. Hogan, K. T., Shimojo, N., Walk, S. F., Engelhard, V. H., Maloy, W. L., Coligan, J. E., et al. (1988) Mutations in the α-helix of HLA-A2 affect presentation but do not inhibit binding of influenza virus matrix peptide. *J. Exp. Med.* **168,** 725–736.

40

Human Antibody Subclass ELISA

Pierre Druilhe and Hasnaa Bouharoun-Tayoun

1. Introduction

Our results have emphasized the potential involvement of the antibody-dependent cellular inhibition (ADCI) mechanism in the protective effect of antibodies mediating acquired immunity to *Plasmodium falciparum* malaria *(1)*. The ADCI mechanism consists of a cooperation between monocytes and cytophilic antibodies able to bind to the Fc receptors present on the monocyte surface. Thus human IgG1 and IgG3 are the main isotypes effective in ADCI, whereas IgG2, IgG4, and IgM are ineffective and could rather have a blocking effect if they are directed to parasite antigens that are targets of effective antibodies *(2–4)*. Therefore, in order to assess the potential role of a given parasite antigen in triggering protective immune responses, it is of interest to characterize the isotype distribution of antibodies directed to this antigen and elicited in protected versus nonprotected individuals.

The enzyme-linked immunosorbent assay (ELISA) is suitable for measuring the amount of antibodies from each isotype that binds to the antigen. The major steps of this protocol are as follows:

1. Capture of the target antigen on plastic wells.
2. Blocking of unbound plastic sites.
3. Incubation with the serum to be tested.
4. Incubation with monoclonal antibodies directed to one of the Human IgG subclasses or IgM.
5. Incubation with anti-mouse Ig conjugated to peroxidase.
6. Colorimetric reaction of peroxidase in the presence of its substrate.
7. Quantitative determination of the reaction intensity by measuring the absorbance.

2. Materials

1. Maxisorb 96-well microplates (Nunc).
2. Coating Buffer (*see* **Note 1**): Phosphate-buffered saline (PBS) pH 7.4.
3. Blocking buffer: PBS, pH 7.4, 2.5% nonfat milk (Regilait).
4. Washing buffer: PBS, pH 7.4, 0.2% Tween-20.
5. Buffer for serum dilution: PBS, pH 7.4, 0.05% Tween-20, 1.25% nonfat milk.
6. Control sera from healthy blood donors with no history of malaria.
7. Murine monoclonal antibodies directed to human isotypes, purchased from Sigma (*see* **Note 2**):
   a. Anti-IgG1: NL16.
   b. Anti-IgG2: HP6002.

c. Anti-IgG3: ZG4.
d. Anti-IgG4: RJ4 (or GB7B).
e. Anti-IgM: MB11.
8. Peroxidase conjugated anti-mouse antibodies (Biosys).
9. Citrate buffer, 1 M, pH 5.5.
10. *Ortho*-phenylene diamine (OPD) as a peroxidase substrate.
11. Hydrogen peroxide (H_2O_2), 30%.
12. ELISA washer.
13. ELISA reader, with 492-nm filter.

3. Methods

1. Dilute the antigen, usually at 1–10 μg/mL, in coating buffer (*see* **Note 1**).
2. Aliquot 100 μL of antigen solution per well, in a 96-well microplate plate. For each serum to be tested and for each control serum, 10 wells are required. Cover the microplate.
3. Incubate overnight at 4°C.
4. Wash the plate three times with washing buffer.
5. Add 300 μL of blocking buffer per well, to block the free plastic sites and avoid nonspecific binding of antibodies.
6. Incubate for 2 h at room temperature.
7. For each serum, prepare 1.2 mL of a 1/100 dilution, using the dilution buffer.
8. Wash the plate five times with washing buffer.
9. Add 100 μL per well of diluted serum, 10 wells for each serum. In each plate include one negative control serum.
10. Incubate for 1 h at room temperature.
11. Wash the plate five times with washing buffer.
12. Add 100 μL per well, in duplicate, of the antiisotype monoclonal antibodies diluted as follows (*see* **Note 3**):
    a. NL16 (anti-IgG1): 1/2000.
    b. HP6002 (anti-IgG2): 1/10,000.
    c. ZG4 (anti-IgG3): 1/10,000.
    d. RJ4 (anti-IgG4): 1/30,000.
    e. MB11 (anti-IgM): 1/30,000.
13. Incubate for 1 h at room temperature.
14. Wash the plate five times with washing buffer.
15. Add 100 μL per well of peroxidase conjugated anti-mouse antibodies diluted 1/4,000.
16. Incubate for 1 h at room temperature.
17. At 5 min before the end of the incubation, prepare the peroxidase substrate solution: for each microplate, combine 11 μL of citrate buffer, 10 g of OPD, and 11 mL of 30% H_2O_2. Protect the solution from light.
18. Wash the plate five times with washing buffer.
19. Add 100 μL per well of the substrate solution.
20. Incubate for 30 min away from light.
21. Immediately read the optical density using a 492-nm filter.
22. For each isotype, calculate the ratio OD sample/OD control.

4. Notes

1. We have used PBS, pH 7.4, for the coating of several malarial antigens (FIRA, RESA, LSA, MSP3, and others). Nevertheless, the optimal conditions for coating (buffer, antigen concentration, incubation time, and temperature) should be determined for each antigen.
2. The choice of monoclonal antibodies (MAb) specific for human immunoglobulin isotypes is based on the result of an international meeting *(5)*, and on their reactivity with defined

human myeloma proteins, purified from white, black, and Asian patients. We had to reject some MAbs because of their specificity for allotypic determinants unequally distributed in human populations. For example, the MAb SH21 (Bioyeda) specific for the IgG2 allotype Gm is unable to recognize IgG2 from black Africans that do not express this allotype.

3. To determine the concentration of each MAb to be used, different myeloma proteins are adjusted to the same concentration, coated onto microplate wells and revealed by a range of antiisotype MAb dilutions. The working concentrations are determined in order to obtain comparable OD for all isotypes when comparable concentration of the different myelomas are used, and to avoid crossreaction of a given antiisotype MAb with different isotypes (which occur at high concentrations, i.e., the MAbs are specific only within a given concentration range).

References

1. Bouharoun-Tayoun, H., Attanah, P., Sabchareon, A., Chongsuphajaisiddhi, T., and Druilhe, P. (1990) Antibodies that protect humans against *Plasmodium falciparum* blood stages do not on their own inhibit parasite growth and invasion in vitro, but act in cooperation with blood monocytes. *J. Exp. Med.* **172,** 1633–1641.
2. Bouharoun-Tayoun, H. and Druilhe, P. (1992) *P. falciparum* malaria. Evidence for an isotype imbalance which may be responsible for the delayed acquisition of protective immunity. *Infect. Immun.* **60,** 1473–1481.
3. Bouharoun-Tayoun, H., Oeuvray, C., Lunel, F., and Druilhe, P. (1995) Mechanisms underlying the monocyte-mediated antibody-dependent killing of *Plasmodium falciparum* asexual blood stages. *J. Exp. Med.* **182,** 409–418.
4. Oeuvray, C., Bouharoun, T. H., Gras, M. H., Bottius, E., Kaidoh, T., Aikawa, M., et al. (1994). Merozoite surface protein-3: a malaria protein inducing antibodies that promote *Plasmodium falciparum* killing by cooperation with blood monocytes. *Blood* **84,** 1594–1602.
5. Jefferis, R., Reimer, C. B., Skvaril, F., de Lange, G., Ling, N. R., Lowe, J., et al. (1985). Evaluation of monoclonal antibodies having specificity for human IgG subclasses: results of an IUIS/WHO collaborative study. *Immunol. Lett.* **10,** 223–248.

41

Systemic Nitric Oxide Production in Human Malaria

I. Analysis of NO Metabolites in Biological Fluids

Nicholas M. Anstey, Craig S. Boutlis, and Jocelyn R. Saunders

1. Introduction
1.1. Nitric Oxide Generalities

Nitric oxide (NO) and the related species nitrosothiols, have multiple important physiological and pathological roles in health and disease that have been extensively reviewed elsewhere *(1–6)*. In brief, NO is involved in modulating or mediating host resistance to tumors and microbes, regulation of blood pressure and vascular tone, neurotransmission, learning, neurotoxicity, and control of cellular growth, and differentiation.

NO is synthesised from the amino acid L-arginine by the actions of a family of enzymes, the NO synthases (NOS), each isoform of which is encoded by a separate gene *(see* Chapter 43). Two NOS isoforms, NOS1 (nNOS) and NOS3 (eNOS), are constitutively expressed and produce low-level NO production. NOS2 (inducible NOS or iNOS) is transcriptionally induced by proinflammatory cytokines (such as tumor necrosis factor [TNF] and interferon-γ [IFN-γ]) and microbial products (e.g., lipoplysaccharide [LPS]). Human mononuclear cell NOS2 expression is increased in African children with asymptomatic malaria.

Immunologically, NO has both beneficial and toxic effects. Because of the large body of evidence implicating proinflammatory cytokines in both protective and pathological immune responses in malaria, and because NO is a downstream mediator and modulator of the cytokine network, there has been much interest in the role of NO in the immune response to malaria and in the pathophysiology of malaria. In studies to date NO appears to be an important mediator/modulator of the protective immune response to all stages of *Plasmodium* infections *(7)*. NO-related activity against liver stages and sexual blood stages is antiparasitic in vitro. In rodent studies, NO production has been associated with both antidisease host-protective effects as well as the direct antiplasmodial effects found in vitro *(7)*. However studies attempting to measure NO production in human malaria have on the whole been more difficult to interpret, and have illustrated the range of methodological difficulties involved in clinical studies of NO biology, and the need for careful control for confounding variables. Recent studies in African children that have controlled for diet and renal impairment have shown an inverse association between disease severity and systemic NO production (as measured by corrected plasma and urine levels of NO metabolites [NOx]) and mononuclear cell inducible nitric oxide synthase (NOS2) *(7)*. Because host immune responses and malaria disease phenotype in humans are influenced by many variables, it will be important to validate the association between NO production and disease-protection in other age groups and in areas with different malaria epidemiology. Such clinical studies must control for potential confounding variables if NOx levels are used as markers of NO production.

In this chapter, we describe a method for the measurement of NO metabolites in bodily fluids, and in the following chapter (*see* Chapter 42), we describe measurement of mononuclear cell inducible nitric oxide synthase (NOS2) expression. These are all measures of systemic NO production that do not measure organ-specific NO production. Local expression of NOS isoforms can be detected by tissue immunohistochemistry and their activity by tissue arginine to citrulline conversion. Details of these tissue assays, however, are beyond the scope of these chapters.

1.2. NO Metabolites in Malaria

In the presence of oxygen, NO is rapidly converted to the stable metabolites, nitrite and nitrate *(8,9)*. Measurement of nitrate + nitrite (NOx) in plasma and urine is a valid marker of systemic NO production in rodents and humans in a variety of disease states *(7)*, provided that there is adequate control for the potential confounding effects of dietary nitrate ingestion and nitrate retention in renal impairment. Recent studies in African children controlling for diet and renal impairment have shown an inverse association between disease severity and NO production as measured by corrected plasma and urine NOx levels. Several early studies described plasma NOx levels in human malaria *(10–15)*. However, it has been difficult to extrapolate NO production from the NOx levels reported in these studies because of absent or insufficient numbers of disease-free control subjects and inadequate control for the potential confounding effects of dietary nitrate ingestion *(16–18)*, renal impairment *(19,20)* decreased fractional excretion of NOx *(7)* and altered volume of distribution of NOx in malaria. All of these confounders will act to increase plasma NOx levels in malaria without reflecting increased NO production *(16,17)*. For these reasons, early studies from Papua New Guinea (PNG), Brazil, Vietnam, and Gabon *(11,13–15)* used high uncorrected plasma nitrate levels to extrapolate increased NO production in severe and cerebral malaria, particularly those with a fatal outcome *(11)*. More recently, better controlled studies have shown that while uncorrected plasma NOx levels are higher in fatal compared with nonfatal severe and cerebral malaria, this difference disappears when individual NOx levels are corrected for renal impairment *(7)*, and at least in coastal African children are markedly lower in cerebral malaria than in fasting malaria-exposed control children *(7)*. Details of sample collection and NOx analysis that will avoid or correct for these confounding variables are included in this chapter. The following method describes the measurement of the NO metabolites, nitrate and nitrite. It relies on the conversion of nitrate to nitrite using *Aspergillus* nitrate reductase, then quantitation of total nitrite (the sum of preexisting nitrite and nitrite derived from nitrate) using the Griess reagents.

2. Materials

2.1. Nitrate and Nitrite Stock Standards

Prepare a 1 M stock of sodium nitrate (NaNO3) (Sigma S-8170) and sodium nitrite. Prepare working stocks of 1 mM by making a 1:1000 dilution of each. Dilute working stocks to 160 μM for the first standard, then serially dilute to 2.5 μM by transferring 0.5 mL into successive vials containing 0.5 mL of dH_2O. Include 0 μM in standard curve since there is usually low-level background in most samples. Store standards at 4°C.

2.2. Stock Solutions and Storage

1. 1 M Tris-HCl, pH 7.5–8.0.
2. 0.2 mM NADPH (Sigma). Store 110-μL aliquots at –20°C.
3. 50 mM Glucose 6-phosphate (G6P) (Boehringer Manneheim). Store 220-μL aliquots at –20°C.
4. 100 U/mL Glucose 6-phosphate dehydrogenase (G6PDH) (Boehringer Manneheim). Store 40-μL aliquots at –20°C.
5. *Aspergillus* nitrate reductase (NR) (Boehringer Manneheim; 20 U/vial). Add 2 mL autoclaved distilled, deionized water = 10 U/mL. Working quickly with vial in ice, freeze 110-μL aliquots of 10 U/mL in 1.5-mL sterile polypropylene freezing tubes at –20°C without delay. Use one such tube for each microtiter plate.

2.3. Premixed Griess (1% Sulfanilamide and 0.1% of Naphthylethylenediamine)

To 97.7 mL of distilled, deionized water and 2.3 mL of concentrated (85%) phosphoric acid (orthophosphoric acid), add 1.0 g of sulfanilamide (Sigma S-9251), and 0.1 g of naphthylethylenediamine (Sigma N-9125). Store in the dark, at 4°C. Discard if the solution turns pink.

3. Methods

3.1. Controlling for Exogenous Nitrate Prior to Sample Collection

This is essential to ensure that endogenous NO production is not attributed to exogenous sources of nitrate such as: (a) ingested nitrate in food and water; (b) inhaled nitrogen oxides (NOx) from air pollution and cigarettes; and (c) contamination of samples with extraneous NOx (e.g., vials) during collection and processing *(17) (see* **Note 1**).

3.1.1. Diet

Nitrate in the human diet comes predominantly from cured meats and vegetables, especially stems, roots, leafy plants, and berry fruits *(16)*. Dietary nitrate is renally excreted and has a half-life of 6–8 h. We have found in both adults and children that a protocol of a low nitrate dinner plus an overnight fast, followed by fasting morning spot collections of urine and plasma, gives NOx levels comparable to those in 24-h collections found after a 24- or 48-h low nitrate diet *(17)*. Low nitrate dinners have been described for use in American *(17)* and African *(7)* settings. In most malaria-endemic areas, a readily available low nitrate dinner is a quarter or half baked chicken served with rice boiled in distilled water or local water that is known to be low in nitrate *(see below) (7)*. Unrestricted nitrate-free water is allowed overnight. The following morning the first void urine is discarded, followed by 1–3 glasses (depending on age) of distilled/nitrate-free water and collection of the second-void urine (along with venous blood) prior to breakfast. In anorexic patients unwell with malaria, an alternative approach is to determine the time since the last significant ingestion of food, which is usually sufficiently long ago (>18 h) in those who are comatose or otherwise severely unwell to exclude a confounding effect of dietary nitrate.

3.1.2. Water

Some environments have high nitrate concentrations in water, often from nitrate fertilizers leaching into water supplies. Although nitrates are usually removed by water treatment facilities in developed countries, this may not be the case in malarious areas. It is recommended that nitrate concentrations be measured in a variety of local community and clinic water supplies prior to commencement of sample collections. Provision of distilled water may be required if more than trivial concentrations of NOx are detected. For studies in infants, we have found that common infant milk formulas (e.g., Enfalac™ and Olac™) and oral rehydration electrolyte preparations (e.g., Gastrolyte™) have undetectable NOx levels (<2 μM) when prepared according to the manufacturer's instructions).

3.1.3. Exogenous NOx in Vials and Plasticware

Corrections must be made to adjust for the contaminating nitrate present in anticoagulant tubes and ultrafilters. Test for NOx contamination of vacutainer tubes and filters prior to sample collection and processing (*see* **Note 2**).

3.2. Sample Collection

3.2.1. Urine

Spot urine samples are collected in sterile pots in adults or from bag urine samples in infants. These contain isopropanol to prevent reduction of nitrate to more reduced unmeasurable compounds, for example, ammonia by denitrifying bacteria. The ratio of urine to isopropanol in the final collection should be between 5:1 and 20:1. Alternatively, in the field, a freshly collected urine specimen may be immediately sterilized by filtering through a 0.22-μm pore size syringe filter into a sterile plastic tube and stored at room temperature. This latter strategy is especially useful for collection and storage of spot urine samples (*see* **Note 3**). NOx collected in isopropanol is stable at room temperature, and although urine creatinine is stable for at least 1 wk at 4°C, the urine is best frozen for long-term storage.

3.2.2. Blood

Venous blood is collected as serum, or plasma in heparin anticoagulant. EDTA will inhibit nitrate reductase used in the method below. Lithium heparin is thus the preferred anticoagulant. NOx is stable in sterile blood at room temperature. While time to centrifugation may alter the ratio of nitrite to nitrate, total NOx is not affected. Early centrifugation may, however, be required for cytokine determination (*see* **Note 4**).

3.3. Sample Preparation

1. Thaw urine and plasma sample and vortex thoroughly.
2. For urine, centrifuge at 13,000g for 3 min; dilute 1 in 10 with distilled water to a volume of 150–250 μL.
3. Plasma is assayed undiluted, but must be deproteinated prior to the assay. Ultrafiltration of 200–250 μL of plasma through Millipore Ultrafree-MC 10,000 NMWL regenerated cellulose filters for 90 min at 15,000g at 15°C results in approx 100 μL of ultrafiltrate (*see* **Note 5**).

3.4. Measurement of NO Metabolites

1. Add 50 μL of sample or standard in duplicate to each well of a 96-well microtitre plate.
2. Immediately prior to each assay, make 1:10 dilutions of the stock solution aliquots of

NADPH (990 μL of sterile distilled deionized water to 110 μL of 0.2-m*M* aliquot), G6P (1980 μL of sterile distilled deionized water to 220 μL of 50 m*M* aliquot), and G6PDH (360 μL sterile deionized water to 40 μL of 100 U/mL aliquot). Mix 2200 μL of 5 m*M* G6P with 330 μL of 10.0 U/mL G6PDH. Thaw nitrate reductase (NR) last, *immediately before use*: Prepare NR at 1 U/mL by adding 990 μL of ddH$_2$O to each 110-μL aliquot of 10 U/mL. Never refreeze and reuse stocks. The volumes given here yield sufficient reaction mix for one plate of 96 reactions.
3. Add the following reagents to each well, **in the order listed**: 7 μL of 1 *M* Tris-HCl (pH 7.5–8.0), 10 μL of 0.02 m*M* NADPH, 23 μL of G6P/ G6PDH mixed as in **Subheading 3.4., step 2**, and 10 μL of 1.0 U/mL nitrate reductase.
4. Total 50 μL of sample/standard plus 50 μL of reagents, to give a final volume of 100 μL per well.
5. Mix, then incubate in the dark at room temperature for 30 min.
6. To each well of a second microtiter plate, add 75 μL of Griess reagent mix just prior to the end of the incubation period.
7. Transfer 75 μL per well of reaction mix from the first microtiter plate to the plate containing Griess reagent (from **step 5**).
8. Mix, then incubate for 10 min at room temperature.
9. Read absorbance at 540–550 nm.

3.5. Result Analysis, Interpretation, and Correction For Renal Impairment

1. Total nitrite + nitrate is indicated by a pink color. NOx concentrations are calculated by comparing to the *nitrate* standard curve, and multiplying by the dilution factor. The standard curve should be linear between 5 μ*M* and 160 μ*M*. If concentrations are outside these ranges, repeat assay using a lesser or greater dilution (*see* **Notes 6** and **7**).
2. The adequacy of nitrate reduction can be checked by comparing nitrite and nitrate standard curves: check stocks of NR and repeat assay if <90% of nitrate is converted to nitrite.
3. Spot urine NOx concentrations are normalized for differences in hydration, and expressed as spot urine NOx:creatinine ratios. Creatinine concentration in urine and plasma can be measured using an Ektachem autoanalyzer (Eastman Kodak Co., Rochester, NY) or using kits from Sigma.
4. Because 60–73% of nitrate is renally excreted, nitrate is retained and plasma NOx elevated in otherwise healthy humans with renal impairment. Plasma NOx results are therefore corrected for renal dysfunction and expressed as plasma NOx:creatinine ratios.
5. Although suggestive, altered production of NOx during inflammation compared to a control group does not prove that altered NOS2 activity is responsible for the change.
6. We have used quantitative measurement of NOS2 in blood mononuclear cells by immunoblotting with specific monoclonal antibody (*see* Chapter 43) to correlate changes in NOx concentration with NOS2 protein to provide evidence that NOS2 activity is the source of altered nitric oxide synthesis in malaria.

4. Notes

1. NOx results are difficult to interpret if dietary nitrate ingestion and renal impairment are not controlled for.
2. While contaminating NOx was undetectable in Becton-Dickinson (B-D) sodium heparin vacutainers, we have found small amounts present in lithium heparin and EDTA vacutainers, with lot to lot variation. Insignificant contamination can become significant if only small volumes of blood are collected *(17)*. Using capillary electrophoresis, we have found considerable variation in NOx contamination of ultrafiltration units. Centrifugation of 70 μL of deionized water through units resulted in mean contaminating NOx

levels in ultrafiltrates that ranged from 3.9 µM and 6.27 µM for two different lot numbers of Millipore ultrafree-MC regenerated cellulose filters through to 182 µM for Gelman Sciences nanospin 10,000 MWCO filters. If it is not possible to use uncontaminated tubes and filters, use low NOx tubes/filters with the same lot number, collect consistent volumes of blood into tubes, add consistent volumes of plasma/serum to filtration units, and include an ultrafiltrate from 200 µL of dH_2O in each assay. A standard correction for resultant contamination can then be applied. We found that washing filters (by vigorous squirting of deionized water onto both sides of filters three times or spinning deionized water through filters at least four times) did not reduce contaminating NOx levels in the ultrafiltrate. Finally, certain brands of gloves, particularly latex, are contaminated with significant amounts of nitrate *(21)* and this is another potential source of extraneous NOx in the laboratory.

3. Thiomersal, an alternative urinary sterilizing agent inhibits nitrate reductase and is thus not recommended.
4. To measure NOx in EDTA plasma, supplementary calcium and magnesium needs to be added at the time of the nitrate reductase assay. Alternatively NOx can be quantitated by capillary electrophoresis.
5. Ultrafiltrate in a refrigerated centrifuge to avoid evaporation.
6. Urine from children with cerebral malaria and adults with severe malaria does not inhibit nitrate reductase at 1:19 and 1:10 dilutions respectively, although pediatric cerebral malaria urine at a 1:5 dilution does. Avoid NOx measurements in urine diluted <1:10.
7. Although this method can be used to measure NOx in cerebrospinal fluid (CSF), interpretation of CSF NOx is difficult: levels are much lower than in plasma (low micromolar range). Intravenous infusion of nitrate in rabbits is known to increase CSF NOx. It is not known how much is derived from choroid plexus filtration of plasma NOx and how much from NO production in adjacent brain parenchyma.

References

1. Bogdan, C. (1998) The multiplex functions of nitric oxide in (auto)immunity. *J. Exp. Med.* **187,** 1361–1365.
2. Fang, F. C. (1997) Mechanisms of nitric oxide-related antimicrobial activity. *J. Clin. Invest.* **99,** 2818–2825.
3. Moncada, S. and Higgs, A. (1993) The L-arginine-nitric oxide pathway. *N. Engl. J. Med.* **329,** 2002–2012.
4. Moncada, S. and Higgs, E. A. (1995) Molecular mechanisms and therapeutic strategies related to nitric oxide [review]. *FASEB J.* **9,** 1319–1330.
5. Nathan, C. and Xie, Q. W. (1994) Nitric oxide synthases—Roles, tolls, and controls. *Cell* **78,** 915–918.
6. Weinberg, J. (1998) Nitric oxide production and nitric synthase type 2 expression by human mononuclear phagocytes: a review. *Mol. Med.* **4,** 557–591.
7. Anstey, N. M., Weinberg, J. B., Hassanali, M. Y., Mwaikambo, E. D., Manyenga, D., Misukonis, M. A., et al. (1996) Nitric oxide in Tanzanian children with malaria: inverse relationship between malaria severity and nitric oxide production/nitric oxide synthase type 2 expression. *J. Exp. Med.* **184,** 557–567.
8. Kosaka, H. K., Imaizumi, K., Imai, K., and Tyuma, I. (1979) Stoichiometry of the reaction of oxyhemoglobin with nitrite. *Biochim. Biophys. Acta.* **581,** 184–188.
9. Westfelt, U. N., Benthin, G., Lundin, S., Stenqvist, O., and Wennmalm, A. (1995) Conversion of inhaled nitric oxide to nitrate in man. *Br. J. Pharmacol.* **114,** 1621–1624.
10. Agbenyega, T., Angus, B., Bedu-Addo, G., Baffoe-Bonnie, B., Griffin, G., Vallance, P., et al. (1997) Plasma nitrogen oxides and blood lactate concentrations in Ghanaian children with malaria. *Trans. R. Soc. Trop. Med. Hyg.* **91,** 298–302.
11. Al Yaman, F. M., Mokela, D., Genton, B., Rockett, K. A., Alpers, M. P., and Clark, I. A. (1996) Association between serum levels of reactive nitrogen intermediates and coma in children with cerebral malaria in Papua New Guinea. *Trans. R. Soc. Trop. Med. Hyg.* **90,** 270–273.
12. Cot, S., Ringwald, P., Mulder, B., Miailhes, P., Yap-Yap, J., Nussler, A. K., et al. (1994) Nitric oxide in cerebral malaria. *J. Infect. Dis.* **169,** 1417,1418.
13. Kremsner, P. G., Winkler, S., Wildling, E., Prada, J., Bienzle, U., Graninger, W., and Nussler, A. K. (1996) High plasma levels of nitrogen oxides are associated with severe disease and correlate with

rapid parasitological and clinical cure in *Plasmodium falciparum* malaria. *Trans. R. Soc. Trop. Med. Hyg.* **90,** 44–47.
14. Nussler, A. K., Eling, W., and Kremsner, P. G. (1994) Patients with *Plasmodium falciparum* malaria and *Plasmodium vivax* malaria show increased nitrite and nitrate plasma levels. *J. Infect. Dis.* **169,** 1418,1419.
15. Prada, J. and Kremsner, P. G. (1995) Enhanced production of reactive nitrogen intermediates in human and murine malaria. *Parasitol. Today* **11,** 409,410.
16. Anonymous (1981) Nitrate, nitrite and nitrogen oxides: environmental distribution and exposure of humans, in *The Health Effects of Nitrate, Nitrite, and N-Nitroso Compounds*, vol. 1. (Peter, F. M., ed.), National Academy Press, Washington, DC, pp. 3–52.
17. Granger, D., Anstey, N., Miller, W., and Weinberg, J. (1999) Measuring nitric oxide production in human clinical studies. *Meth. Enzymol.* **301,** 49–61.
18. Mitchell, H. H., Shonle, H. A., and Grindley, H. S. (1916) The origin of nitrates in the urine. *J. Biol. Chem.* **24,** 461–490.
19. Anstey, N. M., Granger, D. L., and Weinberg, J. B. (1997) Nitrate levels in malaria. *Trans. R. Soc. Trop. Med. Hyg.* **91,** 238–240.
20. Mackenzie, I. M. J., Ekangaki, A., Young, J. D., and Garrard, C. S. (1996) Effect of renal function on serum nitrogen oxide concentrations. *Clin. Chem.* **42,** 440–444.
21. Makela, S., Yazdanpanah, M., Adatja, I., and Ellis, G. (1997) Disposable surgical gloves and Pasteur (Transfer) pipettes as potential sources of contamination in nitrite and nitrate assays. *Clin. Chem.* **43,** 2418.

42

Systemic Nitric Oxide Production in Human Malaria

II. Analysis of Mononuclear Cell Nitric Oxide Synthase Type 2 Antigen Expression

Jocelyn R. Saunders, Mary A. Misukonis, J. Brice Weinberg, and Nicholas M. Anstey

1. Introduction

As described in Chapter 42, niric oxide (NO) is synthesized from the amino acid L-arginine by the actions of a family of enymes, the NO synthases (NOS), each isoform of which is encoded by a separate gene. Two NOS isoforms are calcium-dependent and constitutively expressed and produce low levels of NO: NOS1 (neuronal NOS or nNOS), which is found mostly in neurons and skeletal muscle, and NOS3 (endothelial NOS or eNOS), which is found mostly in endothelial cells. NOS1 is critical for neurotransmission and learning, and NOS3 regulates vascular tone and adhesion of circulating cells. Inducible NOS (iNOS or NOS2) is transcriptionally induced by proinflammatory cytokines (such as tumor necrosis factor-α [TNF-α] and interferon-γ [IFN-γ]) and microbial products (e.g., lipoplysaccharide [LPS]). iNOS is calcium-independent, expressed by many cell types (especially mononuclear phagocytes, hepatocytes, chondrocytes and smooth muscle cells) and is responsible for high output NO production *(1–3)*. While initial studies showed that iNOS expression within mouse macrophages resulted in high-output NO production, until recently there was doubt as to whether human macrophages were capable of producing NO. There is now clear evidence however that human monocytes and tissue macrophages *can* express iNOS and produce NO both in vitro and in vivo *(3)*, including evidence from malaria-exposed Tanzanian children *(4)*.

We have recently found in African children an inverse association between malaria disease severity and iNOS expression by peripheral blood mononuclear cells (PBMC) *(4)*. Plasma and urine concentrations of NO metabolites were correlated with PBMC iNOS expression. We describe here an immunoblot method to allow these associations to be examined in other populations and other age groups, in settings with different malaria epidemiology.

2. Materials

2.1. Cell Culture of Positive and Negative Controls

See **Subheading 2.5.**

1. Cells of the mouse macrophage cell line J774: grow in RPMI 1640 with glutamine, 10% fetal calf serum (FCS), 100 U/mL penicillin, 100 µg/mL streptomycin and 10 mM HEPES at 37°C with 5% CO_2 (*see* **Note 1**).
2. Cells of the human colon carcinoma cell line DLD-1: Grow in Dulbecco's *minimum essential medium* (DMEM) with glutamine, 10% FCS, 100 U/mL penicillin, 100 µg/mL streptomycin, and 10 mM HEPES at 37°C with 5% CO_2.
3. Trypsin ethylenediaminetetraacetic acid (EDTA): 0.05% trypsin, 0.53 mM EDTA, for harvesting cells.
4. Cell lysis reagents: Cocktail of double-distilled water containing Leupeptin (final concentration, 10 µg/mL), PMSF (final concentration, 1 µg/mL; prepare 100 µM stock in dimethyl sulfoxide [DMSO]), Antipain (final concentration, 10 µg/mL), and Aprotinin (final concentration, 10 µg/mL).

2.2. Protein Estimation

Bio-Rad protein assay reagents (cat. no. 500-0002.): Use according to manufacturer's instructions.

2.3. Sodium Dodecyl Sulfate-Polyacrylamide Gel Electrophoresis and Western Blotting

Sodium dodecyl sulfate-polyacrylamide gel electrophoresis (SDS-PAGE) minigel reagents: Prepare using standard methods *(5,6)*.

1. SDS sample buffer: 4% SDS, 10 mM Tris-HCL, pH 7.4, 5.0 mM EDTA, pH 8.0. Prepare 100 mL of 1X solution by mixing 4 g of SDS, 1 mL of 1 M Tris-HCl stock, pH 7.4, and 1 mL of 0.5 M EDTA stock, pH 8.0. Make up to 100 mL with ddH_2O.
2. Sample loading buffer: 125 mM Tris-HCl pH 6.8, 25% glycerol, 6% SDS, dithiothreitol (DTT), β-mercaptoethanol. Prepare 20 mL of a 4X stock solution by mixing 5 mL of 0.5 M Tris-HCl, pH 6.8, 5 mL of glycerol (100%), 1.2 g of SDS, and 0.3 g of DTT. Add ddH_2O to 20 mL final volume. Freeze at –20°C in 1-mL aliquots. On thawing, add 20 µL of β-mercaptoethanol to give a final concentration of 2%.
3. Gel running buffer: 25 mM Tris-base, 192 mM glycine, 0.1% SDS. Prepare 1 L of 10X stock solution by mixing 30.27 g of Tris-base, 144.1 g of glycine, and 10 g of SDS. Make up to 1 L final volume with ddH_2O, adjusting pH to 8.0. Prior to use, dilute stock 1/10 with ddH_2O.
4. Transfer buffer: 20 mM Tris-base, 150 mM glycine. Prepare 1 L of 10X stock solution by mixing 24.2 g of Tris-base and 112.6 g of glycine. Make up to 1 L with ddH_2O. Adjust pH to 8.0 with concentrated HCl. Prior to use, dilute stock 1/10 with 200 mL of methanol and ddH_2O.
5. Stripping buffer:
    a. 2% SDS, 62.5 mM Tris-HCl, pH 6.8, 100 mM β-mercaptoethanol: Prepare 100 mL of buffer by mixing 2 g of SDS, 12.5 mL of 0.5 M Tris-base, pH 6.8, and 700 µL of β-mercaptoethanol. Make up to 100 mL with ddH_2O. Incubate blot in the buffer for 30 min at 50–70°C. Rinse with wash buffer and repeat blocking followed by antibody incubations.
    b. 0.1 M glycine, pH 2.9: Prepare 100 mL of buffer by dissolving 0.75 g of glycine in ddH_2O to 100 mL final volume, adjusting pH to 2.9 with concentrated HCl. Rinse blot in H_2O followed by 20 min in strip buffer at room temperature. Rinse with wash buffer and repeat blocking followed by antibody incubations.

2.4. Immunoprobing and Detection

1. Wash buffer triethanolamine-buffered saline (TBS)-Tween: 10 mM Tris-HCl, pH 8.0, 100 mM NaCl, 1% Tween-20, pH 8.0. Prepare 1 L of 10X stock solution by mixing 90.0 g of NaCl and 12.1 g of Tris-base. Make up volume to 1 L with ddH_2O, adjusting pH to 8.0 with concentrated HCl. Store at 4°C. Prior to use, dilute stock 1/10 with ddH_2O and add 1 mL of Tween-20 per 1 L to give a final concentration of 0.1% (*see* **Note 2**).
2. Blocking and diluent, "blocking buffer": 5% dry milk powder in TBS-Tween. Prepare fresh and pH to 8.0 with NaOH (*see* **Notes 2** and **3**).
3. Primary antibodies:
   a. Anti-iNOS monoclonal antibody, (cat. no. N32020, Transduction Laboratories, Lexington, KY).
   b. Polyclonal rabbit antiactin antibody (cat. no. A2066, Sigma).
4. Secondary antibodies:
   a. Horseradish peroxidase (HRP) conjugated polyclonal anti-mouse IgG (cat. no. 115-035-062, Jackson ImmunoResearch Laboratories, West Grove, PA).
   b. Horseradish peroxidase (HRP) conjugated polyclonal anti-rabbit IgG (cat. no. R14745, Transduction Laboratories).
5. Detection by chemiluminescence: ECL Western blotting chemiluminescence detection reagents (cat. no. RPN 2106, Amersham Pharmacia Biotech).

2.5. Positive and Negative Controls

For known negative and positive NOS2 control cellular extracts, we use cells from the murine macrophage cell line J774 (American Type Culture Collection) (*see* **Note 1**) and cells from the human colon cell line DLD-1 (American Type Culture Collection), untreated and treated with LPS and cytokines as follows:

2.5.1. iNOS Stimulation of Cells

1. J774-1 cells plated at $3–5 \times 10^4$ cells/mL are cultured with rIFN-γ (100 to 200 U/mL) and LPS (200 ng/mL) for 3 d before harvesting.
2. DLD-1 cells grown to complete confluence and then cultured with human TNF-α (100 U/mL), IL-6 (200 U/mL), rIFN-γ (100 U/mL) and IL-1β (0.5 ng/mL) for 2 to 3 d, depending on viability of the cells before harvesting (*see* **Note 4**).

2.5.2. Lysis Preparation

1. Harvest stimulated cells (*see* **Note 5**), wash twice in 15 mL of PBS by centrifugation at approx 200g.
2. Transfer cells to a 1.5-mL microcentrifuge tube, centrifuge and remove all remaining buffer.
3. Lyse cells by addition of protease inhibitor cocktail, outlined in **Subheading 2.1.**, approx 150 µL for the harvested contents of each 75-cm^2 tissue culture flask (*see* **Note 6**). Mix well with a pipette.
4. Freeze mix once in liquid nitrogen and thaw on ice.
5. Centrifuge tube at approx 10,000g in a 1.5-mL microcentrifuge tube at 4°C for 10 min to remove cell debris.
6. Centrifuge the supernatant an additional two times at approx 10,000g at 4°C for 10 min to remove any residual cell debris.
7. Quantitate protein using the Bio-Rad protein assay.
8. Aliquot preparation at 5 µg per aliquot (at approx 1 µg/µL) for immunoblots and 15 µg for arginine to citrulline conversion assay (*see* **Notes 6** and **7**) and immediately store at –70°C.

3. Methods

3.1. Sample Preparation

1. Purify peripheral blood mononuclear cells (PBMCs) from venous blood by Ficoll-Hypaque gradient centrifugation. The volume of blood required for the assay will depend on the mononuclear cell count. PBMCs from 5 to 10 mL in healthy malaria exposed adults and 3–5 mL in malaria-exposed children (in whom monocyte counts are usually higher) is usually adequate for a NOS2 immunoblot (and usually enough for a repeat assay if required). Smaller volumes may be adequate in those with a monocytosis (e.g., such as that often seen in acute malaria). Usually at least 3 million to 5 million cells are required.
2. Wash PBMCs in 15 mL of PBS by centrifugation and count the cell number using trypan blue exclusion.
3. Transfer the PBMCs to a 1-mL cryovial, centrifuge and remove all remaining buffer.
4. Freeze cell pellets at –70°C until use.
5. To analyze, lyse frozen cell pellet in the protease inhibitors mix (*see* **Subheading 2.1.**). The volume added will depend on PBMC numbers/protein content of pellet. Add approx 10 µL per million PBMC. Add 50 µL of protease inhibitor cocktail (e.g., for 5 million cells). *See* **Notes 6** and **7** if cell lysate is also to be used for measuring NOS2 activity by arginine to citrulline conversion. Leave on ice for 15 min.
6. Centrifuge sample in the cryovial at approx 10,000g at 4°C for 10 min to remove cell debris.
7. Centrifuge supernatant once more at approx 10,000g at 4°C for 10 min.
8. Remove 1 µL for protein estimation using the Biorad protein assay.
9. Add 50 µL of SDS sample buffer (4% SDS, 10 mM Tris-HCl, pH 7.4, 5.0 mM EDTA, pH 8.0) to remaining sample and leave on ice for 15min.
10. Aliquot cell lysate either at 30–50 mg per aliquot, and store cell preparation immediately at –70°C.

3.2. General Procedure for Analysis of NOS2 Expression

3.2.1. Protein Separation by SDS-PAGE

1. Prepare a standard 8% mini Laemmli SDS-PAGE denaturing discontinuous polyacrylamide gel.
2. Dilute controls and samples 1 in 4 with loading buffer (see reagents).
3. Centrifuge briefly.
4. Heat to 95°C for 2 min followed by ice for 2 min.
5. Centrifuge briefly.
6. Load positive controls J774 and DLD-1 at 5 µg and 10 µg per well, respectively.
7. Load human samples at 30 µg per well for minigels and 50 µg for large gels (*see* **Note 8**).
8. Run gel at 30 to 40 mamp per gel.

3.2.2. Western Transfer

1. Separated proteins are electrophoretically transferred overnight at 4°C using 50 mamp onto a polyvinylidene difluoride (PVDF) membrane.
2. Monitor transfer by staining membrane in Ponceau S for 10 min, followed by PBS rinse (*see* **Notes 9** and **10**).
3. Coomassie stain the gel after transfer.
4. Improved transfer may be achieved by increasing transfer time and milliamperes used

3.2.3. Immunoprobing of the Membrane

1. Block membrane in 10–20 mL of 5% fresh blocking solution, pH 8.0, for 2 h at room temperature with rocking in a small tray (approx 8 × 8 cm, a large weigh tray is ideal) (*see* **Notes 11** and **12**).

Analysis of NOS2 Antigen Expression

2. Place the membrane and specific primary antibody (mouse anti-iNOS monoclonal antibody), diluted at 1/500 in 5 mL of blocking solution, in a 50-mL Falcon tube (*see* Notes **2** and **3**). Incubate for 1 hr at room temperature or overnight at 4°C on a roller mixer. Alternatively incubate the membrane in primary mouse anti-NOS antibody diluted at 1/500 in 10 mL of blocking solution, in a small tray with rocking, making sure the membrane in covered by the blocking solution.
3. Rinse blot in the small tray and wash with 10 to 20 mL of TBS-Tween with rocking over approx 1 h changing buffer every 15 min. NOTE: never leave the membrane in wash buffer overnight as this will strip the antibodies from the blot.
4. Incubate blot in secondary anti mouse HRP antibody diluted at 1/5,000 in 10 mL of blocking solution for 1 h at room temperature or overnight at 4°C (*see* **Note 13**). Again incubate in either a 50-mL Falcon tube or a small tray. The supplier's product description sheet gives an indication of the appropriate dilution, but optimization is required, investigating several different dilutions.
5. Rinse blot and wash with TBS–Tween with rocking more than 1 h, changing buffer every 15 min.
6. Develop blot in a total of 8 mL of chemiluminescence reagent prepared according to the manufacturer's instructions.

3.2.4. Quantitation and Interpretation

NOS2 levels in samples can be quantified by densitometry, expressing a ratio of the NOS2 band density to either that of a known positive control or that of an actin band (*see* **Notes 14** and **15**). Densitometry of actin, with a mol wt of approx 43, can be determined either by coincubation of the membrane with an antiactin antibody and appropriate secondary antibody or by stripping the membrane after iNOS analysis and reprobing with the actin antibody as listed in reagents **Subheading 2.1.3.**

1. After blocking the membrane as detailed in **Subheading 3.2.3.** incubate with both primary antibodies, the anti-actin antibody at a 1/10,000 dilution in blocking solution. Incubate for 1 h at room temperature with shaking.
2. After washing as detailed in **Subheading 3.2.3.** The membrane can be cut across approximately at the 73-kDa marker. The top half of the membrane can then be incubated with anti-mouse HRP-conjugated antibody, and the bottom half with anti-rabbit HRP-conjugated antibody diluted at a 1/5000. Alternatively, the membrane can be stripped after the analysis of iNOS and reprobed with the antiactin antibody.
3. Wash the membrane as described in **Subheading 3.2.3.** and develop with chemiluminescence reagents as described in the manufacturer's instructions.

4. Notes

1. An alternative positive control cell line is the murine macrophage cells raw 264.7 plated at 1×10^5 cells/mL stimulated with rIFN-γ (200 U/mL) and LPS (1 µg/mL) for 3 d before harvesting, but often a doublet is seen with these cells on Western blot.
2. Antibody suppliers product description sheets should give an indication of the appropriate dilution but optimization of primary and secondary antibodies is essential with the several dilutions of antibody tested.
3. Primary and secondary antibodies vary in quality. Once a pair of primary and secondary antibodies are performing well, it is advised to store reserves of identical lot numbers.
4. The amount of cytokines used to induce iNOS in the DLD-1 colon cancer cell line may be doubled if a larger increase in iNOS is required.
5. Good lysis of J774 is essential and is dependent on harvesting the cells with a low trypsin incubation (0.05% with EDTA), to prevent clumping, followed by scraping the cells loose.

6. Note the importance of the volume of protease cocktail used to lyse the cells, adding too much lysis mix will reduce the protein concentration, preventing the loading of 5 µg on the gel. Optimization of this volume may be required.
7. The same positive control and PBMC samples may also be used in the arginine to citrulline conversion assay. Following sample preparation as detailed in **Subheading 3.1., step 5**, increase the volume of protease inhibitors mix added to the sample to 75 µL. Leave on ice for 15 min and centrifuge as described in **Subheading 3.1.6.** Before adding the SDS buffer in **Subheading 3.1., step 9** remove 25 µL of the sample and freeze in three 8-µL aliquots. These volumes may need to be adjusted should the protein concentrations be too low.
8. For minigels and maxigels, 30 µg and 50 µg of the human PBMC protein respectively loaded on the SDS-PAGE is recommended. Therefore the lysed preparation can be aliquoted to this amount per tube. For blood volume, it has been found that 5–10 mL of blood from an adult with approx 1×10^7 cells gives reliable results. With children, who in general have a higher mononuclear cell count, 4 mL is adequate in most cases.
9. It is important to note that the high mol wt of NOS2 (130) means it may transfer inefficiently. It is therefore important to monitor the transfer as detailed in **Subheading 3.2.2.**
10. Improved transfer may be achieved by increasing transfer time and mAmps.
11. High background can be reduced by increasing the percentage of milk powder in the blocking solution up to 8% and/or increasing the percentage of Tween-20 present in the wash buffer up to 0.3%.
12. Optimal binding of the antibodies requires that TBS-Tween and blocking solution should have a pH of 8.0. The blocking buffer should be fresh and adjusted to pH 8.0 just before use. Do not leave the membrane in blocking buffer alone overnight.
13. If problems with sensitivity occur, the antibody can be incubated with the blot at 37°C for 1–2 h.
14. Quantitation of NOS2 levels in reference to actin can also be achieved by stripping the membrane after NOS2 densitometry and reprobing with an antiactin antibody. The NOS2 levels can then be expressed as a ratio of the sample NOS2 band density to actin band density over the control NOS2 band density to actin band density.
15. The procedure can also be used for analysis of tissue NOS2 content. Tissue samples for analysis once harvested should be frozen prior to homogenization. The tissue should homogenized on ice, using a tissue grinder with 5 vol (v/w of tissue) of ice-cold double-distilled water containing the protease inhibitors described in reagents. The samples are then centrifuged before a protein estimation is made on the supernatant. The supernatant is aliquoted and frozen at –70°C *(4)*.

References

1. Bogdan, C. (1998) The multiplex functions of nitric oxide in (auto)immunity. *J. Exp. Med.* **187**, 1361–1365.
2. Moncada, S. and Higgs, A. (1993) The L-arginine-nitric oxide pathway. *N. Engl. J. Med.* **329**, 2002–2012.
3. Weinberg, J. (1998) Nitric oxide production and nitric synthase type 2 expression by human mononuclear phagocytes: a review. *Mol. Med.* **4**, 557–591.
4. Anstey, N. M., Weinberg, J. B., Hassanali, M. Y., Mwaikambo, E. D., Manyenga, D., Misukonis, M. A., et al. (1996) Nitric oxide in Tanzanian children with malaria: inverse relationship between malaria severity and nitric oxide production/nitric oxide synthase type 2 expression. *J. Exp. Med.* **184**, 557–567.
5. Bollag, D., Rozycki, M., and Edelstein, S. (1996) *Protein Methods.* Wiley, New York.
6. Harlow, E. and Lane, D. (1988) *Antibodies. A Laboratory Manual.* Cold Spring Harbor Press, Cold Spring Harbor, NY.

VI

CELL BIOLOGY TECHNIQUES

43

In Vitro Culture of *Plasmodium* Parasites

James B. Jensen

1. Introduction

Because it has been 25 yr since the successful cultivation of *Plasmodium falciparum* (1), most researchers do not remember how difficult it was to work with malaria parasites, especially in vitro. Before the development of current methods, malaria parasite cultures were always short-term, lasting only a few days with decreasing numbers until the parasites died out. Not only was it extremely inconvenient to constantly begin new cultures, but investigators worked with parasites that were abnormal in the sense that the overall population was dying. Sixty-four years passed between the initial studies of Bass and Johns (2) and the successful development of continuous cultures of *P. falciparum*. During that time, many reported advances were able to extend the time of cultivation by only a few days but always with the inevitable terminal results. Most of the early investigators on in vitro cultivation used the bird malaria parasite *P. lophurae* or the simian parasite *P. knowlesi*, but the first successful cultivation of any malarial parasite was *P. falciparum*, the most important of the human malaria species. Using essentially the same procedures, other species of malarial parasites now have been cultured, including *P. fragile*, *P. inui*, and *P. cynomolgi*, but these species offer no real advantages over the use of *P. falciparum*. Methods for the cultivation of *P. vivax* have been reported but these cultures apparently require a continuous source of human reticulocytes which limits the suitability and presents a nearly insurmountable barrier to all but a few laboratories (3). A recent review of the impact of continuous cultures of *P. falciparum* underscores the tremendous contributions of this technique on malaria research (4). Initially, two methods for cultivation were reported: the Petri dish, candle-jar technique and the continuous flow method (5). However, the latter procedure requires some sophisticated equipment and is expensive to maintain in terms of blood, medium, and serum and thus will not be discussed here. Present popular cultivation methods remain essentially the same as initially reported by Trager and Jensen (1), with moderate refinements outlined in this chapter.

2. Materials

1. Erythrocytes: Human red blood cells are stored at 4°C in the blood preservative, CPDA-1 which maintain the suitability of erythrocytes for malaria cultures for up to 35 d, although fresher cells generally give better results. The blood type does not matter, except that the

cells need to be compatible with the serum source used (*see* **Note 1**). To prepare CPDA-1, combine 26.3 g of sodium citrate, 3.27 g of citric acid, 31.9 g of dextrose, 2.22 g of sodium monobasic phosphate [$NaH_2PO4·H_2O$], 0.275 g of adenine, and 1000 g of water. Add CPDA-1 at a rate of 14 mL per each 100 mL of whole blood.

2. Parasites: *P. falciparum* parasites grow well once they have become adapted to culture. It would be difficult to determine just how many parasite isolates have become established in culture, but Professor David Walliker (Institute of Cell, Animal, and Population Biology, Edinburgh University, Kings Buildings, West Mains Road, Edinburgh, EH9 3JT Scotland) maintains a WHO-sponsored repository of *P. falciparum* strains suitable for cultivation. Many other investigators using malaria parasite cultures can provide frozen stabilates as well. Parasites obtained directly from human infections will generally grow well for a few cycles, after which the numbers drop off significantly and usually require weeks of multiple subcultures before they become permanently stabilized. However, it is far better to conduct experiments with established strains or clones. Most established strains of parasites do not produce gametocytes, although a few are excellent long-term producers of these sexual stages.

3. Gas mixtures: Since the parasites grow best in reduced O_2, and 5% CO_2, some means of providing an appropriate environment must be undertaken. The simplest means of achieving a suitable gas mixture is to use a "candle-jar," that is, a plastic or glass desiccator with a stopcock (*6*). Although adequate for most cultures, the candle-jar does not provide the optimal gas mixture and some strains do not grow well in its relatively higher oxygen environment (15–17%). Therefore, cylinders of premixed gasses (1% O_2, 5% CO_2, and 94% N_2) are used to replace air in of the culture system.

4. Culture medium: RPMI 1640, supplemented with 25 m*M* *N*-2-hydroxthylpiperazine-*N*'-2-ethane-sulfonic acid (HEPES) buffer with L-glutamine, without sodium bicarbonate is supplied in both powdered (1-L packages or 50-L bulk) and liquid form (Gibco-BRL, Life Technologies, Grand Island, NY). The liquid medium is more expensive and generally less suitable, probably due to instability of essential components (*see* **Note 2**). Because RPMI 1640 without $NaHCO_3$ (RPMI-incomplete medium, RP-I) stores longer, being more stable, and is used as a basic medium for washing erythroctes, making up serum supplements, diluting plasma-expanders, etc., it is widely used for many procedures outlined in this chapter. Moreover, complete medium (RP-C) is less stable, and because it contains serum or serum-replacements and often antibiotics, it is far more expensive and used only when the added components are necessary.

5. Serum supplementation: The culture medium will not support parasite development without further supplementation with serum or a serum replacement. In early studies, culture medium was supplemented with 15% AB human serum because it was compatible with all blood types from which original parasites were cultured. However, as noted above (*see* **Note 1**), the serum for stock parasite cultures only needs to be compatible with the erythrocytes used for continuous cultures. Human A+ serum is generally available from United States blood banks and is compatible with the widely available A and O erythrocytes. If new cultures are being established from human infections, type AB is normally used since the donor erythrocyte type cannot be controlled (*see* **Note 3**). Fortunately, commercially available lipid-rich supplemented bovine albumin-based serum replacements, Albumax I and Albumax II (Gibco-BRL) work as well or better than human serum. For several years now, we and others have used Albumax I or Albumax II in parasite culture medium.

6. Plasma-expanders Plasmagel or Physiogel: Theses are used to concentrate mature parasites from culture (*11*). Erythrocytes suspended in plasma, or plasma-expanders containing dextrans or modified gelatin, spontaneously induce longitudinal arrays or stacks of erythrocytes known as rouleaux. Since some dextrans are harmful to mature parasites and

In Vitro Culture

since these reagents are often difficult to procure in the United States, plasma expanders made from gelatin (widely used in Europe, such as Plasmagel, Roger Bellon Laboratories, Neuilly, France; USA distributor: HTI Corp., Buffalo, NY) *(11)* or gelatin dissolved in RPMI can induce rouleaux of uninfected and ring-infected erythrocytes from parasite cultures and are an adequate substitute *(7)*. These agglutinated cells rapidly settle, leaving trophozoite- and schizont-infected erythrocytes in suspension (*see* **Note 4**). To prepare a 2X stock solution of plasma-expanders, dissolve 1 g of 300 bloom gelatin (Sigma, St. Louis, MO) in 96 mL of RP-I at 60°C. Filter sterilize while warm, then cool to room temperature. Add 4 mL of 5% $NaHCO_3$ and store frozen $-20°C$.

7. Rowe's cryoprotectant: To prepare, add 70 mL of anhydrous glycerol to 180 mL of 4.2% sorbitol in 0.85% NaCl. Filter sterilize and store at 4°C *(8)*.
8. Giemsa stain: Parasite cultures are monitored by preparing thin films from the culture and staining them with Giemsa stain (Fisher Scientific, Pittsburgh, PA) (*see* **Note 5**).
9. High quality water: Either triple-distilled or 18-ohm, endotoxin-free water such as produced by a Nanopure ultrafiltration unit (Barnstead/Thermodyn, Dubuque, IA) or equivalent (*see* **Note 6**).
10. Rainin EDP-2 battery-operated motorized pipetors: 25 µL, 100 µL, 1000 µL (Rainin Instrument, Emeryville, CA).
11. Culture-ware: It is not necessary to use special cell-culture plastic-ware for parasite cultures. Standard polystyrene Petri dishes, 33- 60-, and 100-mm as well as 6-, 24-, 48-, and 96-well culture plates are used. Some investigators prefer to use various sizes of tissue-culture flasks (T-flasks) for culturing, but again, this is a personal preference, the pros and cons of which will be discussed below. Other labware, pipets, media bottles, filter sterilization materials, sterile hoods, pipetting aides, and so on are standard laboratory supplies and equipment.

3. Methods

3.1. Preparation of Culture Medium

1. Prepare incomplete medium (RP-I) by dissolving 16.2 g of powdered RPMI 1640 (supplemented with L-glutamine and 25 m*M* HEPES buffer but without sodium bicarbonate; *see* **Notes 2** and **7**) in 900 mL of high quality culture grade water (*see* **Note 6**) in a 1000-mL graduated cylinder on a magnetic stirrer (takes about 20–30 min). Once medium is completely dissolved, add water to a final volume of 960 mL. Filter sterilize through a 0.22-µm filter. Aliquot in 100 mL or 500 mL vol into sterile tight-sealing, screw-capped medium bottles. Store RP-I at 4°C for up to 30–60 d.
2. Prepare a stock solution of 5% sodium bicarbonate by adding 5 g of $NaHCO_3$ to 100 mL of high-quality water. Filter sterilize through a 0.22-µm filter and store at 4°C in tightly capped bottles.
3. Prepare a stock solution of 8% (v/v) Albumax, 0.05% hypoxanthine in RPMI 1640 by adding 8 g of Albumax I or II and 50 mg of hypoxanthine to 100 mL of RPMI 1640. Stir gently for 1–2 h until clear. Filter sterilize through a 0.22-µm filter, aliquot in 5 mL and 25 mL vol, and store at $-20°C$.
4. Prepare complete medium (RP-C) by adding 4.2 mL of 5% sodium bicarbonate solution and 5 mL of 8% Albumax stock solution per 100 mL of RP-I (*see* **Notes 2** and **7**) (final concentration of 0.4% Albumax).
5. Depending upon experience with sterile technique, cultures may be protected from bacterial contamination by adding penicillin-streptomycin solution (penicillin G, 10,000 U/mL; streptomycin sulfate, 10,000 µg/mL; Gibco-BRL) to give final concentrations of 200 U/mL penicillin and 200 µg/mL streptomycin. Penicillin-streptomycin is cheaper and more

effective than the widely used gentamicin sulfate. However, antibiotics are no substitute for good sterile technique because sooner or later poor technique will lead to contaminated cultures—if not with bacteria, then certainly with yeast.

3.2. Preparation of the Parasites

Parasites will remain viable indefinitely when stored in liquid N_2. Hence, parasite cultures are usually initiated from frozen stabilates of established cultures. Many different formulations of cryoprotectants can be used to store the parasites. The procedure described here works well and is widely used.

1. To prepare frozen parasites for culture, quickly thaw cryovials in a 37°C water bath, transfer contents to sterile centrifuge tube, and spin at 300–400g for 5 min. Discard the cryoprotectant and add 300 µL of hypertonic saline (sterile 3.5% NaCl) to draw the glycerol out of the cells and initiate the return of parasites to culture conditions. After 1–2 min, remove the 3.5% NaCl by centrifugation and wash the cells twice in 300 µL of RP-C, and then once in 10 vol of RP-C.
2. To return parasites to frozen storage, snap-freeze in Rowe's cryoprotectant, as follows: Harvest parasites from culture, and wash once in 10 vol of RP-C by gentle centrifugation at 300g for 5 min. Discard supernatant containing erythrocyte ghosts, parasite pigment granules an accumulated debris. Suspend washed parasites at a ratio of 1:1 in Rowe's cryoprotectant, sit for 5 min at room temperature, then disperse 500 µL into 2-mL cryovials, and quickly freeze by adding directly to liquid N_2, or in a dry-ice-absolute ethyl alcohol bath (*see* **Note 8**).
3. Mix thawed and washed parasites at a ratio of 1:4 with freshly washed erythrocytes and place into culture as described below. Generally, we have observed that Giemsa-stained thin film of these newly prepared parasites should show a few normal rings; if so, the parasites will require about 1 wk in vitro with 1–2 subcultures before they return to normal growth rates.
4. To start primary cultures, wash the patient's blood free of plasma and buffy coat, then add the parasites to the culture conditions, as described in **Subheading 3.3.** (*see* **Note 9** for transport of parasites).

3.3. Culture Procedure

First, a general description of the procedure: Parasites are grown in human erythrocytes in a settled layer of cells in RP-C at 37°C in flat-bottomed containers, under a low O_2, ~5% CO_2, balance N_2 atmosphere with 24-h medium changes, requiring subculture by addition of fresh erythrocytes every 4–5 d. Well-established cultures should increase three- to sixfold every 48 h, depending on the parasitemia. If cultures do not increase at this rate, or if they die, something is wrong with the medium, culture ware, gas mixture, and so on. As I have told my students many times: "Properly handled parasites do not die, you have to kill them."

1. Obtain parasites for culture from frozen stocks (*see above*), from ongoing cultures, or from patients. Since the erythrocytes do not remain suitable for parasite development for more than 5–6 d in vitro, fresh cells must be added every 4–5 d, regardless of parasitemia. We have often cloned parasites by dilution until there was only one or less parasites per well in a 96-well plate. We have blindly subcultured these parasites by adding erythrocytes to a ratio of 1:4 every 96 h—generally parasites are not seen on Giemsa thin films until at least 20 d have passed. The key to success is to add fresh cells every 4–5 d. Attempts to encourage a rise in parasitemia by withholding fresh erythrocytes will not

In Vitro Culture

work. One means of raising large numbers of parasites over a short period of time is to split the cultures every 2 d. To do this, remove the medium and divide the cellular contents of the plate between the old and a new 100-mm Petri dish, and add 100 µL of freshly washed erythrocytes and 12 mL of RP-C to each dish. Within a few days of this procedure, you will have too many plates to handle and the parasitemia will be above 12%.

2. Prepare fresh erythrocytes for culture by washing the cells free of CPDA-1 plasma. Transfer cells to a 15-mL centrifuge tube, spin at 300–400g for 5 min, discard plasma and buffy coat, wash cells twice in 10 vol of RP-I, and finally suspend at a ratio of 1:1 in RP-C. Since erythrocytes stored in CPDA-1 for more than a few days have no viable leukocytes or platelets, the "buffy coat" is usually composed of micro-blood clots and cell debris that are of no consequence to the cultures but look alarmingly like mold contamination.
3. Add parasites from cultures to the freshly washed erythrocytes to give a starting parasitemia between 0.1–1.0%, depending upon the experimental design.
4. Transfer cultures to appropriate flat-bottomed containers and add sufficient RP-C to provide approx 3–5% cell suspension (*see* **Note 10**).
5. Provide appropriate atmosphere for the cultures, using the candle-jar method or by gassing with 1% O_2, 5% CO_2, and 94% N_2. If Petri dishes or multiwell plates are used, containers such as a modular incubator chamber (Billups-Rothenberg, Del Mar, CA) can be used for retaining the gas mixture. If one uses screw-capped cell culture flasks (T-flasks), a separate gas chamber is not necessary, but these flasks must be flushed at least 5 min with the gas mixture through a sterile 9-in. cotton-plugged Pasteur pipet before incubating the parasite cultures (*see* **Note 11**).
6. Optimally, cultures are maintained in a stable 37°C incubator—recirculating refrigerated/heating or water-jacket incubators work best.
7. To use the candle-jar: Place the culture vessels in the desiccator, light a short candle and place within, and seal the lid with the stopcock open. Once the candle burns out (2–3 min), close the stopcock to seal in the atmosphere, and place the chamber in a 37°C incubator. This technique is especially useful in developing countries or where a supply of specially mixed gases cannot be procured.
8. Replace culture medium daily, by transferring culture vessels to a sterile hood, gently tipping the vessels without stirring the cells into suspension, and aspirating off the medium through a sterile pipet. Add fresh RP-C (prewarmed to 37°C in a water bath) using a pipet, suspend the settled cells by gentle swirling, then gas the parasite cultures and return them to the incubator (*see* **Note 10**).
9. Generally, provided starting parasitemias are between 0.1 and 1.0%, the parasites should increase about eight times during the first 48 h in culture, and three- to fourfold over the second 48 h. Cultures rarely rise above 10–12% unless the medium is refreshed every 12 h.
10. After 4–5 d of culturing, the parasitemia should reach levels where daily refreshing of medium will not be sufficient to sustain the parasitemia. If higher parasitemias are required for a particular experiment, the media can be changed every 12 h. This is not only a tedious task, but eventually the parasitemia will rise to the point that it cannot be sustained. For this purpose, as well as because of the natural instability of the erythrocytes in vitro, the parasites need to be subcultured every 4 d (*see* **Note 10**).

3.4. Synchronization of Cultures

For many applications, it might be desirable to synchronize the cultures so that all parasites are in the same stage of development. Typically, the cultures contain a mixture of ring, trophozoite, and schizont stages. As the parasite develops from ring through trophozoite to schizont over 44–48 h, major changes in the host-erythrocyte membrane permeability are induced. Between 16 and 20 h after invasion, the infected erythrocyte becomes permeable to many agents that would not normally enter these cells. One such agent is D-sorbitol. Thus, washing a culture with aqueous D-sorbitol *(9,10)* will lyse all trophozoite- and schizont-infected erythrocytes, leaving only uninfected and ring-infected erythrocytes. Many synchrony techniques are based on this principle. Several investigators have demonstrated that the younger parasite (rings and early trophozoites) are resistant to 40°C temperatures, but that the more mature parasites are killed at this temperature. Thus, alternating the cultures between 37° and 40°C can induce synchrony (this procedure is detailed in Chapter 44).

1. Wash asynchronous cultures twice with 5% aqueous sorbitol at 37°C and once with RP-C. Some strains require somewhat more vigorous treatment as outlined in **item 2**.
2. Suspend parasites from culture in 3–4 vol of 15% D-sorbitol at 37°C for 5 min, then dilute with 7 vol of 0.1% (v/v) glucose to give a final concentration of 5% sorbitol. Vortex, incubate another 5 min, and centrifuge. Then wash the cultures twice with RP-C, leaving only ring-stage parasites.
3. Sorbitol-induced synchrony results in a 0–18-h window of parasites (some rings will be newly invaded, whereas others will be 18-h postinvasion), which initially look synchronous, but within a few hours in vitro, will be a mixture of rings and trophozoites, and after 24–30 h, will again be a mixture of all stages. Thus, a second sorbitol treatment 27 h later will create a culture of young rings, more tightly synchronous than achieved by a single sorbitol treatment *(10)*. The drawback to this technique is that the resultant parasitemia is very low, and invasion into doubly sorbitol-treated erythrocytes is generally lower than with untreated erythrocytes. This latter observation may be due to the fact that the erythrocytes have been in vitro for several days.
4. For many experiments, the sorbitol treatment by itself will not achieve a suitable synchrony, especially if the experiment requires higher parasitemias. To obtain higher parasitemias of largely synchronous cultures, a combination of sorbitol treatment and gelatin flotation is needed (*see* **Subheading 2., item 6**).
5. To concentrate parasites to high parasitemia and eliminate ringinfected erythrocytes, wash cultures once in RP-I, then suspend either in 4 vol of gelatin solution (*see* **Subheading 2.6.**) mixed 1:1 with RP-C, or in 35% plasmagel. RP-I in narrow sterile test tubes such as 12 × 75 mm polystyrene snap-cap tubes.
6. Uninfected and ring-infected erythrocytes will settle in 20–30 min at 37°C, leaving in suspension a high concentration of the more mature parasites.
7. Pipet the supernatant from the settled cells and recover parasitized erythrocytes by centrifugation at 300–400*g*. A second gelatin treatment of the settled cells will produce more trophozoite- and schizont-infected erythrocytes, but these will be of lower parasitemia than achieved on the initial separation. However, two gelatin separations does permit the removal of nearly all mature-stage parasites from the culture and allows the rings to be used to initiate a new culture.
8. Both settled (mostly ring-infected) and suspended (mostly trophozoite- and schizont-infected) parasites can be returned to culture, if desired.
9. To achieve highly synchronous cultures, wash the cells in sorbitol (as described in **Subheading 3.4., items 1** and **2**) and then monitor the development of the parasites until they reach schizont stage. Separate these schizonts using gelatin sedimentation (*see* **Subheading 3.4., items 5–7**) and return the schizont-rich portion to culture. Care should be used to culture the highly concentrated schizonts in excess medium, in case they consume all the glucose in the medium.
10. When the separated schizonts begin to release merozoites, add freshly washed erythrocytes (warmed to 37°C in a water bath) to the concentrated schizonts. After the appropriate "window" of synchrony is achieved (0–4 h, 0–6 h old, respectively), wash the culture

In Vitro Culture

with sorbitol to remove all remaining schizonts, leaving a highly synchronous culture of young rings which will all be trophozoites after 24 h of culture, or schizonts after 36 h of culture.

11. To achieve relatively highly parasitemias of highly synchronous cultures, add large numbers of schizonts to lesser numbers of fresh erythrocytes. If high concentrations of trophozoite- or schizont-infected cells are desired, these can be attained by culturing the rings for 24–36 h, then separating the mature-stage parasites by gelatin flotation.

3.5. Microcultures and Radiolabeling

For many applications such as drug studies, growth-inhibition assays (*see* Chapter 51), and parasite metabolic studies using radiolabeled compounds, the cultures need to be set up in 96-well or other multiwell culture plates. This procedure is simply a matter of downsizing the methods mentioned above. Moreover, the incorporation of radiolabeled metabolic precursors provides an objective index of the number of living parasites since counting stained thin films is tedious, subjective and inherently inaccurate. Furthermore, experimental drugs and some expensive reagents are generally acquired only in small quantities. Since 96-well plates are most widely used for this procedure, their use is detailed here.

1. Dilute parasites from stock cultures, maintained as described in **Subheading 3.3.**, to approx 0.5% parasitemia with freshly washed erythrocytes and suspend at a ratio of 1:1 in RP-C.
2. Add 10 µL of parasite suspension, using a 25-µL Rainin EDP-2 electric pipetor, to give 5 µL of parasitized erythrocytes per well. Manual pipetors are far less accurate than the Rainin device.
3. Add 90 µL of RP-C to each well to complete the culture, using a multidispense 1000-µL Rainin EDP-2 pipetor. To examine the impact of antimalarial compounds, specific metabolic inhibitors, antibodies, or other immunoactive agents, etc., titrate the drug, immune sera, radiolabeled reagent, or other reagent of interest, into the RP-C before it is added to the cells. Mixing the drugs, antibodies, radiolabeled reagents, etc. with RP-C before adding them to the wells, greatly reduces pipeting errors. For example, to test for antimalarial activity of a given drug, all wells will contain 10 µL of parasite suspension, control wells will contain parasites in 90 µL of RP-C; and experimental wells will contain 90 µL of RP-C supplemented with specific dilutions of the test drug, generally in triplicate.
4. If the experiment is to last more than 24–30 h, refresh the medium daily by gently aspirating the exhausted culture medium through a 22-gage needle precisely centered in the well. If carefully done, the procedure should leave the parasite layer intact. Add fresh medium using the 1000-µL Rainin EDP-2 electric pipetor set to dispense 90 µL per cycle. Resuspend the cells in the fresh medium by gently finger-tapping the edge of the plate on all sides (takes about 5 min).
5. To assess the parasites, gently remove all media from the culture using a Pasteur pipet, and prepare thin films from the concentrated settled cells. Fix and stain with Giemsa (*see* **Note 5**).
6. Radiolabel as desired. Generally, if serum is used to supplement the RP-C, the label of choice is [^{3}H] hypoxanthine at 1–2 µCi per well for 12–24 h.
7. If Albumax I or II are used in place of serum (highly recommended) these serum replacements are always supplemented with hypoxanthine which complicates the use of [^{3}H] hypoxanthine as a marker of parasite metabolism. We have noted from previous studies that the amino acid phenylalanine is readily taken up from RPMI 1640 even though it is not exogenously required by the parasites *(12)*. Thus, [^{3}H] L-phenylalanine-[ring 2,6-3H (N)] will be readily incorporated into parasite proteins and requires no modifications of the culture medium to ensure efficient uptake of the label. Like [^{3}H]hypoxanthine, [^{3}H]-labeled phenylalanine is used at 1–2 µCi per well.

8. To reduce pipetting errors, it is best to add the radiolabeled directly to the medium before it is added to the cultures for the final 24 h of culture. However, depending upon the experimental protocol, 10 µL of radiolabeled RP-C may be added directly to each well 12–24 h before harvesting the cells (*see* **Note 12**).
9. For some purposes, such as flow cytometry, organelle or enzyme purification, synchronized cultures of nearly 100% parasitemia may be needed. Parasites concentrated from culture using the gelatin-flotation procedure can be further enriched by centrifugation on a discontinuous Percoll gradient. Since it is generally difficult to obtain large quantities of synchronous highly enriched parasites, this procedure uses 1- to 2.0-µL microcentrifuge tubes and a microfuge, but can be scaled up as needed. Construct a gradient by adding 550 µL of 45% Percoll in RP-C to 1.5-mL microcentrifuge tube, and layering onto this 550 µL of 35% Percoll in RP-C. Suspend the parasite-rich pellet from the gelatin separation in an equal volume of RP-C and layer 100–120 µL of this cell suspension over the completed gradient. Centrifuge at 5500*g* for 15 min. Transfer the parasitized cells, concentrated in the middle layer of the gradient, to a 15-mL centrifuge tube and wash twice in RP-I by gentle centrifugation at 300*g* to remove the Percoll. This double-concentration procedure results in a final suspension of erythrocytes containing at least 95% parasites.

4. Notes

1. For general maintenance of parasite cultures, O+ cells are best since they are compatible with all serum types and are readily available from blood banks or volunteers. Once erythrocytes have been stored for 1–2 d in CPDA-1, the leukocytes and platelets disintegrate, leaving only the red blood cells. The chief advantage of culturing parasites in erythrocytes that have been stored a few days in CPDA-1 is that the cultures will only contain two cell types—the parasites and the erythrocytes—and because erythrocytes have limited metabolism and no organelles (no RNA, DNA, mitochondria, lysosomes, and so on) compared to other cells, studies on parasite biochemistry, genetics, physiology, and so on will not be encumbered or contaminated with macromolecules or metabolic processes from platelets or leukocytes, negating complicated procedures for their removal.
2. Several different types of commercially available culture media have been used to grow malaria parasites, including RPMI 1640, RPMI 1630, Ham's nutrient mixture F-12, Dulbecco's modified Eagle medium, M-199, etc. However, none of these media have proven superior to RPMI 1640. Originally, this medium came without HEPES buffer and some sources provide the medium without this buffer added. If such is the case, HEPES can be added to the RPMI at a concentration of 5.94 g/L. We have found that HEPES from some suppliers is toxic to the parasites, but Sigma HEPES works well. Because the parasites consume glucose at a rate 50–70 times higher than the erythrocytes, and produce significant quantities of lactate as a byproduct, some investigators supplement RP-C with additional glucose and additional HEPES. We have found such high-glucose/high HEPES medium to be of no advantage, and do not recommend it. Since the parasites require preformed purines, and since the concentration of free purines is variable and often low in human serum, supplementation of the serum with hypoxanthine (the preferred purine for the parasites) often improves the culture medium (*see* **Note 3**).
3. Obtaining an adequate supply of fresh human serum has always been difficult for laboratories that culture malaria parasites routinely. Human serum is not only expensive, and its use presents potential infection risks from viruses, but over the years, it has become increasingly difficult to obtain, especially in smaller cities where blood-bank supplies are limited. Freshly collected bovine serum and rabbit serum will work, but these animal sera are also not readily obtainable. Moreover, bovine serum contains xanthine oxidase, which breaks down the hypoxanthine, which must then be re-added before the bovine serum will

support cultures. Commercially available fetal bovine serum is generally not suitable, although we have successfully used "Cosmic Calf" from Hyclone (Logan, UT), although it induces some agglutination of the erythrocytes. Some investigators have used outdated human plasma, but the best serum is obtained from freshly clotted human blood, pooled from 20 different donors and kept frozen at –20°C until used. From a 500-mL bag of clotted blood, one usually obtains about 160–200 mL of serum. If this serum is aliquoted at 50 mL per bottle, when full, 1 L bottle will contain sera from 20 donors. Such pooled sera is rich enough to be used at 5% (v/v), otherwise nonpooled serum is used at 10% (v/v).

4. The gelatin flotation technique chiefly separates uninfected erythrocytes and most ring-stage parasites from the trophozoite-schizont-infected cells. Hence this procedure will increase the parasite concentration about 7-8 times initial parasitemia. Thus, a starting parasitemia of 8–10% will result in final parasitemia of 50–80% trophozoite- and schizont-infected erythrocytes after gelatin flotation. Since the technique segregates ring-infected cells from trophozoite- and schizont-infected cells, both portions can be immediately returned to culture. When using gelatin flotation procedure to synchronize cultures, a common mistake is to collect all the mature parasites from say ten 100-mm Petri dishes, obtaining a packed cell volume of 200–300 µL, then adding these concentrated mature parasites to 2–3 new dishes with fresh culture medium and returning them to 37°C, only to discover a few hours later that all the parasites are dead. The cause of the problem is the rate of glucose consumption in the highly concentrated parasites is too high for the volume of medium. Since the schizonts were the major consumers of glucose and they came from 10 dishes containing an initial volume of 140 µL RP-C, returning these to culture in less medium will lead to disaster. Gelatin flotation methods require parasites that produce the classical membrane protrusions known as "knobs" on trophozoite- and schizont-infected erythrocytes. The ability to produce such knobs is slowly lost after several months in vitro, but knobby cultures can be restored by performing a gelatin flotation and returning only the suspended parasites to culture, discarding the settled cells. This simple readjustment will keep the cultures rich in parasites that produce knobs. We have found that, although erythrocytes stored in CPDA-1 up to 35 d will support cultures, erythrocytes stored for more than 10 d do not separate well using this technique.

5. Giemsa stain can be purchased as a dry powder or as a liquid concentrate. Making Giemsa from powder requires some expertise and the stain must be "aged" for months before use. It is more convenient to purchase the liquid concentrate (Fisher Scientific). This liquid stain contains glycerol and methyl alcohol, but no water. Once Giemsa has been diluted to working solution by the addition of pH 7.2 phosphate buffer, it will not last more than an hour or so. Since water deteriorates the stain, the concentrate must be kept water-free. We fill brown serum bottles with Giemsa stock and top them with a rubber""serum cap." To make working stain, 0.2–0.3 mL of stain is drawn by needle through the serum stopper into a 3-mL syringe then filled to 2–3 mL with phosphate buffer, mixed by inverting the syringe and the stain is layered over methanol-fixed thin films for 10 min. The methanol-fixed films should be completely dry before the stain is added, otherwise numerous membrane artifacts will be noted in the erythrocytes. After a suitable staining time, the slides are rinsed with water and dried before microscopic examination. Giemsa stock varies from lot to lot thus ideal staining times may be longer or shorter. The rule of thumb is 10% Giemsa for 10 min. Making thin-films from culture is initially a challenge because of the diluted cells. After removing the culture medium, a Pasteur pipet is used to take only concentrated cells from the settled layer. Care must be taken to assure that little or no clear medium is pipetted into the slide. If only concentrated cells taken from the settled layer are placed on the microslide, a typical "feathered edge" thin film is achieved. These thin-films should be quickly dried and fixed 30 s in absolute methanol before staining with Giemsa.

6. Water is a key ingredient in all media, additives, and reagents used in cultures. In many laboratories where we have had difficulties establishing parasite cultures, water has often been the problem. "Distilled" water has different meaning to investigators from different backgrounds. Initially, the water we used was "house distilled," meaning it had been produced in a large ion-exchanger on the roof of our building (often the case). We took this house "distilled" water and passed it through a tandem-linked pair of water distillation units to obtain "triple-distilled" water. Some""ion exchange" water purification systems exchange many small molecular weight inorganic contaminants for a few large molecular weight organic molecules. Although the conductivity of this water is greatly reduced, some of the organic ions added to the water in the "exchange" process are toxic. Modern water purification systems such as Millipore and Barnstead/Thermodyn use bound resins and filters to remove contaminants and these purification systems produce water suitable for cultures. In our African field laboratory, we laboriously triple distill water using a single glass still—kept running continuously for 15 yr to redistill water three times. Notwithstanding the effort, the results have been more than satisfactory.

7. Like most culture medium, RPMI contains the pH indicator phenol red which alleviates the need for using a meter for pH adjustment. Generally, this medium is orange-red when the pH is 7.4; yellow when acidic, and purple when alkaline. However, the presence of HEPES buffer and serum will tend to shift the color from orange/red to yellow/orange. RP-I is yellow due to the presence of HEPES and the lack of bicarbonate buffer. Upon the addition of $NaHCO_3$, the medium will change to orange over 30–60 s. If the $NaHCO_3$ solution has outgassed and lost much of its CO_2, a color shift from yellow to pink will be immediately noticeable when it is added to the RP-I. Such bicarbonate solutions should be discarded when this color shift occurs rapidly. The orangish medium does not need to be tested with a pH meter. Since Albumax is "more yellow" than serum, it imparts a marked yellow color to RP-C; thus RP-C containing Albumax as a serum supplement appears to be somewhat acidic. However, if carefully made up according to instructions, such yellow/orange medium is fine for the cultures. RP-C that appears pink/purple is not suitable for culture but can be used to wash erythrocytes. Complete culture medium (RP-C) is stable at 4°C for about 1–2 wk, depending on how much CO_2 is retained from the $NaHCO_3$.

8. In theory, cryopreservation of erythrocytes depends upon a relationship between glycerol concentration and the rate of freezing. For example, if untreated erythrocytes were to be sprayed by aerosol directly into liquid N_2, they would not lyse because the freezing would be instantaneous. Conversely, some cryoprotectants contain as much as 35–40% glycerol and the rate of freezing can be slow. In fact, when using high-glycerol cryoprotectants, the cryovials are often placed in polystyrene foam blocks and frozen over 2 d at –20°C then stored at –70°C of liquid N_2. The method outlined in this chapter contains a moderate glycerol concentration and works well when cells are frozen as quickly as possible. We simply place 0.5 mL of 50% cells and cryoprotectant into a 2.0-mL cryovial, shake to coat the sides with cells, clip onto a freezing cane, and drop it immediately into the liquid N_2 so that it snap-freezes. The success of this method is only realized upon thawing. To reclaim parasites from frozen storage, the vials are taken from the liquid N_2, the caps loosened to vent any nitrogen in the vials, and thawed as quickly as possible in a 37°C water bath, taking care not to allow the circulating water to contaminate the cap of the vial. After thawing, contents of the cryovial are transferred with sterile Pasteur pipet to a 15-mL plastic centrifuge tube and centrifuged at 300–400g for 5–6 min. The cherry red cryoprotectant supernatant is discarded and replaced with 300 µL of sterile 3.5% NaCl, and the centrifugation repeated. Again, the supernatant will appear quite red and the cell pellet somewhat reduced, but this is normal. The 300-µL wash is repeated twice using RP-C to remove the concentrated NaCl and return the cells to isotonicity. With each wash, the

supernatant will appear less red. If the cells were excessively damaged during freezing, or if they were inadvertently thawed and refrozen, the hemolysis will be extensive with each wash until no cell pellet is left. Although parasites reclaimed from cryopreservation might contain some trophozoites and schizonts, generally only ring-stage parasites will survive this procedure. Thus, high parasitemia cultures that are proportionally rich in rings are the best candidates for freezing. Parasites reclaimed from frozen storage will require at least two subcultures over 6–9 d before normal parasite growth is restored.

9. Often parasites are collected far from the laboratory where they will be placed into culture. Initially, we transported *Plasmodium*-infected cells on wet ice. Later, when moving established cultures from laboratory to laboratory sometimes from continent to continent, we carried them chilled on wet ice. Now, we know this is a poor practice since the parasites "travel" better at room temperatures. Thus, parasites are best kept at room temperatures when not in culture. Cooler temperatures seems to "stun" them and they take much longer to return to normal growth rates when placed at 37°C. Generally, we add 1.0 mL infected cells to 13 mL of RP-C in a sterile 15-mL tightly capped centrifuge tube and carry them in a shirt pocket, or in a polystyrene container with polystyrene chips to prevent temperature shocks. In fact, highly diluted parasites can be held in RP-C at room temperature for more than 1 wk and will quickly return to normal development when provided with fresh medium, appropriate gas atmosphere, and 37°C.

10. Generally, the cultures are maintained at 3–5% cell suspension, but this proportion is somewhat misleading, since the critical factor is the thickness of the settled layer of cells. Years of experience have taught us that for good culture conditions, 1.0 µL of packed infected erythrocytes should occupy about 32-mm^2 of flat surface. Thus, a 100-mm Petri dish should contain 200–250 µL of packed cells and sufficient RP-C to give 3–5% hematocrit. In practical terms, 1.0 mL of packed erythrocytes is split equally into four 100-mm Petri dishes. When the parasitemia is low, such as when parasites are initially subcultured, 10 mL of RP-C is added to each dish. After 2 d, when the parasitemia has risen, the volume of freshly added medium is raised to 12 mL per dish and on d 3 raised to 14 mL. Thus, cultures are maintained on a sliding scale of medium, depending on the parasitemia. If initial parasitemias are below 1% this sliding scale of cells to medium will produce excellent parasite development using less medium. Since Petri dishes usually come in 100-, 60-, or 35-mm diameters, they can culture about 250, 90, or 30 µL of packed cells per dish. The ratio of 1.0 µL/32 mm^2 can be applied to any type of flat-bottomed vessel. A critical factor, often overlooked is the rate of glucose consumption which is 6–7 times higher for schizonts than for rings, the trophozoites lying somewhere in between. For example, a synchronized culture composed of schizont-staged parasites will consume glucose at a rate nearly 6–7 times higher than the same culture when it was in the ring stage.

11. Regardless of which method is used to provide the appropriate gas environment for the parasites, flushed cell-culture flasks, or a modular incubator chamber, the ideal gas mixture is rarely obtained because some residual air often remains. Nonetheless, the parasites grow best in deoxygenated erythrocytes (being microaerophilic) that take on a dark purple hue after an hour or so in the incubator. If the cultures remain bright red, this generally means that there is too much oxygen overlaying the cells.

12. We use a "dry scintillation" counter linked to an automatic cell harvester (Inotech Biosystems, Rockville, MD) that is extremely convenient and allows us to forego the use of scintillation fluid and the problems associated with the disposal of radioactive organic solvents, but classical liquid scintillation also works well. Since the 96-well plates can hold 200 µL of medium per well, but generally only 100 µL is used, cultures grown for 1–2 d in the 96-well plates can be labeled over the final 24 h of culture, by adding 100 µL of radiolabeled medium on top of the initial 100 µL of culture medium, mix by gently tapping the sides of the plate to suspend the cells and return the plate to culture conditions.

References

1. Trager, W. and Jensen, J. B. (1976) Human malaria parasites in continuous culture. *Science* **193**, 673–675.
2. Bass, C. C. and Johns, F. M. (1912) The cultivation of malarial plasmodia (*Plasmodium vivax* and *Plasmodium falciparum*) in vitro. *J. Exp. Med.* **16**, 567–579.
3. Golenda, C. F., Li, J., and Rosenberg, R. (1997) Continuous *in vitro* propagation of the malaria parasite *Plasmodium vivax*. *Proc. Natl. Acad. Sci. USA* **94**, 6786–6791.
4. Trager, W. and Jensen, J. B. (1997) Continuous culture of *Plasmodium falciparum*. Its impact on malaria research. *Int. J. Parasitol.* **27**, 989–1006.
5. Trager, W. and Jensen, J. B. (1978) Cultivation of malarial parasites, a review. *Nature* **273**, 621,622.
6. Jensen, J. B. and Trager, W. (1977) *Plasmodium falciparum* in culture: use of outdated erythrocytes and description of the candle jar method. *J. Parasitol.* **63**, 883–886.
7. Jensen, J. B. (1978) Concentration of trophozoite- and schizont-infected erythrocytes of *Plasmodium falciparum* from continuous cultures. Am. J. Trop. Med. Hyg. 27, 1274-1276.
8. Rowe, A. W., Eyster, E., and Kellner, A. (1968) Liquid nitrogen preservation of red blood cells for transfusion. *Cryobiology* **5**, 119–128.
9. Lambros, C. and Vanderberg, J. P. (1979) Synchronization of *Plasmodium falciparum* erythrocyte stages in cultures. *J. Parasitol.* **65**, 418–420.
10. Vernes, A., Haynes, J. D., Tapchaisri, P., Williams, J. L., Dutoit, E., and Diggs, C. (1984) *Plasmodium falciparum* strain-specific human antibody inhibits merozoite invasion of erythrocytes. *Am. J. Trop. Med. Hyg.* **33**, 197–203.
11. Pasvol, G., Wilson, R. J. M., Smalley, M. E., and Brown, J. (1978) Separation of viable schizont-infected red cells of *Plasmodium falciparum* from human blood. *Ann. Trop. Med. Parasitol.* **72**, 87,88.
12. Divo, A. A., Geary, T. G., Davis, N. L., and Jensen, J. B. (1985) Nutritional requirements of *Plasmodium falciparum* in culture. I: exogenously supplied dialyzable components necessary for continuous growth. *J. Protozool.* **32**, 59–64.

44

Automated Synchronization of *Plasmodium falciparum* Parasites by Culture in a Temperature-Cycling Incubator

J. David Haynes and J. Kathleen Moch

1. Introduction

Continuously automated synchronization of *Plasmodium falciparum* malaria parasites can be achieved by culturing in a custom temperature-cycling incubator (TCI) whose process controller is programmed to periodically change between three different temperatures:

1. Shortly after invasion of merozoites into erythrocytes at about 37°C, the culture temperature is lowered to about 17°C, and the maturation of ring forms ($Rg_{early} \rightarrow Rg_{late}$) is suspended (and perhaps some remaining schizonts are killed). The time ramping down to and holding at 17°C, generally between 2.5 and 7.25 h, is determined by adding times for the next three steps below and subtracting from 48 hr. Thus, the total cycle length for any given parasite isolate is adjusted to 48 h, even though various free-cycle times at 37°C for different isolates range from about 38 to 45 h.
2. When the culture temperature is again raised to about 37°C, ring stages mature normally to trophozoites with pigment. Depending on the characteristics of the parasite isolate and its adaptation to culture, we allow between 12.75 and 18.5 h for this (including time for ramping to temperature).
3. When the culture temperature is raised to about 40°C, trophozoites (Tr) continue to mature up to a point ($Tr_{early} \rightarrow Tr_{late}$), but then further maturation to schizonts ($Tr_{late} \rightarrow Sz$) is blocked, so that parasites at earlier stages catch up to those at later stages. We allow between 9.5 and 10.5 h for this.
4. When the culture temperature is lowered again to about 37°C, maturation to schizonts ($Tr_{late} \rightarrow Sz$) proceeds, and merozoites are released and invade erythrocytes. We allow between 18.5 and 22.5 h for this.

The total time allowed for **steps 1–4** is exactly 48 h, creating a cycle repeating in synchronization with a human work schedule. Further adjustments in timing of different batches of parasites can be achieved for particular purposes (e.g., obtaining both early and late schizonts in the same morning) by subjecting one batch to a longer time and another batch for a shorter time at the preceding 17°C stage.

Examples of schedules are given for two *P. falciparum* isolates commonly used in vaccine studies, FVO and 3D7. In order to speed up the adaptation to temperature-cycling synchronization, an additional method of synchronization (sorbitol or alanine treatment) is used several times in the first several weeks when parasites are thawed or moved from 37°C incubators into the temperature-cycling incubators, and occasionally thereafter to improve synchrony by eliminating leading and lagging parasites. Synchrony is also improved by movement to create gentle suspension.

1.1. Background

In 1989, Dominic Kwiatkowski *(1)* reported that culturing at febrile temperatures (40°C) for 24 h would create synchrony in the growth of the malaria parasite *P. falciparum* by disrupting the second half of the 2-d erythrocytic cycle. After reproducing his results, we found that shorter times near 40°C would achieve better growth rates and could be cyclically repeated in order to maintain the synchrony. Automation of the temperature changes seemed desirable and several generations of custom incubators with programmable process control of cooling as well as heating were designed in collaboration with ThermoQuest-Forma Scientific, based first on their platelet incubator and then on their smallest environmental chamber. In addition, based on the common practice of retarding the growth of parasites by holding them at room temperature, we incorporated a 22°C or 17°C step in order to control the total length of the cycle. Synchronization by elevated temperatures or "heat-shock" had been described previously for several organisms, including the protozoan *Tetrahymena pyriformis* in the 1950s *(2)*. Gravenor and Kwiatkowski *(3)* showed in 1998 that a stepwise febrile temperature curve over a single 24-h exposure (6-h periods of 37.7, 38.9, 39.4, and 38.8°C) could give partial synchrony of cultured *P. falciparum* parasites, and used the data in a mathematical model that may help explain the observed periodicities of fever and parasitemias as well as febrile-induced limitations on parasitemia in vivo.

2. Materials

2.1. Equipment

1. Biological safety cabinet for sterile manipulations (*see* **Note 1**).
2. Incubator, Temperature-Cycling, ThermoQuest-Forma 3851/3911M custom 11 cubic foot incubator, with programmable Omega CN3802 process controller for heating and refrigeration, with heated glass door (also controlled by Omega CN3802), with solid door cover, with two internal 110V outlets boxes, and without humidification (ThermoQuest, Forma Scientific Division, Attn: Rocky Lada, Custom Products, Box 649, Marietta OH 45750-0649, 1-800-848-3080).
3. Gas cylinder: 5% O_2, 5% CO_2, balance nitrogen (e.g., Air Products), fitted with regulator and 1/4 in. ID Tygon tubing.
4. Labware and supplies for cultures and staining:
    a. Culture flasks, 25 cm^2 and 150 cm^2, with plug seal caps (e.g., Corning 430168).
    b. Centrifuge tubes: 1.5-mL, 15-mL and 500mL.
    c. Pipets, disposable: 1-mL, 5-mL, 10-mL, 25-mL.
    d. Micropipetor: 1–20 µL, 20–200 µL, with disposable tips.
    e. Sterile 9 in. pasteur pipets, with and without cotton plugs.
    f. Tubes: 5-mL, polypropylene, snap-top.
    g. Filter units, 0.2 µm PES membrane; e.g., 150-mL, 250-mL, 500-mL and 1000-mL sizes in the MF75™ Series, extra receiver containers also available (Nalgene, 75 Panorama Creek Drive, PO Box 20365, Rochester, NY 14602-0365, 1-800-625-4327, http://nalgenunc.com).
    h. Filters, small for small volumes: e.g., Acrodisc® syringe filter 0.2 µm HT Tuffryn membrane low protein binding, sterile (Ref 4192, Pall Gelman Laboratories).

i. Slides, 1-mm-thick frosted glass.
j. Jars, Coplin staining.
k. Pail, large, with lid for transporting samples stained for flow cytometry.

5. Centrifuges:
   a. Large, preferably with heating (e.g., Jouan CT 4 22 centrifuge, which also has very little vibration and pellets the parasites gently) for centrifuging cultures at 37°C in a prewarmed centrifuge.
   b. Microfuge, for 1.5-mL tubes.

6. Warming platform and insulating sheets: a thin heated platform (e.g., microplate warming station, bench warmer, 120 V, 3 W, 9.5 × 13.5 × 1 cm, Thermolyne TWW100) is useful for keeping flasks at 34–38°C while being manipulated in the biological safety cabinet (*see* **Note 1**). For larger volumes of culture and media, heat loss to the metal work surface can be reduced by using 1/4-in.-thick closed cell foam plastic sheets (e.g., clean sheets work surface, XLPE, 26.7 × 30.5 cm; Nalge Nunc, cat. no. 6281-1012).

7. Rotating platform: 11 in. × 14 in. platform, 1 in. orbit, adjustable RPM, digital RPM readout is ideal (e.g., Adjustable Rotating Orbital Shaker, AROS 160, Barnstead/Thermolyne, 2555 Kerper Blvd. Dubuque, IA 52001-1478, 1-800-446-6060, http://www.barnstead.com). Topped by a 46 × 38=cm insulating platform under a 46 × 38-cm air circulation filter, either flat (for 25-cm^2 flasks), or propped at about a 4° angle (for 150-cm^2 flasks).

8. Insulating platforms, 46 × 38 cm, made from foam board insulation, 1/2-in.-thick, used to enlarge the platform and to insulate culture flasks from the heat of the rotating platform motor in incubators, cut to size with knife (e.g., Super Tuff-R insulating sheathing, a semirigid polyisocyanurate foam board insulation with poly/aluminum foil facers on both sides, Celotex Corporation, at local building supply stores).

9. Air circulation filter, 46 × 38 cm, air filters, natural organic fiber, 3/4-in.-thick, placed immediately under flasks (on top of insulating platform which is on top of rotating platform) for allowing air flow under culture flasks, cut to size with scissors, place cloth screen side up on foam board insulation on rotating platforms in incubators (from various manufacturers at local building supply stores—do not use fiberglass filters, their glass fibers are dangerous when cut to size).

10. Angle supports, to give a 4° tilt, made from 1/2-in.-thick foam board insulation (*see* **item 8**) cut to size and taped together to fit into one end of the rotating platform (*see* **item 7**). The two layers of foam board (both 29 cm front-to-back) are taped together front-to-back, with the smaller (5.7 cm left-to-right) on the bottom, and the larger (22.8 cm left-to-right) piece of foam board on the top. When fitted into the left side of the AROS rotating platform (inside dimensions 34.7 cm left-to-right × 29.1 cm front-to-back) the larger (22.8 cm) top piece of foam board tilts down to contact the floor of the rotating platform about 12.9 cm from the outer rim (0.3 cm high, 0.65 cm wide) of the rotating platform on the right, giving about a 4° angle tilt from the horizontal for the next piece of foam board, the insulating platform (*see* **item 8**). These are used only for the large 150-cm^2 flask cultures, the 25-cm^2 flask cultures are placed on flat insulating and air circulating platforms without an angle.

11. Thermometer, NIST traceable calibrated, for checking the actual temperatures in the temperature-cycling incubator (e.g., Ertco-Eutechnics 4400-2i.2.5 high precision digital thermometer with Micro Probe and RS-232-C Option, [Fisher, cat. no. 15-060-383] with AC adapter 15-060-385).

12. Table, heavy duty, adjustable height.

2.2. Reagents

1. *P. falciparum* culture adapted isolates: standard isolates such as 3D7 (clone of NF54) and FVO are ideal. Use any standard malaria culture media and technique (*see* Chapters 43 and 51).

2. 0.3 M Alanine, 10 mM HEPES, made to pH 7.5 with NaOH, 5 N (use tissue culture grade or highest grade reagents available, e.g., from Sigma), filter sterilize and store at 4°C.

3. Methods
3.1. Set Up the Temperature-Cycling Incubator

1. Unpack the TCI, Forma custom 3851/3911M incubator with Omega CN3802 RTD process (temperature) controller—**Caution:** Get professional movers with a fork-lift, this incubator weighs about 300 lbs! Place on table, level, install two shelves, and let the TCI sit for 1 d after moving. Be sure to remove any packing, for example, around the compressor cooling fan in upper compartment (front top swings open with some tugging).
2. Turn on the main power. Be sure the toggle switch is set to the Omega Controller.
3. Initial set up for the Omega Controller (note, for example, that the Enter key appears as [ENT], and the six main menus as 1), etc.):

    Press [Func] to get to 2) Exec menu, set to manual ([MAN]) (with arrow keys and press [ENT]) to allow later 6) Initial data entry (may have to first back out of a program by using arrow keys to display [RST] and pressing [ENT]).

    Press [RVS] and [Func] to get to 1) output menu and reset output to 35% using arrow keys and press [ENT] (sets heater manually for 35% of each 3-s cycle to allow a temperature of about 30°C while performing the steps below, otherwise only cooling will be on).

    Press [Func] until see 6) Initial Data [Entry AVLBL] (then check or reset the following): Press [File] until see PV filter: 3 (use arrow keys and press [ENT] to set to 3), PV Bias: 0.0 C (initially at least, may need to adjust later, pressing [ENT] twice moves to PV Bias, then use arrow keys and press [ENT] to set it).

    Press [File] until see R/D Action: [R] (unchanged)
    Cyc Time: 3 s (unchanged)
    Press [Item] until see Out = SSR,   T1 = NON
    Com = NON,   T2 = NON
    Press [File] until see Unit: [°C]
    RTD Type: [Pt]
    Press [File] until see Pt Range: [7] (change to 5 from 7 using arrow keys and press [ENT], so that range [5] is selected).
    0.00° — 50.00°C
    Press [File] until see Power On Mode: [Normal]; (change to [GUA Run] using arrow keys and press [ENT], otherwise will not restart when power fails).
    Press [RVS] then [Func] until see 2) Exec Key: [KEY], select [FIX] with arrow keys and press [ENT])
    Press [File] until see FIX SV: 37.00°C (change to 37°C if not there already)
    PID_:1, Alarm_:1
4. Tape onto the shelf a probe from a NIST traceable calibrated thermometer (reading to 0.1°C or better, for example, a Ertco-Eutechnics 4400 series digital thermometer with microprobe, available from Fisher Scientific).
5. Equilibrate TCI at 37°C several hours.
6. Auto-Tune the PID parameters of the Omega Temperature Controller:

    [RVS] then [Func] until see 2) Exec Key: [KEY], (then use arrow keys until see [AT], and press [ENT]; the PID_1 will start out at the default values of 5%, 300s, 100s, and, after the small red AT on the left of the display logs off and [FIX] lights instead, check the PID_1 values by pressing [Func] until see 4) Control Data and press [File] untill see, for example, PID_1, P: 2.2%, I: 39s, S:13s [typical values]).
7. Adjust the temperature Offset Bias by using the NIST traceable thermometer:

    After another hour of equilibration, compare the Omega CN3802 temperature (PV) reading with the NIST traceable thermometer on the shelf. If the temperature readings differ by more than 0.1°C, go back to **step 3** above in order to adjust the PV Bias (for example, to 0.3°C if the Omega PV temperature is reading 0.3°C higher than the NIST traceable thermometer on the shelf). After 1 h, the two temperatures should agree within 0.1°C, if not, repeat **step 7**.

3.2. Place the Rotating Platforms

In the middle of each shelf of the TCI, about 15 cm back from the front of the shelf, place a rotating platform with insulating and air circulation platforms as described in **Subheading 2.1., item 7–10**.

3.3. Program the Omega Controller for Temperature-Cycling

Press [Func] until 3) Program is displayed, then [File] until the first temperature setting is displayed, then using the arrow and enter keys, set the temperature and time (it will be a dwell if the previous step had the same temperature, and a ramp if the previous temperature was different). Press [Item] to go on to the next step (temperature and time).

Examples of temperature program timings for specific strains of *P. falciparum* parasites (*see* **Notes 2–6**):

1. FVO timings: i19:30, 22(2:22), 37(13:42), 39.8(9:30), 37.2(21:26), 38.2(1:00).
2. 3D7 timings: i18:30, 17(7:15), 37(11:40), 39.8(10:35), 36.8(17:30), 38.2(1:00).
3. Dd2 timings: i19:30, 17(4:45), 37(13:10), 39.8(10:35), 36.8(18:30), 38.2(1:00).

Example of interpretation: Invasion by FVO should end about 19:30 (7:30 PM), at which time the temperature-cycling incubator (TCI) begins cooling down to 22°C; the ramp down plus dwell time at 22°C equal a duration of 2 h 22 min, during which the rings are in suspended animation. Then the TCI begins warming up to 37°C; the ramp plus dwell time at 37°C equal 13 h 42 min, during which the early rings develop into late rings. During the ramp to and dwell at 39.8°C (totaling 9 h 30 min) the late rings develop into late trophozoites. When the TCI ramps down to and dwells at 37.2°C, the inhibition on the development of trophozoites into schizonts is released, schizogony proceeds, and invasion begins (21 h 26 min total). After about 3–5 h of invasion, during the last 1 h of invasion, there is a slow ramp up from 37.2 to 38.2°C, intended partly to signal the end of invasion and partly to speed it up. The total cycle length is exactly 2 d (48 h, 0 min). This program takes 9 steps (5 ramps and 4 dwells), the maximum available without linking programs, before repeating (999 times means indefinitely).

3.4. Prepare the Culture Flasks and Place on the Rotators

Dilute the cultures from culture stocks (*see* **Notes 7** and **8**) or freshly thawed cyropreserved parasites so that, if the parasites are already partially synchronized, the parasites are at about the same part of their cycle that the TCI is currently programmed to handle (the TCI timing can be shifted 1 d to accommodate this). Place the culture flasks on the rotators (*see* **Notes 9** and **10**) with insulating and air circulating platforms, flat at 44 RPM (1-in. orbit) for small flasks (25-cm^2 Corning culture flasks with 7.5 mL culture), and at a 4° angle to the horizontal and at 40 RPM for large flasks (150-cm^2 Corning culture flasks with 45 mL culture).

3.5. Hasten the Achievement of Synchrony, or Tighten it, with Alanine Synchronization

Two or three times within the first 2 wk in the TCI, synchronize the cultures with alanine treatments *(4)* so that the parasites will become synchronized more quickly. Sometimes alanine treatments can improve (tighten) synchrony after the first 2 wk also.

Centrifuge the culture (300g, 5–10 min, 37°C), aspirate the supernatant, add 1.5 packed cell volumes of prewarmed alanine solution (0.3 M alanine, 10 mM HEPES, pH 7.5) to the culture pellet, mix, incubate at 37°C for 10 min, add about 10 vol of warm culture media, centrifuge, aspirate supernatant fluid, add desired amount of culture medium and continue culturing (*see* **Note 11**).

3.6. Continue Temperature-Cycling Synchronization

Depending on the desired schedule, every 2 or 4 d early in the morning at the ring stage before the increase in temperature to 39.8°C, sample the culture flasks to make and stain slides for determining parasitemias. Prepare and warm new flasks with fresh media and erythrocytes, and, before the increase in temperature to 39.8°C, make dilutions from the old cultures to achieve the desired initial parasitemias for another 1 to 2 cycles (2–4 d) of growth. Cultures can be expanded from 7.5 mL in 25 cm^2 into several 150-cm^2 flasks (45 mL each) if the parasites are allowed to grow for two cycles after expansion before use at ~2–4% final parasitemia. If desired, the last two cycles can be in a constant 37°C incubator to diminish any possible effect of temperature changes on the biology of the parasites, keeping in mind that there will be shifts in the time of invasion and subsequent stages for each following cycle at 37°C. We estimate that free-wheeling at 37°C, cultured 3D7 has about a 38- to 40-h cycle, while FVO has about a 44- to 46-h cycle.

3.7. Shift the Time of Appearance of Mature Schizonts, or Other Stages, If So Desired

If, after terminal expansion into large flasks, for example, mature 3D7 schizonts are desired for an experiment earlier in the day than they would appear on the usual schedule, during the preceding cycle, those flasks can be placed in another TCI whose program has been changed to decrease the time at the low temperature, say by 4 h, thereby advancing the appearance of schizonts by about 4 h (*see* **Note 12**). In order to support the more advanced growth of these parasites during the last night of growth, we find it desirable in these cases to change the culture media (by 300g centrifugation for 10 min in 50-mL tubes at 37°C, aspiration of supernatant, pouring in warm complete culture media, returning to flask, gassing, and reincubating) after 8 PM on those flasks.

3.8. Some Future Directions For Improving Synchrony

Even with temperature cycling and occasional alanine synchronization, about 90% of the invasion by *P. falciparum* occurs over about a 4- to 6-h period, probably in part because the parasites grow more quickly in young erythrocytes than in old erythrocytes. In preliminary experiments if the erythrocytes are separated into more uniform ranges by age (by density using silicone fluids), invasion can be reduced to about a 1-h period or window. Differences between batches of erythrocytes may also affect the degree of synchrony.

The maximum rate of air temperature change capable by this temperature-cycling incubator is used for the changes in temperature, however, the rates of temperature changes in both the air and the liquid flask culture may not be optimal (*see* **Note 13**). The synchronization of both *Tetrahymena* (*2*) and *Amoeba* (*5*) is better with more rapid temperature changes.

Parasite Synchronization 495

4. Notes

1. Avoidance of both infection of the worker and contamination of the malaria cultures requires universal precautions and standard sterile techniques using sterile disposable labware and pipetting aids in a laminar flow biological safety cabinet. Contamination can be avoided by meticulous technique without either flaming or antibiotics — one accidental touch and the item is discarded (or filtered again if possible).
2. General information about programming the Omega CN3802 process (temperature) controller: Along the top and the top left side of the Omega controller face are indicators as to the current state of the controller, e.g.,
    a. RUN (when lit, means a program is running);
    b. HLD (when lit, means the program is running but is on hold or pause until released);
    c. ADV (lights momentarily when advancing a program to the next step or a given time);
    d. FIX (when lit, means the set value temperature is kept at the value selected for fix temperature);
    e. MAN (when lit means that the heater output is fixed manually, though the temperature may vary);
    f. AT (when lit means Auto-Tune; the controller is adjusting its control, PID, parameters).
    g. The eight buttons along the bottom of the Omega face are used to navigate the menus and enter the temperature, time, and other control parameters.
3. These timings are for RPMI 1640 culture media containing 10% heat inactivated pooled normal human sera. The conditions for temperature-cycling synchronization have not been worked out for Albumax containing culture media, in which these parasites grow more slowly than in human sera.
4. Some choices of nonintegral temperatures are to enable the observer to more easily determine at a glance which parasite and which part of the cycle.
5. Other strains of *P. falciparum* that have been synchronized with similar programs include FCR-3, Camp, FCB-2, and, not as well, 7G8 (which seemed to be more susceptible to killing at high temperatures, one reason that the times and temperatures were decreased— even less than 10 h and less than 39.8°C might be optimal for 7G8).
6. The actual times set for the ramps are probably not very important now since the current temperature-cycling incubator cannot heat or cool very quickly, giving rather long times before the equilibration of temperature (up to 1 h for the incubator and even longer for the flask culture fluid). Thus, setting a ramp for 0 min (there has to be a ramp step with both temperature and time values), which tells the incubator to change temperature as fast as it can, would be OK, however, in order have a program that reminds us that these temperature changes do not happen instantly, I usually program in 5 min for the small ramps and 15 min for the big ramps (except of course the last slow 1 h ramp up to 38.2, which is there to encourage any laggards to invade, and to remind us that invasion is coming to a close).
7. We recommend 4% Hct human erythrocytes in 10% serum containing culture media with ring form parasitemias of about 0.015% for 3D7 and 0.03% for FVO for initiating two cycles (4 d) of culture without changing the media. This results in final parasitemias of about 1.5–4% rings for 3D7, which under these conditions grows at about eightfold to 15-fold per cycle, and in about 1–2% for FVO, which has lower growth rates, and also does not seem to tolerate high parasitemias without even worse growth the next cycle. Cultures that will go only 2 d without media changes before subculturing can be initiated at higher parasitemias, e.g., 0.15% for 3D7 and 0.25% for FVO, with the same limits on final parasitemias. Outside of the temperature-cycling incubator, in a constant 37°C incubator, both strains can increase their growth rates by about 25–100% per cycle, which may require even further dilutions of initial parasitemias in order to avoid excessively high

final parasitemias that would interfere with synchrony or parasite health; though more frequent media changes can help achieve higher parasitemias, they also can desynchronize the parasites. Growth rates increase and cycle times decrease after 3 mo in culture, at which time the cultures are reinitiated from cryopreserved stocks.

8. Gassing, e.g., with 5% CO_2, 5% O_2, 90% N_2, can be accomplished in about 10 s per flask: small flasks are gassed (5 L/min) through a sterile cotton-plugged Pasteur pipet, and large flasks are gassed through a sterile cotton-plugged 5-mL pipet directly into the culture media to produce bubbles that reach just up to the base of the neck of the flask before sealing the cap.

9. Parasites in the bottom layers of static cultures are relatively nutrient starved compared with those on top, thus, synchrony can be improved by gentle suspension cultures, which provide more uniform distribution of nutrients and gases. Suspension has the added benefit of reducing the incidence of multiple invasion of erythrocytes where crowded parasites compete for nutrients and space. As noted, cultures of different volumes and in different sized flasks require different motions and angles to the horizontal for gentle suspension. Too vigorous suspension may damage the parasites, for example, by precipitating residual serum fibrinogen which can then mechanically damage erythrocytes.

10. Both the RPM and the angle and to the horizontal must be adjusted for the flask size and shape as well as the culture volume per flask, in order to allow adequate suspension while minimizing trauma to the culture—motion that is too vigorous can inhibit growth and even cause precipitation of residual fibrinogen from the serum, leading to floating masses of fibrous material.

11. Alternatively, treat with 1 packed cell volume of alanine solution, add 25–50 vol of culture media, and return to culture. We have found alanine treatments to be just as effective as sorbitol at killing trophozoites and schizonts but result in better growth of the surviving rings in the FVO isolate. The timing of the treatments is important. At least one of the alanine treatments should be about 30 min before the expected end of invasion in order to kill lagging schizonts (at about 7 PM for FVO and 6 PM for 3D7). Another of the alanine treatments should be in the morning close to the beginning of the 39.8°C temperature, in order to kill leading trophozoites.

12. In the absence of another TCI to reprogram, it may be possible to shift the cycle timing by moving flasks by hand at the appropriate time from the TCI with the low temperature into an incubator at 37°C. Of course, many other options are open, including advancing the timing by performing the final two cycles at constant 37°C, or retarding the timing by adding time at the low temperature (either in a TCI, or at room temperature, or possibly even in the refrigerator).

13. The use of a process controller temperature probe in the liquid of a simulated culture can speed up the rates of change slightly by creating over and under temperatures in the air, however, this introduces problems of some temperature instability and a sensitivity to placement of the temperature probe flask. Alternatively, we have left the process temperature probe in the air stream in the wall of incubator, but, based on independent measurements of the temperature in the liquid in a simulated culture flask, introduced slight temporary overshoots and undershoots in the air temperatures during the ramps to the new temperature in order to speed equilibration in the liquid temperature. However, because the changes in air temperature generally seem to be the limiting factor (the changes from 37–40°C generally occur within about 10 min, but it can take about 1 h for the air temperature to go between 17–37°C), we do not always used these slight over and under shoots in air temperature. We await the development of a temperature-cycling incubator with less thermal mass or more powerful heating and cooling in order to address better these temperature change speed issues. Nevertheless, the existing equipment and protocols maintain relatively good synchronization.

Acknowledgments

We thank Dr. Stephen Hoffman and Dr. Jeffrey Lyon for support, advice, and encouragement; and Dr. Gordon Langsley, Dr. Catherine Braun-Breton, and Dr. Hagai Ginsburg for advice on alanine synchronization. The views of the authors do not purport to reflect the position of the Department of Defense.

References

1. Kwiatkowski, D. (1989) Febrile temperatures can synchronize the growth of *Plasmodium falciparum* in vitro. *J. Exp. Med.* **169**, 357–361.
2. Zeuthen, E. (1964) The Temperature-induced division synchrony in *Tetrahymena*, in *Synchrony in Cell Division and Growth* (Zeuthen, E., ed.), Wiley Interscience, New York, pp. 99–157.
3. Gravenor, M. B. and Kwiatkowski, D. (1998) An analysis of the temperature effects of fever on the intra-host population dynamics of *Plasmodium falciparum*. *Parasitology* **117,** 97–105.
4. Ginsburg, H., Nissani, E., and Krugliak, M. (1989) Alkalinization of the food vacuole of malaria parasites by quinoline drugs and alkylamines is not correlated with their antimalarial activity. *Biochem. Pharmacol.* **38,** 2645–2654.
5. Neff, R. J. and Neff, R. H. (1964) Induction of synchronous division in *Amoebae*, in *Synchrony in Cell Division and Growth* (Zeuthen, E., ed.), Wiley Interscience, New York, pp. 213–246.

45

Hepatic Portal Branch Inoculation

John B. Sacci, Jr.

1. Introduction

The study of plasmodial liver stage parasites, in vivo, has been hampered by the relatively small number of liver stage schizonts that can be generated in an animal. This has been true even when large numbers of sporozoites are injected. The number of liver stage parasites can be increased, by increasing the parasite inoculum, but the liver stage schizonts plateau at a sporozoite dose of $3–6 \times 10^6$ *(1)*. It is not clear why this occurs, but the result is that the number of infected hepatocytes is small, relative to the noninfected hepatocytes.

Several authors have utilized a surgical technique to inoculate sporozoites directly into the portal vein *(1,2)*. This places the sporozoites into the liver without having to move through the whole circulatory system, as is the case when parasites are injected intravenously. When this technique is coupled with partial ligation of the liver, the parasites can be directed to a smaller amount of liver tissue and thus increase the ratio of infected hepatocytes to non-infected hepatocytes. Hepatic portal branch inoculation (HPBI) has demonstrated the capacity to significantly increase the number of liver stage parasites in both rats and mice, making it a valuable tool for the dissection of this stage in the parasite's life cycle *(3)*.

2. Materials

2.1. General Reagents

1. 1-mL syringes.
2. 26- and 30-gage needles.
3. Sterile surgical instruments (forceps, scissors, and hemostats).
4. Sterile surgical gloves.
5. 10-mm Microvascular clamp (Roboz Surgical Instrument Co., Washington, DC).
6. Baby serrefine hemostatic clamp.
7. Gelfoam™ sponge (Upjohn, Kalamazoo, MI).
8. 4-0 chromic gut suture (Ethicon, Somerville, NJ).
9. Michel wound clips (Roboz).
10. Sterile 2×2 gauze pads.

2.2. Solutions

1. Avertin.
2. 0.75% Providine iodine.
3. 70% Ethyl alcohol.
4. Sterile normal saline.
5. Buprenorphine (Reckitt and Colman Pharmaceuticals, Richmond, VA).

3. Method

1. Anesthetize the mouse with Avertin (0.02 mL/g body weight), by intraperitoneal injection.
2. Once a deep plane of anesthesia is achieved, remove the hair on the abdomen by shaving.
3. Swab the abdomen with a disinfectant (0.75% providine iodine), then rinse with 70% alcohol and wipe with sterile gauze.
4. Make a midline incision through the skin and muscle layer, changing scissors in between incisions.
5. Expose the portal vein and its branches by reflecting the liver up and retracting the intestines laterally.
6. Wrap the intestines in saline-soaked gauze to prevent them from drying during the procedure (*see* **Note 1**).
7. Apply a 10-mm microvascular clamp to the portal vein, approx 1-cm distal to the bifurcation. This serves to prevent the retrograde flow of sporozoites after inoculation (*see* **Note 2**).
8. Place a baby serrefine hemostatic clamp on the portal vein at the bifurcation of the median and left lateral lobes. This reduces the patient blood flow to the smaller right and caudate lobes. The occlusion of the lobes produces immediate blanching, while the nonclamped lobes remain a deep red color.
9. Inject sporozoites into the portal vein, in a volume of 0.2 mL or less, using a 30-gage needle with the tip bent at 45° (*see* **Notes 3** and **4**).
10. Place a small piece of Gelfoam (2 × 2 mm) over the injection site using a pair of fine forceps, and remove the needle from the portal vein. The Gelfoam promotes homeostasis, reducing the chances of hemorrhage from the injection.
11. Remove the hepatic portal clamp, upstream from the injection site, 15 min after injection, and remove the lobar clamp 60 min after injection.
12. Suture the abdominal muscle layer using 4-0 chromic gut suture and close the skin with Michel wound clips.
13. Administer buprenorphine (0.03 mg/kg) as a single dose, subcutaneously for analgesia.
14. Inject normal saline (0.5 mL subcutaneously into each flank) to offset any dehydration that may have occurred during the surgery.
15. Keep the mouse warm and observe until the anesthetic has worn off, before returning it to its cage.
16. The right and caudate lobes can subsequently be removed, at various time points, for histological or molecular studies.

4. Notes

1. The liver and intestines are very fragile; therefore it is essential that care be exercised in handling them during the surgery. Additionally, they should be kept moist throughout the procedure.
2. Clamping the vessels is a delicate part of the surgery and is greatly facilitated by the use of a magnifying loupe.
3. Injections of sporozoites should be done using the smallest practical volume.
4. Sporozoites can be isolated by gland dissection or gradient and collected in medium with 5% normal mouse serum.

References

1. Meis, J. F. G. M., Verhave, J. P., Jap, P. H. K., Sinden, R. E., and Meuwissen, J. H. E. T. (1983) Ultrastructural observations on the infection of rat liver by *Plasmodium berghei* sporozoites in vivo. *J. Protozool.* **30,** 361–366.
2. Scheller, L. F., Stump, K. C., and Azad, A. F. (1995) *Plasmodium berghei*: production and quantitation of hepatic stages derived from irradiated sporozoites in rats and mice. *J. Parasitol.* **8,** 58–62.
3. Scheller, L. F. and Azad, A. F. (1995) Maintenance of protective immunity against malaria by persistent hepatic parasites derived from irradiated sporozoites. *Proc. Natl. Acad. Sci. USA* **25,** 4066–4068.

46

Hepatocyte Perfusion, Isolation, and Culture

John B. Sacci, Jr.

1. Introduction

The isolation of intact functional hepatocytes is a relatively recent innovation in cell biology, as compared to the isolation of cells from other organs. First successfully accomplished in the late 1960s, isolation of viable hepatocytes has become an increasingly utilized technique for metabolic studies and host–pathogen experiments.

It had been thought that isolation of intact hepatocytes would be impossible because of the firm cell–cell connections between cells. Advances in the understanding of junctional complexes and the extracellular matrix surrounding hepatocytes, however, have led to the successful experimental isolation of hepatocytes. Hepatocytes were first successfully isolated by Howard ct al. (1), although the yields of viable hepatocytes were very low. The initial isolations laid the ground work for subsequent studies by Seglen (2), in which he utilized a two-step process. The first step was a pre-perfusion, using media without Ca^{2+}, which opened up the desmosomes of the junctional complex. The second step utilized the enzymatic separation of the hepatocytes with a collagenase solution. This basic technique has been used for many years, with modifications, and routinely produces large numbers of viable hepatocytes. The two-step method, for the isolation of hepatocytes is described in this chapter.

2. Materials

2.1. Equipment

1. Water bath, set at 37°C.
2. Peristaltic pump.
3. Laminar flow hood.
4. 4 ft. of Sterile latex tubing (3 mm internal diameter).
5. 10% Povidone iodine solution.
6. Sterile surgical instruments (forceps, scissors).
7. 23-gage Butterfly infusion set.
8. 3-0 Sterile suture.
9. Sterile petri dish.
10. Sterile gauze.
11. Sterile latex gloves.
12. Sterile disposable 1-, 2-, and 10-mL pipets.
13. 8-well Lab-tek™ chamber slides.

2.2. Solutions

1. Hanks' balanced salt solution (HBSS): combine 8 g of NaCl, 0.4 g of KCl, 0.2 g of $MgSO_4$ $7H_2O$, 0.06 g of KH_2PO_4, 1.5 g of glucose, 0.5 mg of insulin, 4.9 g of HEPES, and 10 mL of penicillin/streptomycin (10,000 U/mL), and make up volume to 1 L, pH 7.4. Filter-sterilize through a 0.22-µm filter and store at 4°C.
2. HBSS supplemented with 0.075% $CaCl_2$ and 0.025% collagenase (*see* **Note 1**).
3. Earle's minimum essential medium (EMEM) stock: to 500 mL of EMEM, add 1 g of bovine serum albumin (BSA), 1.1 g of $NaHCO_3$, and 5 mg of insulin. Filter-sterilize through a 0.22-µm filter and store at 4°C.
4. 10X HBSS: combine 8 g of NaCl, 0.4 g of KCl, 0.2 g of $MgSO_4·7H_2O$, 0.06 g of KH_2PO_4, 1.5 g of glucose, 0.5 mg of insulin, 4.9 g of HEPES, and 10 mL penicillin/streptomycin (10,000 U/mL), and make up volume to 1 L, pH 7.4. Filter-sterilize through a 0.22-µm filter and store at 4°C.
5. Percoll®.
6. EMEM Complete: supplement EMEM stock with 10% fetal calf serum (FCS), 2% penicillin/streptomycin (10,000 U/mL), 1% glutamine, and 1% nonessential amino acids. Filter-sterilize through a 0.22-µm filter and store at 4°C.
7. EMEM complete with dexamethasone: supplement EMEM stock with 10% FCS, 2% penicillin/streptomycin (10,000 U/mL), 1% glutamine, 1% nonessential amino acids, and 50 mL of dexamethasone (1 mg/mL). Filter-sterilize through a 0.22-µm filter and store at 4°C.
8. 70% EtOH.
9. 0.075% $CaCl_2$ and 0.025% collagenase in HBSS.

3. Methods

1. Set up the perfusion apparatus as described. Maintain the perfusion media in the laminar flow hood, and place the pump and 37°C water bath adjacent to the hood on a portable cart or bench top. Run the tubing from the medium in the hood through the pump and waterbath and finally to a waste bottle. Attach a 23-gage butterfly infusion set to the end of the tubing.
2. Run approx 50 mL of 70% alcohol through the tubing, to waste, at a flow rate of 10 mL/min. Subsequently, run 75 mL of 1X HBSS through the tubing.
3. Anesthetize the mouse by intraperitoneal injection of pentobarbital (50 mg/kg mouse body weight). After the mouse achieves a deep plain of anesthesia, secure it to a platform (a 15-cm × 25-cm styrofoam board) with the abdomen facing up, and place the platform into a tray.
4. Wipe the abdomen with iodine solution and make midline incision through the skin from the sternum to pubis, using sterile forceps and scissors. Then make a horizontal incision n the skin on either side of the vertical incision at the midpoint and reflect the skin flaps out of the way. Using a new set of sterile forceps and scissors, make a similar incision through the abdominal muscle wall.
5. Reflect the intestines to the right out of the abdominal cavity and gently reflect the two large lobes of the liver up to expose the portal vein. Using a fine pair of curved forceps, thread a 3-0 suture around the vein and tie a loose knot around the portal vein 1 cm from where it enters the liver. Then cannulate the portal vein by introducing the 23-gage butterfly needle such that the tip of the needle is 0.5 cm from where the portal vein enters the liver (*see* **Note 2**). A small amount of blood should appear in the butterfly hub at this point; this will confirm that the needle is in the vein. Pull the suture tight to prevent the needle from slipping out of the vein.
6. Turn the pump on and set to deliver 10 mL/min. Puncture the diaphragm of the mouse and cut the aorta. The liver should blanch at this point and its dark red color should be replaced by a yellow-cream color. Perfusion with 1X HBSS should be maintained for 10 min. Then,

turn the pump off and take the inlet tubing from the HBSS and place it into a container of 0.075% $CaCl_2$, 0.025% collagenase in HBSS. Then, turn the pump back on and resume the perfusion at 10 mL/min.
7. Continue the perfusion until fissures develop in the liver either spontaneously or as the result of slight pressure with a pair of forceps.
8. Place approx 10 mL of cold 1X HBSS into a sterile 100-mm Petri dish. Carefully cut the liver away from the stomach and duodenum with sterile scissors and place in the Petri dish. Then place the dish in the laminar flow hood and tease the liver apart using a pair of 1-mL pipets. The liver should be so softened that it will disintegrate with a small amount of mechanical manipulation. Place a sterile 4-in. × 4-in. gauze pad over a 50-mL centrifuge tube and filter the cell suspension into the tube. Repeat the process of adding HBSS to the Petri dish and filtering the cells until there is approx a 40-mL vol in the tube.
9. Centrifuge the hepatocyte suspension at 200 rpm for 3 min at 4°C to pellet the cells. While the cells are being centrifuged, prepare the isomotic Percoll® by mixing 45 mL of Percoll® with 5 mL of 10X HBSS. After centrifugation, gently resuspend the cell pellet with 25 mL of EMEM stock and layer over 20 mL of isomotic Percoll®. Mix the Percoll®-cell suspension by inversion and centrifuge at 1000 rpm for 10 min at 4°C (*see* **Note 3**).
10. After the Percoll® centrifugation, the viable cells will pellet and the nonviable cells will float to the top of the gradient. Aspirate off the non-viable cells and the gradient and resuspend the pellet in 30 mL of EMEM complete. Centrifuge at 200 rpm for 3 min at 4°C.
11. Resuspend the washed cell pellet in 10 mL of complete media and count, using a hemocytometer. Dilute the cells and dispense into 8-well permanox Lab-Tek® chamber slides at a concentration of 9.5×10^5 hepatocytes/0.3 mL. Place the slides into a humidified 37°C CO_2 incubator overnight.
12. By the following morning, the cells should have attached and spread to form a monolayer. Aspirate off the media and replace with EMEM complete containing dexamethasone (*see* **Note 4**).

4. Notes

1. The quality of collagenase varies from supplier to supplier. It is important to test a lot to determine its suitability for liver perfusion. The collagenase can only be tested by doing a liver perfusion. Once a suitable lot is found, a supply should be ordered that will be enough to last at least 1 yr.
2. The cannulation is the most important aspect of the procedure. Ensure that there are no air bubbles in the tubing or butterfly. An air bubble introduced into the liver will obstruct the flow of buffer and result in a large decrease in the recovery of hepatocytes.
3. The centrifugation step with Percoll® removes any damaged cells. Percoll® is a colloidal silica particle suspension coated with polyvinylpyrrolidine. The particles are unable to enter intact or damaged cells. The damaged cells have a lower density than the Percoll® solution and float, while the intact cells have a higher density and pellet. Additionally, nonparenchymal cells will not pellet at the concentration of Percoll® used in this isolation and will be removed at this step.
4. The addition of dexamethasone will promote the maintenance of differentiated hepatocyte morphology and function.

References

1. Howard, R. B., Christensen, A. K., Gibbs, F. A., and Pesch, L. A. (1967) The enzymatic preparation of isolated intact parenchymal cells from rat liver. *J. Cell Biol.* **35,** 675–684.
2. Seglen, P. O. (1976) Preparation of isolated rat liver cells, in: *Methods in Cell Biology*, vol. 13, (Prescott, D. M., ed.), Academic Press, New York, pp. 29–83.

47

Inhibition of Sporozoite Invasion

The Double-Staining Assay

Laurent Rénia, Ana Margarida Vigário, and Elodie Belnoue

1. Introduction

During the life cycle of the malaria parasites in mammals, sporozoites invade hepatocytes, and give rise to thousands of merozoites that are able to infect red blood cells. Invasion of the liver cells is thus a critical step by which infection becomes established in the host, and against which an antiparasite attack might be effective. The development of hepatic cell culture systems *(1–7)* has made it possible to study several aspects of the liver stage infection, including penetration of the hepatocyte by the sporozoite *(8–13)*.

Hollingdale et al. *(14–16)* first used the human embryonic lung cell line WI-38, and later the hepatoma cell line HepG2, to develop the inhibition of sporozoite invasion (ISI) assay, a test of the capacity of drugs or antibodies to prevent invasion of *Plasmodium falciparum* or *P. berghei* sporozoites. They claimed to distinguished microscopically between sporozoites that have invaded and those that have not because only the former cause a sausage shaped-area of staining brought about by the release of CSP in the parasitophorous vacuole.

Mazier et al. *(17)* used primary cultures of human hepatocytes to study the effect of anti-CSP antibodies on sporozoite invasion. Unlike Hollingdale et al., they used a count of trophozoites as a measure of successful invasion by sporozoites. But some are still not transformed even after 5 d in cultures because the development of sporozoites in vitro is heterogeneous. Thus, counting trophozoites as a measure of sporozoites that have successfully invaded may lead to an underestimate of parasite invasion.

To eliminate subjectivity, Zavala et al. *(18)* have proposed using an immunoradiometric assay to measure the amount of internalized CSP in human hepatoma hepG2. But this approach, unlike microscopic observation, does not indicate the precise location of the parasite. For this reason, we have developed a method, the double staining assay, to distinguish in vitro whether a sporozoite is outside the hepatocyte membrane or inside *(19)*. This technique has been successfully used to assess the effect of drugs or antibodies on host–parasite interaction *(13,20)*. It has made it possible to study not only sporozoite invasion but also the transformation of sporozoites into trophozoites and maturation of trophozoites into schizonts. Variants of the double staining assay involving the use of different secondary reagents have recently been developed *(21,22)*.

The double-staining assay relies on sequential staining using a single monoclonal antibody specific for the CSP that is revealed by two different secondary antibodies; the first is coupled to an enzyme, the second to a fluorochrome. Briefly, hepatocyte cultures are inoculated with infective sporozoites. The cultures are fixed with paraformaldehyde at appropriated times after inoculation. This makes the hepatocyte membrane impermeable, and so impedes the penetration of antibodies. External sporozoites are then detected with an anti-CSP monoclonal antibody (MAb) and revealed by peroxidase coupled-secondary antibodies. The cultures are then incubated overnight and fixed in ethanol, which renders the hepatocyte membrane permeable, so allowing internal parasites (both sporozoites and trophozoites) to be detected using the same anti-CSP MAb and an fluoroscein isothiocyanate (FITC)-coupled-secondary antibody. This method therefore discriminates between external or membrane-attached sporozoites, which appear brown, and internal parasites, which appear fluorescent.

The following sections provide a detailed description of the procedures for the culturing hepatocytes, and staining the parasite.

2. Materials

2.1. Hepatocyte Isolation and Culture

1. Hepatocyte perfusion and isolation reagents: see Chapter 46 for details.
2. Female mice, 7- to 12-wk-old.
3. Eight-chamber plastic Lab-Tek slides (Nunc Inc., IL).
4. Dissecting materials: tweezers, scissors.
5. Complete medium: Williams' medium (Life Technologies, Edinburgh, Scotland) supplemented with 10% fetal calf serum (FCS) (Dutscher, Paris, France; aliquot and store at −20°C), 1% penicillin/streptomycin (Life Technologies; 100X stock solution, aliquot and store at −20°C,) and 0.5 µg/mL of fungizone (Sigma). Prepare fresh and store at 4°C.

2.2. Mosquito Dissection

1. *Plasmodium yoelii*-infected mosquitoes.
2. Neubauer or Malassez counting chambers.
3. Sterilin tubes.
4. 1-mL syringes.
5. 26-gage needles.
6. Tissue grinder (10-mL) and pestle (Thomas, Philadelphia, PA).
7. Sterile microscope slides
8. Sterile tips for 200-µL pipet.
9. Small Petri dishes: diameter, 60 mm (Corning).
10. Ethanol, 70%.
11. Washing medium A: Williams' medium containing 5 µg/mL of Fungizone (Sigma)
12. Washing medium B: Williams' medium containing 2% penicillin/streptomycin.

2.3. Sera

1. Immunogen-immunized and control mice.
2. Pasteur pipets.
3. 1.5-mL Eppendorf tubes.
4. 1-mL syringes.
5. 0.22-µm filter (Costar).

2.4. Parasite Staining

1. Moist chamber.
2. Phosphate-buffered saline (PBS), 1X (pH 7.5), diluted 1:10 with dH_2O from 10X PBS stock (Life Technologies). Store at 4°C.
3. Paraformaldehyde fixative: add 40 g of paraformaldehyde powder per 100 mL of dH_2O to obtain a 40% (w/v) solution. To completely dissolve the paraformaldehyde, add one drop of 1 N NaOH, stir the suspension with a magnetic bar, and boil. Add 9 vol of PBS to the cooled 40% stock solution to give a 4% paraformaldehyde working solution. Aliquot and store at –20°C. Aliquots should be used only once (*see* **Note 1**). Extreme care should be taken when boiling the paraformaldehyde solution, as its vapors are highly toxic. Use gloves and a mask when manipulating the solution, and do so under a chemical hood.
4. Mouse MAb NYS1 against *P. yoelii* circumsporozoite protein (1 mg/mL in PBS) *(23)*. Store at –20°C.
5. Peroxidase-conjugated goat anti-mouse antibody (lyophilized; Biosys, Compiègne, France). Store lyophilized product at 4°C. Reconstitute with sterile dH_2O at a concentration of 1 mg/mL, aliquot, and store at –20°C.
6. 3,3-Diaminobenzidine tetrahydrochloride powder (DAB, Sigma). **Caution:** Extreme care should be taken when handling DAB, as this chemical is thought to be carcinogenic. Use gloves and a mask when manipulating the product in powder form. Store at –20°C.
7. H_2O_2 stock solution 30% (Sigma).
8. PBS-azide: 1X PBS, pH 7.5 supplemented with 0.001% (w/v). sodium azide (Sigma). **Caution:** Extreme care should be taken when handling sodium azide, as this chemical is highly toxic. Use glove and mask when manipulating the product in powder form. Store at 4°C.
9. Absolute ethanol. Store at room temperature in a special cabinet. Store a 10-mL aliquot at –20°C for parasite staining.
10. FITC-labeled goat anti-mouse IgG (lyophilized, Biosys). Store lyophilized product at 4°C. If reconstituted with sterile dH_2O at a concentration of 1 mg/mL, aliquots should be made and stored at –20°C.
11. Evans' Blue (Sigma): prepare a 100X stock of 5% (w/v). Store at 4°C.
12. PBS/glycerol: prepare a 1:1 solution of PBS-azide and glycerol (1:1). Store at 4°C.
13. Cover slips, 20–60 mm.

3. Methods
3.1. Hepatocyte Isolation and Culture

See Chapter 46 and **Notes 1–6** for additional details.

1. Prepare hepatocytes by perfusing a liver biopsy from BALB/c mice.
2. Seed hepatocytes at a concentration of 60×10^3 cells in complete medium, in Lab-Tek wells.
3. Incubate at 37°C for 24 h in 3.5% CO_2 before use. This time is normally sufficient to allow the hepatocytes to adhere to the Lab-Tek slides and become confluent.

3.2. Isolation of Sporozoites

1. Collect mosquitoes with sporozoites in their salivary glands in a Sterilin vial.
2. Shake the vial vigorously, to stun the mosquitoes.
3. Catch the mosquitoes with tweezers, then immerse in 70% ethanol.
4. Transfer the mosquitoes to a small Petri dish containing 3 mL of washing solution A. Next, wash twice in two other Petri dishes containing the same medium, washing solution A. Then transfer to two more petri dishes containing washing solution B.
5. Place the mosquitoes on a sterile slide and dissect under a microscope using sterile 24-gage needles. Cut the heads of mosquitoes and remove the salivary glands by making a

light pressure on the thorax. Pull away the salivary glands, which appear translucid, and immerse in a drop of medium (50 µL) at the opposite end of the dissecting slide. Care must be taken not to take the legs or other parts of the mosquito's body.
6. At the end of the dissection of a series of mosquitoes (10–15 per slide), collect the salivary glands using a 200-µL pipet with a sterile tip, and transfer into a small tissue grinder which is kept on ice.
7. When all the glands are collected, homogenize them and dilute the resulting sporozoite suspension in complete medium. Then, count the sporozoites by loading Neubauer or Malassez counting chambers with an aliquot of the sporozoite suspension and allowing to stand for 5–0 min. Sporozoites sink slowly, especially in a small volume and in a thick suspension, and counting them before they have sunk is difficult and inaccurate. Phase-contrast illumination facilitates rapid and accurate counts. The most convenient magnification for counting sporozoites is ×300.

3.3. Preparation of Immune Serum

See **Note 7**.

1. Collect blood from the retro-ocular vein of immunized or control mice using a Pasteur pipet and transfer to a 1.5-mL Eppendorf tube. Do not use heparin.
2. Keep tubes at 4°C for at least for 1 h, then centrifuge at 800g for 15 min.
3. Transfer the serum, which is in the supernatant, to another 1.5-mL Eppendorf tube. Dilute serum (minimum volume of 100 µL) to 1:5 in Williams' medium.
4. Inactivate complement by heating the tube at 56°C for 30 min.
5. Filter the decomplemented serum using a 0.22-µm filter, and keep on ice or at 4°C. Prepare aliquots, and store at –20°C if the serum is not to be used the same day.

3.4. Addition of Immune Sera and Sporozoites to Hepatocyte Cultures

1. Remove the medium from the culture chambers and add 50 µL of the immune serum to the Lab-Tek well. Test the serum in triplicate. Fill two wells in each Lab-Tek with 50 µL of complete medium to serve as control.
2. Add 6×10^4 sporozoites in 50 µL of complete medium to each well.
3. Incubate cultures at 37°C in 3.5% CO_2.

3.5. Parasite Staining

1. At 3 h later, wash the cultures three times with PBS using a 1-mL sterile pipet tip. Washes are necessary to remove the sporozoites which have not completed their entry. For all washing steps, remove the contents of each well using a tip placed in the corner, and add the new solution slowly. Handle the Lab-Teks carefully to avoid damaging the hepatocytes.
2. Fix the cultures by adding 200 µL of 4% paraformaldehyde to each well and incubating at room temperature for 10 min. Take care to thaw the paraformaldehyde solution before use to dissolve any paraformaldehyde aggregates (see **Note 8**).
3. Wash the cultures three times with PBS.
4. Add NYS1 MAb (diluted 1:100 in PBS) to the cultures (see **Note 9**).
5. Incubate for 30 min at 37°C in a moist chamber.
6. Wash cultures twice with PBS
7. Incubate with a peroxidase-conjugated goat anti-mouse antibody diluted 1:100 in PBS at 37°C for 30 min.
8. Wash cultures twice with PBS.
9. Add 200 µL of a solution of freshly prepared DAB (1 mg/mL) and 0.03% H_2O_2 in PBS to the wells. Protect cultures from the light for at least 10 min to allow the development of a brownish precipitate.

10. When precipitates are observed, stop the reaction by washing the wells twice with PBS-azide.
11. Keep the Lab-Teks in a moist chamber and incubate overnight at 37°C in PBS-azide.
12. The following day, wash the cultures twice with PBS-azide.
13. Add cold absolute ethanol (0.3 mL) to each well and incubate Lab-Teks at room temperature for 10 min.
14. Remove the ethanol and wash the cultures twice with PBS-azide.
15. Add NYS1 MAb (diluted 1:100 in PBS) to the cultures.
16. Incubate for 30 min at 37°C in a moist chamber.
17. At the end of the incubation, wash the cultures twice with PBS-azide.
18. Incubate for 30 min at 37°C with FITC-labeled anti-mouse antibody diluted 1:100 in PBS-azide with Evans' blue (0.5%). The solution should appear light blue; if it is dark blue, then too much Evan's Blue has been added.
19. Wash the Lab-Teks twice with PBS-azide and remove the wells.
20. Mount the remaining slides with PBS-glycerol and keep at 4°C before being read (*see* **Note 10**).
21. Read the slides with an epifluorescence microscope under ×250 or ×400 magnification. If slides are not seen when illuminated, it is advisable to shield them from the incoming light (*see* **Note 11**).

4. Notes

1. The rate of sporozoite development in vitro is usually low when mouse hepatocytes are used. In a good experiment, 50 to 500 sporozoites per well may be obtained. But the number of sporozoites is frequently lower than this, diminishing the significance of the results.
2. A limiting factor when dealing with murine malaria parasites is the number and quality of sporozoites. These parasite forms are obtained after dissection of infected mosquitoes. The development of the parasite inside its vector is sensitive to very small variation in heat or humidity, so only a limited number of questions can be answered in any particular experiment.
3. Primary hepatocyte cultures need to be produced for each experiment and this requires the expertise of trained personnel. Also, cultures are frequently contaminated by bacteria or fungi present in the solution containing the sporozoites, despite the presence of antibiotics and antimycotics in the culture medium.
4. This assay has been adapted for human parasites. Because the development of sporozoites of human parasite into schizonts takes longer, hepatocyte cultures are stopped 24 h after sporozoite inoculation instead of 3 h. Nevertheless, the medium is changed after 3 h, and cultures are thoroughly washed to prevent bacterial or fungal contamination. Simian (for *P. vivax and P. cynomolgi*) and human (for *P. falciparum* and *P. vivax*) hepatocytes have been used successfully *(12,19,24–28)*. For safety reasons, the culturing of human hepatocyte and the handling of infected mosquitoes (especially of *P. falciparum*) should be carried out in a P2 laboratory.
5. For practical reasons, the tumoral hepatoma cell line, HepG2-A16 *(15)* has been used more often than the more problematic human hepatocyte to assess the effect of antibodies (principally to *P. falciparum*). In spite of the logistical complications, several critical differences make it is imperative to use primary human hepatocytes if the data obtained are to be reliable. In practical terms the lack of *P. falciparum* development in HepG2 cells directly influences the outcome of invasion inhibition studies. Moreover, HepG2 cells clearly differ functionally from host hepatocytes, as evidenced by the totally artificial receptivity of HepG2 cells for *P. berghei* contrasting with its inability to permit intrahepatocytic development of *P. falciparum (2,15)*. This reveals critical differences that may also affect susceptibility to invasion. Indeed, many comparative studies have revealed major discrepancies in the inhibitory effect of antibodies to sporozoite molecules, this

frequently being near total for HepG2 cells but dramatically weaker for hepatocytes *(17,29–31)*. In one study, antibodies found to significantly inhibit *P. falciparum* sporozoite invasion of HepG2 cells gave the opposite result, i.e., an increase in the number of invaded sporozoites *(31)*. Furthermore, data from three human vaccine trials has shown a lack of correlation between levels of inhibition of sporozoite invasion (ISI) in HepG2 cells and degree of protection conferred *(32–34)*. In support of the relevance of the primary hepatocyte assay, a close correlation has been observed between the levels of antisporozoite immunity in vivo and the in vitro inhibition found with the relevant antisporozoite antibodies, for both *P. falciparum* and *P. yoelii*, *(35–37)*. These findings lend weight to the assertion that data obtained using human hepatocytes will more faithfully reflect the true capacity of these antibodies to limit sporozoite invasion in vivo.

6. The double staining assay could also be used to determine the effect of antibodies or drugs on the developing liver stage. For rodent parasites, cultures are stopped 24 or 48 h after sporozoite inoculation; and for human parasites, cultures are stopped 3 and 5 d after sporozoite inoculation. Anti-Pfhsp1 antibodies are used together with the anti-CSP MAb to detect *P. yoelii* schizonts *(38)*. This technique allows the precise quantification of internal sporozoites, trophozoites and schizonts, the evaluation of transformation rate of sporozoites into trophozoites, and of maturation rate of trophozoites into schizonts *(13,39)*. This test is more reliable than the inhibition of liver-stage development assay (ILSDA) *(40)* which is based on the count of late schizonts but not of invaded sporozoites and trophozoites.

7. Different forms of antibody preparation can be tested in this assay. It is preferable to use purified monoclonal or polyclonal forms *(20,23,28,41,42)* because we have observed that certain molecules present in ascites or in the sera of normal or vaccinated animals may inhibit sporozoite invasion *(39,43,44)*. However, the quantity of serum from immunized mice may be limited. To overcome this problem, sera could be pooled and the antibodies purified. However, it is sometime of interest to correlate the level of antibodies in the serum of individual immunized animals with the outcome of the experiment (e.g., protection). Sera are obtained from mice bled without the use of heparin. This is known to inhibit sporozoite invasion in vitro *(21)*. Diluted immune sera must be then decomplemented because it has been reported that complement may be toxic for the hepatocytes. This effect is frequently seen when serum from one species (e.g., rabbit) is tested with hepatocytes from a different species (e.g., human). Sera must also be filtrated to prevent fungal and bacterial contamination.

8. The first fixation step uses paraformaldehyde to make hepatocyte membrane impermeable and to allow the staining of external but not internal sporozoites. Paraformaldehyde is easy to prepare and to use, but it is unstable and aliquots must be used only once. Frozen aliquot are brought to room temperature and should be used only when the solution is transparent and any aggregates have been fully dissolved. Only paraformaldehyde, and not formaldehyde should be use. Formaldehyde, a 39% paraformaldehyde solution contains 10% methanol stabilize it, and may create holes in the hepatocyte membrane, so allowing antibodies to diffuse into the interior of the infected cells.

9. Anti-CSP mAbs are used as detecting antibodies and have always given reliable results for all parasite species tested so far (*P. yoelii*, *P. berghei*, *P. vivax*, and *P. falciparum*) *(45–48)*. Antibodies to new antigens such as SSP2/TRAP, STARP, and SALSA *(49–54)* may be of interest but this remains to be tested. This assay has been simplified by directly coupling the MAb to peroxidase or to FITC, so making secondary antibodies unnecessary *(27)*. The appropriate dilution of the detecting antibody should be determined before it is used in the assay. To avoid antibody denaturation, care should be taken to avoid repeating the freezing and thawing cycle.

10. If immunofluorescence (IF) detection is performed immediately after ethanol fixation, the fluorescent signal may be very weak. This is caused by the presence of oxygen in the

culture medium, produced by the peroxidase-mediated catalytic transformation of H_2O_2 into H_2O and O_2. Oxygen inhibits intensity of fluorescence by colliding with the fluorophore *(55)*. To avoid this phenomenon, a minimum of 8 h at 37 °C is necessary to ensure the total disappearance of reactive oxygen in the medium. Only then should the IF staining be commenced. Evans' blue is used as a counterstain. It stains the hepatocytes red and allows better discrimination of the schizonts, which appear green. If too much Evans' blue is added, the fluorescence is quenched. If this happens, the coverslips should be removed by immersion in PBS-azide. The slides can be dipped into methanol for 2 min, and washed in PBS-azide.

11. Assessment of the inhibitory activity in infected hepatocytes can be time-consuming. Reading the slides is the longest part of the technique. A trained scientist takes 25–40 min per well. Slides should therefore be examined before they accumulate and their fluorescence fades.

Acknowledgments

This work was made possible in part by financial support from Institut Electricité et Santé, UNDP/World Bank/WHO Special Program for Research and Training in Tropical Diseases (TDR), European Community programs INCO-DC, Fondation pour la Recherche Médicale, and Junta Nacional de Investigação Cientifica e Tecnologica (JNICT). Ana Margarida Vigário held a fellowship from JNICT, Portugal. She received support from Centro de Malaria e Outras Doenças Tropicais, Lisbon, Portugal (Dir: Virgilio do Rosario). Elodie Belnoue is supported by a predoctoral fellowship from the Ministère de l'Education Nationale, de La Recherche et de la Technologie. We thank Irène Landau for providing infected mosquitoes, and Yupin Charoenvit and Stephen Hoffman for providing the NYS1 antibody. We also thank to Dominique Mazier, Myriam Marussig, Sylviane Pied, and Pierre Druilhe for helpful discussions. We thank Dr Jean-Gérard Guillet for his active support. The English text was edited by Dr. Geoff Watts.

References

1. Hollingdale, M. R., Leef, J. L., McCullough, M., and Beaudoin, R. L. (1981) *In vitro* cultivation of the exoerythrocytic stage of *Plasmodium berghei*. *Science* **213**, 1021–1022.
2. Hollingdale, M. R., Leland, P., and Schwartz, A. L. (1983) *In vitro* cultivation of the exoerythrocytic stage of *Plasmodium berghei* in a hepatoma cell line. *Am. J. Trop. Med. Hyg.* **32**, 682–684.
3. Mazier, D., Landau, I., Miltgen, F., Druilhe, P., Lambiotte, M., Baccam, D., and Gentilini, M. (1982) Infestation *in vitro* d'hépatocytes de *Thamnomys* adulte par des sporozoites de *Plasmodium yoelii*: schizogonie et libération de mérozoites infestants. *C. R. Acad. Sci. Paris.* **294**, 963–965.
4. Mazier, D., Landau, I., Druilhe, P., Miltgen, F., Guguen-Guillouzo, C., Baccam, D., et al. (1984) Cultivation of the liver forms of *Plasmodium vivax* in human hepatocytes. *Nature* **307**, 367–369.
5. Mazier, D., Beaudoin, R. L., Mellouk, S., Druilhe, P., Texier, J., Trosper, J., et al. (1985) Complete development of hepatic stages of *Plasmodium falciparum in vitro*. *Science* **227**, 440–442.
6. Millet, P., Collins, W. E., Fisk, T. L., and Ngyuen-Dinh, P. (1988) *In vitro* cultivation of exoerythrocytic stages of the human malaria parasite *Plasmodium malariae*. *Am. J. Trop. Med. Hyg.* **38**, 470–473.
7. Millet, P., Fisk, T. L., Collins, W. E., Broderson, J. R., and Ngyuen-Dinh, P. (1988) Cultivation of exoerythrocytic stages of *Plasmodium cynomolgi, P. knowlesi, P. coatneyi*, and *P. inui* in *Macaca mulatta* hepatocytes. *Am. J. Trop. Med. Hyg.* **39**, 529–534.
8. Cerami, C., Frevert, U., Sinnis, P., Takacs, B., Clavijo, P., Santos, M. J., and Nussenzweig, V. (1992) The basolateral domain of the hepatocyte plasma membrane bears receptors for the circumsporozoite protein of *Plasmodium falciparum* sporozoites. *Cell* **70**, 1021–1033.
9. Frevert, U., Sinnis, P., Cerami, C., Shreffler, W., Takacs, B., and Nussenzweig, V. (1993) Malaria circumsporozoite protein binds to heparan sulfate proteoglycans associated with the surface membrane of hepatocytes. *J. Exp. Med.* **177**, 1287–1298.
10. Sinnis, P., Clavijo, P., Fenyo, D., Chait, B. T., Cerami, C., and Nussenzweig, V. (1994) Structural and functional properties of region II-plus of the malaria circumsporozoite protein. *J. Exp. Med.* **180**, 297–306.

11. Müller, H.-M., Reckmann, I., Hollingdale, M. R., Bujard, H., Robson, K. J. H., and Crisanti, A. (1993) Thrombospondin related anonymous protein (TRAP) of *Plasmodium falciparum* binds specifically to sulfated glycoconjugates and to HepG2 Hepatoma cells suggesting a role for this molecule in sporozoite invasion of hepatocytes. *EMBO J.* **12**, 2881–2889.
12. Pancake, S. J., Holt, G. D., Mellouk, S., and Hoffman, S. L. (1992) Malaria sporozoites and circumsporozoite proteins bind specifically to sulfated glycoconjugates. *J. Cell Biol.* **117,** 1351–1357.
13. Goma, J., Rénia, L., Miltgen, F., and Mazier, D. (1996) Iron overload increases hepatic development of *Plasmodium yoelii* in mice. *Parasitology* **112,** 165–168.
14. Hollingdale, M. R., Zavala, F., Nussenzweig, R. S., and Nussenzweig, V. (1982) Antibodies to the protective antigen of *Plasmodium berghei* sporozoites prevent entry into cultured cells. *J. Immunol.* **128,** 1929,1930.
15. Hollingdale, M. R., Nardin, E. H., Tharavanij, S., Schwartz, A. L., and Nussenzweig, R. S. (1984) Inhibition of entry of *Plasmodium falciparum* and *P. vivax* sporozoites into cultured cells: an *in vitro* assay of protective antibodies. *J. Immunol.* **132,** 909–913.
16. Leland, P., Sigler, C. I., Danforth, H. D., and Hollingdale, M. R. (1984) Inhibition of *Plasmodium berghei* sporozoite invasion of cultured hepatoma cells. *Trans. R. Soc. Trop. Med. Hyg.* **78,** 639–640.
17. Mazier, D., Mellouk, S., Beaudoin, R. L., Texier, B., Druilhe, P., Hockmeyer, W. T., et al. (1986) Effect of antibodies to recombinant and synthetic peptides on *Plasmodium falciparum* sporozoites *in vitro. Science* **231,** 156–159.
18. Zavala, F., Hollingdale, M. R., Schwartz, A. L., Nussenzweig, R. S., and Nussenzweig, V. (1985) Immunoradiometric assay to measure the *in vitro* penetration of sporozoites of malaria parasites into hepatoma cells. *J. Immunol.* **134,** 1202–1205.
19. Rénia, L., Miltgen, F., Charoenvit, Y., Ponnudurai, T. V., Verhave, J. P., Collins, W. E., and Mazier, D. (1988) Malaria sporozoite penetration: a new approach by double staining. *J. Immunol. Methods*, **112,** 201–205.
20. Nudelman, S., Rénia, L., Charoenvit, Y., Yuan, L., Miltgen, F., Beaudoin, R. L., and Mazier, D. (1989) Dual action of anti-sporozoite antibodies *in vitro. J. Immunol.* **143,** 996–1000.
21. Frevert, U., Sinnis, P., Esko, J. D., and Nussenzweig, V. (1996) Cell surface glycosaminoglycans are not obligatory for *Plasmodium berghei* sporozoite invasion *in vitro. Mol. Biochem. Parasitol.* **76,** 257–266.
22. Sinnis, P. (1998) An immunoradiometric assay for the quantification of *Plasmodium* sporozoite invasion of HepG2 cells. *J. Immunol. Methods* **221,** 17–23.
23. Charoenvit, Y., Leef, M. F., Yuan, L. F., Sedegah, M., and Beaudoin, R. L. (1987) Characterization of *Plasmodium yoelii* monoclonal antibodies directed against stage-specific sporozoite antigens. *Infect. Immun.* **55,** 604–608.
24. Millet, P., Collins, W. E., Herman, L., and Cochrane, A. H. (1989) *Plasmodium vivax*:*In vitro* development of exoerythrocytic stages in squirrel monkey hepatocytes and inhibition by an anti-*P. cynomolgi* monoclonal antibody. *Exp. Parasitol.* **69,** 61–93.
25. Millet, P., Kamboj, K. K., Cochrane, A. H., Collins, W. E., Broderson, J. R., Brown, B. G., et al. (1989) *In-vitro* exoerythrocytic development of *Plasmodium cynomolgi bastianellii*: inhibitory activity of monoclonal antibodies against sporozoites of different *P. cynomolgi* strains and of *P. knowlesi. Parasite Immunol.* **11,** 223–230.
26. Millet, P., Collins, W. E., Broderson, J. R., Bathurst, I., Nardin, E. H., and Nussenzweig, R. S. (1991) Inhibitory activity against *Plasmodium vivax* sporozoites induced by plasma from *Saimiri* monkeys immunized with circumsporozoite recombinant proteins or irradiated sporozoites. *Am. J. Trop. Med. Hyg.* **45,** 44–48.
27. Etlinger, H. M., Rénia, L., Matile, H., Manneberg, M., Mazier, D., Trzeciak, A., and Gillessen, D. (1991) Antibody response to a synthetic peptide-based malaria vaccine candidate: influence of sequence variants of the peptide. *Eur. J. Immunol.* **21,** 1505–1511.
28. Fidock, D. A., Pasquetto, V., Gras, H., Badell, E., Eling, W. M. C., Ballou, W. R., et al. (1997) *Plasmodium falciparum* sporozoite invasion is inhibited by naturally acquired or experimentally induced polyclonal antibodies to the STARP antigen. *Eur. J. Immunol.* **27,** 2502–2513.
29. Hollingdale, M. R., Ballou, W. R., Aley, S. B., Young, J. F., Pancake, S., Miller, L. H., et al. (1987) *Plasmodium falciparum*: elicitation by peptides and recombinant circumsporozoite proteins of circulating mouse antibodies inhibiting sporozoite invasion of hepatoma cells. *Exp. Parasitol.* **63,** 345–351.
30. Moelans, I. I., Cohen, J., Marchand, M., Molitor, C., de Wilde, P., Van pelt, J. F., et al. (1995) Induction of Plasmodium falciparum sporozoite-neutralizing antibodies upon vaccination with recombinant Pfs16 vaccinia virus and/or recombinant Pfs16 protein produced in yeast. *Mol. Biochem. Parasitol.* **72,** 179–192.
31. Hollingdale, M. R., Appiah, A., Leland, P., Do Rosario, V. E., Mazier, D., Pied, S., et al. (1990) Activity of human volunteer sera to candidate *Plasmodium falciparum* circumsporozoite vaccines in the inhibition of sporozoite invasion (ISI) of human hepatoma cells and hepatocytes assays. *Trans. R. Soc. Trop. Med. Hyg.* **84,** 325–329.

32. Fries, L. F., Gordon, D. M., Richards, R. L., Egan, J. E., Hollingdale, M. R., Gross, M., et al. (1992) Liposomal malaria vaccine in humans : a safe and potent adjuvant strategy. *Proc. Natl. Acad. Sci. USA* **89,** 358–362.
33. Egan, J. E., Hoffman, S. L., Haynes, J. D., Sadoff, J. C., Schneider, I., Grau, G. E., et al. (1993) Humoral Immune Responses in Volunteers Immunized with Irradiated *Plasmodium falciparum* Sporozoites. *Am. J. Trop. Med. Hyg.* **49,** 166–173.
34. Brown, A. E., Singharaj, P., Webster, H. K., Pipithkul, J., Gordon, D. M., Boslego, J. W., et al. (1994) Safety, Immunogenicity and Limited Efficacy Study of a Recombinant *Plasmodium falciparum* Circumsporozoite Vaccine in Thai Soldiers. *Vaccine* **12,** 102–108.
35. Mellouk, S., Mazier, D., Druilhe, P., Berbiguier, N., and Danis, M. (1986) *In vitro* and *in vivo* results suggest that antisporozoite antibodies do not totally block *Plasmodium falciparum* sporozoite infectivity. *N. Engl. J. Med.* **315,** 648.
36. Mellouk, S., Lunel, F., Sedegah, M., Beaudoin, R. L., and Druilhe, P. (1990) Protection against malaria induced by irradiated sporozoites. *Lancet* **335,** 721.
37. Charoenvit, Y., Mellouk, S., Cole, C., Bechara, R., Leef, M. F., Sedegah, M., et al. (1991) Monoclonal, but not polyclonal, antibodies protect against *Plasmodium yoelii* sporozoites. *J. Immunol.* **146,** 1020–1025.
38. Rénia, L., Mattei, D., Goma, J., Pied, S., Dubois, P., Miltgen, F., Nussler, A., et al. (1990) A malaria heat shock like protein epitope expressed on the infected hepatocyte surface is the target of antibody-dependent cell-mediated cytotoxic mechanisms by non-parechymal liver cells. *Eur. J. Immunol.* **20,** 1445–1449.
39. Nüssler, A., Pied, S., Pontet, M., Miltgen, F., Rénia, L., Gentilini, M., et al. (1991) Inflammatory status and pre-erythrocytic stages of malaria. Role of the C-reactive protein. *Exp. Parasitol.* **72,** 1–7.
40. Mellouk, S., Berbiguier, N., Druilhe, P., Sedegah, M., Galey, B., Yuan, L., et al. (1990) Evaluation of an in vitro assay aimed at measuring protective antibodies against sporozoites. *Bull. World Hlth. Organ,* **68 Suppl,** 52–59.
41. Brahimi, K., Perignon, J. L., Bossus, M., Gras, H., Tartar, A., and Druilhe, P. (1993) Fast immunopurification of small amounts of specific antibodies on peptides bound to ELISA plates. *J. Immunol. Methods* **162,** 69–75.
42. Del Giudice, G., Tougne, C., Rénia, L., Ponnudurai, T., Corradin, G., Pessi, A., et al. (1991) Characterization of murine monoclonal antibodies against a repetitive synthetic peptide from the circumsporozoite protein of the human malaria parasite, *Plasmodium falciparum*. *Mol. Immunol.* **28,** 1003–1009.
43. Pied, S., Nussler, A., Pontet, M., Miltgen, F., Matile, H., Lambert, P. H., and Mazier, D. (1989) C-reactive protein protects against pre-erythrocytic stages of malaria. *Infect. Immun.* **57,** 278–282.
44. Pied, S., Tabone, M. D., Chatellier, G., Marussig, M., Jardel, C., Nosten, F., and Mazier, D. (1995) Non specific resistance against malaria pre-erythrocytic stages: involvement of acute phase proteins. *Parasite* **2,** 263–268.
45. Boulanger, N., Matile, H., and Betschart, B. (1988) Formation of the circumsporozoite protein of *Plasmodium falciparum* in *Anopheles stephensi*. *Acta Tropica* **45,** 55–65.
46. Verhave, J. P., Leeuwenberg, A. D., Ponnudurai, T., Meuwissen, J. H., and van Druten, J. A. (1988) The biotin-streptavidin system in a two-site ELISA for the detection of plasmodial sporozoite antigen in mosquitoes. *Parasite Immunol.* **10,** 17–31.
47. Wirtz, R. A., Zavala, F., Charoenvit, Y., Campbell, G. H., Burkot, T. R., Schneider, I., et al. (1987) Comparative testing of monoclonal antibodies against *Plasmodium falciparum* sporozoites for ELISA development. *Bull. World Hlth. Organ.* **65,** 39–45.
48. Yoshida, N., Nussenzweig, R. S., Potocnjak, P., Nussenzweig, V., and Aikawa, M. (1980) Hybridoma produces protective antibodies directed against the sporozoite stage of malaria parasite. *Science* **207,** 71–73.
49. Fidock, D. A., Bottius, E., Brahimi, K., Moelans, I. I., Aikawa, M., Konings, R. N., et al. (1994) Cloning and characterization of a novel *Plasmodium falciparum* sporozoite surface antigen, STARP. *Mol. Biochem. Parasitol.* **64,** 219–232.
50. Bottius, E., BenMohamed, L., Brahimi, K., Gras, H., Lepers, J. P., Raharimalala, L., et al. (1996) A novel *Plasmodium falciparum* sporozoite and liver stage antigen (SALSA) defines major B, T helper, and CTL epitopes. *J. Immunol.* **156,** 2874–2884.
51. Robson, K. J. H., Hall, J. R. S., Jennings, M. W., Harris, T. J. R., Marsh, K., Newbold, C. I., et al. (1988) A highly conserved amino-acid sequence in thrombospondin, properdin and in proteins from sporozoites and blood stages of a human malaria parasite. *Nature* **335,** 79–82.
52. Rogers, W. O., Malik, A., Mellouk, S., Nakamura, K., Rogers, M. D., Szarfman, A., et al. (1992) Characterization of *Plasmodium falciparum* sporozoite surface protein 2. *Proc. Natl. Acad. Sci. USA* **89,** 9176–9180.
53. Rogers, W. O., Rogers, M. D., Hedstrom, R. C., and Hoffman, S. L. (1992) Characterization of the gene encoding sporozoite surface protein 2, a protective *Plasmodium yoelii* sporozoite antigen. *Mol. Biochem. Parasitol.* **53,** 45–51.

54. Charoenvit, Y., Fallarme, V., Rogers, W. O., Sacci, J. B., Jr., Kaur, M., Aguiar, J. C., et al. (1997) Development of two monoclonal antibodies against *Plasmodium falciparum* sporozoite surface protein 2 and mapping of B-cell epitopes. *Infect. Immun.* **65,** 3430–3437.
55. Lakowitz, J. R. and Weber, G. (1973) Quenching of fluorescence by oxygen: a probe for structural fluctuation in macromolecules. *Biochemistry* **21,** 4161–4167.

48

Inhibition of Liver-Stage Development Assay

John B. Sacci, Jr.

1. Introduction

In vitro infection of hepatocyte monolayer cultures with *plasmodial* sporozoites has become a valuable tool for the dissection of host–cell parasite interactions *(1–4)*. This methodology has facilitated the study of how antibodies, cytokines, drugs, and immune effector cells can influence the infection of hepatocytes with *plasmodial* parasites. In vitro cultures have the advantage of allowing an investigator to isolate hepatocytes from nonparenchymal cells, thus avoiding any confounding of the results that might occur due to an interaction of the test conditions with nonparenchymal cells (kupffer cells, endothelial cells, lymphocytes, etc.). Several different types of inhibition of liver-stage development assay (ILSDA) procedures can be utilized. They fall into two different categories: blocking entry of the parasite into the hepatocyte or inhibiting development of the liver stage schizont after infection. The first procedure is used to determine if immune sera can prevent the entry of parasites into the hepatocytes, while the other can be used to assess the capacity of sera, immune effector cells or drugs to affect the development of the intracellular liver stage parasite. Because these assays are assessing different aspects of the host-parasite relationship, the practical aspects of these tests are slightly different.

Isolation of sporozoites from infected mosquitoes can be done by several methods. Salivary glands can be hand-dissected from mosquitoes and triturated in a ground glass homogenizer to release the parasites. This is laborious, requires the use of a dissecting microscope and does not lend itself to the rapid isolation of parasites from hundreds of mosquitoes. Two other methods have been utilized for a quicker less laborious isolation of sporozoites: gradient separation *(5)*, and centrifugation through glass wool *(6)*. Both methods will produce viable sporozoites. While the gradient method produces a parasite suspension with less mosquito debris, the Ozaki using glass wool is faster. Because of the quicker isolation, the Ozaki method is usually utilized for in vitro infections and will be described below.

2. Materials

2.1. General Reagents

1. *Plasmodium yoelii*-infected mosquitoes.
2. Murine hepatocyte monolayers on 8-well Lab-Tek® chamber slides.
3. 1-mL syringes with 26-gage needles.

4. Sterile Ozaki tubes (*see* **Note 1**).
5. 6-Well tissue culture plate.
6. Microliter pipetors.
7. Sterile pipet tips.
8. Sterile 3 in. × 2 in. glass slides.
9. Fine forceps.
10. Sterile siliconized microfuge tubes.
11. Sterile 12 mm × 75 mm polypropylene round bottom tubes.

2.2. Solutions

1. 70% ethanol.
2. Earle's minimum essential medium (EMEM) stock: To 500 mL of EMEM add 1 g of bovine serum albumin, 1.1 g of $NaHCO_3$, and 5 mg of insulin. Filter-sterilize through a 0.22-µm filter and store at 4°C.
3. EMEM Complete with dexamethasone: Supplement EMEM stock with 10% fetal calf serum (FCS), 2% penicillin/streptomycin (10,000 U/mL), 1% glutamine, 1% nonessential amino acids, and 50 mL of dexamethasone (1 mg/mL). Filter-sterilize through a 0.22-µm filter and store at 4°C.
4. EMEM stock with 4% penicillin/streptomycin (10,000 U/mL).
5. EMEM stock with 2% Fungizone®.
6. Phosphate-buffered saline (PBS).
7. Absolute methanol.
8. Monoclonal anti-*P. yoelii* liver-stage antibody (NYLS3).
9. Fluorescein-conjugated goat anti-mouse IgG.

3. Methods

1. Aspirate mosquitoes from their cage to a 50-mL centrifuge tube and place at –20°C for 5 min.
2. Transfer the mosquitoes to a well of a 6-well plate containing 70% EtOH. Immerse them in the 70% EtOH and then transfer, using fine forceps, to a well containing EMEM stock with 2% Fungizone®, and then to a well containing EMEM stock with 4% penicillin/streptomycin (10,000 U/mL).
3. Remove the mosquitoes from the final wash and place onto a sterile glass slide. Using the tips of 26-gage needles attached to 1-mL syringes, separate the abdomens from the heads and thoraxes, and discard the abdomens. Then, using the beveled needles, separate the heads and thoraxes and place both into the small Ozaki tube. Combine a total of 50 heads and thoraxes in each Ozaki tube and add 50 µL of EMEM complete media. Repeat these steps until all mosquitoes have been cut-up and placed into a series of tubes (*see* **Note 2**).
4. Microfuge the Ozaki tubes at 15,000 rpm for 2 min at room temperature. Collect the medium containing the sporozoites from the bottom tube and transfer to a 12 mm × 75 mm polypropylene tube. Add 50 mL of media to the top of the Ozaki tubes containing the mosquito debris and microfuge again. Collect the filtrate and add to the 12 mm × 75 mm polypropylene tube.
5. Mix the sporozoite suspension well and dilute a small sample (5 µL) 1:10 with media for counting. Count the sample using a phase-contrast hemocytometer to determine the number of sporozoites present.
6. Murine hepatocyte cultures, started the previous day, are utilized for in vitro infections. Aspirate the media from the culture wells and add the diluted parasites to each well.
7. Depending upon the type of assay being done, the infection of the monolayers is done in one of two ways. If sera are being tested for the capacity to inhibit invasion, add 2X concentrations of the diluted sera to triplicate wells in a volume of 25 µL per well. Then add sporozoites (75,000 per well) to each well in a volume of 25 µL per well. Incubate the

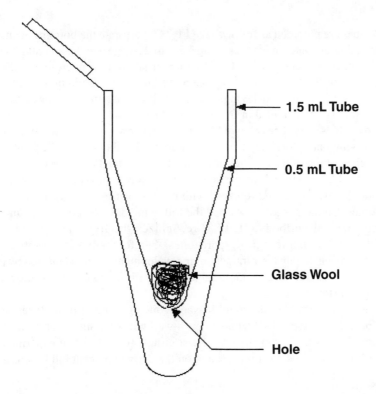

Fig. 1. Diagram of an Ozaki tube used for the isolation of sporozoites. A 0.5-mL microfuge tube, with a small hole at its base containing a plug of glass wool, sits inside a 1.5-mL microfuge tube. Separated heads and thoraxes of mosquitoes are placed in the small tube and centrifuged to release the sporozoites.

cultures for 3 h in a CO_2 incubator. Then wash the cultures 3 or 4 times with fresh media and return to the incubator overnight (*see* **Note 3**). If the effect of antibodies, effector cells or drugs postinvasion is being assessed, infect the cultures with sporozoites (75,000 per well) in a volume of 50 µL per well, incubate for 3 h, wash several times with fresh media, and culture with the diluted serum or drug overnight. At 24 h after infection, change the media on all cultures, adding the test serum or drug (diluted in media) to assess postinvasion effects of the compound. If effector cells are being utilized, add the cells at 24 h after infection.
8. Stop the cultures at 48 h postinfection. Wash the slides three times with PBS and fix with ice cold absolute methanol for 10 min. Wash the fixed slides three times with PBS and visualize the parasites by immunofluorescent staining.
9. Incubate the chamber slides with 100 µL of NYLS3, a *P. yoelii*-specific monoclonal antibody (7) (*see* **Note 4**), for 30 min at 37°C. Wash the slides three times with PBS and incubate for 30 min at 37°C with a fluorescein-conjugated goat anti-mouse IgG (Kirkegard and Perry Laboratories, Gaithersburg, MD) diluted 1:40 in 0.02% Evans' blue.
10. Wash the slides three times with PBS, mount with Vectashield® (Vector laboratories, Burlingame, CA) (*see* **Note 5**), and view using an epifluorescent microscope.
11. Count the numbers of schizonts present in each well and calculate the percent inhibition (*see* **Note 6**).

$$\text{Percent inhibition} = \frac{\text{control} - \text{experimental} \times 100}{\text{control}}$$

4. Notes

1. Ozaki tubes are prepared as follows (see **Fig. 1**): perforate the bottoms of small (0.65 mL) siliconized microcentrifuge tubes using a 27- to 23-gage needle, and plug the bottom with glass wool. Insert each small tube into a larger siliconized microtube (1.5-mL). Cut the snap top off of the large tube. Wrap the nested tubes in aluminum foil, place into a container and sterilize by autoclaving. The glass slides that are used for cutting the mosquitoes should also be wrapped in foil and autoclaved.
2. The mosquitoes should be removed from the EMEM wash in small groups to reduce their risk of drying out before they have been processed and placed in the Ozaki tubes.
3. Thorough washing of the cultures after infection is crucial for the removal of bacteria that may cause contamination. The parasite suspension is removed from the cultures, 3 h after infection, by aspiration and replaced with 0.3 mL of fresh media. This wash is repeated three times. Cultures are then washed the following day and given fresh media.
4. The monoclonal antibody, NYLS3, recognizes the liver- and blood-stage antigen PyHEP17. It does not recognize sporozoites, therefore only parasites that have successfully invaded and begun their intracellular development can be identified by this antibody.
5. Vectashield® contains an antifading agent that retards photobleaching, allowing prolonged viewing and storage.
6. Because of the variable infectivity of sporozoites generated from different blood-stage infections, the number of schizonts in each well will vary significantly from infection to infection. Thus, direct comparisons of test groups from different experiments cannot be done. All comparisons must be based upon the derived percent inhibition.

References

1. Franke, E. D., Hoffman, S. L., Sacci Jr., J. B., Wang, R., Charoenvit, Y., Appella, E., et al. (1999) Pan DR binding sequence provides T cell help for induction of protective antibodies against *Plasmodium yoelii* sporozoites. *Vaccine* **17,** 1201–1205.
2. Renia, L., Grillot, D., Marussig, M., Corradin, G., Miltgen, F., Lambert, P., et al. (1993) Effector functions of circumsporozoite peptide-primed $CD4^+$ T cell clones against *Plasmodium yoelii* liver stages. *J. Immunol.* **150,** 1471–1478.
3. Nussler, A., Pied, S., Goma, J., Renia, L., Miltgen, M., Gentilini, G., et al. (1991) TNF inhibits malaria hepatic stages *in vitro* via IL-6 liver synthesis. *Int. Immunol.* **3,** 317–322.
4. Hoffman, S. L., Weiss, W., Mellouk, S., and Sedegah, M. (1990) Irradiated sporozoite vaccine induces cytotoxic T lymphocytes that recognize malaria antigens on the surface of infected hepatocytes. *Immunol. Lett.* **25,** 33–38.
5. Pacheco, N. D., Strome, C. P. A., Mitchell, F., Bawden, M. P., and Beaudoin, R. L. (1979) Rapid, large-scale isolation of *Plasmodium berghei* sporozoites from infected mosquitoes. *J. Parsitol.* **65,** 414–417.
6. Ozaki, L. S., Gwadz, R. W., and Godson, G. N. (1984) Simple centrifugation method for rapid separation of sporozoites from mosquitoes. *J. Parasit.* **70,** 831–833.
7. Charoenvit, Y., Mellouk, S., Sedegah, M., Toyoshima, T., Leef, M. F., De la Vega, P., et al. (1995) *Plasmodium yoelii*: 17-kDa hepatic and erythrocytic stage protein is the target of an inhibitory monoclonal antibody. *Exp. Parasitol.* **80,** 419–429.

49

T-Cell Mediated Inhibition of Liver-Stage Development Assay

Laurent Rénia, Elodie Belnoue, and Ana Margarida Vigário

1. Introduction

Most studies of the T-cell response to *Plasmodium*-infected hepatocytes have been done with the rodent parasites, *P. yoelii* and *P. berghei*. Early experiments showed that mice immunized with irradiated sporozoites of *P. berghei* and devoid of antibodies after a long-term treatment with anti-IgM antibodies resisted a live sporozoite challenge *(1)*, suggesting that T cells by themselves were able to confer protection. Further transfer of both CD8+ and CD4+ T-cell clones specific for T epitopes from *P. yoelii* or *P. berghei* circumsporozoite or SSP-2 protein confirmed that these T cell populations could confer protection against a sporozoite challenge *(2–6)*. However, these experiments provided indirect evidence that T cells inhibit parasite development in hepatocytes.

The advent of hepatocyte culture of malaria parasites (*see* Chapter 46) has made it possible to develop specific assays to measure the activity of drugs *(7)*, and antibodies (inhibition of sporozoite invasion assay, ISI or double staining assay; *see* Chapter 47) *(8–10)*. However, the role of T cells against in vitro infected hepatocytes remained unstudied because there was no suitable assay. The classical chromium release assay could not be adapted to hepatocytes because these cells incorporate chromium poorly and spontaneously release high levels of this radioactive compound.

To prove that the infected hepatocytes are targets for T-cells, a new assay, the T-cell-mediated liver-stage development assay (TILSA), has been developed *(4,11–14)*. It is based on the coculture of infected primary hepatocytes and parasite-specific T cells. T-cell clones, T-cell lines, or T-cell-enriched preparations from the spleen or lymph nodes of mice immunized with various immunogens (peptides, proteins, DNA or irradiated parasites) can be used. These T cells are added to hepatocyte cultures 3 h after sporozoite inoculation, and the resulting cocultures are kept for 45 h, the time required for rodent malaria parasites to develop in vitro. Counting schizonts in experimental and control wells gives an assessment of T-cell-mediated inhibition, expressed as a percentage reduction in schizont number.

TILSA is now being used to elucidate the mechanisms by which T cells eliminate liver schizonts. The involvement of lymphokines has been studied using antibodies to anti-interleukins *(4,11,13)*. Antigen processing of malaria epitopes can also be studied by adding processing inhibitors (e.g., Brefeldin A). Although developed for rodent malaria, TILSA could also be use for human malaria parasites. So far, this has not be done because of the difficulty of obtaining *P. falciparum* sporozoites, human hepatocytes, and human *P. falciparum*-specific T-cell lines or clones of the same human lymphocyte antigen (HLA) on a regular basis and in the same time-frame.

TILSA relies on three procedures: (a) the preparation of effector T cells from peptide-immunized mice; (b) the coculture of infected hepatocyte with T cells; and (c) the detection of liver schizonts by immunofluorescence staining. These three components will be described in this chapter.

2. Materials
2.1. T-Cell Preparation

1. Cell strainer (Falcon).
2. 75-cm^2 culture flasks.
3. Tissue grinder (15-mL) and pestle (Kontes Glass, Vineland, NJ).
4. Complete and incomplete Freund's adjuvant (CFA, IFA; Sigma, St. Quentin L'Arbresles, France). Store at 4°C.
5. Peptides (Neosystem, Strasbourg, France): dissolve desiccated peptide in H$_2$O at a concentration of 10 mg/mL, aliquot, and store at –20°C.
6. Complete medium: Williams' medium (Life Technologies, Edinburgh, Scotland) supplemented with 10% fetal calf serum (FCS) (Dutscher, Paris, France; aliquot and store at –20°C), 1% penicillin/streptomycin (Life Technologies; 100X stock solution, aliquot and store at –20°C,) and 0.5 µg/mL of Fungizone (Sigma). Prepare fresh and store at 4°C.
7. Trypan blue, 0.4% (w/v) stock solution in dH$_2$O (Sigma). Prepare 0.1% working solution by diluting stock 1:4 with saline.
8. Goat anti-mouse immunoglobulin (lyophilized; Biosys, Compiègne, France). Reconstitute with sterile dH$_2$O at a concentration of 1 mg/mL, aliquot and store at –20°C.
9. Phosphate-buffered saline (PBS), 1X (pH 7.5), diluted 1:10 with dH$_2$O from 10X PBS stock (Life Technologies). Store at 4°C.

2.2. Cocultures of Infected Hepatocytes and Lymphocytes
2.2.1. Hepatocytes

1. Hepatocyte perfusion and isolation reagents: see Chapter 46 for details.
2. Female BALB/c mice, 7- to 12-wk-old (Harlan Olac, Leicester, UK).
3. Eight-chamber plastic Lab-Tek slides (Nunc Inc, IL).
4. Dissecting materials: tweezers, scissors.

2.2.2. Mosquitoes

1. *Plasmodium*-infected mosquitoes.
2. Neubauer or Malassez chamber.
3. Sterilin tubes.
4. 1-mL syringes.
5. 26-gage needles.
6. Teflon/glass homogenizer (10-mL) (Thomas, Philadelphia, PA).
7. Sterile microscope slides.
8. 200-µL pipet and sterile tips.
9. Small Petri dishes (60 mm diameter; Corning).
10. Ethanol, 70%.
11. Washing medium A: Williams' medium containing 5 µg/mL Fungizone (Sigma).
12. Washing medium B: Williams' medium containing 2% penicillin/streptomycin.

2.3. Immunofluorescent Staining

1. Absolute methanol. Care should be taken when handling as this chemical is toxic; use of gloves is recommended. Store at room temperature in a special cabinet. A 20-mL aliquot should be placed at –20°C before doing parasite staining.
2. PBS-azide: 1X PBS, pH 7.5, supplemented with 0.001% (w/v) sodium azide (Sigma). Extreme care should be taken when handling sodium azide, as this chemical is highly toxic. Use gloves and a mask when manipulating the product in powder form. Store at 4°C.
3. Mouse serum anti-Pfhsp70.1 *(15)*. Aliquot and store at –20°C.
4. FITC-labeled goat anti-mouse IgG (lyophilized, Biosys). Store lyophilized product at 4°C. Reconstitute with sterile dH$_2$O at a concentration of 1 mg/mL, aliquot, and store at –20°C.
5. Evans' blue (Sigma): prepare a 100X stock solution of 5% (w/v). Store at 4°C.
6. Coverslips, 20 × 60 mm.
7. PBS-glycerol: prepare a 1:1 solution of PBS-azide and glycerol (1:1). Store at 4°C.

3. Methods
3.1. Preparation of T Cells
3.1.1. Peptide Immunization

1. Make an emulsion of peptide and CFA by mixing 30 µg of peptide, diluted in 50 µL of PBS, with 50 µL of CFA in a syringe. Firmly attach a 24-gage needle and secure both sides of the syringe with adhesive tapes. Fix the syringe on the top of a stirring-rod and stir until the emulsion looks milky. Test the quality of the emulsion by placing a drop of the emulsion onto the surface of ice-water in a beaker; the emulsion should not disperse.

    Note that it is necessary to always prepare at least 20% more volume than required for the number of animals (at 100 µL per animal), to compensate for losses that occur during the preparation. Two to five mice are generally used per group.
2. Immunize mice subcutaneously at the base of the tail with the 100 µL of emulsified solution. Care should be taken to avoid injecting the emulsion in the tail vein, since this will lead to death of the mouse.
3. Boost mice twice at 2–3 wk after the priming and then 1 wk later, with 100 µL of the emulsion containing 30 µg of peptide emulsified in IFA (*see* **step 1**, but substitute CFA for IFA), injected at the base of the tail.

3.1.2. Purification of T Cells

See **Note 1**.

1. At 1 wk after the last boost, sacrifice control and immunized mice, and harvest the periaortic, mesenteric, and inguinal lymph nodes into a tube containing 5 mL of complete medium.
2. Crush the lymph nodes in a tissue grinder, and pass the resulting tissue suspension through a cell strainer to remove fibrous tissue remains.
3. Centrifuge the cells at 450*g* for 7 min at room temperature, and discard the supernatant.
4. Resuspend the pellet in 5 mL of complete medium

3.1.3. Enrichment of T Cells

1. In order to eliminate adherent cells, resuspend cells (obtained as described above) in complete medium containing 20% FCS and incubate overnight in 75-cm^2 culture flasks.
2. Recover nonadherent cells by rinsing the flask four times with complete medium.
3. Remove B cells by panning.
4. Add recovered cells to flasks coated with goat anti-mouse immunoglobulins, and incubate at 4°C for 1 h. To prepare these flasks, add 2 mL of goat anti-mouse immunoglobulin

(50 μg/mL in sterile PBS), incubate at 37°C for 1 h and then wash with complete medium to remove unbound goat immunoglobulins.
5. Recover nonadherent cells by washing the flask three times with complete medium.
6. Centrifuge the cell suspension at 450g for 7 min at room temperature, and discard the supernatant.
7. Resuspend the cell pellet in 1 mL of complete medium.
8. Determine the cell count and viability by trypan blue exclusion, by mixing 50 μL of the cell suspension with 50 μL of a 0.1% trypan blue solution.
9. Verify the efficacy of cellular subset depletion by flow cytometry, using antibodies specific for B and T cells (*see* Chapter 36). Efficiency of depletion must be >90%.

3.2. Coculture of Infected Hepatocytes and T Cells

3.2.1. Isolation of Hepatocytes

See Chapter 46 for additional details (*see* **Notes 2–4**).

1. Prepare hepatocytes by perfusion of a liver biopsy from BALB/c mice.
2. Seed hepatocytes at a concentration of 60×10^3 cells in 5 ml of complete medium, in Lab-Tek wells.
3. Incubate for 24 h at 37°C in 3.5% CO_2 before use. This time is normally sufficient to allow hepatocytes to adhere to the Lab-Tek slides and became confluent.

3.2.2. Sporozoite Isolation

1. Collect infected mosquitoes with *Plasmodium* sporozoites in their salivary glands in a Sterilin vial.
2. Vigorously shake the vial to knock out the mosquitoes.
3. Catch the stunned mosquitoes with tweezers, and wash them in 70% ethanol.
4. Transfer to a small Petri dish containing 3 mL of washing medium A, and then twice again in two other Petri dishes containing the same medium.
5. Transfer to a small Petri dish containing 3 mL of washing medium B and wash twice more.
6. Place mosquitoes on a slide and dissect them under a microscope using sterile 24-gage needles. Cut off the head of mosquitoes; make a light pressure on the thorax to release the salivary glands (which appear translucid), pull the salivary glands away and immerse them in a drop of medium (50 μL) on the opposite side of the dissecting slide. Take care to avoid bringing the legs or other part of the mosquito's body into contact with the salivary glands.
7. At the end of the dissection of a series of mosquitoes (10–15 per slide), collect salivary glands with a 200-μL pipet using a sterile tip and transfer into a small sterile tissue grinder thatis kept on ice.
8. Homogenize the glands and dilute the resulting sporozoite suspension with complete medium in preparation for counting.
9. Place a suspension of sporozoites in a Neubauer or Malassez counting chamber and allow to stand for 5–10 min before counting. The sporozoites sink slowly, especially in a small volume and in a thick suspension, so counting them before they have sunk will be difficult and inaccurate. Phase-contrast illumination facilitates rapid, accurate counting. The most convenient magnification for counting sporozoites is ×300.

3.2.3. Infection of Hepatocytes

1. After removal of medium from the culture chamber, add 6×10^4 sporozoites in 100 μL of fresh medium to Lab-Tek wells.
2. Incubate at 37°C in 3.5% CO_2.
3. After 3 h, wash cultures three times with complete medium to remove any sporozoites which have not completed their entry. These washes are also necessary to remove fungi or

bacteria present in the sporozoite suspension, and omission of these steps may lead to contamination of the culture. For all washing steps, the contents of each well should be removed using a pipet tip placed in the corner of each well and the new solution should be added slowly. Lab-Teks should be handled carefully to avoid damaging the hepatocytes.
4. After the last wash, add 0.3 mL of complete medium containing the T cells to the wells, at different ratios of T cells to infected hepatocytes (30:1, 10:1, and 1:1).
5. Change 50 µL of the culture medium at 21–24 h after sporozoite inoculation.
6. Incubate the cultures for an additional 24 h.

3.3. Parasite Staining

See **Note 5**.

1. At 45–48 h after sporozoite inoculation, stop the cultures by completely removing the medium, using a 1-mL pipet tip.
2. Carefully add 0.5 mL of PBS-azide to the wells and then remove. Repeat this procedure three times to remove the majority of nonadherent cells.
3. Add 0.3 mL of cold methanol to each well and incubate the Lab-Tek slides at room temperature for 10 min.
4. Remove the methanol and wash the culture twice with PBS-azide.
5. Add anti-Pfhsp70.1 serum diluted 1:100 in PBS-azide (for the dilution, *see* **Note 3**) to the hepatocyte culture, and incubate at 37°C for 30 min.
6. At the end of the incubation, wash the cultures twice with PBS-azide.
7. Add fluoroscein isothiocyanae (FITC)-labeled anti-mouse antibody diluted 1:100 in PBS-azide with 0.5% Evans' blue, and incubate at 37°C for 30 min. The solution should appear light blue and not dark blue, which would indicate that too much Evans' blue was added (*see* **Note 5**).
8. Wash Lab-Teks twice with PBS-azide and remove the wells.
9. Mount the remaining slides with PBS-azide–glycerol and keep at 4°C before observation.
10. Read slides with an epifluorescence microscope using ×250 or ×400 magnification (*see* **Note 6**).

4. Notes

1. T cells can be obtained from mice immunized with peptides, proteins, parasite preparations (e.g., irradiated sporozoites), or other immunogens. These immunogens may induce $CD4^+$ or $CD8^+$ T cells that could be recovered from the blood, the spleen, the lymph nodes, or the liver. Thus, in some experiments, it could be of interest to further enriched T-cell preparation in $CD8^+$ or $CD4^+$ T-cell population by depletion with anti-CD4 or anti-CD8 monoclonal antibodies and complement, or using antibody-coated beads and magnetic separation. In most other organs, the proportion of T cells is lower than in lymph nodes and may required extensive purification in order to prevent a nonspecific toxic effect against the parasite. We have observed that spleen cell preparations from normal mice significantly inhibit liver stage parasite development. This nonspecific effect could be reduced by purifying T cells to more than 90% or by reducing the incubation time of the spleen cells with the infected hepatocytes. In the latter case, spleen cells should be added 24 h after sporozoite inoculation and incubated for 24 h with infected hepatocytes. TILSA could also be performed with T-cell clones or lines. These cells induce a greater inhibition of parasite development. The assay should be carried out with a lower effector to target ratio.
2. When dealing with murine malaria parasites, the main limiting factors are the number and the quality of sporozoites. These parasite forms are obtained after dissection of infected mosquitoes. Development of the parasites inside its vector is sensitive to very small variation in heat or humidity. So only a limited number of questions can be asked in any experiment.

3. The rate of sporozoite development in vitro is usually low when mouse hepatocytes are used. In a good experiment, 50–300 schizonts per well may be obtained. But the number of schizonts is frequently lower than this, diminishing the significance of the results.
4. Primary hepatocyte cultures infected with malaria sporozoites need to be produced for each experiment and this requires the expertise of trained personnel. Cultures frequently become contaminated by bacteria or fungi present in the solution containing the sporozoites despite the presence of antibiotics and antimycotics in the culture medium.
5. The parasite can be detected using mouse serum containing antibodies, which react to other liver stage antigens. Only a limited number of antigens expressed by the liver parasite have been identified, and this limits the range of antibodies suitable for detect it. Antibodies to Pfhsp70.1 *(4)* are useful because malaria hsp is highly conserved, they can recognize all malaria liver-stage parasites. Antibodies against antigens such as MSP1 *(16,17)*, 17-kDa *(18)*, and SSP2 *(19)* can also be used but are mostly species-restricted. The reactivity of the antiliver-stage antibodies should be checked regularly to determine the appropriate dilution to be used. To avoid antibody denaturation, care should be taken to avoid repeating the cycle of freezing and thawing. Evans' blue is used as a counterstain. It stains the hepatocytes red and allows better discrimination of the schizonts, which appear green. If too much Evans' blue is added, the fluorescence is quenched. If this happens, the cover slip should be removed as described above, and the slide should be dipped in methanol for 2 min, washed in PBS and mounted.
6. Assessment of the T-cell activity in infected hepatocytes can be time-consuming. Immunization of mice can take 1–2 mo. TILSA usually takes 1 wk. Reading the slides is also time-consuming. A trained scientist takes 15–30 min per well. So slides should be examined before they accumulate and their fluorescence fades. It is worth noting that slides can be restained after removal of the cover slips, by dipping them in Petri dishes containing PBS.

Acknowledgments

This work was made possible in part by financial support from Institut Electricité et Santé, UNDP/World Bank/WHO Special Program for Research and Training in Tropical Diseases (TDR), European Community programs INCO-DC, Fondation pour la Recherche Médicale, and Junta Nacional de Investigação Cientifica e Tecnologica (JNICT). Ana Margarida Vigário held a fellowship from JNICT, Portugal. She received support from Centro de Malaria e Outras Doenças Tropicais, Lisbon, Portugal (Dir: Virgilio do Rosario). Elodie Belnoue is supported by a pre-doctoral fellowship from the Ministère de l'Education Nationale, de La Recherche et de la Technologie. We thank Irène Landau for regularly providing infected mosquitoes and Denise Mattei for providing the recombinant protein R44 used to produce the serum anti-Pfhsp70.1. We also thank to Dominique Mazier, Sylviane Pied and Geoffrey Targett for helpful discussion. We like to acknowledge the active support of Jean-Gérard Guillet. Dr. Owen Parkes edited the English text.

References

1. Chen, D. H., Tigelaar, R. E., and Weinbaum, F. I. (1977) Immunity to sporozoite-induced malaria infection in mice. 1. The effect of Immunization of T and B cell-deficient mice. *J. Immunol.* **118,** 1322–1327.
2. Romero, P. J., Maryanski, J. L., Corradin, G., Nussenzweig, R. S., Nussenzweig, V., and Zavala, F. (1989) Cloned cytotoxic T cells recognize an epitope on the circumsporozoite protein and protect against malaria. *Nature* **341,** 323–325.
3. Rodrigues, M. M., Cordey, A. S., Arreaza, G., Corradin, G., Romero, P., Maryanski, J. L., et al. (1991) CD8+ cytolytic T cell clones derived against the *Plasmodium yoelii* circumsporozoite protein protect against malaria. *Int. Immunol.* **3,** 579–585.

4. Rénia, L., Grillot, D., Marussig, M., Corradin, G., Miltgen, F., Lambert, P.-H., Mazier, D., and Del Giudice, G. (1993) Effector functions of circumsporozoite peptide-primed CD4+ T cell clones against *Plasmodium yoelii* liver stages. *J. Immunol.*, **150**, 1471–1478.
5. Weiss, W. R., Berzofsky, J. A., Houghten, R. A., Sedegah, M., Hollindale, M. R., and Hoffman, S. L. (1992) A T cell clone directed at the circumsporozoite protein which protects mice against both *Plasmodium yoelii* and *Plasmodium berghei*. *J. Immunol.* **149**, 2103–2109.
6. Khusmith, S., Sedegah, M., and Hoffman, S. L. (1994) Complete protection against *Plasmodium yoelii* by adoptive transfer of a CD8+ cytotoxic T-cell clone recognizing sporozoite surface protein 2. *Infect. Immun.* **62**, 2979–2983.
7. Millet, P., Landau, I., Baccam, D., Miltgen, F., Mazier, D., and Peters, W. (1985) Mise au point d'un modèle expérimental "rongeur" pour l'étude *in vitro* des schizonticides exo-érythrocytaires. *Ann. Parasitol. Hum. Comp.* **60**, 211–212.
8. Hollingdale, M. R., Nardin, E. H., Tharavanij, S., Schwartz, A. L., and Nussenzweig, R. S. (1984) Inhibition of entry of *Plasmodium falciparum* and *P. vivax* sporozoites into cultured cells: an *in vitro* assay of protective antibodies. *J. Immunol.* **132**, 909–913.
9. Mazier, D., Mellouk, S., Beaudoin, R. L., Texier, B., Druilhe, P., Hockmeyer, W. T., et al. (1986) Effect of antibodies to recombinant and synthetic peptides on *Plasmodium falciparum* sporozoites *in vitro*. *Science* **231**, 156–159.
10. Rénia, L., Miltgen, F., Charoenvit, Y., Ponnudurai, T., Verhave, J. P., Collins, W. E., and Mazier, D. (1988) Malaria sporozoite penetration : a new approach by double staining. *J. Immunol. Methods* **112**, 201–205.
11. Hoffman, S. L., Isenbarger, D., Long, G. W., Sedegah, M., Szarfman, A., Waters, L., et al. (1989) Sporozoite vaccine induces genetically restricted T cell elimination of malaria from hepatocytes. *Science* **244**, 1078–1081.
12. Weiss, W. R., Mellouk, S., Houghten, R. A., Sedegah, M., Kumar, S., Good, M. F., et al. (1990) Cytotoxic T cells recognize a peptide from the circumsporozoite protein on malaria-infected hepatocytes. *J. Exp. Med.* **171**, 763–773.
13. Rénia, L., Salone-Marussig, M., Grillot, D., Corradin, G., Miltgen, F., Del Giudice, G.,et al. (1991) *In vitro* activity of CD4+ and CD8+ T lymphocytes from mice immunized with a malaria synthetic peptide. *Proc. Natl. Acad. Sci. USA* **88**, 7963–7967.
14. Grillot, D., Michel, M., Muller, I., Tougne, C., Rénia, L., Mazier, D., et al. (1990) Immune responses to defined epitopes on the circumsporozoite protein of the murine malaria parasite, *Plasmodium yoelii*. *Eur. J. Immunol.* **20**, 1215–1222.
15. Rénia, L., Mattei, D., Goma, J., Pied, S., Dubois, P., Miltgen, F., Nussler, A., et al. (1990) A malaria heat shock like protein epitope expressed on the infected hepatocyte surface is the target of antibody-dependent cell-mediated cytotoxic mechanisms by non-parechymal liver cells. *Eur. J. Immunol.* **20**, 1445–1449.
16. Suhrbier, A., Holder, A. A., Wiser, M. F., Nicholas, J., and Sinden, R. E. (1989) Expression of the precursor of the major merozoite surface antigens during the hepatic stage of malaria. *Am. J. Trop. Med. Hyg.* **40**, 19–23.
17. Rénia, L., Ling, I. T., Marussig, M., Miltgen, F., Holder, A. A., and Mazier. D. (1997) Immunization with a recombinant C-terminal fragment of *Plasmodium yoelii* merozoite surface protein 1 protects mice against homologous but not heterologous *P. yoelii* sporozoite challenge. *Infect. Immun.* **65**, 4419–4423
18. Charoenvit, Y., Mellouk, S., Sedegah, M., Toyoshima, T., Leef, M. F., De la Vega, P., et al. (1995) *Plasmodium yoelii*: 17-kDa hepatic and erythrocytic stage protein is the target of an inhibitory monoclonal antibody. *Exp. Parasitol.* **80**, 419–429.
19. Rogers, W. O., Malik, A., Mellouk, S., Nakamura, K., Rogers, M. D., Szarfman, A., et al. (1992) Characterization of *Plasmodium falciparum* sporozoite surface protein 2. *Proc. Natl. Acad. Sci. USA* **89**, 9176–9180.

50

Antibody-Dependent Cellular Inhibition Assay

Pierre Druilhe and Hasnaa Bouharoun-Tayoun

1. Introduction

The antibody-dependent cellular inhibition (ADCI) assay is designed to assess the capability of antibodies to inhibit the in vitro growth of *Plasmodium falciparum* in the presence of monocytes. Our studies have shown that antibodies that proved protective against *P. falciparum* blood stages by passive transfer in humans are unable to inhibit the parasite in vitro unless they are able to cooperate with blood monocytes *(1,2)*. It was also shown that antibodies that were not protective in vivo had no effect on *P. falciparum* growth in the ADCI assay *(3,4)*. The ADCI is therefore an in vitro assay the results of which reflect the protective effect of antimalarial antibodies observed under in vivo conditions in humans.

The antibodies able to cooperate with monocytes should be obviously cytophilic: IgG1 and IgG3 isotypes are efficient in ADCI while IgG2, IgG4, and IgM are not efficient. This is consistent with our findings that in sera from protected individuals, cytophilic anti-*P. falciparum* antibodies are predominant, although in nonprotected patients the antibodies produced against the parasite are mostly noncytophilic *(5)*.

Our results suggest that ADCI likely involves the following succession of events: at the time of schizonts rupture, the contact between some merozoite surface component and cytophilic antibodies bound to monocytes via their Fc fragment triggers the release of soluble mediators which diffuse in the culture medium and block the division of surrounding intraerythrocytic parasites *(6)*.

The major steps involved in the ADCI protocol are as follows:

1. Serum IgG preparation using ion exchange chromatography.
2. Monocyte isolation from a healthy blood donor.
3. Preparation of *P. falciparum* parasites including synchronization and schizont enrichment.
4. Parasite culture, for 96 h, in the presence of antibodies and monocytes.
5. Inhibition effect assessed by microscopic observation and parasite counting.

2. Materials

2.1. IgG Preparation

1. Tris buffer: 0.025 M Tris-HCl, 0.035 M NaCl, pH 8.8.
2. Phosphate-buffered saline (PBS), pH 7.4.
3. GF-05-Trisacryl filtration column (IBF, Biothecnics, Villeneuve La Garenne, France).

4. DEAE-Trisacryl ion exchange chromatography column (IBF).
5. G25 Filtration column.
6. Amicon filters and tubes for protein concentration (mol wt cutoff: 50,000).
7. Sterile Millex filters, 0.22-μm pore size (Millipore Continental Water Systems, Bedford, MA).
8. Spectrophotometer equipped with ultraviolet lamp.

2.2. Monocyte Preparation

1. Heparinized blood collected from a healthy donor, 20–40 mL vol.
2. Ficoll-Hypaque density gradient (Pharmacia LKB, Uppsala, Sweden).
3. Hanks' solution supplemented with $NaHCO_3$, pH 7.0.
4. RPMI 1640 culture medium supplemented with 35 mM HEPES and 23 mM $NaHCO_3$; prepare with mineral water; store at 4°C.
5. Reagents for nonspecific esterase (NSE) staining *(7)*: fixing solution, nitrite, dye, buffer, and substrate
6. 96-Well sterile plastic plates (TPP, Switzerland).
7. Refrigerated centrifuge.
8. CO_2 incubator.
9. Inverted microscope.

2.3. Parasite Preparation

1. RPMI 1640 culture medium (*see* **Subheading 2.2., item 4**).
2. 10% Albumax stock solution; store at 4°C for up to 1 mo.
3. 5% Sorbitol for parasite synchronization.
4. Plasmagel for schizont enrichment.
5. Reagents for fixing and staining of thin smears: methanol, eosine, methylene blue.

3. Methods

3.1. IgG Preparation

IgGs are extracted from human sera (*see* **Note 1**) as follows:

1. Dilute the serum at a ratio of 1 to 3 in Tris-HCl buffer.
2. Filter the diluted serum through a GF-05 Trisacryl gel filtration column previously equilibrated in the Tris-HCl buffer. Ensure that the ratio of serum to filtration gel is 1 vol of undiluted serum to 4 vol of GF-05 gel.
3. Pool the protein-containing fractions
4. Load over a diethylaminoethanol (DEAE)-Trisacryl ion exchange chromatography column previously equilibrated with Tris buffer. Ensure that the ratio of serum to filtration gel is 1 vol of undiluted serum to 4 vol of DEAE gel.
5. Collect fractions of 1 mL volume.
6. Measure the optical density (OD) of each fraction using a 280-nm filter.
7. Calculate the IgG concentration as follows:

$$\text{IgG concentration (mg/mL)} = \frac{\text{OD 280 nm}}{1.4}$$

8. Pool the fractions containing IgGs.
9. Concentrate the IgG solution using Amicon filters. Amicon filters are first soaked in distilled water for 1 h and than adapted to special tubes in which the IgG solution is added.
10. Centrifuge the tubes at 876g for 2 h at 4°C. This usually leads to a 25-fold concentration.
11. Perform a final step of gel filtration using a G25 column previously equilibrated in RPMI culture medium.

12. Collect the IgG fractions in RPMI.
13. Measure the optical density (OD) of each fraction using a 280-nm filter.
14. Calculate the IgG concentration.
15. Pool the fractions containing IgGs.
16. Sterilize the IgG fractions by filtration through 0.22-µm pore size filters.
17. Store the sterile IgG solution at 4°C for up to 1 mo (or add Albumax for longer storage—but this is not recommended).

3.2. Monocyte Preparation

The procedure for monocyte preparation is based on that described by Boyum *(8)* and includes the following steps:

1. Dilute the heparinized blood threefold in Hanks' solution.
2. Carefully layer 2 vol of diluted blood onto 1 vol of Ficoll-Hypaque (maximum volume of 20 mL of diluted blood per tube).
3. Centrifuge at 560*g* for 20 min at 20°C.
4. Remove the mononuclear cell layer at the Ficoll-Hypaque/plasma interface.
5. Add 45 mL of Hanks' solution to the mononuclear cell suspension.
6. Centrifuge at 1000*g* for 15 min at 20°C.
7. Carefully resuspend the pelleted cells in 45 mL of Hanks' solution.
8. Centrifuge again at 1000*g* for 15 min at 20°C. Repeat this washing step twice more.
9. Finally, centrifuge at 180*g* for 6 min at 20°C, to remove any platelets that remain in the supernatant.
10. Resuspend the mononoclear cells in 2 mL of RPMI.
11. Calculate the mononuclear cell concentration (i.e., lymphocytes plus monocytes) in the cell suspension: dilute a 20-µL aliquot of the cell suspension threefold in RPMI and count cell numbers using a hemocytometer (e.g., Malassez type).
12. Determine the number of monocytes using the nonspecific esterase (NSE) staining technique:
    a. In microtube A, add 40 µL of mononuclear cell suspension to 40 µL of fixing solution.
    b. In microtube B, mix the NSE staining reagents in the following order: 60 µL of nitrite, 60 µL of dye, 180 µL of buffer, and 30 µL of substrate
    c. Add the mixture in microtube B to the cells in microtube A.
    d. Take a 20-µL sample of the stained cells and measure the proportion of monocytes: lymphocytes: monocytes will be colored in brown, whereas the lymphocytes will be uncolored. Usually the proportion of monocytes is 10–20% of the total mononuclear cells.
13. Adjust the cell suspension to a concentration of 2×10^5 monocytes per 100 µL, with RPMI.
14. Aliquot the cell suspension in a 96-well plate at 100 µL per well.
15. Incubate for 90 min at 37°C, 5% CO_2. During this incubation, monocytes will adhere to the plastic.
16. Remove the nonadherent cells and wash the monocytes by adding, and thoroughly removing, 200 µL of RPMI in each well.
17. Repeat this washing procedure three times in order to remove all the nonadherent cells.
18. At least 95% of the recovered cells will be monocytes. Control for the cell appearance and the relative homogeneity of cell distribution in the different wells by observation using an inverted microscope (*see* **Notes 2–4**).

3.3. Parasite Preparation

P. falciparum strains are cultivated in RPMI 1640 supplemented with 0.5% Albumax.

Parasites are synchronized by Sorbitol treatments *(9)* as follows:

1. Dilute the sorbitol stock to 5% in mineral water.
2. Centrifuge the asynchronous parasite culture suspension at 250g for 10 min at 20°C.
3. Resuspend the pellet in the 5% sorbitol solution. This will lead to the selective lysis of schizont infected RBC without any effect on the rings and young trophozoites.

When required, schizonts are enriched by flotation on plasmagel (*10*) as follows:

1. Centrifuge cultures containing asynchronous parasites at 250g for 10 min at 20°C
2. Resuspend the pellet at a final concentration of 20% red blood cells (RBC), 30% RPMI, 50% plasmagel.
3. Incubate at 37°C for 30 min. Schizont-infected RBC will remain in the supernatant, whereas young trophozoite-infected and -uninfected RBC will sediment.
4. Collect carefully the supernatant, by centrifugation at 250g for 10 min at 20°C.
5. Prepare a thin smear from the pelleted cells, stain, and determine the parasitemia by microscopic examination.
6. Usually, using this method, synchronous schizont-infected RBC are recovered at ~70% parasitemia.

For the ADCI assay, synchronized early schizont parasites are used. Usually the parasitemia is 0.5–1.0% and the hematocrit 4%.

3.4. The ADCI Assay

1. After the last washing step, add in each monocyte containing well:
   a. 40 µL of RPMI supplemented with 0.5% Albumax (culture medium).
   b. 10 µL of the antibody solution to be tested. Usually the IgGs are used at 10% of their original concentration in the serum (~20 mg/mL for adults from hyperendemic areas, and ~12 mg/mL for children from endemic area and primary attack patients) (*see* **Note 5**).
   c. 50 µL of parasite culture, at 0.5% parasitemia and 4% hematocrit.
2. Control wells consist of the following:
   a. Monocytes (MN) and parasites with normal IgG (N IgG) prepared from the serum of a donor with no history of malaria.
   b. Parasite culture with IgG to be tested without MN.
3. Maintain the culture at 37°C for 96 h in a candle-jar (or a low O_2, 5% CO_2 incubator).
4. Add 50 µL of culture medium to each well after 48 and 72 h.
5. Remove the supernatant after 96 h. Prepare thin smears from each well, stain, and determine the parasitemia by microscopic examination. In order to ensure a relative precision in the parasite counting, a minimum of 50,000 RBC should be counted and the percentage of infected RBC calculated (*see* **Notes 6** and **7**).
6. Calculate the specific growth inhibitory index (SGI), taking into account the possible inhibition induced by monocytes or antibodies alone:

$$SGI = 100 \times 1 - \frac{\text{percent parasitemia with MN and Abs/percent parasitemia with Abs}}{\text{percent parasitemia with MN + N IgG/percent parasitemia with N IgG}}$$

4. Notes

1. IgG preparation from sera to be tested is an essential step because we have frequently observed a nonantibody-dependent inhibition of parasite growth when unfractionated sera were used, probably due to oxidized lipids.
2. Monocyte (MN) function in ADCI is dependent upon several factors such as water used to prepare RPMI 1640. Highly purified water, such as Millipore water, although adequate for parasite culturing, leads to a poor yield in the number of MN recovered after adherence

to the plastic wells. On the other hand, water which contains traces of minerals, such as commercially available Volvic water, or glass-distilled water, provide consistently a good monocyte function.
3. Improved monocyte adherence can be obtained by coating the culture wells with fibronectin, that is, coating with autologous plasma from the MN donor, followed by washing with RPMI 1640, prior to incubation with mononuclear cells.
4. MN from subjects with a viral infection (e.g., influenza) are frequently able to induce a non-IgG-dependent inhibition of parasite growth. This nonspecific inhibition effect could prevent the observation of the IgG-dependent inhibition in ADCI. Therefore, MN donors suspected of having a viral infection, or who have had fever in the past 8 d, should be avoided. The results from ADCI are not reliable when the direct effect of MN alone is greater than 50% inhibition. The preparation of MN in medium containing heterologous serum, such as FCS, results in the differentiation of MN, their progressive transformation into macrophages which have lost their ADCI-promoting effect.
5. If required, murine IgG can be tested in ADCI with human MN. The IgG2a isotype is able to bind to the human Fc γ receptor II present on monocytes shown to be involved in the ADCI mechanism *(6)*.
6. A possible variation of the ADCI assay is the assessment of a competition effect between protective cytophilic antibodies (adults from hyperendemic area) directed to the merozoite surface antigens, and nonprotective antibodies (children from endemic area and primary attack patients) which recognize the same antigens but are not able to trigger the monocyte activation because they do not bind to Fc gamma receptors. Therefore noncytophilc Ig directed to the "critical" antigens may block the ADCI effect of protective antibodies. Each IgG fraction should be used at 10% of its original concentration in the serum.
7. The ADCI assay protocol can be modified and performed as a two-step ADCI with short-term activation of monocytes according to the following procedure:
    a. Incubate MN for 12–18 h with test Ig and synchronous mature schizonts infected RBC, at 5–10% parasitemia. During this first culture time, infected RBC rupture occurs and merozoites are released.
    b. Collect supernatants from each well and centrifuge them at 7000g.
    c. Distribute the supernatants in a 96-well plate, at 100 µL per well.
    d. Add to each well 100 µL of *P. falciparum* asynchronous culture containing fresh medium, at 0.5–1% parasitemia, 5% hematocrit (particular care is taken to reduce to a minimum the leukocyte contamination of the RBC preparation used for this second culture).
    e. At 36 h of culture, add 1 mCi of [^{3}H]hypoxanthine to each well.
    f. At 48 h of culture, harvest cells and estimate [^{3}H] uptake by counting in a liquid scintillation counter.

References

1. Khusmith, S. and Druilhe, P. (1983). Cooperation between antibodies and monocytes that inhibit *in vitro* proliferation of *Plasmodium falciparum*. *Infect. Immun.* **41,** 219–223.
2. Lunel, F. and Druilhe, P. (1989). Effector cells involved in nonspecific and antibody-dependent mechanisms directed against *Plasmodium falciparum* blood stages *in vitro*. *Infect. Immun.* **57,** 2043–2049.
3. Sabchareon, A., Burnouf, T., Ouattara, D., Attanah, T., Bouharoun-Tayoun, H., Chantavanich, P., et al. (1991) Parasitological and clinical response to immunoglobulin administration in *P. falciparum* malaria. *Am. J. Trop. Med. Hyg.* **45,** 297–308.
4. Bouharoun-Tayoun, H., Attanah, P., Sabchareon, A., Chongsuphajaisiddhi, T. and Druilhe, P. (1990) Antibodies that protect humans against *Plasmodium falciparum* blood stages do not on their own inhibit parasite growth and invasion in vitro, but act in cooperation with blood monocytes. *J. Exp. Med.* **172,** 1633–1641.
5. Bouharoun-Tayoun, H. and Druilhe, P. (1992) *P. falciparum* malaria. Evidence for an isotype imbalance which may be responsible for the delayed acquisition of protective immunity. *Infect. Immun.* **60,** 1473–1481.

6. Bouharoun-Tayoun, H., Oeuvray, C., Lunel, F., and Druilhe, P. (1995) Mechanisms underlying the monocyte-mediated antibody-dependent killing of *Plasmodium falciparum* asexual blood stages. *J. Exp. Med.* **182,** 409–418.
7. Tucker, S. B., Pierre, R. V., and Jordon, R. E. (1977) Rapid identification of monocytes in a mixed mononuclear cell preparation. *J. Immunol. Methods* **14,** 267–269.
8. Boyum, A. (1968) Isolation of mononuclear cells and granulocytes from human blood. Isolation of mononuclear cells by centrifugation, and of granulocytes by combining centrifugation and sedimentation at 1g. *Scand. J. Clin. Lab. Invest.* **21,** 77–89.
9. Lambros, C. and Vanderberg, J. P. (1979) Synchronisation of *P. falciparum* erythrocytic stages in culture. *J. Parasitol.* **65,** 418–420.
10. Reese, R. T., Langreth, S. G., and Trager, W. (1979) Isolation of stages of the human parasite *P. falciparum* from culture and from animal blood. *Bull. WHO.* **57,** 53–67.

51

Erythrocytic Malaria Growth or Invasion Inhibition Assays with Emphasis on Suspension Culture GIA

J. David Haynes, J. Kathleen Moch, and Douglas S. Smoot

1. Introduction
1.1. Summary

Erythrocytic cycle malaria parasite growth or invasion inhibition assays (GIA) compare the effects of various test and control substances on malaria parasite growth in erythrocytes or invasion into erythrocytes in vitro. Although inhibitions by antimalarial drugs in vitro correlate well with drug protective levels required in vivo, as yet there are too few data to know how well inhibitions by antibodies in vitro correlate with the types and degrees of immune protection in vivo. Antibody-mediated GIA is frequently complicated by parasite strain-specific inhibitions, as well as nonspecific inhibitory factors generated in sera collected or stored under nonoptimal conditions. In this chapter, we describe methods for collecting and processing sera, for using different strains of parasite, and a simplified method for staining parasite DNA with Hoechst dye 33342 before quantitating parasites using ultraviolet (UV)-excited flow cytometry. We also describe a new type of GIA using suspension cultures in a 48-well plate. Critical to this method is enclosing the plate in a gassed, heat-sealed plastic bag, which, being low mass, can easily be rested at a 13.5° angle on a rotor platform (114 rpm with 1-in. displacement) to produce gentle pulsatile waves of media in each well. The suspension GIA, which, relative to the static GIA, increased inhibition by one antibody and decreased inhibition by another (**Table 1**), may better simulate in vivo blood flow and may thus better predict in vivo efficacy.

1.2. Background

The pathologies and the resulting illnesses in people with malaria are caused by the erythrocytic malaria parasite cycle. By the stage of their life cycle that these protozoan parasites are easily detectable in blood, at about 10 parasites per microliter, a person with 6 L of blood has 6×10^7 parasitized erythrocytes. Continued growth and invasion by the most common and virulent malaria parasite species, *Plasmodium falciparum*, can produce more than 6×10^{11} parasitized erythrocytes per person (>100,000 per microliter), but toxic effects, including anemia, occur at lower levels even in semiimmune people. One concern about these increasing numbers of parasites in the blood is that they might adsorb inhibitory antibodies more rapidly than the secondary

Table 1
An Example of GIA Data[a]

	Inhibition per cycle			
	Suspension		Static	
Isolate	Anti-A (%)	Anti-B (%)	Anti-A (%)	Anti-B (%)
FVO	**52**[a]	−7	**43**	**32**
3D7	**12**	3	4	**26**

[a]Inhibition per cycle (I_{PC}) compared in suspension GIA and static GIA for two inhibitory antibodies. One set of GIA data (from four 48-well plates, two-cycle cultures, 6% Hct, 0.5 (anti-A) or 1 (anti-B) mg/mL IgG, 150 mL per well) were analyzed by **Eq. 3b**, $I_{PC} = 100\% \times [1 - (P_{TF}/P_{CF})^{1/2}]$, where P_{TF} is the final parasitemia in the presence of test antibody and P_{CF} is the final parasitemia in the presence of control antibody. Parasitemias (infected erythrocytes per 100 erythrocytes) were determined by UV flow cytometry of 40,000 Hoechst dye 33342-stained erythrocytes. Initial parasitemias for FVO and 3D7 were 0.04% and 0.02% schizonts, respectively. In the suspension cultures the final control parasitemias for FVO and 3D7 were about 1.6% and 5.3% ring forms, respectively (some variation depending on the source of the control antibody), and in the static cultures the final control parasitemias were about 0.65% and 0.77%, respectively. Standard errors of the mean for the final parasitemias in triplicate wells averaged 0.03% for all but the static 3D7 GIA, which average 0.08%. An inhibition of 52% per cycle corresponded to an inhibition of 77% for the entire GIA. Inhibitions in **bold** were considered significant and were associated with a $p < 0.05$ in a two-tailed Students' t-test when comparing the final parasitemias in the triplicate wells with the test antibody to those with the control antibody. A negative number indicates stimulation—an increased final parasitemia in test compared with control wells.

Interpretation: Apparently anti-A inhibited by a mechanism that was as effective, or more, in suspension cultures as it was in static cultures, but was relatively isolate specific. By contrast, anti-B inhibited both isolates about equally well, but inhibited by a mechanism that was effective in the static cultures only.

immune response can replace them and thus effectively neutralize a moderate immunity. By contrast, sporozoites injected by mosquitoes infect only 1 to 100 liver cells with exoerythrocytic parasites, and the merozoites they release do not reinfect the liver.

Malaria merozoites invade erythrocytes, grow inside erythrocytes as ring forms, and then, as larger amoeboid trophozoites, divide to form schizonts and destroy the erythrocytes, releasing many new merozoites to invade more erythrocytes. In each cycle (about 2 d long for *P. falciparum*) about 16 to 24 merozoites are produced per schizont. More erythrocytes are invaded in a geometric progression of about 12- to 15-fold increase every 2 d in a nonimmune person *(1)*. In vitro, the published growth rates per cycle are much lower (usually between two- and sixfold per cycle), though under optimal conditions we have seen up to 22-fold per cycle, about equal to the number of merozoites per schizont. The in vivo inhibition of erythrocytic malaria parasite growth or invasion is clearly desirable if it can decrease the malaria growth and invasion in the person's blood enough to interrupt the cycle or at least protect the person from serious illness and death. If an in vitro growth or invasion inhibition assay (GIA) were to predict in vivo inhibition, it should be helpful in the rational development of malaria vaccines. It may help select the best antigens, adjuvants, and combinations of immune mechanisms that will decrease the growth rate per cycle to less than 1—at which rate the parasitemia would become less each cycle and the parasites would eventually disappear. The level of inhibition needed to produce a growth rate of less than 1 depends on the uninhibited growth rate of the parasite. If the uninhibited growth rate is 10 or more, as it seems to be in vivo *(1)*, then an inhibition of greater than 90% per cycle will be needed to eliminate the parasite.

In this chapter, the term GIA will refer to a GIA mediated by antimalarial antibodies alone. The traditional antibody-mediated GIA is performed in settled cell layers in 96-well plates (static GIA) but, because in vivo malaria parasites grow and invade erythrocytes in suspension in flowing blood, we have developed and describe here a GIA in suspension cultures in 48-well plates (suspension GIA). Although there have been some reports of antibodies inhibiting growth (as opposed to invasion), all the intracellular growth inhibitions that we have seen in our laboratory were traced to nonspecifically toxic factors. Thus, we often use the term inhibition per cycle to mean inhibition per invasion (a part of a cycle). Another erythrocytic malaria in vitro inhibition assay requires mononuclear leukocytes as well as antimalarial antibodies. This antibody-dependent cellular inhibition (ADCI) assay is described in Chapter 50. Perhaps in part because the ADCI assay seems to exhibit less parasite strain specificity than does the antibody GIA without mononuclear leukocytes, ADCI appears to correlate better with the parasite clearance seen in children given pooled antimalarial antibodies (IgG) from adults immune to malaria.

1.3. Strain Specificity

Several publications have demonstrated that antibodies from immune people or monkeys are often strain-specific, inhibiting one strain or isolate of malaria but not another in vitro. It is also known that in vivo protection from blood-stage malaria is often isolate-specific; for example, from early work using blood stage malaria infections to treat neurosyphilis *(2)*. *P. falciparum* isolate-specific GIA was first reported using parasites and convalescent sera from one village in Africa *(3)*. Similar results were obtained with immune sera from *Aotus* monkeys immunized by repeated infection and drug cure. Hyperimmune sera from monkeys immunized with the Camp strain of *P. falciparum* inhibited the Camp strain much more than the FCR-3 strain, whereas hyperimmune sera from monkeys immunized with the FCR-3 strain inhibited the FCR-3 strain much more than the Camp strain *(4)*. The inhibitory sera also produced immune clusters of merozoites (ICM) by crosslinking merozoite surface antigens *(5,6)*. Miller et al. *(7)* showed that crosslinking *P. knowlesi* merozoites to each other inhibited invasion. Similarly, immune sera from some Cambodian refugees produced ICM and inhibited invasion by an isolate (Camp) from Southeast Asia, but not one from Africa (FCR-3) *(8)*. Probably contributory to these serotypic differences found in GIA and ICM, the Camp, and FCR-3 strains have different amino acid sequences and immunodominant epitopes in the merozoite surface protein MSP-1 *(9,10)*. Of course, it would be helpful to develop vaccines that also elicited inhibitory antibodies that were not strain-specific in GIA.

All the above GIA were static cultures. Although the suspension GIA, described below may better mimic in vivo conditions, suspension GIA is too new to have accumulated much data.

1.4. Definitions and Equations

If we are to eventually and successfully correlate in vitro GIA data and in vivo protection data, it is important not only to use a GIA that mimics the in vivo situation in biologically relevant ways, but also to interpret the GIA data in biologically relevant

Table 2
GIA Inhibitions and Residual Growth Rates Calculated by Different Equations, Given the Same Fraction Surviving Per Cycle, $F_{SPC}{}^a$

	Pi	GR	Fraction surviving per cycle, $F_{SPC} = (P_{TF}/P_{CF})^{1/n} =$										
			1.0	0.9	0.8	0.7	0.6	0.5	0.4	0.3	0.2	0.1	0.05
Inhibitions, standard equations[b]							Calculated % inhibitions						
I_{PC} (per cycle, **Eq. 1b**)	All	All	0	10	20	30	40	50	60	70	80	90	95
I_E (entire, e.g., two cycles, **Eq. 3b**)	All	All	0	19	36	51	64	75	84	91	96	99	99.8
Inhibitions, Pi subtracted (Equation 4)[c]													
I_{ESub} (two cycles, P_{CF} = 3.9%)	0.02%	14	0	19	36	51	64	75	84	91	96	100	100
I_{ESub} (two cycles, P_{CF} = 2.0%)	0.02%	10	0	19	36	52	65	76	85	92	97	100	101
I_{ESub} (two cycles, P_{CF} = 1.8%)	0.20%	3	0	21	41	57	72	84	95	102	108	111	112
I_{ESub} (two cycles, P_{CF} = 1.2%)	0.30%	2	0	25	48	68	85	100	112	121	128	132	133
							Calculated residual growth rates (GR_R)						
Growth rate per cycle, residual (test compared with control conditions) GR_R for given control GR and $F_{SPC}{}^d$		14	14.0	12.6	11.2	9.8	8.4	7.0	5.6	4.2	2.8	1.4	0.7
		10	10.0	9.0	8.0	7.0	6.0	5.0	4.0	3.0	2.0	1.0	0.5
		3	3.0	2.7	2.4	2.1	1.8	1.5	1.2	0.9	0.6	0.3	0.2
		2	2.0	1.8	1.6	1.4	1.2	1.0	0.8	0.6	0.4	0.2	0.1

last column continuing: 0, 100, 100, 101, 101, 113, 133, 0.0, 0.0, 0.0, 0.0, 0.0

[a]The standard equations are robust (**Eqs. 1b** and **3b**), however, subtracting the initial parasitemia, Pi, (Equation 4) is unacceptable.
[b]The standard calculations for inhibition per cycle (I_{PC}) and inhibition for two cycles (I_E) are robust; they are not dependent on the growth rates, the results are the same at all Pi and GR (and meaningful if the biology of the parasite in the GIA is relevant). The inhibition per cycle (I_{PC}) is the most useful, and determines residual growth (GR_R) in the presence of the test substance, e.g., for a GR of 10 in the control, an I_{PC} of 60% gives a GR_R of 4.0.
[c]In contrast, when initial parasitemias are subtracted, the calculations for inhibitions are very sensitive to changes in control growth rates per cycle. Such calculations overestimate the inhibitions at low growth rates and are biologically misleading. The P_{CF} and P_{TF} were calculated from the Pi, GR, and F_{SPC} then substituted into Equation 4 for I_{ESub}: For example, in a two-cycle GIA with Pi = 0.3, GR = 2, and F_{SPC} = 0.4 (60% inhibition per cycle), P_{CF} = 1.2% = 0.3 × 2² and P_{TF} = 0.19% = 0.3 × (2 × 0.4)². Then, by Equation 4, I_{ESub} = 112% = 100% × [1 − (0.19% − 0.3%)/(1.2% − 0.3%)]. (continued)

ways. In order to calculate inhibitions and to discuss GIA data, we use the following definitions and equations (*see also* **Table 2**).

A percentage parasitemia, P, is expressed as the number of parasitized erythrocytes per 100 erythrocytes (*see* **Note 1**).

Inhibition over the entire time of culture, I_E, is commonly expressed as a percentage, and is calculated by subtracting the final test parasitemia in the presence of the tested substance or antibody (P_{TF}) from the final control parasitemia in the presence of a control substance or antibody (P_{CF}) and dividing the result by P_{CF} as measured at the end of the GIA (**Eq. 1a**).

F_S, the final fraction of parasites surviving (or detected) after culture in the presence of the test substance as compared with the control substance, $F_S = P_{TF}/P_{CF}$, is used in **Eqs. 1b** and **c**.

$$I_E = 100\% \times [(P_{CF} - P_{TF})/P_{CF}] \tag{1a}$$

or its equivalent,

$$I_E = 100\% \times [1 - (P_{TF}/P_{CF})] = 100\% \times (1 - F_S) \tag{1b}$$

and, solving for the surviving fraction,

$$F_S = P_{TF}/P_{CF} = (1 - I_E) \tag{1c}$$

If the GIA involved only one cycle (one invasion), then the entire inhibition, I_E, is also the inhibition per cycle, I_{PC}, where PC means per cycle, and F_{SPC} is the fraction surviving per cycle in the test condition relative to the control:

for one cycle

$$I_E = I_{PC} = 100\% \times (1 - F_{SPC}) \tag{1d}$$

On the other hand, for GIA involving more than one cycle, since the inhibitions in each cycle are cumulative, the total inhibition, I_E, in a two-cycle GIA (two invasions) cannot be directly compared with I_E from a one-cycle GIA unless recalculated as the I_{PC}. The basic equation for this is **Eq. 2a**, in which n = number of cycles, and F_{SPC} = the fraction surviving per cycle. Only the parasites surviving (or multiplying) in a cycle can survive (or not) in the next cycle, making it convenient to calculate the ultimate fraction of survivors ($F_S = P_{TF}/P_{CF}$) in test conditions (relative to control conditions) after n cycles, as:

$$F_S = (F_{SPC})^n = P_{TF}/P_{CF} \tag{2a}$$

Table 2 footnotes *(continued)*
Pi = Parasitemia, Initial (time 0, % RBC Parasitized)
GR = Growth Rate per cycle, control
n = Number of cycles (invasions)
P_{CF} = Pi × $(GR)^n$ = parasitemia, control, final
P_{TF} = Pi × $(GR \times F_{SPC})^n$ = parasitemia, test, final
$F_S = P_{TF}/P_{CF}$ = fraction surviving (test condition/control)
$F_{SPC} = (P_{TF}/P_{CF})^{1/n} = 1 - I_{PC}$ = fraction surviving per cycle (test condition/control)
$GR_R = GR \times F_{SPC}$ = growth rate per cycle, residual, test conditions
Eq. 1b: $I_E = 100\% \times [1 - (P_{TF}/P_{CF})]$ = inhibition, entire GIA
Eq. 3b: $I_{PC} = 100\% \times [1 - (P_{TF}/P_{CF})^{1/n}] = 100\% \times (1 - F_{SPC})$ = inhibition per cycle
Equation 4: $I_{ESub} = 100\% \times [1 - (P_{TF} - Pi)/(P_{CF} - Pi)]$ = inhibition, entire GIA, subtracting Pi
[d]Note that the higher control GR are more relevant to the in vivo situation in the absence of other inhibitions.

or, taking the $1/n$ root of both sides and rearranging to get the fraction surviving per cycle,

$$F_{SCP} = (P_{TF}/P_{CF})^{1/n} \tag{2b}$$

Substituting **Eq. 2a** into **Eq. 1b** for I_E, inhibition over the entire GIA gives

$$I_E = 100\% \times [1 - (F_{SPC})^n] \tag{3a}$$

and, more importantly, I_{PC}, inhibition per cycle, by substituting **Eq. 2b** into **Eq. 1d** gives

$$I_{PC} = 100\% \times [1 - (P_{TF}/P_{CF})^{1/n}] \tag{3b}$$

Or, if using previously calculated total inhibitions (I_E)

$$I_{PC} = 100\% \times [1 - (1 - I_E)^{1/n}] \tag{3c}$$

1.5. Interpreting Inhibitions

For example, an $I_{PC} = I_E = 60\%$ in a one-cycle assay ($F_S = 0.4$ of control, by **Eq. 1b**) is the equivalent of $I_E = 84\%$ in a two-cycle assay, as can seen easily seen using **Eq. 2a**, $(F_{SPC})^n = F_S$, $(0.4 \times 0.4) = 0.16$ of control surviving, thus an 84% inhibition (**Eq. 3a**). Working the other way, the I_{PC}, can be determined from the fraction surviving after n cycles, P_{TF}/P_{CF}, if the fraction surviving per cycle is assumed to be the same each cycle, by using **Eq. 3b**. The validity of **Eqs. 3a** and **b** were experimentally verified for at least one immune serum *(11)* and for one MAb (J. D. Haynes, et al., unpublished observations). Further examples of the relationships between these variables can be seen in **Table 2**.

We are not aware of any inhibitions greater than 70% per cycle by purified antibodies as calculated by **Eq. 3b**. In a one-cycle GIA (one invasion) at an initial parasitemia of 0.3% and with a control growth rate of 10, an $I_E = I_{PC} = 70\%$ would reduce the increase in parasitemia from 3% in control to 0.9% in test samples. However, this is still a residual growth of threefold per cycle, high enough to let the parasite continue multiplying and to overwhelm the host immune system in vivo (unless this 70% inhibition is combined with other additive or synergistic immune mechanisms) *(12)*. Over two cycles the residual growth would be $3 \times 3 =$ ninefold, even though by **Eq. 3a** the entire inhibition, $I_E = 100\% - (1 - 70\%)^2 = 91\%$.

Greater than 90% inhibition per cycle is needed to achieve no net growth per cycle in vivo when GR = 10. Reports of greater than 90% have been achieved by using the entire inhibition in a two-cycle GIA, or by using a different method of calculating that involves subtracting the initial parasitemias from the final parasitemias before using the change in parasitemias in an equation similar to **Eq. 1b** to give I_{ESub} (**Table 2**). However, subtracting the initial parasitemia can give the wrong impression, for example, the inhibition is apparently more than complete ($I_{ESub} = 112\%$) in a two-cycle GIA when $F_{SPC} = 0.4$, the growth rate is low (GR = 2) and the true inhibition per cycle is still moderate ($I_{PC} = 60\%$; *see* **Table 2**). The estimated in vivo residual growth rate (GR_R) is key; in this case with $I_{ESub} = 112\%$, if the unimpeded in vivo GR were 10, the GR_R would be fourfold per cycle.

1.6. Nonspecific Inhibition

Although longer GIA are more sensitive, they risk amplifying nonspecific inhibitions by nonspecifically toxic factors that frequently appear in sera or antibodies not optimally collected, prepared, or stored. Suggestions for minimizing nonspecific inhibitions are given in the serum collection and preparation protocols and notes. Although GIA are often performed with immune and control sera, particularly when screening many samples, critically important GIA are often performed with purified antibodies. Specificity controls are also useful, for example, showing that an immune serum inhibits one parasite isolate but not another, or showing that antiserum against an antigen can be neutralized by adding back the antigen (both types of controls have been successful in our laboratory). Perhaps eventually, methods for removing nonspecifically toxic factors from serum without purifying antibodies will be described, validated, and accepted.

In brief, nonspecific toxicities can be minimized in sera by careful collection, storage, and treatment. Blood is collected sterilely in serum-separator-type tubes, clotted briefly, centrifuged to separate serum, and serum is frozen as soon as possible. Before use the serum is also heat-inactivated, adsorbed against the erythrocytes being used in the GIA, and dialyzed against culture medium adjusted to pH 7.4 with NaOH (no $NaHCO_3$) using a 10,000 mol wt cutoff dialysis membrane treated to be nontoxic. Up to 20% (v/v) dialyzed serum or purified antibody can be included in a GIA. In addition, the final concentration of normal heat-inactivated pooled human serum is kept at 10%.

1.7. Overview of Method

Static GIA in 96-well plates is performed essentially as described by Chulay et al. *(11)* and Vernes et al. *(8)*, using late-stage parasites (synchronized by sorbitol or alanine treatments, and, more recently, by methods using temperature-cycling incubators) at 1.5–4% hematocrit (Hct) and, depending on the Hct and strain of parasite, an initial parasitemia (Pi) of about 0.3–0.5% for one-cycle GIA or 0.04–0.1% for two-cycle GIA. Static GIA are also performed in 48-well plates similar to suspension GIA in 48-well plates, but kept flat and still. Final parasites (P_{TF} and P_{TC}) are enumerated using UV laser flow cytometry (**Fig. 1**).

Suspension GIA in 48-well plates are set up at 150 µL per well, 4% Hct and about 0.3% initial parasitemia (*P*i) for one-cycle GIA and 0.015–0.03% parasitemia for two-cycle GIA. The initial GIA parameters (primarily *P*i) must be determined for each parasite isolate in order to avoid the plateau and excess regions in which decreasing the initial number of parasites does not change, or even paradoxically increases, the final number of parasites (**Fig. 2**). Rather than using a bulky gas-tight box, each plate is placed in a plastic bag that has been gassed and heat sealed before placing on a support angled at about 13° from the horizontal, which, in turn, is placed on an orbital rotor at 114 rpm. This results in a gentle wave action that keeps the erythrocytes suspended.

At the end of the GIA when most of the parasites are rings or trophozoites, parasite DNA is stained with Hoechst dye 33342. Then 1 vol of culture (4% hematocrit) is mixed with 3 vol of the dye solution—450 µL of 1.33 µg/mL H33342 in phosphate-buffered saline (PBS), pH 7.4 with 10 m*M* EDTA and 10 m*M* glucose is added directly to the 48-well plate, which is sealed, vortexed, incubated with mixing at 37°C for 45 min, and subsequently kept refrigerated for up to several days. Shortly before performing UV laser-excited flow cytometry on 40,000 total cells, the plate is vortexed and 1:67 dilutions are made into PBS–EDTA–glucose. Parasites with 1 to 4 nuclei (ring forms and later trophozoites) can be distinguished from those with 5 to 24 nuclei (schizonts) by their fluorescence intensities, and parasites not associated with an erythrocyte can

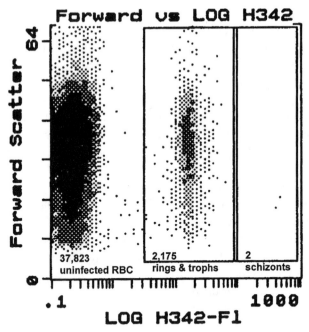

Fig. 1. A typical flow cytometer scatter plot of Hoechst dye 33342 fluorescence vs forward scatter, with gates marked to distinguish uninfected erythrocytes from erythrocytes containing 1–4 times the haploid DNA (rings and trophozoites), and from erythrocytes with 5–25 nuclei (schizonts). In this sample, a total of 40,000 cells were counted, of which there were 2175 (5.44%) erythrocytes infected with rings or trophozoites, 2 (0.005%) erythrocytes infected with schizonts, and 37,823 (94.56%) uninfected erythrocytes.

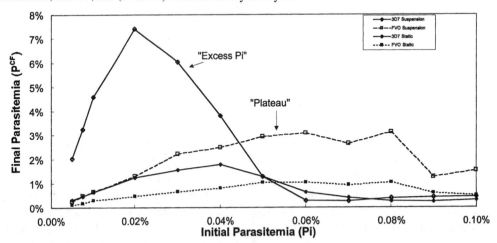

Fig. 2. An example of an experiment for determining the initial conditions for GIA. After examining a range of GIA conditions for each parasite strain, a relatively linear portion of the relationship initial and final parasitemias can be chosen for the GIA, avoiding the "Excess Pi" and "Plateau" regions. This experiment demonstrates that higher initial parasitemias sometimes result in lower final parasitemias in control wells without immune antibodies. This example shows a dramatic effect on 3D7 at 6% hematocrit over two cycles of growth, with a steep decline in final parasitemia for initial parasitemias greater than 0.02% ("Excess Pi"). Based on this graph using these conditions, 0.015%, or at most, 0.02%, would be a reasonable initial

be gated out by their smaller forward-scattering signals. Inhibitions (I_E and I_{PC}) are calculated as described by **Eqs. 1b** and **3b**.

Parasite strains or clones of isolates are selected for study in GIA based on the origin and intended use of the immune sera or antibodies. Each isolate may require different handling and initial parasite concentrations depending on its growth characteristics (**Fig. 2**). When feasible, we run samples with two parasite strains with both static and suspension GIA in parallel (four GIA plates). We currently use clone 3D7 because it is used for sporozoite challenge of human volunteers in vaccine efficacy trials and for producing vaccine candidate antigens, and also use FVO because it is used to immunize and challenge *Aotus* monkeys. We used to routinely perform a two-cycle GIA in order to achieve greater sensitivity, converting I_E to I_{PC} using **Eq. 3b** in order to better determine the likely biological importance of the inhibition. However, lately, we have preferred the one-cycle GIA because it seems more robust, giving a higher percentage of usable assays in which the positive (inhibitory) control antibody is good. It is important to include relevant negative control sera or control antibodies obtained and handled similarly to the test samples in order to accurately determine the appropriate control parasitemia.

1.8. Example of GIA Data

One antibody that we have tested extensively gives about 60% inhibition per cycle against FVO and 25% against 3D7 in the suspension GIA but is slightly less inhibitory in the static GIA (**Table 1**), possibly because in the static GIA the merozoites are released already in contact with adjacent erythrocytes, allowing less time for the antibody to bind and inhibit. In marked contrast, another antibody gives about 25% inhibition against both FVO and 3D7 in the static GIA but no inhibition against either strain in the suspension GIA (**Table 1**). In both cases, purified IgG has been used, and inhibitions have been reversed by preincubating antibodies with the corresponding antigen. We plan to examine possible mechanisms for these differences and will also try additional methods for more gentle suspension because it is possible that the shear forces in the currently used suspension GIA are disrupting agglutinated immune clusters of merozoites (ICM) *(5,6)*. We have no estimate as yet for the shear stress in our suspension cultures. A published estimate for the average shear stress to which malaria *P. falciparum* schizonts are subjected in the brain microvasculature is about 1 dyne/cm^2 *(13)*.

Fig. 1. *(continued)* parasitemia for a two-cycle 3D7 suspension GIA. A higher initial parasitemia, 0.03%, before the "Plateau" would be reasonable for the other GIA (3D7 static, and FVO suspension and static). However, using these conditions for a while, we found too much variability from GIA to GIA that resulted in 30% or more of unusable data because the positive controls were not inhibitory, or were even stimulatory (e.g., in effect pushing 3D7 up the biphasic growth curve from right to left by decreasing the effective parasitemia early on). Small changes in growth rates or small deviations in initial parasitemias can make big differences over two cycles of growth and multiplication. Similar experiments could yield graphs for choosing the initial parasitemias at other hematocrits, for one-cycle GIA, and for other parasite isolates. Based on similar experiments, we have chosen 4% Hct, preferably in 1-cycle GIA, as being more robust and reliable (see text for protocols).

1.9. Perspective

While admitting that GIA immune mechanisms and levels required for protection against malaria are probably complex and interacting, we hope that gathering more data with several standardized GIA methods will help clarify the situation and lead to the general acceptance of an antibody-dependent GIA as one of the in vitro correlates of in vivo immune protection. An urgent need remains the validation and distribution of a reproducibly inhibitory antibody that can be used by everyone as a positive control in GIA. This would allow better comparisons between laboratories and methods.

Eventually, it is likely that solid protection in vivo will involve combining several antigen specificities and immune mechanisms that will work together either additively or synergistically (more than additively). It is hoped that antibody-mediated GIA will help analyze some of these in vitro. We have preliminary evidence for synergism. It is beyond the scope of this chapter to further discuss possible mechanisms for additive or synergistic immune interactions, whereby subeffective doses of two or more immune specificities or mechanisms might combine to produce very effective inhibition, thus rendering a multicomponent vaccine highly protective.

2. Materials
2.1. Equipment

1. Biological safety cabinet for sterile manipulations (*see* **Note 2**).
2. Incubator, 37°C: Humidification and CO_2 are not needed or even desirable, but an incubator with very good temperature uniformity throughout (using horizontal semilaminar air flow or water-jacketing) is ideal; otherwise, uneven temperatures will cause uneven concentration of culture media by evaporation from one part of the 48-well culture plate and condensation onto another part in the gassed sealed bag.
3. Gas cylinder: 5% O_2, 5% CO_2, balance nitrogen (e.g., Air Products), fitted with regulator and 1/4-in. ID Tygon tubing.
4. Labware and supplies for cultures and staining:
    a. Culture flasks, 25 cm^2, with plug seal caps (e.g., Corning 430168, http://www.scienceproducts.corning.com/).
    b. Wide-mouth jars for dialysis (e.g., 1000 mL Straight-Side , Nalgene Gray Polycarbonate, NNI no. 2119-1000).
    c. Centrifuge tubes: 1.5-mL, 15-mL and 500mL.
    d. Pipets, disposable: 1-mL, 5-mL, 10-mL, 25-mL.
    e. Micropipetor: 1–20 µL, 20–200 µL, with disposable tips.
    f. Sterile 9-in. Pasteur pipets, with and without cotton plugs.
    g. Tubes: 5-mL, polypropylene, snap-top.
    h. Filter units, 0.2 µm PES membrane; e.g., 150-mL, 250-mL, 500-mL, and 1000-mL sizes in the MF75™ Series, extra receiver containers also available (Nalgene, Rochester NY, http://nalgenunc.com).
    i. Filters, small for small volumes: e.g., Acrodisc® Syringe Filter 0.2 µm HT Tuffryn membrane low protein binding, sterile (Ref 4192, Pall Gelman Laboratories).
    j. Slides, 1-mm thick frosted glass.
    k. Jars, Coplin staining.
    l. Pail, large, with lid for transporting samples stained for flow cytometry.
5. Centrifuges:
    a. Large, preferably with heating (e.g., Jouan CT 4 22 Centrifuge, which also has very little vibration and pellets the parasites gently) for centrifuging cultures at 37°C in a

Growth Inhibition Assays

   prewarmed centrifuge.
   b. Microfuge, for 1.5-mL tubes, capable of 10,000g.
6. Dialysis chambers: Slide-A-Lyzer dialysis cassettes, 10 mol wt cutoff, in sample sizes 0.5-mL to 3.0-mL, and 0.1- to 0.5-mL (Pierce).
7. Pipetting devices:
   a. Repeating pipetter, continuously adjustable: e.g., Gilson Distriman, with 1.25-mL and 12.5-mL disposable tips (Gilson, Middleton, WI, http://www.gilson.com/).
   b. Multichannel pipet, capable of pipetting 120 µL vol (e.g., Labsystems Finnpipette 4510, 50–300 µL).
8. Microculture plates, 48-well: must have 11.3-mm inner diameter wells (e.g., Costar Corporation, cat. no. 3548); the slightly smaller diameter wells in some 48-well plates do not allow gentle suspension.
9. Warming platform and insulating sheets: a thin heated platform (e.g., microplate warming station, bench warmer, 120 V, 3 W, 9.5 × 13.5 × 1 cm, Thermolyne TWW100) is useful for keeping plates and flasks at 34–38°C while being manipulated in the biological safety cabinet (*see* **Note 3**). For larger volumes of culture and media, heat loss to the metal work surface can be reduced by using 1/4-in.-thick closed cell foam plastic sheets (e.g., clean sheets work surface, XLPE, 26.7 × 30.5 cm; Nalge Nunc, cat. no. 6281-1012).
10. Tubular roll stock: Roll of polyester plastic tubing for making gas-tight bags with heat sealer, 9.5-in. (24-cm) wide, 2-mm thickness tubular roll stock (no. 5 Scotchpak, Kapak, 5305 Parkdale Drive, Minneapolis, MN).
11. Heat sealer for making gas-tight bags: Large 12-in. impulse sealer (e.g., AIE-300, American International Electric, 380 W, set to 3.5 out of 8; Polyheat Bag Sealer 12 in.; PGC Scientific, cat. no. 2-3360-12).
12. Rotating platform: 11-in. × 14-in. platform, 1-in. orbit, adjustable RPM (range at least 30–120 rpm), digital RPM readout is ideal (e.g., adjustable rotating orbital shaker, AROS 160, Barnstead/Thermolyne, Dubuque, IA, http://www.barnstead.com). Fitted with an angled platform at about 13° from horizontal, for supporting 48-well plates sealed in gas-tight bags. An empty closed 3-ring binder (nominal 3-in.-thick, effective thickness about 2.5 in., with a 10.75-in. hypotenuse = about 13.5°) works well, or lightweight angled holders with two levels holding a total of four plates can be made of hardware cloth, ties, and bracing (plans available by email).
13. Plate sealer: Adhesive film, 6 in. × 3.25 in., and roller (e.g., Titer Tops, TTOPS, Diversified Biotech, Boston MA).
14. Paper, thick filter or blotter, 10.8 cm × 14.5 cm (e.g., from one-half of blotter paper, 14.5 cm × 21.5 cm [50], cat. no. 80620730 (Hoefer No. TE46), Hoefer Pharmacia Biotech, San Francisco, CA).
15. Flow cytometer with UV excitation: Instrument capable of exciting between 325 and 375 nm in the UV region and reading the emitted light between 475 and 535 nm in the visible region (we use a Coulter Epics Elite with HeCad laser emitting at 325 nm, reading with emission bandpass filter at 525 nm. The argon 488 nm laser emission is used only for forward scatter and is spatially and temporally separated from the HeCad laser). An electrical power conditioner or UPS should be used on the flow cytometer laser and electronics because any electrical noise will decrease the separation of unparasitized from parasitized erythrocyte fluorescence (the DNA in one parasite genome is only 1% the size of a human cell genome).

2.2. Reagents

1. *P. falciparum* culture adapted isolates: Standard isolates such as 3D7 (clone of NF54) and FVO are ideal. Use any standard malaria culture technique, but well-synchronized cultures are ideal (*see* Chapters 43 and Chapter 44).

2. Human blood: Every 2 wk 1 U (or less) of O positive blood collect in CPD-A1, depleted of leukocytes by standard blood bank filtration within 2 h, tested negative for pathogens, and kept at 4–6°C as a source of erythrocytes for malaria cultures. Some commercial blood banks can supply these units.
3. Heat inactivated pooled normal human sera (10% [v/v]) is included in all cultures: Collect units of sera with care to remove all traces of antiseptic, from 6 to 15 people in 1 d. After heat inactivation and testing small pools of several units each, pool all those supporting good growth for use over the next 6 mo or longer. At a blood bank (e.g., Interstate Blood Bank, Memphis, TN), collect each unit of normal human A positive blood in a blood bag without anticoagulant, clot about 1 h at room temperature, cool to 4°C, centrifuge, express the serum into an empty blood bag, and then seal. Within 24 h, pack the units of serum on wet ice in an insulated shipping box, and send by overnight express to the malaria culture laboratory. When the bags of sera arrive from the blood bank, warm a large water bath to 60°C, place the cold serum bags into the water bath, turn its temperature down to 56°C, monitor the temperature until it reaches 56°C, and allow bags to sit at 56°C another 45 min. Remove the bags from the water bath and dry thoroughly. Using sterile technique and transfer sets (Fenwal), remove sera from bags and pool in a 2-L roller bottle. Aliquot sera in 45 mL total volume into 50-mL polypropylene conical tubes. Place tubes into –70°C freezer until use (for up to 2 yr). Thaw in a 37°C water bath prior to use.
4. Water: Tissue culture grade water (e.g., 18 MOhm, endotoxin-free, Milli-Q Synthesis water, Millipore Corporation, http://www.millipore.com/).
5. Modified RPMI 1640 culture media, powdered: Foil-sealed packets (sufficient for 1 L each) of modified custom RPMI 1640 powder formulation without *p*-aminobenzoic acid (PABA), with 1 mg/L phenol red (1/5 normal), with 1.36 mg/L hypoxanthine, with 2 mM glutamine, with 5.94 g HEPES acid /L, and without sodium bicarbonate (Life Technologies, custom cat. no. 98-5079EC; 9800 Medical Center Drive, Rockville, MD, http://www.lifetech.com).
6. Sodium bicarbonate: 7.5% $NaHCO_3$ w/v solution (Life Technologies), best stored at 4°C in sealed containers.
7. Sodium hydroxide: Using 5 N commercial stock solution, make a 0.6 N working solution by diluting 24 mL of 5 N NaOH into 176 mL of water; store in sealed container at room temperature for up to 6 mo.
8. Control and test sera (human or animal) or purified IgG. Use a 70% alcohol pad to remove all traces of antiseptic and collect blood sterilely in serum separator gel-type tubes containing clot activators and a plug of inert gel (that is more dense than serum but less dense than the clotted cells). All brands tested worked well. Mix the tube of blood end-over-end five times and let clot at room temperature for 45 min. If desired, place the tube of blood on ice or at 4°C to let clot retract for another 1–4 h (will result in a slight increase in yield of serum compared with immediate centrifugation). Centrifuge at 1500g for 10 min to separate serum; remove, aliquot, and freeze the sera as soon as possible. Heat inactivate sera for 45 min at 56°C, preferably before freezing, although heat inactivation can be done after thawing. IgG can be purified using commercially available protein G columns and buffers.
9. PBS–EDTA–glucose: PBS, pH 7.4 with 10 mM EDTA and 10 mM glucose (without calcium or magnesium). Prepare using 10X PBS, pH 7.4 (e.g, cat. no. 3400-1010; Digene Diagnostics, Beltsville, MD); 0.5 M EDTA, pH 8.0 (cat. no. 351-027-100; Quality Biological, Gaithersburg MD), and 500 mM glucose. Adjust to pH 7.4 with 0.6 N NaOH, filter sterilize, and store in receiving flask at 4°C for up to 6 mo.
10. Hoechst dye 33342 (Sigma or Calbiochem): Dissolve at 2 mg/mL in Milli-Q water by warming at 37°C and store stock aliquots at –20°C indefinitely. Do not filter. Before use,

dilute the 2 mg/mL stock 1:1500 (v/v) in PBS-EDTA-glucose, to 1.33 µg/mL—this working solution can be frozen and thawed several times as needed. Hoechst dye 33258 does not work in this protocol.

3. Methods

3.1. Preparation of Media

1. RPMI 1640 culture media without bicarbonate: Modified RPMI 1640 powdered media is sealed in foil packets. To make RPMI 1640 from four media packets, using plastic labware dedicated to tissue culture, collect Milli-Q water using a 2-L graduated cylinder and measure 3 L water into a 5-L beaker or pail. Open the tops of four media packets with scissors and place all powder into the 3 L of water. Use a final 840 mL of Milli-Q water in the cylinder to wash each media packet carefully to remove all remaining powder, being careful not to spill any water or have any go over the side of the bag. Rinse each packet two times and combine with the remaining water in the beaker. Using a magnetic stir bar, stir the media vigorously at room temperature for 30 mins and then filter in 1-L vol using 0.2-µm membrane filter units (PES, Nalgene) (*see* **Note 2**). Rinse the cylinder and pail thoroughly with Milli-Q water and air dry. Store RPMI 1640 culture media without bicarbonate in the sterile receiving vessel at 4°C for up to 2 mo. The liquid culture media is most stable without added sodium bicarbonate; add the sodium bicarbonate shortly before use (*see* **Note 4**).
2. Complete culture media with 10% pooled normal serum (CMS): To prepare CMS, combine the following reagents in the top of a 150-mL 0.2-µm PES membrane filter unit (larger volumes can also be made with larger filter units and extra receiving containers): 130.2 mL of RPMI-1640, 4.8 mL of sodium bicarbonate solution (7.5% w/v), and 15 mL of heat inactivated pooled normal human sera. Apply gentle vacuum to pull solution through into the receiving container. Store CMS in gassed (5% O_2, 5% CO_2), sealed receiving containers at 4°C until use (within 1 wk) or at 37°C up to 24 h before use. After the container is opened and some media removed, gas the remaining media with 5% CO_2, 5% O_2 before resealing the cap. Gassing is performed with a sterile cotton plugged 9-in. glass Pasteur pipet attached to 1/4-in. tygon tubing from the gas cylinder regulator at about 5 psi. Crimp the tubing with the hand holding the pipet to stop the gas flow while entering the flask, then relax the crimp to allow gas flow directed toward the flask wall for about 10 s, crimp tubing, remove the pipet from the flask, and tightly seal the cap.
3. Wash media without serum is prepared and gassed as above.
4. RPMI–NaOH for dialysis: allow RPMI 1640 culture media without sodium bicarbonate (*see* **Note 4**) to equilibrate at about 25°C and place in large flask or beaker. Using a calibrated pH meter, adjust pH to 7.4 by adding 0.6 *N* NaOH dropwise to the RPMI 1640 in receiver flask with gentle mixing (takes about 1.64 mL 0.6 *N* NaOH for every 100 mL of RPMI 1640). Filter using a 0.2-µm PES membrane filter unit. The RPMI-NaOH will be slightly pink (alkaline) at 4°C and orange (neutral) at 37°C because the dominant HEPES buffer has a negative temperature-pK_a coefficient. Store at 4°C until use.

3.2. Dialyzing Test and Control Sera or IgG

1. In a cold room, place 600 mL of RPMI-NaOH into dedicated wide-mouth jars large enough to hold Slide-A-Lyzer® dialysis cassettes and allow for volume.
2. Remove Slide-A-Lyzer cassette from pouch and slip into groove on float buoy. Immerse buoy-cassette in 600 mL of RPMI-NaOH for 1–2 min and remove each cassette as you are ready to add sample.
3. Place a 21-gage needle onto an appropriate sized syringe with sample (3-mL used most often). With bevel of needle sideways, insert the tip of the needle a short distance through one syringe port (located on the corner of the cassette), taking care not to puncture the membrane with needle tip, and inject sample slowly.

4. After injecting sample, tip the cassette so that needle is in the area of remaining air bubble. Remove the air bubble by pulling on syringe plunger. If all the air is not removed once piston is withdrawn fully, use another port to remove the remaining air. Do not insert needle more than once into the same port because multiple insertions can cause leaking and loss of sample.
5. Place cassette containing sample back into 600 mL of RPMI-NaOH solution and allow to dialyze overnight with gentle magnetic stirring. Depending on the size and number of samples, dialysis medium may need to be changed after about 2 hrs. The final ratio of sample to dialysis volume should be at least 1:300 (e.g., 1 or 2 mL dialyzed against 600 mL once; or 2×3 mL dialyzed against 600 mL twice ($1:100 \times 1:100 = 1:10,000$ final ratio).
6. Remove sample from cassette, first filling syringe full of air by pulling back plunger. Note which ports have been used and during removal of sample use a new port. Insert needle into cassette with bevel sideways. Depress plunger slowly, filling cassette with air. Rotate cassette so that the corner where the syringe is located contains the sample on the bottom. Pull back plunger to aspirate sample from the cassette. Place sample into sterile microfuge tube.
7. Spin tube at 10,000g for 3 min to pellet any bacterial contamination.
8. Remove all but the last 10 µL from the tube and place into screw top Nunc tube for storage. Use samples within 48 h or store at –70°C until use.

3.3. Preparing Washed 50% Hematocrit Human O-Positive Erythrocytes

1. Using Fenwal transfer sets, aliquot 25 mL vol from blood bags into 50-mL polystyrene or polypropylene tubes. Place caps on tubes and turn part way, do not tighten caps. Store tubes at 4°C until use. Resuspend tubes three times a week by gentle agitation.
2. To wash a tube of blood, remove plasma supernatant using gentle vacuum or pipetter. Fill tube to 50 mL with wash media. Centrifuge for 10 min at 2,000g. Remove supernatant with gentle vacuum or pipetor. Fill again to 50 mL vol with wash media, mix, and centrifuge again. Remove supernatant with gentle vacuum or pipetor.
3. Estimate packed cell volume. Add an equivalent volume of CMS (minus about 1 mL to account for remaining supernatant fluid) to bring to 50% hematocrit. Resuspend with gentle agitation. Store at 4°C with caps vented until use (up to 2 d).

3.4. Preparing Malaria Cultures in Flasks

1. We recommend growing synchronous parasites in suspension cultures in flasks preparatory to GIA so that most erythrocytes contain only one parasite of a defined stage and so that growth conditions are optimal. Gentle suspensions can be achieved at 4% Hct in 7.5 mL culture volumes in 25-cm^2 flasks (Corning 430168) on an orbital rotating platform at 44 rpm with 1-in. circular displacement. Synchronization can be achieved by a variety of methods, including two treatments about 27 h apart with isotonic sorbitol or alanine, or automated temperature cycling incubators (*see* Chapters 43 and 44).
2. Flask cultures can be diluted the day before the GIA so that the desired parasitemia and stage of parasite are present at the start of the GIA, or diluted the day of the GIA. We routinely begin with schizont stages at a parasitemia that will result in about 1–6% parasitemia after one cycle of invasion, for example, about 0.2% for 3D7 and 0.3% for FVO. A two-cycle GIA is begun with a lower initial parasitemia, for example, about 0.015% for 3D7 and 0.03% for FVO. FVO cultures can be kept static for the last day of culture to increase the recovery of mature stages (FVO mature parasites tend to adhere to the bottom of the flasks in suspension cultures and then not all are added to the GIA).

3.5. Choosing the Initial Parasitemia

Preparatory to using a parasite isolate in actual GIA, analyze its growth characteristics under GIA conditions but without test sera, so that the initial parasitemia chosen

Growth Inhibition Assays

for the actual GIA with each parasite isolate will be well within the region of linear response between initial and final parasitemias (**Fig. 2**). This is essential, otherwise, too high a parasitemia may be unknowingly chosen, so that inhibiting growth or invasion shortly after the beginning of the GIA (equivalent to decreasing the initial parasitemia) will result in no change in final parasitemia, or even paradoxically increase the final parasitemia.

3.6. Adsorbing Test and Control Sera with Human Erythrocytes

Sera are usually adsorbed in 0.5 mL aliquots (therefore, a 1-mL sample would require two sets of tubes). Adsorb immune and control serum samples and sometimes purified animal IgG; purified MAb and human IgG are usually not adsorbed because antierythrocyte antibodies are usually IgM.

1. Prepare washed erythrocytes at 50% hematocrit.
2. Pipet 1.0 mL of 50% erythrocytes into each sterile 1.5-mL tube.
3. Centrifuge at 5000g for 2 min. Aspirate and discard the supernatant.
4. Add 0.5 mL of test sera and resuspend by vortexing.
5. Place on rotating wheel at about 5–20 rpm for 30 min at room temperature.
6. After 30 min, centrifuge, and remove the 0.5 mL sera, including a small amount of erythrocytes.
7. Repeat steps above twice, for a total of three absorptions. During final removal of sera, be careful not to remove many erythrocytes with sera.
8. Place sera sample in clean microcentrifuge tube and centrifuge.
9. Remove all but the pelleted erythrocytes, and place sera in a clean microcentrifuge tube.
10. Centrifuge for 3 min at 10,000–13,000g to remove possible bacterial contamination.
11. Remove all but the bottom 10 µL and place sera in Nunc tube for storage.
12. Store at 4°C if 48 h or less; store at –70°C if more than 48 h until use.

3.7. Test Mix Preparation and Preincubation in Microculture Plates

1. Prepare a test mix at 5X final concentration by diluting, if necessary, the dialyzed test substances (e.g., pretreated immune sera, control sera, or purified IgG) into RPMI-NaOH (also containing 1% heat-inactivated normal human serum if using small concentrations of purified antibodies). The amounts of immune or control sera or IgG to be added in the test mix are calculated on the basis of the final volume of 150 µL. For example, if the final desired concentration of test sera is 20%, then 20% of 150 µL is 30 µL test sera. This 30 mL is the maximum volume allowed in the test mix; larger volumes have given less consistent results. If the desired final concentration is 10%, then the test sera is diluted with an equal volume of RPMI-NaOH. The initial 50% sera in the 30-µL test mix will be diluted to the desired final 10% sera when the 120 µL of malaria culture mix (*see* **Subheading 3.9.**) is added to each well to make the final 150 µL culture volume. For suggestions on calculating total volumes of test mix, *see* **Note 5**.
2. Add the test mix in 30-µL vol with a micropipet each to triplicate wells of microplates.
3. Equilibrate the plate for 1 h at 37°C (flat, static) in a heat-sealed gassed (5% O_2, 5% CO_2) humidified bag (*see* **Subheading 3.8.**).

3.8. Heat-Sealed Gassed Bags

A heat-sealed gassed bag is used for each GIA plate in place of bulkier gas-tight chambers because the suspension cultures require motion that is hindered by the size and mass of chambers (bags are also more convenient). It helps to make a small platform (e.g., of a cardboard box) to hold the bag and plate up to the level of the heat sealer. After incubation, plates are removed by cutting one end of the bag with scissors. After the 1-h preincubation of the plates with the 30-µL test mix, the same bags and blotting papers are reused for the final incubation with the culture mix.

1. For each plate, cut about a 25-cm-length of the 24-cm-wide tubular roll stock and use the heat-sealer to seal one end of the bag. To ensure each seal, repeat it a few millimeters away.
2. Place in the bag a 10.8 cm × 14.5 cm piece of blotter paper, wet with Milli-Q water, and pour off the excess.
3. Insert the 48-well plate with lid (containing the 5X test mix for preincubation or the final GIA culture) on top of the wet blotting paper, heat seal most of the open end of the bag, leaving about 2 cm unsealed at one side.
4. Insert a 5-mL cotton-plugged pipet attached to tubing coming from the 5% CO_2 5% O_2 gas tank and gently inflate the bag twice, each time expressing the gas by hand pressure, then after another inflation, withdraw the pipet and rapidly heat seal the final opening.

3.9. Malaria Culture Mix

Because each 120 µL of malaria culture mix will be diluted 4:5 when added to the 30-µL test mix in each well, the malaria culture concentrate is made up at 5% Hct (5/4 times the final 4% Hct for the GIA) and at 12.5% normal pooled serum (5/4 the final 10% normal serum); the parasitemia does not change. After the 120-µL malaria culture mix is added to each well for GIA (*see* **Subheading 3.10.**), the plate is placed in a bag which is gassed and sealed (*see* **Subheading 3.8.**) and incubated at 37°C for either static culture (flat and still), or for suspension culture (13° angle, 114 rpm; *see* **Subheading 3.12.**).

A spreadsheet is available from the authors by email for calculating the volumes for preparing the malaria culture mix. For example, to prepare enough 0.2% parasitemia malaria culture mix for 48 wells from a flask culture at 4% Hct and 2.0% parasitemia (*see* **Table 3**):

1. Place 10 mL of warm CMS in a 50-mL centrifuge tube, and add 1.19 mL of the culture (4% Hct, 2.0% parasitemia). Using a 50-mL tube allows later withdrawing aliquots with the repeating pipet; using 10 mL of warm CMS helps recover the small volume of culture from the pipet and keep it warm.
2. Centrifuge at 300*g* for 5 min at 37°C.
3. Remove (and discard) the supernatant with a pipet.
4. Resuspend the pellet in 8.39 mL of warm CMS, then add 0.86 mL of 50% Hct (pipetting up and down a few times to recover the cells) and 0.25 mL of normal pooled serum, and gently mix. This is now the malaria culture mix with 5% Hct, 0.2% parasitemia, and 12.5% normal pooled serum, which, after adding 120 µL to each well containing preincubated 30-µL test mix, will achieve 4% Hct, 0.2% parasitemia, and 10% normal pooled sera for a one-cycle GIA culture.

3.10. Adding the Malaria Culture Mix to the Wells

1. Remove the preincubated 48-well plate from the incubator, use scissors to open the gassed bag, and place the plate on one warming platform in the biological safety cabinet and the collapsed bag with the wet blotter paper on another warming platform. Remove the lid of the plate only when necessary to reduce evaporation and gas loss.
2a. For less than 48 wells receiving the same malaria culture mix, it is convenient to use a 1.25-mL repeating pipet set to deliver 120 µL each stroke to each well.

Table 3
Snapshot of a Spreadsheet for Calculating Malaria Culture Mix[a–c]

Inputs, with examples	Steps	Outputs, with examples[d]	Checks, with examples (round to 2 or 3 sig. fig.)
Pc = 2.00%	1	Vol_Cult =1.194 mL	
Pi = 0.20%	2	Cent. in 50-mL tube, asp. Sup.	
No. of wells = 48	3	Vol_CMS = 8.39 mL	
Cult_Hct = 4.0%	4	Vol_50_Hct = 0.86 mL	12.50%% serum
GIA_Hct = 4.0%	5	Vol_Serum = 0.25 mL	5.76 mL Mix in wells
		Add in order, mix all, now = Mix	3.79 mL extra Mix
	Other:	Vol_Mix = 9.55 mL	9.55 mL total Mix
		Vol_PCV = 0.48	0.48 mL_PCV
		Vol_m_PCV = 9.07	
		Hct_of_Mix = 5.0%	5.76 mL Mix for wells

[a] See **Subheading 3.9.**
[b] GIA setup calculator2: Mix = culture mix for 4% Hct final constants:
    150 = final_µL_per_well
    120 = µL_Mix_per_well added to 30 µL sample
    12.5% = nl_serum_Pc_Mix so final = 10%   final_nl_serum
    5% = extra_percent for losses
    3.5 = extra_ml for losses   (using 50-ml tube for Mix)
[c] Pc, parasitemia of culture; Pi, desired initial parasitemia of GIA.
[d] Output equations:
Vol_Mix = extra_mL + wells*(µL_Mix_per_well/1000)*(1 + extra_percent)
Hct_of_Mix = GIA_Hct*(final_µL_per_well/µL_Mix_per_well)
Vol_Cult = Vol_Mix*(Pi*Hct_of_Mix)/(Pc*Cult_Hct)
Vol_50_Hct = ((Vol_Mix*Hct_of_Mix)-(Vol_Cult*Cult_Hct))/50%
Vol_CMS = (((1-nl_serum_Pc_Mix)*(Vol_Mix-Vol_PCV))/(1-10%))-50%*Vol_50_Hct
Vol_Serum = Vol_Mix—Vol_PCV-Vol_CMS-50%*Vol_50_Hct
Vol_PCV = Vol_Mix*Hct_of_Mix
Vol_m_PCV = Vol_Mix-Vol_PCV

2b. For 48 or more wells receiving the same malaria culture mix, it is convenient to use a multichannel pipet to deliver 120 µL to each well.
3. Reinsert the covered plate over the wet blotter paper in the bag, gas and seal as described in **Subheading 3.8.**
4. Plates for static GIA can simply be placed in a 37°C incubator; for suspension GIA, *see* **Subheading 3.11.**

3.11. Suspension GIA Cultures in 48-Well Plates

The ideal total volume is 150 µL per well in wells with 11.3 mm average diameter (Corning 3548 plates); other 48-well plates have smaller diameter wells and do not work well; neither do volumes other than 150 µL work well. Once the plate is in a gassed, heat sealed bag (*see* **Subheading 3.8.**), place the bag containing the plate on an angled support (angled at about 13° from the horizontal), which is on an orbital platform at 114 rpm with a 1-in. circular displacement (*see* **Subheading 2.1., item 12**), all in a 37°C incubator.

3.12. Static GIA Cultures in 48-Well Plates

All conditions for static GIA cultures are as above for suspension GIA cultures, except the culture plate is placed in its gassed bag on a flat, level, stationary surface in

37°C. However, if not comparing directly with suspension GIA, 96-well plates may also be used, and perhaps also a higher starting parasitemia and lower Hct (e.g., 0.4% parasitemia, 1.5% Hct), which are more typical if using a 96-well plate. Note that combined invasion and growth in control wells tends to be two- to threefold less per cycle in static as compared with suspension GIA cultures.

3.13. Staining with Hoechst Dye 33342 in 48-Well Plates

1. At the end of the GIA culture (e.g., about 20 h or less for one-cycle GIA begun with schizonts and about 56 h for two-cycle GIA begun with schizonts), open the gassed bag with scissors and remove the plate.
2. To each well with 150 µL of 4% hematocrit culture add 450 µL of Hoechst dye 33342 (1.33 µg/mL in PBS-EDTA-glucose, warmed to 37°C) (modified from Franklin et al. *[14]*).
3. Seal the top of the plate with adhesive film, and briefly mix by orbital motions with hand or on a plate vortexer.
4. Place the plate in a 37°C incubator for 45 min, on an orbital platform rotator, as for suspension culture (suspension will not be complete with this large volume of fluid in each well, but the motion seems to increase the uniformity of staining).
5. Remove the sealed plate, vortex once more, and place at 4°C. At 4°C in PBS–EDTA–glucose, the staining of ring forms seems to be stable for at least 5 d and of later stages for about 2 d. The Hoechst dye is not as sensitive to photobleaching as some dyes, but it is good practice to limit exposure of the plate to light during storage.
6. To spot-check selected wells, small aliquots for thin smears and Giemsa staining can be removed before (30 µL) or after (120 µL) Hoechst dye staining, centrifuged in 1.5-mL tubes, the pellets resuspended to about 30% Hct, and 3 µL used to make smears.

3.14. UV Flow Cytometry

Perform UV flow cytometry within 2 d of staining as detailed in **Subheading 3.13.** Excitation with the 488-nm laser must be spatially and temporally separated from the UV excitation, otherwise the 488 nm will interfere with reading the emitted blue light from the UV excitation.

1. Between 1 and 4 h before flow cytometry, prepare labeled 5-mL polypropylene tubes with 1 mL of PBS–EDTA–glucose. Remove the Hoechst dye-stained GIA plate from the refrigerator and mix the contents using a plate vortexer.
2. Remove the adhesive film (or use a pipet tip to pierce the plate sealing film twice, including making a small hole for air), and remove 15 µL of the stained culture (about 1% hematocrit and 1 µg/mL Hoechst dye), and place the 15 µL into the appropriately labeled tube containing 1 mL of PBS–EDTA–glucose to make a 1:67 dilution, and gently vortex mix. At this 1:67 dilution, the residual dye in solution does not significantly interfere with flow cytometry (proportionately dilute a 3% Hct culture 1:50 and a 6% Hct culture 1:100).
3. Transport the diluted stained samples surrounded by absorbent diapers in a sealed 10-L bucket (e.g., Nalgene 13-308-2) at room temperature to the flow cytometer.
4. Gently vortex mix tubes one at a time and sample in the flow cytometer.
5. Set a gate for forward (small angle) scatter of the 488-nm laser, in order to count only events associated with (inside or attached to) cells the size of erythrocytes. Normal erythrocyte controls are useful for this.
6. Read intensity of emitted blue light at 530 nm (between 475 and 535 nm) after excitation with 325-nm UV light (between 325 and 375 nm). For the forward-scatter gated cells, set three other gates, one to count relatively nonfluorescent erythrocytes (uninfected), one to count erythrocytes associated with 1–4 parasite genome units of fluorescence (rings and

trophozoites), and another to count erythrocytes associated with 5–30 parasite genome units of fluorescence (schizonts) (*see* **Fig. 1** of fluorescence vs forward scatter with gates). Routinely, 40,000 total erythrocytes are counted, with events accumulating in one of the three bins (total, rings and trophozoites, and schizonts).
7. Analyze data for inhibitions using **Eqs. 1b** and **3b** to determine I_E and I_{PC}. If an unexpectedly large number of schizonts are present, suspect some nonspecifically toxic interference with the GIA. Also spot-check Giemsa-stained thin smears.

4. Notes

1. Of course, more than 100 erythrocytes are examined to determine this percentage parasitemia and the number will depend upon the precision desired. To attain the same precision, fewer total erythrocytes will have to be examined when there is a higher parasitemia. When 40,000 erythrocytes are examined per sample by flow cytometry, the standard error of the mean for triplicate samples is reasonably small, even when the parasitemia is less than 1%; for example, in a recent experiment with a mean parasitemia of 0.75%, the average SEM was 0.04% (range 0.02–0.08%).
2. Avoidance of both infection of the worker and contamination of the malaria cultures requires standard sterile techniques using sterile disposable labware and pipetting aids in a laminar flow biological safety cabinet. Contamination can be avoided by meticulous technique without either flaming or antibiotics—one accidental touch and the item must be discarded (or filtered again if possible).
3. The exposure of parasites to nonculture conditions (e.g., temperatures, CO_2 and O_2 concentrations, and pH) should be minimized. Although malaria ring forms seem to survive lower temperatures and high pH, schizonts do not seem so hardy, and cultures should be kept close to 37°C during manipulation and centrifugation to avoid differentially affecting the different malaria stages, unless, of course, that is what is desired, as in some types of temperature synchronization (*see* Chapter 44). Because the pK_a of HEPES is highly dependent upon temperature ($\Delta pK_a = -0.015/°C$), even without losing CO_2, the pH of media rises as it is cooled. Use of a warming platform or insulating sheets minimizes heat losses. In addition, the pH of small volumes of culture media buffered by HCO_3^-/CO_2 can rapidly rise above pH 7.6 when exposed to moving room air, thus the lid should be kept on the microculture plate between manipulations in the biological safety cabinet.
4. Media is most stable without bicarbonate when stored long term or used for dialysis. Media with bicarbonate loses CO_2 on exposure to room air, and the resulting high pH (which can rise higher than pH 8.0) accelerates the breakdown of nutrients, for example, glutamine to ammonia, as well as having a direct toxic effect on malaria parasites. The loss of nutrients and the toxic breakdown products remain even when the proper pH is restored by gassing again with CO_2.
5. Example for calculating the total volume of each test mix to be prepared:
For each test mix, if comparing GIA with two isolates, each in suspension and static GIA, requires four plates, thus, with triplicate wells in each plate, $2 \times 2 \times 3 = 12$ wells. At 30-µL test mix per well, 30 µL × 12 wells = 0.36-mL test mix, however, it is best to prepare about 15–20% excess volume to allow for losses during handling, for example, 0.36 × 117% = 0.42-mL test mix (enough for 14 wells if there were no losses). These test mixes can be made up in small tubes (or spare wells in a 48-well plate, which can hold 1.7 mL per well), and added to the appropriate wells with a repeating pipetor.

Acknowledgments

We thank Dr. Stephen Hoffman and Dr. Jeffrey Lyon for support, advice, and encouragement; Dr. Kim Lee Sim and Dr. David Narum for the antimalarial inhibitory antibodies anti-A and anti-B used in **Table 1**.

References

1. Cheng, Q., Lawrence, G., Reed, C., Stowers, A., Ranford-Cartwright, L., Creasey, A., Carter, R., and Saul, A. (1997) Measurement of *Plasmodium falciparum* growth rates *in vivo*: a test of malaria vaccines. *Am. J. Trop. Med. Hyg.* **57,** 495–500. (Their estimations of growth rates per 48 h of 12- to 15-fold would give growth rates of about 9-fold per cycle if the cycle length of their 3D7 clone in vivo is similar to the approx 40 h cycle length of our 3D7 clone in vitro.)
2. Boyd, M. F., Stratman-Thomas, W. K., and Kitchen, S. F. (1936) On acquired immunity to *Plasmodium falciparum*. *Am. J. Trop. Med.* **16,** 139–145.
3. Phillips, R. S., Trigg, P. I., Scott-Finnigan, T. J., and Bartholomew, R. K. (1972) Culture of *Plasmodium falciparum in vitro*: a subculture technique used for demonstrating antiplasmodial activity in serum from some Gambians, resident in an endemic malarious area. *Parasitology* **65,** 525–535.
4. Chulay, J. D., Haynes, J. D., and Diggs, C. L. (1985) Serotypes of *Plasmodium falciparum* defined by immune serum inhibition of *in vitro* growth. *Bull. World Health Organ.* **63,** 317–323.
5. Chulay, J. D., Aikawa, M., Diggs, C. L., and Haynes, J. D. (1981) Inhibitory effects of immune monkey serum on synchronized *Plasmodium falciparum* cultures. *Am. J. Trop. Med. Hyg.* **30,** 12–19.
6. Lyon, J. A., Haynes, J. D., Diggs, C. L., Chulay, J. D., and Pratt-Rossiter, J. M. (1986) *Plasmodium falciparum* antigens synthesized by schizonts and stabilized at the merozoite surface by antibodies when schizonts mature in the presence of growth inhibitory immune serum. *J. Immunol.* **136,** 2252–2258.
7. Miller, L. H., Aikawa, M., and Dvorak, J. A. (1975) Malaria (*Plasmodium knowlesi*) merozoites: immunity and the surface coat. *J. Immunol.* **114,** 1237–1242.
8. Vernes, A., Haynes, J. D., Tapchaisri, P., Williams, J. L., Dutoit, E., and Diggs, C. L. (1984) *Plasmodium falciparum* strain-specific human antibody inhibits merozoite invasion of erythrocytes. *Am. J. Trop. Med. Hyg.* **33,** 197–203.
9. Miller, L. H., Roberts, T., Shahabuddin, M., and McCutchan, T. F. (1993) Analysis of sequence diversity in the Plasmodium falciparum merozoite surface protein-1 (MSP-1). *Mol. Biochem. Parasitol.* **59,** 1–14.
10. Haynes, J. D., Vernes, A., and Lyon, J. A. (1987) Serotypic antigens of *Plasmodium falciparum* recognized by serotype-restricted inhibitory human sera. *Am. J. Trop. Med. Hyg.* **36,** 246–256.
11. Chulay, J. D., Haynes, J. D., and Diggs, C. L. (1981) Inhibition of *in vitro* growth of *Plasmodium falciparum* by immune serum from monkeys. *J. Infect. Dis.* **144,** 270–278.
12. Oaks, S. C., Mitchell, V. S., Pearson, G. W., and Carpenter, C. J. (eds.) (1991) Asexual blood-stage vaccines, in *Malaria: Obstacles and Opportunities*. National Academy Press, Washington DC, pp.182–192.
13. Siano, J. P., Grady, K. K., Millet, P., and Wick, T. M. (1998) Short report: *Plasmodium falciparum*: cytoadherence to alpha(v)beta3 on human microvascular endothelial cells. *Am. J. Trop. Med. Hyg.* **59,** 77–79.
14. Franklin, R. M., Brun, R., Grieder, A. (1986) Microscopic and flow cytophotometric analysis of parasitemia in cultures of Plasmodium falciparum vitally stained with Hoechst 33342—application to studies of antimalarial agents. *Z Parasitenkd* **72,** 201–212.

52

Analysis of CSA-Binding Parasites and Antiadhesion Antibodies

Michal Fried and Patrick E. Duffy

1. Introduction

Unlike other malaria parasites of man, *Plasmodium falciparum* parasites are able to sequester in postcapillary venules *(1)*. This allows mature forms of the parasite to avoid circulating through the spleen, where they could be destroyed. Parasites sequester by adhering to receptors on the endothelial surface; endothelial receptors that support parasite binding in vitro include CD36, thrombospondin, ICAM-1, VCAM-1, ELAM-1, CSA, PECAM-1, the integrin $\alpha_v\beta_3$, and P-selectin *(2–9)*. Electron-dense protrusions, called knobs, appear on the surface of infected red blood cells (IRBCs) and act as points of contact with the endothelium *(10,11)*. Variant and invariant antigens expressed by the parasite on the surface of IRBCs have been identified as binding ligands *(12–14)*.

Naturally occurring antibodies that inhibit parasite adhesion have been associated with protection from specific malaria syndromes *(15)*, suggesting that antiadhesion vaccines could be a general strategy for protection from malaria. Antigenic variation by the parasite may be an obstacle to the development of such a vaccine. A family of variant antigens expressed by the parasite will bind some endothelial receptors in vitro *(12)*, and humoral immunity against variant antigens on the surface of IRBCs is strain-specific *(16)*. For example, sera from repeatedly infected *Aotus* monkeys *(17)* or naturally exposed humans *(18)* block parasite binding to melanoma cells or CD36 in a strain-specific fashion.

Sera from naturally exposed humans will also agglutinate IRBCs in a strain-specific fashion *(16)*. Some studies have demonstrated a significant correlation between agglutination and inhibition of parasite binding, while other studies reported inhibition of parasite binding in the absence of agglutination *(19–21)*. This suggests that parasite agglutination may interfere with parasite adhesion, but that the antigen(s) targeted by agglutinating and antiadhesion antibodies may or may not be the same. Corollary to this, agglutinating antibodies may be a confounding factor when assaying sera for antiadhesion activity.

Maternal malaria afflicts millions of women and their fetuses. Unlike many other infectious organisms, *P. falciparum* is most likely to infect a woman during her first pregnancy. *P. falciparum* parasites are able to adhere specifically to chondroitin sulfate A (CSA) on the surface of placental syncytiotrophoblasts *(6)* and sequester in the placenta. Multigravid women in endemic areas have developed antiadhesion antibodies that block parasite binding to CSA, but primigravid women lack these antibodies, explaining the susceptibility to malaria during first pregnancies *(15)*. Unlike antiadhesion antibodies against CD36-binding parasites, antiadhesion antibodies against CSA-binding parasites are strain-independent. In areas of high malaria transmission, sera from multigravid women uniformly inhibit parasite binding to CSA, but do not commonly agglutinate these parasites. Thus, in this case, binding is inhibited by antibodies that presumably target the parasite binding ligand, and not by antibodies that interfere with adhesion by agglutinating the parasite.

This chapter will describe assays to define parasite binding phenotypes and to measure levels of antiadhesion antibodies. Although we will discuss the use of these assays to study CSA-binding parasites, the same techniques can be used to study adhesion and antiadhesion antibodies against parasites with other binding phenotypes. Because parasite agglutination may interfere with parasite binding in vitro, we will also describe an assay that measures agglutination, and suggest approaches to differentiate agglutinating and antiadhesion activities.

2. Materials

2.1. Reagents

1. Malaria parasites: *see* **Subheading 3.** (*see* **Note 1**).
2. Serum: Serum or plasma samples from individuals living in malaria endemic areas. Control serum obtained from naïve donors living in nonmalarious areas (*see* **Note 2**).
3. Phosphate-buffered saline (PBS): 10 mM PBS, pH 7.4
4. Chondroitin sulfate A: CSA from bovine trachea (Sigma, cat. no. C-8529).
5. Blocking solution: 2% bovine serum albumin (BSA) in PBS.
6. Petri dishes (Falcon, cat. no. 1001).
7. Plasmagel (Cellular Products, Buffalo, NY).
8. Gelatin (Sigma, cat. no. G-2500): 5.5% gelatin in RPMI.
9. RPMI (Gibco-BRL).
10. Giemsa: Giemsa stain (Sigma).
11. Glutaraldehyde.

2.2. Equipment

1. Table-top centrifuge (1800g).
2. Water bath incubator at 37°C.
3. Water-jacketed incubator.
4. Rotator.
5. Laminar flow hood.
6. Vacuum pump.
7. Light and fluorescent microscope.

3. Methods

Malaria parasites bind to different receptors in vitro. For the purpose of this chapter, we will focus on parasite binding to CSA. Full analysis of antiadhesion antibodies requires defining the binding phenotype of the parasite isolate, measuring antiadhesion activity in serum samples, and determining the level of agglutinating activity as a potential confounding factor. The total volume of IRBCs required for 1 d of assays is a function of the number of sera to be tested for antiadhesion and agglutinating activity. For each individual agglutination assay, binding assay, or serum binding-inhibition assay, roughly 1–1.5 µL of pelleted red blood cells (where 2–20% of the red blood cells are parasitized) is needed.

3.1. Parasite Collection

Serum inhibition of parasite binding can be assayed using laboratory-adapted parasites (as long as the parasites have retained their adhesion phenotype) or field isolates collected from malaria-infected individuals (see **Note 1**).

Placentas are highly vascular and can harbor high parasite densities, thus they are often an abundant source of parasites. Placental parasites may be obtained by making a deep incision (about 1 cm) in the placenta from the maternal side and collecting blood samples by suction with a syringe. This method is quick and easy, but yields relatively small volumes of blood. In order to obtain large blood volumes, the tissue can be compressed in a grinder, and the blood evacuated into an open container. The blood is collected into a tube containing citrate phosphate dextrose (CPD), ethylenediaminetetraacetic acid (EDTA), or heparin. Peripheral blood samples obtained by venipuncture are similarly collected into CPD, EDTA, or heparin.

3.2. Parasite Preparation

Adhesion and agglutination assays are performed when parasites are in late trophozoite–early schizont stages (young trophozoites do not adhere to the endothelium). Commonly, the placenta contains mature forms of the parasite, and it is possible to use these samples immediately after collection. Parasites obtained from the peripheral circulation are almost always ring-stage parasites, and require at least 20 h of in vitro culture growth in order to develop an adhesive phenotype. Binding assays are performed using red cell suspensions at 5% hematocrit and parasitemias between 2 and 20%. Parasite preparations can be enriched for the trophozoite/schizont stages using Plasmagel or gelatin gradient.

1. Plasmagel-enrichment of mature parasites (22): Warm the filter-sterilized Plasmagel to 37°C. Add Plasmagel at an equal volume to a sample of parasitized blood (40% hematocrit in RPMI). Uninfected RBCs and ring stage IRBCs will aggregate and sediment by gravity over 15–45 min at 37°C (water bath). Recover trophozoite and schizont-stage IRBCs which will remain suspended in the supernatant by centrifugation at 1260g for 5 min, in a tabletop centrifuge. Wash the parasite pellet three times with RPMI.
2. Gelatin enrichment of mature parasites (23): Mature parasites can be concentrated on a gelatin gradient (5.5% gelatin in RPMI). Warm the filter-sterilized gelatin solution to 37°C. Pellet the sample of parasitized blood, then resuspend in the gelatin solution to a 10% hematocrit. Incubate for 30 min at 37°C in a water bath. The uninfected RBCs and ring-stage IRBCs will sediment, and trophozoite and schizont stage IRBCs will remain in the upper layer. Pellet the trophozoite and schizont stage IRBCs and wash three times with RPMI.

3.3. Adhesion Assay to Immobilized Receptors

1. Draw circles on the back of a Petri dish, indicating sites where receptors will be immobilized. Circles should be large enough to accommodate a sample size of 20–30 µL.
2. Apply 20–30 µL of CSA (10 µg/mL) or control molecule (e.g., BSA at 10 µg/mL in PBS) to the interior surface of the Petri dish at sites designated by marked circles. (Use the same

volume during the different steps of the assay, e.g., the same volume of receptor solution, blocking solution, and parasite suspension).

3. Allow CSA to adsorb to the dish for 3 h at room temperature; overlay the covered dish with wet paper towel during adsorption to avoid evaporation. Dishes can be prepared the day before, and adsorption accomplished by incubating the plate at 4°C in a humid environment.
4. Remove the CSA by suction through a yellow tip, and apply 20–30 µL of blocking solution (2% BSA in PBS).
5. Incubate for 30 min at room temperature.
6. Remove blocking solution by suction, and apply the parasite suspension.
7. Incubate for 1 h at room temperature or in an incubator at 37^0C.
8. Wash three times with PBS. To wash, cover the samples with 20 mL of PBS by addition at the edge of the Petri dish. So as not to dislodge bound IRBCs, avoid adding PBS directly to the parasite suspensions. After the samples are covered with PBS, tilt the plate gently to wash off unbound cells. Collect PBS and dislodged cells by suction at the edge of the dish. Do not allow the dish to dry between washes.
9. Fix bound cells with 0.5% glutaraldehyde in PBS for 10 min.
10. Stain with 1% Giemsa solution for 1 min.
11. Wash repeatedly with tap water and allow to dry.
12. Count the number of bound IRBCs. Counting is performed under ×1000 magnification (high-power field). Survey each test site to identify areas with low background (no uninfected RBCs). Count 20 fields that contain the highest numbers of parasites. Follow the same approach (survey for low-background areas and count fields with highest parasite densities) for both the control (BSA) and test (CSA) sites. Specific binding to CSA should significantly exceed binding to the control molecule.

3.4. Serum Inhibition of Parasite Adhesion

1. Prepare Petri dish and parasites as described in **Subheadings 3.2.** and **3.3.**
2. Preincubate parasites with sera at a final dilution of 1:5.
3. Apply the parasite suspension (in the presence of antibodies) to the dish and allow the parasites to bind.
4. Washing, fixing, and staining the dish is performed as described in **Subheading 3.3., steps 6–11**.
5. Count the number of bound IRBCs as described in **Subheading 3.3., step 12**.
6. Compare the number of bound cells in the presence of tested sera (N_{test}) to the number bound in the presence of control sera (N_{bound}). Calculate the percent inhibition (1- {N_{test}/N_{bound}}) (*see* **Notes 3** and **4**).

3.5. Serum Agglutination Assay

Parasites are used at the late trophozoite and early schizont stages and are enriched as described in **Subheading 3.2.** (*see* **ref.** *16*).

1. Wash parasites three times in PBS.
2. Resuspend parasites in 40 µL of RPMI.
3. Add 5 µL of sera (final serum dilution will be 1:10).
4. Add 5 µL of ethidium bromide (200 µg/mL in PBS).
5. Incubate for 1 h at 37°C on a cell rotator.
6. Apply 10 µL of the blood on a slide, and cover with 18 × 18 mm cover slip. Agglutinate formation is observed using fluorescent light. Agglutinates should contain only parasitized RBCS. Count the number of cells in the agglutinate, and number of agglutinates per high-power field (*see* **Note 4**).

4. Notes

1. Informed consent should be obtained from donors before collecting blood or placentas.
2. When using fresh parasite isolates that have not been adapted to culture, only sera with matching blood type can be used for binding-inhibition assays. Use AB type sera from naïve donors as control sera for all parasite isolates. To overcome the necessity to match sera and red cells for ABO type, parasites can be adapted to blood type O during in vitro culture.
3. Statistical analysis: software packages such as Statview can be used to perform statistical analyses. Differences between sera for their level of inhibition can be analyzed by nonparametric methods, such as Mann-Whitney or Kruskal Wallis analyses. In order to correlate the ability of sera to block binding of different isolates, antiadhesion activities can be analyzed with the Spearman rank test.
4. To determine whether inhibition of parasite adhesion is due to antiadhesion antibodies or agglutinating antibodies, compare the results obtained in these two assays. If a serum sample agglutinates and inhibits binding of all parasite isolates tested, then these two processes cannot be distinguished. In our experience, testing placental parasites and adhesion to CSA, few samples of immune sera obtained from multigravid women were able to agglutinate the parasites, while the same sera had uniformly high antiadhesion activity. In those cases where a serum samples agglutinated some parasite isolates but not others, we compared the level of binding inhibition between agglutinated parasites and nonagglutinated parasites. Using paired sign analysis, we were unable to define a relationship between agglutination and antiadhesion activity, suggesting that, at least for CSA-binding parasites, epitopes targeted by agglutinating antibodies are distinct from those targeted by antiadhesion antibodies. Because agglutinating antibodies are strain-specific, parasites that are not agglutinated by a specific serum sample should be identified if at all possible. These parasites can then be used to define the antiadhesion activity present in the serum sample.

References

1. Riganti, M., Pongponratn, E., Tegoshi, T., Looareesuwan, S., Punpoowong, B., and Aikawa, M. (1990) Human cerebral malaria in Thailand: a clinico-pathological correlation. *Immunol. Lett.* **25,** 199–206.
2. Barnwell, J. W., Asch, A. S. , Nachman, R. L., yamay, M., Aikawa, M., and Ingravallo, P. (1989) A human 88 kD membrane glycoprotein (CD36) functions in vitro as a receptor for cytoadherence ligand on *Plasmodium falciparum* infected erythrocytes. *J. Clin. Invest.* **84,** 765–772.
3. Berendt, A. R., Simmons, D. L., Tansey, J., Newbold, C. I., and Marsh, K. (1989) Intercellular adhesion molecule-1 is an endothelial cell adhesion receptor for *Plasmodium falciparum. Nature* **341,** 57–59.
4. Ockenhouse, C. F., Tegoshi, T., Maeno, Y., Benjamin, C., Ho, M., Kan, K. E., et al. (1992) Human vascular endothelial receptors for *Plasmodium falciparum*-infected erythrocyte: roles for endothelial leukocyte adhesion molecule 1 and vascular cell adhesion molecule 1. *J. Exp. Med.* **176,** 1182–1189.
5. Rogerson, S. J., Chaiyaroj, S. C., Ng, K., Reeder, J. C., and Brown, G. V. (1995) Chondroitin sulfate A is a cell surface receptor for *Plasmodium falciparum*-infected erythrocytes. *J. Exp. Med.* **182,** 15–20.
6. Fried, M. and Duffy, P. E. (1996) Adherence of *Plasmodium falciparum* to chondroitin sulfate A in the human placenta. *Science* **272,** 1502–1504.
7. Treutiger, C. J., Heddini, A., Fernandez, V., Muller, W. A., and Wahlgren, M. (1997) PECAM-1/CD31, an endothelial receptor for binding *Plasmodium falciparum* erythrocytes. *Nat. Med.* **3,** 1405–1408.
8. Siano, J. P., Grady, K. K., Millet, P., and Wick, T. M. (1998) Short report: *Plasmodium falciparum*: cytoadherence to alpha(v)beta3 on human microvascular endothelial cells. *Am. J. Trop. Med. Hyg.* **59,** 77–79.
9. Ho, M., Schollaardt, T., Niu, X., Looareesuwan, S., Patel, K. D., and Kubes, P. (1998) Characterization of *Plasmodium falciparum*-infected erythrocyte and P-selectin interaction under flow conditions. *Blood* **91,** 4803–4809.
10. Trager, W., Rudzinska, M. A., and Bradbury, P. C. (1966) The fine structure of *Plasmodium falciparum* and its host erythrocytes in natural malarial infections in man. *Bull. Wld. Hlth. Org.* **35,** 883–885.
11. Luse, S. A. and Miller, L. H. (1971) *Plasmodium falciparum* malaria. Ultrastructure of parasitized erythrocytes in cardiac vessels. *Am. J. Trop. Med. Hyg.* **20,** 655–660.

12. Baruch, D. I., Pasloke, B. L., Singh, H. B., Taraschi, T. F., and Howard, R. J. (1995) Cloning the *Plasmodium falciparum* gene encoding PfEMP1, a malarial variant antigen and adherence receptor on the surface of parasitized human erythrocytes. *Cell* **82,** 77–87.
13. Ockenhouse, C. F., Klotz, F. W., Tandon, N. N., and Jamieson, G. A. (1991) Sequestrin, a CD36 recognition protein on *Plasmodium falciparum* malaria-infected erythrocytes identified by anti-idiotype antibodies. *Proc. Natl. Acad. Sci. USA* **88,** 3175–3179.
14. Crandall, I., Collins, W. E., Gysin, J., and Sherman, I. W. (1993) Synthetic peptides based on motifs present in human band 3 protein inhibit cytoadherence/sequestration of the malaria parasite *Plasmodium falciparum. Proc. Natl. Acad. Sci. USA* **90,** 4703–4707.
15. Fried, M., Nosten, F., Brockman, A., Brabin, B. J., and Duffy, P. E. (1998) Maternal antibodies block malaria. *Nature* **395,** 851-852.
16. Marsh, K. and Howard, R. J. (1986) Antigens induced on erythrocytes by *P. falciparum*: expression of diverse and conserved determinants. *Science* **231,** 150–152.
17. Udeinya, I. J., Miller, L. H., McGregor, I. A., and Jensen, J. B. (1983) *Plasmodium falciparun* strain-specific antibody blocks binding of infected erythrocytes to amelantoic melanoma cells. *Nature* **303,** 429–451.
18. Singh, B., Ho, M., Looareesuwan, S., Mathai, E., Warrell, D. A., and Hommel, M. (1988) *Plasmodium falciparum*: inhibition/reversal of cytoadherence of Thai isolates to melanoma cells by local immune sera. *Clin. Exp. Immunol.* **72,** 145–150.
19. Forsyth, K. P., Philip, G., Smith T., Kum, E., Southwell, B., and Brown G. V. (1989) Diversity of antigens expressed on the surface of erythrocytes infected with mature *Plasmodium falciparum* parasites in Papua New Guinea. *Am. J. Trop. Med. Hyg.* **41,** 259–265.
20. Van Schravendijk, M. R., Rock, E. P. Ito, Y., Aikawa, M., neequaye, J., Ofori-Adjei, D., et al. (1991) Characterization and localization of *Plasmodium falciparum* surface antigens on infected erythrocytes from west african patients. *Blood* **78,** 226–236.
21. Southwell, B. R., Brown, G. V., Forsyth, K. P., Smith, T., Philip, G., and Anders, R. (1989) Field applications of agglutination and cytoadherence assays with *Plasmodium falciparum* from Papua New Guinea. *Trans. R. Soc. Trop. Med. Hyg.* **83,** 464–469.
22. Pavol, G., Wilson, R. J. M., and Smalley, M. E. (1978) Separation of viable schizont-infected red cells of *Plasmodium falciparum* from human blood. *Ann. Trop. Med. Parasitol.* **72,** 87,88.
23. Jensen, J. B. (1978) Concentration from continuous culture of erythrocytes infected with trophozoites and schizonts of *Plasmodium falciparum. Am. J. Trop. Med. Hyg.* **27,** 1274–1276.

53

Analysis of the Adhesive Properties of *Plasmodium falciparum*-Infected Red Blood Cells Under Conditions of Flow

Brian M. Cooke, Ross L. Coppel, and Gerard B. Nash

1. Introduction

Adhesion assays performed under static conditions reveal a great deal about the molecular mechanisms by which malaria-infected red blood cells adhere to the vascular endothelium. Nevertheless, they do not accurately model the process in vivo as they neglect the influence of physiological flow forces to which red blood cells are exposed in the vascular system. Flow-based assays permit the qualitative and quantitative aspects of adhesive interactions to be studied under physiologically relevant flow conditions. Both adhesion of blood cells from flow and their subsequent detachment by flow can be investigated, and quantitative estimates of the adhesive forces can be made. Adhesive interactions under flow depend not only on the affinity between a ligand and its cognate receptor, but also on the kinetics of the interaction, that is, receptor-ligand on and off rates and on the strength of the interaction between particular pairs of cell-expressed receptors and their ligands. These may have little influence on adhesion under static conditions. Thus, the absolute number of parasitized cells able to adhere to an endothelial cell or to purified receptor in a static assay and the nature of the adhesive interaction may be quite different from those found under conditions of flow. We have previously demonstrated substantial differences in the adhesive behavior of parasitized cells between static and flow-based assays. For example, in static assays parasitized cells bind well to thrombospondin, whereas binding under flow at a level that mimics that in postcapillary venules is low level and unstable *(1)*. Similarly, parasites lacking knob-associated histidine-rich protein (KAHRP) bind well in static assays but very poorly when examined under flow *(2)*. In addition to these quantitative differences in adhesion, different receptors can also mediate qualitatively different forms of adhesion that cannot be detected under static conditions. For example, parasitized cells adhering to CD36 under flow conditions remain stationary, whereas those adhering to ICAM-1 demonstrate a continuous rolling-type adhesion *(1)*.

It may be that to understand fully the pathogenesis of severe malaria, it will be necessary to know the proportions of endothelial and parasitized cells that are able to form adhesive interactions under physiologically relevant flow conditions, and the variations in adhesive strengths in these cell populations. We believe that a full understanding of the mechanisms of cytoadhesion is dependent on the description of the effects of flow on this phenomenon.

Here, we describe in detail our flow-based assay using flat, rectangular glass microcapillary tubes (microslides) as parallel-plate flow chambers (*see* **Note 1**) to visualize and quantify adhesion of malaria-infected red blood cells to endothelial cells or purified, immobilized receptors.

2. Materials

1. Microslides (VitroCom, Mountain Lakes, NJ) are precision-made flat open-ended glass microcapillary tubes with good optical qualities. They are available with a variety of well-defined internal dimensions and a length of either 50 mm or 100 mm. For our assay described here, we recommend use of microslides that are 50 mm long with a rectangular cross section of 0.3 mm × 3.0 mm (**Fig. 1**; VitroCom, cat. no. 3530). Coating the internal surface of microslides with monolayers of endothelial cells or purified receptor is described in the next chapter (*see* Chapter 54).
2. Adhesion buffer: Supplement RPMI 1640 (with sodium bicarbonate and without glutamine; Gibco-BRL, Life Technologies, Grand Island, NY) with gentamicin sulfate (20 µg/mL), glucose (11 mM), glutamine (1 mM), HEPES (25 mM), hypoxanthine (200 µM) and ALBUMAX II (1%) (Gibco-BRL). Degas the media under vacuum for 5 min then adjust the pH to 7.2 (or as required) with NaOH (2 M). The buffer can be filter-sterilized and stored at 4°C for a maximum of 1 mo.
3. Infusion-withdrawal syringe pump: We recommend a Harvard model 906 or equivalent fitted with a 50-mL glass syringe (Harvard Apparatus Inc., South Natick, MS).
4. Electronic 3-way solenoid valve with near-zero dead volume. We recommend LFYA series miniature solenoid valves (The Lee Company, Westbrook, CT, cat. no. LFYA 1226032H) or equivalent, to allow smooth change over between flowing suspensions of blood cells or cell-free adhesion buffer. The cells and buffer should be placed in plastic syringe barrels of 2-mL and 50-mL capacity, respectively, and connected directly to the valve.
5. Inverted microscope with bright field and phase-contrast illumination, long working distance condenser and ×20 and ×40 objective lenses. All high-quality inverted microscopes currently on the market are suitable. We use both Olympus (IMT2; Olympus Optical, Tokyo, Japan) and Leica (DMIRB; Leica Microsystems, Wetzlar, Germany) microscopes. To maximize image quality at ×40 magnification, the objective should have medium long (~2 mm) working distance and the highest numerical aperture possible. Water immersion lenses provide suitably high working distances and numerical apertures but are disadvantageous if the assay is being conducted at 37°C in a microscope-stage incubator where there is constant evaporation of water from the lens. Dry ×40 objectives, however, with adequately long working distances are now available with numerical apertures as high as 0.85 (e.g., Leica PL APO 40X, cat. no. 11506140, Leica Microsystems).
6. Video camera: There are currently many suitable choices of camera. Despite the increasing number of many moderately priced digital cameras now available, we currently recommend a high resolution (753 × 582 pixels, 570 horizontal lines) monochrome single CCD (1/3 in.) camera such as Panasonic's WV-BP310 (Matsushita Electric Industrial Co. Ltd., Osaka, Japan.) or similar.
7. Professional video recorder/player (e.g., JVC model BR-S800E; Victor Company of Japan Ltd) preferably with remote editing control unit (e.g., JVC model RM-G800U).
8. High-resolution (1000 horizontal lines) monochrome video monitor, e.g., Panasonic's model WV-BM1400 (Matsushita Electric Industrial).

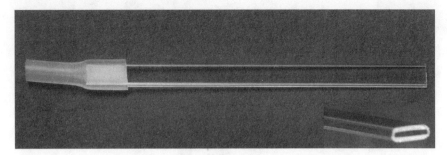

Fig. 1. A microslide with a silicon rubber adaptor attached. The insert (lower right) shows the end-on view of the same microslide.

3. Methods

3.1. Preparation of Parasitized Red Blood Cell Suspensions

Cultures of *P. falciparum*-infected red blood cells in which the parasites are predominantly trophozoites should be used for adhesion assays. For studies of clinical isolates taken directly from individuals with malaria, the ring-infected red cells should be first cultured in vitro for approx 20 h to allow the parasites to mature to trophozoites.

1. Determine the differential parasitemia by counting a minimum of 1000 red blood cells on thin smear stained with Giemsa, using a light microscope and ×100 oil immersion objective lens.
2. Determine accurately the red blood cell count, preferably using an automated analyzer such as a Coulter counter (Beckman Coulter, Fullerton, CA) although a hemocytometer is a suitable, albeit more time consuming, alternative.
3. Centrifuge the red blood cells at 400g for 5 min at room temperature and discard the supernatant.
4. Resuspend the resulting cell pellet in adhesion buffer to give a red cell concentration of 1–1.5×10^8 red blood cells per milliliter (*see* **Note 2**).
5. Use the cell suspension immediately for adhesion assays.

3.2. Setting Up the Flow-Based Microslide Adhesion Assay

The arrangement of the assay system is shown in **Fig. 2A** and **B**. Essentially, cell- or receptor-coated microslides (*see* Chapter 54) are connected to a simple flow control system then mounted on the stage of an inverted microscope as described below. Adhesion assays can be carried out at room temperature, although 37°C is more physiologically relevant. For assays at 37°C, the microscope stage (including the cell suspension and adhesion buffer reservoirs) should be enclosed in a heated air incubator. Suitable stage incubators are commercially available for most inverted microscopes.

1. Prepare cell- or receptor-coated microslides (*see* Chapter 54).
2. Clean the outside surface of the microslide with a tissue soaked in 70% ethanol.
3. Wrap double-sided adhesive tape (3M Ltd, Bracknell, Berkshire, UK), 1-cm long and 0.6-cm wide around each end of the microslide and push a piece of silicon rubber tubing (2-mm bore, 1-mm wall) connected to the flow-control system over each of the taped ends.
4. Secure the microslide in a custom-made yoke mounted on the microscope stage and begin flowing adhesion buffer through the system. Purge the system with adhesion buffer to ensure that any bubbles introduced during the assembly process are removed before beginning the assay.

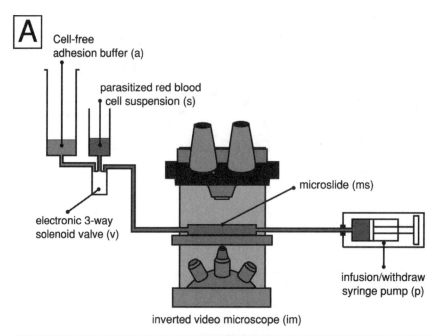
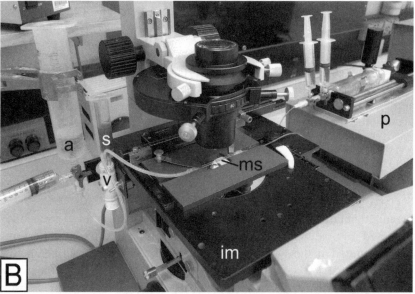

Fig. 2. (**A**) Schematic representation of the microslide adhesion assay. Letters in parentheses correspond to the labels on the photograph in Fig. 2B. (**B**) Photograph of a microslide adhesion assay set up in our laboratory. The microslide (ms) is shown fixed in a yoke on the stage of an inverted microscope (im). Syringe barrels containing parasitized red blood cell suspension (s) and adhesion buffer (a) are connected to an electronic solenoid valve (v). Flow through the microslide is controlled by a Harvard infusion/withdraw syringe pump (p).

3.3. Calculating the Flow Rate to Give the Desired Wall Shear Stress

The power of this assay over conventional static assays is that the hemodynamic flow conditions to which parasitized red blood cells are exposed can be reproduced and, provided that flow is laminar, accurately quantified (*see* **Note 2**). The rate and the nature of the flow of blood, vessel geometry, and therefore wall shear stress, are highly variable in vivo. Wall shear stress in the microcirculation has not been measured directly in humans, but shear stresses have been calculated using animal models, from centerline cell velocities and vessel diameters or from measurement of the pressure drop over a defined length of vessel. Using animal models, wall shear stresses in the range 0.1 to 1.0 Pa have been calculated to exist in postcapillary venules of the microcirculation (*3*). Since parasitized red blood cells in vivo tend to sequester predominantly in postcapillary venules, we recommend that in vitro assays be performed within this range.

The flow rate required to give the desired wall shear stress can be calculated from the dimensions of the microslide and the viscosity of the flowing adhesion buffer. There is no exact, general solution for flow in all rectangular channels, but if the depth of the channel is much less than its width, flow can be defined by

$$Q = (wh\tau)/(6\mu)$$

where w is the width of the microslide, h is its depth, τ is the wall shear stress, and μ is the viscosity of the flowing adhesion buffer. Thus for any given flow rate, the wall shear stress can be calculated or vice versa.

We recommend that the viscosity of the adhesion buffer be determined by viscometry at room temperature (e.g., using a Coulter Viscometer II; Coulter Beckman Inc.), then corrected for ambient temperature using standard tables (*4*). In our system, the viscosity of our adhesion buffer is typically 0.78 mPa·s at 37°C.

As a guide to volumes of red blood cell suspension and flowing buffer required, the flow rate during a typical experiment in our laboratory at 37°C for a wall shear stress of 0.1 Pa would be 0.35 mL/min. A typical 5-min inflow of parasitized cells at 0.1Pa would, therefore, require a minimum of 1.75 mL of cell suspension. At room temperature (21°C), the flow rate would be reduced to 0.25 mL/min to compensate for the increase in viscosity (1.1 mPa·s) at the lower temperature.

3.4 Performing the Adhesion Assay

Two major types of assay can be performed: (a) adhesion from flow, to quantify the ability of parasitized cells to adhere to a target cell or purified receptor over a range of wall shear stresses that mimic those in the vasculature in vivo; and (b) detachment of adherent parasitized cells by stepwise increases in wall shear stress to derive a quantitative estimate of the strength of the interaction between parasitized cells and the cellular target or purified receptor.

3.4.1. Effect of Increasing Shear Stress on Adhesion of Parasitized Cells from Flow

1. Flow the parasitized red blood cell suspension through the microslide for 5 min at the desired wall shear stress.
2. Then flow cell-free adhesion buffer through the microslide at the same stress for 2–5 min, depending on the wall shear stress, to clear nonadherent cells.
3. Quantify adhesion by counting the number of adherent cells while flow continues at the same wall shear stress.
4. Removed adherent cells by increasing the flow rate of the adhesion buffer through the microslide, then repeat the process with the next sample of parasitized cells if required.

3.4.2. Effect of Increasing Stress on Previously Adhered Parasitized Cells

1. Flow the parasitized red blood cell suspension through the microslide at the desired wall shear stress (e.g., 0.1 Pa) for 5 min, followed by cell-free adhesion buffer for 5 min at the same stress to remove nonadherent cells.
2. Count adherent cells in one or more fields of view.
3. Increase the wall shear stress (e.g., to 0.2 Pa) for 2 min followed by cell-free adhesion buffer at the same stress.
4. Recount adherent cells.
5. Repeat this procedure at increasing wall shear stresses (e.g., 0.3, 0.50, and 2.5 Pa or greater) until all adherent cells have been removed.
6. Plot the data as parasitized red cells remaining adherent as a function of wall shear stress (as a percentage of those initially adherent at the initial inflow stress) (*see* **Note 3**).

3.5. Quantifying Adhesion

Quantify adhesion either in real-time by direct microscopic observation or by retrospective analysis of video recordings. Restrict observations of cell adhesion to a region along the centerline of the microslide, avoiding the extreme ends near the inlet and outlet ports where flow may not be laminar.

3.5.1. Direct Microscopic Observation

1. For cell-coated microslides, first confirm the integrity of the cell monolayer using phase-contrast illumination, then count adherent cells.
2. Count adherent cells in 5 to 10 complete microscope fields (each of predetermined area) or a minimum of 100 adherent cells (at higher shear stresses or when levels of adhesion are low). Cells can be most easily visualized and counted under phase-contrast illumination using a ×20 objective lens (**Fig. 3A**); however, an alternative method is to count under bright-field illumination with the field diaphragm stopped down. Use of this optical system allows parasites to be clearly identified within adherent red blood cells (**Fig. 3B**).
3. Express the result as adherent cells per square millimeter or as adherent cells/mm^2/10^7 parasitized cells perfused. The latter accounts for differences in the parasitemia, the red cell count, and the flow rate between individual samples. This can be calculated as:

$$\text{Adherent cells/mm}^2/10^7 \text{ parasitized cells perfused} = N/(CPQt) \times 10^7$$

where N, number of adherent parasitized cells (per mm^2); C, red blood cell concentration (cells/mL); P, parasitemia (expressed as a decimal); Q, flow rate (mL/min); t, time for which the cell suspension was flowed (min).

3.5.2. Quantification by Video Analysis

1. Record assays on videotape then analyze the recordings retrospectively using a professional quality stop-frame video recorder with a high-resolution monochrome video monitor (*see* **Subheading 2.**).
2. Count the number of adherent parasitized cells in several complete monitor fields, each of known area, precalibrated using a stage micrometer.
3. Count a minimum of 100 adherent cells and express the result as adherent cells per square millimeter or as adherent cells/mm^2/10^7 parasitized cells perfused (*see* **Subheading 3.5.1.**).

3.5.3. Quantifying Rolling Adhesion

Quantitation of adhesion of parasitized cells to receptors such as purified CD36 with which parasitized cells form stable and stationary interactions is somewhat less com-

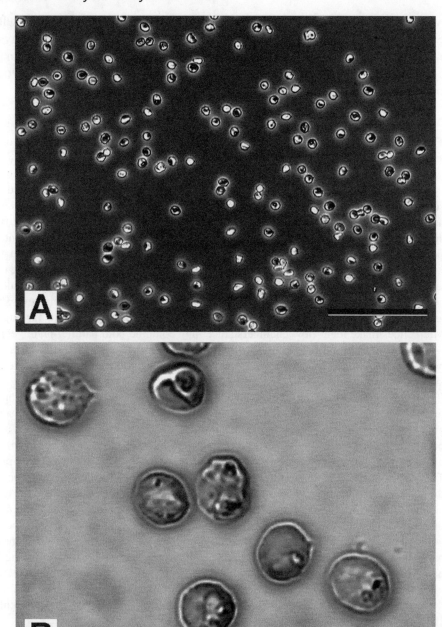

Fig. 3. Digitized phase-contrast images of parasitized red blood cells adhering to purified CD36 in a microslide under flow. (**A**) Typical medium-power (×20 objective) field of view after parasitized cells had been flowed through a CD36-coated microslide and nonadherent had been washed out. Scale bar = 100 µm. (**B**) Higher magnification (×40 objective) of adherent parasitized cells confirms the presence of malaria parasites inside all adherent red blood cells. Scale bar = 10 µm.

plicated to analyze than adhesion to umbilical vein endothelial cells or ICAM-1 on which a large proportion of parasitized cells continuously roll (*1*). In this case, the relative proportions of adherent cells which are stationary or rolling, and the velocity of the rolling cells should be calculated by retrospective computer-assisted analysis of video recordings made during the assay. Most commercially available image analysis packages can be tailored to perform semiautomated cell counts or analyze velocities (distance/time) of rolling cells.

4. Notes

1. The flow chamber must have a well-defined geometry so that flow can be precisely defined and experiments performed under conditions that mimic those in the vasculature in vivo (*see* **Note 2**). Ideally, they should also have good optical qualities to facilitate visualization of adherent cells. A number of studies of adhesion of flowing blood cells (red blood cells, sickle red cells, neutrophils, lymphocytes, and neoplastic cells) to endothelial cells have been carried out in vitro using parallel-plate flow chambers. A variety of different designs have been described, but, in essence, a cell suspension is flowed at a controlled flow rate between a pair of parallel plates, clamped together and separated by a thin gasket (typically 150-µm thick). The lower of the two plates is coated with a confluent monolayer of endothelial cells, and adhesion is observed microscopically after nonadherent cells have been washed out. Cylindrical glass tubes have also been used, however, parallel-plate flow geometry has the advantage that the optical qualities are far superior. Microslides make ideal flow chambers since they combine both parallel-plate geometry with good optical qualities. They also represent a significant simplification of a previously described parallel-plate flow chamber for studying adhesion of parasitized red cells (*5*) that must be clamped together and is altogether more finicky to assemble.

2. If colored dye is injected into the stream of a liquid flowing between two parallel plates, or into a cylindrical tube, the liquid along the central axis of the channel moves with a much greater velocity than that near the wall, and the leading edge of the dye assumes a parabolic profile. Poiseuille was the first to observe that individual red blood cells in flowing blood moved at a range of velocities, now known to be attributable to this velocity profile. The velocity profile arises because the molecules of the liquid are flowing as a series of thin layers (or laminae) with their axes parallel to the sides of the channel, and the fluid in contact with the wall is stationary. Each individual lamina is sliding, or shearing over the ones immediately juxtaposing it, causing viscous friction between adjacent laminae. Each lamina is therefore subjected to a tangential force imposed upon it by the adjacent laminae. The shear stress imposed on any lamina is the force per unit area. The moving laminae directly in contact with the stationary plates in a parallel-plate chamber such as a microslide, or with the wall in a tube with circular cross section, exert a shearing force on the wall that is called the wall shear stress, τ. The unit of shear stress is the pascal (Pa) (1 Pa = 1 N/m^2) (N = Newton).

   The rate of change of velocity between adjacent laminae is termed the shear rate and is measured in inverse seconds (s^{-1}). According to Newton's second law describing the motion of nonaccelerating fluids, the viscous frictional force opposing the forward movement of the laminae represents the constant of proportionality between shear stress and shear rate and is called the viscosity of the flowing liquid; thus viscosity = shear stress/shear rate. The basic unit of viscosity, the pascal-second, is too large to be convenient and the millipascal-second (mPa·s) is more commonly used and is approximately equal to the viscosity of water at room temperature. When flow occurs in such parallel laminae, it is termed laminar flow. There is no transverse mixing between the layers, across the channel or tube and the individual laminae remain distinct and continue to flow side by side.

   For accurate calculation of wall shear stress over a range of shear rates, we recommend that the red blood cell concentration in the parasitized red blood cell suspension under test

is kept low (*see* **Subheading 3.1.**). In this case, the cell suspension can then be assumed to be newtonian, that is, its viscosity is constant regardless of the shear rate. For more concentrated (non-Newtonian) cell suspensions, the relationship between shear rate and viscosity becomes complex so for accurate calculation of wall shear stress, the viscosity must be determined at each level of shear rate.
3. The wall shear stress required to remove 50% of adherent parasitized cells (τ_{50}) can be determined by extrapolation from the resulting wash-off curve and provides a useful and a quantitative measure of the strength of the interaction between parasitized cells and the target.

References

1. Cooke, B. M., Berendt, A. R., Craig, A. G., MacGregor, J., Newbold, C. I., and Nash, G.B. (1994) Rolling and stationary cytoadhesion of red blood cells parasitized by *Plasmodium falciparum*: separate roles for ICAM-1, CD36 and thrombospondin. *Br. J. Haematol.* **87**, 162–170.
2. Crabb, B., Cooke, B. M., Reeder, J. C., Waller, R. F., Caruana, S. R., Davern, K. M., et al. (1997) Targeted gene disruption shows that knobs enable malaria-infected red cells to cytoadhere under physiological shear stress. *Cell* **89**, 287–296.
3. Chen, S. (1969) Blood rheology and its relation to flow resistance and transcapillary exchange with special reference to shock. *Adv. Microcirc.* **2**, 89–103.
4. Weast, R. C. (1975) *Handbook of Chemistry and Physics*. CRC Press, Cleveland, OH, p. F49.
5. Nash, G. B., Cooke, B. M., Marsh, K., Berendt, A., Newbold, C., and Stuart J. (1992) Rheological analysis of the adhesive interactions of red blood cells parasitized by *Plasmodium falciparum*. *Blood* **79**, 798–807.

54

Preparation of Adhesive Targets for Flow-Based Cytoadhesion Assays

Brian M. Cooke, Ross L. Coppel, and Gerard B. Nash

1. Introduction

In the previous chapter (*see* Chapter 53), we described our method for visualizing and quantifying adhesion of malaria-infected red blood cells to endothelial cells or purified receptors using flat, rectangular glass microcapillary tubes (microslides) as flow chambers. In this chapter, we describe methods for producing the receptor- and cell-coated microslides for use in the flow-based adhesion assays. For cell-coated microslides, we have concentrated specifically on human umbilical vein endothelial cells (HUVEC) since these are commonly used for static adhesion assays. For purified receptor, we describe preparation of CD36-coated microslides since this is a receptor to which most parasitized cells (both laboratory-adapted lines and clinical isolates) adhere. However, our methods are not restricted to HUVEC and CD36 and can be modified to suit most other cells or receptors to which parasitized cells adhere. For example, we also describe the use of platelets and C32 amelanotic melanoma cells that provide additional targets for cytoadhesion assays.

2. Materials

1. Microslides (VitroCom Inc., Mountain Lakes, NJ; *see* Chapter 53).
2. 3-Aminopropyltriethoxysilane (APES) (Sigma, St. Louis, MO). Prepare a 4% v/v solution of APES in anhydrous acetone immediately before use. Prepare and work with the solution in polypropylene tubes to avoid unnecessary exposure to glass.
3. Culture medium for HUVEC. Supplement M199 (Earle's modification with sodium bicarbonate; ICN Pharmaceuticals Inc. Costa Mesa, CA, USA) with gentamicin sulfate (28 mg/L), glutamine (2 mM), preservative-free sodium heparin (50 IU/mL), and heat-inactivated human serum (20% v/v). Store at 4°C for a maximum of 1 wk.
4. Collagenase (type IA-S, Sigma, cat. no. C-5894). Reconstitute the lyophyllate in PBS to 10 mg/mL. Filter-sterilize (0.22-µm filter) and store in 0.5 mL aliquots at –20°C. When required, dilute thawed aliquots to 1 mg/mL by adding 4.5 mL of phosphate-buffered saline (PBS). Warm to 37°C and use immediately.
5. Acid citrate dextrose-theophylline (ACD-T) anticoagulant (10X). Prepare an aqueous solution of trisodium citrate (26.5 g/L), citric acid (12.6 g/L), and glucose (25.2 g/L). Adjust the pH to 4.6 using citric acid if necessary. Add theophylline (12.6 g/L) and heat the solution to 70°C in a water bath until the theophylline has completely dissolved. Cool

the solution to room temperature, re-check the pH and adjust to 4.6 if necessary. Dispense 1-mL aliquots into 10-mL sterile tubes and store at –20°C until required.
6. PBS, calcium- and magnesium-free (ICN Pharmaceuticals).
7. Platelet washing buffer (10X). Prepare an aqueous solution of K_2HPO_4 (7.5 g/L), $NaH_2PO_4·2H_2O$ (38 g/L) $Na_2HPO_4·2H_2O$ (7.7 g/L), NaCl (66 g/L), glucose (10 g/L), and theophylline (18 g/L). Heat the solution to 70°C in a water bath until the theophylline has completely dissolved. Check the pH of the solution at room temperature and adjust to 5.6 if required. Dispense into 10-mL aliquots and store at –20°C until required. For use, thaw aliquots and dilute 1:9 with water containing bovine serum albumin (5 mg/mL). Check the pH of the buffer and adjust to 6.4 if necessary.
8. Trypsin (ICN Pharmaceuticals Inc.). Prepare 0.2 mL aliquots of a sterile solution (25 mg/mL) and store at -20°C. When required, dilute a thawed aliquot to 0.5 mg/mL by adding 9.8 mL of PBS.
9. Tyrode's buffer. Prepare an aqueous solution of $NaHCO_3$ (1 g/L), HEPES (2.4 g/L), NaCl (8 g/L), KCl (0.2 g/L) and glucose (1 g/L). Adjust the pH of the solution to 7.4 at room temperature and bottle. Autoclave the solution at 121°C for 15 min. Store at room temperature until required.
10. Microslide cell culture dish. These are not commercially available but can be fabricated in-house by any experienced glass blower by fusing six glass tubing side arms (2 cm long, 1 mm wall, 2 mm bore) through the wall of a 100 mm diameter Pyrex glass Petri dish (**Fig. 1**). Inside the dish, each side arm should be further tapered linearly to an overall external diameter of 2 mm to facilitate connecting the microslides via their silicon tubing adaptor. Connect silicon rubber tubing (2 mm wall thickness, 3 mm bore, ~40 cm long) onto each external arm. Place an autoclavable plastic pinch clip onto each length of tubing (**Fig. 1**).
11. Multichannel peristaltic pump (e.g., model 505S with 308MC pump head and 0.38 mm bore Marprene tubing; Watson-Marlow, Falmouth, UK) connected to a cyclical timing controller (e.g., Watson-Marlow model 501T.)

3. Methods
3.1. Washing and Precoating Microslides

Before microslides can be coated with cell monolayers or purified receptor, they must be acid-washed and precoated. Precoating the internal surfaces of microslides is essential for growth of confluent layers of HUVEC and greatly enhances immobilization of purified receptors.

3.1.1. Acid-Washing Microslides

1. Soak microslides overnight in 50% (Aq) nitric acid. We recommend that microslides are washed in batches of 30 by placing the microslides vertically into a glass 10-mL universal bottle, then slowly fill it with the acid solution using a transfer pipet. This way, the microslides fill rapidly and completely by capillary action without introducing air bubbles that are subsequently difficult to remove.
2. Wash microslides in copious amounts of running tap water for 2 h, then rinse each slide individually with distilled water using a piece of small diameter tubing connected to the distilled water supply.
3. Rinse the microslides twice with 96% ethanol then dry overnight at 37°C. Store the washed microslides in a closed container until required.

3.1.2. Precoating Microslides

The choice of precoating agent depends on the type of cell being cultured or nature of the receptor being immobilized (*see* **Note 1**). We recommend 3-aminopropyl-

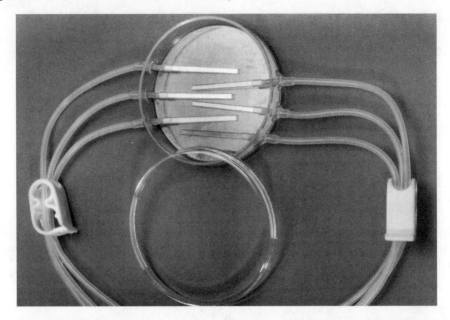

Fig. 1. Specially constructed Pyrex-glass dish used to culture endothelial cells in microslides under conditions of flow. Six microslides are shown connected to the internal side arms. The external silicon rubber tubing would be connected to a peristaltic pump to control flow.

triethoxysilane (APES) as an effective and convenient agent for pretreatment of microslides prior to coating with HUVEC or CD36.

1. Place acid-washed microslides in polystyrene tubes and rinse twice with anhydrous acetone.
2. Immerse microslides in a freshly prepared solution of APES (4% v/v in anhydrous acetone) for 5 min. Remove microslides from the APES then repeat the procedure two more times using a fresh solution of APES each time.
3. Rinse microslides once with anhydrous acetone then allow them to air dry.
4. Wash through each microslide individually with distilled water (as above) then leave to dry at 37°C overnight. Place the microslides into capped glass tubes then dry-heat sterilize. Store at room temperature until required.

3.2. Isolation and Primary Culture of Human Umbilical Vein Endothelial Cells (HUVEC)

All culture techniques must be performed under aseptic conditions in a class 2, laminar airflow cabinet. Cells should be cultured in a humidified CO_2 (5%) incubator maintained at 37°C.

1. Collect human umbilical cords as soon as possible postparturition and place into plastic, disposable pots containing sterile PBS (*see* Note 2). Harvest the endothelial cells lining the lumen of the umbilical vein using the following method based on that of Jaffe et al. (*1*).
2. Precoat 25-cm² tissue culture flasks (3 flasks/cord) with gelatin by adding 2 mL of sterile aqueous gelatin solution (1% w/v; 225 bloom porcine gelatin, Sigma) to each and incubate at 37°C until required.
3. Clean the outside of the cord thoroughly with 70% v/v methanol and transfer the cord onto a surgical instrument tray.

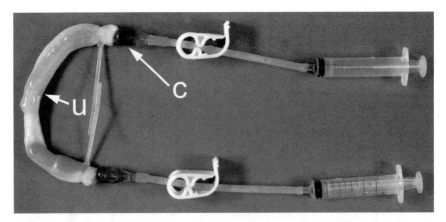

Fig. 2. A section of human umbilical cord (u) after insertion of glass cannulae (c) into the umbilical vein followed by perfusion with collagenase solution to detach the endothelial cells from the vascular intima.

4. Cut off the ends of the cord to give a section 15–20 cm long that is free from any visible signs of damage (including clamp marks, contusion or puncture marks resulting from cord blood sampling).
5. Insert sterile glass cannulae, with silicon rubber tubing attached to one end, into both ends of the vein and secure in place using plastic cable ties (**Fig. 2**).
6. Attach a syringe to the silicon rubber tubing and flush the lumen of the vein with 20 mL of PBS to remove blood and thrombi, and to ensure that the vessel is patent.
7. Inject 5 mL of collagenase solution, prewarmed to 37°C, into the lumen of the vein until it is full and slightly distended.
8. Clamp the silicon rubber tubing attached to the cannulae.
9. Place the entire preparation inside a plastic bag, seal it and incubate at 37°C for 15 min.
10. After incubation, remove the cord and gently massage down its entire length between the thumb and forefinger (to encourage the release of adherent cells from the basement membrane).
11. Flush out the collagenase solution with 30 mL of PBS.
12. Centrifuge the resulting eluate at 640g for 5 min.
13. Remove the supernatant, and resuspend the cell pellet in 2 mL of HUVEC culture medium.
14. Disaggregate any clumps of cells by pipetting the suspension up and down several times using a Pasteur pipet, then add a further 10 mL of culture medium.
15. Remove the previously added gelatin solution from the flasks, rinse once with PBS, and then place 4-mL aliquots of the endothelial cell suspension into each flask.
16. Incubate the flasks at 37°C in a CO_2 incubator.
17. Change the culture medium after 2 h, 24 h, and then every 3 d until the cells are confluent (5–8 d).

3.3. Culturing HUVEC in Microslides

In initial attempts to culture HUVEC in the microslides, mono-dispersed cell suspensions were simply loaded into coated microslides then observed at regular intervals. Although the cells could adhere and spread initially on the lower internal surface of the microslide, after 2 h they reassumed a spherical morphology and became detached from the glass. This occurred because the small volume of medium in a microslide (45 µL) rapidly lost its buffering capacity, and its pH fell when in contact

Targets for Flow-Based Adhesion Assays

with the rapidly metabolizing endothelial cells. However, detachment could be prevented if the medium in the microslide was changed manually every 2 h and the adherent cells continued to grow. Because the cells require 20–24 h of culture before confluency was reached, an automated method to refresh the medium in the microslides at regular intervals is necessary for the flow-based culture method described here (*see* **Note 3**).

1. Remove media from confluent cultures of HUVEC and rinse the cell monolayer twice with 4 mL of PBS containing disodium EDTA (1 m*M*).
2. Add 2 mL of sterile trypsin solution (0.5 mg/mL in PBS) to each flask and incubate at room temperature for 2 to 3 min until the cells become detached.
3. Pipet the suspension up and down several times using a glass Pasteur pipet to disaggregate any clumps of cells and then add 8 mL of HUVEC culture medium.
4. Centrifuge at 400*g* for 5 min then remove and discard the supernatant.
5. Resuspend the cell pellet (containing 2×10^6 cells on average) in 0.5 mL of HUVEC culture medium and disperse them by sucking them in and out of a glass Pasteur pipet several times.
6. Transfer the suspension to one corner of a slightly tilted 35-mm Petri dish and load the cell suspension into the microslides as follows:
7. Attach each of six microslides to an automatic pipetor fitted with a 1-mL sterile tip, via the silicon rubber adaptor attached to the end of the microslide (*see* **Fig.1** and Chapter 53).
8. Aspirate the endothelial cell suspension from the dish into the microslide.
9. Place a sterile, glass microscope slide inside a 100-mm Petri dish and lay the filled microslides across it (to keep them horizontal).
10. Incubate the dish at 37°C for 1 h.
11. Add 50 mL of culture medium to the culture dish (**Fig. 1**). Prime all six lengths of silicon tubing by drawing media from the inside of the dish through each of them in turn then clamp the tubing with the pinch clips.
12. After the 1 h of static incubation, connect the microslides aseptically to the internal side arms of the culture dish, using sterile forceps.
13. Place the dish in a humidified CO_2 incubator and pass the silicon tubing through service ports located in the wall or rear of most incubators.
14. Connect the tubing to the peristaltic pump and set the timer to turn the pump on for 30 s every 2 h at a flow-rate of 0.26 mL/min per microslide. This is equivalent to a wall shear stress of 0.06 Pa (*see* **Subheading 3.3.** and Note 2 in Chapter 53) and a total of approximately three microslide volumes of medium exchanged during the 30-s period. The HUVEC will have reached confluence after 20–24 h. The culture setup is shown schematically in **Fig. 3**.

3.4. Preparation of Platelet-Coated Microslides

Platelet monolayers provide suitable, convenient and easy to prepare targets for studying adhesion of parasitized red blood cells to cell-expressed CD36 *(2)*. Unlike HUVEC, no flow is necessary to produce confluent monolayers of platelets in microslides. In acid washed microslides, platelets will adhere, become activated, and spread to near-confluence within 30 min under static conditions. We have previously shown that adhesion of parasitized cells (both cultured lines and clinical isolates) is ablated by greater than 90% if platelet monolayers are treated with a monoclonal antibody raised against CD36 before parasitized cells are flowed over *(2,3)*.

1. Collect human venous blood with minimum stasis by venipuncture in the antecubital region of the arm and dispense immediately into plastic tubes containing ACD-T (1 mL of ACD-T + 9 mL of whole blood).

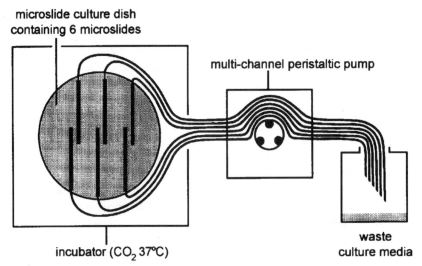

Fig. 3. Schematic representation of the flow-based apparatus used to culture cells in microslides.

2. Centrifuge the anticoagulated blood immediately at 200g for 10 min.
3. Using a plastic transfer pipette, collect the supernatant platelet-rich plasma (PRP), typically containing 2×10^8 platelets/mL, into a clean 10-mL tube, record the volume, and then centrifuge at 2000g for 5 min.
4. Remove the supernatant and carefully resuspend the platelet pellet in 1X platelet washing buffer. Centrifuge for 5 min at 2000g.
5. Remove the supernatant and carefully resuspend the pelleted platelets in Tyrode's buffer (use 2.5 mL of buffer/10 mL of whole blood; equivalent to a final platelet concentration of approx 3×10^8 platelets/mL).
6. Fill microslides immediately by capillary action by dipping one end of the microslide into the suspension of washed platelets.
7. Once filled, place microslides on a glass microscope slide and incubate flat in a humidified chamber at room temperature for 30 min. Platelets adhere to and spread on the lower internal surface of the microslide to form a confluent monolayer.
8. After incubation, rinse microslides through gently with adhesion buffer (*see* **Subheading 2.** and Chapter 53) and use immediately for an adhesion assay.

3.5. Culturing C32 Amelanotic Melanoma Cells in Microslides

The human amelanotic melanoma cell line C32 has also been used extensively as an adhesive target in static adhesion assays. We have made many attempts to culture C32 cells in microslides with varying degrees of success and poor reproducibility. Conditions for growth of C32 cells appeared to be quite different to those for HUVEC. For example, C32 cells will not adhere to and grow well on APES-coated microslides, however, after trying a variety of coatings, gelatin (1% Aq. for 1 h at 37°C followed by a PBS rinse) appeared to give the best result. Like HUVEC, C32 cells would not grow in microslides that had not been pretreated. Better results were also obtained if the number of C32 cells in the loading suspension was reduced to 1×10^6/mL, that is, only one-quarter of the number of cells in the HUVEC suspension. Numbers in excess of this impeded adhesion of C32 cells to the microslide during the 1-h static incubation period. The flow rate of the culture medium was the same as for HUVEC but the frequency of medium changes required was greater (30 s every 30 min), and time to reach confluence was longer (approx 2–3 d). Even under these conditions, growth was unpredictable and the cells did not always reach confluence. The cells also frequently tended to "pile up" leading to perturbation of the laminar flow profile when microslides were used for adhesion assays.

3.6 Use of Fixed Cell-Coated Microslides

Use of fixed cell-coated microslides for adhesion assays is technically advantageous. The choice of fixative is, however, critical and must be chosen with care depending on the target cell or the nature of the receptor and ligand pair under study (*see* **Note 4**). Here we describe the method for fixing cell monolayers with either formalin or glutaraldehyde.

1. Wash the adherent cell monolayer once with PBS.
2. Draw either freshly diluted formalin (1% v/v in PBS) or freshly diluted glutaraldehyde (2% v/v in PBS) into the microslide. Incubate in a humidified chamber at room temperature for 1 h before use or store in the fixative solution at 4°C until required.
3. Before use in an adhesion assay, remove the fixative and wash the cells five times in PBS followed by five washes in adhesion buffer.

3.7. Immobilization of Purified CD36 in Microslides

The example given here is for CD36 although other receptors to which parasitized cells adhere can be used (*see* **Note 1**). For CD36, we recommend using microslides that have been acid washed and pretreated with APES. For other receptors, the choice of the precoating agent must first be determined (*see* **Note 1**) as well as the coating concentration of receptor to give maximal levels of binding.

We routinely purify CD36 in large batches from human platelets using the method of Kronenberg et al. *(4)* and store the protein in aliquots at –70°C until required. We use outdated packs of platelets for transfusion obtained from our local hospital hematology department.

1. Fill APES-coated microslides with 50 µL of CD36 (1 µg/mL) and incubate overnight in a humidified box at 4°C.
2. The following day, incubate the microslides for 1 h at room temperature, and then remove the CD36 and replace with PBS containing 1% bovine serum albumin for 2 h at room temperature to block nonprotein-coated sites.
3. Use the microslides immediately for adhesion assays.

3.8 Adhesion Blocking Experiments with Antibody or Other Soluble Inhibitory Agents

Microslides coated with cells or purified receptor can be pretreated with antibody or compounds with putative antiadhesive activity and provide insights into the molecular mechanisms of cytoadhesion. Pretreatment requires only 50 µL of antibody per microslide at a typical working dilution of antibody. We draw antibody diluted in adhesion buffer into a microslide then incubate the microslide in a humidified chamber at room temperature for 30 min. The microslide can then be connected directly into the flow system and the adhesion assay performed. Alternatively, compounds with putative anti-adhesive properties can be added to the parasitized red blood cell suspension during inflow or added to the cell-free adhesion buffer during the wash-off stage to investigate the ability of compounds to reverse already established adhesion *(5)*.

4. Notes

1. HUVEC will not adhere to and form confluent monolayers in microslides that have not been pretreated. We have compared a number of different reagents that are commonly used in tissue culture to precoat glass and plastic in order to enhance adhesion and growth of cultured cells. Of these, we found APES to be far superior and the most convenient. APES coating also greatly enhances immobilization of platelets, CD36, and ICAM-1. It is not, however, suitable for all cells and receptors. For example, C32 melanoma cells or transfected U293 cells do not adhere to and grow well on APES. In this case, gelatin (1% aq for 1 h at 37°C followed by a PBS rinse) appears to give good results.

    In some cases, modification of the receptor itself is advantageous. For example, chondroitin sulfate A (CSA) is a recently described receptor for parasitized cells but adsorbs poorly to microslides treated with any of our usual armament of precoating agents. When CSA was linked to phosphatidylethanolamine (PE), however, the resulting CSA–PE conjugate adheres well to microslides treated with poly-L-lysine without any interference or alteration of binding of parasitized cells to the CSA moiety itself.

    To coat microslides with poly-L-lysine we prepare a 5 mg/mL solution of poly-L-lysine hydrobromide (mol wt 70,000–150,000; Sigma, cat. no. P1274) in distilled water and store at –20°C until required. To coat, fill acid-washed microslides with the poly-L-lysine solution and incubate at room temperature for 1 h in a humidified box. Wash each slide individually with distilled water and dry at 37°C overnight. The coated slides can be stored in a capped container at 4°C for up to 1 mo.

2. It is advisable to establish a collaboration and appropriate ethics approval with a maternity hospital close to your laboratory and organize to collect umbilical cords on a regular basis. Once established, we have had no problems obtaining a regular supply of fresh umbilical cords. We have also successfully established primary cultures of endothelial cells from umbilical cords that are up to 2 d postparturition.

3. Endothelial cells are cultured in microslides under conditions of flow that can be varied, in contrast to static adhesion assays, in which the endothelial cells are cultured under static conditions. Endothelial cells that have been cultured under flow may be preferable for cytoadhesion studies since the endothelium in vivo is continuously exposed to shear forces exerted by flowing blood.

4. Fixation allows multiple cell-coated microslides to be prepared from a single umbilical cord or batch of platelets form a single donor, which can then be fixed and stored until required. Removal of adherent parasitized cells from an unfixed monolayer of endothelial cells can sometimes cause disruption of the cell monolayer, however, a single fixed target can be used several times over for comparative assays performed on a given day or over a period of time. Fixation not only facilitates and improves the reproducibility of comparative experiments, but also makes adhesion targets available for transport and use in field studies of clinical isolates of malaria parasites in endemic areas *(3,6)*. The fixative, however, must be carefully chosen, depending on the cell type of the monolayer. For example, parasitized cells adhere to both unfixed and fixed HUVEC. Glutaraldehyde fixation does not have a marked affect on the strength of the adhesive interaction between parasitized cells and HUVEC, but does markedly increase the numbers of parasitized cells that adhere. The rolling phenomenon, however, may explain this finding, since adhesion to glutaraldehyde-fixed HUVEC is stationary in nature and rolling is rarely observed. Because parasitized cells continuously roll along and off unfixed HUVEC during inflow and wash out, the total number making adherent contacts is underestimated when counts are finally made. In fact, if all parasitized cells that adhere while the cells are being flowed through and washed out of the microslide are counted, the total number of parasitized cells adhering to unfixed HUVEC and to glutaraldehyde-fixed HUVEC is similar. In contrast to glutaralde-

hyde-fixed HUVEC, adhesion to formalin-fixed HUVEC mirror adhesion to unfixed HUVEC. Thus, if fixed endothelial cells are required for adhesion studies, formalin should be the fixative of choice. Furthermore, adhesion of parasitized cells to formalin-fixed platelets (mediated via CD36) completely mirrors that of unfixed platelets. In contrast, fixation of platelets with glutaraldehyde completely ablates adhesion.

In terms of time of storage of fixed cell-coated microslides, we have stored both glutaraldehyde- and formalin-fixed HUVEC for more than 2 yr without a noticeable decrease in their adhesive properties mediated via ICAM-1. By contrast, the adhesive properties of formalin fixed platelets (mediated via CD36) do decrease over time (detected as a steady decrease in the level of adhesion of parasitized cells over time). We would recommend that microslides coated with formalin-fixed platelets are stored only for a maximum of 1 mo.

References

1. Jaffe, E. A., Nachman, R. L., Becker, C. G., and Minick, C. R. (1973) Culture of human endothelial cells derived from umbilical veins. Identification by morphologic and immunologic criteria. *J. Clin. Invest.* **52,** 2745–2756.
2. Cooke, B. M. and Nash, G. B. (1995) *Plasmodium falciparum*: characterization of adhesion of flowing parasitized red blood cells to platelets. *Exp. Parasitol.* **80,** 116–123.
3. Cooke, B. M., Morris-Jones, S., Greenwood, B. M., and Nash, G. B. (1995) Mechanisms of cytoadhesion of flowing, parasitized red blood cells from Gambian children with falciparum malaria. *Am. J. Trop. Med. Hyg.* **53,** 29–35.
4. Kronenberg, A., Grahl, H., and Kehrel, B. (1998) Human platelet CD36 (GPIIIb, GPIV) binds to cholesteryl-hemisuccinate and can be purified by a simple two-step method making use of this property. *Thromb. Haemost.* **79,** 1021–1024.
5. Cooke, B. M., Nicoll, C. L., Baruch, D. I., and Coppel, R. L. (1998) A recombinant peptide based on PfEMP-1 blocks and reverses adhesion of malaria-infected red blood cells to CD36 under flow. *Mol. Microbiol.* **30,** 83–90.
6. Cooke, B. M., Morris-Jones, S., Greenwood, B. M., and Nash, G.B. (1993) Adhesion of parasitized red blood cells to cultured endothelial cells: a flow-based study of isolates from Gambian children with falciparum malaria. *Parasitology* **107,** 359–368.

55

Triton X-114 Phase Partitioning for Antigen Characterization

Lina Wang and Ross L. Coppel

1. Introduction

In 1981, Bordier *(1)* first demonstrated that Triton X-114 could be exploited as a means of separating hydrophilic and integral membrane proteins. Since then, Triton X-114 phase partitioning has been extensively used for identification and isolation of membrane proteins and has proved to be a convenient method for obtaining enriched preparation of membrane proteins that retain biological activity *(2)*. By making some modifications to the original methodology, Smythe et al. *(3)* have improved the procedure to make it more suitable for isolation of integral membrane proteins from *Plasmodium falciparum*-infected human erythrocytes. The technique now has been used in the characterization of several integral membrane proteins of plasmodia species, including MSP2, AMA1, MSP4, and MSP5 *(4–8)*.

The Triton X series are nonionic detergents consisting of a hydrophobic *p-tert*-octyl phenol group attached to a hydrophilic polyoxyethylene head group (**Fig. 1**). These detergents form a homogenous solution of micelles when dispersed in water at 0°C, but as the temperature is raised, the micelle size increases, and the solution turns turbid at a characteristic temperature known as the cloud point *(9)*. Above this temperature, phase separation occurs in the solution due to aggregation of detergent micelles, and this results in a detergent-rich phase and a detergent-depleted or aqueous phase. Within the Triton X series, the temperature of the cloud point depends mainly on the number of hydrophobic oxyethylene units in the detergent. Thus, the cloud point of the commonly used detergent Triton X-100 is 64°C, whereas that of Triton X-114 is at a more convenient temperature of 20°C *(9)*. This temperature is suitable for the reversible condensation and separation of the detergent under conditions compatible with the isolation of native proteins *(1)*.

Integral membrane proteins are characterized by a hydrophobic domain that interacts with the hydrophobic core of the lipid bilayer. When a membrane system is solublized in Triton X-114, the nonionic detergent can replace most lipid molecules that interact with the hydrophobic domain of integral membrane proteins and lead to the formation of soluble protein–detergent mixed micelles *(1)*. In contrast to integral membrane proteins that have an amphiphilic structure, hydrophilic proteins show little or no hydrophobic interaction with Triton X-114 and their physicochemical properties are not affected. Under conditions promoting the condensation of Triton X-114, the

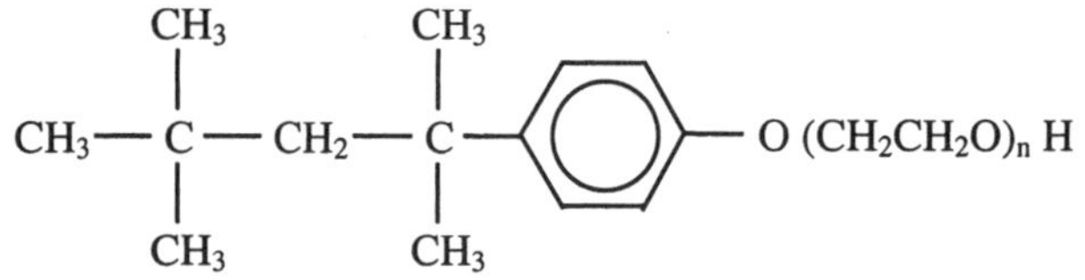

Fig. 1. Structure of the Triton X series. n represents the average number of ethylene oxide units per molecule, which are 9.5, 7.5, and 5.0 for Triton X-100, Triton X-114, and Triton X-45, respectfully.

mixed micelles formed by the membrane proteins and detergent aggregate with the rest of the detergent, while the hydrophilic proteins remain in the aqueous phase.

In the protocol described in this chapter, *Plasmodium*-infected erythrocytes are first solublized in an ice-cold Triton X-114 solution (*see* **Note 1**). The detergent solution mixture is then warmed and separates into two protein-containing phases. The upper aqueous phase contains hydrophilic proteins whereas membrane proteins partition to the lower detergent-enriched phase. Both phases are then analyzed to clarify the group to which the proteins of interest belong. Precondensation of Triton X-114, necessary to remove hydrophilic contaminants before partitioning, is also described.

2. Materials

1. Parasitized erythrocytes, isolated parasites, or other source of parasite proteins.
2. Triton X-114 detergent (Sigma Chemical Co.).
3. Phosphate-buffered saline (PBS): 137 mM NaCl, 2.7 mM KCl, 4.3 mM Na$_2$HPO$_4$, 1.4 mM KH$_2$PO$_4$, pH ~7.3.
4. Sucrose cushion: 6% (w/v) sucrose, 0.06% (w/v) Triton X-114 in PBS.
5. Protease inhibitors (optional).

3. Methods

3.1. Precondensation of Triton X-114

See **Note 2**.

1. Dissolve 1 mL of commercially purchased Triton X-114 in 100 mL of ice-cold PBS with continuous mixing at 4°C for 1 h.
2. Warm the solution to 37°C for 6 h until two phases separate.
3. Remove and discard the upper aqueous phase (~90 mL) and replace with an equal volume of ice-cold PBS.
4. Repeat this condensation procedure two more times.

The final detergent-enriched phase contains ~12% (w/v) detergent and is used as a stock (store at 4°C) to produce all subsequent solutions with PBS.

3.2. Extraction and Phase Separation of Plasmodium Parasites

1. Harvest parasitized erythrocytes and wash three times in PBS. Centrifuge to ~1 mL packed-cell volume (*see* **Notes 3** and **4**).
2. Add 15 mL of 0.5% (w/v) ice-cold Triton X-114 (*see* **Note 5**) and vortex to solubilize the cells. Keep on ice for 90 min and vortex every 10 min.
3. Remove an aliquot (0.1–1.0 mL) of the lysate fraction and freeze at –70°C for later analysis.
4. Centrifuge the remaining lysate at 10,000g for 15 min at 4°C to pellet insoluble materials. Transfer the supernatant to a fresh tube and keep on ice for phase partitioning.

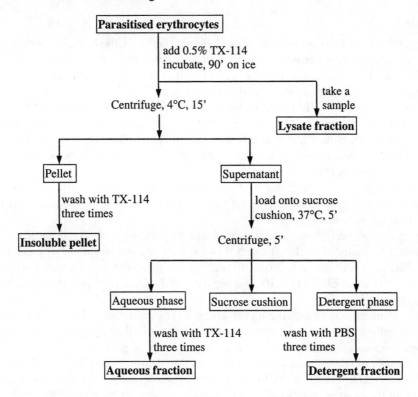

Fig. 2. Flow chart of the Triton X-114 phase partitioning procedure.

5. If the pellet is required for analysis, wash by resuspending in 10 mL of 0.5% (w/v) ice-cold Triton X-114, incubate on ice for 10 min and centrifuge for 15 min at 4°C. Repeat this step three times and freeze the final pellet at –70°C.
6. Layer the supernatant from **step 4** over a 10 mL cushion of chilled sucrose and place in a 37°C water bath for 5 min. The solution will go cloudy at this stage.
7. Centrifuge the solution at 500g for 5 min at 37°C. Carefully remove the upper aqueous layer and store on ice. Then remove the sucrose cushion and discard it. The detergent phase will be evident as an oily droplet at the bottom of the tube (*see* **Note 6**).
8. Resuspend the detergent pellet in 10 mL of ice-cold PBS and redissolve on ice over a 10-min period. Then warm the solution to 37°C for 5 min followed by centrifugation for 5 min. Discard the aqueous phase and retain the detergent pellet.
9. Repeat **step 8** three times to deplete any hydrophilic proteins from the detergent phase. The detergent phase can be resuspended in whatever volume is required with PBS and frozen at –70°C.
10. To obtain an aqueous phase containing only hydrophilic molecules, add 500 µL of the stock Triton X-114 to the aqueous phase from **step 7** and allow to dissolve by placing on ice for 10 min, then incubate at 37°C for 5 min followed by centrifugation for 5 min. Collect the supernatant and discard the detergent pellet.
11. Repeat **step 10** three times and freeze the final aqueous phase at –70°C.
12. Analyze the aqueous and detergent fractions (from **steps 11** and **9**, respectively) as well as the lysate fraction from **step 3** and the insoluble pellet from **step 5** for the protein(s) of interest (*see* **Notes 7–12**).

A diagram of the Triton X-114 phase partitioning procedure is shown in **Fig. 2**.

4. Notes

1. As Triton X-114 is a nondenaturing detergent *(2)*, it is likely that a number of assays dependent on native protein structure will be feasible after phase separation.
2. Commercial preparations of Triton X-114 often contain hydrophilic contaminants and should be precondensed by several rounds of phase separation before use.
3. The starting material described in the protocol is *Plasmodium*-infected erythrocytes. If the protein(s) of interest is associated with the parasite and other proteins of erythrocyte origin cause background in the detection procedure, purified parasites isolated from erythrocytes should be used instead of parasitized erythrocytes. This may be achieved by saponin lysis procedure *(10)*.
4. The starting volume described in the protocol is 1 mL of packed cells. The procedure can be scaled up or down depending on the requirement as long as the ratio of detergent to cells is consistent.
5. If an unexpectedly low yield or degradation of proteins is observed, protease inhibitors should be included in the Triton X-114 solution and the washing PBS. For example, a cocktail of phenylmethylsulfonyl fluoride (1 mM), leupeptin (0.5 µg/mL), trypsin inhibitor (10 µg/mL), EDTA (1 mM), and pepstatin (0.7 µg/mL) can provide protection against many proteases.
6. A convenient way of visualizing the Triton X-114 detergent pellet is by adding a drop of bromophenol blue to the lysate before separation *(3)*. The dye only associates with the detergent micelles and does not affect the properties of protein.
7. A number of methods are available for detecting the proteins of interest in a given phase. Proteins can be visualized by SDS-gel electrophoresis and Coomassie blue or silver staining, or they can be identified by immunoprecipitation or immunoblotting using specific antibodies. In addition, enzymes can be detected by their known activities.
8. A number of analysis and preparative procedures can be performed with the membrane proteins still in the detergent phase. However, if the analysis is complicated by the presence of Triton X-114, proteins can be precipitated from the detergent with acetone and resolublized in PBS *(2)*.
9. Although Triton X-114 phase partitioning is a reliable method for separating membrane proteins and hydrophilic proteins, it should be noted that a single membrane-spanning domain or hydrophobic group might not be sufficient to retain a large protein in the detergent phase. In general, this procedure appears to be most effective for proteins smaller than approx 80 kDa *(11)*.
10. The Triton X-114 partitioning procedure can be used as a means of identifying glycosylphosphatidylinositol (GPI)-anchored membrane proteins. To do that, the detergent phase can be treated with specific enzymes or chemicals to cleave the GPI-anchor and then subjected to a second round of phase separation. Once a GPI-containing protein is released from the lipid component of its anchor, either by GPI-specific enzymatic digestion or chemical cleavage, it will no longer partition into the detergent-enriched phase *(12–14)*. This alteration of partitioning behavior provides a rapid assay for the presence or absence of a GPI anchor and allows identification of GPI-anchored proteins without detailed structural analysis. One caveat is that some preparations of phospolipase are contaminated by proteases, thus it would be prudent to include protease inhibitors in the protocol if this type of analysis is planned.
11. The Triton X-114 phase partitioning can also be used as a highly effective method to remove endotoxin from solutions containing water-soluble proteins *(15)* or DNA used for immunization *(16)*.
12. The Triton X-114 solution should be handled below the cloud point to avoid inadvertent phase separation. When desired, higher cloud points can be achieved by mixing Triton X-

114 with Triton X-100 *(2)*. On the other hand, if the protein of interest is thermolabile and manipulation at a temperature above the cloud point of Triton X-114 is detrimental to its biological activity, the phase separation procedure can be performed at lower temperatures by using mixture of Triton X-114 and Triton X-45 *(2)*.

References

1. Bordier, C. (1981) Phase separation of integral membrane proteins in Triton X-114 solution. *J. Biol. Chem.* **256,** 1604–1607.
2. Brusca, J. S. and Radolf, J. D. (1994) Isolation of integral membrane proteins by phase partitioning with Triton X-114. *Meth. Enzymol.* **228,** 182–193.
3. Smythe, J. A., Murray, P. J., and Anders, R. F. (1990) Improved temperature-dependent phase separation using Triton X-114: Isolation of integral membrane proteins of pathogenic parasites. *J. Meth. Cell. Molec. Biol.* **2,** 133–137.
4. Smythe, J. A., Coppel, R. L., Brown, G. V., Ramasamy, R., Kemp, D. J., and Anders, R. F. (1988) Identification of two integral membrane proteins of *Plasmodium falciparum*. *Proc. Natl. Acad. Sci. USA* **85,** 5195–5199.
5. Marshall, V. M., Silva, A., Foley, M., Cranmer, S., Wang, L., McColl, D. J., Kemp, D. J., and Coppel, R. L. (1997) A sec merozoite surface protein (MSP-4) of *Plasmodium falciparum* that contains an epidermal growth factor-like domain. *Infect. Immun.* **65,** 4460–4467.
6. Crewther, P. E., Culvenor, J. G., Silva, A., Cooper, J. A., and Anders, R. F. (1990) *Plasmodium falciparum*: two antigens of similar size are located in different compartments of the rhoptry. *Exp. Parasitol.* **70,** 193–206.
7. Black, C. G., Wang, L., Hibbs, A., Werner, E., and Coppel, R. L. (1999) Identification of the *Plasmodium chabaudi* homologue of the merozoite surface protein 4 and 5 of *Plasmodium falciparum*. *Infect. Immun.* **67,** 2075–2081.
8. Wu, T., Black, C. G., Wang, L., Hibbs, A. R. and Coppel, R. L. (1999) Lack of sequence diversity in the gene encoding merozoite surface protein 5 (MSP5) of *Plasmodium falciparum*. *Mol. Biochem. Parasitol.* **103,** 243–250.
9. Pryde, J. G. (1986) Triton X-114: a detergent that has come from the cold. *TIBS* **11,** 160–163.
10. Rosenthal, P. J. (1995) *Plasmodium falciparum* - effects of proteinase inhibitors on globin hydrolysis by cultured malaria parasites. *Exp. Parasitol.* **80,** 272–281.
11. Wang, L., Black, C. G., Marshall, V. M., and Coppel, R. L. (1999) Structural and antigenic properties of merozoite surface protein 4 of *Plasmodium falciparum*. *Infect. Immun.* **67,** 2193–2200.
12. Hernandez-Munain, C., Fernandez, M. A., Alcina, A., and Fresno, M. (1991) Characterization of a glycosyl-phosphatidylinositol-anchored membrane protein from *Trypanosoma cruzi*. *Infect. Immun.* **59,** 1409–1416.
13. Bordier, C., Etges, R. J., Ward, J., Turner, M. J., and Cardoso de Almeida, M. L. (1986) *Leishmania* and *Trypanosoma* surface glycoproteins have a common glycophospholipid membrane anchor. *Proc. Natl. Acad. Sci. USA* **83,** 5988–5991.
14. Murray, P. J., Spithill, T. W., and Handman, E. (1989) The PSA-2 glycoprotein complex of *Leishmania major* is a glycosylphosphatidylinositol-linked promastigote surface antigen. *J. Immunol.* **143,** 4221–4226.
15. Aida, Y. and Pabst, M. J. (1990) Removal of endotoxin from protein solutions by phase separation using Triton X-114. *J. Immunol. Methods.* **132,** 191–195.
16. Boyle, J. S., Koniaras, C., and Lew, A. M. (1997) Influence of cellular location of expressed antigen on the efficacy of DNA vaccination: cytotoxic T lymphocyte and antibody responses are suboptimal when antigen is cytoplasmic after intramuscular DNA immunization. *Int. Immunol.* **9,** 1897–1906.

56

Immunoprecipitation for Antigen Localization

John G. T. Menting and Ross L. Coppel

1. Introduction

Immunoprecipitation was introduced by Schwartz and Nathenson *(1)* who used the procedure to isolate radioactively labeled antigens extracted by nonionic detergent. The procedure selectively precipitates an antigen(s) of interest using antibodies as a specific selection component of a precipitating complex. In this way the antigens are isolated and resolution and detection of the antigen by sodium dodecyl sulfate-polyacrylamide gel electrophoresis (SDS-PAGE) is improved by removal of other cellular components. A notable feature of immunoprecipitation protocols used in malaria research is the wide variety of assay conditions used. This reflects the robust nature of this procedure in that antibody to antigen complexes form readily under a variety of conditions. Immunoprecipitation protocols have usually followed the following format: antigens are liberated from parasites by extraction or solubilization with detergent to give a total cellular extract. The soluble extract containing an antigen of interest is incubated with a specific antibody, or perhaps a polyspecific antibody, and a complex of antigen and antibody forms. The complex is sedimented by centrifugation following binding of protein A-Sepharose or fixed *Staphylococcus aureus* cells *(2)* to the Fc portion of the antibody. Alternatively, antigen can be sedimented directly by binding of specific antibody conjugated directly to chromatography beads *(3)*. Precipitated antigens are separated by SDS-PAGE or two-dimensional electrophoresis. Antigens are then detected either by autoradiography for metabolically labeled antigens or conventional immunoblot analysis for non-radioactive antigens. Additional useful procedures and comments are to be found in **refs. 4** and **5**.

There are several key steps in immunoprecipitation.

1. Antigen extraction: The purpose of detergent extraction is to liberate the antigen of interest, whether it is a hydrophilic cytosolic protein or a very aqueous-insoluble integral membrane protein. Following release of the protein into aqueous extraction medium the protein must remain soluble, otherwise the protein may be lost following centrifugation to remove insoluble material. Adequate antigen solubility is also essential to ensure that precipitation is achieved exclusively via the specific antibody. Poor solubility may cause the protein to precipitate spontaneously or to bind nonspecifically to the precipitating reagents. Extraction is generally achieved with a detergent that is only slightly chaotropic or denaturing to proteins (general descriptions of detergent properties and applications are given

in **refs. 6** and **7**). An example of this type of detergent is Nonidet P40, which is essentially the same as Triton X100 *(7)*. Use of this type of detergent minimizes structural and functional alterations of antigen and antibody. For extraction of malaria parasites, extraction buffers usually contain 1% w/v Triton X100, other detergents such as sodium deoxycholate, or a combination of the two (examples are given in **refs. 8–11**). Detergent properties may not be critical where retention of biological function or native antigen structure is unimportant, but the choice of detergent can be critical if there are considerations such as coprecipitation of bound protein partners where the ability to bind must be preserved. We need to consider both the benefits and adverse effects of each detergent. On the one hand, a detergent could decrease the binding between extracted proteins; on the other hand, if the detergent fails to maintain protein solubilities proteins may be precipitated uselessly by nonspecific aggregation. A detergent is required to maintain antigen solubility, minimize nonspecific interactions, maintain three-dimensional structure and preserve native specific interactions. Therefore different experimental requirements determine the need for particular detergent properties. This is exemplified in a comparison of detergents used to extract surface proteins for immunoprecipitation of *Plasmodium knowlesi* surface antigens *(8)*. This study has revealed: (a) detergents solubilize a given antigen to different extents; (b) conditions required to solubilize one antigen component may be inefficient for another; and (c) detergents that efficiently solubilize a membrane component may adversely affect structures recognized by antibodies. It is clear that empirical comparisons of detergents are desirable if not essential. However, in some cases chaotropic detergents must be used to effect solubilization. Such unwanted properties of detergents can be minimized by diluting the detergent following solubilization or extraction *(12)*. With some antigens efficient extraction of antigen (such as an intrinsic protein) can only be achieved with a chaotropic detergent such as sodium dodecyl sulfate (SDS) *(8,13,14)*. In that case, SDS in an initial extraction should be subsequently diluted by addition of a buffer containing a less chaotropic detergent. Addition of a less chaotropic detergent will maintain antigen solubility, prevent antibody denaturation and possibly reverse some secondary or tertiary structure denaturation.
2. Antibody binding: To be useful in immunoprecipitation, an antibody must bind to an antigen with sufficiently great specificity. It is important that an antibody is tested in immunoprecipitation and immunoblot assays to determine whether the level of unwanted extraneous binding is acceptable. Antibodies exhibiting unwanted binding are to be avoided, as a precipitated antigen may only have been caught up in the precipitating complex by low-affinity nonspecific binding or by relatively high-affinity crossreactivity. Extensive use of controls must be made in initial experiments to examine these possibilities. Specific antibody must also bind to antigen sufficiently strongly to resist release during subsequent manipulations, especially the washes. Although antibody-to-antigen binding is usually strong, with dissociation constants in the range 10^{-5}–10^{-10} M *(4)*, it is generally true that the affinity of monoclonal antibodies for a given antigen is less than for a polyclonal antibody simply because polyclonal antibodies will usually contain a mixture of high- and low-affinity species. Because they are monospecific, monoclonal antibody binding is more likely to be adversely affected by alterations in antigen structure following extraction from cells. Polyclonal antibodies are therefore more likely to succeed in an immunoprecipitation, although a mixture of monoclonal antibodies could be just as effective. Lower affinity antibodies may be used if assay conditions can be adjusted to favor the formation of antigen–antibody complexes with weaker-binding antibodies. This can be done by adding excess antibody, by reducing the assay volume to effectively increase reagent concentrations, by incubating at a higher temperature such as 37°C, or by increasing incubation times, overnight if necessary.

3. Precipitation of antigen: While mixtures of antigen and antibody have been used to form precipitating complexes, these aggregates are slow to form and antibody–antigen mole ratios take time to optimize *(2)*. Incubation periods for precipitation are reduced from typically overnight to 1-h incubations by means of a solid-phase precipitating reagent. The precipitating reagent must bind strongly to the chosen antibody to enable the complexed antigens to be isolated and must be readily precipitated by centrifugation. Protein A-Sepharose or protein G-Sepharose are insoluble antibody-binding chromatography beads and used because they bind rapidly and with high affinity to the Fc domain of IgG antibodies *(2)*. Earlier studies also used fixed *Staphylococcus aureus* cells *(2,15,16)* which sediment readily and have protein A expressed on the surface.

There are many applications of immunoprecipitation in research on malarial antigens.

1. Metabolic labeling of parasites with radiochemicals such as [^{35}S]-methionine or [^{3}H]-leucine incorporated into cellular proteins *(11,17)*. This procedure has been used frequently to determine whether a particular antigen is of parasite or host origin (e.g., **ref. 18**). Radiolabeling of parasite proteins can provide vsery high sensitivity of antigen detection and only the antigens, not the added assay components, will be detected. Analysis of immunoprecipitated parasite extracts following pulse-chase labeling of antigens can clearly show the kinetics of immature and mature gene products during the asexual life cycle *(19)*. The major problem encountered with metabolic labeling is that a precipitated antigen detected by autoradiography may not have been the immunodetected protein, but rather a binding partner or a crossreactive antigen. In some cases, however, this may be an advantage where the binding partner is also of interest. The question of which antigen is precipitated by the procedure can be further addressed by carrying out nonradioactive immunoprecipitation with immunoblot detection (*see below*). Specific antigens can then be probed for. Alternatively more aggressive detergents could be tried in order to separate the complexed antigens.

    The merits of different radiolabeled amino acids for biosynthetic incorporation into proteins have been addressed in **refs. 20** and **21**. In general, L-[^{35}S]methionine and L-[4,5-^{3}H]leucine are the radioisotopically-labeled compounds used for metabolic labeling studies in plasmodial species *(9,10,17)*. This is mainly because these compounds have high specific activities and are relatively inexpensive. Other radiolabeled amino acids such as L-[^{3}H]isoleucine *(3)* and L-[^{3}H]histidine *(22)* have also been used. Fluorographic enhancers must be used for detection of [^{3}H]labeled proteins separated by SDS-PAGE, therefore L-[^{35}S]methionine is simpler to use. However, proteins vary greatly in their methionine content with some lacking methionine, necessitating the use of another radiolabel. Radiolabeled antigens can be detected by autoradiography following SDS-PAGE size-dependent separation. Where many metabolically labeled antigens are precipitated, they can be separated into highly resolved patterns by two-dimensional electrophoresis *(23)* that separates according to isoelectric point and size. This type of labeling is useful for examining stage-specific expression and coprecipitation of antigens such as MESA *(24,25)*.

2. Radiolabeled surface proteins: Cell surface proteins can be selectively labeled using [^{125}I] surface-labeling techniques *(26)*. In this way, only the surface protein fraction will be detected in an immunoprecipitation experiment. Antigens precipitated from such a preparation may be separated by SDS-PAGE and detected by autoradiography. However surface radioiodination may alter the isoelectric point of antigens by modifying free amines, thus isoelectric focus analysis or two-dimensional electrophoresis data must be interpreted with caution.

3. Immunoblot analysis of nonradiolabeled antigens; detection of immunoprecipitated antigen on immunoblot by antibody probes. There are two basic types of immunoprecipitation protocol for which different immunoblot detection procedures can be used.

a. The first uses soluble antibody and a solid phase such as protein G-Sepharose to rapidly bring about separation of antigen from the liquid phase. A disadvantage of this method is that antibodies are present in the sample following elution from protein G-Sepharose and for nonradioactive detection of antigen on immunoblot requires the use of an antibody from a different host in order to prevent detection of the first antibody. This problem can be circumvented, if desired, by use of [^{125}I]labeled antibody detection on immunoblots.
b. The second procedure covalently couples the antibody to the solid phase that means that no free antibody will be present. Antigens when removed from precipitated beads, will be free of antibody, although harsh, denaturing elution conditions may cause antibody chains to separate and elute from the beads. This method makes analysis of nonradiolabeled precipitated products easier and reduces background resulting from nonspecific binding of detecting antibodies to precipitating antibody. These immunoblots can be probed with antibodies raised in any species, as there will be no antibodies with which secondary antibody or IgG conjugates will crossreact.

4. Multiprotein complexes: Coprecipitation has been used to identify the coprecipitating antigens in multiprotein complexes such as in the rhoptries *(25,27)*.
5. Immunoprecipitation of functionally radiolabeled antigens: Antigens can be labeled by more specific means than metabolic or surface radiolabel incorporation. For example, immunoprecipitation can be used to identify whether specific antigens have been phosphorylated by endogenous kinases (examples are MESA, RESA, and Pgh-1 *[24,27–29]*) or covalently labeled by specific photoactivatable substrate analogues *(30)*.
6. Immunoprecipitation of biologically active antigen: If the assay conditions are mild, biological activity such as protease activity will be retained *(31)*. This enables us to functionally characterize the immunoprecipitated antigen.
7. Covalent crosslinking of associated antigens: Proteins associating weakly with the antigen of interest can be coprecipitated following covalent crosslinking *(32)*. Care must be taken with this method to minimize nonspecific crosslinking of proteins.
8. Concentration of low-abundance antigens: Immunoprecipitation can be used to concentrate an antigen of interest prior to immunoblot analysis. In whole extracts run on SDS-PAGE the relative amount of an antigen may be so low as to preclude detection or electrophoretic separation of an antigen may be affected by overloading of the gel. Adsorption of an antigen to a solid phase and sedimentation by immunoprecipitation will enable larger amounts of the antigen to be analyzed by SDS-PAGE and/or immunoblot analysis *(33)*.
9. Assessment of polymorphic antigens: Immunoprecipitation has been used to assess the extent of antigenic polymorphisms *(18,34,35)*.
10. Immunomagnetic precipitation: Antibodies can be adsorbed to magnetic beads, Dynabeads® (Dynal, Norway), either directly or via protein A. This procedure is most commonly used for the selection of whole cells rather than soluble antigens. An interesting application of this technique has been to use isolated Plasmodium falciparum-infected erythrocytes as a source of DNA for PCR *(36)*.

Protocols included in this chapter detail facile methods for preparation of metabolically labeled parasite antigens, purification of antibodies, covalent coupling of antibody to Sepharose beads and immunoprecipitation protocols. It is assumed that operators intend to use saponin-purified parasites; however, the procedures are readily adapted for intact infected erythrocytes. Two immunoprecipitation protocols are included. The first uses soluble antibody and protein G-Sepharose solid phase to rapidly bring about separation of antigen from the liquid phase. This procedure has the

Immunoprecipitation

advantage that it can be carried out using unpurified antibodies in antisera, hybridomal cell culture supernatant, or ascites fluid. The procedure is therefore easy and is a good method to try first. For the second procedure antibodies must be purified and covalently coupled to the solid phase. This method has the disadvantage that more work is required to produce the reagents and antibody function and specificity could be compromised by the purification and coupling processes. However, once the reagents are made, the method is fast as only one immunoprecipitation reagent is required.

2. Materials
2.1. Preparation of Parasites and Parasite Lysates
2.1.1. Equipment for Parasite Preparation

1. Culture facilities at an appropriate level of safety will need to be available or parasite material may be obtained from collaborators. A detailed method for in vitro culture of *Plasmodial* species is given in Chapter 44.
2. Benchtop centrifuge capable of holding 10-mL centrifuge tubes and delivering up to $1500g$.
3. Microcentrifuge delivering up to $13,000g$.
4. Side-arm flask and filtration apparatus.
5. Scintillation counter.

2.1.2. Reagents and Disposable Materials for Parasite Preparation

1. L-[^{35}S]methionine, >1000 Ci/mmol^{-1} (Amersham, UK).
2. Giemsa stain.
3. Filters, 0.2 µm (Millipore).
4. Filters, 0.45 µm HA (Millipore).
5. Scintillation fluid (Sigma).

2.1.3. Solutions for Parasite Preparation

1. PBS: Prepare from tablets (Sigma).
2. Filter-sterilized 5% sorbitol. Filter the solution through a 0.2-µm filter under aseptic conditions in a laminar flow or Class II Biohazard cabinet.
3. TCA, 5 and 10% (*see* **Note 1**).
4. Carrier solution for L-[^{35}S]methionine experiments. Add 50 mg of bovine serum albumin (BSA) (or 100 mL of fetal calf serum), 40 mg of L-methionine, 50 µL of 20% sodium azide to 15 mL of PBS, dissolve and make the volume up to 20 mL, storing at 4°C. For labeling with other radioactive amino acids, use the appropriate "cold" amino acid instead of L-methionine. For [^{125}I]labeled samples use potassium iodide (6 mg in 20 mL).
5. Autoclaved or filter-sterilized 0.15% saponin (Sigma). Autoclave a solution of 0.15% saponin in PBS or filter the solution through a 0.2-µm filter under aseptic conditions in a laminar flow hood or Biohazard hood.
6. Filter-sterilized methionine-free culture medium.

2.2. Purification of Antibodies and Preparation of Antibody Immobilized on Sepharose Beads
2.2.1. Equipment for Antibody Preparation

1. Rotating mixer capable of 5 to 100 rpm (Ratek, Australia).
2. Benchtop centrifuge capable of holding 10-mL and 50-mL centrifuge tubes and delivering $1500g$.

3. 15-mm Diameter Econo-Pac column (Bio-Rad) clamped vertically 10–15 cm above the bench.
4. 8 mm, 2 mL Poly-Prep column (Bio-Rad).
5. Ultrafree-15 centrifugal concentrator (Millipore).
6. UV spectrophotometer for OD280 measurements.

2.2.2. Reagents and Disposable Material for Antibody Preparation

1. PBS-washed GammaBind G-Sepharose. Uniformly suspend a 50% slurry of GammaBind G-Sepharose (supplied by Pharmacia, Sweden in PBS plus 20% ethanol) by inversion and shaking (*see* **Note 2**) and add twice the required settled volume of beads to a 50-mL microcentrifuge tube (*see* **Note 3**). Collect the beads by centrifugation for 10 min at 1500g and remove most of the supernatant. Add five times the settled volume of PBS buffer and completely suspend the beads by inversion and shaking. Repeat the centrifugation and discard the supernatant. Wash the beads three times in this way and suspend the pelleted beads as a 50% suspension with one settled volume of PBS.
2. CNBr-activated Sepharose beads (Pharmacia, Sweden).
3. Dialysis tubing (e.g., Spectra/Por, Spectrum).

2.2.3. Solutions for Antibody Preparation

1. PBS (Sigma).
2. 200 mM acetic acid–NaOH, pH 2.8.
3. 1 M Tris-HCl, pH 8.0.
4. 200 mM sodium acetate-acetic acid, pH 4.0.
5. 0.1 M NaHCO$_3$, 0.5 M NaCl, pH 7.9; the pH should not need to be adjusted.
6. 1 M glycine–NaOH, pH 8.0.
7. PBS containing 0.05% sodium azide.

2.3. Immunoprecipitation Protocols

2.3.1. Equipment for Immunoprecipitation

1. Microcentrifuge capable of delivering approx 13,000g.

2.3.2. Reagents for Immunoprecipitation

1. Washed protein G–Sepharose, a 50% slurry: To prepare, uniformly suspend a 50% slurry of GammaBind G–Sepharose (supplied by Pharmacia in PBS plus 20% ethanol) by inversion and shaking (*see* **Note 2**) and add the required settled volume of beads to a 1.5-mL microcentrifuge tube (*see* **Note 3**). Collect the beads by centrifugation for 30 s at maximum speed in a microcentrifuge and remove most of the supernatant. Add 10X settled volume of HNET-PI buffer and completely suspend the beads by flick mixing and inversion. Repeat the centrifugation and discard the supernatant. Wash the beads three times in this manner and suspend the pelleted beads as a 50% suspension with 1X settled volume of HNET-PI.

2.3.3. Solutions for Immunoprecipitation

1. PBS (Sigma).
2. HNET buffer: 25 mM HEPES-NaOH, 150 mM NaCl, 1 mM EDTA, 0.5% Triton X100, pH 7.4.
3. HNET-PI buffer: HNET plus 1:1000 (v/v) stocks of protease inhibitors: pepstatin A (0.7 mg/mL in dimethyl sulfoxide [DMSO]), leupeptin (0.5 mg/mL), aprotinin (1 mg/mL), egg white trypsin inhibitor (10 mg/mL), benzamidine (1 M), and phenylmethylsulfonyl

Immunoprecipitation

fluoride (200 mM in propan-2-ol). Pepstatin A, leupeptin, and aprotinin are obtained from ICN; the remainder are obtained from Sigma.

4. 200 mM acetic acid-NaOH, pH 2.8.
5. 1 M Tris-HCl, pH 8.0.
6. Concentrated 3X nonreducing SDS-PAGE sample loading buffer. Mix 1.875 mL of 1.0 M Tris-HCl, pH 6.8, 3.0 mL of glycerol, 1.0 mL of 0.05% (w/v) bromophenol blue, and 0.60 g of SDS. Add pure water to bring the total volume to 10 mL (after the SDS has dissolved). Use 1 vol of buffer to 2 vol of sample and heat at 90–100°C for about 5 min.

2.4. Analysis of Precipitated Antigens by SDS-PAGE, Autoradiography, and Immunoblot Analysis

2.4.1. Equipment for Antigen Analysis

1. Mini-Protein II apparatus (Bio-Rad) and power supply for running small SDS-PAGE.
2. Mini-Transblot apparatus (Bio-Rad,) and suitable power supply.
3. Vacuum gel dryer (e.g., Bio-Rad).
4. X-ray film (e.g., Fuji, Japan).
5. X-ray film cassette (NEN).

2.4.2. Reagents and Disposables for Antigen Analysis

1. Radioactive protein molecular weight markers from NEN or Amersham.
2. Prestained protein molecular weight markers (e.g., SeeBlue markers, Novex).
3. PVDF membrane (e.g., Polyscreen PVDF membrane NEN).
4. Western transfer buffer.
5. Whatman 3MM paper (Whatman International, England).
6. Anti-(primary antibody host-species) immunoglobulin HRP conjugate (Silenus, Australia; or Sigma, USA).
7. Chemiluminescent detection kit (e.g., Renaissance, NEN).
8. Tracker tape (Amersham).

2.4.3. Solutions for Antigen Analysis

1. Concentrated 3X nonreducing SDS-PAGE sample loading buffer. *See* **Subheading 2.3.3.**
2. PBS (Sigma).
3. PBS-T: PBS containing 0.05% Tween-20.
4. Blocking solution (5S-PBS-T): PBS-T containing 5% skim milk powder.

3. Methods

3.1. Preparation of Parasites

Parasites obtained from field isolates or from cultured parasites can be used. A detailed method for in vitro culture of plasmodial species is given in Chapter 44.

3.1.1. Synchronization of Plasmodium falciparum with Sorbitol

Sorbitol lyses trophozoites and schizonts because these cells are osmotically intolerant, however, ring-stage parasites are left intact. This procedure can convert asynchronous or mixed-asexual stage cultures into synchronous or single-stage cultures *(37)*. The parasites remain viable; however, the period of the subsequent cycle may be longer than usual as the cells recover from the treatment. This procedure must be carried out under aseptic conditions where parasites are to be maintained in continuous culture. To obtain well-synchronized cultures the parasites should be sorbitol-treated again after another growth cycle.

1. Examine Giemsa-stained smears of *Plasmodium falciparum* cultures and select cultures when they contain the most ring-stage parasites.
2. Collect the cells by centrifugation at 500–750g and discard the supernatant.
3. Suspend the cell pellet in 5 vol of 5% sorbitol prewarmed to 37°C.
4. Incubate at 37°C for 5 min.
5. Centrifuge as in **step 2** to pellet the cells and discard the supernatant.
6. Return the cells to culture by suspending them in complete culture medium and add to a fresh culture flask.

3.1.2 Metabolic Labeling of In Vitro Cultured Plasmodium falciparum by Radioactive L-[^{35}S]Methionine Incorporation

Plasmodium falciparum proteins are usually biosynthetically labeled with radioactively labeled amino acids such as L-[^{35}S]methionine. Incorporation of the radiolabel is achieved by incubation with L-[^{35}S]methionine for 2–6 h for asynchronous cultures and 0.25–2 h for synchronous cultures. In order to bring about the highest degree of L-[^{35}S]methionine incorporation possible the culture medium must be deficient in nonradioactive L-methionine. However, the presence of nonradioactive L-methionine in serum, albumax or erythrocytes does not appear to present a problem as many protocols used for metabolic labeling do not include extensive dialysis of serum or erythrocytes to ensure that all nonradioactive L-methionine is absent. Preparation of metabolically labeled proteins in different asexual parasite stages requires the preparation of synchronized cultures as detailed in **Subheading 3.1.1**. All of the above procedures must be conducted under aseptic conditions (*see* **Note 4**).

1. Prepare a 10 mL culture of *Plasmodium falciparum* grown to 5% parasitemia in 2–3% hematocrit. If stage-specific labeling is required, the culture must be synchronous (*see* **Subheading 3.1.1.**). Monitor growth and parasite stages by microscopic examination of Giemsa-stained smears.
2. Collect the cells by centrifugation in a 10-mL plastic centrifuge tube at 500–750g for 5 min. Wash cells three times by suspending cells by slow inversion in 10 mL methionine-free medium and centrifuge as before to sediment cells. Discard the supernatant.
3. Transfer the cells to a fresh flask containing methionine-free medium supplemented with L-[^{35}S]methionine (*see* NOTE 5).
4a. For metabolic-labeling of asynchronous cultures, grow the parasites for 2 h in the presence of 0.2 mCi/mL L-[^{35}S]methionine.
4b. For stage-specific metabolic-labeling, take aliquots of a synchronous culture at various intervals. Asexual stage development can be determined by examination of Giemsa-stained smears. Time 0 is either the time when the parasite culture was synchronized or when invasion of erythrocytes commences. Incubate cells in culture with 60 µCi/mL of L-[^{35}S]methionine (*see* **Note 6**) for 1 h at the required interval or once the required asexual stage has developed.
5. Wash the cells three times in serum-free and methionine-free medium as in **step 2**.
6. Collect the cell pellet and store at –70°C.
7. If purified parasites are required, use the protocol detailed in **Subheading 3.1.4**.

3.1.3. Measurement of Radioactive Amino Acid Incorporation by TCA Precipitation

Comparison of immunoprecipitated metabolically labeled antigens in asexual stages is carried out on similar amounts of radioactively incorporated parasite material. Incorporated counts are measured by TCA precipitation of labeled parasites in the presence of carrier protein (*see* **Note 1**).

Immunoprecipitation

1. Pipet duplicate 1-μL samples of radiolabeled parasites in 0.5-mL carrier solution, and vortex.
2. Add 0.5 mL of 10% TCA, and vortex.
3. Incubate at room temperature for 30 min.
4. Filter through a Millipore HA 0.45-μm filter (*see* **Note 7**).
5. Rinse the filter with 10 mL 5% TCA.
6. Dry the membrane.
7. Place the dry membrane in 3 mL of scintillation fluid and count in a beta counter (*see* **Note 8**).

3.1.4. Purification of Parasites by Saponin Lysis

In vitro-cultured parasites can be removed intact from red blood cells using a detergent, saponin, which lyses erythrocyte membranes. Erythrocyte membranes collapse around parasites but saponin leaves parasite membranes intact (*see* **Note 9**). As a rough estimate, 1 mL of packed cells (total 1×10^{10} infected and uninfected cells) at 5% parasitemia will contain approx 5×10^8 parasitized erythrocytes and will yield 50 μL of packed purified parasites. It must be noted that this pellet will contain variable amounts of contaminating erythrocyte membrane and cytoskeleton.

1. Collect red blood cells from parasite cultures by centrifugation at 500–750g for 5 min at room temperature.
2. After removal of the medium, resuspend the cell pellet in 10 vol of 0.15% saponin in PBS (*see* **Note 10**). Incubate for 5 min at room temperature to lyse erythrocytes.
3. Centrifuge at 1000–1500g for 10 min and discard the supernatant; the supernatant will be dark red if hemolysis is successful.
4. Suspend the pellet in PBS to the volume used in **step 2**. It may be necessary to vortex fairly vigorously to obtain a uniform suspension. The pellet will be very much smaller than the original cell pellet.
5. Repeat the PBS washes twice (a total of three washes). After the second wash the pellet will be even smaller and will be black due to hemozoin sequestered in trophozoites and schizonts (*see* **Note 11**). The pellet will be almost completely free of hemoglobin.
6. Mix the parasites with a fine micropipette tip to uniformly mix them in any residual supernatant fluid and quickly dispense 1- to 5-μL aliquots into screw-capped microcentrifuge tubes and store at –70°C. Do not thaw the parasites and refreeze them prior to use, as proteases will be liberated and can degrade the protein fraction.

3.2. Purification of Antibodies and Preparation of Antibody Immobilized on Sepharose Beads

Antibody purification is time consuming but relatively easy. Antibody purification removes unwanted proteins from crude antibody preparations. **Subheading 3.2.1.** gives details of IgG monoclonal antibody purification from hybridomal supernatant. Purified monoclonal antibodies can be coupled to Sepharose as in **Subheading 3.2.3.** Alternatively if polyclonal antibodies are available as antisera, they can be affinity-purified as in **Subheading 3.2.2.** and coupled to Sepharose as in **Subheading 3.2.3.** Affinity purification of antibodies will give an antibody subset of the original population, lacking antibodies that are unstable under the elution conditions and lacking antibodies that bind to the affinity resin so tightly as to remain bound under the elution conditions. Purified antibodies may therefore have lower binding affinity and different reactivity compared to the original antiserum.

3.2.1. Purification of Monoclonal Antibodies

This protocol is a batch-column procedure that combines rapid and convenient batch-binding with efficient column washing and elution. Note that GammaBind G-Sepharose binds well to IgG subisotypes from all host species but not to IgA, IgE, or IgM. It is best to repeat the purification on an aliquot of culture supernatant to ensure highest possible yields; however, it is not essential and a single round of purification will usually result in good yields.

1. Mix culture supernatant (250 mL), which usually contains 2–25 mg monoclonal antibody, at room temperature with 5 mL of PBS-washed GammaBind G-Sepharose beads in a suitable bottle (*see* **Note 12**).
2. Maintain the beads in suspension by slow rotation at room temperature for 10 min.
3. Let the bottle stand to settle the beads under gravity for 10–15 min and collect in a 16-mm diameter column with an open top; adding the culture supernatant with a 10- or 25-mL graduated pipet. When only a little supernatant remains, suspend the beads and add them also (*see* **Note 13**).
4. Save the used supernatant for a second purification.
5. Wash the beads with 10 mL of PBS and discard the wash fraction.
6. Elute antibody with 8 mL of 200 mM acetic acid-NaOH, pH 2.8. The pH is immediately neutralized by collecting the eluate in a tube containing an equal volume of 1 M Tris-HCl, pH 8.0 and mixing.
7. Regenerate the column with 3 mL of 1.0 M Tris-HCl, pH 8.0, then 8 mL of PBS.
8. Seal the column drain and add 1 bed volume of the used culture supernatant or PBS. Suspend the beads by aspiration with a Pasteur pipet. Add the suspended beads to the used supernatant and repeat the purification (*see* **Note 14**).
9. Combine the antibody eluates and concentrate to 2 mL in a 30 kDa cutoff Ultrafree-15 centrifugal concentrator (2000g in a benchtop centrifuge).
10. The buffer can be replaced with PBS as follows: Leave the concentrated protein in the Ultrafree-15 concentrator and dilute to 15 mL with PBS. Mix by aspiration with a Pasteur pipet and concentrate to 2 mL. Carrying out this procedure three times will effectively replace the buffer.
11. Measure the amount of protein by OD280 measurement (*see* **Note 15**): Switch on a UV spectrophotometer and allow it to warm up for at least 10 min. Place 1 mL of antibody solution in a quartz cuvette and measure the absorbance against an appropriate buffer blank. A 1 mg/mL solution of IgG will have an absorbance of 1.43 *(38)*.

3.2.2. Affinity Purification of Rabbit Polyclonal Antibodies

All chromatography procedures may be conveniently carried out by flowing buffers through an antigen-Sepharose column under gravity.

1. Prepare approx 1 mL of antigen-Sepharose beads using the method detailed in **Subheading 3.2.3.** (*see* **Note 16**).
2. Suspend the beads in PBS and add to an 8-mm Poly-Prep column, removing the bottom cap to allow the buffer to drain.
3. Wash antigen-Sepharose beads three times with 3 mL of PBS.
4. Dilute 2.5 mL of a cognate rabbit immune serum 1 in 2 with PBS and flow through the antigen-Sepharose column under gravity to bind specific antibodies.
5. Wash the column with 4 mL of PBS.
6. Elute antibodies with 1 mL of 200 mM sodium acetate-acetic acid, pH 4.0 and collect into

Immunoprecipitation

      4 mL 1 M Tris-HCl, pH 8.0 to neutralize pH (*see* **Note 17**).
7. Elute with 3 mL of acetic acid-NaOH, pH 2.8, and combine with pH-neutralized eluate from **step 5**.
8. Regenerate the column with 1 mL of 1.0 M Tris-HCl, pH 8.0, then 4 mL PBS.
9. Repeat the purification (*see* **Note 14**).
10. Combine eluates and concentrate to 1 mL in a 30 kDa cutoff Ultrafree-15 centrifugal concentrator.
11. Exchange buffer for PBS as detailed in **Subheading 3.2.1., step 10**, or by dialysis (**Subheading 3.2.3.**).
12. Measure the antibody protein concentration as detailed in **Subheading 3.2.1., step 11**.

3.2.3. Covalent Coupling of Proteinaceous Antigen or Antibody to CnBr-Activated Sepharose Beads

1. Soak a 10- to 15-cm length of dry 10-mm wide dialysis tubing for 10 min in PBS (*see* **Note 18**).
2. Gently rub the membrane to separate the sides (wear gloves) and place a dialysis tubing closure on one end (*see* **Note 19**).
3. Add 2 mg of protein in 2 mL to the dialysis tubing with a Pasteur pipet taking care not to tear the membrane.
4. Seal the dialysis tubing with another closure removing as much air as possible.
5. Stir the sample gently in 2 L of 0.1 M NaHCO$_3$, 0.5 M NaCl, pH 7.9 for 1 h at 21°C. Replace the buffer twice (i.e., a total of three dialyses) incubating each time for 1 h (*see* **Note 20**).
6. Weigh 0.3 g of CNBr-activated Sepharose into a 10-mL plastic centrifuge tube.
7. Add 9 mL of 1 mM HCl and hydrate for 30 min at 21°C with occasional mixing.
8. Pellet beads at 3000 rpm for 10 min in a benchtop centrifuge. Remove the supernatant.
9. Wash the beads three times by adding 9 mL of 1 mM HCl, mixing, centrifuging as in the previous step to collect the beads and remove the supernatant.
10. Suspend the beads as a 50% slurry by adding 1 mL of 0.1 M NaHCO$_3$, 0.5 M NaCl, pH 7.9.
11. Add 2 mg of dialyzed protein from **step 5** and incubate with slow mixing on a rotating mixer for 1 h at 21°C (*see* **Note 21**).
12. Add 1.0 mL of 1 M glycine-NaOH, pH 8.0, and mix overnight at 4°C to block unreacted CNBr groups.
13. Collect the beads by centrifugation as detailed in **step 8** and discard the supernatant.
14. Wash the beads sequentially with 9 mL of each of the following buffers using centrifugation as in **step 8** to collect the beads and discard the supernatant:
    a. 0.1 M NaHCO$_3$, 0.5 M NaCl, pH 7.9.
    b. sodium acetate–acetic acid, 200 mM, pH 4.0.
    c. 0.1 M NaHCO$_3$, 0.5 M NaCl, pH 7.9.
    d. three times with PBS, 0.05% sodium azide (*see* **Note 22**).
15. Store the beads in PBS containing 0.05% sodium azide.

3.3. Immunoprecipitation Protocols

If antigens are to be detected by immunoblot analysis, try immunoprecipitation with 1 µL of saponin-purified parasites prepared as in **Subheading 3.1.4.**, corresponding to 1×10^6 to 1×10^7 parasites. In our experience, this quantity of parasites allows the detection of soluble and membrane-bound antigens expressed at relatively low levels. For radiolabeled parasite preparations, try 1×10^6 dpm per immunoprecipitation as a starting point. For metabolically labeled parasites at different asexual stages, the numbers of parasites and amount of parasite-specific antigens may vary. One could assume that rates of amino-acid incorporation are similar throughout the asexual life cycle, in which case immunoprecipitation of stage-specific samples containing equal amounts of incorporated counts can be equated to equal amounts of parasite protein. However, this is probably not true, and it is probably better to try to work with labeled material derived from equal numbers of parasites. This section details procedures for extraction of parasites and two immunoprecipitation procedures.

3.3.1. Detergent Extraction of Parasites

1. Add 30 µL of HNET-PI to 1 µL (packed cell volume) of purified parasites in a screw-capped 1.5-mL microcentrifuge tube (*see* **Note 23**).
2. Aspirate thoroughly using a micropipetor to break up the parasite pellet and lyse for 5 min at 37°C. Save 3 µL of whole lysate for further analysis, mix with 9 µL of HNET-PI (sample 1, *see* **Note 24**) and incubate under the same conditions as the immunoprecipitated sample.
3. Sediment the black insoluble material by centrifugation at full-speed (13,000*g*) for 5 min in a microcentrifuge and carefully remove the soluble supernatant fraction using a micropipettor (*see* **Note 25**).
4. To 25 µL of supernatant add 175 µL of HNET-PI (*see* **Notes 26** and **27**). Save 12 µL of the supernatant for detection of antigen by immunoblot (sample 2) and incubate under the same conditions as the immunoprecipitated sample.

3.3.2. Immunoprecipitation Using Soluble, Specific Antibody

This method works best with metabolically labeled cells. With immunoblot detection of antigens, the detecting antibodies must be carefully chosen to ensure that binding to antibody chains (which are coeluted with antigen and are therefore also present on blots) does not occur.

1. Aliquot 180 µL of extract from **Subheading 3.3.1., step 4** into a 1.5-mL screw-capped centrifuge tube.
2. Preadsorb GammaBind G-Sepharose binding antigens by addition of 10 µL of washed 50% slurry of GammaBind G-Sepharose and maintain the beads in uniform suspension by slow mixing on a rotating wheel for 30 min. Remove the beads by centrifugation for 30 s at maximum speed in a microcentrifuge and collect the supernatant for use in **step 3** (*see* **Note 28**).
3. Add 20 µL of a 1 mg/mL solution of antigen-specific monoclonal antibody or polyclonal antibody in PBS (*see* **Notes 29–31**).
4. Incubate for 4 h at room temperature, 21°C (*see* **Note 32**).
5. Add 100 mL of 50% washed GammaBind G-Sepharose (*see* **Note 33**) and incubate for 30 min at 21°C on a rotating mixer (*see* **Notes 30** and **31**).
6. Collect the beads by centrifugation for 30 s at maximum speed in a microcentrifuge and remove and discard most of the supernatant.
7. Add 1 mL of HNET-PI and suspend the beads by slow inversion. Collect the beads by centrifugation for 30 s at maximum speed in a microcentrifuge and remove and discard most of the supernatant. Wash five times with HNET-PI and save the beads (*see* **Note 34**). Remove as much supernatant as possible and wash once with 1 mL of PBS (*see* **Notes 35** and **36**).
8. Extract beads by acid elution (*see* **Note 37**). Add 10 µL of 200 m*M* acetic acid-NaOH, pH 2.8. Incubate 5 min at room temperature, centrifuge 30 s at maximum speed in a microcentrifuge and remove the supernatant from the beads carefully so as not to take up any beads. Add 3 µL of 1 *M* Tris-HCl, pH 8.0 to neutralize the pH. Repeat the extraction

of beads and combine the samples. Add 10 µL three times concentrated nonreducing SDS-PAGE sample loading buffer (*see* **Note 38**) to the combined neutralized eluate fractions. Heat at 95–100°C for 5–10 min. Samples can be stored indefinitely at –20°C.

3.3.3. Immunoprecipitation Using Insoluble-Specific Antibody

1. Aliquot 180 µL of extract from **Subheading 3.3.1., step 4**, into a 1.5-mL screw-capped centrifuge tube and add 10 µL of HNET-PI.
2. Add 10 µL of a 50% slurry of insoluble antibody (prepared as in **Subheading 3.2.3.**) and maintain the beads in uniform suspension by slow mixing on a rotating wheel for 3 h at 21°C (*see* **Notes 32** and **39**).
3. Collect the beads by centrifugation for 30 s at maximum speed in a microcentrifuge and remove and discard most of the supernatant.
4. Add 0.5 mL of HNET-PI and suspend the beads by slow inversion. Collect the beads by centrifugation for 30 s at maximum speed in a microcentrifuge and remove and discard most of the supernatant. Wash three times and save the beads. Remove as much supernatant as possible and wash once with 0.5 mL of PBS (*see* **Note 35**).
5. Extract beads by acid elution (*see* **Note 37**). Add 10 µL of 200 mM acetic acid-NaOH, pH 2.8. Incubate 5 min at room temperature, centrifuge 30 s at maximum speed in a microcentrifuge and remove the supernatant from the beads carefully so as not to take up any beads. Add 3 µL of 1 M Tris-HCl, pH 8.0 to neutralize the pH. Repeat the extraction of beads and combine the samples. Add 10 µL three times concentrated nonreducing SDS-PAGE sample loading buffer (*see* **Note 38**) to the combined neutralized eluate fractions. Heat at 95–100°C for 5–10 min. Samples can be stored indefinitely at –20°C.

3.4. Analysis of Precipitated Antigens by SDS-PAGE, Autoradiography, and Immunoblot Analysis

3.4.1. SDS-PAGE Separation of Proteins

1. Add 5 µL of concentrated nonreducing sample loading buffer to samples 1 and 2 (**Subheading 3.3.1.**) and adjust the volumes to 20 µL with PBS or pure water.
2. Prepare a suitable percentage polyacrylamide gel for SDS-PAGE (use 12% or a gradient if uncertain of the antigen size).
3. Load 20 µL of immunoprecipitated samples and control samples 1 and 2 into the wells. If analysis of antigens is by immunoblot, load another 20 µL of each immunoprecipitated sample onto a duplicate gel for subsequent immunoblot analysis with pre-immune or other control antibody. Separate proteins by SDS-PAGE at 20 mA per gel, running for approx 1 h until the bromophenol blue dye front has just left the gel.
4. Prepare suitable molecular weight markers: (a) for immunoblots: 5 µL of SeeBlue markers plus 5 µL sample loading buffer plus 10 µL of PBS or water. (b) For autoradiograms: Use 5 µL radioactive protein molecular weight markers.

3.4.2. Immunoblot Analysis of Immunoprecipitated Antigens

1. Cut a piece of PVDF membrane (NEN) slightly larger than the size of the gel. Immerse for 10 s in 10 mL of 100% methanol, wash for 2 min in 20 mL of pure water, and three times for 5 min in 20 mL Western transfer buffer with agitation.
2. Transfer proteins to PVDF using a mini-Transblot apparatus (Bio-Rad) at 100 V, 60 min (*see* **Note 40**).
3. Block the filter for at least 1 h with 5S-PBS-T with constant agitation at room temperature, or overnight at 4°C.
4. Probe the filter with a primary antibody (*see* **Note 41**) dissolved in 15 mL of 5S-PBS-T for

1 h at room temperature.
5. Wash three times with 30 mL of 5S-PBS-T.
6. Probe with an anti-(primary antibody host-species) immunoglobulin HRP conjugate diluted according to the manufacturer's recommendation, in 15 mL of 5S-PBS-T.
7. Wash twice with 30 mL of 5S-PBS-T.
8. Wash twice with 30 mL of PBS.
9. Use chemiluminescent detection to visualize protein bands, according to the manufacturer's recommendations.
10. Seal the filter in plastic wrap, place fluorescent tracker tape on the corners to enable orientation of the filter and expose to X-ray film.

3.4.3. Autoradiographic Detection of Protein Bands

1. Fix the gel. This can be done by placing it in 25:65:10 (v/v/v) isopropanol:water:acetic acid for 30 min with agitation. Then soak it in water for 10 min. Alternatively, stain with Coomassie blue, which also fixes proteins (*see* **Note 42**).
2. Place the gel on a slightly larger piece of Whatman 3MM paper, ensuring that the gel is not distorted.
3. Dry on a gel dryer at 60–80°C.
4. Use autoradiographic detection at room temperature to locate bands. Protein sizes can be matched to radioactive protein markers.

4. Notes

1. TCA is highly corrosive and suitable precautions must be taken to protect the eyes and skin.
2. Beads should never be stirred with a magnetic stirrer/stir-bar as this may damage the beads.
3. When collecting or dispensing beads by micropipetor cut approx 4 mm of the pipet tip off with scissors else the beads may not be aspirated.
4. Appropriate safety precautions for biohazardous materials must be taken when working with material obtained from malarial pathogens grown in human sera and/or erythrocytes.
5. Suitable precautions should be taken when using radioactive material, such as wearing disposable gloves and using radiation shields.
6. L-[^{35}S]methionine is used in the range 10–100 µCi/mL for stage-specific labeling.
7. Filtration may be carried out under vacuum or by application of the sample by syringe. However, if the latter is to be used take great care and use luer-lock fittings to prevent sprays of TCA in the event of a leak.
8. For [^{35}S] or [^{3}H] place the filter in a suitable scintillant and count in a beta counter. [^{125}I] can be counted directly in a gamma counter.
9. Once liberated from erythrocytes, parasites are no longer viable and cannot be cultured further, so these procedures do not need to be carried out under aseptic conditions.
10. The ratio of saponin to cell pellet volume can be lower; however, this results in high hemoglobin concentrations in the supernatant, which can make it difficult to distinguish the pellet.
11. Usually the volume of purified parasites is roughly one-twentieth of the original packed cell volume, for a culture of 5% trophozoites.
12. The capacity of GammaBind G-Sepharose beads for antibody is about 6 mg murine antibody per milliliter of gel.
13. Beads can be collected rapidly by loading the upper bead-free supernatant first as the column flow-rate decreases greatly once the beads are collected in the column.
14. A further round of purification may yield additional antibody and a comparison of the amounts of protein eluted in the first two rounds will indicate whether a further purification is warranted.
15. OD280 measurements can be confounded by interference due to absorption at 280 nm due

Immunoprecipitation

to buffer components. It is best to measure the OD280 on antibodies in PBS and use a buffer blank to correct for any contaminants. A more reliable way to measure protein is to use the Lowry protein assay that can be obtained in kit form from several commercial sources such as Bio-Rad or Pierce.

16. As a very rough approximation, if a 50 kDa antigen is covalently crosslinked to CNBr-Sepharose at 2 mg antigen per milliliter of beads we can expect the antibody-binding capacity of antigen-Sepharose to be approx 6 mg monoclonal antibody per milliliter of beads and greater than 6 mg for polyclonal antibody because of the multiple epitopes. However, steric factors, in particular, will affect the final antibody yields.
17. The initial elution will elute weaker-binding antibodies and will help to preserve binding activities that may be adversely affected by elution at lower pH.
18. Dialysis membrane with 12–14 kDa cutoff should be used for proteins down to 30 kDa. Use 30 kDa cutoff membrane for antibodies.
19. The tubing can be knotted but knots stretch the pores in the membrane.
20. This procedure is adequate for proteins in dilute buffers; however, if the protein is in a high concentration of Tris-HCl, glycine, or another buffer which will react with CNBr-activated Sepharose, then the number of buffer changes may be increased to four or more. Dialysis should be carried out at 4°C for labile proteins, but the rate of diffusion will be slower, so incubation periods should be extended at 4°C.
21. If a short incubation period and little protein are used then crosslinking and distortion of protein structure will be minimized. Proteins have many free amines that can react with CNBr-reactive groups on the matrix, so extended reaction times may extensively crosslink and distort protein structure.
22. Sodium azide is very toxic, wear appropriate protective clothing, especially gloves.
23. Twenty microliters of unpurified parasites at 5% parasitemia will contain a similar number of parasites to 1 µL of saponin-purified parasites. A theoretical estimation of parasite numbers is 5×10^5 parasites per mL of packed unpurified cells at 5% parasitemia. Packed purified parasites should contain 1×10^6 to 1×10^7 parasites per mL. If unpurified parasites are used, increase the assay volume to dilute the red blood cell components by 20- to 50-fold, preferably using a 1-mL assay vol and include an uninfected red blood cell control.
24. Comparison of samples 1 and 2 will indicate whether or not the antigen or antigens of interest have been extracted and solubilized.
25. Triton X100 insoluble material can be solubilized in a small volume of SDS-containing buffer and then SDS diluted with a buffer containing a nonchaotropic detergent. SDS may irreversibly denature proteins.
26. The assay volume can be altered. In many cases a volume of 1 mL is convenient and will enable the use of larger volumes of dilute antibodies (e.g., hybridomal culture supernatant) to be used. Where antibody affinity is low, it may be necessary to reduce the assay volume to 100 µL. Volumes less than 100 µL are much less satisfactory to work with.
27. BSA can be added to immunoprecipitation buffer as a means to stabilize the antigen of interest. Concentrations used vary from 1 to 10% *(8,26)*, but are usually about 1%. BSA may also help to reduce proteolysis and can reduce background on immunoblots. BSA may be included in wash buffers to reduce background labeling *(10,11)*.
28. This step need not be done if appropriate controls are used. For example, include a sample in which the specific antibody is replaced by an irrelevant antibody.
29. Usually the antigen-containing sample is incubated with antibody prior to a second incubation with solid phase precipitant. However, the solid phase can be incubated with the antibody, washed and then incubated with antigen *(3)*. This will ensure that no excess antibody will remain in solution and that only antibodies to which the solid phase can adsorb will bind to the antigen.
30. Optimization of the amount of antibody used must be done by a series of experiments with

increasing amounts of antibody. Comparing the amount of recovered antigen with a known volume of the detergent-extracted antigen on immunoblot will provide a comparison and estimate of percent antigen recovered. It is also important to know how much antibody protein is being used and whether sufficient protein G-Sepharose is being used. Refer to the manufacturer's specifications to estimate the capacity of antibody that can be bound. Use a slight excess of protein G-Sepharose, twice that required.

31. If unpurified antibodies are to be used, the assay should be carried out in a larger volume, 1 mL, to provide sufficient volume for dilution of culture and serum components. For untested antibodies try 200 µL of hybridomal supernatant or 20–50 µL of polyclonal antiserum. Once an immunoprecipitated product is observed, different volumes of unpurified antibodies should be tried.

32. The binding of antigen to antibody, while frequently a high-affinity interaction, is not always rapid and may involve a relatively slow high-affinity interaction (39). The interaction may be modulated by interference from other proteins and buffer components and therefore it is worth trying longer incubation periods such as 4–8 h or even overnight. Incubate a sample of extracted antigen in parallel to determine whether the antigen survives extended incubation. Binding will be more rapid at higher temperatures, room temperature or 37°C. At higher temperatures; however, rates of proteolysis will increase and stability may decline. Prolonged incubations are best carried out at 4°C.

33. Either protein G Sepharose or protein A-Sepharose can be used. The capacity of protein G-Sepharose for human IgG is about 20 mg per mL beads, and 6 mg of mouse IgG. The IgG-binding capacities of protein A-Sepharose and protein G-Sepharose are about the same. Protein A does not bind well to some IgG isotypes but protein G binds to most Ig classes with higher affinity than protein A and is therefore the better reagent. GammaBind G-Sepharose beads have a recombinant form of protein G covalently coupled which binds less serum albumin than native protein G.

34. If the antibody has weak binding activity reduce the number of washes to three to minimize antigen losses during washing.

35. Remove most of the detergent prior to SDS-PAGE. If detergent remains, the binding of SDS may be reduced and antigen migration on SDS-PAGE will be affected.

36. Virtually all of the supernatant can be removed if desired by an additional brief centrifugation and removal of the supernatant with a fine gel-loading micropipetor tip. Remember that most of the bead volume will consist of buffer within the beads.

37. Alternatively beads can be extracted directly with non-reducing SDS-PAGE sample loading buffer as follows. Extract beads by addition of 5 µL three times nonreducing SDS-PAGE sample loading buffer (see **Note 38**) plus 15 mL of PBS or pure water. Heat at 95–100°C (5–10 min), cool briefly, centrifuge 30 s at maximum speed in a microcentrifuge, vortex vigorously and centrifuge for 30 s at maximum speed in a microcentrifuge. Remove the supernatant with a fine gel-loading micropipet tip to recover as much as possible. Reextract the beads as before and combine the supernatants as the eluted fraction (40 mL total). Samples can be stored indefinitely at –20°C.

38. Nonreducing conditions are best if antigens are to be detected by immunoblot analysis. Reduced antibodies can cause background or nonspecific binding on immunoblots.

39. Use an irrelevant antibody-Sepharose as control for antigen binding to the matrix.

40. If gel percentage is greater than 12% transfer for 75 min to ensure transfer of large proteins.

41. (a) Note that the samples obtained with the soluble antibody protocol will contain antibodies. For example, if the specific antibody was from a rabbit, then the antibody used on the immunoblot must be from another host to prevent detection of the specific antibody on the blot. (b) If the insoluble antibody procedure is used, any antibody can be used as the sample will not contain antibodies.

42. Coomassie staining will not interfere with the detection of [^{35}S]-labeled bands, however,

it may cause quenching of the [^{3}H] signal.

References

1. Schwartz, B. D. and Nathenson, S. G. (1971) Isolation of H-2 alloantigens solubilized by the detergent NP-40. *J. Immunol.* **107,** 1363–1367.
2. Kessler, S. W. (1975) Rapid isolation of antigens from cells with a staphylococcal protein A-antibody adsorbent: parameters of the interaction of antibody-antigen complexes with protein A. *J. Immunol.* **115,** 1617–1624.
3. Chulay, J. D., Lyon, J. A., Haynes, J. D., Meierovics, A. I., Atkinson, C. T., and Aikawa, M. (1987) Monoclonal antibody characterization of *Plasmodium falciparum* antigens in immune complexes formed when schizonts rupture in the presence of immune serum. *J. Immunol.* **139,** 2768–2774.
4. Goding, J. W. (1983) in *Monoclonal Antibodies: Principles and Practice. Production and Application of Monoclonal antibodies in Cell Biology, Biochemistry and Immunology,* Academic Press, London.
5. (1991) Isolation and analysis of proteins in *Current Protocols in Immunology,* Wiley, New York, pp. 8.0.1–8.12.9.
6. Hjelmeland, L. M. and Chrambach, A. (1984) Solubilization of functional membrane proteins. *Methods Enzymol.* **104,** 305–318.
7. Neugebauer, J. (1988) A guide to the properties and uses of deterÖînts in biology and biochemistry.
8. Howard, R. J. and Barnwell, J. W. (1984) Solubilization and immunoprecipitation of ^{125}I-labelled antigens from schizont-infected erythrocytes using non-ionic, anionic and zwitterionic detergents. *Parasitol.* **88,** 27–36.
9. Freeman, R. R. and Holder, A. A. (1983) Surface antigens of malaria merozoites. A high molecular weight precursor is processed to an 83,000 molecular weight form expressed on the surface of Plasmodium falciparum merozoites. *J. Exp. Med.* **158,** 1647–1653.
10. Holder, A. A. and Freeman, R. R. (1982) Biosynthesis and processing of a *Plasmodium falciparum* schizont antigen recognized by immune serum and a *Plasmodium knowlesi* monoclonal antibody. *J. Exp. Med.* **156,** 1528–1538.
11. Pirson, P. J. and Perkins, M. E. (1985) Characterization with monoclonal antibodies of a surface antigen of *Plasmodium falciparum* merozoites. *J. Immunol.* **134,** 1946–1951.
12. Brown, G. V., Coppel, R. L., Vrbova, H., Grumont, R. J., and Anders, R. F. (1982) Plasmodium falciparum: comparative analysis of erythrocyte stage-dependent protein antigens. *Exp. Parasitol.* **53,** 279–284.
13. Foley, M., Tilley, L., and Anders, R. F. (1991) The ring-infected erythrocyte surface antigen of *Plasmodium falciparum* associates with spectrin in the erythrocyte membrane. *Mol. Biochem. Parasitol.* **46,** 137–148.
14. van Schravendijk, M. R., Rock, E. P., Marsh, K., Ito, Y., Aikawa, M., Neequaye, J., et al. (1991) Characterization and localization of *Plasmodium falciparum* surface antigens on infected erythrocytes from west African patients. *Blood* **78,** 226–236.
15. Brown, G. V., Anders, R. F., Mitchell, G. F., and Heywood, P. F. (1982) Target antigens of purified human immunoglobulins which inhibit growth of *Plasmodium falciparum* in vitro. *Nature* **297,** 591–593.
16. Shantz, E. M. (1983) Pansorbin. Staphylococcus aureus cells: Review and bibliography of the immunological applications of fixed Protein A-bearing *Staphylococcus aureus* cells.
17. Knopf, P. M., Brown, G. V., Howard, R. J., and Mitchell, G. F. (1979) Immunoprecipitation of biosynthetically-labelled products in the identification of antigens of murine red cells infected with the protozoan parasite, *Plasmodium berghei. Aust. J. Exp. Biol. Med. Sci.* **57,** 603–615.
18. Anders, R. F., Brown, G. V., and Edwards, A. (1983) Characterization of an S-antigen synthesized by several isolates of *Plasmodium falciparum. Proc. Natl. Acad. Sci. USA* **80,** 6652–6656.
19. McColl, D. J., Silva, A., Foley, M., Kun, J. F., Favaloro, J. M., Thompson, J. K., et al. (1994) Molecular variation in a novel polymorphic antigen associated with *Plasmodium falciparum* merozoites. *Mol. Biochem. Parasitol.* **68,** 53–67.
20. Coligan, J. E. and Kindt, T. J. (1981) Determination of protein primary structure by radiochemical techniques. *J. Immunol. Meth.* **47,** 1–11.
21. Vitetta, E. S., Capra, J. D., Klapper, D. G., Klein, J., and Uhr, J. W. (1976) The partial amino-acid sequence of an H-2K molecule. *Proc. Natl. Acad. Sci. USA* **73,** 905–909.
22. Schofield, L., Bushell, G. R., Cooper, J. A., Saul, A. J., Upcroft, J. A., and Kidson, C. (1986) A rhoptry antigen of *Plasmodium falciparum* contains conserved and variable epitopes recognized by inhibitory monoclonal antibodies. *Mol. Biochem. Parasitol.* **18,** 183–195.
23. Brown, G. V., Anders, R. F., and Knowles, G. (1983) Differential effect of immunoglobulin on the in vitro growth of several isolates of *Plasmodium falciparum. Infect. Immun.* **39,** 1228–1235.
24. Coppel, R. L., Lustigman, S., Murray, L., and Anders, R. F. (1988) MESA is *a Plasmodium falciparum* phosphoprotein associated with the erythrocyte membrane skeleton. *Mol. Biochem. Parasitol.* **31,** 223–231.
25. Lustigman, S., Anders, R. F., Brown, G. V. and Coppel, R. L. (1988) A component of an antigenic

rhoptry complex of *Plasmodium falciparum* is modified after merozoite invasion. *Mol. Biochem. Parasitol.* **30,** 217–224.
26. Howard, R. J., Barnwell, J. W., Kao, V., Daniel, W. A., and Aley, S. B. (1982) Radioiodination of new protein antigens on the surface of *Plasmodium knowlesi* schizont-infected erythrocytes. *Mol. Biochem. Parasitol.,* **6,** 343–367.
27. Lustigman, S., Anders, R. F., Brown, G. V., and Coppel, R. L. (1990) The mature-parasite-infected erythrocyte surface antigen (MESA) of *Plasmodium falciparum* associates with the erythrocyte membrane skeletal protein, band 4. 1. *Mol. Biochem. Parasitol.* **38,** 261–270.
28. Foley, M., Murray, L. J., and Anders, R. F. (1990) The ring-infected erythrocyte surface antigen protein of *Plasmodium falciparum* is phosphorylated upon association with the host cell membrane. *Mol. Biochem. Parasitol.* **38,** 69–76.
29. Lim, A. S. Y. and Cowman, A. F. (1993) Phosphyorylation of a P-glycoprotein homologue in *Plasmodium falciparum. Mol. Biochem. Parasitol.* **62,** 293–302.
30. Menting, J. G., Tilley, L., Deady, L. W., Ng, K., Simpson, R. J., Cowman, A. F. and Foley, M. (1997) The antimalarial drug, chloroquine, interacts with lactate dehydrogenase from Pasmodium falciparum. *Mol. Biochem. Parasitol.* **88,** 215–224.
31. Roggwiller, E., Fricaud, A. C., Blisnick, T., and Braun-Breton, C. (1997) Host urokinase-type plasminogen activator participates in the release of malaria merozoites from infected erythrocytes. *Mol. Biochem. Parasitol.* **86,** 49–59.
32. Perkins, M. E. and Rocco, L. J. (1990) Chemical cross-linking of *Plasmodium falciparum* glycoprotein, Pf200 (190-205 kDa), to the S-antigen at the merozoite surface. *Exp. Parasitol.* **70,** 207–216.
33. Wiser, M. F. and Schweiger, H. G. (1986) Increased sensitivity in antigen detection during immunoblot analysis resulting from antigen enrichment via immunoprecipitation. *Anal. Biochem.* **155,** 71–77.
34. David, P. H., Hudson, D. E., Hadley, T. J., Klotz, F. W., and Miller, L. H. (1985) Immunization of monkeys with a 140 kilodalton merozoite surface protein of Plasmodium knowlesi malaria: appearance of alternate forms of this protein. *J. Immunol.* **134,** 4146–4152.
35. McBride, J. S., Newbold, C. I. and Anand, R. (1985) Polymorphism of a high molecular weight schizont antigen of the human malaria parasite *Plasmodium falciparum. J. Exp. Med.* **161,** 160–180.
36. Seesod, N., Lunderberg, J., Hedrum, A., Aslund, L., Holder, A., Thaithong, S., and Uhlen, M. (1993) Immunomagnetic purification to facilitate DNA diagnosis of *Plasmodium falciparum. J. Clin. Microbiol.* **31,** 2715–2719.
37. Lambros, C. and Vanderberg, J. P. (1979) Synchronization of *Plasmodium falciparum* erythrocytic stages in culture. *J. Parasitol.* **65,** 418–420.
38. (1991) Antibody detection and preparation, in *Current Protocols in Immunology*, Wiley, New York, pp. 2.0.1–2.10.4.
39. Scopes, R. K. (1994) in *Protein Purification, Principles and Practice*, Springer-Verlag, New York.

VII

TESTING INTERVENTIONS FOR MALARIA CONTROL

57

Field Trials

Pedro L. Alonso and John J. Aponte

1. Introduction

Current malaria control strategies are based on interventions aimed at either reducing exposure to infectious bites or at chemosupressing the parasite while in the host. In brief, the most commonly used methods to reduce exposure have relied on trying to reduce the number of adult vectors with the use of insecticides, often through residual indoor spraying, or aiming to reduce human vector contact through bednets, repellents, or insecticide-treated materials. Measures aimed against the parasite are based on the use of drugs either as treatment or as prophylaxis. The emergence and spread of resistance against the most commonly used drugs as well as the limitations in the use of insecticides, coupled with the technological developments of the last 30 years, have accelerated the impetus to develop and test new drugs and vaccines against malaria.

The development and testing of new interventions for application in public health requires a sequence of trials that is similar regardless of the tool itself. Drugs, vector control measures, and new vaccines need to generate information on the safety and acceptability of the measure and then undergo the pivotal trial(s) that determine whether this new tool is efficacious and deserves further development toward its application as a public health tool.

This chapter attempts to establish a wide framework for the testing of new interventions, but concentrates on the testing of vaccines. We do so for two major reasons. First of all, there is already an established methodology and expertise in testing new drugs and insecticides. Second, we are still in the early days of development and testing of malaria vaccines, and there are a number of particularities and uncertainties involved with which the international scientific and public health community has little accumulated experience. Although this chapter concentrates primarily on phase III trials of malaria vaccines, it specifically avoids the necessary speculation regarding the ways different vaccines might be used in different situations and target groups in order to help define its role for malaria control (*1,2*). Particularly important and problematic are the unanswered questions regarding the kind and duration of protection likely to be produced with the kind of vaccines likely to become available.

Why do we need phase III trials for new interventions? These trials provide evidence on the efficacy of new control tools, including vaccines (*see* **Note 1**). This is particularly important in the case of malaria for three crucial reasons:

1. We still have a rather limited and incomplete understanding of immunity to malaria that prevents us from being able to predict from any given human immune responses (*see* **Note 2**) whether someone is protected from malaria infection or its clinical consequences.
2. With perhaps the exception of membrane feeding experiments to establish whether a mosquito can become infectious following a bloodmeal containing gametocytes, we have very limited in vitro assays that help us predict the outcome of an infection.
3. We have still a rather limited capacity to predict how does the impact on one end point translates into other end points. This high level of uncertainty heavily conditions the choice of end points and the interpretation and comparability of results.

2. Materials

2.1. Information on the Control Strategy

In other chapters, the different steps involved in the development of a new tool until it reaches the stage when an efficacy trial can be contemplated have been reviewed. A first and necessary step is to review all that information *(3)* (*see* **Note 3**).

2.2. Study Area and Characterization

The need for an adequate selection of the study area cannot be overemphasized. There are two central components: operational criteria and epidemiological criteria.

2.2.1. Operational Criteria

1. Firm and informed commitment of the national and local authorities, as well as that of the study population and the associated investigators.
2. Involvement of national research institutions and investigators, with the highly desirable consequence of building up the local research capacity to conduct such trials.
3. Reasonable expectations of social and political stability.

2.2.2. Epidemiological Criteria

1. Accurate baseline data. It is necessary to conduct a trial where the incidence of malaria is sufficient to allow a precise assessment of efficacy, and where the intervention is likely to be relevant for malaria control.
2. Well-characterized site, ideally including:
   a. Availability of maps and demographic data.
   b. Levels of drug resistance and a reasonable understanding of the health services structure, the quality of its performance, case management practices, and consumption patterns of antimalarials.
   c. Knowledge on the vector composition and seasonal variations, estimate of the entomological inoculation rate and use of vector control or man vector reduction methods.
   d. Precise estimates of the incidence of the different potential end points, its seasonal variation and its age dependence among the possible target groups.

2.3. Development and Approval of a Study Protocol

A detailed protocol must be developed by the investigators and submitted to the relevant national authorities and institutional review boards for its consideration and approval prior to the initiation of any activities.

2.4. Data and Safety Monitoring Board

It is highly desirable that an independent board that can monitor and testify that a trial protocol is adequately being followed should be constituted before the trial starts. The Board should include members with adequate and recognized expertise in different fields relevant to malaria control and the intervention to be tested, including a clinical monitor. Its functions should be defined and regulated by well-defined and agreed-upon standard operating procedures.

3. Methods
3.1. Objectives and End Points

The specification of the trial's objectives essentially relates to the identification of the end points to be measured.

3.1.1. Primary Objective

The primary objective, which also determines the size of the sample, must aim to estimate the protective efficacy of the intervention on a single end point. A full discussion of the criteria and constraints to select such an end point is beyond the scope of this article. In brief, **Fig. 1** aims to represent schematically the possible relations between different interventions and the likely outcomes and endpoints of phase III trials, and the possible roles of additional cofactors (*see* **Note 4**). Many (but clearly not all) would probably agree that the primary end point for a phase III trial should be as close as possible to the biological target of the intervention. In this case, incidence of infection would be the clear choice of end point for the evaluation of vector control measures and man vector reduction methods as well as preerythrocytic and transmission blocking vaccines. Incidence of mild or uncomplicated disease may be the best choice for asexual-stage vaccines and antitoxin vaccines. The more relevant public health end points may need to wait for later phase III trials should this be acceptable to the national and research review committees or for phase IV trials and program implementation and evaluation (*see* **Note 5** and **Fig. 2**).

3.1.2. Secondary Objectives

Secondary objectives include the following:

1. Other efficacy end points that may be chosen (if the trial size allows for enough power) among other primary efficacy end points; or other secondary endpoints such as parasite prevalence and density, prevalence and intensity of anemia, infectivity of humans, or multiplicity of infections.
2. Immunogenicity (*see* **Note 2**), including dynamics of the immune response and whether natural boosting takes place, as well as correlation between immune responses elicited by the vaccine and presence or absence of protection.
3. Safety ands its careful assessment must continue as the number of individuals receiving the intervention increases and the number possibility of detecting rare side effects increases.

3.2. Target Groups

This must be clearly defined, and, as with the choice of end points, it is often an arbitrary decision. Ideally the group in which the trial is carried out should be the same as the one that is the target of the future control or vaccination program. In areas of

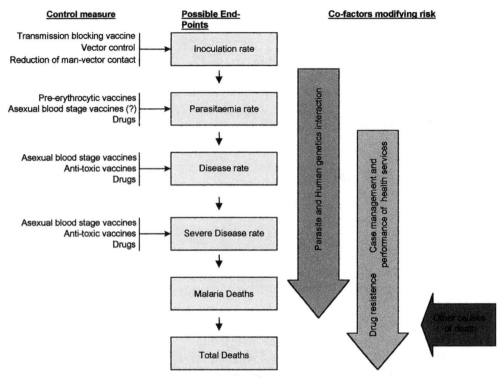

Fig. 1. Schematic of the relationship between different interventions, the likely biological targets and their end points, and the potential role of cofactors that modify the risk of malaria. The figure also aims to provide a framework of the relation between different end points for its selection in phase III trials.

very low malaria transmission, and where the population rarely develops significant levels of immunity, as is the case in most of Latin America or Asia, this is likely to mean the whole population *(4)*. On the other hand, for most of sub-Saharan Africa, this is likely to mean young infants receiving the vaccine through the Expanded Programme of Immunization (EPI) scheme *(5,6)*. However, a presumably less fragile group, for example, children aged 1–4 yr, may be a more reasonable way to start *(7,8)*. Pregnant women are an important group at risk and a major target for protection using malaria control tools. Given the particularities of malaria and its impact in this group *(9)*, as well as the implications of testing a new drug or vaccine in pregnant women, one anticipates that the new intervention will be tested in this target group only after it has been shown to be efficacious in other groups.

3.3. Study Design

The only adequate and acceptable study design requires a double-blind, randomized, controlled comparison between the test intervention (vaccinated) and an nonintervened group (unvaccinated) *(10)*.

When possible, the unit of randomization should be the individual, as it minimizes the risk of accidental bias. There are, however, situations when individual randomization may not be possible. Interventions where the effect is likely to operate beyond the individual may require randomization by transmission unit. Vector control measures, including insecticide-treated bednets, transmission-blocking vaccines, preerythrocytic vaccines, or drugs that reduce transmissibility or viability of gametocytes may require such randomization.

The choice of control during phase III trials is heavily dependent on the type of intervention. For vaccines, the best control is a placebo, and the best one is probably one that contains all the vaccine ingredients with the exception of the antigen. The vaccine's carrier and adjuvant would thus be included in the placebo. There is, however, growing debate as to the use of such placebos *(11)*, and recent trials have been controlled with other vaccines, that are not thought to modify the risk of malaria such as oral poliovirus vaccine (OPV) *(12)* or hepatitis B vaccine (HBV) *(13)*, but that are likely to provide some benefit to the participants.

3.4. Sample Size

The number of individuals recruited into a trial, and that complete the follow-up should be sufficient to provide reliable estimates of the efficacy of an intervention against the outcome established in the primary objective. The size of the trial depends on the frequency of the primary end point in the control group, the size of the difference that we want to detect, the precision of the estimate, the power, and the level of significance *(14,15)* (*see* **Note 6**).

3.5. Measuring End Points

This is a critical aspect of the trial that should be carefully considered in the preparatory phases. A detailed description of the different methods is beyond the scope of this chapter (*see* **Note 7**). All field, clinical and laboratory procedures should be reviewed and specific standard operating procedures (SOP) drawn and adhered to. Particular attention must be drawn to establishing normal laboratory values in populations where there are often no reference values (e.g., LFT among infants), procedures for quality control and monitoring of the trial, reproducibility of measurements within and between observers and over time, sensitivity and specificity of diagnostic methods, or case definitions for the chosen end points and procedures for recording and validating data. All procedures should satisfy the *ICH Guideline for Good Clinical Practice* *(12)*.

Measurement of the endpoints is done through either one or a combination of methods. Cross-sectional surveys are often used to estimate the prevalence of malariometric parameters (parasitemia, splenomegaly, anemia, etc.) in a given population or cohort, and are often carried out before and after the intervention. Repeated cross-sectional surveys at defined time intervals have been used to determine incidence of parasitemia. Clinical episodes may be detected by active case detection, implying the search of suspect cases, often including regular home visits to a cohort of children, or through passive case detection, which relies on cases presenting at health facilities. Detection and investigation of all deaths is required in all trials, even if mortality is not the primary efficacy end point. The most appropriate method for determining mortality will be dependent on local circumstances, but it may often include a combination of passive and active surveillance *(15)*.

3.6. Implementation

A phase III trial classically proceeds in three periods:

1. A preparatory period of about 1 year when the protocol is developed and approved, final baseline data are collected, personnel are recruited and trained, data collection methods are established, standardized and validated and intense community preparation is carried out.
2. Intervention and evaluation period.
3. Final analysis period, which starts after the files have been locked and the code has been broken, and should produce the final report *(16)*.

Among all the critical aspects for the success of a trial present in every one of the periods, involvement and preparation of the community is worth highlighting. The purpose of the trial and the methods to be used should be discussed in detail with representatives of the communities in which the trial is planned.

3.7. Data Management and Analysis

Data management should be preferably done at the trial site in accordance with GCP guidelines *(16)*. Detailed quality control procedures should be established for the programs to be used, data entry, cleaning, and storage of all data both in electronic form and paper.

In order to ensure that the results do not influence the choice of primary end point, the investigators should prepare an analytical plan prior to breaking the code. This plan specifies the primary analysis of efficacy, the inclusion criteria, the case definition to be used and the methods of data analysis.

Vaccine efficacy is usually estimated as (1-relative risk)% where the relative risk may be estimated as the relative rate (r_1/r_0) using the incidence rate in vaccine (r_1) and placebo group (r_0), or the odds ratio ($odds_1/odds_0$), using the odds of being a case in the vaccine group ($odds_1$) and placebo group ($odds_0$), or the hazard ratio ($h(t)_1/h(t)_0$) using the hazard in vaccine group ($h(t)_1$) and placebo group ($h(t)_0$). The appropriate method of analysis depends on the trial design and on the outcome. For an efficacy trial it may be preferable to include in the analysis only the individuals who complete the vaccination schedule, whereas for an effectiveness trial it may be preferred an *intention to treat* analysis. If the morbidity surveillance is continuos, the incidence rate for each group is calculated as the number of cases by the total person-time at risk. Individuals who are lost of follow-up, withdraw or die, should be included up to the date of loss, withdrawal, or death. Poisson regression models may be used to obtain a confidence interval for the vaccine efficacy and if indicated, to adjust for important imbalances. If the incidences change within each group by age or by season, these factors should be included in the regression models. If there is interest in the time to and event, survival techniques including Cox regression models may be used. When multiple episodes of the same individual are included in the analysis, the individual should be removed both from numerator and denominator for a period of time (usually 30 d) until he is at risk again. However, for this analysis it is necessary to take into account the heterogeneity in susceptibility. It is very likely that once a subject has experienced one or more episodes, he or she is more or less likely to experience another episode than those who have experienced no episodes up to that time in the same group.

3.8. Ethical Considerations

The design and implementation of the trial should conform to both national regulations and requirements and international ethical standards as outlines in *ICH Guideline for Good Clinical Practice (12)* and in the guidelines for biomedical research involving human subjects *(17)*. There must be witnessed informed consent from those who participate in the trial. Researchers should check adequate understanding of the implications of participation, as well as procedures involved through key questions to participants, or its parents or guardians in the case of children. There should be no pressure to participate, and no difference of services offered to those who agree or do not agree.

4. Notes

1. Some older guidelines and papers have often distinguished between phase IIb and phase III trials *(18,19)*. This nomenclature is slightly unfortunate and was based on the size of the trial. In this chapter, we do not distinguish one from the other as both of them have determining efficacy as primary objective, and therefore the size of the trial will be determined by the conditions and calculations involved in determining the sample size.
2. In this chapter, the terms immune and immunity will indicate resistance to infection or some of its clinical consequences acquired following previous infection or vaccination. Immune responses describe the host's humoral or cellular immune responses that follow exposure to antigens (either through infection or vaccination), but they do not necessarily reflect or correlate with a state of protection against infection or its adverse effects. Finally, a vaccine's immunogenicity refers to its capacity to generate immune responses that may or may not correlate with protection.
3. The information that we need from previous phases includes
   a. A detailed product definition.
   b. Information from preclinical research.
   c. Information from phase I studies, including safety, acceptability, and immunogenicity.
   d. Information from phase IIa (artificial challenge) studies where applicable, including efficacy and relation to challenge dose, duration, and boostability, and preferably an evaluation of immunological tests as indicators of protection.
4. The selection of a phase III trial primary end point depends on several components:
   a. Although necessarily oversimplified, **Fig. 1** aims to illustrate the uncertainty of extrapolation introduced by the existence of cofactors that modify the risk of a given end point. Given a certain impact on an endpoint (e.g., reduction in the incidence of parasitemia) it is difficult to predict what will be the impact on downstream end points (e.g., reduction of deaths). Locally operating cofactors such as human—parasite interactions, quality, and availability of case management, or the interaction between other causes of death, could profoundly modify that relation.
   b. The Public Health importance of the endpoints increases as one moves down toward the more severe end points. It has often been considered desirable to choose, at an early stage, the more severe end points such as severe malaria or overall mortality. On the other hand, the more downstream an endpoint is, the further away it is from the biological target, and the more it is dependent on the quantity and effect of other operating cofactors. A choice of this endpoint increases the public health relevance of the result, makes it locally more relevant, but decreases the generalizability and the comparability between trials (*see* **Fig. 2**).
   c. Ethical constraints may imply that once efficacy against even a mild endpoint has been established, it may be ethically questionable to conduct randomized controlled trials (RCT) against more severe end points.
5. The experience accumulated over the past years with the development and testing of insecticide-treated bednets, is a good example on how the perception and drive to develop a strategy is changed when the intervention is shown to have an impact on the more public health relevant end points, such as overall mortality *(20)*. Initial trials with incidence of infection or mild disease as the primary endpoint had yielded conflicting results, and in

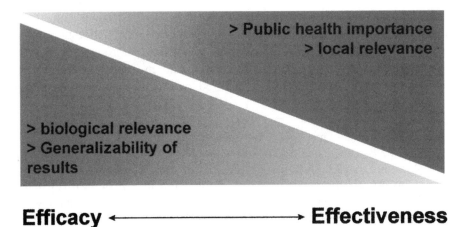

Fig. 2. Schematic of the relationship between efficacy and effectiveness trials, to illustrate that in field trials there is never a "pure" efficacy or effectiveness situation, but rather a mixture of one and another in different proportions. As one moves toward an effectiveness-type trial, this is more influenced by local cofactors, including the functioning of the health services. In this case, the results are of greater local and public health relevance, and tend to use "downstream" end points. On the contrary, as one moves to the left, the trial should be less dependent on local conditions and factors, therefore aims to provide more "biological evidence" that should be generalizable. The immediate public health and local relevance diminishes in a directly proportional way. The end points chosen tend to be those closer to the biological target of the intervention.

the most optimistic scenarios only had a mild impact. Insecticide-treated bednets were close to being abandoned as a potential strategy. The efficacy of this tool against different endpoints is also a good example of the uncertainty of extrapolation of a given impact on an end point, to what the impact on other downstream end points may be *(22)*.

6. For individually randomized trials, where the protective efficacy (VE) is estimated as $100 \times (1 - r_1/r_0)$, the person-years required per group to estimate the vaccine efficacy with a 90% power at 5% significance level can be obtained using the approximate formula:

$$y = 10.5(r_0 + r_1)/(r_0 - r_1)^2$$

For a sample size based on the precision on the estimation rather than on the power, the person-years at risk required in each group can be obtained using the formula:

$$y = (1.96/\log_e f)^2 (1/r_0 + 1/r_1)$$

where f is a factor used to evaluate the width of the confidence interval which extend from VE/f to VExf. For trials where the randomization unit is the community, the number of communities required in each group can be estimated using the formula

$$c = 1 + 10.5[(r_0 + r_1)/y + k^2(r_0^2 + r_1^2)]/(r_0 - r_1)^2$$

where y is the person-years of observation in each community and k is the coefficient of variation of the (true) incidence rate among the communities in each group *(14)*.

7. Incidence of disease is the most frequently used end point in phase III trials. The diagnosis of clinical malaria rests on clinical suspicion, often fever above a certain threshold, for example, 37.5°C, in the presence of asexual parasitaemia of a given density. In areas of low transmission (and hence low immunity), asymptomatic parasitemia is rare, while in areas of high transmission it is common and must be taken into account when determining case definitions. An appropriate case definition is one with high specificity (few cases caused by nonmalarial illness but with a concurrent parasitemia are included) and with a high sensitivity (few true malaria cases are omitted). Variations in the sensitivity of a case definition modify the power of the study. Loss of specificity will bias results toward 0 and will therefore yield underestimates of true efficacy. Trials should determine and evaluate case definitions, based on a single entry point and by appropriate procedures *(22,23)*.

References

1. World Health Organization (1997) *Guidelines for the Evaluation of* Plasmodium falciparum *Asexual Blood-Stage Vaccines in Populations Exposed to Natural Infection.* Document TDR/MAL/VAC/97.
2. Alonso, P. L., Molyneaux, M. E., and Smith, T. (1995) Design and methodology of field-based intervention trials of malaria vaccines. *Parasitol. Today* **11,** 197–200.
3. World Health Organization (1992b) Good manufacturing practices for biological products. Annex 1. *WHO Tech. Rep. Ser. 822.*
4. Valero, M. V., Amador, L. R., Galindo, C., Figueroa, J., Bello, M. S., Murillo, L. A., et al. (1993) Vaccination with SPf66, a chemically synthesised vaccine, against *P. falciparum* malaria in Colombia. *Lancet* **341,** 705–710.
5. Acosta, C. J., Galindo, C., Schellenberg, D. M., Kahigwa, E., Urassa, H., Aponte, J. J., Armstrong-Schellenberg, J. R., et al. (1998) Evaluation of the SPF66 vaccine for malaria control when delivered through the EPI scheme in Tanzania *Trop. Med. Int. Health* **4,** 368–376.
6. World Health Organization (1993) *Framework for Evaluating a Vaccine for the EPI.* WHO/EPI/GEN/93.5.
7. Alonso, P. L., Tanner, M., Smith, T., Hayes, R. J., Armstrong-Schellenberg, J. R., Lopez, M. C., et al. (1994) A Trial of the synthetic malaria vaccine SPF66 in Tanzania: rationale and design. *Vaccine* **12,** 181–186.
8. Alonso, P. L., Smith, T., Armstrong-Schellenberg, J. R., Masanja, H., Mwankusye, S., Urassa, H., et al. (1994) Randomised trial of efficacy of SPF66 vaccine against *Plasmodium falciparum* malaria in children in southern Tanzania. *Lancet* **344,** 1175–1181.
9. Menendez, C. (1995) Malaria during pregnancy: a priority area of malaria research and control. *Parasitol. Today* **11,** 178-183.
10. Smith P. G. and Hayes, R .J. (1991) Design and conduct of field trials of malaria vaccines, in *Malaria: Waiting for the Vaccine.* (Targett, G. A. T., ed.), Wiley, New York, pp. 199–215.
11. Stoute, J. A., Slaoui, M., Heppner, G., Momin, P., Kester, K. E., Desmons, P., et al. (1997) A preliminary evalaution of a recombinant circumsporozoite protein vaccine against *Plasmodium falciparum* malaria. *N. Engl. J. Med.* **336,** 86–91.
12. D'Alessandro, U., et al. (1995) Efficacy trial of malaria vaccine SPf66 in Gambian infants. *Lancet* **346,** 462–467.
13. Nosten, F., et al. (1996) Randomized double blind placebo controlled trial of SPf66 malaria vaccine in children in North Western Thailand. *Lancet* **348,** 701–707.
14. World Health Organization (1986) *Sample Size Determination. A User's Manual. WHO Epidemiology and Statistical Methodology Unit.* Document WHO/HST/ESM86.1.
15. Smith, P. G. and Morrow, R. H. (1996) *Methods for Field Trials of Interventions Against Tropical*
16. ICH (1997) *Harmonised Tripartite Guideline for Good Clinical Practice*, Bookwood medical publications Ltd, Second Edition.
17. Council of International Organizations of Medical Sciences (1993) *International Ethical Guidelines for Biomedical Research Involving Human Subjects.* CIOMS, Geneva.
18. World Health Organization (1989) *Guidelines for the Evaluation of* Plasmodium falciparum *Asexual Blood-Stage Vaccines in Populations Exposed to Natural Infection.* Document TDR/MAP/SVE/PF/89.5.
19. World Health Organization (1992) *Guidelines for Community-Based Trials of Vaccines Against the Sexual Stages of Malaria Parasites.* Document TDR/CTD/TBV/92.
20. Alonso, P. L., Lindsay, S.W., Armstrong, J. R. M., Conteh, M., Hill, A. G., David P. H., et al. (1991) The effect of insecticide-treated bednets on mortality of Gambian children. *Lancet* **337,** 1499–1502.
21. Alonso, P. L., Lindsay, S. W., Armstrong-Schellenberg, J. R., Keita, K., Gomez, P., Shenton, F. C., et al. (1993) The impact of the interventions on mortality and morbidity from malaria *Trans. Roy. Soc. Trop. Med. Hyg.* **87,** 37–44.

22. Smith, T., Armstrong, J., and Hayes, R. (1997) Attributable fraction estimates and case definitions for malaria in endemic areas *Stat. Med.* **13,** 2345–2358.
23. Armstrong, J., Smith, T., Alonso, P. L., and Hayes, R. (1994) What is clinical malaria? Finding case definition for field research in highly endemic areas *Parasitol. Today* **10,** 439–442.

Index

A

ACD, *see* Active case detection
Active case detection (ACD), risk assessment, 17, 18
ADCI, *see* Antibody-dependent cellular inhibition
Adhesion, *see* Chondroitin sulfate A; Flow-based cytoadhesion assays
AES, *see* Average enlarged spleen
Alanine,
    brain analysis, 72–74
    synchronization of cultures, 493, 494, 496
Annual parasite incidence (API), risk assessment, 17
Anopheles,
    blood-feeding stage identification, 98
    Christopher's stages, ovarian development, 98, 99
    criteria for malaria vector, 3
    field handling, 96, 97
    parity status determination, 98
    polymerase chain reaction for species identification,
        amplification reaction, 100
        DNA extraction, 100
        overview, 99–101
        primers, 100, 101
    specimen processing, assembly-line approaches, 97
    taxonomic identification, 97, 98, 100, 101
    vector incrimination, *see* Entomological inoculation rate
Antibody-dependent cellular inhibition (ADCI),
    assay,
        antibodies, 529
        calculations, 532
        immunoglobulin preparation, 529–532
        incubation conditions, 532, 533
        materials, 529, 530
        monocyte preparation, 530–533
        overview, 529
        parasite preparation, 530–532
        strain specificity, 537
    malaria defense, 457, 529
API, *see* Annual parasite incidence
Attributable fraction, calculation, 21
Attributable risk, calculation, 20, 21
Average enlarged spleen (AES), scoring, 15

B

Blood, sensitivity of microscopic examination for parasitemia, 189
Blood-feeding stage, identification, 98
Brefeldin A, inhibition of cytokine secretion, 414, 420, 429, 434

C

CD36, immobilization on microslides, 577, 578
Cell adhesion, *see* Chondroitin sulfate A; Flow-based cytoadhesion assays
Chloroquine, primate model treatment, 90, 91
Chondroitin sulfate A (CSA),
    antiadhesion antibody assay,
        immobilized receptor adhesion, 557, 558
        materials, 555, 559
        parasite collection and preparation, 557, 559
        serum agglutination assay, 558, 559
        serum inhibition of adhesion, 558, 559
    pregnancy expression and parasite adhesion, 555, 556
Christopher's stages, ovarian development, 98, 99
Chromium release assay, cytotoxic T lymphocytes, 439, 442, 453–455
Chromosome purification, *see* Plasmodium falciparum
Circumsporozoite protein (CSP),
    adenoviral vector vaccine and enzyme-linked immunospot assay of T cell response,
        advantages, 361, 362
        CD4+ cell assay, 363, 366
        EL4 cell supernatant preparation, 362, 364
        immunization of mice, 364, 366
        interferon-γ CD8+ T cell assay, 363, 365, 366
        interleukin-5 CD8+ T cell assay, 363, 366

lymphocyte purification,
  liver, 362–365
  spleen, 363, 365
  materials, 362, 363
 antibodies and risk assessment, 16
 enzyme-linked immunosorbent assay, 6, 7
Clinical trials, *see* Field trials, malaria interventions
Complementary DNA library, *Plasmodium,*
  gene representation probabilities, 241, 242, 245, 246, 265
  lambda library preparation,
    adapter ligation, 266, 270
    amplification of library, 272–274
    blunt end generation, 266, 269, 270
    complementary DNA synthesis, 266, 269, 273, 274
    kinase reaction, 266, 270
    ligation of DNA into vector, 267, 271
    materials, 266, 267, 273
    packaging, 271, 272
    poly(A) RNA purification, 266, 268, 269, 273
    restriction digestion, 266, 270, 271
    RNA isolation and quantification, 266–268
    size fractionation of DNA, 267, 271
    titration of packaged library, 272
  mung bean nuclease digest library,
    amplification and storage, 262, 263
    blunt end generation, 260, 261, 263
    cleavage specificity modification, 253
    digestion, 258, 263
    DNA preparation,
      cesium chloride centrifugation for separation of parasite and host DNA, 257, 258
      extraction, 256, 257, 263
      isolation from gels, 258, 259
    ligation,
      adapters, 261
      DNA fragments, 261–263
      materials, 255, 256, 263
      overview, 253, 255
      screening, 262
      size of mung bean fragments versus size of coding regions in *Plasmodium,* 253, 254
    transformation, 262
  random-sheared DNA library preparation,
    adapter ligation, 247, 248
    blunt end generation, 247

cesium chloride centrifugation for separation of parasite and host DNA, 243, 244, 250
    DNA isolation, 242, 243
    DNase treatment for linear DNA removal, 248
    gene representation in random libraries, 241, 242, 245, 246
    insert–vector ligation, 248, 251
    materials, 242, 243, 250
    plating of library, 248, 249
    screening, 249, 250
    shearing of DNA,
      pulsed-field gel electrophoresis-purified DNA, 246, 247, 251
      total genomic DNA, 246, 250
    size selection, 247
    vector DNA preparation, 246
  stage-specific Plasmodium falciparum library construction and subtractive hybridization,
    antisense-strand DNA preparation, 280, 282–284, 288
    applications, 277
    library construction, 279, 281, 282, 287
    materials, 279–281
    overview, 277, 279
    phage DNA purification, 280, 282, 288
    sense-strand complementary RNA preparation, 280, 284, 285, 288
    subtracted library,
      probe preparation for screening, 287, 289
      production, 281, 286–288
    subtractive hybridization, 280, 281, 285, 286, 288
CSA, *see* Chondroitin sulfate A
CSP, *see* Circumsporozoite protein
Cumulative incidence, calculation, 19
Cytokine, *see also specific cytokines,*
  definition, 423
  dynamics in measurement, 424, 425
  flow cytometry of T cell responses,
    activation agents and conditions, 428, 429, 433, 434
    advantages, 410, 425, 426
    analysis, 416–418
    antibody labeling, 414, 415, 420
    Brefeldin A inhibition of secretion, 414, 420, 429, 434
    cell culture, 427, 433

Index

data acquisition, 415, 416, 420
Fc receptor blocking, 429
fixation, 430
intracellular staining, 410, 411, 418, 419, 430, 431, 434
materials, 426, 432
media, 426–428
permeabilization, 430
surface labeling for phenotype determination, 429, 430
nomenclature, 423, 424
overview of assays, 424
Th1 versus Th2 response in malaria, 369, 380, 425

D

DNA extraction, *see Plasmodium* DNA extraction
DNA vaccination,
advantages and versatility, 347, 348
immune responses, 347
mouse immunization,
intradermal immunization, 357, 358
intramuscular, 356–358
preparation,
equipment, 348, 349
expression analysis,
cell culture of UM449 cells, 354
transient transfection, 348–351, 354, 355, 358
Western blot, 350–352, 355, 356, 358
plasmid DNA production and purification,
cesium chloride gradient centrifugation, 353, 354, 358
large-scale cell culture, 352
lysate preparation, 352, 353
materials, 348–351, 358
reagents, 349, 350
stock solutions, 350–352, 358

E

EIR, *see* Entomological inoculation rate
Electroporation,
episomal transformation, 310
Plasmodium berghei gene targeting, 320, 326, 329
ELISA, *see* Enzyme-linked immunosorbent assay
ELISPOT, *see* Enzyme-linked immunospot assay
Entomological inoculation rate (EIR),
bloodmeal enzyme-linked immunosorbent assay, 7, 8
definition, 3
determination,
calculation, 8–10
landing/biting catches of mosquitoes, 5, 6, 10
materials, 4
sampling stations and site selection, 4, 5
malaria epidemiology relationship, 9
sporozoite rate determination,
enzyme-linked immunosorbent assay, 6, 7
salivary gland dissection and examination, 6, 10
Enzyme-linked immunosorbent assay (ELISA),
antibody subclass assay,
incubation conditions and reading, 458, 459
materials, 457–459
overview, 457
bloodmeal, 7, 8
sporozoite rate determination, 6, 7
Enzyme-linked immunospot assay (ELISPOT),
adenoviral vector vaccine and T cell response,
advantages, 361, 362
CD4+ cell assay, 363, 366
EL4 cell supernatant preparation, 362, 364
immunization of mice, 364, 366
interferon-γ CD8+ T cell assay, 363, 365, 366
interleukin-5 CD8+ T cell assay, 363, 366
lymphocyte purification,
liver, 362–365
spleen, 363, 365
materials, 362, 363
interferon-γ CD8+ T cell assay, 363, 365, 366, 439, 440, 442, 443
Epidemiology, malaria,
active surveillance,
active case detection, 17, 18
active case survey, 18
gametocyte rate, 18
period prevalence, 18, 19
endemic grading, 13, 14
entomological factors, 14
entomological inoculation rate relationship, 9
environmental factors, 14
incidence of infection,
attributable fraction, 21
attributable risk, 20, 21
cumulative incidence, 19
density, 20

passive surveillance,
   annual parasite incidence, 17
   passive case detection, 16, 17
quality control in molecular epidemiological studies, 120
serology, 16
spleen enlargement, 14, 15
triad, 13
Episomal transformation, see Plasmodium berghei
Erythrocytic malaria,
   abundance of infected erythrocytes, 535, 536
   growth or invasion inhibition assay, see Growth or invasion inhibition assay
   mouse models,
      anemia detection,
         flow cytometry of erythrocytes, 70
         hematocrit, 70
         hemoglobin determination in whole blood, 70, 71
      body temperature measurement, 69
      body weight measurement, 70
      brain measurements,
         alanine, 72–74
         density, 71, 72
         Evans' blue leakage, 72
         histology processing, 72
         lactate, 72–74
         pyruvate, 73, 74
         weight, 71
      cerebral malaria pathogenesis, 61–63
      clinical examination, 69
      Giemsa staining of blood films, 68, 74
      glucose measurement in blood, 71
      immune response, 61, 64, 65
      infection,
         blood injection, 69
         mosquito feeding, 69, 74
         sporozoite injection, 69
      materials, 66–68
      parasite frozen stock preparation and thawing, 68, 74
      Plasmodium species and host strains, 58–60
      Plasmodium yoelii, 61, 73, 74

F

Field trials, malaria interventions,
   approaches for intervention, 607
   clinical trial phases, 608, 613
   data management and analysis, 612
   data and safety monitoring board, 609
   end point measurement, 611
   ethics, 612, 613
   implementation of phase III trial, 611, 612
   information review of control strategy, 608, 613
   objectives,
      primary objective, 609, 613, 614
      secondary objectives, 609, 613
   sample size, 611, 614
   study area,
      epidemiological criteria, 608
      operational criteria, 608
   study design, 610, 611
   study protocol development and approval, 608
   target groups, 609, 610
Flow-based cytoadhesion assays,
   factors influencing adhesion, 561
   inhibitor experiments, 577
   microslide preparation,
      acid washing, 572
      CD36 immobilization, 577, 578
      fixed cell-coated microslides, 577–579
      human umbilical vein endothelial cells,
         culture on microslides, 574, 578
         isolation, 573, 574
      materials, 571, 572
      melanoma cell culture on microslides, 576, 577
      overview, 571
      platelet coating, 575, 576
      precoating, 572, 573, 578
   parasite agglutination interference with adhesion, 555
   parasite receptors, 555
   Plasmodium falciparum-infected red blood cell microslide assay,
      flow chamber, 562, 568
      flow rate calculation, 564, 565, 569
      materials, 562
      quantitative analysis,
         direct microscopy, 566
         rolling adhesion, 566–568
         video analysis, 566
      red blood cell preparation, 563, 568, 569
      setup, 563
      shear stress modification studies, 565, 566
Flow cytometry,
   CD4+ T cell tracing after adoptive transfer in mice, 406
   erythrocytes, 70
   growth or invasion inhibition assay, 552, 553
   limiting dilution analysis, 392, 393

T cell antigen-specific proliferation,
    advantages, 410
    analysis,
        cytokine production, 416–418
        phenotype, 416–418
        total cell division within mixed cell population, 416
    antigen-presenting cell preparation, 413, 419
    cell thawing, 413
    data acquisition, 415, 416, 420
    dyes, 409, 410
    lymphocyte separation from peripheral blood, 410, 412
    materials, 410, 411, 418, 419
    overview, 409–412
    PKH26 labeling, 410–413, 419
    proliferation culture setup, 413, 414, 419, 420
    surface labeling for phenotype determination, 411, 415
T cell cytokine responses,
    activation agents and conditions, 428, 429, 433, 434
    advantages, 410, 425, 426
    analysis, 416–418
    antibody labeling, 414, 415, 420
    Brefeldin A inhibition of secretion, 414, 420, 429, 434
    cell culture, 427, 433
    data acquisition, 415, 416, 420
    Fc receptor blocking, 429
    fixation, 430
    intracellular staining, 410, 411, 418, 419, 430, 431, 434
    materials, 426, 432
    media, 426–428
    permeabilization, 430
    surface labeling for phenotype determination, 429, 430

G

Gametocyte, asexual-free gametocyte purification, 29, 30, 37, 38
Gametocyte rate, risk assessment, 18
Gene expression analysis, *see* Northern blot; Reverse transcriptase–polymerase chain reaction; RNA hybridization, *in situ*
Genomic library, *see* Complementary DNA library, *Plasmodium*; Yeast artificial chromosome
GIA, *see* Growth or invasion inhibition assay

Giemsa stain, blood films, 27, 68, 74
Growth or invasion inhibition assay (GIA),
    antibody-mediated inhibition assay,
        adsorbing test and control sera, 549
        data example, 543
        dialyzing test and control sera or immunoglobulin, 547, 548
        equipment, 544, 545, 553
        erythrocyte preparation, 548
        flow cytometry, 552, 553
        heat-sealed gassed bags, 549, 550
        Hoechst 33342 staining, 552
        initial parasitemia, 548, 549
        malaria culture mix and well addition, 550, 551
        media preparation, 547, 553
        overview, 541, 543
        parasite culture, 548
        reagents, 545–547
        static cultures, 551, 552
        suspension cultures in microtiter plates, 551
        test mix preparation and preincubation, 549, 553
    applications, 536, 537
    equations, 537–540
    inhibition per cycle, 535–537
    inhibitors, 535
    interpretation, 540
    nonspecific inhibition, 540, 541
    percentage parasitemia, 539, 553
    prospects, 544
    strain specificity, 537
    survival fraction, 539, 540

H

Hepatocyte, isolation,
    culture, 505, 506
    equipment, 503
    historical perspectives, 503
    inhibition of sporozoite invasion assay, 508, 509, 511, 512
    mouse cell isolation, 44, 45, 47
    Percoll gradient centrifugation, 505, 506
    perfusion, 504–506
    solutions, 504, 506
    T cell inhibition of liver stage development assay, 524–526
High-performance liquid chromatography (HPLC), peptide vaccines, 336, 338, 343
HPLC, *see* High-performance liquid chromatography

I

IFN-γ, *see* Interferon-γ
IL-5, *see* Interleukin-5
ILSDA, *see* Inhibition of liver stage development assay,
Immunoprecipitation, *see Plasmodium* antigens
Inducible nitric oxide synthase, *see* Nitric oxide
Inhibition of liver stage development assay (ILSDA), *see also* T cell inhibition of liver stage development assay,
   approaches, 517
   calculations, 519
   hepatocyte culture infection, 518, 519
   immunostaining, 519
   materials, 517–519
   mosquito dissection, 518, 519
Inhibition of sporozoite invasion (ISI) assay,
   development, 507
   double-staining assay,
      epifluorescence microscopy, 511, 513
      hepatocyte isolation and culture, 508, 509, 511, 512
      immune serum,
         incubation, 510
         preparation, 508, 510
      materials, 508, 509
      mosquito dissection for sporozoite isolation, 508–510
      overview, 507, 508
      parasite staining, 509–511
Inoculation, *see* Sporozoite; Vaccination
In situ RNA hybridization, *see* RNA hybridization, *in situ*
Interferon-γ (IFN-γ), ELISPOT, 363, 365, 366, 439, 440, 442, 443
Interleukin-5 (IL-5), ELISPOT, 363, 366
ISI, *see* Inhibition of sporozoite invasion

L

Lactate, brain analysis, 72–74
Lambda library, *see* Complementary DNA library, *Plasmodium*
Limiting dilution analysis, *see* T cell

M

Malaria,
   animal models, see also Mouse models, malaria; Plasmodium berghei; Primate models, malaria
   epidemiology, *see* Epidemiology, malaria
   host range and parasites in models,
      primates, 77–84
      rodents, 25, 26, 42, 57
   vector species, 25, 26
Mass spectrometry (MS), peptide vaccines, 336, 338, 339, 343
Merozoite surface protein (MSP), *see* Polymerase chain reaction
Metabolic radiolabeling, *see Plasmodium* antigens
Microsatellite analysis, *see Plasmodium falciparum*
Molecular fingerprinting, *see* Restriction fragment length polymorphism
Mosquito, *see Anopheles*
Mouse models, malaria,
   erythrocytic stage malaria,
      anemia detection,
         flow cytometry of erythrocytes, 70
         hematocrit, 70
         hemoglobin determination in whole blood, 70, 71
      body temperature measurement, 69
      body weight measurement, 70
      brain measurements,
         alanine, 72–74
         density, 71, 72
         Evans' blue leakage, 72
         histology processing, 72
         lactate, 72–74
         pyruvate, 73, 74
         weight, 71
      cerebral malaria pathogenesis, 61–63
      clinical examination, 69
      Giemsa staining of blood films, 68, 74
      glucose measurement in blood, 71
      immune response, 61, 64, 65
      infection,
         blood injection, 69
         mosquito feeding, 69, 74
         sporozoite injection, 69
      materials, 66–68
      parasite frozen stock preparation and thawing, 68, 74
      Plasmodium species and host strains, 58–60
      Plasmodium yoelii, 61, 73, 74
   overview, 41, 57
   Plasmodium berghei maintenance, *see Plasmodium berghei*
   pre-erythrocytic stage malaria,
      applications, 42
      hepatocyte isolation, 44, 45, 47

historical perspective, 41, 42
hosts, 43, 48, 49
liver stage detection with histology, 43–47, 49, 50
materials, 43–45
mosquito dissection, 43, 49
parasite staining, 45, 48
sporozoite,
    infection by injection, 43, 45, 48
    irradiation for immunization, 44, 47, 50
    isolation, 45, 47
MS, *see* Mass spectrometry
MSP, *see* Merozoite surface protein
Mung bean nuclease digest library, *see* Complementary DNA library, *Plasmodium*

N

Nested polymerase chain reaction, *see* Polymerase chain reaction
Nitric oxide (NO),
immunological effects, 461
inducible nitric oxide synthase expression,
    Bio-Rad assay for protein, 470
    control cells,
        culture, 470
        lysate preparation, 471, 474
        stimulation, 471, 473
    malaria response, 469
    materials for assay, 470, 471, 473, 474
    sample preparation, 472, 474
    Western blot analysis, 470–474
malaria effects on levels, 461, 462
metabolites, 462
nitrate and nitrite assay,
    exogenous nitrate control before sample collection,
        contamination prevention, 464–466
        diet, 463
        water, 464
    Griess assay, 464, 465
    materials, 462, 463
    sample collection,
        blood, 464, 466
        urine, 464, 466
    sample preparation, 464
synthesis, 461, 469
NO, *see* Nitric oxide
Northern blot,
limitations in gene expression analysis, 213
Plasmodium RNA, 155

O

Oocyst, staining with mercurochrome, 27, 28, 35
Ookinete,
culture, 36, 37, 39
enrichment, 30, 38
staining, 35
Ovarian development, Christopher's stages, 98, 99

P

Parity status, determination, 98
Passive case detection (PCD), risk assessment, 16, 17
PCD, *see* Passive case detection
PCR, *see* Polymerase chain reaction
Peptide vaccination,
immunogenicity limitations, 335
synthetic peptide vaccine preparation,
    absorption spectroscopy, 336, 337, 339, 343
    adjuvant preparation,
        CFA, 340, 343
        IFA, 340, 343
        Montanide ISA 51, 340, 341
        Montanide ISA 720, 341
        OM-174, 341, 343
        QS-21, 341
    aluminum hydroxide salt preparation, 341–343
    high-performance liquid chromatography, 336, 338, 343
    mass spectrometry, 336, 338, 339, 343
    materials, 335–337
    mouse injection, 337, 342, 343
    size-exclusion chromatography, 336–338
    solubilization of peptides, 339, 340, 343
    storage, 339
    end point immunological analysis, 342, 343
Period prevalence, risk assessment, 18, 19
PFGE, *see* Pulsed-field gel electrophoresis
Phenylhydrazine, reticulocytosis induction in rodent models, 27, 37
Plasmodium antigens,
immunoprecipitation for localization,
    antibody binding, 588
    antibody preparation,
        materials, 591, 592, 600
        monoclonal antibody purification, 595, 596, 600, 601
        polyclonal antibody purification, 595–595, 601
        Sepharose bead coupling, 597
    applications, 589, 590
    approaches, 587

extraction of antigen, 587, 588
immunoprecipitation,
   insoluble-specific antibody
      immunoprecipitation, 599, 602
   materials, 592, 593, 600
   parasite detergent extraction, 598, 601
   soluble-specific antibody
      immunoprecipitation, 598, 599, 601, 602
parasite preparation,
   materials, 591, 600
   metabolic radiolabeling with sulfur-35 methionine, 594, 600
   radiolabel incorporation quantification, 594, 595, 600
   saponin lysis, 595, 600
   synchronization of cultures with sorbitol, 593, 594
precipitation reaction, 589
Western blot analysis,
   autoradiography, 600, 603
   blotting and immunostaining, 599, 600, 602
   gel electrophoresis, 599
   materials, 593
Triton X-114 phase partitioning,
   applications, 581
   extraction and phase separation of parasites, 582–585
   materials, 582
   membrane protein features, 581, 582
   precondensation of detergent, 582, 584
   principles, 581
   principles, 581, 582
Plasmodium berghei,
cryopreservation,
   blood-stages, 31, 38
   sporozoites, 31
episomal transformation,
   cloning of parasites, 311, 312
   cryopreservation of parent populations, 311
   curing episomes, 312–314
   efficiency, 305, 306
   electroporation, 310
   equipment, 307
   overview, 305, 306
   plasmid rescue, 312
   pyrimethamine treatment and selection of resistant parasites, 310, 311
   rats,
      handling, 308
      infection, 308
      phenylhydrazine treatment, 308
   reagents, 307, 308
   resistant parasite transfer to naive mice, 311
   schizont targeting, 306
   schizont,
      culture, 308, 309, 313, 314
      injection into rats, 310, 314
      purification, 309, 314
   selectable markers, 306
   time requirements, 308
   transfection construct preparation and purification, 309, 310
gene expression analysis, *see* RNA hybridization, *in situ*
gene targeting by homologous recombination,
   advantages over *Plasmodium falciparum* transformation system, 317
   cloning of parasites, 321, 327, 328, 330
   electroporation, 320, 326, 329
   gene inactivation,
      insertion plasmids, 323, 329, 330
      replacement plasmids, 323, 329
   gene modification,
      insertion plasmids, 324, 330
      replacement plasmids, 323, 324, 330
   genomic DNA preparation and analysis, 321, 328
   materials, 319–321, 329
   phenotypic analysis of recombinant clones, 328, 329
   recipient parasite preparation, 319, 320, 325, 326, 329, 330
   selectable marker, 317
   selection of parasites, 320, 321, 326, 327, 330
   targeting plasmid construction, 319, 321, 323, 324
maintenance in mosquito,
   infection by direct feeds on mice, 33, 34, 39
   membrane feeders and feeding, 27, 34, 35
   oocyst staining with mercurochrome, 27, 28, 35
   ookinete staining, 35
   sporozoite analysis, 35
maintenance in vitro,
   exoerythrocytic stage culture,
      blood-stage culture, 36, 39
      fixation and staining, 28
      host cells, 36
      ookinete culture, 36, 37, 39
      parasites, 36
   overview, 35

Plasmodium yoelii comparison, 42, 43
pre-erythrocytic stage, 42, 43
rodent malaria models,
    advantages, 25
    anesthesia, 26, 27
    asexual stage purification, 28, 29
    drug cure of asexual infections, 33
    Giemsa staining of blood films, 27
    host strains, 32
    infection of mice, 32, 33, 38, 39
    life-cycle stage purification,
        asexual-free gametocyte purification, 29, 30, 37, 38
        mixed asexual blood stages, 37
        ookinete enrichment, 30, 38
        schizont purification, 29, 37, 39
        sporozoite purification, 30, 38
    materials, 26–30
    media, 28
    mosquito feed, 27
    parasite strains, 32
    parasitemia monitoring, 33, 39
    phenylhydrazine for reticulocytosis induction, 27, 37
Plasmodium brasilianum, strains for primate models, 83
Plasmodium chabaudi, CD4+ T cell limiting dilution analysis,
    advantages, 385, 386
    antigen preparation, 390, 397
    antigen-presenting cell preparation,
        spleen collection and processing, 389, 390, 397
        T cell depletion, 390
    CD4+ T cell preparation,
        negative selection, 389
        positive selection, 389, 397
        spleen collection and processing, 388
    enzyme-linked immunosorbent assay,
        cytokines, 394
        malaria antibodies, 394, 395
    flow cytometry analysis, 392, 393
    interleukin-2 assay, 393
    interleukin-3 assay, 393
    materials, 386–388, 396, 397
    precursor frequency calculations, 395
    principles, 386, 388
    restimulation, 391
    setup, 390, 391
    tritiated thymidine pulsing, 393
Plasmodium coatneyi, Hackeri strain for primate models, 82
Plasmodium culture,

automated synchronization of *Plasmodium falciparum* cultures,
    alanine synchronization, 493, 494, 496
    equipment, 490, 491, 495
    flask preparation, 493, 495, 496
    historical perspective, 490
    incubator setup, 492, 493
    Omega controller programming, 493, 495
    prospects, 494, 496
    reagents, 491, 492
    rotating platform placement, 493
    schizont appearance time modification, 494, 496
    temperature cycling, 489, 490, 494
cryopreservation of stocks, 479, 480, 486
historical perspective, 477
incubation, 480, 481, 487
materials, 477–479, 484, 485
medium, 479, 484–486
microculture and radiolabeling, 482–484, 487
synchronization, 481, 482, 593, 584
Plasmodium cynomolgi, strains for primate models, 82
Plasmodium DNA extraction,
    equipment, 160
    extraction and pelleting, 161, 162
    genomic DNA preparation, 293, 294, 297, 302, 303
    overview, 159, 160
    quality criteria, 159
    reagents, 160–162
    RNase treatment, 163
    Southern blotting, *see* Southern blot, parasite DNA
    storage, 161, 163
Plasmodium falciparum,
    adhesion, *see* Chondroitin sulfate A; Flow-based cytoadhesion assays
    chromosome purification,
        equipment, 235
        parasite preparation, 236–238, 240
        pulsed-field gel electrophoresis, 237–240
        reagents, 235, 236
    gene expression analysis, *see* RNA hybridization, *in situ*
    genome sequencing, 235, 246
    genomic library, *see* Complementary DNA library, *Plasmodium*; Yeast artificial chromosome
    genotyping with polymerase chain reaction–restriction fragment length polymorphism analysis,
        amplification reactions, 119, 120, 127

consecutive sample analysis from same
    patient, 125, 128
cross-contamination prevention, 120
DdeI digestion, 124, 125
DNA extraction, 118, 126, 127
equipment, 118
gel electrophoresis of products, 121
*Hinf*I digestion and analysis, 121, 123,
    124, 128
merozoite surface protein 2
    polymorphisms, 117
quality control in epidemiological
    studies, 120
reagents, 118, 119, 127
life cycle, 536
microsatellite analysis,
  applications, 132, 133
  equipment, 133
  fluorescence labeling of polymerase
      chain reaction products, 135
  radioactive labeling of polymerase chain
      reaction products, 134, 135
  reagents, 133, 134
  simple sequence length polymorphisms
      and advantages in analysis, 131
primate model strains,
  INDO-1, 79
  Malayan IV, 79
  miscellaneous strains, 79
  Santa Lucia, 78
  Uganda Palo Alto, 78, 79
  Vietnam Oak Knoll, 78
Southern blot analysis, *see* Southern blot,
    parasite DNA
synchronization of cultures, *see*
    Plasmodium culture
Western blotting of proteins,
  applications, 177
  chemiluminescent immunodetection,
      180, 183–188
  electroblotting, 179, 182–186
  electrophoresis, 179, 181, 182, 184, 186
  materials, 178–180, 184, 185
  overview,
    chemiluminescence detection, 178
    denaturing gel electrophoresis, 177, 178
    electroblotting, 178
  protein sample preparation, 178–181, 186
Plasmodium fieldi, strains for primate models,
    83
Plasmodium fragile, strains for primate
    models, 82
Plasmodium genomic library, *see*
    Complementary DNA library,
    Plasmodium
Plasmodium genotyping, *see* Polymerase chain
    reaction
Plasmodium gonderi, Mandril strain for
    primate models, 83, 84
Plasmodium inui, strains for primate models,
    82, 83
Plasmodium knowlesi, strains for primate
    models, 81, 82
Plasmodium malariae, primate model strains,
  Uganda I, 81
  China I, 81
Plasmodium ovale, Nigeria I strain for primate
    models, 81
Plasmodium RNA extraction,
  agarose gel electrophoresis, 154, 157
  applications, 151, 155, 156
  DNase treatment, 154, 155
  extraction and pelleting, 153, 156
  formamide removal, 154, 157
  materials, 152, 156
  Northern blot, 155
  overview, 151
  parasite harvesting, 152, 153, 156
  poly(A) RNA isolation, 156
  spectrophotometric quantification, 154, 157
  storage of RNA, 154, 157
Plasmodium simiovale, strains for primate
    models, 83
Plasmodium simium, strains for primate
    models, 83
Plasmodium vivax, primate model strains,
  AMRU-1, 80
  Brazil I, 80
  Chesson, 80
  Indonesia I, 80
  Indonesia XIX, 80
  Mauritana strains, 80
  miscellaneous strains, 81
  North Korean, 81
  Salvador I, 79, 80
  Salvador II, 80
  Thai III, 80
  Vietnam Nam, 81
  Vietnam Ong, 81
  Vietnam Palo Alto, 80
Polymerase chain reaction (PCR),
  Anopheles species identification,
    amplification reaction, 100
    DNA extraction, 100
    overview, 99–101
    primers, 100, 101
  liver-stage parasite quantification with
      TaqMan® automated system,

complementary DNA synthesis, 138
equipment, 137
interpretation of results, 139
rationale, 137
reagents, 137, 138
real-time quantification, 139
RNA extraction and isolation, 138, 139
Plasmodium detection and genotyping with nested polymerase chain reaction,
advantages, 103, 189, 190
amplification reactions, 106–109, 112–114, 192–195, 198, 199
equipment, 104, 112, 190, 197
facilities, 104, 111, 190, 196, 197
gel electrophoresis of products, 109, 110, 114, 195, 196, 200
genetic marker criteria, 103, 104
interpretation, 110, 111, 196
primers, 106, 107, 192
reagents, 104, 105, 112, 190, 191, 197, 198
sample collection, 105, 191
sensitivity, 189
single-base polymorphisms, 104
template preparation, 105, 106, 191, 192
Plasmodium falciparum genotyping with polymerase chain reaction–restriction fragment length polymorphism analysis,
amplification reactions, 119, 120, 127
consecutive sample analysis from same patient, 125, 128
cross-contamination prevention, 120
DdeI digestion, 124, 125
DNA extraction, 118, 126, 127
equipment, 118
gel electrophoresis of products, 121
*Hin*fI digestion and analysis, 121, 123, 124, 128
merozoite surface protein 2 polymorphisms, 117
quality control in epidemiological studies, 120
reagents, 118, 119, 127
Plasmodium falciparum microsatellite analysis,
applications, 132, 133
equipment, 133
fluorescence labeling of polymerase chain reaction products, 135
radioactive labeling of polymerase chain reaction products, 134, 135
reagents, 133, 134
simple sequence length polymorphisms and advantages in analysis, 131

RNA, *see* Reverse transcriptase–polymerase chain reaction
Primate models, malaria,
animals,
care and ethics, 77, 78, 84
species, 77, 78, 85
blood collection, 88, 89
chloroquine treatment, 90, 91
erythrocyte parasitized stocks,
freezing, 89
infection, 87
thawing, 87
materials, 85, 86, 90
mosquito infection, 89
parasitemia monitoring, 87, 88, 90
Plasmodium brasilianum strains, 83
Plasmodium coatneyi Hackeri strain, 82
Plasmodium cynomolgi strains, 82
Plasmodium falciparum strains,
INDO-1, 79
Malayan IV, 79
miscellaneous strains, 79
Santa Lucia, 78
Uganda Palo Alto, 78, 79
Vietnam Oak Knoll, 78
Plasmodium fieldi strains, 83
Plasmodium fragile strains, 82
Plasmodium gonderi Mandril strain, 83, 84
Plasmodium inui strains, 82, 83
Plasmodium knowlesi strains, 81, 82
Plasmodium malariae strains,
China I, 81
Uganda I, 81
Plasmodium ovale Nigeria I strain, 81
Plasmodium simiovale strain, 83
Plasmodium simium strains, 83
Plasmodium vivax strains,
AMRU-1, 80
Brazil I, 80
Chesson, 80
Indonesia I, 80
Indonesia XIX, 80
Mauritana strains, 80
miscellaneous strains, 81
North Korean, 81
Salvador I, 79, 80
Salvador II, 80
Thai III, 80
Vietnam Nam, 81
Vietnam Ong, 81
Vietnam Palo Alto, 80
splenectomy, 88, 90
sporozoite transmission, 89, 90
Pulsed-field gel electrophoresis (PFGE),

chromosome purification from *Plasmodium falciparum*, 237–240
shearing of purified DNA for library preparation, 246, 247, 251
Southern blotting, 171
yeast artificial chromosome libraries,
　clone analysis, 296, 297, 302
　size fractionation of ligation reaction, 295, 299
Pyruvate, brain analysis, 73, 74

R

Restriction fragment length polymorphism (RFLP),
　parasite characterization,
　　agarose gel electrophoresis, 208, 211
　　DNA isolation, 207, 208, 211
　　equipment, 206, 207
　　hybridization, 210, 211
　　probe labeling, 209, 210, 212
　　reagents, 207
　　restriction digest, 208, 211
　　Southern blotting, 208, 209, 211, 212
　Plasmodium falciparum genotyping with polymerase chain reaction–restriction fragment length polymorphism analysis,
　　amplification reactions, 119, 120, 127
　　consecutive sample analysis from same patient, 125, 128
　　cross-contamination prevention, 120
　　*Dde*I digestion, 124, 125
　　DNA extraction, 118, 126, 127
　　equipment, 118
　　gel electrophoresis of products, 121
　　*Hinf*I digestion and analysis, 121, 123, 124, 128
　　merozoite surface protein 2 polymorphisms, 117
　　quality control in epidemiological studies, 120
　　reagents, 118, 119, 127
　principles, 205, 206
Reverse transcriptase–polymerase chain reaction (RT-PCR),
　gene expression analysis in parasites,
　　advantages, 213, 214
　　amplification reactions,
　　　first polymerase chain reaction, 218, 223
　　　nested polymerase chain reaction, 219, 223
　　cell preparation, 215, 221, 222
　　contamination precautions, 219, 220
　　gel electrophoresis of products, 219, 223
　　interpretation of results, 219, 224
　　materials, 214, 215, 221
　　overview, 213–215
　　reverse transcription,
　　　gene-specific primers, 218, 223
　　　random primers, 217, 222, 223
　　RNA extraction, 215–217, 222
　liver-stage parasite quantification with competitive reverse transcriptase–polymerase chain reaction,
　　advantages, 141
　　competitive amplification reactions, 146, 147
　　competitor plasmid construction, 142, 143
　　complementary DNA synthesis, 146
　　materials, 144, 145
　　principles, 141–144
　　quantitative imaging, 147, 148
　　RNA isolation from liver, 145, 147
　liver-stage parasite quantification with TaqMan® automated system,
　　complementary DNA synthesis, 138
　　equipment, 137
　　interpretation of results, 139
　　rationale, 137
　　reagents, 137, 138
　　real-time quantification, 139
　　RNA extraction and isolation, 138, 139
RFLP, *see* Restriction fragment length polymorphism
Risk assessment, *see* Entomological inoculation rate; Epidemiology, malaria
RNA extraction, *see* Plasmodium RNA extraction
RNA hybridization, *in situ*,
　applications, 225
　blood-stage *Plasmodium*,
　　alkaline phosphatase-conjugated antibody detection, 230, 231
　　fluorochrome-conjugated antibody detection, 231
　　hybridization, 230
　　smear preparation, 230
　buffers, 227, 228
　controls, 232
　equipment, 226, 227
　gene expression analysis advantages, 225
　probe preparation,
　　labeling systems, 229, 232, 233
　　materials, 227, 228
　　selection of probe, 228, 229

transcription of labeled antisense RNA
        probes, 229, 230
    signal detection, 228
    sporozoites, 232
    whole mount hybridization and detection,
        228, 231–233
RT-PCR, *see* Reverse transcriptase–
    polymerase chain reaction

S

Schizont,
    episomal transformation, *see*
        Plasmodium berghei
    purification, 29, 37, 39
Sorbitol, synchronization of cultures, 593, 594
Southern blot, parasite DNA,
    agarose gel electrophoresis, 168, 172
    hybridization and washing, 171, 174,
        175
    materials, 166, 167, 172
    nylon membrane transfer,
        capillary transfer, 170, 173
        gel preparation, 168, 173
        vacuum transfer, 168–170, 173
    principles, 165, 166
    probe radiolabeling, 170, 171, 173, 174
    pulsed-field gels, 171
    restriction digestion of genomic DNA,
        167, 168, 172
    restriction fragment length
        polymorphism for parasite
        characterization,
            blotting, 208, 209, 211, 212
            hybridization, 210, 211
            probe labeling, 209, 210, 212
    stripping of membrane, 171, 175
    time requirements, 172
Spleen, enlargement and risk assessment, 14,
    15
Sporozoite,
    cryopreservation, 31
    cytotoxic T lymphocyte epitopes, 438
    hepatic portal branch inoculation,
        materials, 499, 500
        overview, 499
        technique, 500
    inhibition assay, *see* Inhibition of
        sporozoite invasion assay
    pre-erythrocytic stage malaria mouse
        model,
            infection by injection, 43, 45, 48
            irradiation for immunization, 44, 47, 50
            isolation, 45, 47

primate model transmission, 89, 90
purification, 30, 38
rate determination,
    enzyme-linked immunosorbent assay,
        6, 7
    salivary gland dissection and
        examination, 6, 10
RNA hybridization, *in situ*, 232
Subtractive hybridization,
    stage-specific *Plasmodium falciparum* library,
        antisense-strand DNA preparation, 280,
            282–284, 288
        applications, 277
        library construction, 279, 281, 282, 287
        materials, 279–281
        overview, 277, 279
        phage DNA purification, 280, 282, 288
        sense-strand complementary RNA
            preparation, 280, 284, 285, 288
        subtracted library,
            probe preparation for screening, 287, 289
            production, 281, 286–288
        subtractive hybridization, 280, 281, 285,
            286, 288
    yeast artificial chromosome library, *see*
        Yeast artificial chromosome
Synchronization of cultures, *see* Plasmodium
    culture

T

TaqMan®, *see* Polymerase chain reaction
T cell,
    adenoviral vector vaccine and enzyme-linked
        immunospot assay of response,
            advantages, 361, 362
            CD4+ cell assay, 363, 366
            EL4 cell supernatant preparation, 362, 364
            immunization of mice, 364, 366
            interferon-γ CD8+ T cell assay, 363,
                365, 366
            interleukin-5 CD8+ T cell assay, 363, 366
            lymphocyte purification,
                liver, 362–365
                spleen, 363, 365
            materials, 362, 363
    antigenic peptide identification, 370–372
    antigen presentation, 369, 370
    CD4+ T cell limiting dilution analysis,
        advantages, 385, 386
        antigen preparation, 390, 397
        antigen-presenting cell preparation,
            spleen collection and processing, 389,
                390, 397

T cell depletion, 390
CD4+ T cell preparation,
    negative selection, 389
    positive selection, 389, 397
    spleen collection and processing, 388
enzyme-linked immunosorbent assay,
    cytokines, 394
    malaria antibodies, 394, 395
flow cytometry analysis, 392, 393
interleukin-2 assay, 393
interleukin-3 assay, 393
materials, 386–388, 396, 397
precursor frequency calculations, 395
principles, 386, 388
restimulation, 391
setup, 390, 391
tritiated thymidine pulsing, 393
CD4+ T cell proliferation assays,
    human assays,
        antigen-reactive clone profunction, 378
        antigens, 373
        calculations, 377, 378
        cryopreservation and thawing of
            cells, 377, 381
        epitope-specific clone generation,
            378–380
        medium, 373
        overview, 375
        peptides, 373, 380
        peripheral blood mononuclear cell
            preparation, 376
        proliferation assay, 376, 377, 381
        subjects, 374
        venipuncture, 374
    mouse assays,
        immunization, 372, 373, 380
        lymph node proliferation assay, 374,
            375, 380
        materials, 372, 373, 380
CD4+ T cell tracing after adoptive transfer
    in mice,
    adoptive transfer, 403, 404
    5-(and 6-)carboxyfluorescein diacetate
        succinimidyl ester labeling of T
        cells, 403, 406
    fluorescence-activated cell sorting, 406
    materials, 401, 402
    organ collection,
        bone marrow, 405
        liver, 404, 405
        lung, 405
        lymph nodes, 404, 405
        spleen, 404, 405
    parasite challenge of mice, 404

peripheral blood collection, 404
T cell line generation,
    cell line growth and maintenance, 403
    immunization, 402, 406
    lymph node cell collection, 402, 403, 406
CD8+ T cell assays,
    chromium release assay, 439, 442, 453–455
    interferon-γ ELISPOT, 439, 440, 442, 443
    liver cell preparation and culture, 439,
        441, 442
    lymph node cells,
        preparation and culture, 438, 439, 443
        restimulation, 440, 441, 443
    lymphoblastoid cell lymphoma
        generation, 450
    materials, 438–440, 445–447, 454
    mitomycin C treatment of cells, 451
    peptide toxicity assessment, 451
    peptide-specific cytotoxic T lymphocyte
        generation, 451, 452
    peripheral blood mononuclear cells,
        freezing, 449, 454
        isolation, 447, 448, 454
        thawing, 449, 450
    phytohemagglutinin P-activated
        lymphoblast generation, 450
    spleen cells,
        preparation and culture, 439, 440, 443
        restimulation, 440, 441, 443
    virally stimulated cytotoxic T lymphocyte
        generation, 452, 453, 455
epitope classification, 371, 373
flow cytometry of antigen-specific
        proliferation,
    advantages, 410
    analysis,
        cytokine production, 416–418
        phenotype, 416–418
        total cell division within mixed cell
            population, 416
    antigen-presenting cell preparation, 413, 419
    cell thawing, 413
    data acquisition, 415, 416, 420
    dyes, 409, 410
    lymphocyte separation from peripheral
        blood, 410, 412
    materials, 410, 411, 418, 419
    overview, 409–412
    PKH26 labeling, 410–413, 419
    proliferation culture setup, 413, 414,
        419, 420
    surface labeling for phenotype
        determination, 411, 415
flow cytometry of cytokine responses,

activation agents and conditions, 428, 429, 433, 434
advantages, 410, 425, 426
analysis, 416–418
antibody labeling, 414, 415, 420
Brefeldin A inhibition of secretion, 414, 420, 429, 434
cell culture, 427, 433
data acquisition, 415, 416, 420
Fc receptor blocking, 429
fixation, 430
intracellular staining, 410, 411, 418, 419, 430, 431, 434
materials, 426, 432
media, 426–428
permeabilization, 430
surface labeling for phenotype determination, 429, 430
malaria immune response, 369, 385, 401, 409, 437, 438, 445
Th1 versus Th2 response in malaria, 369, 380, 425
T cell inhibition of liver stage development assay (TILSA),
development, 521
hepatocyte,
infection, 524, 525
isolation, 524–526
materials, 522, 523
overview, 521, 522
parasite staining and fluorescence microscopy, 525, 526
sporozoite isolation, 524
T cell preparation,
enrichment, 523, 524
peptide immunization, 523
purification, 523, 525
TILSA, see T cell inhibition of liver stage development assay
Transformation, see *Plasmodium berghei*
Triton X-114, see *Plasmodium* antigens

V

Vaccination,
antiadhesion vaccines, 555
clinical trials, see Field trials, malaria interventions
DNA, see DNA vaccination
peptides, see Peptide vaccination
viral vector vaccine, see Circumsporozoite protein

W

Western blot,
DNA vaccine expression analysis, 350–352, 355, 356, 358
inducible nitric oxide synthase, 470–474
Plasmodium antigen localization,
autoradiography, 600, 603
blotting and immunostaining, 599, 600, 602
gel electrophoresis, 599
materials, 593
Plasmodium falciparum proteins,
applications, 177
chemiluminescent immunodetection, 180, 183–188
electroblotting, 179, 182–186
electrophoresis, 179, 181, 182, 184, 186
materials, 178–180, 184, 185
overview,
chemiluminescence detection, 178
denaturing gel electrophoresis, 177, 178
electroblotting, 178
protein sample preparation, 178–181, 186

Y

YAC, see Yeast artificial chromosome
Yeast artificial chromosome (YAC),
applications, 291, 292
insert size capability, 291
library construction overview, 292, 293
Plasmodium falciparum library preparation,
filter,
hybridization, 296, 301, 302
preparation, 296, 301, 303
genomic DNA,
partial digestion, 294, 297, 298
preparation, 293, 294, 297, 302, 303
ligation, 294, 295, 298, 299
materials, 293–297
pulsed-field gel electrophoresis,
clone analysis, 296, 297, 302
size fractionation of ligation reaction, 295, 299
replication, 296, 300
spheroplast preparation and transformation, 295, 296, 299, 300, 303
storage, 296, 301
vector preparation, 294, 298
Plasmodium libraries, 291, 292